FIFTH EDITION

Introduction to
Technical Mathematics

FIFTH EDITION

Introduction to
Technical Mathematics

Allyn J. Washington
Dutchess Community College

Mario F. Triola
Dutchess Community College

Ellena E. Reda
Dutchess Community College

Boston San Francisco New York
London Toronto Sydney Tokyo Singapore Madrid
Mexico City Munich Paris Cape Town Hong Kong Montreal

Publisher	Greg Tobin
Editor in Chief	Deirdre Lynch
Project Editor	Joanne Ha
Editorial Assistant	Susan Whalen
Associate Managing Editor	Jeffrey Holcomb
Senior Designer	Joyce Cosentino Wells
Photo Research	Beth Anderson and Christina Gleason
Digital Asset Manager	Marianne Groth
Media Producer	Sharon Tomasulo
Software Development	Bob Carroll and Mary Durnwald
Marketing Manager	Jay Jenkins
Marketing Assistant	Alexandra Waibel
Senior Author Support/ Technology Specialist	Joe Vetere
Senior Prepress Supervisor	Caroline Fell
Senior Manufacturing Buyer	Evelyn Beaton
Project Manager	Jennifer Albanese
Cover Design	Night & Day Design
Text Design	Susan Carsten Raymond
Production Coordination	Shelley Creager, Techbooks, Inc.
Composition and Illustrations	Techbooks, Inc.

Cover photos: © George Kavanagh/Getty Images: Wind tunnel. INSET: © Gregor Schuster/Getty Images: Scans of human brain. ©Mechanik/Shutterstock. © Gregor Schuster/Getty Images: Keyboard.

Page 1 © Corbis Page 45 © NASA Page 71 © Photolibrary.Com Page 111 © Steve Gray/istockphoto.com Page 155 © Getty/Stockbyte Silver Page 199 © Sean Sexton Collection/Corbis (image courtesy of Addison-Wesley) Page 233 © Corbis Page 269 © AP/Wide World Photos Page 295 © Sean McBride/istockphoto.com Page 331 © Marek Pawluczuk/istockphoto.com Page 365 © Joseph Helfenberger/istockphoto.com Page 397 © Getty/Digital Vision Page 429 © Alinari / Art Resource, NY Page 483 © JupiterImages Corporation Page 539 © Getty/Blend Images Page 585 © Corbis Page 609 © Getty/Photodisk Green

Many of the designations used by manufacturers and sellers to distinguish their products are claimed as trademarks. Where those designations appear in this book, and Addison-Wesley was aware of a trademark claim, the designations have been printed in initial caps or all caps.

This interior of this book was composed in QuarkXpress 4.11.

Library of Congress Catologing-in-Publication Data

Washington, Allyn J.
   Introduction to technical mathematics / Allyn J. Washington, Mario F. Triola, Ellena Reda. — 5th ed.
      p. cm.
   ISBN-13: 978-0-321-37417-2   ISBN-10: 0-321-37417-7
   1. Mathematics. I. Triola, Mario F. II. Reda, Ellena. III. Title.

QA39.2.W374 2006
510—dc22

2006057735

23 18

PEARSON

To my loving wife, Millie
~ AJW

To Ginny
~ Marty

To my wonderful husband, Tony.
Thanks for all your support.
~ Ellena

Preface

Welcome

We are pleased to bring you the fifth edition of *Introduction to Technical Mathematics*. This book is intended for students enrolled in technical programs at two-year and four-year colleges and technical institutions. It is designed to give students a basic, working knowledge of mathematics. The textbook has a clean, uncluttered, and modern look. The material is presented in such a way as to accommodate those students who may have struggled with mathematics in the past. Explanations are written in clear, simple language and try to provide students with the background behind why a particular process or method works. The concepts developed in this textbook make it an appropriate choice for students who may be moving on to more advanced work in mathematics or students for whom this is a terminal mathematics course.

The examples and exercises in this textbook include applications from a variety of technical fields. These applications are intended to help illustrate the uses of mathematics and provide students with examples of the applicability of the concepts they are learning. The technical material itself is not developed and no prior technical knowledge is assumed or required to work through these applications.

Although any scientific calculator is appropriate for this course and can be used with this textbook, specific references to the TI-83/84 Plus graphing calculator are made throughout the book.

Changes to the Fifth Edition

The fifth edition has undergone a substantial reorganization and includes extensive revision of some topics. The first thing users of previous editions will notice is that the Arithmetic Review has been moved to the Appendix. Consequently, Chapter 1 begins with basic algebraic concepts. This chapter will serve as a review for many students. Because good critical thinking and problem-solving skills are essential, a new section on problem-solving strategies is included in Chapter 1 as well. The new Chapter 3 (Introduction to Algebra) combines information presented in Chapters 4 and 7 of the previous edition to form a more comprehensive introduction to basic algebra skills. Chapter 5 (Graphs), previously Chapter 13, has been moved up in this edition in order to allow for the introduction of the concept of functions earlier in the text and to provide for the possibility of graphical solutions throughout the remainder of the textbook, when appropriate. Chapter 6 (Introduction to Geometry) now includes a section on volume, previously located in Chapter 15 in the fourth edition. This places all of the geometry together and allows for the development of this topic in one, all-inclusive chapter. In Chapter 8 (Factoring), Section 8.3 of the previous edition, which dealt with factoring trinomials, has been separated into two sections. Section 8.2 focuses on factoring trinomials and lets students build their skills before moving on to Section 8.3, factoring general trinomials. In Chapter 11 (Quadratic Equations) two topics have been added, completing the square and graphing a quadratic function. Interpolation and logarithmic tables have been removed from Chapter 12 (Exponential and Logarithmic Functions). New to this chapter is the introduction of exponential functions. Two sections focusing on solving exponential and logarithmic equations have been added to this chapter as well. A section introducing log-log and semi-log graphs has also been added to this chapter. Chapter 14 (Oblique Triangles and Vectors) combines information from Chapters 17 and 18 in the previous edition. Chapter 15 (Graphs of Trigonometric Functions) contains a new section that focuses on the sine function as a function of time. Finally, a new Chapter 17 (Introduction to Data Analysis) has been added to include introductory material on data analysis. This chapter is not meant to be a comprehensive unit on statistics, but rather introduces students to some basic concepts they are likely to encounter in other courses that they are taking.

Chapter Openers

All of the chapter openers have been rewritten. Whereas some of the chapter openers feature a real-world application of mathematics that are tied to concepts presented within the chapter, others are meant to be more informative in nature.

Section Objectives

The section begins with learning objectives that are clearly stated and written in behavioral terms. This provides a quick overview of what will be covered in each section.

New Problems and Updated Exercises

Many new and updated examples and exercises are included throughout this textbook. The applied problems include numerous applications from a variety of technical fields. These applications are intended to illustrate the many uses of mathematics and provide students with examples of the applicability of the concepts they are learning.

Now Try It Problems

These problems are designed to provide students with an opportunity to practice a concept or skill presented within a given section. They are intended to be used in class as a mechanism for assessing students' understanding of the material presented. They can be assigned as group or individual work. Answers to the "Now Try It" problems are placed at the end of each section just after the section exercises.

Using Technology

A scientific calculator should be a minimum requirement for this textbook. The use of a graphing calculator can increase learning and, when used appropriately, often helps students understand basic concepts. Students who are visual learners will benefit from the use of the calculator as a way to picture some of the concepts presented. Specific references to the TI-83/84 Plus graphing calculator are made throughout the book. In addition to numerous screenshots, step-by-step keystroke instructions are illustrated in the optional Using Technology feature.

Cautions and Notes

Found in the margin of the text, these boxes warn the student about common errors, emphasize a particular idea, or clarify an important concept.

Chapter Summaries

An extensive chapter summary can be found at the end of each chapter. Because understanding the vocabulary of mathematics is crucial, the summary provides a list of terms introduced in that chapter. The chapter summary also includes key concepts and important formulas.

For Students

Student's Solutions Manual

- By Roxane Barrows and Cheryl Mansky, *Hocking College.*
- Provides detailed solutions to every odd-numbered section exercise.
- ISBN-13: 978-0-321-37419-6
- ISBN-10: 0-321-37419-3

Graphing Calculator Manual

- By Sandra DeGuzman, *Dutchess Community College.*
- Provides an overview with detailed instruction on the use of the TI-83/84+.
- ISBN-13: 978-0-321-45062-3
- ISBN-10: 0-321-45062-0

Addison-Wesley Math Tutor Center

- Provides tutoring through a registration number that can be packaged with a new textbook or purchased separately.
- Staffed by qualified college mathematics instructors.
- Accessible via toll-free telephone, toll-free fax, e-mail, and the Internet.
- *www.aw-bc.com/tutorcenter*

For Instructors

Instructor's Solutions Manual

- By Roxane Barrows and Cheryl Mansky, *Hocking College.*
- Contains detailed solutions to every section exercise.
- ISBN-13: 978-0-321-37418-9
- ISBN-10: 0-321-37418-5

Printed Test Bank

- By David Harris, *Trident Technical College.*
- Contains two short-answer test forms for every chapter. Answer keys are included.
- ISBN-13: 978-0-321-45066-1
- ISBN-10: 0-321-45066-3

TestGen®

- Enables instructors to build, edit, print, and administer tests.
- Features a computerized bank of questions developed to cover all text objectives.
- Available on a dual-platform Windows/Macintosh CD-ROM.
- ISBN-13: 978-0-321-45414-0
- ISBN-10: 0-321-45414-6

Chapter Tests

A practice test follows each set of chapter review exercises. These questions are written to highlight the main topics presented in each chapter.

Addison-Wesley Math Adjunct Support Center

The Addison-Wesley Math Adjunct Support Center is staffed by qualified mathematics instructors with over 50 years of combined experience at both the community college and university level. Assistance is provided for faculty in the following areas:

- Syllabus consultation
- Tips on using materials packaged with your book
- Book-specific content assistance
- Teaching suggestions, including advice on classroom strategies

For more information, visit *www.aw-bc.com/tutorcenter/math-adjunct.html.*

MyMathLab®: *www.mymathlab.com*

MyMathLab® is a series of text-specific, easily customizable online courses for Addison-Wesley textbooks in mathematics and statistics. Powered by CourseCompass™ (Pearson Education's online teaching and learning environment) and MathXL® (our online homework, tutorial, and assessment system), MyMathLab gives you the tools you need to deliver all or a portion of your course online, whether your students are in a lab setting or working from home. MyMathLab provides a rich and flexible set of course materials, featuring free-response exercises that are algorithmically generated for unlimited practice and mastery. Students can also use online tools to improve their understanding and performance independently. Instructors can use MyMathLab's homework and test managers to select and assign online exercises correlated directly with the textbook, and they can also create and assign their own online exercises and import TestGen tests for added flexibility. MyMathLab's online grade book—designed specifically for mathematics and statistics—automatically tracks students' homework and test results and gives the instructor control over how to calculate final grades. Instructors can also add offline (paper-and-pencil) grades to the gradebook. MyMathLab is available to qualified adopters. For more information, visit our Web site at *www.mymathlab.com* or contact your Addison-Wesley sales representative.

MathXL®: *www.mathxl.com*

MathXL® is a powerful online homework, tutorial, and assessment system that accompanies Addison-Wesley textbooks in mathematics or statistics. With MathXL, instructors can create, edit, and assign online homework and tests using algorithmically generated exercises correlated with the textbook at the objective level. They can also create and assign their own online exercises and import TestGen tests for added flexibility. All student work is tracked in MathXL's online grade-book. Students can take chapter tests in MathXL and receive personalized study plans based on their test results. The study plan diagnoses weaknesses and links students directly to tutorial exercises for the objectives they need to study and retest. MathXL is available to qualified adopters. For more information, visit our Web site at *www.mathxl.com* or contact your Addison-Wesley sales representative.

InterAct Math Tutorial Web site: *www.interactmath.com*

Get practice and tutorial help online! This interactive tutorial Web site provides algorithmically generated practice exercises that correlate directly with the exercises in the textbook. Students can retry an exercise as many times as they like, with new values each time, for unlimited practice and mastery. Every exercise is accompanied by an interactive guided solution that provides helpful feedback for incorrect answers, and students can also view a worked-out sample problem that steps them through an exercise similar to the one they're working on.

Acknowledgments

I am grateful for the many helpful suggestions made by reviewers of the fifth edition. I would like to thank the following individuals:

Stan Adamski, Owens Community College, Toledo Campus

Douglas Carbone, Central Maine Community College

Cloyd Payne, Owens Community College

Saverio Perugini, Gateway Community College

Hank Regis, Valencia Community College

Jamal Salahat, Owens Community College

Dina Spain, Horry Georgetown Technical College

Other helpful suggestions came from the following people, who checked the manuscript for accuracy: Lori Booze, Alicia Schlintz, and Patricia Nelson.

I would like to thank Allyn and Marty for giving me the opportunity to work on this edition with them, and everyone at Addison-Wesley who helped out with the revision: Carter Fenton, who brought me on board for this revision; Deirdre Lynch, who worked with me throughout the project's completion; and Joanne Ha, Jennifer Albanese, Susan Whalen, Jeff Holcomb, Jay Jenkins, Alexandra Waibel, Joyce Wells, Christina Gleason, and Sharon Tomasulo. Thanks as well to everyone at Techbooks, Inc. for their producing this edition: Peggy Hood, Shelley Creager and the entire production team.

~ Ellena Reda

To the Student

Welcome to *Introduction to Technical Mathematics*. No matter what your previous experiences and successes have been in mathematics, this is an opportunity to explore mathematics from a different perspective. We hope that this text will help you to recognize the importance of mathematics in many technical fields.

The most effective way to learn just about anything is to practice and reinforce the skill you hope to acquire. Think of some of the things that you do well or of some special talent that you have developed. To reach a certain level of achievement you had to practice. Very few of us have a natural talent that allows us to be successful the first time we try to learn a new task. Mathematics is no different. To be successful in a mathematics class you need to make a commitment to your course work right from the beginning. The algebra skills you develop in this course will serve as the foundation for your success in future mathematics courses.

How To Be a Successful Mathematics Student

Researchers have identified three variables affecting academic achievement. These are cognitive entry skills (how much math a person knows before entering the course), the quality of instruction (classroom atmosphere, teaching style, lab instruction, textbook content and format), and affective characteristics (academic self-concept, attitude, anxiety, and study habits). As a student, you have the most control over your affective characteristics. Your academic self-concept is a significant predictor of mathematics achievement.

No matter what your mathematics ability, it is always possible to improve your approach to your learning. You may not need to make sweeping changes to the way you study and learn, but by thinking about how you learn, and what motivates you, you will continue to grow as a learner.

There are steps that you can take to ensure that your experience in this mathematics course is a successful one.

1. **A positive attitude** is the most important quality that you can bring with you to any class you take. It will ultimately affect your success. Henry Ford is credited with saying "If you think you can do a thing, or think you can't do a thing, you're right." Put aside any negative feelings you might have about mathematics based on past experiences. This is a chance to start fresh. Decide that you will do well in the course. Have confidence in yourself and believe that you can be successful.

2. Be an **active learner.** Attend class regularly. This may seem obvious but this is particularly important when taking a mathematics class. Each topic builds on previous material. If you miss a class you will have a more difficult time following the lecture the next time you come to class. If you need to miss a class, take the time to contact your instructor or another student enrolled in the class to find out what you missed.

3. **Come to class prepared.** Bring your textbook, calculator, notebook, and pen or pencil to every class. Take notes and don't be afraid to ask questions when you don't understand something.

4. **Be on time** for each class and stay for the whole class. Arriving on time insures that you will not miss any announcements that the instructor will make at the beginning of class. Staying until the end of class is courteous and respectful. As the class wraps up, this is the perfect opportunity to ask any questions you might have about material presented that day. Assignments are often given out at the end of class.

5. **Learn your instructor's** name, office location, office hours, and how to get in touch with him or her.

6. **Set aside regular study time.** Even if you have only a few moments, go over your notes from the lecture after class. The repetition will help you remember the material. Remembering and understanding everything you heard during class can be difficult when you are not able to pause, and think about the material being presented. In a study on recall after listening to a lecture, students forgot more than 90% of the points from the lecture after just 2 weeks. **Schedule adequate time during the week at regular intervals to review material presented in class.** It is much more useful to study an hour a night each night than to try and cram in 6–7 hours of studying in one night. The mind needs time to absorb mathematics and one night of cramming cannot replace several nights of diligence.

7. **Read your textbook.** Many students believe that they cannot read and understand their math textbooks. However, great care has been taken in writing this textbook to present the material in an understandable way. Take some time to read through the chapter opener, section introductions, and each section before it is covered in class. If you have read the section, then the lecture can build on and clarify what you have read.

8. **Don't be afraid to get extra help.** If you struggle with a topic, do not get discouraged. Be proactive. Make sure that you take advantage of all the resources available to you. If you feel like you are getting lost in the course, be sure to ask your instructor for some help. If your school has a tutoring center make use of it.

9. **Learn how math fits into your education, life, and career goals.** You are taking this course for a reason. Determine why it is important for you to be successful in this course. Use these reasons to keep yourself motivated.

Contents

Signed Numbers

Many of us have heard meteorologists use the term "wind chill" to describe how the air temperature feels on a cold and blustery winter day. The effect of air temperature (in degrees Fahrenheit) and wind speed (in miles per hour) on how cold we feel is called the **wind chill factor.** A winter day with a strong wind can seem much colder than one with only a mild wind, though the air temperature may be exactly the same. Wind chill is a measure of how fast energy is lost from an area of exposed skin when the wind blows. So although the actual air temperature is not as low as the wind chill makes us feel, the risk of frostbite from low wind chill indices makes this a winter weather hazard.

The Wind Chill chart in Figure 1.1 shows how wind chill can make a fairly moderate winter day seem much colder. For example, a temperature of 30° Fahrenheit might not seem so bad, but combined with a 15 mile per hour (mph) wind, it can feel like it is only 19° Fahrenheit. A temperature of 10°F

Wind Chill Chart

Temperature (°F)

Calm	40	35	30	25	20	15	10	5	0	–5	–10	–15	–20	–25	–30	–35	–40	–45
5	36	31	25	19	13	7	1	–5	–11	–16	–22	–28	–34	–63	–46	–52	–63	–63
10	34	27	21	15	9	3	–4	–10	–16	–22	–28	–35	–41	–63	–53	–59	–72	–72
15	32	25	19	13	6	0	–7	–13	–19	–26	–32	–39	–45	–63	–58	–64	–77	–77
20	30	24	17	11	4	–2	–9	–15	–22	–29	–35	–42	–48	–63	–61	–68	–81	–81
25	29	23	16	9	3	–4	–11	–17	–24	–31	–37	–44	–51	–63	–64	–71	–84	–84
30	28	22	15	8	1	–5	–12	–19	–26	–33	–39	–46	–53	–63	–67	–73	–87	–87
35	28	21	14	7	0	–7	–14	–21	–27	–34	–41	–48	–55	–63	–69	–76	–89	–89
40	27	20	13	6	–1	–8	–15	–22	–29	–36	–43	–50	–57	–63	–71	–78	–91	–91
45	26	19	12	5	–2	–9	–16	–23	–30	–37	–44	–51	–58	–63	–72	–79	–93	–93
50	26	19	12	4	–3	–10	–17	–24	–31	–38	–45	–52	–60	–63	–74	–95	–95	–95
55	25	18	11	4	–3	–11	–18	–25	–32	–39	–46	–54	–61	–63	–75	–97	–97	–97
60	25	17	10	3	–4	–11	–19	–26	–33	–40	–46	–55	–62	–76	–76	–98	–98	–98

Wind (mph)

Frostbite occurs in 15 minutes or less

Figure 1.1

might seem chilly enough but when you add winds of 20 miles per hour, it can feel like −9°F (or 9 degrees below zero).

1.1   Signed Numbers

In the real world it is relatively easy to find examples of both positive and negative numbers. Temperatures, changes in the stock market and in stock prices, checkbook balances, and elevations below sea level are just a few of the more common examples often cited. In electronics, voltage dropping below a given reference level is measured in negative volts. In this chapter we will study signed numbers and the principles of arithmetic that apply to them.

Section Objectives
- Learn basic language for sets of numbers
- Compare integers
- Find opposites of integers
- Find the absolute value of a number

Integers are the set of whole numbers and their opposites. The number line can be used to give a visual representation of the set of integers. Refer to the number line (Figure 1.2) below as you read the definitions that follow.

Section Definitions
- Whole numbers greater than zero are called **positive integers** or natural numbers. These numbers are located to the right of zero on the number line.

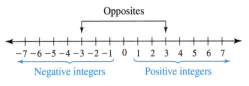

Figure 1.2

- Whole numbers less than zero are called **negative integers.** These numbers are located to the left of zero on the number line.
- The integer **zero** is neutral (not positive or negative).
- Two integers are considered **opposites** if they are the same distance away from zero, but are located on opposite sides of the number line. Using the number line in Figure 1.2 we notice that $+3$ and -3 are opposites.

A negative number is a number less than zero. The symbols for addition ($+$) and subtraction ($-$) are also used to indicate positive and negative numbers, respectively. For example, $+5$ (positive 5) is a positive number and -5 (negative 5) is a negative number. Because positive numbers correspond directly to the numbers we have been using (we made no attempt to associate a sign with them, using only $+$ and $-$ to designate addition and subtraction), the plus sign can be omitted before a positive number if there is no danger of confusion. However, we shall never omit the negative sign before a negative number.

Another name given to the natural numbers is **positive integers.** *The negative counterparts of the positive integers are called* **negative integers.** *If we combine the positive integers, zero (which is neither positive nor negative), and the negative integers, we get the set of all* **integers.** In the following sections, we will sometimes refer to the integers.

---

natural numbers (or positive integers):
$1, 2, 3, \ldots$
negative integers:
$-1, -2, -3, \ldots$
integers: $\ldots -3, -2, -1, 0, 1, 2, 3, \ldots$

Example 1

The numbers 7 (which equals $+7$) and $+7$ are positive integers. The number -7 is a negative integer. The number 0 is an integer, but it is neither positive nor negative. The number $\frac{2}{3}$ is not an integer, because it is not a natural number or the negative of a natural number.

Example 2

We can use integers to describe each of the following situations:

Situation	Integer representation
10 degrees Fahrenheit above zero	$+10°$
a loss of 16 dollars	$-\$16$
a gain of 15 yards in a football game	$+15$ yd
250 feet below sea level	-250 ft

In this discussion, we see that *the plus and the minus signs are used in two senses. One is to indicate the operation of addition or subtraction; the other is to designate a positive or negative number.* Consider the following example.

Example 3

The expression $3 + 6$ means "add the number 6 to the number 3." Here the plus sign indicates the operation of addition, and the numbers are **unsigned numbers,** *which are positive numbers.*

The expression $(+3) + (+6)$ means "add the number +6 to the number +3." Here the middle plus sign indicates addition, whereas the other plus signs denote signed numbers. Of course, the result would be the same as $3 + 6$. See the diagram at left.

The expression $3 - 6$ means "subtract the number 6 from the number 3." Here the minus sign indicates subtraction, and the numbers are unsigned.

The expression $(+3) - (+6)$ means "subtract the number +6 from the number +3." Note that the positive signs designate only signed numbers.

The expression $(+3) - (-6)$ means "subtract the number −6 from the number +3." Here the first minus sign indicates subtraction, whereas the second designates the negative number. See the diagram at left.

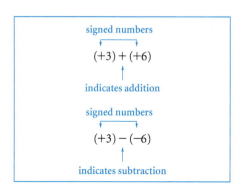

All positive and negative numbers can be located on a number line as seen in Figure 1.3. As we see, every positive number has a corresponding opposite negative number.

When numbers are marked positive or negative, they are called **signed numbers** (or *directed numbers*) to indicate their opposite qualities, as shown on the scale. Because positive numbers are directed to the right of the origin, it follows that numbers increase from left to right. Therefore: *Any number on the scale is smaller than any number located farther to the right. The symbols $>$ and $<$ mean* **"greater than"** *and* **"less than,"** *respectively.* Their use is shown in the following example.

Figure 1.3

Example 4

The expression $2 > 1$ means "the number 2 is greater than the number 1." It also means that on the scale in Figure 1.3 we would find 2 to the right of 1. See the diagram at left.

The expression $0 < 4$ means "0 is less that 4." This means that 0 is to the left of 4. See the diagram at left.

The expression $-3 < 0$ means "−3 is less than 0." The number −3 lies to the left of 0 on the scale.

The expression $-3 > -5$ means "−3 is greater than −5." The number −3 is to the right of the number −5 on the scale.

In general, larger numbers are farther to the right whereas smaller numbers are farther to the left on the standard number line.

Although the positive direction is conveniently taken to the right of the origin on the number scale, we can select any direction as the positive direction so long as we take the opposite direction to be negative. Consider the illustrations in Example 5.

Example
5

Units, such as feet, will be spelled out in this chapter. Unit symbols will be used when units of measure are introduced in Chapter 2, and in all subsequent chapters.

Sea level is used as the reference point by which land elevation is measured. The land elevation at sea level is 0 feet. Elevations above sea level are measured by positive numbers whereas elevations below sea level are measured by negative numbers. Mt. McKinley has the highest land elevation in North America at 20,320 feet (or +20,320). The lowest land elevation in North America is 282 feet below sea level (or −282) and is found in Death Valley. See Figure 1.4a.

If a temperature above zero (an arbitrarily chosen reference level is called positive, a temperature below zero would be called negative. See Figure 1.4b.

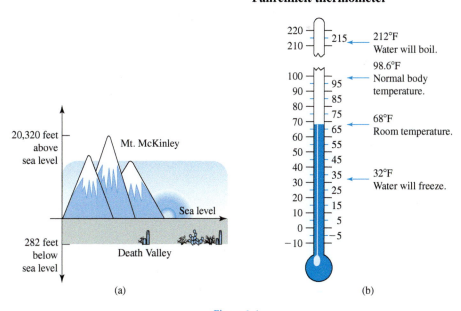

Figure 1.4

The value of a number without its sign is called its **absolute value.** That is, if we disregard the signs of +5 and −5, the value would be the same. This is equivalent

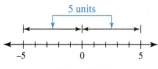

Figure 1.5

to saying that the distances from the origin to the points $+5$ and -5 on the number scale are equal. (See Figure 1.5). The absolute value of a number is indicated by the symbol $|\ |$. The number $+5$ does not equal the number -5, but $|+5| = |-5|$. Absolute values can be used in performing arithmetic operations with signed numbers. Consider the following example.

Example

6

The absolute value of $+8$ is 8. The absolute value of -2 is 2. The absolute value of $-\frac{3}{2}$ is $\frac{3}{2}$. We can write these equalities as $|+8| = 8, |-2| = 2$, and $\left|-\frac{3}{2}\right| = \frac{3}{2}$. Also,

$$|-7.3| = 7.3 \qquad\qquad |9.2| = 9.2$$

$$|-22.6| = 22.6 \qquad\qquad |524| = 524$$

1.1 Exercises

In Exercises 1 through 8, locate the approximate positions of the given numbers on a number scale such as that in Figure 1.2 or Figure 1.3.

1. 5; -5

2. -0.5; 0.5

3. -2.3; $+3.2$

4. $+1.43$; -3.14

5. $+\sqrt{3}$; $-\sqrt{2}$;

6. $-\frac{17}{6}$; $+\frac{3}{5}$

7. $+\frac{13}{22}$; $-\frac{22}{13}$

8. $-\frac{13}{4}$; $+\frac{28}{5}$

In Exercises 9 through 12, determine which of the given numbers is the largest.

9. 2; -9; 0

10. -3; -6; 0

11. -3; -1; -5

12. -7; -8; -9

In Exercises 13 through 24, insert the proper sign ($>$ or $<$ or $=$) between the given numbers.

13. 6   2

14. 8 -3

15. 0   4

16. -3 0

17. -3 -7

18. -9 -8

19. -7 -5

20. -1 -5

21. $\sqrt{5}$ 2.2

22. $\sqrt{10}$ -3.2

23. $|6|$ $|-6|$

24. $|-3|$ $-|-3|$

In Exercises 25 through 28, find the absolute value of each of the given numbers.

25. $+6$; -6

26. -5; 5

27. $-\frac{6}{7}$; $\frac{8}{5}$

28. 2.4; -0.1

In Exercises 29 through 41, certain applications of signed numbers are indicated.

29. In writing a computer program for a bank, the programmer represents a $50 deposit as $+50$. What signed number would represent a $30 withdrawal?

30. In an electronic device, the voltage at a certain point is 0.2 volt (V) above that of the reference and is designated as +0.2 volt (V). How is the voltage 0.5 volt (V) below the reference designated?

31. If a surveyor represents the altitude of a point 12 meters above sea level by +12, what signed number represents an altitude that is 8 meters below sea level?

32. The amounts (in millions of dollars) of indebtedness of two corporations are represented by −3 and −6. Which corporation has the smaller debt?

33. If the number of years between the current year and the year 2000 is represented by a positive number, what number represents the number of years between the current year and 1995?

34. The image of an object formed by a lens can be formed either to the right or to the left of the lens, depending on the position of the object. If an image is formed 10 centimeters to the left and its distance is designated as −10 centimeters, what would be the designation of an image formed 4.5 centimeters to the right?

35. Use the proper sign (> or <) to show which temperature, −30°C or −5°C, is higher.

36. Use the proper sign (> or <) to show which elevation is higher: −80 meters or −60 meters.

37. Use the proper sign (> or <) to show which voltage reading is higher: −2 volts or −5 volts.

38. A quality control specialist analyzes production data by calculating correlation coefficients. Use the proper sign (> or <) to show which coefficient is the larger number: −0.94 or −0.58.

39. The Navy was comparing the depths of two submarines. One was 890 feet below sea level, and the other was 1425 feet below sea level.
  a. Represent each depth as a negative value.
  b. Which submarine was at the greater depth?

40. Use the Wind Chill Chart (Figure 1.1) presented at the beginning of the chapter to answer the following questions.
  a. Some combinations of temperatures and wind speeds give the same wind chill factor as other combinations. For example, (20°F and 10 miles per hour) and (25°F and 25 miles per hour) both give a wind chill factor of 9°F. Find three combinations that result in a wind chill of −16°F.
  b. When does it seem colder (have a lower wind chill factor), at 20°F with a wind of 20 miles per hour or at 0°F with the wind at 5 miles per hour?
  c. Approximately how much does the wind chill factor drop with an increase of wind speed from 5 to 10 miles per hour for each of the temperatures given?
  d. One cold, blustery day a −22°F wind chill factor is reported on the news. Using the table, give four approximate temperature/wind speed combinations that would produce this wind chill factor.

41. Alternating current (ac) voltages change rapidly between positive and negative values. If a voltage of 100 volts decreases to −200 volts
  a. Which of these voltages is greater?
  b. Which of these voltages has the greater absolute value?

1.2 Addition and Subtraction of Signed Numbers

Section Objectives
• Add integers with like signs
• Add integers with unlike signs
• Subtract integers with like signs
• Subtract integers with unlike signs

Section 1.1 introduced signed numbers and ways to express them. This section presents the method for performing the operations of addition and subtraction with signed numbers. We want our procedures for the addition and subtraction of signed numbers to conform to the way things generally work in the real world. We have already established that positive numbers are equal to their equivalent unsigned numbers.

Many of us already have some familiarity with how to add positive and negative numbers.

Example 1

A good example of adding signed numbers comes from the game of football. Think of moving up field as a gain in yardage and down field as a loss in yardage. Consider the following examples:

a. A 7 yard loss and a 3 yard gain would result in a net loss of 4 yards.

b. A 6 yard gain and a 3 yard gain results in a net gain of 9 yards.

c. A 4 yard loss and a 6 yard loss results in a net loss of 10 yards.

d. A 1 yard gain and an 8 yard loss results in a net loss of 7 yards.

Each of the statements in Example 1 can be written using signed numbers. This can be seen in Example 2. A gain will be indicated by a positive number and a loss will be indicated by a negative number.

Example 2

a. $(-7) + 3 = -4$

b. $6 + 3 = 9$

c. $(-4) + (-6) = -10$

d. $1 + (-8) = -7$

Procedure for Adding Signed Numbers

- When adding two numbers that have **like** (the same) **signs** (both positive or both negative), add the two numbers and keep their common sign.
- When adding two numbers that have **unlike** (or different) **signs** (one positive and one negative), subtract the two numbers and then keep the sign of the number with the larger absolute value.

Example
3

a. $2 + 3 = 5$ the sum of two positive numbers is always positive

b. $(-2) + (-3) = -5$ the sum of two negative numbers is always negative

c. $7 + (-4) = 3$ the $+7$ has the larger absolute value so the result is positive

d. $(-7) + 4 = -3$ the (-7) has the larger absolute value so the result is negative

All calculator screenshots shown in this text are for the TI-83/84 Plus graphing calculator. Calculator keystrokes will appear throughout the book as needed. Sample screenshots will accompany various examples. It is important to note that screenshots and instructions for other calculators may differ.

 *Using Technology*

Use the TI-83/84 plus calculator to check your results for parts (b)–(d) in Example 3. Compare your results with those shown in Figure 1.6.

Here's How

To enter a negative number into the calculator press the $(-)$ key before entering the number. This key should be located on the bottom row of the calculator. **Do not use this key to indicate subtraction.** The key used for the operation of subtraction is located in the right-hand column of your calculator. The ENTER key acts like an equal sign. Press ENTER to calculate a numerical result.

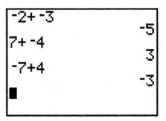

Figure 1.6

Subtraction of signed numbers takes a little more concentration and work. The method presented here is a two-step process that involves *subtraction by adding a number's opposite.* The advantage to using this method is that it allows us to use the procedure for the addition of signed numbers.

Procedure for Subtracting Signed Numbers

To subtract two signed numbers

- Change the sign of the number being subtracted (the second number) to its opposite sign
- Change the subtraction sign to addition
- Now follow the procedure for the addition of signed numbers

Example
4

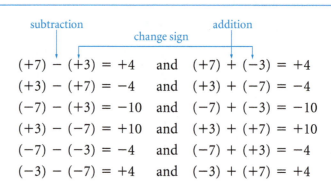

$$(+7) - (+3) = +4 \quad \text{and} \quad (+7) + (-3) = +4$$
$$(+3) - (+7) = -4 \quad \text{and} \quad (+3) + (-7) = -4$$
$$(-7) - (+3) = -10 \quad \text{and} \quad (-7) + (-3) = -10$$
$$(+3) - (-7) = +10 \quad \text{and} \quad (+3) + (+7) = +10$$
$$(-7) - (-3) = -4 \quad \text{and} \quad (-7) + (+3) = -4$$
$$(-3) - (-7) = +4 \quad \text{and} \quad (-3) + (+7) = +4$$

Several important conclusions can be drawn from Example 4. The first is that:

To subtract one signed number from another, we change the sign of the subtrahend (the number after the subtraction symbol) and then add it to the minuend (the number before the subtraction symbol), using the rules for the addition of signed numbers.

Also, from the first two illustrations, we see that:

Subtracting a positive number is equivalent to adding a negative number of the same absolute value.
Subtracting a negative number is equivalent to adding a positive number of the same absolute value.

This last point can be thought of as removing (subtracting) a loss (a negative number), which results in a gain.

Example
5

$$(+8) - (+5) = (+8) + (-5) = +3$$
$$(+5) - (+8) = (+5) + (-8) = -3$$
$$(+8) - (-5) = (+8) + (+5) = +13$$
$$(+5) - (-8) = (+5) + (+8) = +13$$

Thus we see that $-(+5)$ and $+(-5)$ are the same and $-(-5)$ and $+(+5)$ are the same.

Now Try It!

Perform the indicated operations:

1. $4 + (-7)$
2. $-9 + 16$
3. $2 - (-5)$
4. $10 - 14$
5. $(-8) + (-17)$
6. $(-5) - 7$

When you become more familiar with these operations, you will be able to do them without actually writing in the step where a negative number is changed to a positive number. This can be shown immediately with signed numbers. When this is done we must remember that ***adding a negative number is equivalent to subtracting an unsigned number and subtracting a negative number is equivalent to adding an unsigned number.***

Example
6

adding a negative number
$$(+7) + (-3) = 7 - 3 = 4$$

subtracting a negative number
$$(+7) - (-3) = 7 + 3 = 10$$
$$(-7) - (-3) = -7 + 3 = -4$$
$$(-7) + (-3) = -7 - 3 = -10$$
$$(-4) + (-2) - (-7) = -4 - 2 + 7 = -6 + 7 = 1$$
$$(-5) - (-2) - (+9) = -5 + 2 - 9 = -3 - 9 = -12$$

The following examples illustrate the applied use of addition and subtraction of signed numbers.

Example
7

An oil well that is 30 m below ground level is drilled 40 m deeper. The total depth (in meters) is represented as

$$(-30) + (-40) = -70.$$

Example
8

The temperature of an engine coolant is −8°C. If the temperature increases by 12°C, its new value is represented (in degrees Celsius) by

$$(-8) + (+12) = +4.$$

1.2 Exercises

In Exercises 1 through 32, perform the indicated operations on the given signed numbers.

1. $2 + 9$

2. $3 + 8$

3. $(-6) + (-9)$

4. $(-6) + (-5)$

5. $(-3) + 6$

6. $(-1) + 8$

7. $2 + (-10)$

8. $3 + (-11)$

9. $12 - 5$

10. $14 - 6$

11. $7 - 10$

12. $2 - 15$

13. $(-6) - 4$

14. $(-3) - 8$

15. $7 - (-9)$

16. $14 - (-2)$

17. $(-6) - (-7)$

18. $(-9) - (-5)$

19. $1 + (-5) + (-2)$ **20.** $6 + 5 + (-3)$

21. $2 + (-8) - 2$ **22.** $(-4) - 8 + (-9)$

23. $5 - (-3) - 7$ **24.** $(-7) - (-1) - 6$

25. $(-7) - (-15) + (-2)$

26. $(-6) + (-13) - (-11)$

27. $(-9) - (-5) + (-8) - 5$

28. $(-17) - (-5) + (-4) - 6$

29. $3 - (-7) - 9 + (-3)$

30. $6 + (-11) - (-5) - 8$

31. $(-6) - 9 - (-12) + (-4) - (-1)$

32. $(-5) + 6 - 3 - (-14) + (-3)$

In Exercises 33 through 37, set up the given problems in terms of the addition or subtraction of signed numbers and then perform the operations. (The operation of addition or subtraction that should be used in setting up *the problem is indicated in each case.)*

33. The temperature of a fluid is $-10°C$. (a) If the temperature increases (use addition) by $5°C$, what is the resulting temperature? (b) If the temperature decreases (use subtraction) by $5°C$, what is the resulting temperature?

34. A fluid boils at $120°C$ and freezes at $-30°C$. How much higher is the boiling point than the freezing point? (Use subtraction.)

35. The current in an electric circuit changes from 12 amperes to -5 amperes. What is the difference between a current of 12 amperes and a current of -5 amperes? (Use subtraction.)

36. In testing an aircraft altimeter, technicians use air pressure to simulate an altitude of 6000 feet, and an error of -150 feet is noted. What is the altimeter reading at this simulated altitude? (Use addition.)

37. In four successive years, the population of a certain city changes as follows: increases by 5000; increases by 2000; decreases by 3000; increases by 1000. Find the sum (use addition) of these changes.

In Exercises 38 through 45, use a scientific or graphing calculator to perform the indicated operations on the given signed numbers.

38. $(-2) + (-5)$ **39.** $(-8) + (3)$

40. $(+7) - (-2)$ **41.** $(+3) - (-8)$

42. $(-4) - (-3) + (-2)$

43. $(+8) - (-5) + (-1)$

44. $(-2) - (-5) - (-4)$

45. $(+12) - (-20) - (-15) + (-3)$

Now Try It! Answers

1. -3 **2.** 7 **3.** 7 **4.** -4 **5.** -25 **6.** -12

1.3 **Multiplication and Division of Signed Numbers**

Section Objectives
• Multiplication with two integers
• Multiplication with zero
• Division with two integers
• Division with zero

In developing the rules for the multiplication of signed numbers, we want to preserve the definitions already presented. We know that multiplication is a process of repeated addition. We now extend this same concept to signed numbers so that the multiplication of signed numbers is the process of repeated addition, where equal numbers are added together a specified number of times.

The numbers to be multiplied are called **factors** *and the result is the* **product.** The process of multiplication has certain basic properties:

> **Properties of Multiplication**
>
> **Commutative Property**
> $a \times b = b \times a$ for all real numbers
>
> **Associative Property**
> $a \times (b \times c) = (a \times b) \times c$ for all real numbers

1. The **commutative law** for multiplication tells us that *the order in which we multiply two numbers does not matter.*

2. The **associative law** deals with the multiplication of more than two factors. It tells us that *the grouping of the numbers being multiplied (the factors) does not matter.*

With these ideas in mind, consider the following example.

Example 1

Since $3 \times 4 = 12$ and since $3 = +3, 4 = +4$, and $12 = +12$, we have $(+3) \times (+4) = +12$.

Since $3 \times (-4)$ means $(-4) + (-4) + (-4)$, and $(-4) + (-4) + (-4) = -12$, and since $3 = +3$, we have $(+3) \times (-4) = -12$.

Also, since multiplication is commutative, we have $(-4) \times (+3) = (+3) \times (-4) = -12$.

We see from Example 1 that the product of two positive numbers is positive. Also, the product of a positive number and a negative number is negative. From this, we get the following general principle:

Changing the sign of one factor changes the sign of the product.

Applying this principle, we can determine the product of two negative numbers.

Example 2

From Example 1, we have seen that

$$(-4) \times (+3) = -12.$$

Applying the principle that changing the sign of one factor changes the sign of the product, we change the factor $+3$ to -3 and the result changes from -12 to $+12$. This leads to

$$(-4) \times (-3) = +12.$$

In general,

The product of two negative numbers is positive.

This last illustration can be thought of as subtracting -3 four times or $-(-3) - (-3) - (-3) - (-3) = +12$. We can also think of this as negating four \$3 debts, which would result in a gain of \$12. Note that this example is consistent with the fact that subtracting a negative number is equivalent to adding a positive number of the same absolute value.

From the preceding discussion, we can state the following results:

Procedure for Multiplying Signed Numbers

- When multiplying two numbers that have the same sign (both positive or both negative) their product will always be positive.

- When multiplying two numbers that have different signs the answer is always negative.

If there are three or more factors, we can use the associative and commutative laws and multiply two factors at a time, in any order, observing each time the sign of the product. We can then conclude that:

1. The product is positive if all factors are positive or there is an even number of negative factors.

2. The product is negative if there is an odd number of negative factors.

If any factor is zero, it follows that the product is zero because zero times any number is zero.

Example 3

like signs

$(+4) \times (+7) = +28$

$(-4) \times (-7) = +28$

unlike signs

$(+4) \times (-7) = -28$

$(-4) \times (+7) = -28$

$(+2) \times (+3) \times (+5) = +30$ no negative factor

$(-2) \times (+3) \times (+5) = -30$ one negative factor

$(-2) \times (-3) \times (+5) = +30$ two negative factors

$(-2) \times (-3) \times (-5) = -30$ three negative factors

$(-1) \times (-2) \times (-3) \times (-5) = +30$ four negative factors

$(-1) \times (0) \times (-3) \times (-5) = 0$ zero is a factor

In any product where the factors are signed numbers, the multiplication of these signed numbers can be indicated by writing them, in parentheses, adjacent to each other. This eliminates the need for using $\times$ or $\cdot$ as signs of multiplication.

Example 4

We can write $(-3) \times (+4)$ as $(-3)(+4)$ because we realize that this form indicates the multiplication of -3 and $+4$.
Therefore $(-2)(-3)(+4) = +24$, since

$$(-2)(-3)(+4) = (-2) \times (-3) \times (+4).$$

Example 5

While flying an instrument approach to an airport, a pilot at 4000 ft descends at the rate of 500 ft/min for 3.0 min and then climbs at the rate of 800 ft/min for 5.0 min.

To determine the resulting altitude, we note that the altitude decreases in a descent and increases in a climb. We therefore represent lost altitude with a negative number and gained altitude with a positive number. Combining the changes, we get

$$\underset{\text{dist.} = \text{rate} \times \text{time}}{(+4000) + \overset{\text{start} \qquad \text{descend} \qquad \text{climb}}{(-500)(+3.0) + (+800)(+5.0)}}$$

$$= 4000 - 1500 + 4000$$
$$= 6500 \text{ ft.}$$

The plane will be at an altitude of 6500 ft.

After having discussed addition, subtraction, and multiplication, we now consider the division of positive and negative numbers. We want to continue to make our definitions consistent with those already given. We have seen that division is the process of determining how many times one number is contained in another. Here, *the number being divided is called the* **dividend,** *the number being divided into the dividend is the* **divisor,** *and the result is the* **quotient.**

Division is the inverse process of multiplication, because the product of the quotient and the divisor equals the dividend. We can now establish the principles used in the division of signed numbers from those we have already established for the multiplication of signed numbers. The basis for these principles is seen in Example 6.

Now Try It! *1*

Multiply the following:

1. $4(-7)$
2. $-9(-6)$
3. $-2(5)$
4. $10(14)$
5. $0(-17)$

Example 6

$$(+15) \div (+5) = +3 \quad \text{since} \quad (+5) \times (+3) = +15$$
$$(+15) \div (-5) = -3 \quad \text{since} \quad (-5) \times (-3) = +15$$
$$(-15) \div (+5) = -3 \quad \text{since} \quad (+5) \times (-3) = -15$$
$$(-15) \div (-5) = +3 \quad \text{since} \quad (-5) \times (+3) = -15$$

We see from Example 6 that:

Procedure for Dividing Signed Numbers

- When dividing two numbers that have the same sign (both positive or both negative) the quotient will always be positive.

- When dividing two numbers that have different signs the quotient will always be negative.

We should note that the division of two signed numbers follows the same principles as the multiplication of two signed numbers.

The division of one signed number by another is often expressed in the form of a fraction, as illustrated in Example 7.

Example 7

$(+21) \div (-3)$ in fractional form is $\dfrac{+21}{-3}$ so that

unlike signs $\dfrac{+21}{-3} = -7$ unlike signs $\dfrac{-21}{+3} = -7$ like signs $\dfrac{-21}{-3} = +7$.

Now Try It! 2

Divide the following:

1. $42 \div (-7)$
2. $-39 \div (-3)$
3. $-72 \div 9$
4. $\dfrac{108}{4}$
5. $\dfrac{-119}{-7}$
6. $\dfrac{0}{-17}$

If either the numerator or the denominator (or both) of a fraction is the product of two or more signed numbers, this product can be found and then the value of the fraction determined by the appropriate division. However, because the same basic principles apply for the multiplication and division of signed numbers, we need determine only the number of negative factors of the numerator and denominator combined. *If the total number of negative factors is odd, the quotient is negative. Otherwise, it is positive.* The reduction of the fraction to lowest terms then follows as with any arithmetic fraction. *Also, factors common to both the numerator and the denominator can be divided out.*

Example
8

two negative factors

$$\frac{(-2)(+12)}{-4} = +\frac{(2)(12)}{4} = \frac{(1)(12)}{2} = \frac{6}{1} = 6$$

The sign of the result is positive because the original fraction has two negative factors (-2 and -4) in the numerator and denominator. The resulting fraction is then simplified by dividing out two factors of 2 from both the numerator and the denominator. Because the resulting denominator is 1, it does not have to be written.

In a similar manner, we have

three negative factors

$$\frac{(-3)(+6)(-5)}{(-1)(+10)} = -\frac{(3)(6)(5)}{(1)(10)} = -\frac{(3)(6)(1)}{(1)(2)} = -\frac{(3)(3)(1)}{(1)(1)} = -9$$

The result is negative because the original fraction had three negative factors. The fraction was then simplified by dividing out a factor of 5 and 2.

Example
9

Six partners form a company that suffers a total loss of $7200. Express the loss for each partner as a single value.

It is common to express gains as positive amounts and losses as negative amounts so that the loss for each partner is

$$\frac{-\$7200}{6} = -\$1200.$$

Zero is the only number that cannot be used as a divisor. Because division is the inverse of multiplication, we can say that $0 \div$ (any number) $= 0$ because (any number) $\times 0 = 0$. This means that zero divided by any number (except zero) is equal to zero. However, if zero is a divisor as in (any number) $\div 0$, we cannot determine the quotient that will equal the *number* when multiplied by zero. (Any quotient that may be tried gives a product of zero when multiplied by zero and does not give the *number*.) In the case of $0 \div 0$, any positive or negative number can be a quotient, because $0 \times$ (*any number*) $= 0$. This result is said to be indeterminate.

Because *division by zero* yields either no solution or an indeterminate solution:

Division by zero is undefined at all times.

We must remember, however, that all other operations with zero are perfectly valid.

Example
10

1. $\dfrac{0}{4} = 0$, since $0 \times 4 = 0$.

2. $\dfrac{-5}{0}$ is undefined since (no number) $\times 0 = -5$.

3. $3 + 0 = 3$ and $0 - (-3) = 0 + 3 = 3$.

▶ Note that **division by zero is undefined.**

When we introduced subtraction where the result was not positive, we in turn introduced negative numbers. This was necessary because the results could not be expressed as positive numbers. When we divide one integer by another, it often happens that the results are not integers. *The name* **rational number** *is given to any number that can be expressed as the division of one integer by another (not zero).* For example, $\frac{2}{3}$, $\frac{8}{4}$, and $\frac{1097}{431}$ are rational numbers. We shall later learn of numbers that cannot be expressed in this way (although we have already met a few: $\sqrt{2}$, for example).

1.3   Exercises

In Exercises 1 through 14, find the product. Pay careful attention to the sign of your answer.

1. $(7)(-9)$ **2.** $(-4)(11)$

3. $(-7)(12)$ **4.** $15(-4)$

5. $(-2)(-15)$ **6.** $(-9)(-8)$

7. $3(0)$ **8.** $-8(7)$

9. $(4)(-2)(7)$ **10.** $(-3)(6)(10)(-1)$

11. $(-1)(-3)(-5)(-2)$ **12.** $8(0)(-4)(-1)$

13. $3(-1)(4)(-7)(-2)$ **14.** $(3)(-5)(-2)(0)(-9)$

In Exercises 15 through 26, find the quotient. Pay careful attention to the sign of your answer.

15. $32 \div (-4)$ **16.** $27 \div (-3)$

17. $(-48) \div 16$ **18.** $(-35) \div (-5)$

19. $\dfrac{-22}{2}$ **20.** $\dfrac{20}{-4}$

21. $\dfrac{-60}{-4}$ **22.** $\dfrac{-52}{-13}$

23. $\dfrac{0}{-6}$ **24.** $\dfrac{-16}{0}$

25. $\dfrac{-36}{0}$ **26.** $\dfrac{0}{13}$

In Exercises 27 through 34, perform the indicated operation. Pay careful attention to the sign of your answer.

27. $\dfrac{8(-4)}{2}$ **28.** $\dfrac{(-6)(3)}{-9}$

29. $\dfrac{(-5)(15)}{25(-1)}$ **30.** $\dfrac{(-36)(12)}{(-4)(-27)}$

31. $\dfrac{(-3)(-24)}{(-2)(-12)}$ **32.** $\dfrac{(-6)(4)}{3(8)}$

33. $\dfrac{10(-6)(-14)}{8(-21)}$ **34.** $\dfrac{(-22)(8)(18)}{6(-44)}$

In Exercises 35 through 41 read each statement carefully and determine whether the answer would be positive, negative, or zero.

35. Multiplying three negative integers.

36. Subtracting a negative integer from a positive integer.

37. Dividing a positive integer by a negative integer.

38. Multiplying a positive integer by a positive integer.

39. Dividing a positive integer by a negative integer, and then multiplying the result by a negative integer.

40. Adding two negative integers and then dividing the result by a negative integer.

41. Dividing a negative integer by a negative integer, subtracting that result from a positive integer, and finally multiplying your result by zero.

In Exercises 42 through 46 set up the given problems as either a multiplication or division problem using signed numbers. Then solve each problem.

42. Metal expands when it is heated and contracts when cooled. The length of a particular metal rod changes by 2 millimeter for each 1°C change in temperature. What is the total change in the length of the rod as the temperature changes from 0°C to −25°C?

43. An integrated circuit voltage of 0.6 volt decreases by 0.3 volts per second. How long does it take to reach a voltage of −0.9 volts?

44. Cisco (CSCO) is the leading supplier of networking equipment and network management for the Internet. Suppose the price of a share of this stock falls $3.00 each day for four consecutive days. Find the net change in the price of the stock during this period.

45. The temperature of a coolant used in a nuclear reactor is dropping by 2°C each day. At this rate, how long will it take for the temperature to drop by 10°C?

46. The voltage in a certain electric circuit is decreasing at the rate of 2 millivolts per second. At a given time the voltage is 220 millivolts. What was the voltage 10 seconds earlier?

In Exercises 47 through 58 use a scientific or graphing calculator to evaluate the given expressions.

47. $(-3)(-6)$ **48.** $14482 \div (-26)$

49. $(-7)(8)$ **50.** $(-5)(-4)$

51. $(-63) \div (-3)$ **52.** $4(-2)$

53. $12 \div 0$ **54.** $(-17)(-23)$

55. $\dfrac{-108}{12}$ **56.** $\dfrac{26(-12)}{13(-3)}$

57. $\dfrac{2(3)(4)}{(-2)(-6)}$ **58.** $\dfrac{(-8)(-2)(0)}{(-7)(3)}$

Answers

Now Try It! 1
1. −28   **2.** 54   **3.** −10   **4.** 140   **5.** 0

Now Try It! 2
1. −6   **2.** 13   **3.** −8   **4.** 27   **5.** 17   **6.** 0

1.4 **Powers and Roots**

Section Objectives
• Identify the base and exponent of an expression
• Find the root of a number

An **exponent** provides us with a simpler way to write an expression that includes repeated multiplication.

In many situations we encounter a number that is to be multiplied by itself several times. Rather than writing the number over and over repeatedly, we can use a notation in which *we write the number once and write the number of times it is a factor as an* **exponent.**

Repeated Multiplication	*Exponential Form*
$7 \times 7 \times 7 \times 7 \times 7$	7^5
$(2x)(2x)(2x)(2x)$	$(2x)^4$

The notation a^n tells us we need to multiply a times itself n times. This means that $\underbrace{a^n = a \times a \times a \times \cdots \times a}_{n \text{ factors}}$ where n **is called the exponent** and

x **is called the base.**

Example
1

2 factors of 3
$3 \times 3 = 3^2$
3 factors of 7
$7 \times 7 \times 7 = 7^3$

Instead of writing 3×3, we write 3^2. Here the 2 is the exponent and the 3 is the base. The expression is read as "3 to the second power" or "3 squared."

Rather than writing $7 \times 7 \times 7$, we write 7^3. Here the 3 is the exponent and the 7 is the base. The expression is read as "7 to the third power" or "7 cubed."

The product $5 \times 5 \times 5 \times 5 \times 5 \times 5$ is written as 5^6. Here 6 is the exponent and 5 is the base. The expression is read as "the sixth power of 5" or "5 to the sixth power."

If a number is raised to the first power—that is, if it appears only once in a product—the exponent 1 is not usually written. Also, if we are given a number written in terms of an exponent, we can write the number in ordinary notation by performing the indicated multiplication. Consider the illustrations in Example 2.

Example
2

$5^1 = 5$ (normally the 1 would not appear)

$4^5 = 4 \times 4 \times 4 \times 4 \times 4 = 1024$

$(2^3)(3^2) = (2 \times 2 \times 2)(3 \times 3) = (8) \times (9) = 72$

Example 3

It is important that you be able to correctly identify the **base** and the **exponent.**

In the expression $(3)^5$ we identify the base as 3 and the exponent as 5.

In the expression $(-4)^2$ we identify the base as -4 and the exponent as 2.

In the expression -4^2 we identify the base as 4 and the exponent as 2.

Caution
$(-4)^2$ is not the same as -4^2.

We must note the use of parentheses in raising a signed number to a power. The meaning of $(-4)^2$ and -4^2 is vastly different. In $(-4)^2$ the base of -4 is being raised to the 2nd power. Because parentheses are not placed around the -4 in the expression -4^2, only the 4 is being raised to the 2nd power. We can think of -4^2 as being the same as $-(4^2)$.

 *Using Technology*

Let's look at how the calculator handles $(-4)^2$ and -4^2. You will notice that you get two different results depending on whether or not parentheses are used.

Here's How

- To raise a number to the second power enter the number and press the x^2 key located in the left-hand column. Then press ENTER. See Figure 1.7.

- To raise a number to any other power enter the number, press the $\wedge$ key (called the carat key), followed by the appropriate power. Then press ENTER. See Figure 1.8.

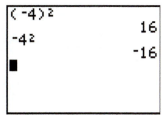

Figure 1.7

Figure 1.8

Now that we have discussed the meaning of a power of a number, we consider the concept of the roots of a number. Essentially, finding the root of a number is the reverse of raising a number to a power. *The* **square root** *of a given number is that number which when squared equals the given number.* We place the **radical sign** $\sqrt{}$ over the given number to indicate its square root.

Example 4

The square root of 9 is 3, since $3^2 = 9$. We write this as $\sqrt{9} = 3$.

The square root of 64 is 8, since $8^2 = 64$. We write this as $\sqrt{64} = 8$.

$\sqrt{25}$ (square root of 25) $= 5$, since $5^2 = 25$.

$\sqrt{144}$ (square root of 144) $= 12$, since $12^2 = 144$.

Example 5

The time (in seconds) it takes an object to fall 576 feet is found by evaluating

$$\sqrt{\frac{576}{16}}.$$

After first dividing 576 by 16 to get a quotient of 36, we proceed to evaluate $\sqrt{36}$.

$\sqrt{36} = 6$, since $6^2 = 36$.

The time is 6 seconds.

Square roots are extremely important in basic mathematics. In addition to square roots, other roots are also used in many applications. We will now briefly consider some of the other roots.

The **cube root** *of a given number is that number which when cubed equals the given number. The* **fourth root** *of a given number is that positive number which when raised to the fourth power equals the given number. The fifth root, sixth root, and so on are defined in a similar manner. The notation* $\sqrt[3]{}$ *is used for a cube root,* $\sqrt[4]{}$ *is used for the fourth root, and so on.*

> If no number appears on the radical, it means square root $\sqrt{}$.

Example 6

The cube root of 27 is 3, since $3^3 = 3 \times 3 \times 3 = 27$. This cube root is written as $\sqrt[3]{27} = 3$.

The fourth root of 625 is 5, since $5^4 = 5 \times 5 \times 5 \times 5 = 625$. This fourth root is written as $\sqrt[4]{625} = 5$.

$\sqrt[4]{81}$ (fourth root of 81) $= 3$, since $3^4 = 81$.

$\sqrt[3]{125}$ (cube root of 125) $= 5$, since $5^3 = 125$.

$\sqrt[3]{-8} = -2$, since $(-2)^3 = 8$.

$\sqrt[3]{-125} = -5$, since $(-5)^3 = -125$.

So far we have determined only powers and roots of whole numbers. We can also determine powers and roots of any decimal. In doing so, we must carefully note the position of the decimal point.

Example
7

In this section we have used only positive exponents, but exponents can also be negative or zero. Such exponents will be considered in later sections.

Squaring 0.6, we have $0.6^2 = 0.6 \times 0.6 = 0.36$.

In the same way, $0.6^3 = 0.6 \times 0.6 \times 0.6 = 0.216$.

Also, $2.7^2 = 7.29$ and $2.73^2 = 7.4529$.

$\sqrt{0.36}$ (square root of 0.36) $= 0.6$, since $0.6^2 = 0.36$.

$\sqrt{1.44} = 1.2$, since $1.2^2 = 1.44$

1.4 Exercises

In Exercises 1 through 8, rewrite the given expressions by using the appropriate base and exponent. (Don't evaluate.)

1. $8 \times 8 \times 8$

2. $4 \times 4 \times 4 \times 4 \times 4$

3. $2 \times 2 \times 2 \times 2$

4. $5 \times 5 \times 5$

5. $3 \times 3 \times 3 \times 3 \times 3$

6. $12 \times 12 \times 12$

7. $10 \times 10 \times 10 \times 10 \times 10$

8. $6 \times 6 \times 6 \times 6 \times 6 \times 6 \times 6 \times 6$

In Exercises 9 through 16, write the given expressions as products.

9. 8^2

10. $(-2)^5$

11. -3^6

12. -4^3

13. 7^8

14. 10^4

15. 5^6

16. 6^5

In Exercises 17 through 44, evaluate the given expressions.

17. -3^5

18. 2^7

19. 4^3

20. -6^4

21. 0.3^2

22. -12^2

23. 3.5^3

24. 0.08^3

25. $\sqrt{16}$

26. $\sqrt{81}$

27. $\sqrt{121}$

28. $\sqrt{400}$

29. $\sqrt[3]{-64}$

30. $\sqrt[4]{16}$

31. $\sqrt[5]{32}$

32. $\sqrt[5]{-243}$

33. $\sqrt{0.16}$

34. $\sqrt{0.81}$

35. $\sqrt{0.09}$

36. $\sqrt{1.21}$

37. $(3^3)(2^2)$

38. $5^2(-4)^3$

39. $(-5)^3(6)^2$

40. $(7^3)(10^4)$

41. $(\sqrt{121})(3^4)$

42. $(\sqrt{144})(4^3)$

43. $(0.7^2)(\sqrt{1.69})$

44. $(3.5^2)(\sqrt{0.04})$

In Exercises 45 through 56, solve the given problems.

45. The amount of electrical energy required to heat one cup of coffee is about $7 \times (10^4)$ joule. Write this number in ordinary notation.

46. In UHF television broadcasting, the signal may have a frequency of $2 \times (10^9)$ cycles per second. Write this number in ordinary notation.

47. The time (in seconds) required for an object to fall 225 feet due to gravity can be found by evaluating

$$\frac{\sqrt{225}}{4}$$

Find the time required.

48. In calculating the diameter (in inches) of a vent pipe for a commercial oil burner, a technician must evaluate $\sqrt{484}$. Find that value.

49. Suppose you save 1¢ on the first day of January, 2¢ on the second day, 4¢ on the third, and you continue to

double your deposit each day. The number of cents required for the last day of January is 2^{30}. What is the value of 2^{30} cents?

50. An investment of \$2000 at a compound interest rate of $4\frac{1}{4}\%$ left in the bank for $5\frac{1}{2}$ years will be worth $(2000)(1.0425)^{5.5}$ dollars. Find this amount, correct to the nearest cent.

51. The distance traveled by an object falling (from rest) under the force of gravity depends upon how long the fall is. The distance (in feet) fallen after a time of t seconds is found by $D = 16t^2$. If an object falls for 5.5 seconds, the distance covered by this object is $16(5.5)^2$. How far did this object fall?

52. The volume of a cube can be found by using the formula $V = s^3$ where s represents the length of one side of the cube. Find the volume of a cube that has a side that measures 8 inches in length.

53. When current flows through a resistance, electrical energy is converted into heat. The **power** output of a lamp, resistor, or other component, is defined as the rate of change of electrical energy to heat, light, or some other form of energy. The formula for the power dissipated in a resistance is given by $P = I^2R$, where I represents current and R represents resistance. Find the power (P) when $R = 365$ ohm and $I = 0.586$ amps. Round your result to three decimal places.

54. The resistance in an amplifier circuit can be found by evaluating the formula $\sqrt{Z^2 - X^2}$. Find the resistance for $Z = 5.4$ ohms and $X = 2.8$ ohms. Round your result to one decimal place.

55. A TV screen is 20.4 inches wide and 17.5 inches high. The length of the diagonal is the dimension used to describe the size of the TV screen and is found by evaluating $\sqrt{w^2 + h^2}$ where w represents width and h represents height. Find the diagonal correct to the nearest whole number. What size is this TV screen?

56. A car costs \$32,000 new and is worth \$20,480 after two years. The annual rate of depreciation is found using the formula $100(1 - \sqrt{V/C})$ where C is the original cost of the car and V is its value after two years. Find the annual rate of depreciation and express your answer as a percent.

1.5   Order of Operations

Section Objectives
- Simplify expressions using order of operation
- Recognize and use different grouping symbols

The preceding sections involved the basic operations with signed numbers. Section 1.2 covered addition and subtraction and Section 1.3 covered multiplication and division of signed numbers. In this section, we will consider expressions involving combinations of those operations. In evaluating such expressions, it is extremely important to know that *there is an order in which the operations are performed.* Given a problem such as $6 + 5 \times 4$, it is possible to arrive at two distinctly different values.

If I add $6 + 5$ to get 11 and then multiply that result by 4, I get an answer of 44.

If I multiply 5×4 to get 20 and then add 6 to my product, I get an answer of 26.

Depending on how you approach the arithmetic operations specified within the problem you can get different results. Yet we all know that problems such as this can have only one correct solution. So which one is right, 44 or 26?

In this section we will look at the rules for **order of operation** in order to evaluate an expression such as $6 + 5 \times 4$. The order of operation we follow is a *convention* that we all use so that everyone always arrives at the same result. This means that at some point in time mathematicians decided on an order in which operations should be performed. Order of operation tells us which operations should be performed in what order.

Order of Operation

Given a particular problem that contains more than one arithmetic operation, we use the following priorities for addition, subtraction, multiplication, division, and exponents:

1. Evaluate expressions contained within **parentheses** or other grouping symbols.

2. Simplify all values that contain **exponents.**

3. Evaluate all **multiplications or divisions,** performing whichever operation comes first when reading from left to right.

4. Evaluate all **additions or subtractions,** performing whichever operation comes first when reading from left to right.

To help you to remember the order of operations you can use the popular sentence

Please **E**xcuse **M**y **D**ear **A**unt **S**ally, or the acnronym PEMDAS:

Parentheses

Exponents

Multiplication and **D**ivision, from left to right

Addition and **S**ubtraction, from left to right

Example
1

To correctly evaluate $(-12) \div (3) - (-2)(5)$, *we must first perform the division and then the multiplication.* **We do not perform the subtraction until these other operations are completed.** After dividing and multiplying we get

$$(-4) - (-10).$$

We can now do the subtraction by changing to unsigned numbers as follows:

$$(-4) + (10) = -4 + 10 = 6.$$

Example 1 had no powers or expressions enclosed within parentheses so the order of operations simply required that the division and multiplication preceded the subtraction. The next example includes powers.

Example 2

To correctly evaluate $2 + (2)^3 - (6)^2 \div (3)$, we note that the expression involves addition, subtraction, two powers, and division. *The order of operations indicates that the powers be evaluated first,* so we get

$$(2) + (8) - (36) \div (3).$$

The operation of division is the next priority; dividing leads to

$$(2) + (8) - (12).$$

Finally, we perform the addition and subtraction by converting to unsigned numbers as follows:

$$(2) + (8) - (12) = 2 + 8 - 12 = -2.$$

Example 3

Note that this example has parentheses nested within other grouping symbols (brackets). When this occurs, we should begin with the innermost set of parentheses and then proceed to work outward. In the evaluation of $[(2) - (5)] + (3)^2(4)$ *our order of operations requires that we begin with the expression enclosed within* **brackets.** We evaluate $(2) - (5)$ to get $2 - 5 = -3$, and the problem is now reduced to

$$(-3) + (3)^2(4).$$

After expressions enclosed within grouping symbols, the next priority involves powers, so the expression is further simplified to

$$(-3) + (9)(4).$$

The third priority here is the multiplication which, when performed, gives

$$(-3) + (36).$$

Finally, we add to get $-3 + 36 = 33$.

The greatest number of errors in evaluating expressions probably occurs when dealing with minus signs. Some of these common difficulties are illustrated in Examples 4–7.

Example 4

In the expression

$$-(-7) - (-2)^4 - (-6)(2)$$

▶ *there are sometimes difficulties with the $-(-7)$ because there is nothing before it.* We can look at this in either one of two ways. First, we could simply place a zero before it, and we see that it amounts to the subtraction of -7. The other way is to treat the minus sign in front as -1. It then becomes a factor in a multiplication. We therefore get

$$-(-7) = 0 - (-7) \qquad \text{or} \qquad -(-7) = (-1)(-7).$$

Another difficulty arises when it is noticed that there is a minus sign before
▶ $(-2)^4$. *There is a tendency to make $-(-2)^4$ equal to $+(+2)^4$, but this is not true.* Remember, we must evaluate the $(-2)^4$ first, and it equals $+16$. Thus we have the subtraction of 16, not the addition of 16 as $+(+2)^4$ would suggest. We therefore get:

$$-(-7) - (-2)^4 - (-6)(2) = 7 - (16) - (-12)$$
$$= 7 - 16 + 12$$
$$= 3.$$

We have already seen that parentheses group numbers and those numbers must be considered first. In addition to numbers enclosed within parentheses or other grouping symbols, we must also give top priority to numbers that are grouped as parts of a fraction as in Example 5. In effect, the fraction bar serves as a grouping symbol.

Example 5

In evaluating the expression

$$\frac{5}{2-3} + \frac{(-4)^2}{-2}$$

we see that we must first subtract 3 from 2 in the first denominator before we can perform the division. This does not violate our basic order of operations because the $2 - 3$ is grouped very specifically by being the denominator of the fraction. It is a step that must be made prior to performing the basic division.

▶ As for $\dfrac{(-4)^2}{-2}$, *there is often a tendency to make this* $\dfrac{(+4)^2}{+2}$, *but again this is not true.* We must remember the meaning of $(-4)^2$ as $(-4)(-4)$. Therefore, we should evaluate the power before we try to evaluate the fraction. Thus,

$$\frac{5}{2-3} + \frac{(-4)^2}{-2} = \frac{5}{-1} + \frac{(-4)^2}{-2} = \frac{5}{-1} + \frac{+16}{-2} = -5 + (-8)$$
$$= -5 - 8$$
$$= -13.$$

Example 6 further illustrates the order of operations that has been outlined in this section.

Example **6**

multiplication

$$(7) - (-3)(4) - \frac{(-6)(1)}{-2} = (7) - (-12) - (3)$$

multiplication and division
$$= 7 + 12 - 3$$
$$= 16$$

power

$$3(7) - (-5)^2 - \frac{12}{(-2)(-1)} = 3(7) - (25) - \frac{12}{(-2)(-1)}$$

multiplication | multiplication and division
$$= (21) - (25) - (6)$$
$$= 21 - 25 - 6$$
$$= -10$$

grouping

$$-2(3 - 7) - \frac{(-3)^2}{-9} = -2(-4) - \frac{(-3)^2}{-9}$$

power

$$= -2(-4) - \frac{+9}{-9}$$

multiplication and division
$$= +8 - (-1)$$
$$= 8 + 1$$
$$= 9$$

Now Try It!

Use the rules for order of operation to evaluate each of the following.

1. $5 - 8(2 - 7)$
2. $4 \div 2(2)$
3. $8 - 2 + 2$
4. $8 - (2 + 2)$
5. $(-2)^3(2 + 3^2)$
6. $\dfrac{2(-5 - 4) - 3(7 - 5)}{-3 - 4 - 1}$

We note in the second example that $3(7)$ was treated as $(3)(7)$, and in the third example $-2(-4)$ was treated as $(-2)(-4)$. Also, note in the third example that we performed the subtraction within the parentheses before multiplying.

Example

7

During a certain chemical reaction, the temperature of a mixture starts at 20°C, then drops 5°C/h for 3 h, and then rises 4°C/h for 7 h. Representing decreasing temperatures by negative numbers and increasing temperatures by positive numbers, we can express the resulting temperature as

$$20°C + (-5°C/h)(3 \text{ h}) + (4°C/h)(7 \text{ h})$$

start   temperature drop   temperature rise

$$= 20°C - 15°C + 28°C = 33°C$$

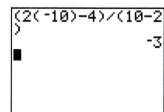

Using Technology

Your calculator is **order of operation specific**. This means it has been programmed with the rules for order of operation. Parentheses are very important when using a scientific or graphing calculator. Let's evaluate $\dfrac{2(-10) - 4}{10 - 2}$.

When using a scientific or graphing calculator it is important to use parentheses where appropriate. In this problem we want to use parentheses around the expression in the numerator and around the expression in the denominator. Compare your result with Figure 1.9. If you did not get the same result be sure to check your parentheses.

Figure 1.9

1.5 *Exercises*

In Exercises 1 through 28, evaluate the given expressions.

1. $(4) - (-8)(2)$

2. $(-6) + (-3)(-9)$

3. $(-9) + \dfrac{-10}{2}$

4. $(-1) + \dfrac{18}{-9}$

5. $(-1)(7) + (-6)(7)$

6. $(-5)(8) - \dfrac{15}{-3}$

7. $\dfrac{-22}{2} - (5)(-2)$

8. $(3)(-10) + (-12)(5)$

9. $(-8)(9) - \dfrac{-9}{3}$

10. $\dfrac{-60}{4} + (-1)(-9)$

11. $\dfrac{-18}{2} - \dfrac{24}{-6}$

12. $\dfrac{40}{-5} - \dfrac{-90}{15}$

13. $(-7)(15) - (-1)^2$

14. $(-6)(4) - (-2)^2$

15. $4^2 - 2(3) \div (4 - 2)$

16. $-(-9) + (-5)^3$

17. $-2(-9) - (-1)^4 + \dfrac{-18}{6}$

18. $-3(7) + (-3)^3 + \dfrac{-48}{4}$

19. $\dfrac{6 + 7 \times 2}{10 - 7}$

20. $(-3 + 5)^2 + 3(-4) - 5$

21. $-20 \div (-5) + 36 \div (-4)$

22. $2^2(5^2) + (-3 + 1)(2)$

23. $120 - 8(3) + 12$

24. $\sqrt{5^2 + (-12)^2}$

25. $\dfrac{6(14 - 20)}{3(5) - 6}$

26. $3(4^2 + 1) - 30 \div 3$

27. $26 + 8 \div 2^3 - 3(-3)$

28. $\dfrac{4(6 + 2) + (-8 + 3)}{6(2 - 4) - 2^2}$

In Exercises 29 through 36, set up the given problems in terms of operations on signed numbers and then solve. Unless otherwise indicated, all numbers are approximate.

29. A plane descends 1000 feet per minute for 4 minutes and then ascends 500 feet per minute for 6 minutes. Find its net change in altitude during this time.

30. In a certain city, the temperature rises 2°F per hour for 5 h, and then falls 3°F per hour for 10 h. Find the net change in temperature during this time.

31. The change in altitude of an object falling under gravity is determined by finding the product of -9.75 and the square of the time of the fall, and then dividing by 2. Determine the change in altitude (in meters) of an object that has fallen for 4.00 seconds.

32. The formula for converting a certain temperature from the Celsius scale to the Fahrenheit scale, is given by

$$F = \frac{9C}{5} + 32.$$ Convert 27°C to °F.

33. When drilling for oil, a geologist calculates the amount of water that must be used to lighten mud. The amount of water (in gallons) is

$$\frac{102(18.05 - 15.02)}{15.02 - 8.33}$$

Find that amount.

34. A solar engineer needs to compute the total roof area of an industrial plant with an irregular shape. The area (in square feet) can be found from

$$(15)(20) + (32)^2 + (32 - 6)(10)$$

Find the area.

35. A holding tank containing 2000 gallons of oil is drained at the rate of 120 gallons per hour for 6 hours and then filled at the rate of 300 gallons per hour for 4 hours. How much oil is in the tank now?

36. An electric current of 3.25 milliamperes decreases at the rate of 0.15 milliamperes per second for 12 seconds and then increases by 0.05 milliamperes per second for 6 seconds. What is the resulting current?

In Exercises 37 through 41, use a scientific or graphing calculator to evaluate the given expressions.

37. $\dfrac{(15)}{10 - 7} - \dfrac{-2}{-1}$

38. $\dfrac{(-3)^4}{8 - 2} - \dfrac{36}{-4}$

39. $\dfrac{(5)^4}{6 - 8} + \dfrac{9 - 4}{(3)^2}$

40. $\dfrac{2 - 5}{6 - 18} - \dfrac{(4)^3 - (3)^4}{53 - 70}$

41. $(-2)^6 - (3)^6 - \dfrac{(-2)^5 - (2)^3}{(-2)(-2)(5)}$

Now Try It! Answers

1. 45   **2.** 1   **3.** 8   **4.** 4   **5.** -88 **6.** 3

1.6 Scientific Notation

Section Objectives
- Change from ordinary notation to scientific notation
- Change from scientific notation to ordinary notation
- Add, subtract, multiply, and divide numbers that are in scientific notation

In technical and scientific work we often use numbers that are extremely large or extremely small in magnitude. For example, a light-year is equivalent to a distance of 9,460,530,000,000,000 meters. In addition to simply representing very large and very small numbers, we must often perform calculations with them. For example, the frequency (in hertz) in a certain electric circuit is found by evaluating 4,260,000 ÷ 0.0009425. In this section we begin with the representation of such numbers, and then proceed to consider calculations involving them.

Example
1

The following are just a few examples of where we might encounter very large or very small numbers:

1. The distance of the earth from the sun is about 149,400,000,000 kilometers.

2. In June, 2005 the U.S. national debt was approximately $78,000,000,000,000.

3. The mass of a single proton is 0.00000000000000000000000001673 kilograms.

4. Light travels at about 300,000,000 meters per second (m/s).

5. Some of the world's fastest computers can perform a single calculation in about 0.000000000000 26 second.

Expressing such numbers with conventional notation is often cumbersome. The notation referred to as **scientific notation** is normally used to represent these numbers of extreme magnitudes.

Scientific Notation

A number is said to be in **scientific notation** when it is expressed as the product of a number between 1 and 10 and a power of 10.

$$p \times 10^k \text{ where } 1 \leq p < 10 \text{ and } k \text{ is any integer}$$
$$1.65 \times 10^{10} \quad \Rightarrow \quad 16{,}500{,}000{,}000$$

number   power of 10

continued on next page

continues from previous page
A positive power of 10 indicates that we multiply by the power of 10.

A negative power of 10 indicates that we divide by the power of 10.

A **googol** is 10^{100}, which is 1 followed by 100 zeros, a very large number. This number has no real meaning in mathematics and is often used to help explain the concept of a very large number. The popular search engine Google takes its name from this number to reflect the infinite amount of information that can be found on the Internet.

The procedure for converting a number from ordinary notation to scientific notation is summarized by the following steps:

1. Move the decimal point to the right of the first nonzero digit.
2. Multiply this new number by a power of 10 whose exponent k is determined as follows:
   a. If the original number is greater than 1, k is the number of places the decimal point was moved to the left.
   b. If the original number is between 0 and 1, k is negative and corresponds to the number of places the decimal point was moved to the right.

Example **2**

Example 2 illustrates how numbers are written in scientific notation.

Consider the examples presented in Example 1. Keep in mind that a positive power of 10 indicates multiplication by the power of 10 whereas a negative power of 10 indicates division by the power of 10.

1. The distance of the earth from the sun is about 149,400,000,000 kilometers.

   $149{,}400{,}000{,}000 = 1.494 \times 10^{11}$ kilometers

2. In June, 2005 the U.S. national debt was approximately \$78,000,000,000,000.

   $78{,}000{,}000{,}000{,}000 = 7.8 \times 10^{13}$

3. The mass of a single proton is 0.00000000000000000000000000001673 kilograms.

   $0.00000000000000000000000000001673$

   $= 1.673 \times \dfrac{1}{1{,}000{,}000{,}000{,}000{,}000{,}000{,}000{,}000{,}000}$

   $= 1.673 \times \dfrac{1}{10^{27}}$

   $= 1.673 \times 10^{-27}$ kilograms

Here we see that
$1/1{,}000{,}000{,}000{,}000{,}000{,}000{,}000{,}000{,}000$
is the same as $\dfrac{1}{10^{27}}$

this verifies that $10^{-27} = \dfrac{1}{10^{27}}$

4. Light travels at about 300,000,000 meters per second (m/s).

   $300{,}000{,}000 = 3.0 \times 10^{8}$ meters per second

5. Some of the world's fastest computers can perform a single calculation in about 0.00000000000026 second.

   $0.00000000000026 = 2.6 \times \dfrac{1}{10{,}000{,}000{,}000{,}000}$

   $= 2.6 \times \dfrac{1}{10^{13}} = 2.6 \times 10^{-13}$ seconds

Example 3

To change 85,000 to scientific notation, first move the decimal point to the right of the first nonzero digit, which in this case is 8. We get 8.5. Now multiply 8.5 by 10^k where $k = 4$ since 85,000 is greater than 1 and we moved the decimal point 4 places to the left. Our final result is 8.5×10^4.

(Note that $10^4 = 10,000$ and $8.5 \times 10,000 = 85,000$ so that the value of the original number does not change.)

Example 4

To change 0.0000824 to scientific notation, first move the decimal point to the right of the first nonzero digit to get 8.24. Now multiply this result by 10^k where $k = -5$ since the original number is between 0 and 1 and the decimal point was moved 5 places to the right. We get 8.24×10^{-5}.

In Example 5 we see a clear pattern evolving. Shifting the decimal point to the left gives us a positive exponent whereas a shift to the right yields a negative exponent.

Example 5

$$56,000 = 5.6 \times 10^4$$
4 places to left

$$0.143 = 1.43 \times 10^{-1}$$
1 place to right

$$0.000804 = 8.04 \times 10^{-4}$$
4 places to right

$$2.97 = 2.97 \times 10^0$$
0 places

Now Try It! 1

Convert each of the numbers in Example 1 from ordinary notation to scientific notation.

To change a number from scientific notation to ordinary notation, the procedure described above is reversed. Example 6 illustrates the method of changing from scientific notation to ordinary notation.

Example 6

To change 2.9×10^3 to ordinary notation, we must move the decimal point 3 places to the right. Therefore, additional zeros must be included for the proper location of the decimal point. We get

$$2.9 \times 10^3 = 2900.$$

This is reasonable when we consider that $10^3 = 1000$ so that 2.9×10^3 is really 2.9×1000, or 2900.

To change 7.36×10^{-2} to ordinary notation, we must move the decimal point 2 places to the left. Again we must include additional zeros so that the decimal point can be properly located. We get

$$7.36 \times 10^{-2} = 0.0736.$$

Calculations with very large or very small numbers can be made by first expressing all numbers in scientific notation. Then the actual calculation can be performed on numbers between 1 and 10, using the laws of exponents to find the proper power of 10 for the result. It is acceptable to leave the result in scientific notation.

Basic Calculations using Scientific Notation

Adding and Subtracting Numbers in Scientific Notation

The key to adding or subtracting numbers in scientific notation is to *make sure that the powers of 10 (the exponents) are the same.*

- If the powers of 10 are the same, add or subtract the decimal number and keep the same power of 10.

- If the powers of 10 are different, the problem must be rewritten so that the powers are the same before you do any addition or subtraction. This can be done by shifting the decimal place right or left, as appropriate, and adjusting the exponent to reflect that shift.

Multiplying Numbers in Scientific Notation

- Multiply the decimal values.

- Add the exponents of the powers of 10.

Dividing Numbers in Scientific Notation

- Divide the decimal values.

- Subtract the exponent in the denominator from the exponent in the numerator.

Example 7

To add $(2.2 \times 10^3) + (3.0 \times 10^3)$ we notice that the powers of 10 are the same. Therefore, we just add the 2.2 + 3.0 to get 5.2. We keep the same power of 10 so that our final result is 5.2×10^3.

To subtract $(4.8 \times 10^7) - (6.2 \times 10^5)$ we notice that the powers of 10 are different. The problem needs to be rewritten so that the exponents are the same. We rewrite 4.8×10^7 as 480×10^5. Now the problem becomes $(480 \times 10^5) - (6.2 \times 10^5)$. Since the powers of 10 are now the same, we can subtract. The result of our subtraction is (473.8×10^5), which can be written as 4.738×10^7 in scientific notation.

$$(4.8 \times 10^7) - (6.2 \times 10^5) =$$
$$(480 \times 10^5) - (6.2 \times 10^5) = 473.8 \times 10^5 = 4.738 \times 10^7$$

Example 8

To multiply $(4.7 \times 10^6) \times (3.2 \times 10^{-2})$ we multiply the decimal values and add the exponents.

$$(4.7 \times 10^6) \times (3.2 \times 10^{-2}) = (4.7)(3.2) \times 10^{6+(-2)} =$$
$$15.04 \times 10^4 = 1.504 \times 10^5$$

Example 9

To evaluate $9,620,000,000,000 \div 0.0000000000328$, we may set up the calculation as

$$\frac{9.62 \times 10^{12}}{3.28 \times 10^{-11}} = \left(\frac{9.62}{3.28}\right) \times 10^{\overset{12-(-11)}{23}} = 2.93 \times 10^{23}.$$

The power of 10 in this result is large enough so that we would normally leave it in this form. Even on most calculators with the scientific notation feature, it would be necessary to express the numerator and denominator in scientific notation before performing the calculation.

Now Try It! 2

Perform the indicated operations. The results should be written in scientific notation.

1. $(10.3 \times 10^5) + (7.1 \times 10^5)$
2. $(3.6 \times 10^4) + (4.7 \times 10^6)$
3. $(0.0056 \times 10^2) - (2.8 \times 10^{-3})$
4. $(3.3 \times 10^7)(2.5 \times 10^{-12})$
5. $\dfrac{4.5 \times 10^{16}}{1.2 \times 10^{-8}}$

Example 10

In Example 1 we noted that some of the world's fastest computers can perform a single calculation in about 0.00000000000026 second. How long would it take for one of these computers to do 5 million calculations?

First we begin by changing each of the numbers to scientific notation.

$$0.00000000000026 = 2.6 \times 10^{-13}$$
$$5 \text{ million} = 5,000,000 = 5 \times 10^6$$

Our problem then becomes

$$(2.6 \times 10^{-13})(5 \times 10^6) = (2.6)(5) \times 10^{(-13+6)} = 13 \times 10^{-7}$$
$$= 1.3 \times 10^{-6} \text{ seconds.}$$

Using Technology

If we were to multiply two very large numbers together on the TI-83/84 Plus calculator our result might be too large to fit on the display so the calculator will convert your result to scientific notation. We can also enter numbers given in scientific notation into our calculator.

Here's How

- Enter the expression 128,000 × 865,000 into your calculator and press ENTER. Compare your result with Figure 1.10.

    The calculator display returned **1.1072E11.** The *E* before the last two digits indicates the power of 11.

    1.1072E11 is equivalent to **1.1072 × 10^{11}**

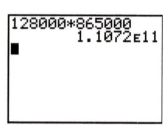

Figure 1.10

- You can also enter numbers that are already in scientific notation. Type in the decimal portion of the number given above. Locate **EE** on your calculator (found above the comma key). Press **2nd EE.** An E will appear in the display. Type in the appropriate **power of 10.** Enter 3.23 × 10^6 in the calculator by following the instructions above. Compare your result with Figure 1.11.

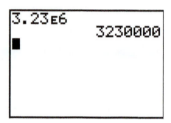

Figure 1.11

1.6 *Exercises*

In Exercises 1 through 8, change the given numbers from scientific notation to ordinary notation. Using the process outlined on p 31.

1. 4×10^6

2. 3.8×10^9

3. 8×10^{-2}

4. 7.03×10^{-11}

5. 2.17×10^0

6. 7.93×10^{-1}

7. 3.65×10^{-3}

8. 8.04×10^3

In Exercises 9 through 20, change the given numbers from ordinary notation to scientific notation. Using the process outlined on p 32.

9. 3000

10. 420,000

11. 0.076

12. 0.0029

13. 0.704

14. 0.0108

15. 9.21

16. 10.3

17. 0.000053

18. 1,006,000

19. 2,010,000,000

20. 0.0004923

In Exercises 21 through 33, perform the indicated calculations by first expressing all numbers in scientific notation.

21. (6700)(23,200)

22. 48,510 + 9700

23. (0.0153)(0.608)

24. (79,500)(0.00854)

25. 4950 − 231

26. $\dfrac{3740}{80,500,000}$

27. $\dfrac{0.0186}{0.0000665}$

28. 0.0064 − 4.2

29. 103,700 + 8364

30. 0.00005820 ÷ 8635

31. 0.0000385 ÷ 0.000000903

32. $\dfrac{(6.80)(8,040,000)}{4,200,000}$

33. $\dfrac{(0.0753)(73,900)}{0.0000811}$

In Exercises 34 through 46, change numbers in ordinary notation to scientific notation and change numbers in scientific notation to ordinary notation.

34. The speed of a communications satellite is about 17,500 miles per hour.

35. The mass of an electron is 0.00000000000000000000000000091 gram.

36. The wavelength of red light is about 0.00000065 meter.

37. It takes about 5×10^4 pounds of water to grow one bushel of corn.

38. Some computers perform addition in 1.5×10^{-9} second.

39. The area of the oceans of the earth is about 360,000,000 square kilometers.

40. A typical capacitor has a capacitance of 0.00005 F.

41. The power of a radio signal received from a laser beam probe is 1.6×10^{-12} watts.

42. The planet Mercury is an average of 36,000,000 miles away from the sun.

43. The half-life of uranium 235 is 710,000,000 years.

44. The faintest sound that can be heard has an intensity of about 10^{-12} watts per square meter.

45. One electron volt is approximately 0.0000000000000000006 joule.

46. A 40-gigabyte hard drive has 4×10^{10} bytes of memory.

In Exercises 47 through 54, perform the indicated calculations by first expressing all numbers in scientific notation. Use your calculator when appropriate.

47. The transmitting frequency of a television signal is given by the formula $f = \dfrac{v}{\lambda}$, where v is the velocity of the signal and λ (the lowercase Greek lambda) is the wavelength. Determine f (in hertz) if $v = 3.00 \times 10^{10}$ centimeters per second and $\lambda = 4.95 \times 10^2$ centimeters.

48. When resistors are wired in series, the total resistance is the sum of the individual resistors. Given three resistors wired in series with individual resistances of 4.83×10^3 ohms, 2.47×10^4 ohms, and 3.77×10^5 ohms, find the total resistance. Express your answer in scientific notation.

49. Find the power dissipated in a resistor if a current of 3.75×10^{-3} amps produces a voltage drop of 7.24×10^{-4} volts across the resistor using the formula $P = IV$. Express your answer in scientific notation.

50. If a computer can do a single calculation in 0.0000000002 seconds, how long, in seconds, will it take to do 8 trillion (8,000,000,000,000) calculations. Using your calculator to solve this problem, express your answer in scientific notation.

51. The shortest blip of light produced at the Center of Laser Studies in California lasts 0.00000000000020 seconds. If the speed of light is 300,000,000 meters per second, how many meters does the light travel in that time? Express your answer in scientific notation.

52. A radio signal travels 300,000 kilometers per second. How long does it take for a radio signal to travel 5000 kilometers? Express your result in scientific notation.

53. A report indicates that people in the United States drink an average of 26 gallons of coffee per person annually. Determine the total amount of coffee consumed if the coffee-drinking population of the United States is 250 million people. Express your answer in both ordinary and scientific notation.

54. In 1998 the United States produced more than 1,073,500,000,000 kilowatt hours of electricity. If 1 kilowatt hour is equivalent to 0.076 gallons of oil, how many gallons of oil would it take to produce that much electricity? Express your answer in ordinary and scientific notation.

Answers

Now Try It! 1
1. 1.494×10^8 km   **2.** 7.8×10^{13} **3.** 1.673×10^{-27} **4.** 3×10^8 m/s   **5.** 2.6×10^{-13}

Now Try It! 2
1. 17.4×10^5 **2.** 4.736×10^6 **3.** 5.572×10^{-1} **4.** 8.25×10^{-5} **5.** 3.75×10^{24}

1.7 Problem-Solving Strategies

Section Objectives
- Identify techniques that can be applied to general problem solving
- Apply problem-solving strategies

Mathematics was developed in order to solve real world problems, so it seems only natural that we spend some time thinking about, and looking at, some problem-solving techniques. Although you may never come across problems exactly like those presented throughout this textbook, you will have to deal with technical material that is presented to you in written format. In the work world, it is highly unlikely that you will be presented with an equation to solve, and more likely that you will be given a complex problem for which you will have to find a solution. Your job will be to sort through the information presented and come up with some techniques to solve the problem at hand.

Problem solving is more of an art than a science. It requires ingenuity, experience, and a bag of tricks. The most important rule of problem solving is to be flexible. No one strategy works all of the time. Try to stay away from looking for the "one way" to solve a problem. There is generally more than one approach that can be used to solve a problem.

Having said this, most problem solving can be broken down into several basic steps. As you read the steps below remember to think of them as guidelines that will help you stay focused, rather than as a set of rules.

Step 1: Understand the Problem
This may seem like an obvious statement but it is often missed. Students often jump in and attempt to solve a problem that they do not understand. Before you get started consider the following:

- Make sure you understand what the problem is asking.
- Make sure you have some idea as to where you are going with the problem.
- Think about how this problem relates to other problems you may have come across, or other problems in the real world.
- Try listing the given information, thinking about what assumptions you can make.
- Try rewording the problem to help clarify the question.

Step 2: Choose a Strategy for Solving the Problem
Once you understand what the problem is asking, think about how you might go about solving the problem. Often this can be the hardest step for students.

- Make a list of possible strategies and hints that might help in the overall solution of the problem.

- Draw a diagram, if applicable.
- Have some idea of what the answer should look like.
- Estimate that answer.

Step 3: Solve the Problem

This is the step where you actually work through the mathematical details of the problem.

- Use the strategy you selected to work through the problem.
- Work neatly and in an organized manner.

Step 4: Check and Think about your Solution

This step is often overlooked. There is a tendency to feel that once we have a solution we are finished with the problem. However, the solution is meaningless if it is wrong or cannot be explained to others (remember we solve problems in the real world and often need to be able to explain our solutions).

- Make sure your answer is reasonable.
- Does your answer make sense?
- Does your answer agree with the estimate you made earlier?

If you have always had difficulty solving problems in the past, the steps outlined above should help you organize your problem-solving approach. But the only way to truly improve your ability to solve problems is through practice and experience. The more problems you solve, the less daunting they will seem.

Here are some other things to keep in mind.

There may be more than one right answer. How do we stop global warming? There is no single best answer to this question. In mathematics it can also be true that there is no single right answer to a problem.

There may be more than one way to approach the problem. Using the example above concerning global warming, there will be more than one way to approach the problem.

Use appropriate tools. Is a calculator an appropriate tool to solve a particular problem? If you are taking a statistics class, a computer program such as Excel would be an appropriate tool to use for statistical analysis.

Do not spin your wheels. Do not get bogged down with a problem. If you are spending a lot of time on a problem and getting no where with it, try leaving the problem for a while. You will be amazed at what you might be able to do when you come back to the problem later on.

Let's try some of these strategies on Example 1.

Example
1

Suppose we want to find the length of a belt that will fit around two pulleys. Each pulley has a diameter of 6 inches and the centers of the two pulleys are 10 inches apart.

Using the methods outlined in this section *we begin by writing down what we know.*

- each pulley has a diameter of 6 inches
- the pulleys are 10 inches apart, measured from their centers
- the belt goes around both pulleys.

We might decide that we need to *draw a picture* to get a better sense of the problem. See Figure 1.12.

This helps us to visualize the path the belt is taking. We realize that the belt is on each pulley for only half of the distance around the pulley. We can now develop a plan that includes trying to find $\frac{1}{2}$ the distance around each pulley and adding that measurement to the 10 inches pulled on the top and bottom of Figure 1.12.

The distance around a circle (also known as the circumference of a circle) is found by using the formula $\pi(d)$, where d represents the diameter and $\pi \approx 3.14$.

$$\pi d \approx 3.14(6) = 18.84$$

Because we want only $\frac{1}{2}$ of the distance around each circle, we divide 18.84 in half and get 9.42 inches as the distance around half of each pulley.

We now have all the information we need to solve the problem:

Length of belt between pulleys on top $+ \frac{1}{2}$ distance around pulley $+$ length of belt between pulleys on bottom $+ \frac{1}{2}$ distance around pulley = total length of the belt.

10 inches + 9.42 inches + 10 inches + 9.42 inches = 38.84 inches.

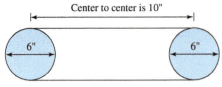

Center to center is 10"

6"   6"

Figure 1.12

> Remember that the best way to become better problem solvers is to solve more problems.

1.7 Exercises

Solve Exercises 1 through 10 to the best of your ability using the problem-solving strategies presented in this section. Some of these problems can be solved algebraically, but many can be solved by thinking through the information using an informal, less mathematical approach. You should be able to give a brief explanation as to how you arrived at your solution.

1. In New York the legal speed limit on Interstate 90 is 65 miles per hour. How long does it take you to travel 390 miles along Interstate 90 if you went the speed limit the whole time?

2. A traffic counter is a device designed to count the number of vehicles passing along a street. They look like a thin black tube stretched across a road and are connected to a "brain box" at the side of the road. The device registers one "count" each time a set of wheels (that is, wheels on a single axle) rolls over the tube. A normal car registers 2 counts—one for the front wheels and one for the rear wheels. A light truck with three axles registers 3 counts. Suppose that during a one-hour period a traffic counter registers 45 counts on a residential street on which only cars and light trucks are allowed. How many cars and light trucks passed over the traffic counter? (There are multiple answers to this problem).

3. The normal breathing rate for most adults is 18 to 20 breaths per minute. The average person inhales, or takes in, about 0.5 liters of air with each breath. Assuming a breathing rate is 20 breaths per minute, how much air does the average person inhale each hour?

4. It takes you 30 seconds to walk from the first (ground) floor of a building to the third floor. How long will it take to walk from the first floor to the sixth floor (at the same pace assuming all of the floors have the same height)?

5. The average of 10 students' test scores was 69. Three students scored 80, 60, and 40, respectively. Three students scored 90 and two students scored 50. The other two students received the same score. What grade did each student receive?

6. At the beginning of the semester your English Literature professor assigned the classic *War and Peace* to be read by the end of the term. You put off reading the book and suddenly realize that you have only 5 days to read the book before a major test. The copy of the book that you have contains 1232 pages. You estimate that you can read about 11 pages per hour. Assuming that you cannot read around the clock, will you be able to finish the book before the exam?

7. A 150-foot rope is suspended at its two ends from the tops of two 100-foot flagpoles. The lowest point of the rope is 25 feet from the ground. What is the distance between the flagpoles?

8. A woman bought a horse for $600 and then sold it for $700. She bought it back for $800 and then sold it for $900. How much money did she profit or lose after these two transactions?

9. In a community of 416 people, each person owns either a dog, a cat, or both a dog and a cat. If there are 316 dog owners and 280 cat owners, how many of the dog owners do not also own a cat?

10. You recently won the lottery and received a check for $20,000. You decide to place part of your money in a savings account paying 8% per year and part in a certificate of deposit paying 7% per year. If you wish to obtain an overall return of $1500 per year, how much would you place in each investment?

Chapter Summary

Key Terms

integer	dividend	square root
negative integer	divisor	radical sign
positive integer	quotient	order of operation
factor	exponent	scientific notation
product	base	

Key Concepts

• Integers are whole numbers and zero as well as positive and negative values. A positive integer is greater than zero and a negative integer is less than zero. The symbol for "greater than" is $>$ and the symbol for "less than" is $<$.

• When **adding two numbers that have the same sign** (both positive or both negative) add the two numbers without their signs. The sign of the sum is the same as the sign of the two numbers being added. When **adding two numbers that have the different signs** (one positive, one negative), subtract the two numbers without their signs. The sign of the sum is the same as the sign of the larger number.

• The operation of **subtraction of signed numbers** involves two steps: first change the subtraction sign to an addition sign and then change the number following the subtraction sign to its opposite. Now follow the rules for the addition of signed numbers.

• When **multiplying and dividing two signed numbers,** if both numbers have the same sign (both positive or both negative), the answer is always positive. If both numbers have different signs (one positive, one negative), the answer is always negative.

• Division by zero is not allowed.

• The **order of operations** for algebra provides a systematic way to evaluate expressions that involve more than one operation. The order is as follows: parentheses, exponents, multiplication and division (in the order in which they appear), and finally, addition and subtraction (in the order in which they appear).

• A number is said to be in *scientific notation* when it is written as a number between one and ten multiplied by ten raised to an integer power. The number takes the form $p \times 10^n$ *where* $1 \leq p < 10$ *and n is any integer.*

• **Problem-solving strategies** include understanding the problem, choosing a strategy to use to solve the problem, doing the work necessary to find a solution, and checking your solution.

Review Exercises

In Exercises 1 through 32, evaluate the given expressions by performing the indicated operations.

1. $(4) + (-6)$

2. $(-6) + (-2)$

3. $(-5) - (8)$

4. $(-3) - (-7)$

5. $(-9)(7)$

6. $(-8)(-12)$

7. $(-63) \div (-7)$

8. $(-72) \div (9)$

9. $(-3) - (-9) + (4)$

10. $(10) + (-4) - (-7)$

11. $(+9) - (2) - (-12)$

12. $(-6) - (5) + (-8)$

13. $(-2)(-6)(8)$

14. $(-1)(4)(7)$

15. $(5)(3)(-2)(2)$

16. $(-10)(6)(-2)(-3)$

17. $\dfrac{(-5)(-6)}{3}$

18. $\dfrac{(-8)(7)}{-14}$

19. $\dfrac{(-18)(4)}{(+6)(2)}$

20. $\dfrac{(8)(-9)}{(-4)(-3)}$

21. $(-4)^4$

22. $(-7)^3$

23. $\dfrac{(-1)(-5)(45)}{(9)(5)(1)}$

24. $\dfrac{(-8)(-16)(3)}{(-2)(6)(-4)}$

25. $(-2)(4) - \dfrac{(-6)}{2} - (-5)$

26. $-(-2)(-5) + \dfrac{(-9)}{(3)} - \dfrac{(-16)}{(4)}$

27. $\dfrac{(-2)(-3)}{-6} + (-4)(2) - \dfrac{-8}{4}$

28. $(-5) - (-2)(-5) + \dfrac{4}{-2}$

29. $-(-7) - (-6)(2) - (-3)^2$

30. $(-4) - \dfrac{-8}{2} - (-2)^4$

31. $-\dfrac{(-2)(-3)}{-6} + (-4)(2) - 3^2$

32. $(-5) + \dfrac{-14}{-2} - 2^4 - (-1)^2$

In Exercises 33 through 37, use order of operation to evaluate the following expressions.

33. $3^2 - 13$ **34.** $5 - (-4)^3$ **35.** $\dfrac{6 + 7(2)}{10 - 7}$

36. $(-3 + 5)^2 + 3(-4) - 5$

37. $-20 \div (-5) + 36 \div (-4)$

In Exercises 38 through 43, change each number to scientific notation.

38. 205,000

39. 9,805,000,000

40. 4005

41. 0.00035

42. 0.00000075

43. 0.707

In Exercises 44 through 49, change each number to ordinary notation.

44. 4.7×10^4

45. 3.02×10^8

46. 83.19×10^1

47. 1.87×10^{-3}

48. 7.7×10^{-9}

49. 9.1×10^{-12}

In Exercises 50 through 54, perform the indicated operation by first expressing all numbers in scientific notation.

50. $(19.3 \times 10^5) + (17.1 \times 10^5)$

51. $(3.06 \times 10^4) + (4.07 \times 10^6)$

52. $(0.0056 \times 10^2) - (2.8 \times 10^{-3})$

53. $(7.3 \times 10^7)(3.5 \times 10^{-11})$ **54.** $\dfrac{14.5 \times 10^9}{1.25 \times 10^{-7}}$

In Exercises 55 through 66, solve the given problems. Use a scientific or graphing calculator when appropriate.

55. A car starts with 6 quarts of oil. If it consumes 3 quarts of oil and 2 quarts are added, how much oil is left?

56. An astronomer writes a program and represents the year 2000 CE as +2000. How would she represent the year that is 5000 years before 2000 CE?

57. The lowest floor of a house is 4 feet below ground level, and the rooftop is 22 feet above ground level. Find the distance from the lowest floor to the rooftop.

58. In a golf tournament, a golfer is 2 strokes over par in the first round, 1 stroke under par in the second round, 3 strokes under par in the third round, and 2 strokes under par for the fourth round. What is his net score relative to par for the tournament?

59. In three successive months in a certain area, the cost of living decreases 0.1, increases 0.3, and decreases 0.4 percentage points. What is the net change in the cost of living for these three months?

60. An employee has an expense account of $800. If she spends $45 each week for 8 weeks, what will her balance be?

61. A cylinder is used to compress water. The pressure in pounds per square inch required to compress 600 cubic inches of water by 3.0 cubic inches can be found by evaluating

$$\frac{(-330000)(-3.0)}{600}$$

Find that pressure.

62. The current (in amperes) in a certain electric circuit is found by evaluating

$$\frac{(+5.96) - (-0.795)(8.04)}{8.04 + 3.85}$$

Find the current.

63. In analyzing the weather, a meteorologist notes that on one day the temperature started at $-8°C$. The temperature dropped 2°C each day for 4 days and it

then rose 3°C each day for 3 days. What was the temperature at the end of the time period?

64. An electron moves at the rate of 300,000,000 meters per second. How many meters will it move in $\frac{1}{4}$ (0.25) of a second?

65. Light travels at an approximate speed of 180,000 miles per second. The minimum distance from the earth to the sun is 87,600,000 miles. How long does it take for the sun's light to reach the earth in seconds?

66. A hydrogen atom has a mass of 1.6735×10^{-24} grams, whereas an oxygen atom has a mass of 2.6561×10^{-23} grams. Which atom has the smaller mass? How much smaller?

Chapter Test ▸◂▸◂▸•▸◂▸◂▸•▸◂▸◂▸•▸◂▸◂▸•▸◂▸◂

In Exercises 1 through 12, perform the indicated operations:

1. $-3 + (-5)$

2. $-2 + 7$

3. $-8 - 3$

4. $-2(3)(-4)$

5. $7 - (4 - 2) - 1$

6. $(-1/2)^3$

7. $-18 \div -6$

8. $12 - 8(4 + 1)$

9. $-3 + 2^3 - 7$

10. $\dfrac{4.55^2 - \sqrt{27}}{3.2^3}$ rounded to the nearest hundredth

11. $(3.5^2 + 5.3^{3.5})^3$ rounded to the nearest tenth.

12. $\sqrt{3.4^3 + 100}$ rounded to the nearest tenth.

13. Express 0.0000000000017 in scientific notation.

14. Express 5,670,000 in scientific notation.

15. Express 0.0035×10^{-5} in ordinary notation.

16. Express 12.4×10^3 in ordinary notation.

17. Add 4.5×10^{-4} to 5.9×10^{-2}.

18. Subtract 3.21×10^3 from 4.92×10^4.

19. Multiply $(2.46 \times 10^{-6})(5.9 \times 10^{-3})$.

20. Divide $\dfrac{2.29 \times 10^7}{7.328 \times 10^{-2}}$.

In Exercises 21 through 23, solve the following problems.

21. In the last 10 plays of a Giants' football game, the following yards were gained or lost: lost 5 yards, gained 7 yards, gained 9 yards, lost 5 yards, gained 15 yards, gained 4 yards, lost 6 yards, gained 20 yards, lost 1 yard, and gained 5 yards. What was the net total yardage for these 10 plays?

22. One coulomb is the amount of electric charge carried by 1 ampere of current for one second. There are approximately 6.2415×10^{18} electrons in one coulomb. How many coulombs do 4.1601×10^{20} electrons represent?

23. A quality assurance technician intern is paid at the rate of $12.50 for the first 40 hours with time and a half for overtime (anything over 40 hours). If the intern recently brought home a paycheck for $725, how many hours of overtime did she work? Assume that no deductions were made from her paycheck.

2 Units of Measurement and Approximate Numbers

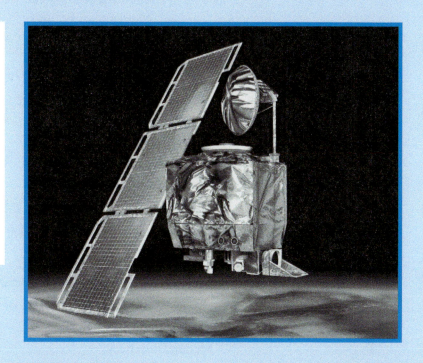

In December 1998 the Mars Orbiter was launched. Its mission was to explore the climate history of the planet Mars. In addition to monitoring the polar ice caps on Mars, the Orbiter was to serve as a relay station for a second mission scheduled for December 1999. However, when the Mars Orbiter hit the Martian atmosphere in September 1999, it immediately broke up. Investigators determined that the cause of the breakup of the Orbiter was human error. Instead of approaching the Martian atmosphere at an altitude of between 87 and 93 miles, the Orbiter came in at 37 miles above the surface of the planet. The minimum survival altitude was 53 miles. What caused this miscalculation in altitude? It was determined that one team used the U.S. Customary units of measurement (e.g., inches, feet, and pounds) whereas the other used metric units for a key spacecraft operation. This information was critical to the maneuvers required to place the spacecraft in the proper Mars orbit. This was a costly mistake because the Mars Orbiter, which cost approximately $357 million dollars to build, launch, and fly, was lost.

In this chapter we will review the U.S. Customary System of measurement and learn about the metric system. We will also learn to convert measurements between both systems. We will also study procedures for changing measurements within a system, as in reducing centimeters to millimeters. We will study the importance of significant digits in calculations and learn the correct procedures for rounding numbers. These topics all have great practical importance for scientists and technicians.

We will also learn to solve many different types of problems. For example, suppose that a pilot estimates that she must take on 12 gallons of avgas (aviation grade gasoline). If she is in Canada, she must convert that amount to liters. How many liters should she request? We will solve many such problems in this chapter.

2.1 Working with Units of Measure

Section Objectives
- Become familiar with the U.S. Customary System of measurement
- Identify units of measurement for length, weight, and capacity
- Introduce unit analysis
- Use unit analysis to convert from one U.S. Customary unit of measurement to another

Most scientific and technical calculations involve numbers that represent a measurement or a count of a specific quantity. *Such numbers are called* **denominate numbers.** *Associated with these denominate numbers are* **units of measure.** A measured distance of 3 feet for example, has the denominate number 3 associated with the foot as the unit of measurement. The United States currently uses two systems of measurements—the **U.S. Customary System** and the **International System of Units** or **SI System,** commonly referred to as the metric system. In this section we will focus on the U.S. Customary System and introduce a systematic way of converting units, called **unit** or **dimensional analysis.**

In the U.S. Customary System the units for measuring length are inch, feet, yard, and mile. The units for measuring weight are ounce, pound, and ton, and the units for measuring volume are cup, pint, quart, and gallon. The following table lists U.S. Customary units and their equivalents for length, weight, and capacity.

There are many occasions in math, science, and technology where we are presented with information in one set of units and need to convert to another set of units. Many students have difficulty determining how to convert from one set of units to another. **Unit** or **dimensional analysis** provides a systematic method for changing from one unit to another. When using **unit analysis** we work to cancel out units appearing in both the numerator and the denominator. In the end, the unit that you are looking for on the right-hand side of the equation is the same as the unit that remains on the left-hand side of the equation after canceling all appropriate units. In this section we will focus on conversions using U.S. Customary units. Later we will learn how to do unit conversions between U.S. Customary units and metric units.

Example 1

Suppose we wanted to know the depth of the Grand Canyon in inches, given that the Grand Canyon is 1.3000 miles deep.

Before we can begin the problem we need to know how many inches there are in a foot and how many feet there are in a mile. From Table 2.1 we know that 12 in = 1 ft and 5280 ft = 1 mile. These relationships can be written in the following format (commonly called **unit fractions**):

$$1 = \frac{12 \text{ in}}{1 \text{ ft}} \quad \text{or} \quad \frac{1 \text{ ft}}{12 \text{ in}} \quad \text{and} \quad 1 = \frac{5280 \text{ ft}}{1 \text{ mile}} \quad \text{or} \quad \frac{1 \text{ mile}}{5280 \text{ ft}}$$

Knowing that we want the units on the right-hand side of the equation to be in inches and that the given information is in miles, we need to come up with a conversion that gets us from miles to inches and allows the cancellation of common units on the left-hand side of the equation. If we set up the problem as follows, we see that we can cancel the common units of miles and feet, leaving us only with inches as our final dimension.

$$\frac{1.3000 \text{ mi}}{1} \times \frac{5280 \text{ ft}}{1 \text{ mi}} \times \frac{12 \text{ in}}{1 \text{ ft}} = 82,368 \text{ in}$$

Table 2.1 U.S. Customary Units

Length: Inch, foot, yard, and mile are units of length.
12 inches (in.) = 1 foot (ft)
3 feet = 1 yard (yd)
5280 feet = 1 mile (mi)
1760 yards = 1 mile

Weight: Ounce, pound, and ton are units of weight.
16 ounces (oz) = 1 pound (lb)
2000 lb = 1 ton (T)

Capacity: Cup, pint, quart, and gallon are used to measure capacity.
8 ounces (oz) = 1 cup (c)
2 cups (c) = 1 pint (pt)
2 pints (pt) = 1 quart (qt)
4 quarts (qt) = 1 gallon (gal)

Example 2

As a security guard for a hazardous waste site you are required to make 2 trips around the site each hour. The perimeter of the hazardous waste site is 4,000 ft around. Find the total distance, in miles, the security guard travels during an 8-hour day.

We know that we want the final units for the right-hand side of the equation to be miles. We also know that 8 *hours* = 1 *work shift*, 2 *trips* are taken in 1 *hour*, the guard travels 4000 *feet* in 1 *trip*, and there are 5280 *feet* in 1 *mile*. Each of these relationships can be set up as a unit fraction as in Example 1.

Using the given information we can set up the following equation, keeping in mind that the method we are using to solve this problem focuses on canceling out units on the left-hand side of the equation in order to match the units on the right-hand side of the equation.

$$\frac{8\ \cancel{hours}}{1\ day} \times \frac{2\ \cancel{trips}}{1\ \cancel{hour}} \times \frac{4000\ \cancel{ft}}{1\ \cancel{trip}} \times \frac{1\ mile}{5280\ \cancel{ft}} \approx 12.12\ \text{mi per shift}$$

To Work Through a Unit Analysis Problem

When using **unit analysis** we work to cancel out units appearing in both the numerator and the denominator.

- Determine what conversions are needed.
- Write the given information as unit fractions with a denominator of 1.
- Set up the necessary conversions as a multiplication problem so that the units to be eliminated can be easily cancelled.
- Cancel unwanted units.
- Perform the numerical calculations.

Example 3

In mixing concrete for a do-it-yourself home project you determine that you will need approximately 2.5 gallons of water to mix with $\frac{1}{2}$ bag of concrete. All you have at hand to add water to the mix is a 20-oz water bottle. How many bottles of water will you need to add to the concrete mix?

Using the steps outlined above we begin by identifying the conversions we need to solve this problem. We know that

20 oz = 1 bottle	2 pints (pt) = 1 quart (qt)
8 ounces (oz) = 1 cup (c)	4 quarts (qt) = 1 gallon (gal)
2 cups (c) = 1 pint (pt)	2.5 gallons are needed to do the job

Using this information we can set up the following multiplication problem so that the units to be eliminated can be easily cancelled out.

$$\frac{1\ bottle}{20\ \cancel{oz}} \times \frac{8\ \cancel{oz}}{1\ \cancel{c}} \times \frac{2\ \cancel{c}}{1\ \cancel{pt}} \times \frac{2\ \cancel{pt}}{1\ \cancel{qt}} \times \frac{4\ \cancel{qt}}{1\ \cancel{gal}} \times \frac{2.5\ \cancel{gal}}{1} = 16\ \text{bottles of water}$$

2.1 Exercises

In Exercises 1 through 12 use unit analysis to make the necessary conversions.

1. Change 8 hours to seconds.

2. Change 3.5 hours to seconds.

3. Change 25 yards to inches.

4. Change 6 miles to inches.

5. Change 85 pounds to ounces.

6. Change 120 ounces to pounds.

7. Change 88 feet per second to miles per hour.

8. Change 440 feet per second to miles per hour.

9. Change 65 miles per hour to feet per second.

10. Change 35 miles per hour to feet per second.

11. Change $45,000 per year to dollars per hour (assume a 40-hour work week and a 52-week year).

12. Change $10.25 per hour to dollars per year (assume a 40-hour work week and a 52-week year).

In Exercises 13 through 16 use problem-solving strategies and unit analysis to solve the following problems.

13. A marathon is a long distance running event that covers a distance of 26.2 miles. If the average runner can run 515 feet in one minute, how many hours will it take to complete the marathon? (Round your result to one decimal place).

14. The cost of a gallon of gas is $3.35 per gallon. If you travel 40 miles a day (back and forth to work) and your gas mileage is 25 miles per gallon, determine how much you will spend on gas each week.

15. A rocket's velocity is generally given in feet per second (ft/s) rather than miles per hour (mph). However, most of us find it easier to conceptualize speed in miles per hour. Find the speed in mph of a rocket traveling at 37,000 ft/s.

16. The amount of blood pumped each minute is measured by multiplying the heart rate per minute by the amount of blood pumped with each beat. At rest this amounts to about 4.22 quarts of blood per minute. Determine the number of gallons of blood pumped by the heart each day.

2.2 Units of Measurement: The Metric System

Section Objectives
- Become familiar with the International System of Units (metric system).
- Become familiar with metric prefixes.
- Identify metric units of measurement for length, mass, capacity, temperature, time and electrical units.
- Become familiar with the symbols used to represent units of measure.

In the United States we use both the metric and U.S. Customary Systems of measurement. Every other major industrial country uses the metric system, but the United States is in a very long state of transition. Some of the major U.S.

The Metric Conversion Act of 1975 designated the metric system as the preferred system of weights and measures for U.S. trade and commerce. It also provided for establishing a U.S. Metric Board to coordinate the "voluntary" transition to the metric system. Some industries jumped on the bandwagon. Most did not. In 1988, the Metric Conversion Act was amended to mandate that all federal agencies implement the metric system in procurement, grants, and other business-related activities. In 1991, President George H. Bush issued Executive Order 12770 mandating the transition to metric measurement for all federal agencies. The United States is in a very long state of transition.

industrial companies have already converted to the metric system, and most others will do so soon. We must be familiar with both systems, because they are often used. Important advantages of the metric system are:

1. There is a definite need for the world to have one standard system of units. Industries not using the worldwide standard will be at a disadvantage.

2. In the metric system, *units of different magnitudes vary by a power of 10, and therefore only a shift of a decimal point is necessary when changing units.* No similar arrangement exists in the U.S. Customary System.

3. When measuring some physical quantity, only one metric unit will apply. Length, for example, is measured in terms of meters. The U.S. Customary System measures lengths in several different units, including inches, feet, yards, miles, and rods.

To better see the advantages of the metric system, consider Example 1.

Example 1

To convert 3 mi to inches in the U.S. Customary System, we get

$$\frac{3 \text{ mi}}{1} \times \frac{5{,}280 \text{ ft}}{1 \text{ mi}} \times \frac{12 \text{ in.}}{1 \text{ ft}} = 190{,}080 \text{ in.}$$

To convert 3 km to centimeters in the metric system, we get

$$\frac{3 \text{ km}}{1} \times \frac{1{,}000 \text{ m}}{1 \text{ km}} \times \frac{100 \text{ cm}}{1 \text{ m}} = 300{,}000 \text{ cm.}$$

Although both of these conversions are similar, the required arithmetic is much simpler in the metric system.

The system now referred to as the metric system is actually the **International System of Units,** or the **SI** system. The metric system uses seven **base units** which are used to measure these quantities: (1) length, (2) mass, (3) time, (4) electric current, (5) temperature, (6) amount of substance, and (7) luminous intensity. Other units, referred to as **derived units,** are given in terms of the units of these quantities.

Example 2

The unit for electric charge, the *coulomb,* is defined as one ampere second (the unit of electric current times the unit of time). Therefore, electric charge is defined in terms of the base units of electric current and time.

The coulomb is a derived unit with a special name, but some derived units are not given special names. Speed, for example, might be in terms of meters per second, but there is no special name given to the units of speed.

In the U.S. Customary System, the base unit of length is the **foot** whereas the **meter** is the base unit of length in the metric system. Both systems use the **second** as the **base unit** of time. A basic problem arises with the base units for mass and force. The weight of an object is the *force* with which it is attracted to the earth, but the *mass* of an object is a measure of the amount of material in it. Mass and weight are therefore different quantities, and it is not strictly correct to convert **pounds** (force) to **kilograms** (mass). However, the kilogram is commonly used for weight even though it is actually a unit of mass. (The metric unit of force, the **newton,** is also used for weight.)

As for the other base units, the **SI** system defines the **ampere** as the base unit of electric current, and this unit can also be used in the U.S. Customary System. As for temperature, **degrees Fahrenheit** are used with the U.S. Customary System and **degrees Celsius** are used with the metric system. (Actually, the **kelvin** is defined as the base unit of thermodynamic temperature, although for temperature intervals, one kelvin equals one degree Celsius.) In the **SI** system, the base unit for the amount of a substance is the mole and the base unit of luminous intensity is the **candela.**

Because the metric system has units of different magnitude varying by powers of 10, the arithmetic is simpler and requires only a shift of the decimal point. No consistent arrangement exists in the U.S. Customary System, so we must contend with a variety of multiplying factors such as 12, 16, 32, 36, and 5,280. Also, the metric system uses a specific set of **prefixes** that denote the particular multiple being used. Table 2.2 shows some of the common prefixes with their symbols and meanings.

Table 2.2 **Metric Prefixes**

Prefix	Symbol	Factor by which unit is multiplied	Common meaning
Tera	T	$1{,}000{,}000{,}000{,}000 = 10^{12}$	trillion
Giga	G	$1{,}000{,}000{,}000 = 10^{9}$	billion
mega	M	$1{,}000{,}000 = 10^{6}$	million
kilo	k	$1000 = 10^{3}$	thousand
centi	c	$0.01 = \dfrac{1}{10^{2}}$	hundredth
milli	m	$0.001 = \dfrac{1}{10^{3}}$	thousandth
micro	μ	$0.000001 = \dfrac{1}{10^{6}}$	millionth
nano	n	$0.000000001 = \dfrac{1}{10^{9}}$	billionth
pico	p	$0.000000000001 = \dfrac{1}{10^{12}}$	trillionth

Example 3

The symbol for the meter is m. A kilometer therefore equals one thousand meters. This is shown as

$$1 \text{ km} = 10^3 \text{ m} \qquad \text{or} \qquad 1 \text{ km} = 1000 \text{ m}$$

represents 1000   represents meter

When changing a quantity from one metric prefix to another, we can use Figure 2.1 to simplify the process of shifting the decimal point. This is illustrated in the following example.

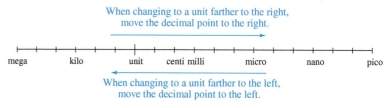

Figure 2.1

Example 4

Figure 2.1 indicates that a change from kilograms to centigrams involves moving the decimal point to the right (toward centi) five places. Therefore, 0.0345 kilograms is equal to 3450 centigrams, which is expressed as 0.0345 kg = 3450 cg.

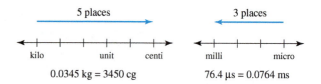

0.0345 kg = 3450 cg   76.4 μs = 0.0764 ms

A change from microseconds to milliseconds requires that the decimal point be moved three places to the left. As a result, 76.4 microseconds becomes 0.0764 milliseconds, or 76.4 μs = 0.0764 ms. Such changes are examples of *reduction,* which will be considered in Section 2.3.

Table 2.3 lists some of the units most commonly used in the metric system, U.S. Customary System, or both. Also shown are the symbols and some of the basic relationships.

The metric prefixes are used with the various units of the metric system as in Example 4. Example 5 further illustrates the use of units, prefixes, and symbols.

Table 2.3 **Metric and U.S. Customary Units**

Quantity	Metric			U.S. Customary		
	Unit	Symbol		Unit	Symbol	
Length	meter	m		foot	ft	
	centimeter	cm	1 cm = 0.01 m	yard	yd	1 yd = 3 ft
	millimeter	mm	1 mm = 0.001 m	inch	in.	1 in. = $\frac{1}{12}$ ft
	kilometer	km	1 km = 1000 m	mile	mi	1 mi = 5280 ft
Mass (SI units) or Force (U.S. Customary units)	kilogram	kg		pound	lb	
	gram	g	1 g = 0.001 kg	ounce	oz	1 oz = $\frac{1}{16}$ lb
	milligram	mg	1 mg = 0.001 g	short ton	t	1 t = 2000 lb
Capacity	liter	L		quart	qt	
	milliliter	mL	1 mL = 0.001 L	pint	pt	1 pt = $\frac{1}{2}$ qt
	kiloliter	kL	1 kL = 1000 L	gallon	gal	1 gal = 4 qt
Temperature	degrees Celsius	°C		degrees Fahrenheit	°F	
Time (both systems)	second	s				
	minute	min	1 min = 60 s			
	hour	h	1 h = 3600 s			
Electrical units (both systems)	ampere	A	current			
	volt	V	potential (voltage)			
	ohm	Ω	resistance			
	coulomb	C	charge			
	watt	W	power			

Note: The SI symbol for liter is 1, but because this symbol may be confused with the numeral 1, the symbol L is recognized for use in the U.S. and Canada. Also, ℓ is recognized in several countries.

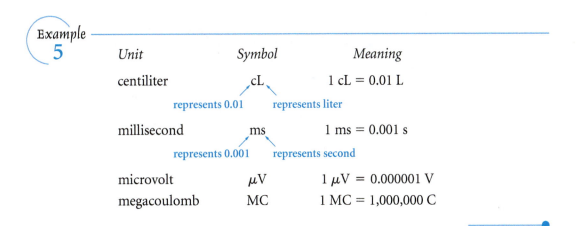

Example 5

Unit	Symbol	Meaning
centiliter	cL	1 cL = 0.01 L
millisecond	ms	1 ms = 0.001 s
microvolt	μV	1 μV = 0.000001 V
megacoulomb	MC	1 MC = 1,000,000 C

As mentioned earlier, the derived units for other quantities are expressible in the base units. For example, the units of area, volume, and speed are expressed in terms of the base units. The units for energy and pressure have special names, but they may be expressed in terms of the base units.

Considering the meaning of exponents given in Section 1.4, we can think of area as being measured in units of length $\times$ length $=$ (length)^2. Therefore, area is measured in square feet (ft^2), square miles (mi^2), square meters (m^2), and so on.

To find the volume of a rectangular solid, we multiply the length by the width by the depth. This means that volume is measured in units of length $\times$ length $\times$ length $=$ (length)^3. Thus we measure volume in cubic feet (ft^3), or cubic centimeters (cm^3), and so on.

Density, which is weight (or mass) per unit volume, is measured in lb/ft^3 or kg/m^3, for example.

If we have the product of two different units, we show that the units are to be multiplied by placing a raised dot between the symbols. To show the unit kilogram meter (simply a space between unit names indicates product), we use kg $\cdot$ m.

In designating units of area or volume, we use square or cubic units and exponents in the designation.

If a unit is used in the SI system or in both systems, we use the SI symbol for the unit. As an example, we use s rather than sec for seconds. Standard U.S. Customary symbols are used for units in that system only. However, in writing numbers we use the notation that is still prevalent. For long numbers, we use commas to separate groupings of three. (The SI system uses spaces, not commas.) For example, we write twenty-five thousand as 25,000 (not as 25 000, which is the SI style). It is expected, however, that the SI style will eventually become universal.

In this section we have considered the meaning and use of units of measurement, along with an introduction to the metric system. This discussion considered the **base units** as well as the derived units directly associated with the base units. In Section 2.3 we shall consider the problem of changing, or converting, one set of units to another.

Now Try It!

Reduce the following:

1. 1 km to centimeters
2. 1 kV to millivolts
3. 4 m to millimeters
4. 60 μs to seconds
5. 250 mm^2 to square meters

2.2 Exercises

In Exercises 1 through 8, give the symbol and meaning of the given unit.

1. milliampere
2. micrometer
3. kilovolt
4. megawatt
5. kilowatt
6. nanosecond
7. megaliter
8. centigram

In Exercises 9 through 16, give the name and meaning for the units whose symbols are given. (See Example 5.)

9. MV
10. kV
11. μs
12. kW
13. cV
14. mΩ
15. nA
16. ps

In Exercises 17 through 20, use only the base units to give the proper symbols (a) in the metric system and (b) in the U.S. Customary System for the units of the given quantities.

17. area   **18.** volume

19. speed (distance per unit of time)

20. acceleration (distance per squared unit of time)

In Exercises 21 through 32, rewrite the given quantities so that all numbers are between 1 and 10. Use the proper metric symbol for the given unit. For example, 0.03 meter would be expressed as 3 cm.

21. 4 seconds   **22.** 3 hours

23. 8 meters   **24.** 2 kilometers

25. 0.04 second   **26.** 0.07 liter

27. 0.000003 second   **28.** 0.06 meter

29. 0.08 gram   **30.** 0.005 meter

31. 4000 grams   **32.** 9000 watts

In Exercises 33 through 44, give the proper symbols of the units of the indicated quantities in terms of the units given.

33. The sides of a rectangle are given in millimeters. What is the symbol for the units of area of the rectangle?

34. The edges of a box are given in inches. What is the symbol for the units of volume of the box?

35. The driver of a car determines the number of miles traveled and the number of gallons of gasoline consumed. What is the symbol for the gasoline consumption of the car?

36. The distance a rocket travels is measured in kilometers, and the time of travel is measured in seconds. What is the symbol for the units of the speed of the rocket?

37. The unit of frequency is the *hertz* (Hz), which has units of the reciprocal of seconds. What is the symbol for the hertz in terms of seconds?

38. The unit of electric capacitance is the *farad* (F), which has units of seconds per ohm. What is the symbol for the farad in terms of seconds and ohms?

39. Pressure can be expressed as the number of pounds per square foot. What is the symbol for the units of pressure?

40. The U.S. Customary unit for power is the *horsepower* (hp), which has units of foot pounds per second. What is the symbol for the units of the horsepower in terms of feet, pounds, and seconds?

41. The metric unit for force is the *newton* (N), which has units of kilogram meters per second squared (only seconds are squared). What is the symbol for the newton in terms of kilograms, meters, and seconds?

42. The metric unit for energy is the *joule* (J), which has units of a newton meter. What is the symbol for the joule in terms of newtons and meters?

43. The unit of electric charge, the *coulomb* (C), has units of an ampere second. What is the symbol of the coulomb in terms of amperes and seconds?

44. The unit for electric potential (voltage) is the *volt* (V), which has units of a joule per ampere second. What is the symbol for the volt in terms of joules, amperes, and seconds?

Now Try It! Answers

1. 100,000 cm   **2.** 1,000,000 mV   **3.** 4000 mm   **4.** 0.00006 s   **5.** 0.00025 m

2.3 **Reduction and Conversion of Units**

Section Objectives
- Understand the difference between reduction and conversion
- Use reduction to go from one metric unit to another
- Use conversion to change between U.S. Customary units of measure and metric units of measure

Recall that denominate numbers represent measurements or counts, and they are associated with units of measurement. When we are working with denominate numbers, it is sometimes necessary to change from one set of units to another. *A change within a system is called a* **reduction.** *A change from one system to another is called a* **conversion.**

To find the cost of floor covering for a room, we must know the area of the room, and this is generally given in square feet. Floor-covering cost is often given in cost per square yard so that it is necessary to reduce square feet to square yards. Here we are changing from square feet to square yards, which are both U.S. Customary units. This is called a reduction because we are changing units within the same system. All of the problems in Section 2.1 are examples of reduction.

Distances in Europe are given in kilometers. To determine equivalent distances in miles, we must convert kilometers to miles. Here we are changing from metric units to U.S. Customary units. This is called a conversion because we are changing units in one system to those in another.

For purposes of changing units, Table 2.4 gives some basic reduction and conversion factors. Other reduction factors are given in Table 2.3.

Table 2.4 **Reduction and Conversion Factors**

Reduction factors		
$144 \text{ in.}^2 = 1 \text{ ft}^2$	(all are exact)	$100 \text{ mm}^2 = 1 \text{ cm}^2$
$9 \text{ ft}^2 = 1 \text{ yd}^2$		$10{,}000 \text{ cm}^2 = 1 \text{ m}^2$
$1728 \text{ in.}^3 = 1 \text{ ft}^3$		$1000 \text{ mm}^3 = 1 \text{ cm}^3$
$27 \text{ ft}^3 = 1 \text{ yd}^3$		$1000 \text{ cm}^3 = 1 \text{ L}$

Conversion factors		
$1 \text{ in.} = 2.54 \text{ cm}$ (exactly)		$1 \text{ ft}^3 = 28.32 \text{ L}$
$1 \text{ m} = 39.37 \text{ in.}$	(all others are approx.)	$1 \text{ pt} = 473.2 \text{ cm}^3$
$1 \text{ mi} = 1.609 \text{ km}$		$1 \text{ L} = 1.057 \text{ qt}$
$1 \text{ lb} = 453.6 \text{ g}$		
$1 \text{ kg} = 2.205 \text{ lb}$		

To change a given number of one set of units to another set of units, we use unit analysis which was introduced in Section 2.2.

Example 1

If we have a number representing feet per second to be multiplied by another number representing seconds per minute, as far as the units are concerned

<table>
<tr><td>

Remember

The convenient way to use the values in the tables is in the form of a **unit fraction.** For example, since 1 in. = 2.54 cm we can write

$$\frac{1 \text{ in.}}{2.54 \text{ cm}} \quad \text{or} \quad \frac{2.54 \text{ cm}}{1 \text{ in.}}.$$

</td></tr>
</table>

we have

$$\frac{\text{ft}}{\text{s}} \times \frac{\text{s}}{\text{min}} = \frac{\text{ft} \times \cancel{\text{s}}}{\cancel{\text{s}} \times \text{min}} = \frac{\text{ft}}{\text{min}}.$$

This means that the final result would be in feet per minute.

In changing a number in one set of units to another set of units, we use reduction and conversion factors from Tables 2.3 and 2.4 and apply the principle illustrated in Example 1 for operating with the units themselves.

Example 2

Convert a distance of 2.800 mi to kilometers.
Since 1 mi = 1.609 km, the fraction

$$\frac{1.609 \text{ km}}{1 \text{ mi}}$$

is equal to 1 and we can therefore multiply 2.800 mi by that fraction to get

$$2.800 \cancel{\text{mi}} \left(\frac{1.609 \text{ km}}{1 \cancel{\text{mi}}} \right) = 4.505 \text{ km}.$$

▶ In this example we know to use the fraction 1.609 km/1 mi because we want the number of kilometers per mile. Also, we see that ***this fraction works because the mi units cancel.*** If we used the fraction 1 mi/1.609 km and tried to set up the conversion as

$$2.800 \text{ mi} \left(\frac{1 \text{ mi}}{1.609 \text{ km}} \right)$$

we see that no cancellation of units occurs. This shows that 1 mi/1.609 km is the wrong form to use.

Example 3

Reduce 30 mi/h to feet per second.

$$30 \text{ mi/h} = \left(30 \frac{\cancel{\text{mi}}}{\cancel{\text{h}}} \right) \left(\frac{5280 \text{ ft}}{1 \cancel{\text{mi}}} \right) \left(\frac{1 \cancel{\text{h}}}{60 \cancel{\text{min}}} \right) \left(\frac{1 \cancel{\text{min}}}{60 \text{ s}} \right)$$

$$= \frac{\overset{1}{\cancel{(30)}} \overset{88}{\cancel{(5280)}} \text{ ft}}{\underset{2}{\cancel{(60)}} \underset{1}{\cancel{(60)}} \text{ s}} = 44 \text{ ft/s}$$

Note that the only units remaining are those that were required.

Example 4

Reduce 575 g/cm³ to kilograms per cubic meter.

$$575\frac{g}{cm^3} = \left(575\frac{g}{cm^3}\right)\left(\frac{100\ cm}{1\ m}\right)^3\left(\frac{1\ kg}{1000\ g}\right)$$

$$= \left(575\frac{g}{cm^3}\right)\left(\frac{\overset{1000}{\cancel{1000000}}\ \cancel{cm^3}}{1\ m^3}\right)\left(\frac{1\ kg}{\underset{1}{\cancel{1000}}\ \cancel{g}}\right)$$

$$= 575 \times 1000\frac{kg}{m^3} = 575{,}000\frac{kg}{m^3}$$

Note that the fraction 100 cm/1 m was cubed so that the cm³ units could then be divided out.

Example 5

Convert 62.80 lb/in.² to kilograms per square meter.

$$62.80\ lb/in.^2 = \left(62.80\frac{lb}{in.^2}\right)\left(\frac{1\ kg}{2.205\ lb}\right)\left(\frac{1\ in.}{254\ cm}\right)^2\left(\frac{100\ cm}{1\ m}\right)^2$$

$$= \left(62.80\frac{\cancel{lb}}{\cancel{in.^2}}\right)\left(\frac{1\ kg}{2.205\ \cancel{lb}}\right)\left(\frac{1\ \cancel{in.^2}}{2.54^2\ \cancel{cm^2}}\right)\left(\frac{10000\ \cancel{cm^2}}{1\ m^2}\right)$$

$$= \frac{(62.80)(10000)}{(2.205)(2.54^2)}\frac{kg}{m^2} = 44{,}150\ kg/m^2$$

Example 6

A certain laptop weighs 8.00 pounds. What is its weight in kilograms?

To begin this problem we need to know that 1 kg = 2.205 lb. We set up a conversion sequence that allows us to cancel common units so that we end up with a result in kilograms.

$$\frac{8.00\ \cancel{lb}}{1} \times \frac{1\ kg}{2.205\ \cancel{lb}} = 3.63\ kg$$

Example 7

Solid copper wire is specified by its gauge. For example, 22-gauge wire is common to electronic equipment and has a diameter of 0.0253 inches. Convert the diameter to millimeters (mm).

To begin this problem we need to know that 2.54 cm = 1 in. and 10 mm = 1 cm. We set up the following conversion sequence:

$$\frac{0.0253 \text{ in.}}{1} \times \frac{2.54 \text{ cm}}{1 \text{ in.}} \times \frac{10 \text{ mm}}{1 \text{ cm}} = 0.643 \text{ mm}$$

Example 8

The average distance from the earth to the moon is 238,857 miles. The speed of a radio signal is the same as the speed of light which is 300 million meters/second. How many seconds does it take for a radio signal to reach the moon from earth?

To begin this problem we need to know how many kilometers (km) there are in a mile as well as the number of meters in a kilometer. We set up the following conversion sequence:

$$\frac{238,857 \text{ miles}}{1} \times \frac{1.609 \text{ km}}{1 \text{ mile}} \times \frac{1000 \text{ meters}}{1 \text{ km}} \times \frac{1 \text{ s}}{300,000,000 \text{ meters}} = 1.28 \text{ s}$$

Now Try It!

Convert the following:

1. 30 mi/h to feet per second
2. 6.50 kg to pounds
3. 30 ns to minutes
4. 575 g/cm^3 to kg/m^3
5. 326 mL to quarts

2.3 Exercises

In Exercises 1 through 16, use only the values given in Tables 2.3 and 2.4. Where applicable, round off to the indicated place value.

1. How many millimeters are there in 1 cm?

2. How many centigrams are there in 1 kg?

3. How many inches are there in 1 mi?

4. How many ounces are there in 1 t?

5. Reduce 2 ft^2 to square inches.

6. Reduce 50 cm^3 to cubic millimeters.

7. Convert 12 in. to centimeters.

8. Convert 8 kg to pounds. (Round off to tenths.)

9. Reduce 55 gal to quarts.

10. Reduce 60 kL to milliliters.

11. Convert 25 qt to liters. (Round off to tenths.)

12. Convert 27 cm to inches. (Round off to tenths.)

13. Reduce 5.2 m^2 to square centimeters.

14. Reduce 0.205 L to cubic centimeters.

15. Convert 256.3 L to cubic feet. (Round off to hundredths.)

16. Convert 967 in.^{3} to liters. (Round off to tenths.)

In Exercises 17 through 30 solve the given problems. Round off to the indicated place value.

17. While driving in France, a sales representative encounters a speed limit sign indicating 80 km/h. Convert that speed limit to mi/h. (Round off to units.)

18. A foreign car's weight is given as 1764 kg. Convert that value to pounds. (Round off to tens.)

19. The manual for a car specifies that the crankcase oil capacity is 4 qt. Convert that capacity to liters. (Round off to tenths.)

20. Near the earth's surface, the acceleration due to gravity is about 980 cm/s². Convert this to feet per second squared. (Round off to tenths.)

21. The speed of sound is about 770 mi/h. Change this speed to feet per second. (Round off to units.)

22. The density of water is about 62.4 lb/ft³. Convert this to kilograms per cubic meter (round off to units).

23. The average density of the earth is about 5.52 g/cm³. Express this in kilograms per cubic meter (round off to units).

24. The amount of water that can be pumped from a well is found to be 72 gal/min. Convert that rate to liters per second. (Round off to hundredths.)

25. The Olympic running track is 400 m. Convert this to miles.

26. The length of a power transmission line is 3,300 ft. How many meters is this?

27. A 100-yd. American football field would have to be specified in meters in England. What is the length of an American football field in meters?

28. The speed of sound is about 1130 ft/s. Change this speed to kilometers per hour.

29. A certain chemical is added to a pool at the rate of 3.47 oz per gallon of water. Convert this to pounds of chemical per cubic foot (ft³) of water.

30. The graph of braking distance versus car speed is part of a parabola and given by the equation $d = kv^2$. The distance, d, has units in meters and velocity, v, has units in meter/second. Find the units of k.

Now Try It! Answers

1. 44 ft/s   **2.** 14.3 lb   **3.** 0.0000000005 s   **4.** 575,000 kg/m³   **5.** 0.345 qt

2.4   Approximate Numbers and Significant Digits

Section Objectives
• Understand the difference between an approximate number and an exact value
• Determine the significant digits of a number

If you used a calculator to work through Example 6 in Section 2.3, your result was 3.628117914 kg. This result is misleading because a ten-digit answer is not necessary. The number of digits in the result determines its accuracy. In this section we consider the proper use of both accuracy and precision, which will be defined soon.

Some numbers are exact whereas others are approximate. When we perform calculations on numbers, we must consider the accuracy of these numbers, because they affect the accuracy of the results obtained. *Most numbers involved in technical and scientific work are approximate, because they represent some measurement. However, certain other numbers are exact, having been arrived at through some definition or counting process.* We can decide if a number is approximate or exact if we know how the number was determined.

Example 1

If a surveyor measures the distance between two benchmarks as 156.2 ft, we know that the 156.2 is **approximate.** A more precise measuring device may cause us to determine the length as 156.18 ft. However, regardless of the method of measurement used, we can never determine this length exactly.

If a voltage shown on a voltmeter is read as 116 V, the 116 is approximate. A more precise voltmeter may show the voltage as 115.7 V. However, this voltage cannot be determined exactly.

Example 2

If a computer counts the number of students majoring in science or a technology and prints this number as 768, this 768 is **exact.** We know the number of students was not 767 or 769. Because 768 was determined through a counting process, it is exact.

When we say that 60 s = 1 min, the 60 is exact, because this is a definition. By this definition there are exactly 60 s in 1 min. As another example illustrating how a number can be exact through definition, the inch is now officially defined so that 1 in. = 2.54 cm exactly.

When we are writing approximate numbers, we must often include some zeros so that the decimal point will be properly located, as in 0.003 or 5,600. However, except for these zeros, all other digits are considered to be **significant digits.** When we make certain computations with approximate numbers, we must know the number of significant digits. Example 3 illustrates how we determine this.

Example 3

All numbers in this example are assumed to be approximate.

34.7 has three significant digits.

8900 has two significant digits. We assume that *the two zeros are placeholders* (unless we have specific knowledge to the contrary). (Note: The overbar symbol (¯) is sometimes used to show that trailing zeros are significant digits. For example, 89,0$\overline{0}$0 has four significant digits.)

706.1 has four significant digits. The zero is not used for the location of the decimal point. It shows specifically the number of tens in the number.

5.90 has three significant digits. The ***zero is not necessary a placeholder and should not be written unless it is significant.***

Determining the Number of Significant Digits

- All non-zero digits are significant.
- Zeros placed between two significant digits are significant.
- Zeros placed to the right of both the decimal point and another significant digit are significant.
- Zeros placed at the beginning of a decimal problem or before other digits are not significant.
- Zeros used as placeholders are not significant.

Examine Table 2.5 which gives additional examples of approximate numbers along with the proper number of significant digits.

Table 2.5

Approximate number	Number of significant digits	Comment
87,000	2	The zeros are placeholders and are not counted as significant digits.
408,000	3	The zero farthest to the left is not a placeholder but the other three zeros are placeholders and are not counted as significant digits.
4.0005	5	The zeros are not placeholders; they do count as significant digits.
0.004	1	All zeros are placeholders and do not count as significant digits.
4.000	4	The zeros are not *required* as placeholders and they do count as significant digits.
0.000503	3	The three digits farthest to the right are significant, but the four zeros farthest to the left are placeholders and they do not count as significant digits.

Note that all nonzero digits are significant. Zeros, other than those used as placeholders for proper positioning of the decimal point, are also significant.

In computations involving approximate numbers, the position of the decimal point as well as the number of significant digits is important. See Figure 2.2.

The **precision** *of a number refers to the decimal position of the last significant digit.*

The **accuracy** *of a number refers to the number of significant digits in the number.*

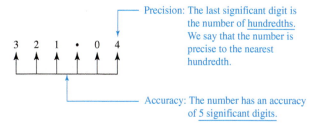

Precision: The last significant digit is the number of hundredths. We say that the number is precise to the nearest hundredth.

Accuracy: The number has an accuracy of 5 significant digits.

Figure 2.2

Example
4

Suppose that you are measuring an electric current with two ammeters. One ammeter reads 0.031 A and the other reads 0.0312 A. The second reading is more precise—the last significant digit is the number of ten-thousandths, whereas the first reading is expressed only to thousandths. The second reading is also more accurate, because it has three significant digits rather than two.

A machine part is measured to be 2.5 cm long. It is coated with a film 0.025 cm thick. The thickness of the film has been measured to a greater precision, although the two measurements have the same accuracy: two significant digits.

A segment of a newly completed highway is 9270 ft long. The concrete surface is 0.8 ft thick. Of these two numbers, 9270 is more accurate, because it contains three significant digits, and 0.8 is more precise, because it is expressed to tenths.

We can formalize the process for determining accuracy as follows: To determine the accuracy of a number, count the number of significant digits by following the rule for the appropriate case, as follows.

Case 1. The number has no digits to the right of the decimal point:

Count the digits from left to right, but stop with the last nonzero digit.

3 0 8 0 0   Accuracy: 3 significant digits

3

Case 2. The number has at least one digit to the right of the decimal point:

Begin with the leftmost nonzero digit and count all digits to the right.

0 . 0 0 2 0 3 0   Accuracy: 4 significant digits

4

The last significant digit of an approximate number is not completely accurate. It has usually been determined by estimation or rounding off. However, we do know that it is in error by at most one-half a unit in its place value.

Example 5

When we measure the distance in Example 1 to be 156.2 ft, we are saying that the length is at least 156.15 ft and no longer than 156.25 ft. Any value between these two, rounded off to tenths, would be expressed as 156.2 ft.

In converting the fraction $\frac{2}{3}$ to the decimal form 0.667, we are saying that the value is between 0.6665 and 0.6675.

The principle of **rounding off** *is to write the closest approximation, with the last significant digit in a specified position, or with a specified number of significant digits.* We shall now formalize the process of rounding off as follows:

Procedure for Rounding Off

- Determine the number of significant digits.
- Consider the digit in the next place to the right.
- If this digit is less than 5, we accept the digit in the last place.
- If this digit is greater than or equal to 5, we increase the digit in the last place by 1 and this result digit becomes the final significant digit in our approximation.

Example 6

70,360 rounded off to three significant digits is 70,400.

third sig. digit   5 or greater

70360

Add 1 to the 3 and replace 6 with 0

70,430 rounded off to three significant digits is 70,400.
187.35 rounded off to four significant digits is 187.4.
71,500 rounded off to two significant digits is 72,000.

Example 7

39.72 rounded off to two significant digits is 40.

▶ When 1 is added to the second significant digit, 9, we get 10, and ***the 1 must be carried to the tens.*** Also, because there are no significant digits in the result to

▶ the right of the decimal point, ***the 7 and 2 are not replaced with zeros.***

> ## Expressing Results From Arithmetic Operations on Approximate Numbers
>
> 1. *When approximate numbers are **added** or **subtracted**, the result is expressed with the precision of the least precise number.*
>
> 2. *When approximate numbers are **multiplied** or **divided**, the result is expressed with the accuracy of the least accurate number. That is, use the smallest number of significant digits.*
>
> 3. *When the root of an approximate number is found, the **accuracy of the result is the same as the accuracy of the original number.** That is, use the same number of significant digits.*

Example
8

Add the approximate numbers 73.2, 8.0627, 93.57, and 66.296.

In addition or subtraction, the number of decimal places in the answer should be the same as the least number of decimal places in any of the numbers being added or subtracted.

In this problem 73.2 has the least number of decimal places. Therefore, we will round off our final result to one decimal place.

$$73.2 + 8.0627 + 93.57 + 66.296 = 241.1287$$

Rounding our result off to one decimal place gives us a final result of 241.1

Example
9

Divide 292.6 by 3.4, where each is an approximate number.

In multiplication or division, the number of significant digits in the answer should be the same as the least number of significant digits in any of the numbers being multiplied or divided.

In this problem 3.4 has the least number of significant digits. It shows 2 significant digits whereas 292.6 shows 4 significant digits. Therefore, the final result will be written to two significant digits.

$$292.6 \div 3.4 = 86.0588$$

Rounding our results off to two significant digits gives us a final result of 86.

2.4 Exercises

In Exercises 1 through 8, determine whether the numbers given are exact or approximate.

1. An engine has 6 cylinders.

2. The speed of sound is 770 mi/h.

3. A copy machine weighs 67 lb.

4. A gear has 36 teeth.

5. The melting point of gold is 1063°C.

6. The 25 math students had a mean test grade of 79.3.

7. A semiconductor 1 cm by 1 mm is priced at $2.80.

8. A calculator has 40 keys and its battery lasts for 987 h of use.

In Exercises 9 through 16, determine the number of significant digits in the given approximate numbers.

9. 563;   4029

10. 46.8;   5200

11. 3799;   2001

12. 0.0025;   0.6237

13. 5.80;   5.08

14. 7000;   7000.0

15. 10060;   403020

16. 9.000;   0.009

In Exercises 17 through 24, determine which of the two numbers in each pair of approximate numbers is (a) more precise and (b) more accurate.

17. 3.764;   2.81

18. 0.041;   7.673

19. 30.8;   0.01

20. 70,370;   50,400

21. 0.1;   78.0

22. 7040;   37.1

23. 7000;   0.004

24. 50.060;   8.914

In Exercises 25 through 36, round off the given approximate numbers (a) to three significant digits and (b) to two significant digits.

25. 5.713

26. 53.72

27. 6.934

28. 27.81

29. 4096

30. 287.4

31. 46792

32. 32768

33. 501.46

34. 7435

35. 0.21505

36. 0.6350

In Exercises 37 through 48, solve the given problems.

37. A surveyor measured the road frontage of a parcel of land and obtained a distance of 128.3 ft. Based on this result, find the lowest possible value and the highest possible value.

38. A micrometer caliper is used to measure the diameter of a bolt and a reading of 0.768 in. is obtained. Based on this measurement, find the smallest possible diameter and the largest possible diameter.

39. An automobile manufacturer claims that the gasoline tank on a certain car holds approximately 82 L. What are the very least and the very greatest capacities?

40. The maximum weight an industrial robot can lift is measured as 46 lb. Based on this measurement, what is the lowest possible value and the highest possible value of this maximum weight?

41. A chemist has a container of 164.0 mL of sulfuric acid. Convert this volume to quarts and use the same degree of accuracy.

42. A design specification calls for a washer that is 0.0625 in. thick. Convert this thickness to millimeters and use the same degree of accuracy.

43. For the 100-yd length of a football field, assume that all three digits are significant. Convert this length to meters and use the same degree of accuracy.

44. An architect specifies that a support beam must be 7.315 m long. Convert that length to feet and use the same degree of accuracy.

45. Three adjacent lots had road frontage measured to be 150.4 ft, 95.66 ft, and 81 ft. What is the total length of the frontages of these three lots? Round off your results to the proper degree of precision.

46. The 2 daytime running lights on the front of a new model car each draw a current of 0.35 A. The 2 front headlights each draw 0.445 A and the 2 taillights each draw 0.457 A. Find the total current drawn by all of these lamps. Round off your results to the proper degree of precision.

47. In an electric circuit, the power (in watts) is found by multiplying the current (in amperes) by the voltage (in volts). Find the power developed in a circuit with a current of 0.0225 A and a voltage of 1.657 V. Round off your results to the proper degree of accuracy.

48. Find the average velocity of an object that falls 259 feet in 4.028 seconds. Round off your results to the proper degree of accuracy.

Chapter Summary

Key Terms

denominate numbers

units of measure

U.S. Customary System

International System of
    Units

unit analysis

unit fraction

base units

reduction

conversion

approximate values

exact values

significant digits

precision

accuracy

rounding off

Key Concepts

• The United States currently uses two systems of measurements—the U.S. Customary System and the International System of Units or SI System, commonly referred to as the metric system. It is important that you be able to work with units in each of these systems.

• Unit analysis provides a systematic method for changing from one unit to another. When using unit analysis we work to cancel out units appearing in both the numerator and the denominator. In the end, the unit that you are looking for on the right-hand side of the equation is the same as the unit that remains on the left-hand side of the equation after canceling all appropriate units.

• In the metric system, units of different magnitudes vary by a power of ten. Therefore, shifting the decimal to the right or to the left is all that is needed when changing units within this system. It is important to understand the metric prefixes and what they represent in order to successfully shift the decimal in the correct direction.

• Use unit analysis as a systematic approach for both reduction and conversion of units. Reduction is used when changing units within the same system. Conversion refers to changing units from one system to another.

• Significant digits are very important in science and technical fields. The significance of a digit has to do with whether it represents a true measurement or not. Any digit that is actually measured or estimated will be considered significant. Placeholders, or digits that have not been measured or estimated, are not considered significant.

Review Exercises

In Exercises 1 through 4, determine the number of significant digits in the given approximate numbers.

1. 6508;   70.43

2. 0.06;   0.60

3. 6070;   6007

4. 10.00;   0.002030

In Exercises 5 through 8, determine which of each pair of approximate numbers is (a) more precise and (b) more accurate.

5. 7.32;   73.2

6. 8000;   80.0

7. 6.49;   207.31

8. 98.568;   0.0021

In Exercises 9 through 16, round off each of the given approximate numbers (a) to three significant digits and (b) to two significant digits.

9. 98.46   **10.** 2.734

11. 60540   **12.** 219500

13. 672.8   **14.** 69005

15. 0.7000   **16.** 4935

In Exercises 17 through 20, give the meaning of each metric unit.

17. μg   **18.** cA

19. kilosecond   **20.** megavolt

In Exercises 21 through 32, the given numbers are approximate. Make the indicated reductions and conversions.

21. Reduce 385 mm^3 to cubic centimeters.

22. Reduce 0.475 m^2 to square centimeters.

23. Convert 5.2 in. to centimeters.

24. Convert 46.5 in. to meters.

25. Reduce 4.452 gal to pints.

26. Reduce 18.5 km to centimeters.

27. Convert 27 ft^3 to liters.

28. Convert 3.206 km to miles.

29. Reduce 0.43 ft^3 to cubic inches.

30. Reduce 28.3 in. to square feet.

31. Convert 2.45 mi/h to meters per second.

32. Convert 52 ft/min to centimeters per hour.

In Exercises 33 through 42 solve the given problems.

33. A machinist mills a template and measures the thickness to be 0.186 in. Based on that measurement, find the lowest value and the highest value of the thickness.

34. A carbon-zinc size D flashlight battery has a voltage measured as 1.5 V. Based on that measurement, find the lowest value and the highest value of the voltage.

35. A chemist combines three solids with masses of 2.841 g, 3.729 g, and 15.27 g. What is the total of these three masses?

36. A wire 4.39 m long is cut into 12 equal sections. What is the length of each section?

37. One meter of steel will increase in length by 0.000012 m for each 1°C increase in temperature. What is the increase in length of a steel girder 38 m long if the temperature increases by 45°C?

38. A board measures 5.25 in. wide after shrinking 12% while drying. What was the original width of the board?

39. If the density of gasoline is 5.6 lb/gal, find its density in kilograms per liter.

40. An FAA flight service station reports that the winds aloft at 10,000 ft are at a speed of 45 knots. Find that speed in miles per hour. Assume that a knot is exactly one nautical mile per hour and that one nautical mile is about 6080 ft.

41. Cross-sectional areas of wires are often measured in square mils, where 1 mil = 0.001 in. exactly. Find the cross-sectional area (in square inches) of a wire with an area of 287 mil^2.

42. A technician records the time required for a robot's arm to swing from the extreme right to the extreme left. A time of 4.3 s is recorded. The time for the return swing is then recorded as 4.75 s. What is the difference between the two times?

Chapter Test

1. Determine the number of significant digits in the following approximate values:
   a. 40,700
   b. 0.0723
   c. 605

2. Round off each of the following approximate values to three significant digits:
   a. 4.361
   b. 0.006155
   c. 105.2

3. Give the name and meaning for the units whose symbols are:
   a. kV
   b. mA
   c. mm

4. Make the indicated reductions or conversions:
   a. Reduce 1 km to centimeters
   b. Convert 5.60 kg to pounds

5. A car's gasoline tank holds 56 L. Convert this capacity to gallons.

6. Water flows from a kitchen faucet at the rate of 8.5 gallons per minute. What is this rate in liters per second?

7. A certain chemical is added to a pool at the rate of 3.47 oz per gallon of water. Convert this to pounds of chemical per cubic foot (ft^3) of water.

8. The height of the Empire State Building is 1414 feet. The height of the Eiffel Tower in Paris is 300.12 meters. Use unit analysis to find the height of the Empire State Building in meters, then determine which building is taller.

9. The formula for converting Celsius temperatures to Fahrenheit is $F = 1.8C + 32$, where C is the temperature in degrees Celsius and F is the temperature in degrees Fahrenheit. If the boiling point of water is 100°C, what is it in degrees Fahrenheit?

10. A commercial jet with 230 passengers on a 2850 km flight from Vancouver to Chicago averaged 765 kilometers per hour and used fuel at a rate of 5650 liters per hour.
    a. What was the fuel consumption in kilometers per liter?
    b. What was the fuel consumption in liters per passenger?

3

Introduction to Algebra

Where am I ever going to use this? Why do I need to learn this? These questions are asked regularly in algebra classes across the country. Most students view algebra as a vague process that involves the manipulation of symbols based on a set of rules they often do not understand. Algebra is a powerful tool for solving real-world problems. It is estimated that over 75% of all jobs require skill in higher-level mathematics. Algebra is used in the "hard" sciences, such as biology, chemistry, and physics; the "soft" sciences, such as economics, psychology, and sociology; engineering fields, such as civil, mechanical, and industrial engineering; and technological fields such as computers and telecommunications.

Many employers require at least a basic knowledge of algebra, because it helps us to solve many problems that would otherwise remain unsolved. One important feature of algebra is its use of letters to represent unknown quantities.

Consider the following two statements that describe the relationship between the Celsius and Fahrenheit temperature scales:

Verbal form: The Celsius temperature equals five-ninths of the Fahrenheit temperature minus thirty-two degrees.

Algebraic form: $C = \frac{5}{9}(F - 32°)$

In addition to being unclear or ambiguous, the verbal statement does not express the relationship with the clarity of the algebraic statement. Calculations can be done with much greater ease if the algebraic expression is used instead of the verbal expression. Also, we shall see that the algebraic form enables us to solve some problems that could never be solved with the verbal form.

In this chapter we explore the meaning of algebra and some of the important terminology associated with it. We also examine the basic algebraic operations.

In addition to the general theory of algebra, this chapter also includes applied problems. For example, because aviation weather reports give some temperatures in the Fahrenheit scale and others in the Celsius scale, a pilot may need to convert a reported temperature of 10°F to a temperature on the Celsius scale. Using the above algebraic form of the equation relating Celsius and Fahrenheit temperatures, we will show how this can be done.

3.1   Working with Formulas

Section Objectives
• Become familiar with basic terminology associated with the study of algebra
• Understand the usefulness of formulas
• Evaluate simple formulas

We begin by defining algebra as a generalization of arithmetic that uses letters to represent numbers.

If we wished to find the cost of a piece of sheet metal, we would multiply the area, say in square feet, by the cost of the metal per square foot. If the piece is rectangular, we find the area by multiplying the length by the width. Consequently, the cost equals the length (in feet) times the width (in feet) times the cost of the sheet metal (per square foot). Rather than writing out such statements as the one just given, we can write

$$C = l \times w \times c$$

where it is specified and understood that C is the total cost, l is the length, w is the width, and c is the cost per unit area of the metal. Note that C and c represent different quantities.

The algebraic equation $C = l \times w \times c$ is an example of a **formula**. A **formula** is nothing more than a rule or equation written in mathematical language.

The letters C, l, w, and c are *literal symbols* or **variables** that are simply letters used to represent numbers. Variables can be assigned any value.

Formulas can also contain letters or symbols that represent **constant** values. These are values that do not vary.

A major advantage of a formula is its general use in many different cases. The formula $C = l \times w \times c$ can be applied to an infinite number of different situations. If we want to find the cost of a rectangular piece of sheet metal with specific dimensions, we need only substitute the appropriate numbers into the formula.

Example 1

If a rectangular piece of sheet metal is 5 ft long and 4 ft wide and costs $3/ft^2, we have

$$l = 5 \text{ ft} \qquad w = 4 \text{ ft} \qquad c = \$3/\text{ft}^2.$$

The formula $C = l \times w \times c$ becomes, in this case,

substitute

$$C = 5 \times 4 \times 3 = \$60.$$

It would cost $60 to purchase that piece. For another rectangular piece with different dimensions and cost we would still substitute into the same formula, because it is a general expression that can be used in many different cases.

Recall that **factors** are *quantities being multiplied.* If numbers represented by symbols are to be multiplied, the expression is written without the signs of multiplication. The symbols are simply placed adjacent to each other. *One of these may be a numerical constant, in which case it is normally written first and is called the* **numerical coefficient** *of the expression.*

Example 2

Instead of writing $C = l \times w \times c$, the expression for the cost of the sheet metal given earlier would be written as

$$C = lwc. \qquad lwc = l \times w \times c$$

The formula for the area of a triangle is given by $A = \frac{1}{2} \times b \times h$ where b represents the length of the base of the triangle and h represents its height. We can rewrite this formula without the signs of multiplication as follows:

$$A = \frac{1}{2} \times b \times h \text{ is the same as } A = \frac{1}{2}bh.$$

In this example, $\frac{1}{2}$ is a numerical coefficient.

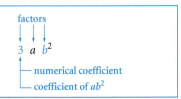

factors

$$3 \quad a \quad b^2$$

— numerical coefficient
— coefficient of ab^2

> In a given product, the quantities being multiplied are called **factors** of the product, just as they are in arithmetic. In general, the quantities multiplying a factor are the **coefficient** of that factor.

Example 3

In the expression lwc, the l, w, and c are factors.

In the expression $3ab^2$, the 3, a, and b^2 are factors.

In the expression $3ab^2$, we know that $b^2 = bb$. We can list the factors of this expression without the use of the exponents: 3, a, b, and b.

Wherever mathematics can be used, we can use symbols to represent the quantities involved. A verbal statement must be translated into an algebraic expression so that it can be used in algebra. The following illustrates several verbal statements and their equivalent algebraic formulas.

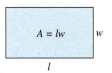

$A = lw$ w

l

Figure 3.1

e

e

e

Figure 3.2

Figure 3.3

Verbal statement	Formula	Meaning of literal symbols
1. The area of a rectangle equals the length times the width.	$A = lw$	A is the area. l is the length. w is the width. (See Figure 3.1.)
2. The volume of a cube equals the cube of the length of one edge.	$V = e^3$	V is the volume. e is the length of an edge. (See Figure 3.2.)
3. The pitch of a screw thread times the number of threads per inch equals 1.	$pN = 1$	p is the pitch. N is the number of threads/in. (See Figure 3.3.)

Verbal statement	Formula	Meaning of literal symbols
4. The distance an object travels equals the average speed multiplied by the time of travel.	$d = rt$	d is the distance. r is the average speed. t is the elapsed time.
5. The simple interest earned on principal equals the principal times the rate of interest times the time the money is invested.	$I = Prt$	I is the interest. P is the principal. r is the rate of interest. t is the time.
6. The voltage across a resistor in an electric circuit equals the current in the circuit times the resistance.	$V = IR$	V is the voltage. I is the current. R is the resistance.

These formulas are valid for the given conditions. As mentioned before, if we wish to determine the result of using any of these formulas for specific values, we substitute these values for the letters in the formula and then calculate the result. Example 4 illustrates this in the case of two of these formulas.

Example **4**

The volume of a cube 3.0 in. on an edge is found by substituting 3.0 for e in the formula $V = e^3$. (See Figure 3.4) We get

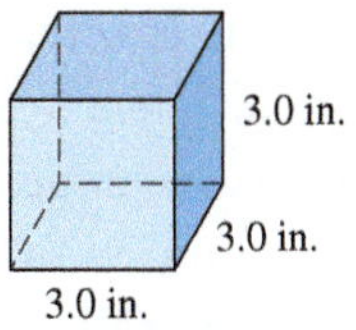

$$V = (3.0)^3 = 27 \text{ in.}^3$$

If the cube had been 3.0 cm on an edge, the volume would have been 27 cm^3.
▶ **We must be careful to attach the proper units to the result.**

3.0 in.
3.0 in.
3.0 in.

Figure 3.4

Example **5**

If the current in a certain electric circuit is 3.0 A, the voltage across a 6.0-Ω resistor in the circuit is found by substituting 3.0 for I and 6.0 for R in the formula $V = IR$. We get

$$V = (3.0)(6.0) = 18 \text{ V}.$$

▶ *Note that the letter V has a different meaning in each of these cases.* In the formula $V = IR$, the literal symbol V represents voltage. But the symbol V in 18 V represents the symbol for volts, a unit of measure. Here we have a unit symbol that is the same as the literal symbol in a formula, and we must be careful to avoid confusion between these two uses.

Example **6**

For paving costs of $3.50 a square yard, find the cost to the nearest dollar of paving a triangular court whose base is 35 yd and whose height is 21 yd.

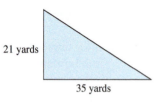

Figure 3.5

As noted in Example 2, the formula for the area of a triangle is $A = \frac{1}{2}bh$. By substituting 35 yd for b and 21 yd for h we get

$$A = \frac{1}{2}(35)(21) = 367.5 \approx 370 \text{ yd}^2.$$

We are not finished though, because the original problem asked for the cost of paving. Since it costs $3.50/yd^2, the final cost will be

$$\$3.50(367.5) = \$1286.25 \approx \$1300.$$

Example 7

What is the circumference of a circular lake 33 m in diameter?

The formula for the circumference of a circle is $C = 2\pi r$ or $C = \pi d$. In this case we substitute 33 for the diameter and 3.14 for π.

$$C = (3.14)(33) = 103.62 \text{ m} \approx 100 \text{ m}.$$

Note that in this example, the Greek letter π (pi) is a symbol that is used to represent a constant value.

When selecting letters to be used as literal symbols, it is often helpful to choose letters that suggest the quantity being represented. The letter V is commonly used for voltage or volume. Also, t frequently represents time, d and s are often used for distance, r for a rate, and so on. It is also common to use the letters x, y, or z to represent unknown variable quantities, whereas the letters a, b, c, and k are often used to represent unknown constant quantities.

3.1 Exercises

In Exercises 1 through 8, identify the individual factors, without exponents, of the given algebraic expressions. (List them in the same way as shown for Exercises 1 through 8.)

1. bc

2. $2ax$

3. $7pqr$

4. $3xyz$

5. i^2R

6. $17a^2b$

7. abc^3

8. $\pi r^2 h$

In Exercises 9 through 16, identify the coefficient of the factor that is listed second by examining the expression listed first.

9. $6x$; x

10. $3s^2$; s^2

11. $2\pi r$; r

12. $8a^2b$; b

13. $4\pi emr$; mr

14. $qBLD$; D

15. mr^2w^2; r^2

16. $36a^2cd$; a^2d

In Exercises 17 through 24, use the literal symbols listed at the end of each problem to translate the given statements into algebraic formulas.

17. A first number x equals four times a second number y. (x, y)

18. A first number equals the square of a second number. (x, y)

19. In a given length, the number of millimeters equals 10 times the number of centimeters. (m, c)

20. The total surface area A of a cube equals six times the square of the length of one of its edges e. (A, e)

21. The volume, in gallons, of a rectangular container equals 7.48 times the length times the width times the depth (these dimensions are measured in feet). (V, l, w, d) See Figure 3.6.

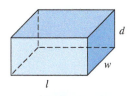

d

w

l

Figure 3.6

22. The heat H developed (per second) in a resistor in an electric circuit equals the product of the resistance R and the square of the current i in the circuit. (H, R, i)

23. The distance d an object falls due to gravity equals $\frac{1}{2}$ times the acceleration g due to gravity times the square of the time of fall t. (d, g, t)

24. The rate of emission R of energy per unit area of the filament of an electric light bulb equals about 0.00002 times the fourth power of the thermodynamic temperature T of the filament. (R, T)

In Exercises 25 through 32, determine the required formula.

25. The number N of feet in a distance of x miles.

26. The area A of a square of side s.

27. The number N of square feet in a rectangular area x yd by y yd.

28. The cost C of renting a masonry drill for t hours if the rental fee is $3/h.

29. The number of bolts N in a shipment if there are n boxes of bolts and each box contains 24 bolts.

30. The number N of bits in a computer with x bytes if each byte consists of 8 bits.

31. The cost C of putting an edge strip around a square piece of wood of side s, if the strip cost c ¢/ft.

32. The total area A of N square tiles if each tile has sides of length s cm.

In Exercises 33 through 44, evaluate the required formula for the given values. Many of these formulas can be found on pages 74–75.

33. A missile travels at the rate of 6,000 mi/h. Find the distance it will travel in 15 min. (Note the units given.)

34. Find the simple interest on $500 at an interest rate of 8% for 3 years.

35. Find the voltage in a circuit in which the current is 0.075 A and the resistance is 20 Ω.

36. Find the volume of a cube for which the edge is 4.3 cm.

37. A holding tank for industrial waste is a rectangular container 4.0 ft by 8.0 ft by 3.0 ft. Find the capacity of this tank in gallons. (See Exercise 21.)

38. The formula for the circumference of a circle is $C = 2\pi r$ and the formula for the area of a circle is $A = \pi r^2$ where $\pi = 3.14$ and r represents the radius of the circle. Find the circumference and area of a circle if the radius is 5 in., 10 in., and 40 in.

39. If the distance d an object falls due to gravity equals $\frac{1}{2}$ times the acceleration g due to gravity times the square of the time of the fall t $(d = \frac{1}{2}gt^2)$, find d if $g = 32.2$ ft/s² and $t = 2.25$ s.

40. The formula for converting Celsius temperatures to Fahrenheit is $F = 1.8C + 32$, where C is the temperature in degrees Celsius and F is the temperature in degrees Fahrenheit. If the boiling point of water is 100°C, what is it in degrees Fahrenheit?

41. Use the formula developed in Exercise 29 to find N if $n = 144$.

42. One bag of fertilizer will cover 500 ft² of lawn. Your front lawn is approximately 105 ft by 50 ft, and your back yard is approximately 130 ft by 100 ft. How many bags of fertilizer will you need to fertilize both your front and back yard?

43. If an edge strip costs $1.35/ft, find the cost of putting an edge strip around a square piece of wood whose side measures 12 ft. Use the formula developed in Exercise 31.

44. When a capacitor C is connected in parallel with an inductor L, the combination is referred to as a **tank circuit** (see Figure 3.7). The circuit has a natural frequency of oscillation given by the formula $f = \sqrt{\dfrac{1}{2\pi\sqrt{LC}}}$ where f is the resonant frequency measured in Hertz, L is

measured in henries (H), and C is measured in farads (F). Determine the frequency for the tank circuit below given $L = 15.2\,\mu\text{H}$ and $C = 82\,\mu\text{F}$.

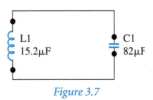

L1
15.2μF

C1
82μF

Figure 3.7

3.2   Basic Algebraic Expressions

Section Objectives
- Recognize algebraic expressions
- Identify terms of an algebraic expression
- Simplify algebraic expressions
- Identify polynomials, monomials, binomials, and trinomials
- Identify the degree of a polynomial expression

In the last section we discussed literal symbols and used them in some simple formulas. The discussion, examples, and exercises of that section used only the multiplication of literal symbols. We must often add, subtract, multiply, divide, and perform other operations on literal symbols. This section presents an introduction to these operations. Here we establish the need for such operations, and also introduce some new terms necessary for later topics.

An **algebraic expression** is used to describe a combination of constants and variables joined by arithmetic operations (addition, subtraction, multiplication, division, roots, and powers).

If an algebraic expression contains parts that are separated by a plus or minus sign, these parts are referred to as **terms.** Those terms that have identical variable parts are called **like terms.**

> *An algebraic expression **does not** contain an equal sign.*

Example
1

Examples of algebraic expressions include $4x^2 - 7x + 3$, $3x + 5$, and $4xy$. The algebraic expression $3x + 5$ has two terms whereas the algebraic expression $4x^2 - 7x + 3$ has three terms and $4xy$ has one term.

Example

2

The algebraic expression $x + 3xy - 2x$ contains three terms: x, $3xy$, and $-2x$. The terms x and $-2x$ are like terms, because the literal parts are the same. The $3xy$ term is not like the x and $-2x$ terms because the *factor y* makes the literal part different.

Remember that it was not enough to say that like terms have the same variables, it was equally important that their variables were raised to the same power.

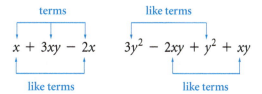

The expression $3y^2 - 2xy + y^2 + xy$ has four terms: $3y^2$, $-2xy$, y^2, and xy. The terms $3y^2$ and y^2 are like terms, and the terms $-2xy$ and xy are like terms.

The algebraic expression $3ab^2 + 4ab + 7b^2a$ has $3ab^2$ and $7b^2a$ as like terms. The term $4ab$ is not like $3ab^2$, but $7b^2a$ is like $3ab^2$, since $b^2a = ab^2$.

A **polynomial** is an algebraic expression consisting of a single term or the sum of terms, each of which is one of the following:

1. A constant.

2. A variable with an exponent that is a positive integer.

3. A product of constants and variables with exponents that are positive integers.

Each of the algebraic expressions presented thus far are examples of polynomials. Specific names have been given to some types of polynomials so that they can be referred to easily.

The following expressions are polynomials: $2x^2 - 3$, $4x^3 - x - 7$, and $x^3y - 4xy^2 - 6$.

The following expressions are not polynomials: $2x^2 + \frac{1}{x}$, $4x^3 - \sqrt{x} - 7$, and $\frac{1}{x^2 + 1}$.

A **monomial** is a polynomial with one term, such as $7x$.

A **binomial** is a polynomial with two terms, such as $5x - 3$.

A **trinomial** is a polynomial with three terms, such as $7x^3 - 2x + 5$.

The **degree of a term** is the sum of the exponents of the variables for that term. The **degree of a polynomial** is the largest degree of all its terms.

Example 3

The term $-3x^2$ has degree 2 because the exponent of the variable is 2.

The term $2x^2y^3$ has degree 5 because the sum of the variable exponents is $2 + 3 = 5$.

The degree of the polynomial $6x^3 + 2x^2 - 9$ is 3 because that is the largest degree of any of the terms of this polynomial.

Now Try It! 1

Determine the degree of each of the following:

a. $3xyz^2$
b. $5a^8$
c. $4x^2y^2 + 7xy^2 - xy$
d. $2x^2 - 4x + 7$

Suppose we want to multiply two algebraic expressions, at least one of which is itself the sum of terms, or the difference of terms, so we must be able to indicate that the sum or difference is to be considered as one of the factors. *For this purpose, we use parentheses to group the terms together.* Example 4 illustrates this use of parentheses.

Example 4

A supplier of electronics components charges p dollars for a certain transformer. If it cost c dollars to make, the profit on the sale of a transformer is $p - c$ dollars. If n transformers are sold, the total profit P is

$p - c$ grouped by parentheses

$$P = n(p - c).$$

▶ **We do not write this as $np - c$;** this arrangement would indicate that n and p are to be multiplied and c is to be subtracted from the product np.

To indicate that the sum of two numbers a and b is to be multiplied by the difference between two other numbers c and d, we write

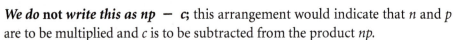

grouped by parentheses

$$(a + b)(c - d).$$

This expression is a *one-term* expression, although the *factors $a + b$ and $c - d$* each have two terms.

Now Try It! 2

Identify each of the following as a monomial, binomial or trinomial.

a. $3xz^2 + 2x$
b. $5a^8$
c. $2x^2 - 4x + 7$
d. $5x - 10$

In algebra, division of one number by another is most commonly designated as a fraction. That is, $a \div b$ would be written as

$$\frac{a}{b}$$

or occasionally as a/b. Here a is the numerator and b is the denominator. The following example illustrates that division is necessary for the algebraic solution of certain problems.

Example 5

The frequency f of a radio wave equals its velocity v divided by its wavelength l. This is written as

$$f = \frac{v}{l}.$$

In a particular type of electric circuitry, the combined resistance C of two resistors r and R is found to be

$$C = \frac{rR}{r + R}.$$

Here we note that the numerator is the product of r and R and the denominator is the sum of r and R.

Another important type of algebraic expression involves the square root of a number. We can define the square root of a number as follows:

$$\sqrt{n} = a \quad \text{if} \quad a^2 = n \quad \text{and } a \text{ is nonnegative}$$

(At this time we are considering only nonnegative numbers for n and a.) Here again we see the use of a concise algebraic expression.

Numerous technical applications in algebra involve the use of square roots. The following example illustrates one of them.

Example 6

One application of square roots is the determination of the time it takes an object to fall. The approximate time (in seconds) that it takes for an object to fall due to the influence of gravity equals the square root of the quotient of the distance (in feet) fallen divided by 16. The formula for this is

$$t = \sqrt{\frac{d}{16}}.$$

As we noted in Section 3.1, an algebraic expression can be evaluated for a specific set of values by substituting the proper values. Example 7 evaluates some expressions given in earlier examples.

Example
7

The conversion from the Celsius temperature scale to the Kelvin temperature scale is given by $K = C + 273$. If $C = 20°$, the temperature in the Kelvin scale can be found as follows:

$$K = C + 273 \qquad \text{general expression}$$
$$= 20 + 273 \qquad \text{substitute 20 for } C$$
$$= 293. \qquad \text{result for specific case of } C = 20°C$$

In Example 4, if $p = \$210$, $c = \$120$, and $n = 6$, the profit on these sales, $P = n(p - c)$, becomes

$$P = 6(210 - 120) = 6(90) = \$540.$$

In Example 6, if an object falls 144 ft, the time of fall,

$$t = \sqrt{\frac{d}{16}}$$

becomes

$$t = \sqrt{\frac{144}{16}} = \sqrt{9.0} = 3.0 \text{ s.}$$

Example
8

The formula $C = \frac{5}{9}(F - 32°)$ relates temperatures on the Celsius and Fahreheit scales. Because aviation weather reports give some temperatures in the Fahrenheit scale and others in the Celsius scale, a pilot may need to convert a reported temperature of 10°F to the Celsius scale. Using the above formula we get

$$C = \frac{5}{9}(10° - 32°) = \frac{5}{9}(-22°) \approx -12°.$$

That is, 10°F is equivalent to $-12°$C (rounded to the nearest degree).

3.2 Exercises

In Exercises 1 through 4, identify the terms of the given algebraic expressions.

1. $x^2 + 4xy - 7x$

2. $a + 2ab - \dfrac{a}{b}$

3. $12 - 5xy + 7x - \dfrac{x}{8}$

4. $3(x + y) - 6a + 3x$

In Exercises 5 through 12, identify the like terms in the given expressions.

5. $3x - 7y + 2x$

6. $9a - 2b + 5b$

7. $3 + x + 5x - 3y$

8. $a + ab + 2a - 3b$

9. $5m^2 - 8mn + m^2n - mn + \dfrac{m}{n}$

10. $6R - bR + 3R^2 - 5bR$

11. $6(x - y) + (x + y) - 3(x - y)$

12. $5(a - b) - 3a + 7b + (a - b)$

In Exercises 13 through 16, indicate the multiplication of the given factors. (Only indicate the multiplication—do not multiply.)

13. $6, a, a - x$

14. $3, x^2, a - x^2$

15. $x^2, a - x, a + x$

16. $3, a^2, a^2 + x^2, a^2 - x^2$

In Exercises 17 through 20, express the indicated divisions as fractions.

17. $2 \div 5a$

18. $x^2 \div a$

19. $6 \div (a - b)$

20. $(3x^2 - 2x + 5) \div (x + 2)$

In Exercises 21 through 24, evaluate the formulas by using the given values.

21. $K = C + 273$, for $C = 55$

22. $P = n(p - c)$, for $n = 8, p = \$250, c = \$140$

23. $t = \sqrt{\dfrac{d}{16}}$, for $d = 64$

24. $C = \dfrac{5}{9}(F - 32°)$, for $F = 68°$

In Exercises 25 through 32, write the required formula.

25. A section x ft long is cut from 50 ft piece of cable. Express the length L of the remaining cable in terms of x.

26. A rectangular place has length l and width w. Write a formula for its perimeter, which is the sum of the lengths of all four sides.

27. A shipping carton is a rectangular box with square ends. If the square ends have sides of length x and the

overall length of the box is l, write a formula for the total area A (of the six sides) of the surface in terms of x and l. See Figure 3.8.

Figure 3.8

28. The voltage V across an electric circuit equals the current I times the resistance in the circuit. If the resistance in a certain circuit is the sum of resistances R and r, write the formula for the voltage.

29. The midrange m of a collection of numbers is one-half the sum of the smallest number s and the largest number l. Express this statement as a formula.

30. The value V of a machine depreciates so that its value after t years is its original value p divided by the sum of t and 1. Express this statement as a formula.

31. The efficiency E of an engine is defined as the difference of the heat input I and heat output P (subtract P from I) divided by the heat input. Express this statement as a formula.

32. The time T for one complete oscillation of a pendulum equals approximately 6.28 times the square root of the quotient of the length l of the pendulum and the acceleration g due to gravity. Find the resulting formula.

In Exercises 33 through 42, evaluate the required formula for the given values.

33. In Exercise 25, find the length of the remaining piece if 27.4 ft is cut off.

34. In Exercise 26, find the perimeter if the length and width are 29.2 cm and 37.9 cm, respectively.

35. In Exercise 27, find the total surface area if $x = 2.91$ ft and $l = 5.23$ ft.

36. In Exercise 28, find the voltage if $I = 0.00427$ A, $R = 82.6\ \Omega$, and $r = 1.08\ \Omega$.

37. What is the value of a $4000 machine after it has depreciated for 3 years? (Use the information from Exercise 30.)

38. In Exercise 32, find the time T (in seconds) if $l = 5.26$ ft and $g = 32.2$ ft/s^2.

39. In Exercise 31, find the efficiency if the heat input is 21,500 J (joules) and the heat output is 7,600 J.

40. Use the formula developed in Exercise 30 to find the value of a computer valued at $5297 after it has depreciated for 4.08 years.

41. Use the formula developed for Exercise 27 to find the total area if $x = 1.238$ in. and $l = 1.348$ in.

42. Use the formula developed in Exercise 28 to determine the voltage across an electric circuit if $I = 0.03672$ A, $R = 82.64$ Ω, and $r = 2.359$ Ω.

Answers

Now Try It! 1
a. 4   **b.** 8   **c.** 4   **d.** 7

Now Try It! 2
a. binomial   **b.** monomial   **c.** trinomial   **d.** binomial

3.3   Addition and Subtraction of Algebraic Expressions

Section Objectives
- Be able to add or subtract algebraic expressions
- Apply the commutative law of addition
- Apply the associative law of addition

In this section we consider the addition and subtraction of algebraic expressions.

If an **algebraic expression** *contains like terms as well as unlike terms, we can* **simplify** *it by combining the like terms.* The like terms are combined by adding or subtracting their numerical coefficients. The result is expressed as a sum of unlike terms. Example 1 illustrates the simplification of a polynomial.

Example 1

In the polynomial

$$3a^2 - 2ab + a^2 - b + 5ab$$

the $3a^2$ and a^2 are like terms and the $-2ab$ and $5ab$ are like terms. Adding $3a^2$ and a^2, we obtain $4a^2$. **(*The coefficient of a^2 is 1, which is not written.*)** When we subtract $2ab$ from $5ab$, we obtain $3ab$. Thus the given polynomial simplifies to

$$4a^2 + 3ab - b.$$

Since there are no like terms in this result, no further simplification is possible and we now know that

$$3a^2 - 2ab + a^2 - b + 5ab = 4a^2 + 3ab - b.$$

When we perform addition, we using the basic **axiom** (*an accepted but unproved statement*) that the order of addition of terms does not matter.

For any numbers a and b the **commutative law for addition** states that

$$a + b = b + a.$$

For any numbers a, b, and c the **associative property of addition** tells us that the order of grouping terms does not matter. That is,

$$a + (b + c) = (a + b) + c.$$

Example 2 illustrates the way we actually use this axiom to simplify a polynomial.

Example 2

The multinomial

$$x^2 + 5x - 3 + 4x + 5$$

is equivalent to

$$x^2 + 5x + 4x - 3 + 5$$

if we interchange the third and fourth terms according to the commutative law. We then combine the like terms $5x$ and $4x$, obtaining

$$x^2 + 9x + 2.$$

In general, we do not have to rewrite such expressions in order to add. However, the fact that we can add like terms is due to this axiom. Thus

$$x^2 + 5x - 3 + 4x + 5 = x^2 + 9x + 2.$$

Now Try It! 1

Combine like terms:

a. $2x - 3y - 5x + 2y$
b. $4y - x + 3z - 7z + 3y - 10x$
c. $3x^3 + 7x - 5 - 3x^2 + 9 + 2x^2 - x^3$

We see that in the addition of polynomials or other algebraic expressions *we can* **combine** *only like terms, but for unlike terms we can only* **indicate** *the sum.*

Example
3

Add $4xy + 3a - x$ and $2xy - 5a + 2x$.

For the sum given above, we write

$$(4xy + 3a - x) + (2xy - 5a + 2x).$$

It is not possible to proceed with the addition until we write an equivalent expression without parentheses. Only then are we able to combine like terms. Removing the parentheses from this expression, we have

$$4xy + 3a - x + 2xy - 5a + 2x.$$

Combining like terms, we have $6xy - 2a + x$, which is the result given above. Thus we can state that

$$(4xy + 3a - x) + (2xy - 5a + 2x) = 4xy + 3a - x + 2xy - 5a + 2x$$
$$= 6xy - 2a + x.$$

Generalizing the results of Example 3, we see that:

When parentheses are preceded by a plus sign, and the parentheses are removed, **the sign of each term within the parentheses is retained.**

Recall from Section 1.2 that *when we subtract a signed number we may change the sign of a number and proceed as in addition.* For example, $8 - 3 = 8 + (-3) = 5$. This same principle is used to subtract one algebraic expression from another. Example 4 illustrates the use of this principle in subtraction of algebraic expressions.

Example
4

Subtract $2xy - 5a + 2x$ from $4xy + 3a - x$.

Using parentheses to indicate that we are subtracting two algebraic expressions, we have

$$(4xy + 3a - x) - (2xy - 5a + 2x).$$

It is necessary to change the sign of each number being subtracted. We remove the second set of parentheses here and also change the sign of *each term* within it. This leads to

$$4xy + 3a - x - 2xy + 5a - 2x = 2xy + 8a - 3x.$$

Generalizing the results of Example 4, we see that:

> When parentheses are preceded by a minus sign, and the parentheses are removed, **the sign of each term within the parentheses is changed.**

Example 5

+ sign before parentheses

1. $3c + (2b - c) = 3c + 2b - c = 2b + 2c$

$2b = +2b$ signs retained

− sign before parentheses

2. $3c - (2b - c) = 3c - 2b + c = -2b + 4c$

$2b = +2b$ signs changed

3. $3c - (-2b + c) = 3c + 2b - c = 2b + 2c$

signs changed

Note in each case that the parentheses are removed and the sign before the parentheses is also removed. Also, in the first two examples we treat $2b$ as $+2b$.

Now Try It! 2

Perform the indicated operations.

a. $(3x + 5) + (6x - 8)$
b. $4a^2 - (2a^2 - 6b + 12)$
c. $(15x^3 + x - 18) - (10x^3 - 3x + 13)$
d. $(3x^2 + 3x - 5) + (x^2 + 4x + 7)$

Example 6

signs retained

1. $4a - (a - 3x + 2y) + (-4x + y) = 4a - a + 3x - 2y - 4x + y$

signs changed

$$= 3a - x - y$$

2. $-2x^2 + 3 + (5x^2 - 2) - (-x^2 + 2x - 1)$

$$= -2x^2 + 3 + 5x^2 - 2 + x^2 - 2x + 1 = 4x^2 - 2x + 2$$

▶ Again we note carefully that **when a minus sign precedes the parentheses, the sign of every term is changed** in removing the parentheses.

Brackets [ ] *and* **braces** { } *are also used to group terms, particularly when a group of terms is contained within another group.* (Another symbol of grouping, the **bar** or **vinculum** ‾, is used, especially for radicals and fractions. The expression $\sqrt{a + b}$ means the square root of the *quantity a + b* and

$$\frac{c}{a + b}$$

means c divided by the quantity $a + b$. In simplifying expressions containing more than one type of grouping symbols, we can remove the symbols one at a time. In general, it is better to *remove the innermost symbols first.* Consider the illustrations in Example 7.

Example
7

Simplify the expression $5a - [2a - (3a + 6)]$.

The innermost grouping symbols are the parentheses, and we begin by removing them. After the parentheses are removed we simplify and then remove the brackets. This leads to

$$5a - [2a - (3a + 6)] = 5a - [2a - 3a - 6] \quad \text{parentheses removed}$$
$$= 5a - [-a - 6] \quad \text{simplify within brackets}$$
$$= 5a + a + 6 \quad \text{brackets removed}$$
$$= 6a + 6.$$

Although a variety of grouping symbols can be used in algebraic expressions, the use of calculators and computers tends to result in parentheses nested within other parentheses. It is usually best to begin with the innermost set of parentheses and work outward.

3.3 Exercises

In Exercises 1 through 6 simplify the given algebraic expression by performing the indicated operation.

1. $x + y + 5x$

2. $3x + 5xy - x$

3. $3a - 5b^2 + 4a + b^2$

4. $2x - y + 5x - y$

5. $2s + 3t - s + 4s$

6. $-5m + 2n - m + 8m$

In Exercises 7 through 22, remove the symbols of grouping and simplify the given expressions.

7. $4a + (3 - 2a)$

8. $6x + (-5x + 4)$

9. $3 - (4x + 7)$

10. $9y - (4y - 8)$

11. $(4s - 9) + (8 - 2s)$

12. $(5x - 2 + 3y) + (4 - 3x)$

13. $(t - 7 + 3y) - (2 - y + t)$

14. $-(6b^3 - 3as + 4x^2) - (2s - 3b^3 - 6x^2)$

15. $4 + [6x - (3 - 4x)]$

16. $3 + (5n - (2 - n))$

17. $3s - (2 - (4 - s))$

18. $7 - [2 - (4x^2 - 2)]$

19. $-(7 - x) - [(2x + 3) - (7x - 2)]$

20. $-(5 - x) - [(5x - 7) - (2 - 3x + b)]$

21. $(t - 5x) - \{[(6p^2 - x) - 9] - (6t + p^2 - x)\}$

22. $8 - \{6xy - [7 - (2xy - 5)] - [6 - (xy - 8)]\}$

In Exercises 23 through 26, perform the indicated operations.

23. Subtract $3a^2 - 7x + 2$ from the sum of $a^2 + x + 1$ and $4a^2 - 5x + 2$.

24. Subtract $5 - 3x$ from the sum of $6s + 5 - 2x$ and $12 + 7x$.

25. Subtract the sum of $4x + 2 - 3t$ and $x - 1 - t$ from the sum of $3x - t + 4t$ and $2x + 8 - 5t$.

26. Subtract the sum of $2a^2 - 7xy + 3x$ and $-a^2 + 10xy$ from the sum of $5a^2 - xy - 2x$ and $-3a^2 + xy + 5x$.

In Exercises 27 through 40 solve the following problems by performing the necessary addition or subtraction and expressing your result in simplest form.

27. One transmitter antenna has a length of $(x - a)$ feet, whereas a second transmitter antenna is measured to be $(3x + 6a)$ feet long. What is the combined length of both transmitter antennae?

28. The local newspaper determines that on Monday through Saturday it prints $5x + 1000$ papers daily, whereas on Sundays it prints $4x - 2000$ papers. What is the sum total of papers printed each week?

29. The length of each side of a triangle is given as follows: $2x + 3, x - 7$, and $5x + 4$. Determine the perimeter of (or distance around) the triangle in Figure 3.9.

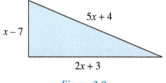

Figure 3.9

30. A small company that assembles gas-operated motors for lawn-care machinery concluded that profits for the month of July can be determined by the algebraic expression $325x - 3500$, and profits for April can be determined by the expression $70x + 1025$. Find a simplified expression that expresses the difference in profit from April to July.

31. A company has two manufacturing plants with daily production levels of $6x + 20$ items and $4x - 8$ items, respectively. How many more items are produced daily by the first plant?

32. A reforesting program has $2x + 1,000$ trees planted in one area and $x - 2,000$ trees planted in a second area. How many more trees are planted in the first area?

33. The resistance (in ohms) in one circuit is given by $5R - 20$ whereas a second circuit has a resistance of $3R - 50$. How much more resistance is in the first circuit?

34. One production method results in a cost (in dollars) of $12x - 400$ whereas a second method results in a cost of $10x - 300$. By how much does the cost of the first production method exceed that of the second production method?

35. In analyzing the forces acting on a beam, we may encounter the expression $4M - (2M + 120) + 80$. Simplify this expression.

36. The interest earned on a certain amount of money deposited in a high-yield mutual fund with a rate of return of 9% can be expressed by $0.09x$. A larger deposit is placed in a certificate of deposit that pays 5% and is represented by the expression $(x + 2500)(0.05)$. Determine an algebraic expression that represents the total interest paid by these two accounts.

37. Determine the difference in altitude between an object dropped from a twelve-story building to the ground 160 feet below, represented by the expression $-16t^2 + 160$ ft and one dropped a height represented by the expression $-16t^2 + 80$ ft.

38. An acoustical engineer calculates the velocity of a certain sound to be $(0.6C + 1089.0) + (10.3 - 0.1C)$. Simplify this expression.

39. The length of one side of a rectangle is $(8 - x)$ and the width is $(20 - 3x)$. The distance around this rectangle is

represented by the expression $2(8 - x) + 2(20 - 3x)$. Simplify this expression.

40. The voltage across one resistor is represented by the expression $7x - 20$ V, and the voltage across a second resistor is represented by the expression $2x - 10$ V. Determine the difference in the voltage across these two resistors.

Answers

Now Try It! 1
a. $-3x - y$ **b.** $-11x + 7y - 4z$ **c.** $2x^3 - x^2 + 7x + 4$

Now Try It! 2
a. $9x - 3$ **b.** $2a^2 + 6b - 12$ **c.** $5x^3 + 4x - 31$ **d.** $4x^2 + 7x + 2$

3.4 Multiplication of Algebraic Expressions

Section Objectives
- Introduce the commutative law of multiplication
- Introduce the associative law of multiplication
- Introduce and apply the distributive law
- Introduce and apply the rules of exponents when multiplying variables
- Be able to multiply algebraic expressions

Let us now consider multiplication. First we will consider the product of two or more monomials. We will then consider the product of a monomial and a polynomial. Before we begin, we introduce the **commutative** and **associative laws for multiplication.** These two laws state that the *order of multiplication and the order of grouping terms does not matter.* Stated algebraically they are shown as:

For any numbers a and b the **commutative law for multiplication** states that

$$ab = ba.$$

For any numbers a, b, and c the **associative property of multiplication** tells us that the order of grouping terms does not matter. That is,

$$a(bc) = (ab)c.$$

The **distributive law** is also a basic law of algebra that is very useful when multiplying algebraic expressions. This law states that given a product of a single term times a sum, multiplying the sum is equivalent to multiplying each term in the sum. Algebraically we write

The **distributive law** states that for any numbers a, b, and c,

$$a(b + c) = ab + ac.$$

An illustration of the distributive property can be seen in the following examples.

Example 1

a. $3(x + y) = 3x + 3y$

b. $-4(x - 4) = -4x + 16$

c. $6(a + 2b - 7c) = 6a + 12b - 42c$

d. $3ab(2a - 5bc) = (3ab)(2a) + (3ab)(-5bc)$
$$= 6aab - 15abbc$$
$$= 6a^2b - 15ab^2c$$

Example 2

Simplify the expression $3s + 2(s + 2t) - s(7 - 2t)$.

First we remove the parentheses by using the distributive law. Then we will combine the like terms.

$$3s + 2(s + 2t) - s(7 - 2t) = 3s + 2s + 4t - 7s + 2st$$
$$= -2s + 4t + 2st$$

As we saw in Example 1(d), multiplying variables can require the use of exponents. Recall from Chapter 1 that an **exponent** provides us with a simpler way to write an expression that includes repeated multiplication.

The notation a^n tells us that we need to multiply a times itself n times. This means that $\underbrace{a \times a \times a \times \cdots \times a}_{n \text{ factors}}$, where **$n$ is called the exponent** and **a is called the base.**

Example 3

$$(a^3)(a^4) = (a \times a \times a)(a \times a \times a \times a) = a^7$$

Also, $a^{3+4} = a^7$. We see that we can multiply a^3 by a^4 as follows:

$$(a^3)(a^4) = a^{3+4} = a^7.$$

Generalizing the results of Example 3, we see that if we multiply a^m by a^n, a^m has m factors of a, a^n has n factors of a, and therefore

$$\boxed{(a^m)(a^n) = a^{m+n}}$$

Sometimes it is necessary to take a number raised to a power, such as a^3, and raise it to a power, as in $(a^3)^4$. Consider the following example.

Example 4

$$(a^3)^4 = (a^3)(a^3)(a^3)(a^3)$$
$$= (a \times a \times a)(a \times a \times a)(a \times a \times a)(a \times a \times a) = a^{12}$$

Since $a^{3 \times 4} = a^{12}$, we see that we can raise a^3 to the fourth power by writing

$$(a^3)^4 = a^{3 \times 4} = a^{12}.$$

Generalizing the results of Example 4, we see that if we raise a^m to the nth power, we get

$$\boxed{(a^m)^n = a^{mn}}$$

Example 5

add exponents

1. $(b^4)(b^5) = b^{4+5} = b^9$

multiply exponents

2. $(b^4)^5 = b^{4 \times 5} = b^{20}$

add exponents

3. $bb^4 = b^{1+4} = b^5$ note that bb^4 is really $b^1 b^4$

has an exponent of 1

To multiply terms containing more than one base, the exponents of like bases are added. Example 6 demonstrates this.

Example
6

$$(a^3b^2)(a^2b^4) = a^3b^2a^2b^4 = a^3a^2b^2b^4 = a^{3+2}b^{2+4} = a^5b^6$$

add exponents of a

add exponents of b

▶ We can indicate the multiplication of a^5 and b^6 only as shown. ***It is not possible to simplify further.***

The power of a product leads to another important result in operating with exponents. Example 7 will allow us to develop a general expression for this case.

Example
7

$$(xy)^4 = (xy)(xy)(xy)(xy) = x^4y^4$$
$$(3c^2d^5)^3 = (3c^2d^5)(3c^2d^5)(3c^2d^5) = 27c^6d^{15}$$

Also, $3^3(c^2)^3(d^5)^3 = 27c^6d^{15}$.

Now Try It! 1

Simplify

a. x^5x^2
b. $y^2y^3y^a$
c. $(a^{-3})^3$
d. $(w^2)^4$
e. $(3x)^3$
f. $(-2x^2y^7)^3$

We see from Example 7 that *if a product of factors is raised to a power* **n,** *the result is the product of each factor raised to the power* **n.** Generalizing this result, we have

$$\boxed{(ab)^n = a^nb^n}$$

Example
8

1. $(3n)^4 = 3^4n^4 = 81n^4$

2. $(a^5b^6)^2 = (a^5)^2(b^6)^2 = a^{10}b^{12}$ — exponents multiplied

3. $(3na^5)^4 = (3)^4(n)^4(a^5)^4 = 3^4n^4a^{20} = 81n^4a^{20}$ — exponents multiplied

When we multiply monomials we use the notation of exponents and the rules for multiplying signed numbers. We first multiply the numerical coefficients and then the literal factors. The product of the new numerical coefficient and the literal factors is the required product.

The rules for integral exponents, along with the distributive law, provide us with all we need in order to successfully multiply algebraic expressions.

The following examples will look at finding the product of

- a monomial times a monomial
- a monomial times a polynomial
- a polynomial times a polynomial

Example 9

$$2ab^3(3a^2bc) = (2 \times 3)(a \times a^2 \times b^3 \times b \times c) = 6a^3b^4c$$
$$-2x^2y^3(4ax^2y) = -8ax^4y^4$$
$$(-p^2q)(3pq^2r)(-2qrs) = 6p^3q^4r^2s$$

Note the difference between expressions such as $(a^2 + b^2)^3$ and $(a^2b^2)^3$. In the first expression we have a *sum* raised to a power, whereas in the second expression we have a *product* raised to a power. $(ab)^n = a^nb^n$ works only when we are multiplying or dividing within the parentheses. We can express $(a^2b^2)^3$ as a^6b^6, but $(a^2 + b^2)^3$ is *not* $a^6 + b^6$.

To multiply a monomial and a polynomial, we use the distributive law. This law indicates that we must *multiply each term of the polynomial by the monomial.*

Example 10

$$2ax(3ax^2 + 5x^3) = 2ax(3ax^2) + 2ax(5x^3)$$
$$= 6a^2x^3 + 10ax^4$$

$$3xy^3(x + 4x^2y + 2y) = 3xy^3(x) + 3xy^3(4x^2y) + 3xy^3(2y)$$
$$= 3x^2y^3 + 12x^3y^4 + 6xy^4$$

Note that in the second part of Example 10, the polynomial contained three terms so that the distributive law was extended. In general, the polynomial may contain any number of terms. The key point to remember is that every one of those terms must be multiplied by the monomial.

One common algebraic error is to multiply the monomial with only the first term of the multinomial. Another common algebraic error is the failure to correctly assign signs to the terms appearing in the result.

Example 10 included only positive signs so that an error in signs is not likely. However, we must be careful with expressions involving negative signs, such as those in Example 11.

Example 11

$$-5x^2y(3x - 2xy + 4y) = -5x^2y(3x) - 5x^2y(-2xy) - 5x^2y(4y)$$
$$= -15x^3y + 10x^3y^2 - 20x^2y^2$$

$$-2ab(-3ab - a^3b + 4b^3) = -2ab(-3ab) - 2ab(-a^3b) - 2ab(4b^3)$$
$$= 6a^2b^2 + 2a^4b^2 - 8ab^4$$

In Examples 10 and 11, one factor was a monomial and the other was a polynomial. We now proceed to consider the multiplication of two polynomials.

To multiply one polynomial by another, we multiply each term of one by each term of the other, again using the rules outlined above. This is a result of the distributive law.

Example 12

$$(x - 2)(x + 7) = x(x) + x(7) + (-2)(x) + (-2)(7)$$
$$= x^2 + 7x - 2x - 14$$
$$= x^2 + 5x - 14$$

$$(x^2 - 2xy)(ab + a^2) = x^2(ab) + x^2(a^2) + (-2xy)(ab) + (-2xy)(a^2)$$
$$= abx^2 + a^2x^2 - 2abxy - 2a^2xy$$

Example 13

The dimensions of a room (in feet) are given as $2x + 1$ and $3x - 2$. Find an expression for the area of that room.

Since area is the product of length and width (the given dimensions), we express the area (in square feet) as

$$(2x + 1)(3x - 2) = 2x(3x) + 2x(-2) + 1(3x) + 1(-2)$$
$$= 6x^2 - 4x + 3x - 2$$
$$= 6x^2 - x - 2$$

Now Try It! 2

Multiply

a. $(7x^4)(4x^7)$

b. $(-2x^2y^3)(4x^2y)$

c. $-6x^2y(2xy - 7x^3y^2 + 3y^2)$

d. $5a^2(-4a^3 + 3a + 6)$

e. $(x^2 + y)(2xy - xy^2)$

f. $(7a^2b + 3c)(8a^2b - 3c)$

Example 14

$$(2a + 3)(2a^2 + 3a - b) = (2a)(2a^2) + (2a)(3a) + (2a)(-b) + (3)(2a^2)$$
$$+ (3)(3a) + (3)(-b)$$
$$= 4a^3 + 6a^2 - 2ab + 6a^2 + 9a - 3b$$
$$= 4a^3 + 12a^2 - 2ab + 9a - 3b$$

Example
15

$$(x - y)(x^2 + xy + y^2) = x(x^2) + x(xy) + x(y^2) - y(x^2) - y(xy) - y(y^2)$$
$$= x^3 + x^2y + xy^2 - x^2y - xy^2 - y^3$$
$$= x^3 - y^3$$

Note that the product includes like terms that can be combined.

Example
16

The expression $(2a + b)^2$ is equivalent to $(2a + b)(2a + b)$. We therefore square $(2a + b)$ as follows:

$$(2a + b)^2 = (2a + b)(2a + b)$$
$$= 2a(2a) + 2a(b) + b(2a) + b(b)$$
$$= 4a^2 + 2ab + 2ab + b^2$$
$$= 4a^2 + 4ab + b^2$$

A common error in such cases is simply to square the individual terms. Here we see that $(2a + b)^2$ is not $4a^2 + b^2$. The result must include the 4ab term.

3.4   *Exercises*

In Exercises 1 through 52, perform the indicated multiplications.

1. x^3x^7

2. n^2n^6

3. y^5y^2y

4. $p^2p^3p^5$

5. $t^3t^5t^2$

6. $(x^3)^8$

7. $(n^2)^7$

8. $(t^3)^5$

9. $(-ax^2b)^2$

10. $(-a^3b)^3$

11. $(-at^2)^5$

12. $(-cx^2y)^6$

13. $(-4rs)(-3st^2)(-7rt)$

14. $(-2axy^3)(7ay^4)(-ax^4y)$

15. $(-2st^3x)^3$

16. $(-3axt^7)^4$

17. $2a(a + 3x)$

18. $3x(2x - 5)$

19. $3a^2(2x - a^2)$

20. $4b^3(3 + 2b^2)$

21. $(-2st)(sx - t^2y)$

22. $(-8y)(8y + t^2)$

23. $(-3xy)(x^2y - 3axy^6)$

24. $(-5y^6)(-uy^7 - hpy)$

25. $(x - 3)(x - 1)$

26. $(t + 5)(t + 2)$

27. $(s - 2)(s + 3)$

28. $(x - 6)(x - 4)$

29. $(x + 1)(2x - 1)$

30. $(3a - 2)(5a + 4)$

31. $(5v + 1)(2v + 3)$

32. $(3v - 2)(4v - 3)$

33. $(a - x)(a - 2x)$

34. $(x + 2y)(x - y)$

35. $(2a - c)(3a + 2c)$

36. $(3s + 4t)(5s + 2t)$

37. $(2x - 5t)(x + 9)$

38. $(4x - 9uy)(2 + 3uy)$

39. $(2a - 9py)(2a + 9py)$

40. $(s - 3xu^2)(-xu^2 - 8s)$

41. $(2a + 1)(a^2 - 3a - 5)$

42. $(3x - 2)(2x^2 + x - 3)$

43. $(a - x)(a + 2xy - 3x)$

44. $(2 - x)(5 + x - x^2)$

45. $(x - 2)^2$

46. $(a + 5)^2$

47. $(x + 2y)^2$

48. $(2a - 3b)^2$

49. $(x - 2)(x + 3)(x - 4)$

50. $(2x - 3)(x + 1)(3x - 4)$

51. $(x + 1)^3$

52. $(2a - x)^3$

In Exercises 53 through 60, identify the equations as true or false. If an equation is false, explain why the equation is not correct.

53. a. $(2x)^3 = 2x^3$
   b. $(x^2y)^3 = x^5y^3$
   c. $(3x^2)^3 = 27x^6$

54. a. $t^5t^2 = t^{10}$
   b. $s^4s^3 = s^7$
   c. $(-x^3)^4 = -x^{12}$

55. a. $(at^3)^4 = a^4t^{12}$
   b. $a^2a^5 = a^7$
   c. $(x^2y^3) = (xy)^6$

56. a. $(3x)^4 = 81x^4$
   b. $x^2x^4 = x^8$
   c. $x^2x^3 = x^5$

57. a. $(a + b)^2 = a^2 + b^2$
   b. $(x - 3)^2 = x^2 - 9$
   c. $(x - 1)^2 = x^2 - 2x + 1$

58. a. $(2x + 3)^2 = 4x^2 + 9$
   b. $(x + y)^2 = x^2 + 2xy + y^2$
   c. $(x + y)^2 = x^2 + y^2$

59. a. $(t + 1)^2 = t^2 + t + 1$
   b. $(t + 1)^2 = t^2 + 2t + 1$
   c. $(t + 1)(t - 1) = t^2 - 1$

60. a. $x^2 - y^2 = (x - y)(x + y)$
   b. $x^2 - y^2 = (x - y)(x - y)$
   c. $x^2 + 9 = (x + 3)(x + 3)$

In Exercises 61 through 68, solve the given problems.

61. The length of a rectangular solar panel is $x + 3$ feet and its width is $x - 2$ feet. Find an expression for its area. See Figure 3.10.

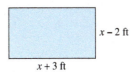

$x - 2$ ft

$x + 3$ ft

Figure 3.10

62. In determining the optimal size of a container, a manufacturer encounters the equation

$$V = 2x(8 - x)(20 - 3x)$$

Multiply out the right member of this equation.

63. The total production cost of x pills used for lowering blood pressure is given by

$$C = 200(x + 3)(x + 5)$$

Multiply out the right member of this equation.

64. By multiplication, show that $(x + y)(x^2 - xy + y^2) = x^3 + y^3$. Then show by substitution that the same value is obtained for each side for the values $x = -2$ and $y = -3$.

65. Use multiplication to show that $(2x + y)(2x - y) = 4x^2 - y^2$. Then substitute 5 for x and 3 for y in both sides of the equation to show that the same value is obtained.

66. A parabolic antenna is filled with a liquid, and the force of the pressure leads to the expression.

$$w(1 - x)(4 - x^2)$$

Multiply out this expression.

67. In determining the focal length of a lens, we use the *lensmaker's equation*

$$\frac{1}{f} = (n - 1)(r_2 - r_1)$$

Multiply out the right side of this equation.

68. The voltage across an electric resistor equals the product of the current and the resistance. The voltage across one resistor is twice that across a second resistor. The current is 3 A in the first resistor and 4 A in the second. Find the resistance (in ohms) if the second resistance is 5 Ω less than the first.

Answers

Now Try It! 1

a. x^7 **b.** y^{5+a} **c.** a^{-9} **d.** w^8 **e.** $27x^3$ **f.** $-8x^6y^{21}$

Now Try It! 2

a. $28x^{11}$ **b.** $-8x^4y^4$ **c.** $-12x^3y^2 + 42x^5y^3 - 18x^2y^3$ **d.** $-20a^5 + 15a^3 + 30a^2$
e. $2x^3y - x^3y^2 + 2xy^2 - xy^3$ **f.** $56a^4b^2 - 21a^2bc + 24a^2bc - 9c^2$

3.5 Division of Algebraic Expressions

Section Objectives

• Introduce and apply the rules of exponents when dividing algebraic expressions
• Be able to divide algebraic expressions

In the division of algebraic expressions the principle to be followed is the same as in the simplification of arithmetic fractions.

> *Factors common to both the numerator and denominator can be divided out.*

The following examples illustrate this basic principle.

Example 1

In simplifying the algebraic fraction

$$\frac{3xy}{ax}$$

we note that both the numerator and denominator have a factor of x. Thus, dividing out this common factor of x, we have

$$\frac{3xy}{ax} = \frac{3x\!\!\!/y}{a\!\!\!/x} = \frac{3y}{a}.$$

In dividing out the factor of x, we are stating that $x/x = 1$, and we know that multiplying the remaining factors by 1 does not change the resulting product.

In the following illustration, the resulting *denominator* is 1 and is not included in the final result.

$$\frac{3xy}{x} = \frac{3x\!\!\!/y}{x\!\!\!/} = \frac{3y}{1} = 3y.$$

However, if the resulting *numerator* is 1, it must be retained, as in the following illustration:

$$\frac{x}{ax} = \frac{\overset{1}{\cancel{x}}}{a\underset{1}{\cancel{x}}} = \frac{1}{a}.$$

When you become more familiar with this operation, you will not have to write in the step where the indicated cancellations are shown.

Example 2 shows additional steps in dividing out factors common to both the numerator and the denominator of a fraction.

Example 2

$$\frac{a}{10abc} = \frac{1}{10bc} \qquad\qquad \text{common factor of } a$$

$$\frac{4x}{6xyz} = \frac{2}{3yz} \qquad\qquad \text{common factor of } 2x$$

$$\frac{3x^2y}{x} = \frac{3xxy}{x} = \frac{3xy}{1} = 3xy \qquad \text{common factor of } x$$

$$\frac{5a^2bc}{10ab} = \frac{5aabc}{2(5)ab} = \frac{ac}{2} \qquad \text{common factor of } 5ab$$

$$\frac{3(x+y)^2}{(x+y)} = 3(x+y) \qquad\qquad \text{common factor of } (x+y)$$

In division, as in multiplication, the use of exponents plays an important role.

Example 3

Divide a^5 by a^2.

This can be rewritten as the fraction $\dfrac{a^5}{a^2}$.

As we saw in Examples 1 and 2, if a fraction has a factor that is common to both the numerator and the denominator, this factor can be divided out. The common factor in this problem is a^2.

$$\frac{a^5}{a^2} = \frac{a^3\cancel{a^2}}{\cancel{a^2}} = a^3 \quad \text{In this problem we divided out the } a^2.$$

Because $a^{5-2} = a^3$ we see that we can divide a^5 by a^2 as follows:

$$\frac{a^5}{a^2} = a^{5-2} = a^3.$$

Let us now divide a^2 by a^5. This is expressed as a fraction, and then the common factor of a^2 is divided out. This leads to

$$\frac{a^2}{a^5} = \frac{a^2}{a^2 a^3} = \frac{1}{a^3}.$$

Here we see that

$$\frac{a^2}{a^5} = \frac{1}{a^{5-2}} = \frac{1}{a^3}.$$

If we divide a^5 by a^5, we get

$$\frac{a^5}{a^5} = 1.$$

Generalizing the results of Example 1, we see that if we divide a^m by a^n, we have

$$\frac{a^m}{a^n} = a^{m-n} \text{ if } m > n \text{ and } a \neq 0$$

$$\frac{a^m}{a^n} = \frac{1}{a^{n-m}} \text{ if } n > m \text{ and } a \neq 0$$

$$\frac{a^m}{a^n} = 1 \text{ if } m = n \text{ and } a \neq 0$$

Example 4

larger exponent

$$\frac{x^4}{x^3} = x^{4-3} = x^1 = x \qquad \frac{x^3}{x^4} = \frac{1}{x^{4-3}} = \frac{1}{x^1} = \frac{1}{x}$$

subtract   larger exponent   subtract

$$\frac{3c^6}{c^2} = 3c^{6-2} = 3c^4 \qquad \frac{c^2}{3c^6} = \frac{1}{3c^{6-2}} = \frac{1}{3c^4}$$

equal exponents $\dfrac{x^3}{x^3} = 1$ $\dfrac{2s^2}{s^2} = 2(1) = 2$

Example 5

When finding the dimensions of a container that will use the least amount of material for a given volume, we obtain the expression

$$\frac{6x^3}{x^2}.$$

Simplify this expression.

Applying $\dfrac{a^m}{a^n} = a^{m-n}$, we get

$$\frac{6x^3}{x^2} = 6x^{3-2} = 6x^1 = 6x.$$

Example 6

$$-16a^2b \div 4a = \frac{-16a^2b}{4a} = -4ab$$

$$36a^2b^3 \div (-12ab) = \frac{36a^2b^3}{-12ab} = -3ab^2$$

$$-18x^3y \div (-12xy^4) = \frac{-18x^3y}{-12xy^4} = \frac{3x^2}{2y^3}$$

$$-8ab^2x^5 \div (-14a^2b^2x) = \frac{-8ab^2x^5}{-14a^2b^2x} = \frac{4x^4}{7a}$$

Now Try It! 1

Divide
a. z^7 by z^3
b. $45a^3$ by $15a$
c. $\dfrac{-18x^2yz^4}{9xy^3}$
d. $\dfrac{20x^3y^3z^3}{4x^3yz^5}$

We now consider the division of a polynomial by a monomial.

In arithmetic, since

$$\frac{2+3}{7} = \frac{2}{7} + \frac{3}{7} \qquad \text{or} \qquad \frac{a+b}{c} = \frac{a}{c} + \frac{b}{c}$$

we see that when the numerator is a sum we must divide each of the numbers of the numerator separately by the number of the denominator.

▶ Since algebraic expressions represent numbers, we have the following method of dividing a polynomial by a monomial: ***Divide each term of the polynomial by the monomial*** and add the resulting terms algebraically to obtain the quotient.

Note that every term of the numerator must be divided by the monomial in the denominator.

Example 7

$$\frac{x^3 + x^2}{x} = \frac{x^3}{x} + \frac{x^2}{x} = x^2 + x$$

> **Note:**
>
> A very **common algebraic error** arises when a monomial is not divided into each term of the numerator. For example,
>
> $$\frac{x^5 + 7}{x}$$
>
> *does* **not** *equal* $x^4 + 7$ since each term of the numerator has not been divided by x. Actually, no further simplification is possible here.

Example
8

$$(2mn^2 - 3n) \div n = \frac{2mn^2 - 3n}{n}$$

$$= \frac{2mn^2}{n} - \frac{3n}{n} = 2mn - 3$$

each term of
numerator
divided by
denominator

$$(4x^2y + 2xy^3) \div 2xy = \frac{4x^2y + 2xy^3}{2xy}$$

$$= \frac{4x^2y}{2xy} + \frac{2xy^3}{2xy} = 2x + y^2$$

$$(7a^3b^2 - 28a^3b^3 + 35a^2b^2) \div (-7ab^2) = \frac{7a^3b^2 - 28a^3b^3 + 35a^2b^2}{-7ab^2}$$

$$= \frac{7a^3b^2}{-7ab^2} + \frac{-28a^3b^3}{-7ab^2} + \frac{35a^2b^2}{-7ab^2}$$

$$= -a^2 + 4a^2b - 5a$$

In dividing by a polynomial, we shall deal exclusively with polynomials.

To divide one polynomial by another, we arrange both the dividend and the divisor in descending powers of the same literal factor. Then we divide the first term of the dividend by the first term of the divisor. The result of this division is the first term of the quotient. We then multiply the entire divisor by the first term of the quotient and subtract this product from the dividend. Now we repeat these operations by dividing the first term of the divisor into the first term of the difference just obtained. The result of this division is the second term of the quotient. We multiply the entire divisor by this second term and subtract the result from the first difference. We repeat this process until the difference is either zero or a quantity whose degree is less than that of the divisor.

Example
9

Divide $2x^2 + 5x - 3$ by $2x - 1$.

Because both the dividend and the divisor are already in descending powers of x, no rearrangement is necessary. Then we set up the division as follows:

$$2x - 1 \overline{)2x^2 + 5x - 3}$$

We now determine that $2x^2$ divided by $2x$ is x. Thus x becomes the first term of the quotient:

$$x \longleftarrow \qquad \frac{2x^2}{2x}$$
$$2x - 1 \overline{)2x^2 + 5x - 3}$$

We now multiply $2x - 1$ by x and place the product below the dividend:

$$\begin{array}{r} x \\ 2x - 1 \overline{)2x^2 + 5x - 3} \\ 2x^2 - x \longleftarrow \quad x(2x - 1) \end{array}$$

▶ The $2x^2 - x$ is now **subtracted** *from the dividend:*

$$\begin{array}{r} x \\ 2x - 1 \overline{)2x^2 + 5x - 3} \\ \underline{2x^2 - x} \\ 6x - 3 \end{array} \qquad -3 - 0 = -3$$

$$2x^2 - 2x^2 = 0 \qquad\qquad +5x - (-x) = 5x + x = 6x$$

We now determine that $2x$, the first term of the divisor, divided into $6x$, the first term of the remainder, is 3. Thus 3 becomes the second term of the quotient:

$$\begin{array}{r} x + 3 \\ 2x - 1 \overline{)2x^2 + 5x - 3} \\ \underline{2x^2 - x} \\ 6x - 3 \end{array} \qquad \frac{6x}{2x}$$

The divisor is now multiplied by 3, and the product is placed below the remainder and subtracted:

$$\begin{array}{r} x + 3 \longleftarrow \quad \text{quotient} \\ 2x - 1 \overline{)2x^2 + 5x - 3} \\ \underline{2x^2 - x} \\ 6x - 3 \\ \underline{6x - 3} \longleftarrow 3(2x - 1) \\ 0 \longleftarrow [(6x - 3) - (6x - 3)] = 0 \end{array}$$

Because the remainder is zero and all terms of the dividend have been used, the division is complete. We conclude that

$$(2x^2 + 5x - 3) \div (2x - 1) = x + 3$$

As a check, we can verify that $(x + 3)(2x - 1) = 2x^2 + 5x - 3$.

Example 10

We can find the efficiency of a certain motor by dividing the output power by the input power. We find an expression for efficiency by dividing $3x^2 - 4x + x^3 - 12$ by $x + 2$.

The dividend is arranged in descending powers of x, and the division proceeds as shown:

$$
\begin{array}{r}
x^2 + x \quad - 6 \quad \text{quotient} \\
x + 2 \overline{)\, x^3 + 3x^2 - 4x - 12} \quad \text{dividend} \\
\underline{x^3 + 2x^2} \quad\quad\quad \text{subtract} \\
x^2 - 4x - 12 \\
\underline{x^2 + 2x} \quad\quad \text{subtract} \\
- 6x - 12 \\
\underline{- 6x - 12} \quad \text{subtract} \\
0 \quad \text{remainder}
\end{array}
$$

divisor

$-4x - (+2x) = -4x - 2x$
$\qquad\qquad\quad = -6x$

This division is exact.

The division can be checked by multiplication. In this case

$$(x + 2)(x^2 + x - 6) = x^3 + x^2 - 6x + 2x^2 + 2x - 12$$
$$= x^3 + 3x^2 - 4x - 12.$$

Because the product equals the dividend, the result checks.

Now Try It! 2

Divide

a. $\dfrac{21a^6 - 18a^3}{3a^2}$

b. $\dfrac{3xy^2 - 3x^2y}{xy}$

c. $2a^2 + 11a + 5$ by $2a + 1$

d. $c^4 + 2c^2 - c + 2$ by $c^2 - c + 1$

Example 11

Divide $4y^3 + 6y^2 + 1$ by $2y - 1$.

Because no first-power term in y appears in the dividend, we insert one with a zero coefficient to as a placeholder:

$$
\begin{array}{r}
2y^2 + 4y \quad + 2 \\
2y - 1 \overline{)\, 4y^3 + 6y^2 + 0(y) + 1} \\
\underline{4y^3 - 2y^2} \quad\quad\quad \text{subtract} \\
8y^2 + 0(y) + 1 \\
\underline{8y^2 - 4y} \quad\quad \text{subtract} \\
4y \quad + 1 \\
\underline{4y \quad - 2} \quad \text{subtract} \\
+ 3 \quad \text{remainder}
\end{array}
$$

$0y - (-4y) = 4y$

The quotient in this case is

$$2y^2 + 4y + 2 + \frac{3}{2y - 1}.$$

Note how the remainder is expressed as part of the quotient.

3.5 Exercises

In Exercises 1 through 44, perform the indicated divisions.

1. $x^7 \div x^4$

2. $x^2 \div x^8$

3. $a^5 \div a^4$

4. $p \div p^9$

5. $\dfrac{8n^4}{2n}$

6. $\dfrac{9m}{6m^5}$

7. $\dfrac{-x^4y^3}{x^2y}$

8. $\dfrac{-4ca^2t^4}{-2c^3a}$

9. $\dfrac{-5x^2yr}{20xyr^3}$

10. $\dfrac{-7yt^3u}{-6yt}$

11. $\dfrac{9abc^4ds}{15bc^4d^4}$

12. $\dfrac{-12a^5b^3c^7}{18a^5b^2c^8}$

13. $(6ab + 5a) \div a$

14. $(2x^2 - 3x) \div x$

15. $(9m^2 - 3m) \div 3m$

16. $(8s^3 + 2s^2) \div 2s$

17. $(a^3x^4 - a^2x^3) \div (-ax^2)$

18. $(-2xy^6 - 4x^2y^5) \div (2xy^3)$

19. $(3xy^4 - 6x^2y^5) \div (3xy^2)$

20. $(-14p^3q^5 + 49p^4q^2) \div (-7p^2q^2)$

21. $\dfrac{a^2b^2c^3 - a^3b^4c^6 - 2a^3bc^2}{a^2bc^2}$

22. $\dfrac{8rs^2t^5 - 18r^3st^4 - 16rst^3}{-2rst^2}$

23. $\dfrac{a^2b^3 - 2a^3b^4 - ab - ab^2}{-ab}$

24. $\dfrac{5m^2n^2y - 30mn^2 + 35m^2n^8}{-5mn^2}$

25. $(x^2 - 2x - 3) \div (x + 1)$

26. $(x^2 + 3x - 10) \div (x - 2)$

27. $(2x^2 - 5x - 3) \div (x - 3)$

28. $(3x^2 + 4x - 5) \div (x + 2)$

29. $(8x^2 + 6x - 7) \div (2x + 3)$

30. $(6x^2 + 7x - 3) \div (3x - 1)$

31. $\dfrac{8x^3 - 18x^2 - 7x + 12}{4x - 3}$

32. $\dfrac{4x^3 - 20x - 2x^2 - 12}{2x + 5}$

33. $(5x - 5x^2 + 2x^3 - 6) \div (x - 2)$

34. $(1 - x^2 + 6x^3) \div (1 + 2x)$

35. $\dfrac{x^4 - 1}{x - 1}$

36. $\dfrac{x^3 - 8}{x - 2}$

37. $\dfrac{5x^2 - 5}{x + 1}$

38. $\dfrac{x^6 - 3x^5 - x^3 - 3x^2 + x - 3}{x - 3}$

39. $\dfrac{6x^4 - 5x^3 + 7x^2 + x - 3}{3x - 1}$

40. $\dfrac{6x^4 + 15x^3 - 25x^2 - 22x + 21}{2x + 7}$

41. $\dfrac{a^2 + ab - 12b^2}{a + 4b}$

42. $\dfrac{2x^2 + xy - 6y^2}{x + 2y}$

43. $\dfrac{x^3 - 5x^2y + 2xy^2 + 2y^3}{x - y}$

44. $\dfrac{4s^3 - 8s^2t + 5st^2 - t^3}{2s - t}$

In Exercises 45 through 50, identify the given equations as true or false. If an equation is false, explain why it is not correct.

45. a. $x^8 \div x^2 = x^4$
    b. $r \div r^3 = r^2$
    c. $x^6 \div x^3 = x^2$

46. a. $\dfrac{6a^2x^3}{2ax} = 3ax^2$
    b. $\dfrac{8x^2y}{8x^2y} = 0$
    c. $\dfrac{ax^{10}y^8}{ax^{12}y^{10}} = \dfrac{x^2}{y^2}$

47. a. $\dfrac{x^2 + y^2}{x} = x + y^2$
    b. $\dfrac{x^2 - ax}{x} = x - a$
    c. $\dfrac{x^2 - 4x + 4}{x - 2} = x - 2$

48. a. $\dfrac{x^2 + 3xy}{x} = x^2 + 3y$
    b. $\dfrac{x^2 + y^2}{x + y} = x + y$
    c. $\dfrac{x^2 + 2xy + y^2}{x + y} = x + y$

49. a. $\dfrac{10t^4 - 5t^2}{-5t} = -2t^3 + t$
    b. $\dfrac{-12x^5 + 8x^2}{-4x} = 3x^4 - 2x$
    c. $\dfrac{6x^2 - 8}{x^2} = -2$

50. a. $\dfrac{8x^3 - 4ax^2}{2ax} = \dfrac{4x^2 - 2ax}{a}$
    b. $\dfrac{x^2 + 4}{x + 2} = x + 2$
    c. $\dfrac{x^2 - 6x + 9}{x - 3} = x - 3$

In Exercises 51 through 54, solve the given problems.

51. The volume of a gas is expanding at a rate (in cubic inches per minute) described by the expression

$$\frac{6r^2 + 8r + 10}{r + 2}$$

Perform the indicated division.

52. In attempting to determine the efficiency of a diesel engine, an automotive engineer encounters the algebraic expression

$$(x^2 + 5x + 6) \div (x + 3)$$

Perform the indicated division.

53. If a satellite used for surveying travels $3x^2 - 8x - 28$ miles in $x + 2$ hours, find an expression for the speed in miles per hour.

54. Under certain conditions, when dealing with the electronics of coils, we may encounter the expression

$$\frac{60r^2}{6r + 1.2}$$

Perform the indicated division.

Answers

Now Try It! 1

a. z^4 **b.** $3a^2$ **c.** $\dfrac{-2xz^4}{y^2}$ **d.** $\dfrac{5y^2}{z^2}$

Now Try It! 2

a. $7a^4 - 6a$ **b.** $3y - 3x$ **c.** $a + 5$ **d.** $c^2 + c + 2$

Chapter Summary

Key Terms

formula	algebraic expression	trinomial
factors	polynomial	degree of a term
coefficient	monomial	degree of a polynomial
like terms	binomial	

Key Concepts

Properties of Real Numbers

Commutative property of addition	$a + b = b + a$
Commutative property of multiplication	$ab = ba$
Associative property of addition	$a + (b + c) = (a + b) + c$
Associative property of multiplication	$a(bc) = (ab)c$
Distributive law	$a(b + c) = ab + ac$

Laws of Exponents

$$(a^m)(a^n) = a^{m+n}$$

$$(a^m)^n = a^{mn}$$

$$(ab)^n = a^n b^n$$

$$\frac{a^m}{a^n} = a^{m-n} \ (m > n, a \neq 0)$$

$$\frac{a^m}{a^n} = \frac{1}{a^{n-m}} \ (n > m, a \neq 0)$$

$$a^{-n} = \frac{1}{a^n} \ a \neq 0$$

• Algebra is a generalization of arithmetic that uses letters, called variables, to represent numbers. An algebraic expression is used to describe a combination of constants and variables joined by arithmetic operations. If an algebraic expression contains parts that are separated by a plus or minus sign, these parts are referred to as terms. Those terms that have identical variable parts are called like terms. An algebraic expression having more than one term is often referred to as a polynomial. The degree of a term is the sum of the exponents of the variables for that term. The degree of a polynomial is the largest degree of all its terms.

• Addition and subtraction of algebraic expressions is done by combining like terms and using the rules for addition and subtraction of signed numbers.

• Multiplication and division of algebraic expressions rely on using the laws of exponents and the rules for multiplying and dividing signed numbers.

Review Exercises

In Exercises 1 through 4, identify the like terms in the given expressions.

1. $6ax - 7a + 5a + 8x$

2. $3ax^2 + 4ax + 5a^2x - 6ax^2$

3. $5(a - b) + 6ab - 7(a - b)$

4. $8x^2 - 2x^2 + 5x - 2$

In Exercises 5 through 66, simplify the given expressions.

5. $a + 3b + 6a - 7b$

6. $6a^2b - a^2b + 5ab - 2a^2b$

7. $6x(2a - 3b)$

8. $-3a(8 - 5b)$

9. $3(x + y) - 2y$

10. $x(2x + 6y) - x^2$

11. $a - (x - 2a)$

12. $x - (3s - 2x)$

13. $2(x - 5y) - (3y - 7x)$

14. $-(y + 2s) - (5s - y)$

15. $-2 + (n + 4) - (6 - n)$

16. $t - (5 + 2t) + (t - 7)$

17. $2x - 3(x - 3y) - (y - x)$

18. $-8x + 4r - 2(-r - 3x)$

19. $(-3a^2b)(2ab^5)$

20. $(-2s^3t^2)(-4st^4)$

21. $(-2xy^2z)^3(-7xy^3z^5)$

22. $(-2x^2y)^2(-3xy^5)$

23. $(3ab^2)^3$

24. $(2a^4c)^4$

25. $(-2x^2yz^3)^4$

26. $(-x^3y^2z^4)^5$

27. $(-8a^2px^4) \div (2apx)$

28. $(-18rs^3t^5) \div (-24r^4st^6)$

29. $\dfrac{15x^2y^3z}{3xy^7z^4}$

30. $\dfrac{48ab^5c^2}{18a^7bc^6}$

31. $5 - [x - (3 - 4x)]$

32. $3y + [(5y - 2) - 6y]$

33. $2x + [(2x - a) - (a - x)]$

34. $-(3xy - y) - 2[2x - (y - xy)]$

35. $2 - \{2 - [3x - (7 - x)]\}$

36. $4a + \{a - 3 + [2a - (3 - 2a)]\}$

37. $-\{2b - [b - (4 - 5b)] + (6 - b)\}$

38. $2x - y - 2\{3x - [y - (3y - 4x)]\}$

39. $2x^2(x^3 - 3x)$

40. $s^3(3s^4 - 2s)$

41. $-2a(a^2x - at)$

42. $3a^2j(-3j^4 + 4a - aj)$

43. $(x - 3)(2x + 7)$

44. $(3x - 2)(x + 5)$

45. $(2a - 5b)(3a + 2b)$

46. $(2x - y)(y - 5x)$

47. $(x + 1)(x^2 - x + 1)$

48. $(x - 3)(2x^2 + x - 2)$

49. $2(2 - x)(x^2 - x - 4)$

50. $-3(xy - q)(x - 2y + 3q)$

51. $(2x^3y^5 - 3x^6y^2) \div (-x^3y^2)$

52. $(-9a^3b^4 + 12ab^5) \div (-3ab^2)$

53. $\dfrac{h^2j^4 - 3hj^6 - 6h^4j^7}{hj^4}$

54. $\dfrac{-18f^3g^2k^2 + 24f^2gk^6 - 36fgk^4}{-6fgk^2}$

55. $(x^2 + x - 12) \div (x - 3)$

56. $(6x^2 - 7x - 5) \div (2x + 1)$

57. $\dfrac{2x^3 + x^2 - x - 8}{2x + 3}$

58. $\dfrac{6x^3 + x^2 - 12x + 7}{3x + 5}$

59. $\dfrac{x^3 + 1}{x + 1}$

60. $\dfrac{4x^3 + x - 1}{2x - 1}$

61. $\dfrac{2x^4 - x^3 - 2x - 8}{x^2 + 2}$

62. $\dfrac{6x^4 + x^3 + 5x^2 + 2}{2x^2 - x + 1}$

63. $\dfrac{3x^2 + 5xy - 2y^2}{3x - y}$

64. $\dfrac{-a^2b - 2b^2 + 3a^4}{a^2 - b}$

65. $\dfrac{x^3 - 1}{x - 1}$

66. $\dfrac{2x^3 - 7x^2 + 5x - 6}{2x^2 - x + 2}$

In Exercises 67 through 72, evaluate the given expression for the indicated values.

67. $x(2x - 1) + 8x^2$ for $x = -2$

68. $2y(y - 3) - y^2$ for $y = -3$

69. $3x^2 - 5y + 2x^2 + 8y$ for $x = 4$ and $y = 6$

70. $2(x - 1) + 7x^3 + x(8 - 7x)$ for $x = -4$

71. $\dfrac{a^2bc^3}{a} + c^2(2abc + 5)$ for $a = 6, b = 7,$ and $c = 12$

72. $\dfrac{a^2(x - y)^2}{(x - y)} - x(a^2 + y) + y(a^2 + 5x)$ for $x = 8$
and $y = -6$

In Exercises 73 through 82, solve the given problems.

73. In analyzing the burning temperature of a plastic, a fire-science specialist encounters the expression

$$(1.8C + 32) - 0.05(1.8C - 32)$$

Simplify this expression.

74. In the analysis of the transfer of heat through two glass surfaces, the expression

$$8(T_1 - T_2) - 6(T_1 - T_2)$$

is encountered. Simplify this expression.

75. When finding the center of mass of a particular area, we may encounter the expression

$$(2x - x^2) - (3x^2 - 6x)$$

Simplify this expression.

76. One factory produced $x + 12$ units in a given day, and a second factory produced $2x + 1$ units. How many more units were produced by the second factory?

77. The total revenue obtained by selling x units of a certain item is given by the expression

$$x(36 - 4x)$$

Perform the indicated multiplication.

78. Under certain conditions, the distance (in feet) that an object is above the ground is given by the expression

$$16t(5 - 2t)$$

Perform the indicated multiplication.

79. Perform the indicated multiplication in the following expression, which is sometimes used in applications involving the expansion of a heated surface.

$$lh(1 + at)^2$$

80. Simplify the following expression, which is sometimes used in applications involving the interference of light from a double source.

$$(2x + d)^2 - (2x - d)^2$$

81. The centripetal force (in pounds) of a car on a curve is found from the equation $F = mv^2/r$. Find an expression for F if $m = 6x + 2, v = x + 5,$ and $r = 3x + 1$. Also, find the force (in pounds) if $x = 10$.

82. The average rate of speed can be found from the equation $r = \dfrac{d}{t}$. Find an expression for r in mi/h if the distance traveled by an automobile is $20x^2 - 202x - 84$ miles in a time of $2x - 21$ hours.

Chapter Test

Identify the like terms in the following expressions.

1. $8ax - 10a + 9a + 2x$

2. $5ax^2 + 10ax - 7a^2x - 16ax^2$

Simplify

3. $2a + 3b + 8a - 7b$

4. $2a^2b - 9a^2b - 12a^2b$

5. $6a(2a - 3ab) - 12ab$

6. $(6a^2b^4)(2ab^2)$

7. $(-2x^2yz^2)^3$

8. $\dfrac{-8s^4d^3}{2ad^4}$

9. $(3x + 4y)(2x - 7y)$

10. $\dfrac{3x^2y^3 + 9xy^6 - 18x^2y^4}{-3xy^2}$

11. $\dfrac{4x^3 + 6x^2 + 1}{2x - 1}$

12. Evaluate $3x^2 - 5y + 2x^2 + 8y$ for $x = 3$ and $y = -3$

13. An analysis of the electrical potential of a conductor includes the expression $2V(r - a) - V(b - a)$. Simplify this expression.

14. A computer analysis of the velocity of a link in an industrial robot leads to the expression $4(t - h) - 2(t + h)^2$. Simplify this expression.

15. When analyzing the motion of a communication satellite, the expression $\dfrac{kr^2 - 2h^2k + h^2rv^2}{k^2r}$ is used. Simplify this expression.

4

Simple Equations and Inequalities

Equations are often used to describe or model the way certain things happen in the real world. Many times solving an equation will provide us with information that we would not otherwise know. Basic electronics involves the use of equations.

For example, a conductor is a material that allows an electric current to pass through it. Resistance (in ohms) is a measure of the degree to which the material used in the conductor opposes the flow of the current. Therefore, a high resistance inhibits the flow of electric current and a low resistance makes electric current flow possible. At a given temperature, the resistance of a material depends on the type, length, and cross section of conductor material and is given by the following literal equation:

$$R = \frac{\rho l}{A}$$

where ρ represents the resistivity of the material. This is a property of the specific material used. l represents the length of material in meters and A is the cross sectional area of the material. Suppose you were provided with a copper wire and the material's resistivity. You are able to measure the length of the wire with a ruler and its resistance with an ohmmeter. How would you determine the wire's cross sectional area? (Your ruler is not accurate enough to measure the wire's diameter.)

In this chapter we will focus on how we might solve equations such as these.

In Chapter 3 we defined an **algebraic expression** as a collection of variables and constants joined by arithmetic operations. We noted early in Section 3.2 that an *algebraic expression **does not** contain an equal sign.*

When two algebraic expressions are set equal to each other they form an **equation**. An equation has two sides and an equal sign.

$$3x^2 - 4x \ = \ 2x + 5$$

left side   right side

equal sign

The **solution** to an equation is that value for the variable that makes the equation a true statement. In other words, it is that value which makes both sides of the equation equal.

> **Important**
> An algebraic expression **does not** contain an equal sign. An algebraic expression can be simplified but cannot be solved.
> An equation **does** contain an equal sign. We can solve an equation.

We begin this chapter with simple equations, and then proceed to consider formulas, inequalities, ratios, proportions, and variations. We will also recall our problem-solving strategies, introduced in Chapter 1, as we solve application problems.

4.1 Solving a Simple Equation

Section Objectives
- Understand the difference between an algebraic expression and an equation
- Translate verbal statements into symbol representations
- Introduce the addition property of equality
- Introduce the multiplication property of equality
- Master techniques for solving simple equations

We have stated that an equation is a mathematical statement that two algebraic expressions are equal. We begin this section with an example of an equation, and then proceed to study methods of solving equations.

Example 1

$3x - 4 = 2x - 1$ is an equation. In this equation it is understood that the variable x represents the unknown. If we substitute $x = 3$ into each side of the equation, we obtain $5 = 5$ Thus $x = 3$ is the solution of the equation. This is often stated as "3 **satisfies** the equation." If we try any other value of x, we will find that the two sides of the equation are not equal. If we substitute zero for x, for example, we get $-4 = -1$ so that $x = 0$ is not a solution of the equation.

This chapter shows methods that can be used to solve equations such as the one in Example 1. At this time we will work only with equations with unknowns raised to the first power called a linear equation. Later chapters show how other types of equations are solved. This chapter also includes brief coverage of simple inequalities.

A large part of algebra is devoted to finding solutions to various equations. In particular, technical and scientific work involves equations of all kinds, and their solutions are needed for finding or confirming scientific information. Example 2 illustrates some statements about numbers that lead to equations.

Example 2

In each case we shall let x be the unknown number.

Statement	Equation
1. A number decreased by seven equals twelve.	$x - 7 = 12$
2. Nine added to a number equals three.	$x + 9 = 3$
3. A number divided by two equals seven.	$\dfrac{x}{2} = 7$

Statement	*Equation*
4. Five times a number equals twenty.	$5x = 20$
5. Three less than twice a number equals one more than the number.	$2x - 3 = 1 + x$
6. Sixteen less three times a number equals two added to four times the number.	$16 - 3x = 2 + 4x$

Because an equation has both sides equal for the appropriate value of *x*, equations have these basic properties:

1. If the same number is added to each side, the two sides are still equal.

2. If the same number is subtracted from each side, the two sides remain equal to each other.

3. If the two sides are multiplied by the same nonzero number, they remain equal.

4. If the two sides are divided by the same number (provided it is not zero), they remain equal.

Addition Property of Equality

Adding (or subtracting) the same number from both sides of an equation does not change the problem but rather produces an equivalent equation.

If $a = b$ then $a + c = b + c$ and $a - c = b - c$.

Multiplication Property of Equality

Multiplying (or dividing) both sides of an equation by the same number (not zero) does not change the problem but rather produces an equivalent equation.

If $a = b$ and c is not equal to 0, then $ac = bc$ and $\dfrac{a}{c} = \dfrac{b}{c}$.

By performing these operations, we can isolate *x* on one side of the equation and the other numerical quantities on the other side. The following examples illustrate the method used to solve the equations of Example 2.

Example 3

Solve $x - 7 = 12$.

We note that when we add 7 to the left side, only x will remain. By adding 7 to both sides, we obtain

$$x - 7 + 7 = 12 + 7$$
$$x = 19$$

because $-7 + 7 = 0$ and $12 + 7 = 19$. The solution is $x = 19$.

Caution

When solving any equation it is important to follow this basic rule of algebra: *Whatever is done to one side of an equation must be done to the other side in order to preserve the equality.*

It is always wise to *check* the solution by following these simple steps:

▶ 1. ***In the original equation,*** substitute the solution for the unknown.

2. Simplify both sides of the equation by performing the indicated operations.

3. If both sides of the equation simplify to the same number, then the solution is correct. (If both sides simplify to different numbers, the solution is wrong.)

The solution of $x = 19$ in Example 3 can be checked by following the preceding steps. Upon substitution of 19 for x in the original equation we get $19 - 7 = 12$, which simplifies to $12 = 12$ so that the solution of $x = 19$ is correct.

Example 4

Solve $x + 9 = 3$.

When we subtract 9 from each side, only x will remain on the left side. Performing this operation, we have

$$x + 9 - 9 = 3 - 9$$
$$x = -6$$

Substitution into the original equation gives $3 = 3$ so that the solution $x = -6$ checks.

Example 5

1. Solve $\dfrac{x}{2} = 7$.

If we multiply the left side by 2, it becomes $2x/2$, or x. Thus, by multiplying both sides by 2, we obtain the solution

$$2\left(\frac{x}{2}\right) = 2(7)$$
$$x = 14$$

Now Try It! 1

Solve

a. $x + 5 = 2$

b. $-2x = 14$

c. $x - 5 = -5$

d. $\dfrac{x}{5} = 14$

e. $5x + 10 = -10$

Substitution of the solution $x = 14$ into the original equation gives $7 = 7$, which means that it checks.

2. Solve $5x = 20$.

If we divide both sides by 5, the left side then becomes x and the solution is

$$\frac{5x}{5} = \frac{20}{5}$$

$$x = 4$$

Substitution of the solution $x = 4$ in the equation gives $20 = 20$.

Each of the equations in Examples 3–5 was solved by one of the basic operations. However, many equations require that we perform several operations to obtain the solution. In solving such equations it is usually a good strategy to follow these steps:

Procedure for Solving Equations

1. Eliminate any fractions by multiplying both sides of the equation by a common denominator, or expressions that eliminate the denominators.

2. Use the distributive law to remove any grouping symbols.

3. Combine like terms on each side of the equation.

4. Perform the same operations on both sides of the equation until $x = result$ is obtained.

5. Check your solution by substituting back in the original equation.

Example 6

Solve $2x - 3 = 1 + x$.

By combining terms containing x on one side and the other terms on the other side, we can isolate x and thereby find the solution. The solution proceeds as follows:

$2x - 3 = 1 + x$	original equation
$2x - 3 - x = 1 + x - x$	subtract x from each side
$x - 3 = 1$	combine like terms
$x - 3 + 3 = 1 + 3$	add 3 to each side
$x = 4$	simplify

Because the last equation gives the desired value of x directly, the solution is complete.

▶   Check: Substituting *in the original equation,* we obtain

$$2(4) - 3 = 1 + 4, \quad \text{or} \quad 5 = 5$$

Example 7

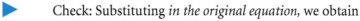

In trying to find the center of gravity of a beam, an engineer needs to solve the equation $16 - 3x = 2 + 4x$. Solve for x.

The solution proceeds as follows:

$16 - 3x = 2 + 4x$	original equation
$16 - 3x + 3x = 2 + 4x + 3x$	add $3x$ to each side
$16 = 2 + 7x$	combine like terms
$16 - 2 = 2 + 7x - 2$	subtract 2 from each side
$14 = 7x$	combine like terms
$\dfrac{14}{7} = \dfrac{7x}{7}$	divide each side by 7
$2 = x \quad \text{or} \quad x = 2$	simplify

Check: Substituting 2 for x in the original equation, we obtain $10 = 10$.

When you are more familiar with solving equations like these, you can do the step of adding or subtracting a term, or dividing by a factor, by inspection. See Examples 8 and 9.

Example 8

Solve $3(2x + 1) = 2x - 9$.

We can proceed as follows:

$3(2x + 1) = 2x - 9$	original equation
$6x + 3 = 2x - 9$	perform multiplication to remove parentheses
$4x + 3 = -9$	subtract $2x$ from each side (by inspection)
$4x = -12$	subtract 3 from each side (by inspection)
$x = -3$	divide each side by 4 (by inspection)

Check: Substituting in the original equation, we obtain

$$3(2(-3) + 1) = 2(-3) - 9$$
$$3(-6 + 1) = -6 - 9 \quad \text{or} \quad -15 = -15$$

Example
9

In solving

$$\frac{x}{3} - \frac{1}{2} = \frac{x}{2} - \frac{5}{2}$$

for x, it would be wise to first multiply each side by 6 so that fractions are no longer involved. In so doing, we must be sure to multiply each term by 6. We get

$2x - 3 = 3x - 15$ multiply each side by 6

$-3 = x - 15$ subtract $2x$ from each side

$12 = x$ or $x = 12$ add 15 to each side

Substituting 12 for x in the original equation, we obtain $\frac{7}{2} = \frac{7}{2}$, so that the solution of $x = 12$ is verified.

Now Try It! 2

Solve for x

a. $4x - 1 = x + 8$

b. $2(x - 1) = 4x - 13$

c. $-3(x - 5) = 2(2x + 1) - 8$

d. $\frac{2x}{3} - 5 = \frac{3x}{2}$

e. $\frac{4x - 2(x - 4)}{3} = -8$

Earlier, we listed steps for a solution strategy, but the order of those steps can vary. No specific order is required and most equations can be solved by any one of several different procedures. In example 10, the solution does not follow the order suggested by the solution strategy listed earlier, but it does yield the correct result.

Example
10

In solving

$$500(2x - 3) = 200(x - 2) + 500$$

for x, we can work with smaller numbers by first dividing both sides by 100 to get

$$5(2x - 3) = 2(x - 2) + 5$$

We can now proceed with the order suggested by the earlier solution strategy.

$10x - 15 = 2x - 4 + 5$ remove parentheses

$10x - 15 = 2x + 1$ combine

$8x - 15 = 1$ subtract $2x$ from each side

$8x = 16$ add 15 to each side

$x = 2$ divide each side by 8

We check the solution by substituting 2 for x in the original equation. Because the result simplifies to $500 = 500$, the solution is verified.

Now that we have had an opportunity to solve a variety of equations we come to several additional points about equations:

- An **identity** *is an equation that is true <u>for all values of x.</u>*
- A **conditional equation** *is true only <u>for specific values of x.</u>* The equations we have solved so far have all been conditional equations.
- A **contradiction** *is an equation that is not true <u>for any value of x.</u>*

Example 11 illustrates an identity and a contradiction.

1. $2x + 2 = 2(x + 1)$ is true for all values of x. Proceeding with the solution, we would arrive at $0 = 0$, which is true for any value of x. This equation is an **identity.**

2. $x + 1 = x + 2$ is an equation, although in attempting to solve it we subtract x from both sides to get $1 = 2$, which *is not true for any value of x.* This equation has no solution and is a **contradiction.**

4.1 Exercises

In Exercises 1 through 32, solve the given equations for x. Check the solution of each equation by substituting the value found in the equation.

1. $x + 3 = 5$

2. $x - 4 = 7$

3. $x + 5 = 8$

4. $x + 6 = 15$

5. $2x = 14$

6. $3x = 21$

7. $\dfrac{x}{7} = 5$

8. $\dfrac{x}{3} = 8$

9. $\dfrac{x}{6} = -3$

10. $\dfrac{x}{5} = -8$

11. $x + 6 = 2$

12. $x - 1 = -7$

13. $2x - 5 = 13$

14. $3x + 20 = 5$

15. $4x + 11 = 3$

16. $3x - 2 = 16$

17. $3 + 6x = 24 - x$

18. $5 - x = 8x - 13$

19. $3x - 6 = x - 18$

20. $14 - 2x = 5x + 7$

21. $2(x - 1) = x - 3$

22. $3(x + 2) = 2x + 11$

23. $\dfrac{x}{2} = x - 4$

24. $\dfrac{x}{3} = 12 - x$

25. $\dfrac{x}{3} + \dfrac{x}{4} = \dfrac{7}{12}$

26. $\dfrac{x}{2} + 2 = \dfrac{x}{5} + 5$

27. $3(x - 1) + x = 2(x - 1)$

28. $3(2 - 3x) = 7(x + 1) - 33$

29. $6x - 1 = 3(x - 2) + 6$

30. $4(2x + 1) = 3(x - 3) - 7$

31. $0.2(3 - x) - 0.3(x - 2) = 0.1x$

32. $0.2(3x - 4) = -0.2(3 - 4x) + 0.7$

In Exercises 33 through 44, set up the appropriate equation from the given statement and then solve.

33. Twice a number is 18.

34. Three more than a number is 11.

35. Half a number is 16.

36. A number less 6 equals 7.

37. Five more than a number is 8.

38. Three times a number is 33.

39. A number divided by 4 is 17.

40. Six more than twice a number equals 4 times the number.

41. Three less than 5 times a number is the number less 11.

42. Half a number equals twice the difference of the number and 12.

43. Five less than 4 times a number equals one-fifth of the number.

44. Seven less than 6 times a number equals the number less 14.

In Exercises 45 through 48, answer the given questions.

45. Distinguish between $x + 5 = x + 7$ and $x + 5 = x + 5$ as to the type of solution.

46. Distinguish between $x + 1 = x - 5$ and $x + 1 = 5 - x$ as to the type of solution.

47. Which of the following are identities?
  a. $2(x + 1) + 1 = 3 + 2x$
  b. $3(x + 1) = 1 + 3x$
  c. $5x = 4(x + 1) - x$
  d. $2x - 3 = 3(x - 1) - x$
  e. $4x - 5 = 4(x - 5)$

48. Which of the following equations are contradictions?
  a. $x - 7 = x - 8$ **b.** $1 - x = 1 + x$
  c. $2x + 2 = 2(x + 1)$ **d.** $3x - 6 = 3(x - 6)$
  e. $2x + 4 = 5 + 2x$

In Exercises 49 through 52, solve the given problem.

49. In finding the maximum temperature T (in °C) for a computer integrated circuit, the equation $1.1 = (T - 76)/40$ is used. Find the value of T.

50. To find the voltage V in a circuit in a TV remote control unit, the equation $1.12V - 0.67(10.5 - V) = 0$ is used. Find V.

51. In blending two different gasolines with different octanes, in order to find the number n of gallons of one octane needed, the equation $0.14n + 0.06(2000 - n) = 0.09(2000)$ is used. Find the value of n.

52. The cost of shipping a FedEx Priority Overnight package weighing one pound or more a distance of 1001 to 1400 miles is given by the equation $c = 2.8p + 21.05$ where c represents the cost and p represents the weight in pounds.
  a. Find the cost of shipping packages that weigh 2 pounds, 5 pounds, and 7 pounds.
  b. If a package costs $239.45 to ship, how much does it weigh?

Answers

Now Try It! 1
a. $x = -3$ **b.** $x = -7$ **c.** $x = 0$ **d.** $x = 70$ **e.** $x = -4$

Now Try It! 2
a. $x = 3$ **b.** $x = 5.5$ **c.** $x = 3$ **d.** $x = -2$ **e.** $x = -16$

4.2 Simple Formulas and Literal Equations

Section Objectives
• To successfully manipulate literal equations and formulas

In the previous section we considered only equations containing the unknown x and other specific numbers. All the formulas included in Chapter 3 are actually equations, because they express equality of algebraic expressions. Most of these formulas contain more than one literal symbol. In this section we extend the methods of solving equations to formulas and other equations containing more than one literal symbol, as in Example 1.

Example 1

In converting temperatures from the Celsius scale to the Kelvin scale, the equation

$$K = C + 273$$

is used. Because it is sometimes necessary to convert temperatures from the Kelvin scale to the Celsius scale, we want an expression for C. Solve for C.

Because the objective is to isolate C, we simply subtract 273 from each side of the equation to get

$$K - 273 = C \text{ or } C = K - 273$$

which is the required solution.

A **literal equation** is simply an equation with more than one variable. **Formulas** are excellent examples of literal equations. Formulas generally describe a relationship between specific variables, for example $y = mx + b$ in mathematics and $V = IR$ in electronics.

In Example 1 our result for C includes the literal symbol K instead of a single specific number. Any equation containing literal symbols other than the unknown, or required symbol, is solved in the same general manner. In each case the result will include literal symbols and not only a specific number. Also, *many of the operations will be in terms of literal symbols,* commonly referred to as variables.

To solve a literal equation for a specific variable, rearrange the equation so that the variable to be solved for is on one side of the equal sign with all other variables and constant terms on the other side.

When rearranging a literal equation, follow the same basic rule of algebra used for solving any equation. Specifically, *whatever is done to one side of an equation must be done to the other side in order to preserve the equality.*

Example 2

In the equation $ay + b = 2c$, solve for b.

Because the object is to isolate b, we subtract ay from each side of the equation. This results in

$$b = 2c - ay$$

which is the required solution.

Example 3

Remember our conductor problem from the start of this chapter. A conductor is a material that permits the flow of electrons. Resistance is a measure of the resistance of a material to that flow of electrons. At a given temperature, the resistance of a material is dependent on the type, length, and cross section of the material and is given by the following literal equation: $R = \dfrac{\rho l}{A}$. Solve this literal equation for A (the wire's cross sectional area).

$$R = \frac{\rho l}{A} \qquad \text{original equation}$$

$$AR = \frac{\rho l}{A}A \qquad \text{multiply both sides of the equation by } A$$

$$\frac{AR}{R} = \frac{\rho l}{R} \qquad \text{divide both sides of the equation by } R$$

$$A = \frac{\rho l}{R} \qquad \text{final result}$$

Example 4

In the study of forces on the wing of an aircraft, the equation $M = R(L - x)$ is used. Solve for L.

$$M = R(L - x) \qquad \text{original equation}$$

$$M = RL - Rx \qquad \text{use distributive law}$$

$$M + Rx = RL \qquad \text{add } Rx \text{ to each side}$$

$$\frac{M + Rx}{R} = L \qquad \text{divide each side by } R$$

or

$$L = \frac{M + Rx}{R}$$

▶ This last equation gives the required solution. Note that **R cannot *be divided out of the numerator and the denominator because it is* not a factor of the entire numerator.**

Example 5

In analyzing the transfer of heat through a wall, a solar engineer must solve the equation

$$q = \frac{KA(B - C)}{L}$$

for B. The solution proceeds as follows.

$$q = \frac{KA(B - C)}{L} \qquad \text{original equation}$$

$$qL = KA(B - C) \qquad \text{multiply each side by } L$$

$$qL = KAB - KAC \qquad \text{multiply to remove parentheses}$$

$$qL + KAC = KAB \qquad \text{add } KAC \text{ to each side}$$

$$\frac{qL + KAC}{KA} = B \qquad \text{divide each side by } KA$$

or

$$B = \frac{qL + KAC}{KA}$$

The last formula is the desired result. Note that KA cannot be divided out of the numerator and denominator because it is not a factor of the entire numerator.

In many problems it is necessary to refer to two or more values of a quantity. For example, we may wish to represent the resistances in several different integrated circuits, representing resistance by the letter R. Instead of choosing a different letter for each resistance, we can use **subscripts** on the letter R for this purpose. Thus R_1 could be the resistance of the first circuit, R_2 the resistance of the second circuit, and so on. It must be remembered that R_1 **and** R_2 **are different literal numbers** just as x and y are different. Example 6 shows the solution of a literal equation involving subscripts.

Example 6

Solve the equation $as_1 + cs_2 = 3a(s_1 + a)$ for s_1.

The solution proceeds as follows:

$$as_1 + cs_2 = 3a(s_1 + a) \qquad \text{original equation}$$

$$as_1 + cs_2 = 3as_1 + 3a^2 \qquad \text{remove parentheses}$$

$$cs_2 = 2as_1 + 3a^2 \qquad \text{subtract } as_1 \text{ from each side}$$

$$cs_2 - 3a^2 = 2as_1 \qquad \text{subtract } 3a^2 \text{ from each side}$$

$$\frac{cs_2 - 3a^2}{2a} = s_1 \qquad \text{divide each side by } 2a$$

or

$$s_1 = \frac{cs_2 - 3a^2}{2a}$$

This last equation gives the required solution.

Now Try It!

a. $f = \dfrac{P}{A}$ solve for A

b. $v = v_0 + at$ solve for a

c. $A = \pi r(2h + r)$ solve for h

d. $E = \dfrac{PL}{ae}$ solve for e.

Example
7

In doing calculations with pulleys, the equation

$$L = 3.14(r_1 + r_2) + 2d$$

must be solved for r_1. The solution proceeds as follows:

$$L = 3.14(r_1 + r_2) + 2d \qquad \text{original equation}$$

$$L = 3.14r_1 + 3.14r_2 + 2d \qquad \text{remove parentheses}$$

$$L - 3.14r_2 - 2d = 3.14r_1 \qquad \text{subtract } 3.14r_2 \text{ and } 2d \text{ from each side}$$

$$\frac{L - 3.14r_2 - 2d}{3.14} = r_1 \qquad \text{divide each side by 3.14}$$

or

$$r_1 = \frac{L - 3.14r_2 - 2d}{3.14}$$

The solution is shown in the last equation.

It often happens that problems are presented in verbal form and must be restated in an algebraic form before the solution is possible. In Section 4.4, we will consider general methods for converting verbal statements into equations, but Example 8 illustrates how a formula can be set up from a statement and then solved for one of its symbols.

Example
8

One missile travels at a speed of v_1 mi/h for 3 h and another missile goes v_2 mi/h for $3 + t$ hours, and the total distance they travel is d. Solve the resulting formula for t.

If the first missile goes v_1 mi/h for 3 h, then it goes a distance of $3v_1$ mi. Similarly, the distance the second missile travels is $v_2(3 + t)$, so we have the total distance d as

$$d = 3v_1 + v_2(3 + t)$$

Solving this formula for t, we get the following:

$$d = 3v_1 + 3v_2 + v_2t \qquad \text{remove parentheses}$$

$$d - 3v_1 - 3v_2 = v_2t \qquad \text{subtract } 3v_1 \text{ and } 3v_2 \text{ from each side}$$

$$t = \frac{d - 3v_1 - 3v_2}{v_2} \qquad \text{divide each side by } v_2 \text{ and then switch sides}$$

The last formula is the desired result.

4.2 Exercises

The formulas in Exercises 1 through 28 are used in the technical areas listed at the right. Solve for the indicated literal number.

1. $N = r(A - s)$, for r (engineering: stress)

2. $D = 2R(C - P)$, for C (business: depreciation)

3. $R_1L_2 = R_2L_1$, for R_2 (electricity)

4. $S = \dfrac{A - B}{A}$, for B (electricity: motors)

5. $v_2 = v_1 + at$, for v_1 (physics: motion)

6. $C = a + bx$, for a (economics: cost analysis)

7. $PV = RT$, for T (chemistry: gas law)

8. $E = IR$, for I (electricity)

9. $I = \dfrac{5300\,CE}{d^2}$, for C (atomic physics)

10. $E = \dfrac{mv^2}{2}$, for m (physics: energy)

11. $l = \dfrac{yd}{mR}$, for R (optics)

12. $P = \dfrac{N + 2}{D_0}$, for N (mechanics: gears)

13. $A = 180 - (B + C)$, for B (geometry)

14. $T_d = 3(T_2 - T_1)$, for T_1 (drilling for oil)

15. $R = \dfrac{CVL}{M}$, for L (water evaporation)

16. $r = \dfrac{g_2 - g_1}{L}$, for g_2 (surveying)

17. $D_p = \dfrac{MD_m}{P}$, for P (aerial photography)

18. $F = \frac{9}{5}C + 32$, for C (temperature conversion)

19. $L = 3.14(r_1 + r_2) + 2d$, for r_2 (pulleys)

20. $Q_1 = P(Q_2 - Q_1)$, for Q_2 (refrigeration)

21. $L = L_0(1 + at)$, for a (temperature expansion)

22. $a = V(k - PV)$, for k (biology)

23. $p - p_a = dg(y_2 - y_1)$, for y_2 (pressure gauges)

24. $F = A_2 - A_1 + P(V_2 - V_1)$, for V_2 (chemistry)

25. $A = \dfrac{n_1p_1 + n_2p_2}{n_1 + n_2}$, for p_1 (economics)

26. $Q = \dfrac{kAT(t_2 - t_1)}{d}$, for t_2 (heat conduction)

27. $f = \dfrac{f_s u}{u + v_2}$, for v_2 (sound)

28. $P = \dfrac{V_1(V_2 - V_1)}{gJ}$, for J (jet engine power)

In Exercises 29 through 44, solve for the indicated literal number.

29. $a = bc + d$, for d

30. $x - 2y = 3t$, for x

31. $ax + 3y = f$, for x

32. $3ay + b = 7q$, for y

33. $2a(x + y) = 3y$, for x

34. $3x(a - b) = 2x$, for a

35. $\dfrac{a}{2} = b + 2$, for a

36. $\dfrac{y}{x} = 2 - a$, for y

37. $x_1 = x_2 + a(3 + b)$, for b

38. $s_1 + s_2 = 2(a - b)$, for a

39. $R_3 = \dfrac{R_1 + R_2}{2}$, for R_2

40. $m + n = \dfrac{2n + 1}{3}$, for n

41. $7a(y + z) = 3(y + 2)$, for z

42. $3x(x + y) = 2(3 - x)$, for y

43. $3(x + a) + a(x + y) = 4x$, for y

44. $2a(a - x) = 3a(a - b) + 2ax$, for b

In Exercises 45 through 52, set up the required formula and solve for the indicated letter.

45. The mean A of three numbers a, b, and c equals their sum divided by 3. Solve for c.

46. Computer output is obtained with three printers. The first printer can do x lines in one minute, the second printer can do $x + 100$ lines in one minute, and the third printer can do $x + 300$ lines per minute. A total of T lines is printed in one minute. Solve for x.

47. The current I in a circuit with a resistor R and a battery with voltage E and internal resistance r equals E divided by the sum of r and R. Solve for R.

48. A vending machine contains x nickels, $x + 5$ dimes, and $x - 6$ quarters. If the total value of these coins is y cents, solve for x.

49. If x dollars are invested at a rate of r_1, and $x + 1000$ dollars are invested at a rate of r_2, the total interest for one year is I. Solve for r_2.

50. One computer does C calculations per second for $t + 8$ seconds. A second computer does $C + 100$ calculations per second for t seconds. Together they do a total of N calculations. Solve for C.

51. A microwave transmitter can handle x telephone connections while seven separate cables can handle y connections each. This combined system can handle C connections. Solve for y.

52. A fire protection unit uses two water pumps to drain a flooded basement. One pump can remove A gal/h whereas the second pump can drain B gal/h. The first pump is run for t hours and the second pump is run

for $t - 2$ hours so that K gallons are removed. Solve for B.

In Exercises 53 through 56, find the indicated values.

53. The pressure p (in kPa) at a depth of h (in m) below the surface of water is given by the formula $p = p_0 + kh$, where p_0 is the atmospheric pressure. Find h when $p = 205$ kPa, $p_0 = 101$ kPa, and $k = 9.80$ kPa/m.

54. A formula relating Fahrenheit temperature F and Celsius temperature C is $F = \frac{9}{5}C + 32$. Find the Celsius temperature that corresponds to 90.2°F.

55. In forestry, a formula used to determine the volume V of a wood log is $V = \frac{1}{2}L(B + b)$ where L is the length of the log and B and b represent the area of the ends of the log (see Figure 4.1).

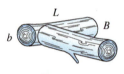

Figure 4.1

Find b (in ft^2) if $V = 38.6$ ft^3, $L = 16.1$ ft, and $B = 2.63$ ft^2.

56. The voltage V_1 across resistance R_1 (in ohms, Ω) is given by $V_1 = \dfrac{VR_1}{R_1 + R_2}$, where V is the voltage across resistances R_1 and R_2 (see Figure 4.2). Find R_2 (in ohms, Ω) if $R_1 = 3.56\ \Omega$, $V_1 = 6.30$ V, and $V = 12.0$ V.

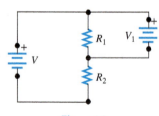

Figure 4.2

Now Try It! Answers

a. $A = \dfrac{P}{f}$ **b.** $a = \dfrac{v - v_0}{t}$ **c.** $h = \dfrac{a - 2\pi r^2}{2\pi r}$ **d.** $e = \dfrac{PL}{Ea}$

4.3 Simple Inequalities

Section Objectives
- Introduce the concept of simple inequalities
- Present properties of inequalities
- Solve simple linear inequalities

In Section 4.1, we defined an equation to be a mathematical statement that two algebraic expressions are equal. In the same manner, *an* **inequality** *is a mathematical statement that one algebraic expression is greater than, or less than, another algebraic expression.*

Equation-solving plays a major role in mathematics and its applications. However, there are also occasions that require the solutions of inequalities. Some of these are shown through the examples and exercises of this section.

In Chapter 1, we introduced the signs of inequality $>$ and $<$. We now review those symbols and introduce two new symbols.

1. The symbol $<$ means "less than" and $a < b$ is interpreted as "a is less than b." For example, $2 < 5$.

2. The symbol $>$ means "greater than" and $a > b$ is interpreted as "a is greater than b." For example, $8 > 3$.

3. The symbol $\leq$ means "less than or equal to." The expression $a \leq b$ is interpreted as "a is less than or equal to b." For example, $3 \leq 7$ and $3 \leq 3$ are both true.

4. The symbol $\geq$ means "greater than or equal to." The expression $a \geq b$ is interpreted as "a is greater than or equal to b." For example, $9 \geq 6$ and $6 \geq 6$ are both true.

These signs define the **sense** *of the inequality, which refers to the direction in which the inequality sign is pointed. The two sides of the inequality are known as* **members** *of the inequality.*

Example 1

same

The inequalities $x + 3 < 6$ and $x + 9 < 12$ have the *same sense* because both inequality symbols are in the same direction.

The inequalities $x + 4 > 1$ and $3 - x < 2$ have *opposite senses* because the inequality symbols are in opposite directions.

*The **solution** of an inequality consists of all values of the variable that satisfy the inequality.* That is, when any such value is substituted into the inequality, it becomes a correct statement mathematically. For most of the inequalities we consider here, the solution will consist of an unlimited number of values bounded by a specific real number.

Example 2

In the inequality $x + 4 > 1$, if we substitute any number greater than -3 for x, we find that the inequality is satisfied. That is, if we substitute $-2, 0, 4, \pi, -\sqrt{2}$, and so on for x, the inequality is satisfied. Thus the solution consists of *all* real numbers that are greater than -3. No value -3 or less satisfies the inequality. We express the solution as $x > -3$.

When finding the solution set of an inequality, we work with each member and do operations that are similar to those used in solving equations.

Properties of Inequalities

- If a, b, and c are real numbers with $a < b$, then $a + c < b + c$.
- If a, b, and c are real numbers with $a < b$, then $a - c < b - c$.
- If a, b, and c are real numbers with $a < b$ and $c > 0$, then $ac < bc$ and $\frac{a}{c} < \frac{b}{c}$.
- If a, b, and c are real numbers with $a < b$ and $c < 0$, then $ac > bc$ and $\frac{a}{c} > \frac{b}{c}$.

These properties are verified in Examples 3 and 4.

Example 3

We know that $2 < 6$. If 3 is added to each member of the inequality, we obtain $2 + 3 < 6 + 3$, or $5 < 9$, which is still a correct inequality.

If 3 is subtracted from each member of the inequality $2 < 6$, we obtain $2 - 3 < 6 - 3$, or $-1 < 3$, which is a correct inequality.

If each member of the inequality $2 < 6$ is multiplied by 3, we obtain $2 \times 3 < 6 \times 3$, or $6 < 18$, which is a correct inequality.

If each member of the inequality $2 < 6$ is divided by 3, we obtain $\frac{2}{3} < \frac{6}{3}$, or $\frac{2}{3} < 2$, which is a correct inequality.

This example illustrates the principle that the sense of an inequality does not change if we add a number to each member, subtract a number from each member, or multiply or divide by a *positive* number.

Example
4

If we multiply each member of the inequality $2 < 6$ by -3, we obtain -6 for the left member and -18 for the right member. However, we know that $-6 > -18$. Thus we obtain $-6 > -18$; the sense of the inequality has been reversed.

If each member of the inequality $2 < 6$ is divided by -2, we obtain $-1 > -3$, and again the sense of the inequality is reversed.

This example illustrates the important principle that multiplication or division by a *negative* number *reverses* the sense of that inequality.

The goal in performing these operations is to isolate the unknown on one side of the inequality and the other numbers on the other side. In this way, we can solve the inequality and determine the values that satisfy it. This is the same as in solving an equation. Examples 5, 6, 7, and 8 illustrate solving inequalities.

Example
5

1. Solve the inequality $x + 6 < 9$.

   By subtracting 6 from each member, we obtain $x < 3$, which means that the values of x less than 3 satisfy the inequality. Thus the solution is $x < 3$.

2. Solve the inequality $x - 6 < 9$.

   By adding 6 to each member, we obtain $x < 15$, which is the solution.

Example
6

1. Solve the inequality $\dfrac{x}{2} > 5$.

   By multiplying each member by 2, we obtain $x > 10$, which is the solution.

2. Solve the inequality $2x > 16$.

   By dividing each member by 2, we obtain $x > 8$, which is the solution.

Example
7

Solve the inequality $6 - 2x < 20$.

By first subtracting 6 from each member, we obtain $-2x < 14$. Then *if each member is divided by -2, we obtain $x > -7$, noting that the sense of the inequality has been reversed.* Thus the solution is $x > -7$.

Caution

If each member of the inequality is multiplied or divided by a negative number, the sign of the inequality is reversed.

Example 8 illustrates the solution of an inequality that requires several basic operations. The operations performed are indicated.

Example
8

Solve the inequality $6(x - 5) \geq 10 + x$.

$$6(x - 5) \geq 10 + x \qquad \text{original inequality}$$
$$6x - 30 \geq 10 + x \qquad \text{use distributive law}$$
$$6x \geq 40 + x \qquad \text{add 30 to each member}$$
$$5x \geq 40 \qquad \text{subtract } x \text{ from each member}$$
$$x \geq 8 \qquad \text{divide each member by 5}$$

Thus the solution is $x \geq 8$.

Example 9 illustrates an applied use of an inequality.

Example
9

A computer center is cooled by an air conditioner that is automatically activated if the temperature equals or exceeds 77°F. Because the Celsius and Fahrenheit scales are related by the equation

$$F = \frac{9}{5}C + 32$$

the air conditioner goes on when

$$\frac{9}{5}C + 32 \geq 77$$

If we wish to find the Celsius temperatures for which the air conditioner operates, we get

$$\frac{9}{5}C + 32 \geq 77 \qquad \text{original inequality}$$
$$9C + 160 \geq 385 \qquad \text{multiply each member by 5}$$
$$9C \geq 225 \qquad \text{subtract 160 from each member}$$
$$C \geq 25 \qquad \text{divide each member by 9}$$

The air conditioner goes on when the temperature equals or exceeds 25° on the Celsius scale.

Now Try It!

Solve

a. $2x + 3 < 7$

b. $-4x > 28$

c. $2x - 5 > x$

d. $10x + 4 < 16 + 12x$

e. $4x - 7 \geq 2x + 9$

Example
10

Another applied use of inequalities occurs with measurements. Measurements are not exact; they are only approximations, and the error associated with a measurement is often given. A surveyor, for example, might describe a particular distance (in meters) as 354.27 ± 0.05. This 0.05 error factor can be included in an inequality by describing the distance x (in meters) by

$$354.22 \leq x \leq 354.32$$

This inequality, which we note has *three* members, is read as "x is greater than or equal to 354.22 m *and* less than or equal to 354.32 m." This means that x is between or equal to these values, which has the same meaning as 354.27 m ± 0.05 m. Such inequalities are sometimes called *double inequalities* because they have two conditions that must be satisfied. We should stress that *both* conditions must be satisfied and that the inequality 354.22 ≤ x ≤ 354.32 really means

$$354.22 \leq x \; and \; x \leq 354.32$$

4.3 Exercises

In Exercises 1 through 20, solve the given inequalities.

1. $x - 7 > 5$

2. $x + 8 < 4$

3. $x + 2 < 7$

4. $x - 2 > 9$

5. $\dfrac{x}{3} > 6$

6. $\dfrac{x}{4} < 5$

7. $4x < -24$

8. $3x > 6$

9. $-2x \geq -8$

10. $-x \leq 5$

11. $2x + 7 \leq 5$

12. $5x - 3 \geq -13$

13. $2 - x > 6$

14. $3 - x < 1$

15. $4x - 1 < x - 7$

16. $2x + 1 > 10 - x$

17. $3(x - 2) > -9$

18. $6(x + 5) < x + 10$

19. $2(2x + 1) \leq 5x - 4$

20. $3(1 - 2x) \geq 1 - 4x$

In Exercises 21 through 30, set up inequalities from the given statements.

21. An electron microscope can magnify an object from 2000 times to 1,000,000 times its normal size. Assuming that these values are exact, express this magnification M as an inequality.

22. A satellite put into orbit near the earth's surface will have an elliptic orbit if its velocity v is between 18,000 mi/h and 25,000 mi/hr. Write this as an inequality in terms of v.

23. The velocity v of an ultrasound wave in soft human tissue can be represented as 1550 ± 60 m/s, where ±60 m/s gives the possible variation in the velocity. Express the possible velocities as an inequality.

24. A voltmeter is designed so that any reading is in error by no more than 1.3 V. If a measurement indicates 14.2 V, write an inequality that describes the true value of the voltage x.

25. A gauge is used to measure the pressure in a pipe used to transport natural gas. This gauge is designed so that any reading is in error by no more than 2.3 lb/in.^{2}. If the gauge indicates 64.9 lb/in.^{2}, write an inequality that describes the true value of the pressure P.

26. The rental cost for a machine is $100 for setup, plus $4 per hour of operation. How many hours of operation will result in total charges that exceed $240?

27. A sales representative earns a salary of $300 per week plus a commission of $6 for each unit sold. What is known about the number of units sold in one week if the total income is greater than $540?

28. A student has grades 80, 92, 86, and 78 on four tests in a certain course. The final exam in the course will count as two test grades. What grade must the student get on the final exam so that her average (arithmetic mean) will fall above an 80 for the course? (The arithmetic mean of several scores is found by adding the scores and dividing that total by the total number of scores).

29. The voltage drop across a resistor is found by multiplying the current i (in A) and the resistance R (in Ω). Find the possible voltage drops across a variable resistor R, if the minimum resistance is 1,600 Ω, the maximum resistance

is 3,600 Ω, and the current is constant at 0.0025 A. Represent your final result as an inequality.

30. In business a profit is realized when the revenue, or money coming into the business, is greater than the costs of operating the business. The weekly cost of manufacturing x computer chips is given by the expression $1500 + 75x$. The revenue from selling these computer chips is given by the algebraic expression $85x - 3300$. What is the minimum number of chips that must be sold each week in order to produce a profit?

Now Try It! Answers

a. $x < 2$ **b.** $x < -7$ **c.** $x > 5$ **d.** $x > -6$ **e.** $x \geq 8$

4.4 Problem Solving Strategies and Word Problems

Section Objectives

• Develop and apply general strategies for solving word problems.

Mathematics is particularly useful in technical areas because it can be used to solve many applied problems. Some of these problems are in formula form and can be solved directly. Many other problems are in verbal form and they must first be set up mathematically before direct methods of solution are possible. This section explains how to solve such problems. We will work primarily on the interpretation of verbal statements. It isn't possible to present specific rules for interpreting verbal statements for mathematical formulation. Certain conditions are *implied,* and these are the ones we must recognize. A very important step to the solution is a careful reading of the statement to make sure that all terms and phrases are understood.

Recall that in Section 1.7 we looked at problem-solving strategies (you may want to go back and review this section). These guidelines were designed to give you some methodology for sorting through the information presented in a problem with an eye on arriving at an appropriate solution. These included understanding the problem, choosing a strategy for solving the problem, solving the problem, and checking your solution.

In this section we will expand upon our problem-solving strategies as we take written information and work to create a mathematical equation that we can solve. In the following examples we will use a process generally used for solving word problems. The process is outlined as follows:

General Procedure for Solving Word Problems

1. Read the problem carefully.
2. Clearly identify the unknown quantities and assign an appropriate letter (or variable) to represent them, stating this choice clearly.
3. If possible, represent all of the unknowns in terms of just one variable.
4. If possible, construct a sketch using the known and unknown quantities.
5. Analyze the statement of the problem and construct the necessary equation. This is the most difficult step, but it does become easier with practice.
6. Solve the equation, clearly stating the solution.
7. Check the solution *with the original statement of the problem.*

Keep in mind that the ***primary goal is to obtain an equation from the written information.*** Once this is accomplished, solving the equation is relatively simple.

Example 1

A rectangular plate is to be the base for a circuit board. Its length is 2 in more than its width and it has a perimeter of 36 in. Find the dimensions.

We must first understand the meaning of all the terms used. Here we must know what is meant by rectangle, perimeter, and dimensions. Once we are sure of the terms, we must recognize what is required. We are told to find the dimensions. This should mean "find the length and the width of the rectangle." Once this is established, we let x (or any appropriate symbol) represent one of these quantities. Thus we write:

Let x = the width, in inches, of the rectangle   Step 2

Next we look to the statement for other key information. The order in which we find it useful may not be the order in which it is presented. For example, here we are told that the rectangle's length is 2 in. more than its width. Since we let x be the width, then

$x + 2$ = the length, in inches, of the rectangle   Step 3

We now have a representation of both required quantities. See Figure 4.3.   Step 4

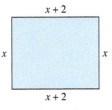

Figure 4.3

Now that we have identified the unknown quantities, we must look for information by which we can establish an equation. We look to the portion of the statement that is as yet unused. That is: "A rectangle . . . has a perimeter of 36 in." Recalling that the perimeter of a rectangle equals twice the length plus twice the width, we can now multiply the width x by 2, add this to twice the length, $x + 2$, and then equate this sum to 36, the known value of the perimeter. This leads to the equation

perimeter (algebraically) ⟶ $2x + 2(x + 2) = 36$ ⟵ perimeter (numerically)   Step 5

which we can now solve as follows:

$$2x + 2x + 4 = 36$$
$$4x = 32$$
$$x = 8$$ Step 6

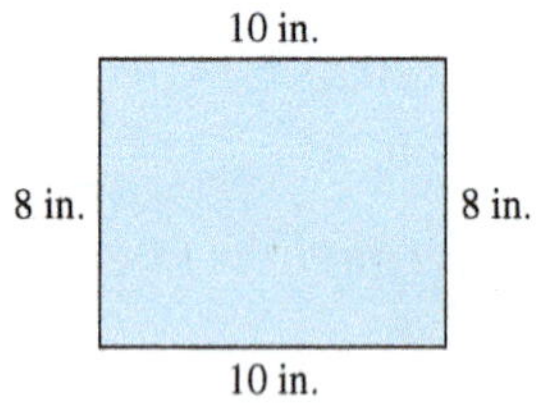

▶ Therefore, the width is 8 in. and the length is 10 in. See Figure 4.4. ***Checking this result with the statement of the problem, not the derived equation,*** we note that a rectangle with these dimensions has a perimeter of 36 in., which means that the solution checks.   Step 7

Our analysis was complete once we established the equation. Probably a great deal of what was just stated seemed quite obvious. However, even more involved problems require a similar technique of reading and interpreting, which is the key to the solution.

10 in.

8 in.   8 in.

10 in.

Figure 4.4

(Example
  2

In designing an electric circuit, it is found that 34 resistors with a total resistance of 56 Ω are required. Two different resistances, 1.5 Ω and 2.0 Ω, are used. How many of each are required in the circuit?

The phrase "how many of each" tells us that the number of each type of resistor is to be determined. We choose our variable to represent one of the resistors as follows:

Let $r =$ the number of 1.5 Ω resistors   Step 2

Since there are 34 resistors in total we know that

$34 - r =$ the number of 2.0 Ω resistors   Step 3

A sketch with 34 resistors may not make much sense here so we will skip step 4. We do know that the total resistance is 56 Ω. This means that

$$\underbrace{1.5r}_{\substack{\text{total resistance} \\ \text{of 1.5 } \Omega \text{ resistors}}} + \underbrace{2.0(34 - r)}_{\substack{\text{total resistance} \\ \text{of 2.0 } \Omega \text{ resistors}}} = 56 \ \Omega \qquad \text{Step 5}$$

$$1.5r + 68 - 2.0r = 56$$
$$-0.5r = -12$$
$$r = 24$$ Step 6

Therefore, there are 24 1.5 Ω resistors and 10 2.0 Ω resistors. The total resistance of these is $24(1.5) + 10(2.0) = 36 + 20 = 56\ \Omega$. We see that this checks with the statement of the problem.   Step 7

Example 3

Several 6-V and 12-V batteries are arranged so that their individual voltages combine to be a power supply of 84 V. How many of each type are present if the total number of batteries is 10?

    The phrase "how many of each type" tells us that the number of 6-V batteries and the number of 12-V batteries are to be determined. Choosing one quantity as x, we write:

Let $x = $ the number of 6-V batteries   Step 2

Then the phrase "the total number of batteries is 10" means that

$10 - x = $ the number of 12-V batteries   Step 3

Since the total voltage is 84 V, by multiplying the number of 6-V batteries (x) by 6 and the number of 12-V batteries ($10 - x$) by 12 and equating this sum to 84 (the total voltage), we get

$$6x + 12(10 - x) = 84$$ Step 5

where the first term $6x$ is the voltage of 6-V batteries (voltage per battery times number of 6-V batteries) and $12(10-x)$ is the voltage of 12-V batteries, with 84 the total voltage.

Solving this equation, we have

$$6x + 120 - 12x = 84 \qquad\qquad -6x = -36$$ Step 6
$$120 - 6x = 84 \qquad\qquad\qquad x = 6$$

There are six 6-V batteries and four 12-V batteries. Checking this, we see that these 10 batteries produce a total voltage of 84 V.   Step 7

Example 4

A machinist made 132 machine parts of two different types. He made 12 more type 1 than type 2 parts. How many of each did he make?

    From the statement of the problem, we see that we are to determine the number of each type. Therefore, we write:

Let $x = $ number of type 1 parts and $132 - x = $ number of type 2 parts

From the statement "he made 12 more type 1 than type 2 parts," we write

$$x = (132 - x) + 12$$

which is the required equation. Solving this equation, we get

$$x = 132 - x + 12$$
$$2x = 144$$
$$x = 72$$

Therefore, he made 72 type 1 parts and 60 type 2 parts. Note that **this checks** with the statement of the problem.

Example 5

A car travels 40 mi/h for 2 h along a certain route. Then a second car **starts along** the same route, traveling 60 mi/h. When will the second car overtake **the first?**

The word "when" means "at what time" or "for what value of t." **We can let t** represent the time either car has been traveling; so, choosing one quantity, **we write:**

Let $t =$ the time, in hours, the first car has traveled

The fact that the first car has been traveling for 2 h when the second **car starts** means that

$t - 2 =$ time, in hours, the second car has traveled

▶ The key to setting up the equation is the word "overtake," which **implies, al-** though it does not state explicitly, that ***the cars will have gone the same distance.*** Because distance equals rate times time, we establish the equation by **equating** the distance the first car travels, $40t$, to the distance the second **car travels,** $60(t - 2)$. Therefore, the equation is

$$\underset{\substack{\text{distance}\\ \text{traveled by}\\ \text{first car}}}{\underbrace{40t}} = \underset{\substack{\text{distance}\\ \text{traveled by}\\ \text{second car}}}{\underbrace{60(t - 2)}}$$

rate × time   rate × time

Solving this equation, we get

$$40t = 60t - 120$$
$$-20t = -120$$
$$t = 6$$

Therefore, the first car travels 6 h and the second car 4 h. Note that **a car travel-** ing 40 mi/h for 6 h goes 240 mi, as does a car traveling 60 mi/h for 4 h. **This con-** firms that the solution checks.

Example 6

A space shuttle maneuvers so that it can "capture" an already-orbiting satellite that is 6000 km ahead of it. If the satellite is moving at 27,000 km/h and the shuttle is moving at 29,500 km/h, how long will it take the shuttle to reach the satellite?

Let t = the time for the shuttle to reach the satellite

Using the same concept as in Example 5 (distance equals rate times time) we set up the following equation:

$$\underbrace{29,500t}_{\substack{\text{distance traveled} \\ \text{by the shuttle}}} = \underbrace{6000}_{\substack{\text{initial distance} \\ \text{between the shuttle} \\ \text{and satellite}}} + \underbrace{27,000t}_{\substack{\text{distance traveled} \\ \text{by the satellite}}}$$

$$2500t = 6000$$
$$t = 2.4 \text{ h}$$

This means that it will take the shuttle 2.4 h to reach the satellite. In 2.4 h, the shuttle will travel 70,800 km and the satellite will travel 64,800 km. We see that the solution checks with the statement of the problem.

Example 7

Suppose 100 kg of a cement-sand mixture is 40% sand. How many kilograms of sand must be added so that the resulting mixture will be 60% sand?

▶   We establish the equation by expressing ***the number of kilograms of sand in the final mixture as a sum of the sand originally present, 40 kg, and what is added,* n kg.** We then equate this to the amount in the final mixture, which is $0.60(100 + n)$. (The final mixture is 60% sand, and there is a total of $(100 + n)$ kg.) Thus

Let n = number of kilograms of sand to be added

$$\overset{\substack{40\% \quad \text{number} \\ \text{sand} \quad \text{of kg}}}{(0.40)(100)} + n = \overset{\substack{60\% \quad \text{number} \\ \text{sand} \quad \text{of kg}}}{0.60(100 + n)}$$

kg of sand kg of   kg of sand
in original sand   in final
mixture   added   mixture

$$40 + n = 0.60(100 + n)$$
$$40 + n = 60 + 0.60n$$
$$0.40n = 20$$
$$n = 50 \text{ kg}$$

Checking this with the original statement, we find that the final mixture will be 150 kg, of which 90 kg will be sand. Since

$$\frac{90}{150} = 0.60$$

the solution checks.

4.4 Exercises

Solve each of the following.

1. When three resistors are connected in series, their resistances are added to produce a total resistance of 970 Ω. One of them has a resistance of 530 Ω, and the others have resistance levels equal to each other. Find the resistance levels of the other two resistors.

2. A power supply has 2 printed circuit boards that contain a combined total of 222 components. One board has 6 more than twice the number of components on the other board. How many components are in each board?

3. An architect determines that if she reduces the dimensions of a square room by 2 ft on each side, the perimeter will be 56 ft. What is the length of the original room before the reduction?

4. The sum of two currents is 200 mA, and the larger current is 30 mA more than the smaller current. Find the value of the smaller current.

5. A 12-ft support stud is cut into two pieces so that one is three times as long as the other. How long is the longer piece?

6. An inlet pipe provides 4 gal/min more than a second pipe and 6 gal/min less than a third pipe. Together they supply 50 gal/min. How much does each provide?

7. A riverfront boat storage area is rectangular with fencing on all sides except the side along the river. If 550 m of fencing is used and the side along the river is 50 m shorter than the two longer sides, find the dimensions of the site.

8. Manufacturers rate print speed in pages per minute (ppm) produced at low resolution. A particular laser printer can print 2.5 times faster than a second laser printer. Together they can print 21 pages per minute. What is the print speed of each printer?

9. One laptop computer has 5.5 times the storage capacity of another. Together they can store 65 Gbytes of information. What is the storage capacity of each laptop computer?

10. Some resistors cost $0.12 each, whereas others cost $1.08 each. Sixty-five resistors cost a total of $22.20. How many of each resistor was purchased?

11. The online cost for an in-home smoke detector and carbon monoxide alarm is $4.29 more than twice the cost of a leading competitor's unit. If the two units are purchased together, their combined cost is $87.36. What is the cost of the more expensive alarm?

12. A company plans to issue 24,500 shares of two different kinds of stock which will have a combined value of $800,000. One of the stocks is worth $100 per share and the other stock is worth $25 per share. How many shares of each stock will be issued?

13. Three different oil storage tanks have a combined capacity of 4,400 gal. The largest tank holds three times as much as the smallest tank and twice as much as the other tank. What is the capacity of each tank?

14. One square cover plate has sides that are 5 mm longer than those of a second square cover plate. If the larger plate has a perimeter of 352 mm, find the perimeter of the smaller cover plate.

15. Two vans used for parts delivery are 420 mi apart when they begin traveling toward each other. One van goes 50 mi/h while the other goes 55 mi/h. How long do they drive before they meet?

16. A courier travels to and from a manufacturing plant in 7 h. His average speed to the plant is 80 km/h, and the return trip averages 20 km/h slower. How long did it take to reach the plant?

17. Two low-flying missiles are launched from the same location at sea and sent in opposite directions. After 5 s they are 33,000 ft apart. If one missile travels 600 ft/s faster than the other, what is the speed of this faster missile?

18. Two gears have a total of 80 teeth, and one gear has 48 fewer teeth than the other. How many teeth does each gear have?

19. If you need 50 lb of an alloy that is 50% nickel, how many pounds of an alloy with 80% nickel must be mixed with an alloy that is 40% nickel?

20. According to Kirchhoff's current law, the sum of the currents into a node equals the current out of the node. The current out of a node is 650 mA and three currents go into it. The largest current is twice the smallest and 100 mA more than the other current. Find all three currents.

21. The base material for a certain roadbed is 80% crushed rock when measured by weight. How many tons of this material must be mixed with another material containing 30% rock to make 150 tons of material that is 40% crushed rock?

22. Approximately 4.5 million wrecked cars are recycled in two consecutive years. There were 700,000 more recycled the second year than in the first year. How many are recycled each year?

23. A person pays $4800 in state and federal income taxes in a year. The federal income tax is five times as great as the state income tax. How much does the person pay on each of these income taxes?

24. A metallurgist melts and mixes 100 g of solder which is 50% tin with another solder that is 10% tin. How many grams of the second type must be used if the result is to be 25% tin?

25. A certain car's cooling system has an 8-qt capacity and is filled with a mixture that is 30% alcohol. How much of this mixture must be drained off and replaced with pure alcohol if the solution is to be 50% alcohol?

26. A company leases petroleum rights to 140 acres of land for $37,000. If part of the land leases for $200 per acre while the remainder leases for $300 per acre, find the amount of land leased at each price.

27. A walkway 3 m wide is constructed along the outside edge of a square courtyard. If the perimeter of the courtyard is 320 m, what is the perimeter of the square formed by the outer edge of the walkway?

28. A company anticipates a budget surplus of $84,000. It is decided that this amount is to be distributed for additional advertising, with radio getting one share, newspapers getting four shares, and direct mail getting two shares. How much is spent for additional newspaper advertising?

29. If $8000 is invested at 7%, how much must be invested at 9% in order for the total interest amount to equal $1550?

30. Two stock investments totalled $15,000. One stock led to a 40% gain, but the other stock resulted in a 10% loss. If the net result is a profit of $2000, how much was invested in each stock?

4.5 Ratio, Proportion and Variation

Section Objectives
- Work with ratios and proportions
- Solve problems using proportions
- Solve problems using direct variations
- Solve problems using indirect variations

This section introduces some of the important terms in many applications of mathematics, including those in science and technology.

> *The **ratio** of one number to another is the first number divided by the second number.* That is, the ratio of a to b is a/b.

(Another notation is $a:b$.) By this definition we use division to compare two numbers.

Example 1

1. The ratio of 12 to 8 is $12 \div 8$, or $\frac{12}{8}$. Simplifying the fraction we obtain $\frac{3}{2}$, which means that the ratio of 12 to 8 is the same as the ratio of 3 to 2. It also means that 12 is $\frac{3}{2}$ as large as 8.

2. The ratio of 2 to 9 is $\frac{2}{9}$.

3. The ratio of 10 to 2 is $\frac{10}{2}$, or $\frac{5}{1}$.

Every measurement is a ratio of the measured magnitude to an accepted unit of measurement. When we say that an object is 4 m long, we are saying that the length of the object is four times as long as the unit of length, the meter. Other examples of ratios are scales of measurements on maps, density (mass/volume), and pressure (force/area). As shown by these examples, ratios may compare quantities of the same kind or they may express divisions of magnitudes of different quantities.

Ratios are often used to compare like quantities. When used for this purpose we usually express each of the quantities in the same units. Example 2 illustrates this use of ratios.

Example 2

1. The length of a laboratory is 18 ft, and the width is 12 ft. See Figure 4.5. Therefore the ratio of the length of the laboratory to the width is 18 ft/12 ft, or $\frac{3}{2}$.

2. If the width of the room is expressed as 4 yd, we have the ratio as 18 ft/4 yd, or 9 ft/2 yd. However, this does not clearly show the ratio. It is preferable and more meaningful to change one of the measurements to the same units as the other. Changing the width from 4 yd to 12 ft, we express the ratio as 18 ft/12 ft or $\frac{3}{2}$, as above. From this ratio we easily see that the length of the room is $\frac{3}{2}$ as long as the width.

3. There are some special cases where ratios involve like quantities but different units. One drainage pipe specification requires a slope of $\frac{1}{4}$ in. for each foot of horizontal distance so that the ratio of vertical distance to horizontal distance is $\frac{1}{4}$ in./ft.

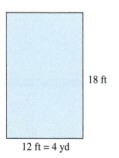

18 ft

12 ft = 4 yd

Figure 4.5

As we have noted, ratios are often used to compare like quantities, but they may also be used to compare unlike quantities. When we do compare unlike quantities, we must attach the proper units to the ratio. This is illustrated in Example 3.

Example 3

If a car travels 80 mi in 2 h and consumes 4 gal of gasoline, the ratio of distance to time is 80 mi/2 h, or 40 mi/h. Also, the ratio of distance to fuel consumed is 80 mi/4 gal, or 20 mi/gal. In each case we must note the units of the ratio.

A *statement of equality of two ratios is a* **proportion.** By this definition,

$$\frac{a}{b} = \frac{c}{d}$$

is a proportion, equating the ratios a/b and c/d.

(Another way of denoting a proportion is $a{:}b = c{:}d$, which is read as "*a* is to *b* as *c* is to *d*.") Thus we see that a proportion is an equation.

Example 4

1. The ratio $\frac{16}{6}$ equals the ratio $\frac{8}{3}$. We may state this equality by writing the proportion $\frac{16}{6} = \frac{8}{3}$.
2. $\frac{18}{12} = \frac{3}{2}$ and $\frac{80}{4} = \frac{20}{1}$ are proportions.
3. $80 \text{ km}/2\text{h} = 40 \text{ km}/1\text{h} = 40 \text{ km/h}$ is a proportion.

In a given proportion, if one ratio is known and one part of the other ratio is also known, then the unknown part can be found. This is done by letting an appropriate literal symbol represent the unknown part and then solving the resulting equation. This is illustrated in Examples 5–8.

Example 5

The ratio of a given number to 3 is the same as the ratio of 16 to 6. Find the number.

First, let $x =$ the number. We then set up the indicated proportion as

$$\frac{x}{3} = \frac{16}{6}$$

Solving for x, we multiply each side by 3:

$$\frac{3x}{3} = \frac{16(3)}{6}$$

which is simplified to

$$x = 8$$

Checking, we see that

$$\frac{8}{3} = \frac{16}{6}$$

Example 6

On a certain blueprint, a measurement of 25.0 ft is represented by 2.00 in. What is the actual distance between two points if they are 5.00 in. apart on the blueprint?

Let $x =$ the required distance and then note that 25.0 is to 2.00 as x is to 5.00. Thus:

actual distances

$$\frac{25.0}{2.00} = \frac{x}{5.00}$$

blueprint distances

By multiplying each side by 5.00, we solve the equation. This leads to

$$\frac{5.00(25.0)}{2.00} = \frac{5.00x}{5.00}$$

or

$$x = 62.5 \text{ ft}$$

Example 7

A certain alloy is 5 parts tin and 2 parts lead. How many kilograms of each are there in 35 kg of the alloy?

First, we let $x =$ the number of kilograms of tin in the given amount of alloy. Next, we note that there are 7 total parts of alloy, of which 5 are tin. Thus 5 is to 7 as x is to 35. This gives us the equation

parts tin $\longrightarrow$ $\dfrac{5}{7} = \dfrac{x}{35}$ $\longleftarrow$ weight tin
total parts $\longrightarrow$ $\phantom{\dfrac{5}{7} = }$ $\longleftarrow$ total weight

Multiplying each side by 35, we get

$$\frac{35(5)}{7} = \frac{35x}{35}$$

or

$$x = 25 \text{ kg}$$

There are 25 kg of tin and 10 kg of lead in the given amount of alloy.

Example 8

Similar triangles are triangles that have the same shape, but not the same size. A number of applications in geometry involve proportions that are based on the property that corresponding sides of similar triangles are proportional.

Suppose you wish to have a tree on your property cut down. The specialist who cuts the tree down must know the height of a tree in order to help determine which way the tree is to fall in order to avoid hitting your house. He uses a 6-ft pole and determines that this pole casts a 4-ft shadow at the same time that the tree casts a 35-ft shadow. He can use this information to determine the height of the tree (see Figure 4.6).

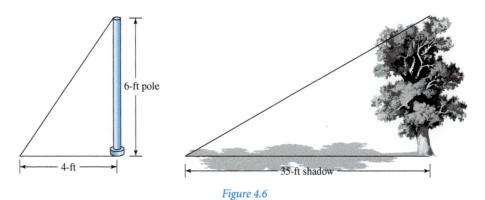

Figure 4.6

The two triangles formed in the picture above are similar triangles. If we let $x =$ height of the tree, we can set up the following proportion:

$$\frac{4}{6} = \frac{35}{x}$$

where 4 and 35 are the lengths of the shadows, and 6 and x are the height of the pole and the tree, respectively.

Using a technique called *cross multiplication* we do the following:

$$\frac{4}{6} \diagup\!\!\!\!\diagdown \frac{35}{x}$$

Cross multiplication is the same as multiplying each side of the proportion by both denominators. Using this technique we can get the same result in just one step by multiplying across the diagonals in a proportion as shown here.

$$4x = 210$$

$$x = 52 \text{ ft tall}$$

Many formulas we use today are derived through observation and experimental evidence. This method is very important in the technologies. For the remainder of this section we will look at illustrations showing how a number of formulas are established using this approach.

The ratio of one quantity to another remains the same in many applied situations. This relationship is the basis for determining many important formulas. Consider the cases in Examples 10–14.

Example 9

1. In a given length, the ratio of the length in inches to the length in feet is 12. This can be written as $\dfrac{i}{f} = 12$

2. Experimentation shows that the ratio of the electric resistance of a wire to the length of the wire is always the same. This can be written as $\dfrac{R}{l} = k$

3. It can be shown that for a given mass of gas, if the pressure remains constant, the ratio of the volume of the gas to the thermodynamic temperature remains constant. This can be written as $\dfrac{V}{T} = k$

In Section 3.1 we noted that *a* **variable** *is a quantity that may take on different values,* but *a* **constant** *takes on only one value during a given discussion.* We can express the fact that a ratio of two variables remains constant by the equation

$$\frac{y}{x} = k$$

where x and y are the variables and k is the constant.

1. In the first illustration of Example 9, the number of inches i and the number of feet f are variables. Each can take on any value so long as their ratio remains 12. This means that when we consider many different distances, and therefore many different values of i and f, the ratio is always 12.

2. In the second illustration of Example 9, the resistance R and the length l are the variables and k is the constant. The value of k may differ for different wires, but *for a given wire k* takes on a specific value, regardless of the length of the wire.

3. In the third illustration of Example 9, the volume V and temperature T are the variables and k is the constant. Again k may differ from one body of gas to another, but it's a specific constant for any one given body of gas (although the volume and temperature can vary).

By convention, letters near the end of the alphabet, such as x, y, and z, are generally used to denote variables. Letters near the beginning of the alphabet, such as *a*, *b*, and *c*, are generally used as constants. The meaning of other letters, such as *k* in this case, are specified in the problem.

If we solve the equation $y/x = k$ for *y*, we get

$$y = kx \qquad \text{direct variation}$$

*This equation is read as "y is **proportional to** x" or "y **varies directly as** x." This type of relationship is known as* **direct variation.**

Given a set of values for *x* and *y*, we can determine the value of the **constant of proportionality** *k*. Then we can substitute this value for *k* to obtain the relationship between *x* and *y*. Then we can find *y* for any other value of *x*. Conversely, we can find *x* for a given value of *y*. Consider the following example.

Example 10

If *y* varies directly as *x*, and $x = 6$ when $y = 18$, find the value of *y* when $x = 5$.

First we express the relationship "*y* varies directly as *x*" as

$$y = kx$$

Next we substitute $x = 6$ and $y = 18$ into the equation and we get

$$18 = 6k$$

▶ or $k = 3$. For this example *the constant of proportionality is 3, and this is substituted into* **y** $=$ **kx,** *giving*

$$y = 3x$$

as the equation between **y** *and* **x.** Now for any given value of *x* we may find the value of *y* by substitution. For $x = 5$, we get

$$y = 3(5) = 15$$

In many problems, one variable will vary directly as another variable raised to some power. Example 11 illustrates this in the case of an applied problem.

Example 11

The distance *d* that an object falls under the influence of gravity is proportional to the square of the time *t* of fall. If an object falls 64.0 ft in 2.00 s, how far does it fall in 6.00 s?

To express the fact that *d* varies directly as the square of *t*, we write

$$d = kt^2 \qquad \textit{directly} \text{ indicates multiplication of constant by variable}$$

Then, using the fact that $d = 64.0$ ft when $t = 2.00$ s, we get

$$64.0 = k(2.00)^2$$

which gives us $k = 16.0$ ft/s^2. In general,

$$d = 16.0t^2$$

We now substitute $t = 6.00$ s and we get

$$d = 16.0(6.00)^2 = 16.0(36.0) = 576 \text{ ft}$$

This means that an object falling under the influence of gravity will fall 576 ft in 6.00 s.

Note that the constant of proportionality in Example 11 has a set of units associated with it. (In this case, ft/s^2.) This will be the case unless the quantities related by k have precisely the same units. *We can determine the units for k by solving the equation for k and noting the units on the other side of the equation.*

Example 12

In Example 11, when we solve for k we can find its units as well as its value if we include the units in the calculation. In this case we have

$$d = kt^2$$
$$64.0 \text{ ft} = k(2.00 \text{ s})^2$$
$$k = \frac{64.0 \text{ ft}}{4.00 \text{ s}^2} = 16.0 \text{ ft/s}^2$$

In the first illustration of Example 9, the 12 has units of inches per foot. In the second illustration, if R is measured in ohms and l in feet, the units for k are ohms per foot. In the third illustration, if V is measured in liters and T in kelvins, the units for k are liters per kelvin. *The units actually used in a given problem will determine the units in which k will be measured for that problem.*

Another important type of variation in **inverse variation,** *which occurs when the product of two variables is constant,* as in $x \times y = k$. This can also be expressed by the equation

$$\boxed{y = \frac{k}{x} \qquad \text{inverse variation}}$$

Here y varies inversely as x (or y is inversely proportional to x), and k is the constant of proportionality. With direct variation in $y = kx$, the variable y increases as x increases and y decreases as x decreases. There is a direct relationship in the way that x and y vary. With the inverse variation of $y = k/x$, note that as x increases, y decreases. Also, if x decreases, y must increase. With inverse variation we have an opposite or inverse relationship between x and y.

Example 13

A business firm found that the volume V of sales of a certain item was inversely proportional to the price p.

This statement is expressed as

$$V = \frac{k}{p} \qquad inversely \text{ indicates division of constant by variable}$$

If we know that $V = 1000$ sales/week for $p = \$5$, we get

$$1000 = \frac{k}{5}$$

or $k = 5000$ dollar-sales/week. This means that the equation relating V and p is

$$V = \frac{5000}{p}$$

For any value of p, we may find the corresponding value of V. For example, if $p = \$4$ we get

$$V = \frac{5000}{4} = 1250 \text{ sales/week}$$

Finally, it is possible to relate one variable to more than one other variable by means of **combined variation.** Let us consider the following example.

Example 14

1. The equation expressing the fact that the force F between two electrically charged particles, with charges q_1 and q_2, varies directly as the product q_1 and q_2 is

   $$F = kq_1q_2 \qquad F \text{ varies directly as } q_1 \times q_2$$

2. The equation expressing the fact that y varies directly as x and inversely as z is

   $$y = \frac{kx}{z} \qquad y \begin{cases} \text{varies directly as } x \\ \text{varies inversely as } z \end{cases}$$

   Note that the word "and" appears in the statement, but the formula contains only products and quotients. The word *"and" **is used only to note that y varies in more than one way; it does not imply addition.***

3. The equation expressing the fact that s varies directly as the square of t and inversely as the cube of v is

$$s = \frac{kt^2}{v^3}$$

Note again that only a product and quotient appear in the formula.

As before, in each case we can determine k by knowing one set of values of the variables. After replacing k by its known value, we can proceed to determine the value of one variable if we know the values of the others.

4.5 Exercises

In Exercises 1 through 8, set up the given ratios as fractions and simplify when possible.

1. 12 to 5; 3 to 23

2. 2 to 11; 13 to 7

3. 21 to 3; 2 to 12

4. 5 to 45; 60 to 15

5. 6 to 9; 12 to 18

6. 20 to 24; 5 to 30

7. 6 to 33; 8 to 28

8. 100 to 10; 72 to 156

In Exercises 9 through 20, find the indicated ratios.

9. 30 A to 8 A

10. 4 mi to 22 mi

11. 9 cm to 30 cm

12. 50 W to 16 W

13. 8 in. to 4 ft

14. 5 cm to 40 mm

15. 80 s to 3 min

16. 500 lb to 3 t

17. 12 m to 6 s

18. 2 in. to 100°C

19. 8 lb to 36 ft^3

20. 25 L to 35 h

In Exercises 21 through 28, solve the given proportions for x.

21. $\dfrac{x}{2} = \dfrac{5}{8}$

22. $\dfrac{x}{3} = \dfrac{7}{12}$

23. $\dfrac{3}{14} = \dfrac{x}{4}$

24. $\dfrac{8}{15} = \dfrac{x}{9}$.

25. $\dfrac{3}{x} = \dfrac{9}{15}$

26. $\dfrac{12}{x} = \dfrac{1}{8}$

27. $\dfrac{4}{3} = \dfrac{12}{x}$

28. $\dfrac{25}{16} = \dfrac{30}{x}$

In Exercises 29 through 46, solve the given problems by first setting up the proper proportion.

29. The ratio of a number to 6 is the same as the ratio of 70 to 30. Find the number.

30. The ratio of a number to 15 is the same as the ratio of 17 to 60. Find the number.

31. The ratio of a number to 40 is the same as the ratio of 7 to 16. Find the number.

32. The ratio of a number to 44 is the same as the ratio of 8 to 33. Find the number.

33. The pitch of a roof is defined as the ratio of the rise to the span. If the rise is 6 ft and the span is 24 ft, find the pitch.

34. A stock is priced at $120 per share and annual earnings average to $8 per share. What is the price-to-earnings ratio?

35. If a substance of 908 g weighs 2.00 lb, how many grams of the substance would weigh 10.0 lb?

36. 60 mi/h = 88 ft/s; what speed in miles per hour is 66 ft/s?

37. A rectangular picture 20 in. by 15 in. is to be enlarged so that the width of the enlargement is 25 in. What should be the length of the enlargement? See Figure 4.7.

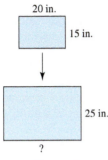

20 in.

15 in.

25 in.

?

Figure 4.7

38. A support beam 15 m long is cut into two pieces, the lengths of which are in the ratio of 2 to 3. What is the length of each piece?

39. A tablet of medication contains two substances in the ratio of 9 to 5. If a tablet contains 280 mg of medication, how much of each type is in the tablet?

40. Five clicks on an adjustment screw cause the inlet valve to change by 0.035 cm. How many clicks are required for a change of 0.021 cm?

41. Sixty gallons of oil flow through a pipe in 8.0 h. How long will it take 280 gal to flow through this same pipe?

42. A truck can go 245 mi on 35.0 gal of diesel fuel. How much diesel fuel would be required to go 1250 mi?

43. A shaft is designed to have a uniform taper. There is a taper of 3.5 mm in 14 cm of its length. How much taper is there in 37 cm of its length?

44. Under standard conditions in a 12 mi/h wind, a sugar pine seed will travel 77 m while a ponderosa pine seed will travel 130 m. In a wind of 19 mi/h the sugar pine seed travels 125 m. How far should the ponderosa pine seed travel in the same 19 mi/h wind?

45. For a hydraulic press, the mechanical advantage is the ratio of the large piston area to the small piston area. Find the mechanical advantage if the large and small pistons have areas of 27 cm^2 and 15 cm^2, respectively.

46. The shunt law, when applied to an electric circuit with parallel resistors, states that currents I_1, I_2, and resistances R_1, R_2 are related by

$$\frac{I_1}{I_2} = \frac{R_2}{R_1}$$

If $I_1 = 3$ A, $I_2 = 4$ A, and $R_1 = 12\Omega$, find R_2.

In Exercises 47 through 58, express the given statements as equations.

47. y varies directly as t.

48. x varies directly as s.

49. y varies directly as the square of s.

50. s varies directly as the cube of t.

51. t varies inversely as y.

52. y varies inversely as the square of x.

53. y varies directly as the product st.

54. s varies directly as the product xyz.

55. y varies directly as s and inversely as t.

56. y varies directly as s and inversely as the square of t.

57. x varies directly as the product yz and inversely as the square of t.

58. v varies directly as the cube of s and inversely as t.

In Exercises 59 through 64, give the equation relating the variables after evaluating the constant of proportionality for the given set of values.

59. y varies directly as s, and $y = 25$ when $s = 5$.

60. y varies inversely as t, and $y = 2$ when $t = 7$.

61. u is inversely proportional to the square of d, and $u = 17$ when $d = 4$.

62. q is inversely proportional to the square root of p, and $q = 5$ when $p = 9$.

63. y is directly proportional to x and inversely proportional to t, and $y = 6$ when $x = 2$ and $t = 3$.

64. t is directly proportional to n and inversely proportional to p, and $t = 21$ when $n = 3$ and $p = 5$.

In Exercises 65 through 71, find the required value by setting up the general equation and then evaluating.

65. Find s when $t = 4$ if s varies directly as t and $s = 20$ when $t = 5$.

66. Find y when $x = 5$ if y varies directly as x and $y = 36$ when $x = 2$.

67. Find q when $p = 5$ if q varies inversely as p and $q = 8$ when $p = 4$.

68. Find y when $x = 6$ if y varies inversely as x and $y = 15$ when $x = 3$.

69. Find s when $p = 75$ and $q = 5$ if s varies directly as p and inversely as the square of q and $s = 100$ when $p = 4$ and $q = 6$.

70. Find s when $v = 9$ and $t = 3$ if s varies directly as the square of v and inversely as t. We know that $s = 27$ when $v = 3$ and $t = 1$.

71. Find z when $x = 12$ and $y = 4$ if z varies inversely as the product xy. We know that $z = 4$ when $x = 2$ and $y = 3$.

In Exercises 72 through 82, solve the applied problems.

72. A fire science specialist studies the motion of an object projected upward from an explosion. The velocity v of the falling object is proportional to the time t of the fall. Find the equation relating v and t if $v = 96$ ft/s when $t = 3.0$ s.

73. An electric circuit has a fixed resistance so that the voltage E varies directly as the current I. If the voltage is 115 V when the current is 5.0 A, find the equation that relates E and I.

74. In chemistry, the general gas law states that the pressure P of a gas varies directly as the thermodynamic temperature T and inversely as the volume V. Express this statement as a formula. The constant of proportionality is called R.

75. According to Coulomb's law, the force F between two charges is directly proportional to the product of the charges (Q_1 and Q_2) and inversely proportional to the square of the distance s between them. Express this relationship as a formula.

76. The intensity of illumination I on a surface varies inversely as the square of the distance d from the source. If $I = 12.0$ footcandles when $d = 12.5$ ft, find I when $d = 10.0$ ft.

77. The horsepower necessary to propel a motorboat is proportional to the cube of the speed of the boat. If 10.5 hp is necessary to go 10.0 mi/h, what power is required for 15.0 mi/h?

78. The heat loss through rock wool insulation is inversely proportional to the thickness of the rock wool. If the loss through 6 in. of rock wool is 3200 Btu/h, find the loss through 2.5 in. of rock wool.

79. The electrical resistance R of a wire varies directly as the length l and inversely as the square of the diameter d. If a certain wire is 100.0 ft long with a diameter of 0.00200 ft, its resistance is 6.50 Ω. Find the resistance if the same material is used for a wire that is 25.0 ft long with a diameter of 0.00750 ft.

80. The kinetic energy E (energy due to motion) of an object varies directly as the square of its velocity v. Given that $E = 5000$ kg $\cdot$ m^2/s^2 when $v = 20$ m/s, find E when $v = 50$ m/s.

81. An industrial robot is being designed, and there is a need for two gears that will mesh. The number N of teeth on the first gear varies directly as r_2 and inversely as r_1, where r_1 is the number of revolutions per minute of the first gear and r_2 is the number of revolutions per minute of the second gear. The first gear turns at 150 r/min while the second gear turns at 200 r/min. Find the number of revolutions per minute for the second gear if the first gear turns at the rate of 180 r/min.

82. A 300.0-m length of aluminum wire contracts in length by 0.230 m when cooled from 90.0°F to 30.0°F. Its decrease in length varies directly as the product of its original length L and the change in temperature $T_2 - T_1$. Find the constant of proportionality.

Chapter Summary

Key Terms

equation

identity

conditional equation

contradiction

literal equation

formula

inequality

solution

ratio

proportion

direct variation

constant of proportionality

inverse variation

combined variation

Key Concepts

Addition Property of Equality

If $a = b$ then $a + c = b + c$ and $a - c = b - c$

Multiplication Property of Equality

If $a = b$ and c is not equal to 0, then

$$ac = bc \text{ and } \frac{a}{c} = \frac{b}{c}$$

- When solving any equation it is important to follow this basic rule of algebra: *Whatever is done to one side of an equation must be done to the other side in order to preserve the equality.*

Procedures for Solving Equations

1. Eliminate any fractions by multiplying both sides of the equation by the lowest common denominator.
2. Use the distributive law to remove any grouping symbols.

3. Combine like terms on each side of the equation.
4. Perform the same operations on both sides of the equations until $x = result$ is obtained.
5. Check your solution by substituting back in the original equation.

Properties of Inequalities

If a, b, and c are real numbers with $a < b$, then $a + c < b + c$.

If a, b, and c are real numbers with $a < b$, then $a - c < b - c$.

If a, b, and c are real numbers with $a < b$ and $c > 0$, then $ac < bc$ and $\frac{a}{c} < \frac{b}{c}$.

If a, b, and c are real numbers with $a < b$ and $c < 0$, then $ac > bc$ and $\frac{a}{c} > \frac{b}{c}$.

General Procedure for Solving Word Problems

1. Read the problem carefully.
2. Clearly identify the unknown quantities and assign an appropriate letter (or variable) to represent them, stating this choice clearly.
3. If possible, represent all of the unknowns in terms of just one variable.
4. If possible, construct a sketch using the known and unknown quantities.
5. Analyze the statement of the problem and construct the necessary equation. This is the most difficult step, but it does become easier with practice.
6. Solve the equation, clearly stating the solution.
7. Check the solution **with the original statement of the problem.**

Formulas

Direct variation $y = kx$

Inverse variation $y = \dfrac{k}{x}$

Review Exercises

In Exercises 1 through 12, find the solution for each equation. Check by substituting the value found in the original equation.

1. $x - 3 = 15$

2. $x + 14 = 11$

3. $3y = 27$

4. $26q = 13$

5. $4x + 21 = 5$

6. $2n - 1 = 17$

7. $3(x + 1) = x + 11$

8. $2(y - 1) = 19 - 5y$

9. $2(1 - 2t) = 11 - t$

10. $19 - 3x = 5(1 - 2x)$

11. $7(s - 1) + 2(s + 2) = 3s$

12. $8(t + 2) = 3(2t + 13)$

In Exercises 13 through 32, solve for the indicated letter in terms of the others. Fields of study in which each formula was developed are in parentheses.

13. $R = R_1 + R_2 + R_3$, for R_3 (electricity)

14. $F = \dfrac{wa}{g}$, for g (physics: force)

15. $r = \dfrac{ms_1}{s_2}$, for s_2 (mathematics: statistics)

16. $P = I^2R$, for R (electricity)

17. $d_m = (n - 1)A$, for n (physics: optics)

18. $T_2w = q(T_2 - T_1)$, for T_1 (chemistry: energy)

19. $M_1V_1 + M_2V_2 = PT$, for M_1 (economics)

20. $f(u + v_s) = f_s u$, for v_s (physics: sound)

21. $R = \dfrac{wL}{H(w + L)}$, for H (interior design)

22. $A = \dfrac{-\mu R_0}{r + R_0}$, for r (electronics)

23. $W = T(S_1 - S_2) - H_1 + H_2$, for T (refrigeration)

24. $2p + dv^2 = 2d(C - W)$, for C (fluid flow)

25. $a(2 + 3x) = 3y + 2ax$, for y

26. $c(ax + c) = b(x + c)$, for a

27. $a(x + b) = b(x + c)$, for c

28. $2(y + a) - 3 = y(2 + a)$, for y

29. $3(2 - x) = a(a + b) - 2x$, for b

30. $r_1(r_1 - r_2) - 3r_3 = r_1r_2 + r_1^2$, for r_3

31. $3a(a + 2x) + a^2 = a(2 + a)$, for x

32. $\dfrac{x(a + 4)}{2} = 3(x + 5)$, for a

In Exercises 33 through 44, solve the given inequalities.

33. $x - 2 < 7$

34. $x + 4 < 2$

35. $2x \geq 10$

36. $6x \geq 18$

37. $3x + 10 < 1$

38. $5x - 1 < 19$

39. $6 - x > 10$

40. $5 - x > 4$

41. $5(x + 1) < x - 3$

42. $4(x + 2) < x - 4$

43. $12 - 2x \geq 3(x - 1)$

44. $9 - x \geq 3(x - 5)$

In Exercises 45 through 48, find the indicated ratios.

45. 2 ft to 36 in.

46. 80 mm to 15 cm

47. 4 min to 40 s

48. 75¢ to $3

In Exercises 49 through 52, solve the given proportions for x.

49. $\dfrac{x}{3} = \dfrac{8}{9}$

50. $\dfrac{x}{4} = \dfrac{17}{12}$

51. $\dfrac{3}{10} = \dfrac{x}{15}$

52. $\dfrac{6}{5} = \dfrac{x}{10}$

In Exercises 53 through 60, find the number.

53. One more than twice a number is nine.

54. Three less than a number is five.

55. Three times a number is eight more than the number.

56. Five more than a number is ten more than three times the number.

57. The ratio of a number to 5 is the same as the ratio of 7 to 15.

58. The ratio of a number to 7 is the same as the ratio of 25 to 28.

59. The ratio of a number to 12 is the same as the ratio of 7 to 8.

60. The ratio of a number to 25 is the same as the ratio of 13 to 15.

In Exercises 61 through 68, solve the given problems by using variation.

61. y varies directly as x; $y = 24$ when $x = 4$; find the resulting equation relating y and x.

62. s varies directly as the square of t; $s = 60$ when $t = 2$; find the resulting equation relating s and t.

63. m is inversely proportional to the square root of r; $m = 5$ when $r = 9$; find the resulting equation relating m and r.

64. v is inversely proportional to the cube of z; $v = 3$ when $z = 2$; find the resulting equation relating v and z.

65. f is directly proportional to m and inversely proportional to p; $f = 8$ when $m = 4$ and $p = 5$; find f when $m = 3$ and $p = 6$.

66. y is directly proportional to the cube of x and inversely proportional to the square of t; $y = 16$ when $x = 2$ and $t = 3$; find y when $x = 3$ and $t = 4$.

67. s varies directly as the product of t and u and inversely as v; $s = 27$ when $t = 3$, $u = 5$, and $v = 6$; find s for $t = 2$, $u = 3$, and $v = 4$.

68. v varies directly as the square root of n and inversely as the product of m and p; $v = 16$ when $n = 4$, $m = 3$, and $p = 6$; find v for $n = 9$, $m = 5$, and $p = 2$.

In Exercises 69 through 84, solve the given problems by first setting up the proper equation.

69. An architect designs a rectangular window such that the width is 18 in. less than the height. If the perimeter

of the window is 180 in., what are its dimensions? See Figure 4.8.

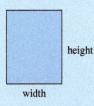

height

width

Figure 4.8

70. A square tract of land is enclosed with fencing and then divided in half by additional fencing parallel to two of the sides. If 75 m of fencing is used, what is the length of one side of the tract? See Figure 4.9.

Figure 4.9

71. The cost of manufacturing one type of computer memory storage device is three times the cost of manufacturing another type. The total cost of producing two of the first type and three of the second type is $450. What is the cost of producing each type?

72. A rectangular security area is enclosed with 120 ft of fencing. If this rectangular area is arranged so that the ratio of the length to the width is $\frac{3}{2}$, find the dimensions.

73. If you get 68 on one test, 73 on a second test, and 84 on a third test, what score must you get on a fourth test in order to have an average (arithmetic mean) of 80?

74. An architect begins with a design of a square room, but she decides to make it rectangular by doubling the lengths of two opposite sides. If the resulting perimeter is 72 ft, what was the perimeter of the original room?

75. A company has three checking accounts for petty cash. One account has $3000 more than that of a second, which has $4500 more than the third account. The three accounts have a total of $21,600. How much is in each account?

76. The electric current in one transistor is three times that in another transistor. If the sum of the currents is 12 mA, what is the current in each?

77. Two jets, 12,000 km apart, start toward each other and meet 3 h later. One is going 300 km/h faster than the other. What is the speed of each?

78. An environmental scientist finds that a certain fish can swim 12 mi/h in still water. This fish swims upstream against a 3.0 mi/h current, then returns to its starting point by swimming downstream. The upstream trip took 0.20 h longer than the downstream trip. How far upstream did the fish swim?

79. How much water must be added to 15 L of alcohol to make a solution which is 60% alcohol?

80. A chemist has 30 mL of a 12% solution of sulfuric acid. How many milliliters of a 40% solution must be added to get a 15% solution?

81. A $12,500 account yields an interest rate of 8.00%. How much must be added to this account so that the total interest yield amounts to $1720?

82. If $6500 is invested at 6.00%, how much more must be invested at 8.00% in order for the total investment to yield an average of 7.50%?

83. The weight w on the end of a spring varies directly as the length x that the spring stretches. If a weight of 8.0 lb stretches the spring 2.3 in., what weight will stretch it 5.7 in.?

84. The period of a pendulum varies directly as the square root of its length. If a pendulum 1.00 ft long has a period of 1.11 s, find the period of a pendulum 10.0 ft long.

Chapter Test

In Exercises 1 through 3, solve the given equations.

1. $3x + 1 = x - 8$

2. $\dfrac{2(n - 4)}{3} = \dfrac{5}{4}$

3. $3t - 2(7 - t) = 5(2t + 1)$

In Exercises 4 through 6, solve for the indicated letter.

4. $R = n^2Z$ *for Z*   **5.** $V = IR + Ir$ *for R*

6. $2(J + 1) = \dfrac{f}{B}$ *for J*

In Exercises 7 and 8, solve the inequalities.

7. $2x - 12 > 0$

8. $-7(x + 1) < 2(x + 4)$

In Exercises 9 and 10, find the indicate ratio.

9. 19 V to 3 V   **10.** 6 min to 45 sec

In Exercise 11, solve the proportion.

11. $\dfrac{5}{8} = \dfrac{x}{120}$

In Exercises 12 through 15, solve the problems.

12. Two computer software programs have a combined cost of $390. If one program costs $129 more than the other program, find the cost of each.

13. A 5-ft piece of tubing weighs 1.7 pounds. What is the weight of a 27-ft piece of the same tubing?

14. An automobile engine produces p cm^3 of carbon monoxide proportional to the time t that it idles. Find the equation relating p and t if $p = 60,000$ cm^3 and $t = 2$ min.

15. The resistance of a conductor is directly proportional to the length of the conductor. If the resistance of 2.6 miles of a coaxial cable is 75 Ω, find the resistance of 75 miles of that same cable line.

5 Graphs

Ask anyone you know what they are paying for car insurance and you will find that car insurance rates vary quite a bit from person to person, region to region, and insurance company to insurance company. The premiums you pay for car insurance are based on a variety of factors, some of which you can control, and others that you cannot control. Some of the more common factors that play a role in what you might pay for your car insurance include:

- Your age and the amount of driving experience you have. Young, inexperienced drivers generally pay more for their car insurance than most adults.

- Where you live. Car insurance companies determine their rates based upon the area in which you live.

- What you intend to use the car for. We all use our cars for different reasons (work, school, pleasure). Therefore, insurance companies have different rates for different uses. Generally, the less you drive, the less you may be charged for car insurance.

- Your driving record and car insurance history. The more accidents and tickets you have, the higher your insurance rates will be.

- The type of car you own. Expensive cars cost more to repair or replace, and so insurance costs more. Sports cars or high-performance cars also carry a higher premium.

We can say that the cost of our car insurance is determined by a number of different variables including age, experience, where you live, primary use for the car, driving record, and type of car.

Many situations in our world involve variables that are related to one another in some way. In mathematics, *functions* are often used to describe the relationship between these variables. We can say that the cost of car insurance is a *function* of several variables.

In this chapter we introduce the concept of a *function* and develop the basic procedures for constructing a *graph*. We will then look at how functions can be described using graphs. Later in the chapter we will explore various relationships as we look at linear and nonlinear functions. We will expand the use of the graphing calculator throughout this chapter by looking at graphical solutions for a variety of functions. Finally, we will look at the graphs of inequalities.

5.1 Functions and Function Notation

Section Objectives
- Understand that a function is a relationship between two variables
- Understand the definition of a function
- Identify the independent and dependent variable
- Find the domain and range of a function
- Use function notation to evaluate a function

In the earlier chapters we established the basic operations of algebra and discussed the solution of certain types of equations. When working with literal equations and formulas, we saw that the various quantities relate to one another in a specific way.

If a scientific experiment were to be performed to determine whether or not a relationship exists between the distance an object falls and the time of its fall, observation of the results should indicate (approximately at least) that $s = 16t^2$, where s is the distance (in feet) and t is the time (in seconds) of the fall. The experiment would show that the distance and the time are related.

In attempting to find a new oil reserve, it is found that the cost of drilling a well can be determined from the depth of that well. The time it takes a particular missile to reach its target can be computed from the distance to the target. The cost of mailing a package is determined by its weight and where it is being sent.

The percentage of chromium in iron-base alloys affects the rate of corrosion of the alloy. In general, as the percentage of chromium in the alloy increases, the rate of corrosion decreases. It is possible to set up an approximate equation or chart to show the relation of the chromium percentage and corrosion rate.

Considerations such as these lead us to the important mathematical concept of a **function.**

A **function** is a relationship between two variables in which any value for the first variable (for example x) is associated with exactly one value of the second variable (for example y). We say that the second variable is a *function* of the first variable.

The first variable is often described as the *input variable* or the **independent variable,** and the second variable is called the *output variable* or the **dependent variable.**

The first variable is referred to as the independent (or input) variable because we can assign any appropriate value we wish to this variable. We control the values of the independent variable. The second variable is called the dependent (or output) variable because its value is determined by our choice for the independent variable.

With $s = 16t^2$ expressing the relationship between distance s and time t, we have the variables s and t related so that for each value of t there is a single value for s. Therefore, this equation describes a function. With $s = 16t^2$ it is easier to substitute values for t and calculate the corresponding values for s. If we assign values to t and calculate the corresponding values of s, then t is called the **independent variable** and s is called the **dependent variable.** Functions require that for each value of the independent variable, the dependent variable will have only one corresponding value.

Example
1

The formula that expresses the area of a circle in terms of its radius is $A = \pi r^2$. Here A is a function of r. The variable r is the independent variable, and A is the dependent variable. We can see that the words "dependent" and "independent" are appropriately chosen. As the formula is written, we choose a value for r, and once this is done, the value of A is determined by that choice. For this example, we say that the area of a circle is a function of the radius of the circle. This formula and other formulas from geometry are introduced in Chapter 6.

Example 2

The stretching force of a steel spring is directly proportional to the elongation of the spring. This can be stated in an equation as $F = kx$. Here x is the independent variable and F is the dependent variable. (k is not a variable; it is a constant.)

We might wonder why the word "function" is introduced, because the formulas we have been dealing with seem sufficient. *In mathematics, the word "function" specifically refers to the operation that is performed on the independent variable to find the dependent variable.* Consider the following example.

Example 3

The distance s (in feet) that an object falls in t seconds is given by $s = 16t^2$. The electric power developed in a 16-Ω resistor by a current of i amperes is given by $P = 16i^2$. Here s is a function of t and P is a function of i. However, the function is the same. That is, to evaluate the dependent variable we square the independent variable and multiply this result by 16. ***Even though the letters are different, the operation is the same.***

A function is often written using notation that indicates the relationship between the independent and dependent variable. In Example 3 we stated that s was a function of t and that P was a function of i. We can use conventional **function notation** to represent this information as follows:

s is a function of t $s = f(t)$

P is a function of i $P = f(i)$

The phrase "y is a function of x" is represented by the notation $y = f(x)$. The notation $f(x)$ is read "f of x"

Using function notation helps to clearly identify the independent and dependent variables. If $s = f(t)$ tells us that s is a function of t, we know that t must be the independent variable and s must be the dependent variable. In general

$$y = f(x)$$

dependent variable   independent variable

Example
4

For the equation $y = x^2 - 5$, we say that $y = f(x)$, where $f(x) = x^2 - 5$. We now have $y = x^2 - 5$ and $f(x) = x^2 - 5$ as different ways of describing the same function.

> It is important to note that $f(x)$ is a special notation and does not mean f times x.

▶ One good use of the functional notation $f(x)$ is to express the value of a function for a specified value of the independent variable. ***Thus "the value of the function $f(x)$ when $x = a$" is expressed as $f(a)$.***

Example
5

If $f(x) = 5 - 2x$, $f(0)$ is the value of the function for $x = 0$ so that

$$f(0) = 5 - 2(0) = 5 \qquad \text{substitute 0 for } x$$

To find $f(-1)$, we have

$$f(-1) = 5 - 2(-1) = 5 + 2 = 7 \qquad \text{substitute } -1 \text{ for } x$$

We now know that $f(0) = 5$ and $f(-1) = 7$. Note that to find the value of the function for a number specified on the left, we substitute this number into the function on the right. Also, if we state that $y = f(x)$, then $y = 5$ for $x = 0$ and $y = 7$ for $x = -1$.

Example
6

If $f(x) = 4x - 2x^2$, find $f(-3)$.
To find $f(-3)$, we substitute -3 for x in the function. This gives us

$$
\begin{aligned}
f(-3) &= 4(-3) - 2(-3)^2 \qquad \text{substitute } -3 \text{ for } x \\
&= -12 - 2(9) \\
&= -12 - 18 \\
&= -30
\end{aligned}
$$

Function notation allows us to show the relationship between the independent and dependent variables in a single notation. For instance, in Example 6 we might say that "when the independent variable, x, is -3, the dependent variable is -30." Using function notation, we can say $f(-3) = -30$.

We know that the cost of mailing a letter is a function of the weight of that letter. The notation $f(1/2) = 0.37$ tells us that the cost of mailing a letter weighing $\frac{1}{2}$ ounce is \$0.37. We know that in order to convert from Celsius to Fahrenheit we use the formula $F = 1.8C + 32$. The notation $f(20) = 68°$ tells us that a temperature of 20°C is equivalent to 68°F.

The substitutions might involve literal quantities rather than numerical values as can be seen in Example 7.

> ### Example 7
>
> If $f(x) = x^2 + 1$, find $f(a)$, $f(-a)$, and $f(a + 1)$.
>
> To find $f(a)$, we substitute a for x in the given equation to get
>
> $$f(a) = a^2 + 1 \qquad \text{substitute } a \text{ for } x$$
>
> We can find $f(-a)$ by substituting $-a$ for x as follows:
>
> $$f(-a) = (-a)^2 + 1 = a^2 + 1 \qquad \text{substitute } -a \text{ for } x$$
>
> Finally, we evaluate $f(a + 1)$ by the same procedure.
>
> $$\text{substitute } a + 1 \text{ for } x$$
> $$f(a + 1) = (a + 1)^2 + 1 = a^2 + 2a + 1 + 1 = a^2 + 2a + 2$$

Now Try It!

Given $f(a) = 5a + 7$

Find:

a. $f(0)$
b. $f(4)$
c. $f(-5)$
d. $f(12)$

It is possible to identify functions from verbal statements. Such functions may be based upon known formulas or a proper interpretation of a given statement.

> ### Example 8
>
> The cost of fencing in a square prison farm is \$20 per foot of perimeter. Express the total cost as a function of the length of a side.
>
> The perimeter of a square is four times the length of a side and the total length of that perimeter is to be multiplied by \$20. This is expressed as
>
> cost per foot of perimeter
> perimeter
> $$C = 20(4s)$$
> $$= 80s$$
>
> The total cost, as a function of the length of a side, is $C = 80s$.

In the definition of a function, it was stipulated that any value for the independent variable must yield only a single value of the dependent variable. This requirement is stressed in more advanced and theoretical courses. *If a value of the independent variable yields more than one value of the dependent variable, the relationship is called a* **relation** *instead of a function.* A relation involves two variables related so that values of the second variable can be determined from values of the first variable. A function is a relation in which each value of the first variable yields only one value of the second. A function is therefore a special type of relation. However, there are relations that are not functions.

Example 9

For $y^2 = x^2$, if $x = 2$, then y can be either $+2$ or -2. Because a value of x yields more than one single value for y, we have a relation, not a function.

Earlier in this section we defined a function as a specific type of relationship between two variables, the independent and the dependent variable. Not all functions are valid for all values of the independent variable, nor are all possible values of the dependent variable possible.

The **domain** of a function is the set of all possible values that can be used for the independent variable.

The **range** of a function is the set of all possible appropriate output values that can be assumed by the dependent variable.

Example 10 illustrates these two definitions.

Example 10

In Example 1, where the area of a circle is given by the function $A = \pi r^2$, there is no real meaning for negative values of the radius or the area. Thus we would restrict values of r and A to zero and greater. That is, $r \geq 0$ and $A \geq 0$.

If we consider only real numbers, $f(x) = \sqrt{x - 1}$ is valid only for values of x greater than or equal to 1. That is, $x \geq 1$. Also, because *the positive square root is indicated,* the values of the function are zero or greater. For $f(x) = \sqrt{x - 1}$, the domain is described by $x \geq 1$ and the range is the collection of all values zero or greater ($y \geq 0$).

If $f(x) = 1/x$, all values of x are possible except $x = 0$, because that value would necessitate division by zero, which is undefined.

> **Caution**
>
> Values that lead to *division by zero*, or a *negative under a square root sign*, should not be included in the domain of a function.

5.1 Exercises

In Exercises 1 through 4, identify the dependent and independent variable of each function.

1. $y = 3x^4$

2. $s = -16t^2$

3. $p = \dfrac{c}{V}$ (*c* is a constant)

4. $v = a(1.05)^t$ (*a* is a constant)

In Exercises 5 through 8, state the basic operation of the function that is to be performed on the independent variable. For example, with $f(x) = x^2$ we "square the value of the independent variable."

5. $f(x) = 3x$

6. $f(y) = y + 3$

7. $f(r) = 2 - r^2$

8. $f(s) = s + s^2$

In Exercises 9 through 12, express each of the following using function notation.

9. $y = 5 - x$

10. $F = 6q^2$

11. $v = t^2 - 3t$

12. $s = 10^{-5r}$

In Exercises 13 through 28, evaluate the given functions whenever possible.

13. $f(x) = x;$ $f(0), f(3)$

14. $f(x) = x + 2;$ $f(0), f(-3)$

15. $f(t) = 2t - 1;$ $f(4), f(-2)$

16. $f(r) = 2 - r;$ $f(3), f(-3)$

17. $f(x) = 3x^2 - 2;$ $f(0), f\left(\frac{1}{2}\right)$

18. $f(z) = z^2 - z;$ $f(2), f(-2)$

19. $f(x) = 3 - x^2;$ $f(2), f(-0.3)$

20. $f(x) = 2x - 3x^2;$ $f(-1), f(-3)$

21. $f(s) = s^3;$ $f(-1), f(2)$

22. $f(p) = p^3 + 2p - 1;$ $f(a)$

23. $f(t) = 3t - t^3;$ $f(1), f(-2)$

24. $f(x) = x^3 - x^4;$ $f(3), f(-3)$

25. $f(q) = \dfrac{q}{q - 3};$ $f(-3), f(3)$

26. $f(v) = v + \dfrac{1}{v};$ $f\left(\dfrac{1}{2}\right), f(0)$

27. $f(x) = x - 2x^2;$ $f(a^2), f\left(\dfrac{1}{a}\right)$

28. $f(x) = 4^x;$ $f(-2), f\left(\dfrac{1}{2}\right)$

In Exercises 29 through 32, determine which of the given relations are also functions.

29. $y > x^3$

30. $y = x^4$

31. $y = \pm\sqrt{x}$

32. $|y| = |x|$

In Exercises 33 through 37, state an appropriate domain for each function.

33. Your salary is based on the number of hours you work each week.

34. The distance traveled in one day while heading off on vacation is a function of the amount of time spent driving if one travels at a constant speed of 65 mph.

35. $f(x) = \sqrt{x + 7}$

36. $f(x) = \dfrac{x + 1}{x - 2}$

37. $f(x) = 2x - \dfrac{1}{x}$

In Exercises 38 through 45, find the indicated functions. Be sure to use appropriate function notation.

38. Express the circumference c of a circle as a function of its radius r.

39. Express the volume V of a cube as a function of an edge e.

40. The number of tons t of air pollutants emitted by burning a certain fuel is 0.02 times the number of tons n of the fuel which is burned.

41. If the sales tax on a drill press is 5%, express the total cost C as a function of the list price P.

42. Express the distance s is kilometers traveled in 3 h as a function of the velocity v in kilometers per hour.

43. The voltage V across an electric resistor varies directly as the current i in the resistor. Given that the voltage is 10 V when the current is 2 A, express the current as a function of the voltage.

44. The stopping distance d of a car varies directly as the square of its speed s. Express d as a function of s, given that $d = 80$ m when $s = 100$ km/h.

45. The cost of renting a power tool is $12 for the first 6 hours and $6 for each additional hour after this.
    a. Use function notation to describe this scenario.
    b. Identify the independent and dependent variables.
    c. Identify a reasonable domain for this function.

46. The equation $c = 3n + 68$ describes the cost per student, c, of a class trip if n students go on the trip.
  a. Identify the independent and dependent variables.
  b. Rewrite the given equation using function notation.
  c. Write the following sentence using function notation "If 12 students go on the trip, the cost per student is $104."

  d. Describe a reasonable domain for this problem.
  e. Find $c(37)$.
  f. Describe what information the following notation gives you $c(24) = 140$.

Now Try It! Answers

a. $f(0) = 7$ **b.** $f(4) = 27$ **c.** $f(-5) = -18$ **d.** $f(12) = 67$

5.2   The Rectangular Coordinate System

Section Objectives
- Introduce the rectangular coordinate system
- Identify quadrants, axes, origins, ordered pairs, and coordinates
- Graph points in rectangular coordinates

> Even though we use the generic letters x and y for our graph, it is important to realize that when working with an application you can use any two variables that are appropriate for your function.

In this section we will discuss the basic system used for representing the graph of a function. Why are graphs of functions important? It is hard to pick up a newspaper, magazine, or journal and not find some type of information presented to us in graphical form. Graphs give us a good visual representation of the relationship between two variables. They can give us a picture of what the function looks like. This gives us a better idea of what we can expect from the function. Later in this chapter we will also see that graphs can be used to find the real solution of an equation.

A function represents two related sets of numbers. Therefore, we need to use a two-dimensional system to represent this relationship. To create this two-dimensional graph we use two number lines that are drawn perpendicular to each other and intersect at the number zero. We call this two-dimensional graph the **rectangular coordinate system.** The point of intersection of the two number lines is called the **origin.** This can be seen in Figure 5.1.

We call the horizontal number line the *x* **axis** and the vertical number line the *y* **axis.** The horizontal axis represents our independent variable and the vertical axis represents our dependent variable. The x axis contains positive values to the right of the origin and negative values to the left of the origin. On the y axis the positive values are above the origin and the negative values are below

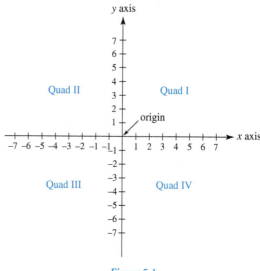

Figure 5.1

the origin. The *x* and *y* axes divide our coordinate system into four areas called **quadrants.** These are traditionally labeled with Roman numerals in a counter-clockwise direction.

Each point in the coordinate plane is uniquely identified by a pair of numbers. That means that it takes two values to locate one point. These numbers are written as an **ordered pair** using the format **(x, y).** The concept of an *ordered* pair is important. The first number in the ordered pair is always the *x* value or the *independent variable.* The second number is always the *y* value or the *dependent variable.* The *x* value is sometimes referred to as the **abscissa** and the *y* value is referred to as the **ordinate.** The values of *x* and *y* are called the **coordinates** of a point.

Figure 5.2 shows a point *P* located in the first quadrant. Its horizontal distance from the origin is 3.5 units, and its vertical distance from the origin is 3 units. We describe the location of point *P* by the ordered pair (3.5, 3). To plot any point *P*(*x, y*), simply count over *x* units on the horizontal and up (or down) *y* units on the vertical.

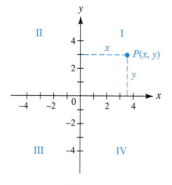

Figure 5.2

Example

1

Locate the points *A*(2, 3) and *B*(−4, −3) on the rectangular coordinate system. (The point *A*(2, 3) is the point *A* which has (2, 3) as its coordinates.)

The coordinates (2, 3) for point *A* means that the point is 2 units to the *right* of the *y* axis and 3 units *above* the *x* axis, as shown in Figure 5.3. The coordinates (−4, −3) for *B* means that the point is 4 units to the *left* of the *y* axis and 3 units *below* the *x* axis, as shown.

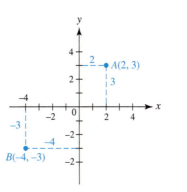

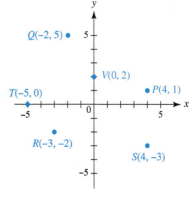

Caution
Even though there are two values in the ordered pair (x, y), these values locate **one point** on the graph.

Figure 5.3   *Figure 5.4*

Example 2

In attempting to analyze the relationship between an index of hardness and an index of color durability for different paint samples, it becomes necessary to plot the points $P(4, 1)$, $Q(-2, 5)$, $R(-3, -2)$, $S(4, -3)$, $T(-5, 0)$, and $V(0, 2)$. Those points are plotted in Figure 5.4.

Example 3

What is the sign of the ratio of the abscissa to the ordinate of a point in the second quadrant?

Because any point in the second quadrant is above the x axis, the y value (ordinate) is positive. Also, since any point in the second quadrant is to the left of the y axis, the x value (abscissa) is negative. The ratio of a negative number to a positive number is negative, which is the required answer. See Figure 5.5.

Example 4

In what quadrant does a point (a, b) lie if $a < 0$ and $b < 0$?

A point for which the abscissa is negative (which is the meaning of $a < 0$) must be in either the second or third quadrant. A point for which the ordinate is negative is in either the third or fourth quadrant. See Figure 5.6. The x coordinate must be in the second or third quadrant, and the y coordinate must be in the third or fourth quadrant. This implies that the point must be in the third quadrant.

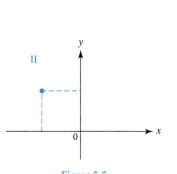

Figure 5.5

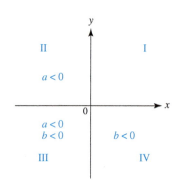

Figure 5.6

5.2 *Exercises*

In Exercises 1 through 8, determine (at least approximately) the coordinates of the points shown in Figure 5.7.

1. *A, B*

2. *C, D*

3. *E, F*

4. *G, H*

5. *I, J*

6. *K, L*

7. *M, N*

8. *P, Q*

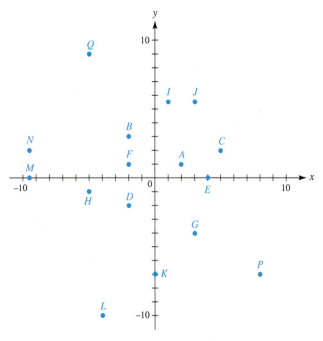

Figure 5.7

In Exercises 9 through 16, plot (at least approximately) the given points.

9. $A(0, 3)$; $B(3, 0)$

10. $A(-3, -6)$; $B(2, -4)$

11. $A(-1, 4)$; $B(2, 5)$

12. $A(0, -6)$; $B(-3, 4)$

13. $A(\frac{1}{2}, 5)$; $B(-\frac{1}{2}, -5)$

14. $A(0, \frac{1}{2})$; $B(\frac{1}{4}, 2)$

15. $A(2.5, -3.5)$; $B(0, -3.4)$

16. $A(-5.5, 8.1)$; $B(\frac{3}{2}, -3)$

In Exercises 17 through 20, identify the quadrant in which each point is located.

17. $(2, 12)$; $(-3, 4)$

18. $(5, -2)$; $(-2, -8)$

19. $(-3, -5)$; $(-2, 1)$

20. $(7, -10)$; $(-9, 2)$

In Exercise 21 through 26, answer the questions.

21. What is the value of x for each point on the y axis?

22. What is the value of y for each point on the x axis?

23. In which quadrants is the ratio of the abscissa to the ordinate positive?

24. In which quadrants is the ratio of the abscissa to the ordinate negative?

25. In what quadrant does a point (a, b) lie if $a > 0$ and $b > 0$?

26. In what quadrant does a point (a, b) lie if $a < 0$ and $b > 0$?

5.3   The Graph of a Function

Section Objectives
- Determine the graph of a function

Having introduced the rectangular coordinate system in the last section, we are now ready to sketch the graph of a function. In this way we can get a "picture" of the function, and this picture allows us to see the behavior and properties of the function. In this section we restrict our attention to functions whose graphs are either **straight lines** or **parabolas** (to be explained soon).

> *The graph of a function consists of all points whose coordinates (x, y) satisfy the functional relationship $y = f(x)$.*

By choosing a specific value for x, we can then find the corresponding value for y by evaluating $f(x)$. In this way we obtain as many points as necessary to plot the graph of the function. Usually we need to find only enough points to get a good approximation to the graph by joining these points with a smooth curve.

Just as there were problem-solving strategies we used in earlier chapters to help set up word problems, there are a few basic steps that we can follow to help build a graph.

Procedure for Creating a Graph

1. Create a table of ordered pairs. This is done by letting x take on several values and calculating corresponding values for y.

2. Plot each ordered pair.

3. Join the points from left to right using a smooth curve.

Here are a few other things to consider when creating a graph: If variables other than x and y are being used, clearly identify your independent and dependent variables and label your axes accordingly. Find an appropriate scale for your graph. Make sure you check your graph to make sure that all of your points lie on the graph you draw.

We shall first consider the graph of a **linear function,** which is a function of the form

$$f(x) = ax + b$$

Here *a* and *b* are constants. *It is called* **linear** *because the graph of such a function is always a straight line.* Example 1 illustrates the basic technique.

Example
1

Graph the function described by $f(x) = 3x + 2$.

Because $f(x) = 3x + 2$, by letting $y = f(x)$ we can write $y = 3x + 2$. Now by substituting numbers for *x*, we can find the corresponding values for *y*. If $x = 0$, then

$$y = f(0) = 3(0) + 2 = 2$$

so that the point $(0, 2)$ is on the graph of $f(x) = 3x + 2$. If $x = 1$, then

$$y = f(1) = 3(1) + 2 = 5$$

This means that the point $(1, 5)$ is on the graph of $f(x) = 3x + 2$. This information will be most helpful when it appears in tabular form. In preparing the table, *list the values of x in order* so that the points indicated can be joined in order. After the table has been set up, the indicated points are plotted on a rectangular coordinate system and then joined. The table and graph for this function are shown in Figure 5.8. Note that the line in the graph is extended beyond the points found in the table. This indicates that we know it continues in each direction.

The table of *x* and *y* values can be constructed either vertically or horizontally. An advantage of the vertical format is that it becomes easier to read the coordinates of the different points.

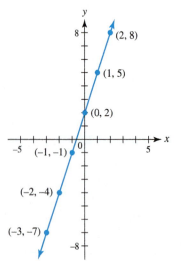

Figure 5.8

$y = 3x + 2$	x	y
$y = 3(-3) + 2 = -7$	-3	-7
$y = 3(-2) + 2 = -4$	-2	-4
$y = 3(-1) + 2 = -1$	-1	-1
$y = 3(0) + 2 = 2$	0	2
$y = 3(1) + 2 = 5$	1	5
$y = 3(2) + 2 = 8$	2	8

We now consider the graph of the **quadratic function.**

$$f(x) = ax^2 + bx + c \quad \text{where } a \neq 0$$

Here a, b, and c are constants. *The **graph** of this function is a **parabola.*** The examples that follow illustrate this function and its graph. This function will be covered in much greater detail in Chapter 11.

Example 2

A parabolic antenna is to be constructed so that its cross section is described by the equation $y = \frac{1}{2}x^2$. Graph the function $y = \frac{1}{2}x^2$.

Choosing values of x and obtaining corresponding values of y, we can determine the coordinates of a set of representative points on the graph. In finding the y values we must *be careful in handling negative values of x*: Remember that the square of a negative number is a positive number. For example, if $x = -3$,

$$y = \frac{1}{2}(-3)^2 = \frac{1}{2}(9) = \frac{9}{2}$$

With these ideas in mind we obtain the following table for $y = \frac{1}{2}x^2$. Note that the points are connected in Figure 5.9 with a smooth curve, not straight-line segments.

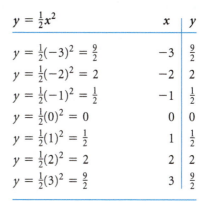

$y = \frac{1}{2}x^2$	x	y
$y = \frac{1}{2}(-3)^2 = \frac{9}{2}$	-3	$\frac{9}{2}$
$y = \frac{1}{2}(-2)^2 = 2$	-2	2
$y = \frac{1}{2}(-1)^2 = \frac{1}{2}$	-1	$\frac{1}{2}$
$y = \frac{1}{2}(0)^2 = 0$	0	0
$y = \frac{1}{2}(1)^2 = \frac{1}{2}$	1	$\frac{1}{2}$
$y = \frac{1}{2}(2)^2 = 2$	2	2
$y = \frac{1}{2}(3)^2 = \frac{9}{2}$	3	$\frac{9}{2}$

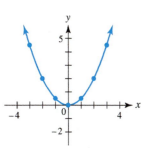

Figure 5.9

Figure 5.9 shows the basic shape of a parabola, which is the graph of a quadratic function. The parabola can be shifted right or left, and up or down, or it can open down. When graphing a parabola, it is important to find where the curve stops falling (or rising) and begins to rise (or fall). *The point at which this change in vertical direction occurs is called the **vertex.*** Instead of randomly plotting points on the parabola, we can use the following equation to find the x coordinate of the turning point.

$$x = \frac{-b}{2a} \quad \text{where} \quad y = ax^2 + bx + c$$

x coordinate of vertex

The y coordinate of the vertex can be found by substituting the x coordinate into the original quadratic function. The points we plot should include the vertex and points to its right and left. This is illustrated in Example 3.

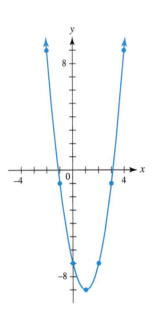

Figure 5.10

Example 3

Graph the function $y = 2x^2 - 4x - 7$.

Instead of selecting points at random, we begin by finding the x coordinate of the vertex. From the given quadratic function we have $a = 2$, $b = -4$, and $c = -7$ so that

$$x = \frac{-b}{2a} = \frac{-(-4)}{2(2)} = 1$$

The x coordinates we select should include $x = 1$ and some values greater than 1 and less than 1. We get the following table which is used to develop the graph shown in Figure 5.10.

$y = 2x^2 - 4x - 7$	x	y
$y = 2(-2)^2 - 4(-2) - 7 = 9$	-2	9
$y = 2(-1)^2 - 4(-1) - 7 = -1$	-1	-1
$y = 2(0)^2 - 4(0) - 7 = -7$	0	-7
$y = 2(1)^2 - 4(1) - 7 = -9$	1	-9
$y = 2(2)^2 - 4(2) - 7 = -7$	2	-7
$y = 2(3)^2 - 4(3) - 7 = -1$	3	-1
$y = 2(4)^2 - 4(4) - 7 = 9$	4	9

In Example 3, if we include more values of x, we may have to adjust the *scale* of the vertical axis. That is, the intervals along the y axis can be changed so that the graph will fit.

We can graph any type of function using the process presented in Examples 1–3.

Example 4

Graph $y = x^3 - 2$

We follow the same process as in Examples 1–3 by creating a table of values as shown below on the left. We plot these points to create the graph in Figure 5.11.

x	y	x	y
-4	-66	1	-1
-3	-29	2	6
-2	-10	3	25
-1	-3	4	62
0	-2		

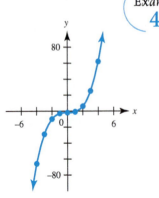

Figure 5.11

The use of graphs in mathematics is extensive. We can use graphs to solve equations. Numerous applications can be modeled by using a graph. Graphing calculators give us a tool that can display graphs quickly, saving time normally spent on evaluating the function and plotting points by hand.

 Using Technology

Graph the function $f(x) = 3x + 2$ (from Example 1) using a graphing calculator. We begin by entering the function.

Here's How

Press the Y = key. Enter the equation $3x + 2$ after the = sign at Y1. Your screen should look like the picture on the right (Figure 5.12a).

Now we need to tell the graphing calculator what range of values we want to use for x and y. This can be done in one of two ways. Try them both.

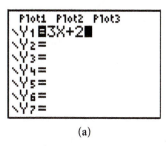

(a)

1. Press ZOOM, then #6. This gives you the standard window in which x and y both take on values from -10 to 10.

2. Press WINDOW. This allows you to enter the range of values you wish to display on the axes of your graph. The scale (scl) tells the calculator how far apart the tick marks on each axis should be. Set your window using the values shown in Figure 5.12b. Now press GRAPH. Compare your graph with Figure 5.12c.

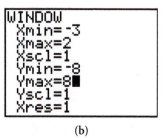

(b)

Try Another One

Repeat this process to create a graph for the parabola found in Example 3. To clear the current function from Y1, position the cursor anywhere in the line and press the CLEAR key.

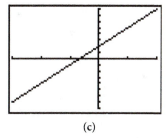

(c)

Figure 5.12

Enter $f(x) = 2x^2 - 4x - 7$ for Y1 (see Figure 5.13a).

Press ZOOM, followed by option #6 (see Figure 5.13b).

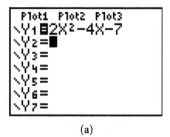

(a)

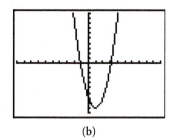

(b)

Figure 5.13

5.3 *Exercises*

In Exercises 1 through 22, graph the given functions.

1. $y = x$

2. $y = -2x$

3. $y = 3x - 5$

4. $y = 2x + 14$

5. $y = 9 - 4x$

6. $y = 8 - 2x$

7. $y = -5x + 3$

8. $y = \frac{1}{3}x - 2$

9. $y = \frac{2 - x}{3}$

10. $y = \frac{4 + 3x}{2}$

11. $y = x^2$

12. $y = x^2 - 4$

13. $y = 2x - x^2$

14. $y = 2 + \frac{1}{3}x^2$

15. $y = 2x^2 - 3x + 2$

16. $y = 3 - x - x^2$

17. $y = x^2 - 3x + 1$

18. $y = 20x - 5x^2$

19. $y = \frac{1}{2}x^2 + 2$

20. $y = -x^2 - 4x + 1$

21. $y = x^4 - 4x^2$

22. $y = x^3 - 2x^2 + x - 3$

In Exercises 23 through 32, graph the functions by plotting the dependent variable along the y axis and the independent variable along the horizontal axis. In the applied problems, be certain to determine whether or not negative values of the variables and the scales on each axis are meaningful.

23. A spring is stretched x in. by a force F. The equation relating x and F is

$$F = kx$$

where k is a constant. A force of 10 lb stretches a given spring 2 in. Plot the graph for F and x.

24. The *mechanical advantage* of an inclined plane is the ratio of the length of the plane to its height. This can be expressed in a formula as

$$M = \frac{L}{h}$$

Suppose that the height of a given plane is 2 m. Plot the graph of the mechanical advantage and length.

25. The electric resistance of wire resistors varies with the temperature according to the relation

$$R_2 = R_1 + R_1\alpha(T_2 - T_1)$$

where R_2 is the resistance at temperature T_2, R_1 is the resistance at temperature T_1, and α (the Greek alpha) is a constant depending on the type of wire. Plot the graph of R_2 and T_2 for a copper wire resistor ($\alpha = 0.004/°C$), given that $R_1 = 20\ \Omega$ and $T_1 = 10°C$.

26. A firm can produce up to 500 units per day of an item that sells for \$2. Fixed costs are \$200 daily and each item costs 50¢ to produce. Find the equation relating the profit p and the number x of units produced daily. Plot the graph.

27. The formula relating the Celsius and Fahrenheit temperature scales is

$$F = \frac{9}{5}C + 32$$

Plot the graph of this function.

28. The distance h (in feet) above the surface of the earth of an object as a function of the time (in seconds) is given by

$$h = 60t - 16t^2$$

if the object is given an initial upward velocity of 60 ft/s. Plot the graph.

29. The mass per unit length m of a bridge at a distance of x meters from the center of the bridge is

$$m = 150 + 0.8x^2$$

Plot the graph of m and x, given that the bridge is 100 m long and m is measured in kilograms per meter.

30. The resistance R (in Ω) of a resistor as a function of the temperature T (in °C) is given by $R = 250(1 + 0.0032T)$. Graph R as a function of T.

31. The height h (in m) of a rocket as a function of the time t (in s) is given by the function $h = 1500t - 4.9t^2$. Graph h as a function of t.

32. A formula used to determine the number N of board feet of lumber that can be cut from a 4-ft section of a log with diameter d (in in.) is $N = 0.22d^2 - 0.71d$. Graph N as a function of d for values of d from 10 in. to 40 in.

In Exercises 33 through 36, use a graphing calculator to obtain graphs of the given functions.

33. $y = 5x - 3$ **34.** $y = -2.7x + 1.4$

35. $y = x^2$ **36.** $y = -2.3x^2 - 1.1x - 2.0$

5.4 Graph of a Linear Function

Section Objectives
- Find the slope of a line using ordered pairs
- Describe the slope of horizontal and vertical lines
- Find x and y intercepts from a linear equation
- Find the slope and y intercept from a linear equation
- Find the equations of horizontal and vertical lines

In Section 5.3 we graphed a linear function by creating a table of ordered pairs. This was done by assigning values to x and calculating corresponding values for y. We then plotted each ordered pair in the rectangular coordinate system and joined the points to create a graph of a line. In this section we will look at how to create the graph of a line using the *slope* of a line. We will also look at the points of intersection between the line and the horizontal and vertical axes. Finally we will look at the equations of horizontal and vertical lines and discuss the slope of these lines.

The **slope** of a line describes the steepness of the line. Slope is measured as a ratio between the vertical change to the horizontal change between any two points on the line. The vertical change between any two points on a line is often referred to as the **rise,** and the horizontal change between the same points is the **run.** Therefore, we can describe the **slope** of a linear function, represented by the letter m, as

$$\text{slope} = m = \frac{\text{vertical change}}{\text{horizontal change}} = \frac{\text{rise}}{\text{run}}$$

Example 1

The straight line in Figure 5.14 has a rise of 8 and a run of 6. Therefore, the slope is

$$\text{slope} = m = \frac{\text{rise}}{\text{run}} = \frac{8}{6}$$

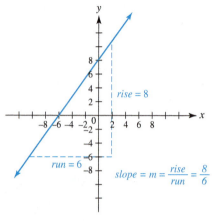

Figure 5.14

A line with a *positive slope* rises as we go from left to right. A line with a *negative slope* falls as we go from left to right.

Example 2

Find the slope of a line that has a vertical change of 30 units as we move 3 units along the positive *x* axis.

In this case the rise is 30 and the run is 3, so the slope is $m = \dfrac{30}{3} = 10$.

We have described the slope of a line as the ratio between the vertical change to the horizontal change between any two points on the line. If (x_1, y_1) and (x_2, y_2) are two points on a line, the slope of the line can be found using the following formula:

$$\text{slope} = m = \frac{y_2 - y_1}{x_2 - x_1}$$

Be careful not to mix up the subscripts when applying the formula for slope.

The slope is $\dfrac{y_2 - y_1}{x_2 - x_1}$ but the slope is

not $\dfrac{y_2 - y_1}{x_1 - x_2}$.

If we put together all of the various ways we have of describing slope, we see that

$$\text{slope} = m = \frac{\text{vertical change}}{\text{horizontal change}} = \frac{\text{rise}}{\text{run}} = \frac{y_2 - y_1}{x_2 - x_1}$$

Example 3

Find the slope of the line connecting the points $(-5, 1)$ and $(3, 5)$.

It does not matter which point we call (x_1, y_1) and which point we call (x_2, y_2). Let $(x_1, y_1) = (-5, 1)$ and $(x_2, y_2) = (3, 5)$

The slope $= m = \dfrac{y_2 - y_1}{x_2 - x_1} = \dfrac{5 - 1}{3 - (-5)} = \dfrac{5 - 1}{3 + 5} = \dfrac{4}{8} = \dfrac{1}{2}$. Therefore, the

slope of the line connecting these two points is $\dfrac{1}{2}$.

Suppose we had chosen to let $(x_1, y_1) = (3, 5)$ and $(x_2, y_2) = (-5, 1)$. Then

$$\text{slope} = m = \frac{y_2 - y_1}{x_2 - x_1} = \frac{1 - 5}{-5 - 3} = \frac{-4}{-8} = \frac{1}{2}$$

We can see that our choice for (x_1, y_1) and (x_2, y_2) does not matter as long as we apply the formula for slope and the rules for adding and subtracting signed numbers correctly. In either case we end up with the same result.

The slopes of horizontal and vertical lines require closer inspection. The y values on a horizontal line are the same everywhere on that line. We can see this in Figure 5.15. The ordered pairs shown on that graph are $(-3, 4)$, $(-2, 4)$, $(0, 4)$, and $(2, 4)$. When we apply the slope formula to any pair of points we see that vertical change will always be 0. Therefore, *the slope of a horizontal line is always 0.*

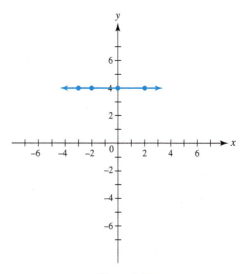

Figure 5.15

The x values on a vertical line are the same everywhere on that line. We see this in Figure 5.16. The ordered pairs shown on that graph are $(2, 5)$, $(2, 4)$, $(2, 3)$, and $(2, 2)$. When we apply the formula for slope we see that the vertical change is always 0. The zero will be in the denominator of our ratio, and division by zero is not defined. For this reason we say that *the slope of a vertical line is always undefined.*

The slope of a linear function is constant. That means that it is the same everywhere on the line. Given three points on the same line, A, B, and C, the slope between A and B is the same as the slope between B and C, which is the same as the slope between A and C.

The knowledge that a function of the form $y = mx + b$ is a straight line can be used to a definite advantage. By finding two points, we can draw the line. Two special points easily found are those where the curve crosses each of the axes. *These points are known as the **intercepts** of the line.* The reason these are easily found is that one of the coordinates of each intercept is zero.

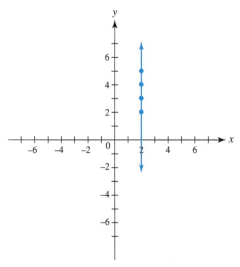

Figure 5.16

The **x-intercept** is the point where the graph crosses the *x* axis. This point always has a *y* coordinate of 0, and is written $(x, 0)$.

The **y-intercept** is the point where the graph crosses the *y* axis. This point always has an *x* coordinate of 0, and is written $(0, y)$.

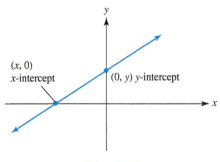

Figure 5.17

To find the *x*-intercept of a linear function we let $y = 0$ and find the corresponding value for *x*. To find the *y*-intercept of a linear function we let $x = 0$ and find the corresponding value for *y*. When graphing a linear function, a third point should also be found as a check.

Example

4

Find the *x* and *y* intercepts for the linear function $y = -2x - 5$.

To find the *x* intercept we let $y = 0$ and find *x*.

$$0 = -2x - 5$$
$$5 = -2x$$
$$-2.5 = x$$

The ordered pair for the *x* intercept is $(-2.5, 0)$.

Similarly, to find the y intercept we let $x = 0$ and find y.

$$y = -2(0) - 5$$
$$y = 0 - 5$$
$$y = -5$$

The ordered pair for the y intercept is $(0, -5)$.

Previously we defined a linear function as having the form $f(x) = ax + b$, where a and b are constants. We modify that definition now to include the concepts of slope and y-intercept.

A linear function can be written in the format $y = mx + b$ where m represents the slope of the line and b represents the y-intercept of the line.

In the equation $y = 2x + 1$, the slope is 2 and the y-intercept is 1.

slope y-intercept

Example 5

Write the equation of a linear function that has a slope of -3 and crosses the y axis at 7.

Because $m = -3$ and $b = 7$, we can substitute these values in the equation $y = mx + b$ to get the required equation. The result is $y = -3x + 7$.

Example 6

Find the slope and y-intercept for the equation $y = -4x - 2$.

By inspection we see that the slope is the coefficient of the x term and the y-intercept is the constant term. Therefore $m = -4$ and $b = -2$.

Example 7

Find the slope and y-intercept of $2y + 4x = 6$.

Before we can identify the slope and y-intercept, we must rewrite the given equation so that it is in $y = mx + b$ form.

$$2y + 4x = 6$$
$$2y = -4x + 6$$
$$y = -2x + 3$$

Now by inspection we can identify the slope, $m = -2$, and the y-intercept, $b = 3$.

Earlier in this section we noted that the slope of a horizontal line is always 0 and that the slope of a vertical line is always undefined. The equation for a horizontal line can be written as $y = b$ where $(0, b)$ is its y-intercept. The equation for a vertical line is $x = a$ where $(a, 0)$ is its x-intercept.

5.4 *Exercises*

In Exercises 1 through 4, find the slope of the line using the given information.

1. rise = 6, run = 3

2. rise = -27, run = -9

3. rise = -5, run = 5

4. rise = 7, run = 0

In Exercises 5 through 16, find the slope of the line passing through the given pairs of points.

5. $(2, 4)$ and $(5, 7)$

6. $(4, 9)$ and $(-3, -7)$

7. $(-2, 5)$ and $(5, -6)$

8. $(3, -3)$ and $(-6, 3)$

9. $(2.5, -3)$ and $(0, -3.4)$

10. $(-3, -8)$ and $(2, -8)$

11. $(1, 3)$ and $(1, 7)$

12. $(0, 3)$ and $(3, 0)$

13. $(5, 4)$ and $(-2, 4)$

14. $(-4, 3)$ and $(-4, 7)$

15. $(0.4, 0.5)$ and $(-0.2, 0.2)$

16. $(-2.8, 3.4)$ and $(1.2, 4.2)$

In Exercises 17 through 24, find the x- and y-intercept of the line with the given equation. Sketch the graph using the x- and y-intercepts. Use your graphing calculator to check your graph.

17. $x + 2y = 4$

18. $y = -3x + 3$

19. $5x - 2y = 20$

20. $-5y = 5 - x$

21. $y = 0.25x + 4.5$

22. $y = 0.08x - 2.4$

23. $y = \dfrac{5}{4}x + 2$

24. $4x - 3y = 12$

In Exercises 25 through 32, find the slope and y-intercept of the line with the given equation. Sketch the graph using the slope and y-intercept. Use your graphing calculator to check your graph.

25. $y = -2x + 1$

26. $y = -4x$

27. $y = x + 4$

28. $5x - 2y = 40$

29. $-2y = 7 + 4x$

30. $24x + 4y = 16$

31. $1.5x + 2.4y = 3.0$

32. $4.0x - 3.5y = 1.5$

5.5 Graphs of Other Functions

Section Objectives
• To introduce the graphs of nonlinear functions

This section introduces the graphs of several other functions.

We use the basic procedure presented in Section 5.3. That is, we select specific values of x and find the corresponding values of y in order to find particular points that lie on the graph of the function. Examples 1–4 plot the graphs of functions that are neither linear nor quadratic.

Example
1

A population growth model is described by the equation $y = 2^x$. Graph the function $y = 2^x$.

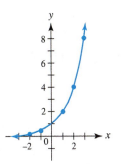

Figure 5.18

$y = 2^x$	x	y
$y = 2^{-2} = \frac{1}{2^2} = \frac{1}{4}$	-2	$\frac{1}{4}$
$y = 2^{-1} = \frac{1}{2}$	-1	$\frac{1}{2}$
$y = 2^0 = 1$	0	1
$y = 2^1 = 2$	1	2
$y = 2^2 = 4$	2	4
$y = 2^3 = 8$	3	8

In order to plot the graph of this function, *we must deal with powers of 2, including both positive and negative powers.* Note the evaluations shown with the table above from which the points are plotted in Figure 5.18.

This is an example of the graph of an **exponential function.**

In general, an exponential function will be of the form $y = b^x$ where b is a positive constant. The shape of any other exponential function will be similar to the graph of Figure 5.18, although it may curve downward instead of upward.

Example
2

Graph the function $y = 2 + \dfrac{1}{x}$.

In finding the points for this graph we must be careful not to set $x = 0$, because division by zero is undefined. To get an accurate graph, however, we must choose values of x near zero. Consequently, *we include fractional values of x* in the following table. For example, if $x = \frac{1}{2}$, then

$$y = 2 + \frac{1}{\frac{1}{2}} = 2 + 2 = 4$$

This graph is known as a **hyperbola** (see Figure 5.19).

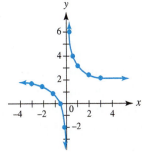

Figure 5.19

x	-3	-2	-1	$-\frac{1}{2}$	$-\frac{1}{4}$	$\frac{1}{4}$	$\frac{1}{2}$	1	2	3
y	$\frac{5}{3}$	$\frac{3}{2}$	1	0	-2	6	4	3	$\frac{5}{2}$	$\frac{7}{3}$

Example 3

Graph the function $y = x^3 - 3x$.

Proper use of signed numbers and their powers is essential in finding the values of y for values of x, *particularly negative values.* For example, if $x = -2$, then

$$y = (-2)^3 - 3(-2) = -8 + 6 = -2$$

The following table gives the points used to plot the graph in Figure 5.20.

x	−3	−2	−1	0	1	2	3
y	−18	−2	2	0	−2	2	18

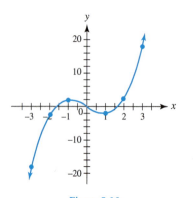

Figure 5.20

This curve is known as a **cubic curve.** Note that *the scale of the y axis is different from the scale of the x axis.* This is done when the range of values used differs considerably from the values in the domain.

Example 4

Graph the function $y = \sqrt{x + 1}$.

In finding the y values of this graph, we must be careful to use only values of x that will give us real values for y. This means that the value under the radical sign must be zero or greater. For example, if $x = -5$, we would have $y = \sqrt{-4}$, which is an imaginary number. Therefore, x cannot be -5. In fact, **we cannot have any value of x that is less than −1.** Also, the values of y will be greater than zero, except for $x = -1$.

This is due to the fact that **y is equal to the principal (positive) square root of x + 1.** Thus we find the following points for the graph:

x	−1	0	1	2	3	4	6	8
y	0	1	1.4	1.7	2	2.2	2.6	3

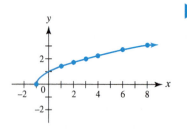

Figure 5.21

The graph is plotted in Figure 5.21. This graph is half of a parabola. All quadratic functions ($y = ax^2 + bx + c$) graph as parabolas that open up or down, but this function graphs as half of a parabola that opens to the right.

Examples 1, 2, 3, and 4 illustrated an exponential function, a hyperbola, a cubic curve, and a parabola, respectively. There are many other functions that do not belong to one of these categories. Some of the following exercises involve functions that we have not illustrated, but the same basic procedure applies. Substitute values for x, find the corresponding values of y, and plot the points that result. Then connect the points with a smooth curve.

In this section we look at several functions that may not be familiar to us.

Let's look at the graph of an **exponential function** on the graphing calculator.

To graph $y = 2^x$ we begin by entering the equation as we did in Section 5.3.

• Press Y= and enter the equation 2^x.

To raise 2 to the x power you will need to use the carat key $\wedge$. Your keystrokes would be $2 \wedge x$.

• Press ZOOM, #6 . This gives you the graph in the standard window as seen in Figure 5.22a.

If we wish to get a better look at the graph, we can change the window settings. We will use the settings that were supplied in Example 1.

• Press WINDOW.

• Change the x and y values to match those found in Example 1. Leave the *Xscl* and *Yscl* at 1.

• See the figure on the right.

• Then press GRAPH to see the graph in Figure 5.22b.

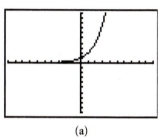

(a)

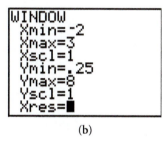

(b)

Figure 5.22

This graph in Figure 5.23 gives us a different perspective of the exponential function.

To graph the square root function found in Example 4, you will have to use the square root symbol. This symbol is located directly above the x^2 key.

• Press $2^{\text{nd}} x^2$ to access the square root key. You will see a $\sqrt{\ }$ (on your screen.

To graph $y = \sqrt{x + 1}$, press Y= and enter the equation by using the following keystrokes:

$2^{\text{nd}} x^2 \ x + 1)$. Then press ZOOM, #6 to see the graph in Figure 5.24.

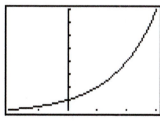

Figure 5.23

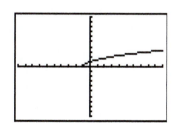

Figure 5.24

5.5 *Exercises*

In Exercises 1 through 18, graph the given functions. Use your graphing calculator to check your graphs.

1. $y = 3^x$

2. $y = 3^{2x}$

3. $y = \dfrac{1}{x}$

4. $y = 1 + \dfrac{2}{x}$

5. $y = \sqrt{x + 4}$

6. $y = \sqrt{1 - x}$

7. $y = \sqrt{4 - x}$

8. $y = x^3$

9. $y = -\dfrac{1}{2}x^3$

10. $y = 2x^3 - 10$

11. $y = 6x - x^3$

12. $y = x^4 - 2x^2$

13. $y = \dfrac{1}{x^2}$

14. $y = \dfrac{1}{x^2 + 1}$

15. $y = \sqrt{x^2 + 1}$

16. $y = \sqrt{25 - x^2}$

17. $y = \dfrac{1}{\sqrt{x}}$

18. $y = \dfrac{1}{\sqrt{1 - x}}$

In Exercises 19 through 26, graph the functions by plotting the dependent variable along the y axis and the independent variable along the x axis. In the applied problems, be certain to determine whether or not negative values of the variables and the scales on each axis are meaningful.

19. The number of bacteria in a certain culture increases by 50% each hour. The number N of bacteria present after t hours is

$$N = 1000\left(\frac{3}{2}\right)^t$$

given that 1000 were originally present. Graph the function.

20. Under the condition of constant temperature, it is found that the pressure p (in kilopascals) and the volume V (in cubic centimeters) in an experiment on air are related by

$$p = \frac{1000}{V}$$

Plot the graph.

21. The electric current I in a circuit with a voltage of 50 V, a constant resistor of 10 Ω, and a variable resistor R is given by the equation

$$I = \frac{50}{10 + R}$$

Plot the graph of I and R.

22. An object is p in. from a lens of a focal length of 5 in. The distance from the lens to the image is given by

$$q = \frac{5p}{p - 5}$$

Plot the graph.

23. The total profit P a manufacturer makes in producing x units of a commodity is given by

$$P = x^3 - 3x^2 - 5x - 150$$

Plot the graph, using values of 0 through 10 for x.

24. Under certain conditions the deflection d of a beam at a distance of x ft from one end is

$$d = 0.05(30x^2 - x^3)$$

where d is measured in inches. Plot the graph of d and x, given that the beam is 10 ft long.

25. After t years, the population of trees planted as a forestry experiment is given by

$$P = 100(1.02)^t$$

Plot the graph.

26. If an amount of P dollars is deposited in a bank, the value V of the account after time t (in years) is given by

$$V = P(1 + r)^t$$

where r is the interest rate in decimal form. Let $P = 100$, $r = 0.05$, and plot the graph.

5.6 Graphical Solutions

Section Objectives
- To accurately read information from a graph
- To be able to solve equations graphically

An important aspect of working with graphs is being able to read information from a graph. This section demonstrates how this is done and also how equations can be solved by the use of graphs.

The procedure for reading values from a graph is essentially the reverse of plotting the coordinates of points. The following examples illustrate this method.

Example 1

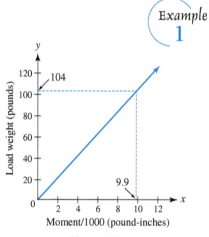

Figure 5.25

When operating a light aircraft, the pilot, passengers, baggage, and fuel must not exceed the maximum allowable safe load. It is also essential that those items are distributed so that they fall within prescribed center-of-gravity limitations. One particular aircraft operating manual provides the loading graph for baggage shown in Figure 5.25. If the baggage weighs 104 lb, find the corresponding value of the moment/1000 (pound-inches).

Knowing that the baggage weighs 104 lb, we locate that weight on the vertical scale and then move horizontally to meet the graph of the line at the point shown. At that point, the horizontal scale has a coordinate of about 9.9 pound-inches. This procedure is repeated for the graphs representing the pilot, passengers, and fuel, and the combined results are then used to determine whether the aircraft is correctly balanced. Every pilot must be able to use this procedure.

Example 2

From the graph shown in Figure 5.26, determine the value of y for $x = 3.5$.

We first locate 3.5 on the x axis and then construct a line (dashed line in the figure) perpendicular to the x axis *until it crosses the curve*. From this point on the curve, we draw another line perpendicular to the y axis *until it crosses the y axis*. The value at which this line crosses is the required answer. Therefore, $y = 1.4$ (approximated from the graph) for $x = 3.5$. In general, when given the x coordinate of a point of a graph, we move vertically (up or down) until we meet the graph and can then approximate the y coordinate. Also, if we know the y coordinate, we can move horizontally (right or left) until we meet the curve. The corresponding x coordinate can then be approximated.

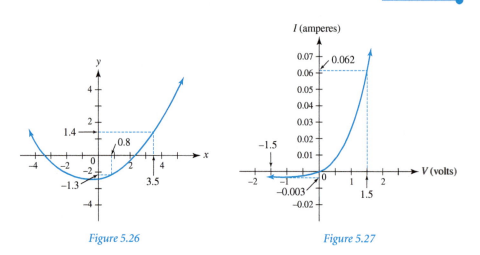

Figure 5.26   Figure 5.27

Example 3

The current as a function of the voltage for a typical type of transistor is shown in Figure 5.27. We can see that for a voltage of -1.5 V (voltage and current can be considered as having direction) the current is -0.003 A. For a voltage of 1.5 V the current is about 0.062 A.

Example 4

A chemical explosion propels an object upward and its distance (in feet) above ground level is related to time t (in seconds) after the explosion by the function $d = -16t^2 + 128t$. From the graph of this function determine how long it takes for the projectile to return to ground level. Also find the maximum height achieved by this projectile.

Because *negative values of t are meaningless* in this situation, our table of values begins with $t = 0$ as follows:

t	0	1	2	3	4	5	6	7	8	9
d	0	112	192	240	256	240	192	112	0	−144

We also ignore the last pair of values (9 and -144) because d cannot be negative. The resulting graph is shown in Figure 5.28. From the graph we see that the projectile returns to ground level ($d = 0$) after 8 s. Also, the graph shows that the maximum value of d is approximately 256 ft.

Noting that the function is quadratic, we can determine the t coordinate of the vertex at $-b/2a$. This gives us $t = -128/(-32) = 4$ s, which confirms our conclusion that $d = 256$ ft is the maximum height. (When $t = 4$ s, $d = 256$ ft.)

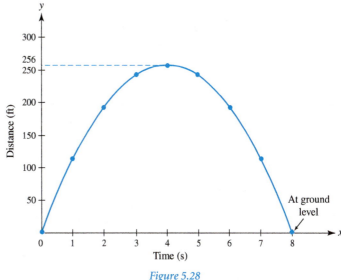

Figure 5.28

We know that the point where the graph touches the *x* axis is called the *x* intercept. This point is also called a **zero of a function.**

Example 5

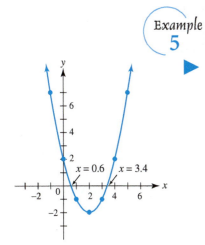

Figure 5.29

Solve the equation $x^2 - 4x + 2 = 0$ graphically.

In solving this equation, we wish to ***find those values of x that make the left side zero.*** By setting $y = x^2 - 4x + 2$ and finding those values of *x* for which *y* is zero, we have found the solutions to the equation. Therefore, we graph

$$y = x^2 - 4x + 2$$

for which the table is

x	−1	0	1	2	3	4	5
y	7	2	−1	−2	−1	2	7

From Figure 5.29 we see that the graph crosses the *x* axis at approximately $x = 0.6$ and $x = 3.4$.

Example 6

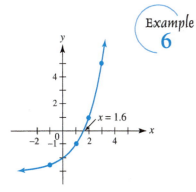

Figure 5.30

Solve the equation $2^x - 3 = 0$ graphically.

First we set $y = 2^x - 3$ and graph this function by constructing the following table:

x	−1	0	1	2	3
y	−2.5	−2	−1	1	5

From the graph in Figure 5.30, we see that the required solution is approximately $x = 1.6$, which is the value for which $y = 0$, or where the curve crosses the *x* axis.

We can use the **TRACE feature** of the graphing calculator to determine coordinates on the graph of a function (as was done in Example 2).

We begin by entering a function into the graphing calculator.

- Press Y = and enter the equation $Y_1 = x^2 + 3x - 5$.
- Then press ZOOM #6 to create a graph in the standard viewing window.
- Now press TRACE. Notice that a blinking cursor appears on the graph. Because we are in the standard window, the cursor initially appears at the point where $x = 0$.
- Notice that the coordinates for this point are displayed on the bottom of the screen and that the equation for this graph is displayed at the top of the screen (see Figure 5.31).

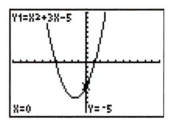

Figure 5.31

- You can move the blinking cursor along the curve by using the left and right arrow keys. Notice that the coordinates at the bottom of the screen change as the cursor moves. These values represent ordered pairs that satisfy the equation. You may want to round these values off.

In Example 5, we found the solution to the given equation. Because this equation was set equal to zero, the solution to the equation is its x-intercepts. These points are also commonly referred to as the **zeros of a function.**

We begin by entering the function used in Example 2 into the graphing calculator.

- Press Y = and enter the equation $Y_1 = x^2 - 4x + 2$.
- Press WINDOW and graph this function so that the Xmin $= -5$, Xmax $= 6$, Ymin $= -5$, and Ymax $= 10$.
- Press GRAPH.
- While in the graph screen, press 2nd TRACE to access the CALC menu. Then press #2 to access the "zero" feature (see Figure 5.32).
- Notice that a blinking cursor appears on the graph and that "Left Bound?" is displayed on the bottom of the graph (see Figure 5.33). You need to move

Figure 5.32

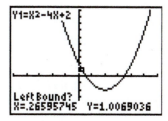

Figure 5.33

the blinking cursor so that it is on the graph *to the left of the x-intercept.* Then press ENTER.

- You are now asked for the "Right Bound?"(see Figure 5.34). Move your cursor so that it is on the graph *to the right of the same x-intercept.* Press ENTER.

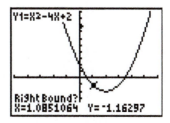

Figure 5.34

- At this point you will see "Guess?" on the bottom of the screen. Just press ENTER once again.

- You should now see the screen shown in Figure 5.35. Notice that your cursor is on the *x* axis and the values for that zero are displayed at the bottom of the screen.

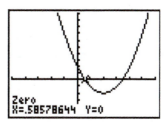

Figure 5.35

- You need to repeat this process to find the value of the second solution to this equation.

- Compare this value to the values found in Example 2.

5.6 Exercises

In Exercises 1 through 4, find the approximate values for y for the indicated values of x from the given figures.

1. Find y for $x = -1$ and $x = 2$ from Figure 5.36(a).

2. Find y for $x = 1.5$ and $x = 4.7$ from Figure 5.36(b).

3. Find y for $x = -2.3$ and $x = 1.8$ from Figure 5.36(c).

4. Find y for $x = -15$ and $x = 37$ from Figure 5.36(d).

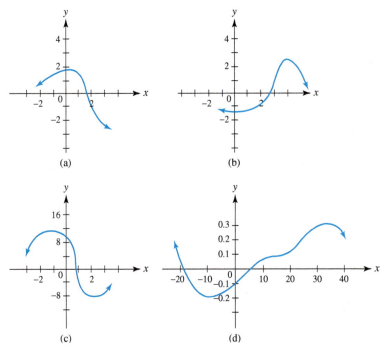

(a) (b) (c) (d)

Figure 5.36

In Exercises 5 through 12, first graph the given functions and then use the graph to determine the approximate values of y for the indicated values of x.

5. $y = 3x + 2;$ $x = 1.5, x = 3.2$

6. $y = -x + 3;$ $x = -1.5, x = 1.2$

7. $y = 8 - 3x;$ $x = -0.4, x = 2.1$

8. $y = 2 - x^2;$ $x = -1.8, x = 1.8$

9. $y = 2x^2 - 5x + 1;$ $x = -1.1, x = 2.7$

10. $y = 6x - x^3;$ $x = -0.7, x = 2.3$

11. $y = \dfrac{3}{x - 3};$ $x = 1.6, x = 3.6$

12. $y = \sqrt{2x + 4};$ $x = 0.8, x = 3.1$

In Exercises 13 through 24, solve the given equations graphically. Use your graphing calculator to verify your results.

13. $7x - 5 = 0$

14. $3x + 13 = 0$

15. $2x^2 - x - 4 = 0$

16. $x^2 + 3x - 5 = 0$

17. $x^3 - 5 = 0$

18. $2x^3 - 5x + 4 = 0$

19. $\sqrt{2x + 9} - x = 0$

20. $2x - \sqrt{x + 6} = 0$

21. $\sqrt[3]{2x + 1} - 2 = 0$ **22.** $2\sqrt{x} + x - 2 = 0$

23. $3^x - 5 = 0$ **24.** $3x - 2^x = 0$

In Exercises 25 through 32, determine the required values from the appropriate graph.

25. Figure 5.37 shows the graph of the charge on a capacitor and the time. Determine the charge on the capacitor at $t = 0.005$ s and $t = 0.050$ s.

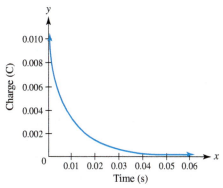

Figure 5.37

26. Figure 5.38 shows the graph of the displacement of a valve and the time. Determine the valve displacement at $t = 0.25$ s and $t = 0.60$ s.

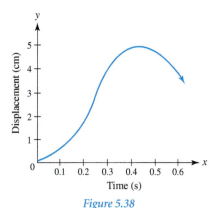

Figure 5.38

27. Figure 5.39 shows the graph of the number of grams of a certain compound that will dissolve in 100 g of water and the temperature of the water. Determine the number of grams that dissolve at 33°C and 58°C.

28. Figure 5.40 shows the graph of milligrams of new mass on a certain plant and the time in days. How much new mass is on the plant after 12 days? After 22 days?

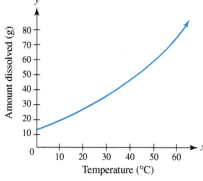

Figure 5.39

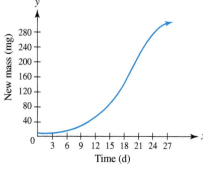

Figure 5.40

In Exercises 29 through 32, solve the given equations graphically.

29. Under certain conditions the force F (in newtons) on an object is found from the equation

$$0.8F - 22 = 0$$

Solve for F.

30. The distance d (in feet) that an object is above the surface of the earth as a function of time is given by

$$d = 85 + 60t - 16t^2$$

When will it hit the ground? [*Hint:* What does d equal when the object is on the ground?]

31. To find the radius (in inches) of a 1-qt container that requires the least amount of material to make, we must solve the equation

$$2\pi r - \frac{57.8}{r^2} = 0$$

Solve for r and thereby determine the required radius.

32. The current i (in amperes) in a particular circuit as a function of time t (in seconds) is given by

$$i = 0.002(1 - e^{-80000t})$$

For what value of t is $i = 0.190$ mA? (Use values of t in microseconds.)

5.7   Graphing Inequalities

Section Objectives
- Successfully graph an inequality
- Identify solutions to inequalities from a graph

In this section we will consider the graphs of inequalities. We introduce the fundamental approach in Example 1.

Example
1

Graph the inequality $x > 3$ on the rectangular coordinate system.

   We begin by graphing $x = 3$ which is a vertical line through $x = 3$ on the x axis. Every point along that line satisfies the equation $x = 3$, regardless of the value of the y coordinate. All points to the right of that line have an x coordinate greater than 3 and therefore satisfy the inequality $x > 3$. In Figure 5.41 we depict our graph by shading the region containing those points. It is customary to **represent the boundary with dashes whenever the boundary itself is not included as part of the solution.** (Solid boundaries are included as part of the solution.) The graph of $x > 3$ is therefore represented by the shaded region of Figure 5.41.

Example
2

Graph the inequality $y \geq -2$ on the rectangular coordinate system.

   Because $y \geq -2$ means y is greater than or equal to -2, we **begin by showing the line $y = -2$ as a solid line because the points on the line itself do satisfy the given inequality.** All points on or above this line also satisfy the original inequality. We therefore shade the region shown in Figure 5.42.

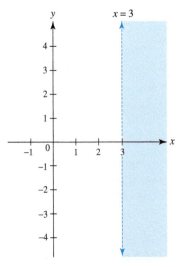

Figure 5.41

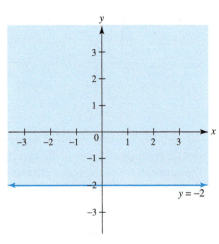

Figure 5.42

Example
3

Graph $y < 3 - x$.

We first draw the line $y = 3 - x$, which has intercepts $(0, 3)$ and $(3, 0)$. Because points on that line will not satisfy the given inequality, the boundary will not be included, so a dashed line is used (see Figure 5.43). We now choose any point on one side of the line as a **test point:** the origin is usually a good choice. Choosing the origin, we substitute zero for x and zero for y in the inequality to get $0 < 3$, *which is true*. Since the coordinates of the origin satisfy the given inequality, *the graph of $y < 3 - x$ is the region on the side of the line that contains the origin*. We therefore get the shaded region shown in Figure 5.43.

If the test point had not satisfied the given inequality, then the region representing the graph of $y < 3 - x$ would have been on the other side of the line.

> For inequalities that contain $<$ or $>$ symbols, we represent the **boundary** of our graph with a **dashed line.**
>
> For inequalities that contain $\leq$ or $\geq$ symbols, we represent the **boundary** of our graph with a **solid line.**

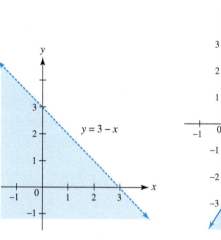

Figure 5.43

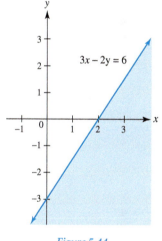

Figure 5.44

Example
4

Graph the inequality $3x - 2y \geq 6$.

We begin by graphing the line $3x - 2y = 6$, which has intercepts $(0, -3)$ and $(2, 0)$, as shown in Figure 5.44. We use a solid line because the points on the line itself do satisfy the given inequality. Using the origin $(0, 0)$ as a test point, we substitute zero for x and zero for y in the given inequality to get $0 \geq 6$, *which is false*. We therefore *shade the side of the line that does not contain the origin* and get the graph shown in Figure 5.44.

Example
5

Graph the inequality $y \geq x^2$.

We first graph the parabola $y = x^2$ by using the following table of values:

x	-3	-2	-1	0	1	2	3
y	9	4	1	0	1	4	9

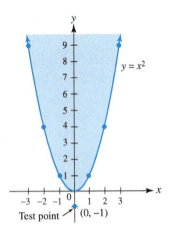

Figure 5.45

We *draw the parabola using a solid curve* (instead of dashes) because the points on the parabola itself do satisfy the given inequality. In this case, we cannot use the origin as a test point because it lies on the parabola; we must *choose a point not on the boundary* itself. Choosing $(0, -1)$, we substitute zero for x and -1 for y in the given inequality to get $-1 \geq 0$, *which is false*. We therefore shade the region not containing the test point of $(0, -1)$ and get the graph shown in Figure 5.45.

Procedure for Graphing Inequalities

1. Graph the inequality by replacing the inequality sign with an equal sign. That is, graph the inequality as if it were an equation.
2. Represent the graph with a dashed line if the inequality is $<$ or $>$. Represent the graph with a solid line if the inequality is $\leq$ or $\geq$.
3. Select a test point not on the graph and substitute the ordered pair into the original inequality.
4. If the inequality holds true for the test point, shade the area that contains the test point; otherwise, shade the area that does not contain the test point.

Example 6

In any given day, a refinery can produce x gallons of gasoline and y gallons of diesel fuel, in any combination. However, the available equipment allows a maximum total output of 1000 gal. Make a graph of the different possible production combinations.

Because the maximum total output is 1000 gal, we get $x + y \leq 1000$. We graph the line $x + y = 1000$ and use the origin as a test point (see Figure 5.46). Substituting zero for both x and y in the inequality, we get $0 \leq 1000$, which is

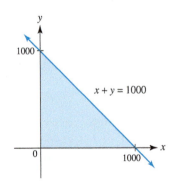

Figure 5.46

true. We therefore shade the region below the line. However, we include only the first quadrant points since $x \geq 0$ and $y \geq 0$. (Negative values for x or y would be meaningless in the context of this problem.) The shaded region therefore represents the graph of all the different possible production combinations.

Example 6 illustrates an application of graphing inequalities. By graphing additional constraints, we can proceed to determine how to best distribute resources. This procedure is used in a branch of applied mathematics called *linear programming*.

 Using Technology

Suppose we want to graph the inequality $3x - 2y \geq 6$ on the graphing calculator.

- Begin by rewriting the inequality for y in terms of x. This becomes $y \leq 1.5x - 3$.

- Press Y = and enter the right-hand side of the above inequality for Y_1.

- Next, choose a shading option for the equation. Use the left arrow key to move the cursor to the left of Y_1 and press ENTER until you see the symbol ⌐, which represents shading below the line. We choose this symbol because y is less than $1.5x - 3$.

- Press ZOOM 6 to draw the graph in the standard viewing window. You will see the graph shown in Figure 5.47.

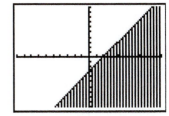

Figure 5.47

Compare your result with Figure 5.47.

5.7 *Exercises*

In Exercises 1 through 28, graph the given inequalities on the rectangular coordinate system. Use your graphing calculator to check your graphs when possible.

1. $x > 0$

2. $x \leq 2$

3. $y \geq 3$

4. $y \leq 0$

5. $x \leq -3$

6. $y \leq -4$

7. $y > -5$

8. $x > -8$

9. $y > 1 - x$

10. $y < x + 1$

11. $y < 2x - 1$

12. $y > 3x + 4$

13. $y \leq 2x + 3$

14. $y \geq 3x - 5$

15. $\frac{1}{2}y < 4$

16. $\frac{1}{3}y < 6$

17. $3y < 8 - 6x$

18. $2y \geq 3 + 2x$

19. $y - 3 > x$

20. $y - 1 \leq 2x$

21. $3x + 4y < 3$

22. $4x + 3y \geq 5$

23. $2x - 5y \geq 10$

24. $6x - y < 6$

25. $y < x^2$

26. $y > x^2 + 1$

27. $y < 3 - x^2$

28. $y - x^3$

In Exercises 29 through 32, use the rectangular coordinate system to graph the inequality suggested by the stated conditions.

29. A company can produce x resistors and y capacitors in 1 h. The number of resistors must be less than or equal to 50. With no constraints on the number of capacitors, graph the possible values of x and y. Exclude from the graph any negative values of x or y.

30. A supplier must provide x gallons of hydrochloric acid and y gallons of sulfuric acid each week. The total amount of acid must be at least 40 gal. Given only the preceding constraints and the fact that neither x nor y can be negative, graph the possible values of x and y.

31. An investor plans to put an amount of money into stock purchases and a savings account. The amount x invested in stocks must be less than or equal to the amount y allocated to the savings account. Given only this constraint and the fact that neither x nor y can be negative, graph the possible values of x and y.

32. A wood products company processes lumber and plywood. Market demand requires that the number of lumber units x must be at least twice the number of plywood units y. Given only this constraint and the fact that neither x nor y can be negative, graph the possible values of x and y.

Chapter Summary

Key Terms

function	x axis	straight lines
domain	y axis	parabolas
range	ordered pair	slope
independent variable	abscissa	x-intercept
dependent variable	ordinate	y-intercept
rectangular coordinate system	coordinates	vertex
quadrants	graph	zeros of a function
function notation relation	origin	test point

Key Concepts

• A function is a relationship between two variables in which any input value is associated with exactly one output value. The domain of a function is the set of all possible input values, and the range is the set of all possible output values.

• The function notation $f(x)$ is read "f of x." The f and x are not being multiplied together when using function notation. To evaluate a function at $x = a$, we write $f(a)$ and substitute a for every value of x in our equation. We then simplify our results.

• Graphs of functions give us good visual representations of the relationship between the two function variables. We use a two-dimensional system called the rectangular coordinate system. Each point in the coordinate plane is uniquely identified by an ordered pair written in the format (x, y).

• To create a graph we create a table of ordered pairs by letting x take on several values and calculating corresponding values for y. Then we plot each ordered pair and join the points from left to right using a smooth curve.

• A quadratic function has the equation $f(x) = ax^2 + bx + c$ where $a \neq 0$. To find the x coordinate of the vertex, or turning point, of the graph of a quadratic function we use the formula $x = \dfrac{-b}{2a}$. The y coordinate is found by substituting the x coordinate into the original function.

• Slope is used to describe the steepness of a line and is sometimes referred to as $\dfrac{rise}{run}$. We can find the slope of the line joining the points (x_1, y_1) and (x_2, y_2) using the formula $m = \dfrac{y_2 - y_1}{x_2 - x_1}$. Lines that rise from left to right have a positive slope, whereas those that drop from left to right have a negative slope. Horizontal lines have a 0 slope. Vertical lines have an undefined slope.

• The x-intercept is the point where the graph touches the x axis and is written $(a, 0)$. The y-intercept is the point where the graph touches the y axis and is written $(0, b)$.

• The general form for a linear function is $y = mx + b$ where m represents the slope of the line and b represents the y-intercept. The equation of a vertical line is $x = a$. The equation of a horizontal line is $y = b$.

• To solve an inequality, graph the equation as if it contained an equal sign. Substitute in a test point to find out which side of the graph to shade. Use a solid graph line for inequalities containing $\leq$ or $\geq$. Use a dashed line for inequalities containing $<$ or $>$.

• The graphing calculator was used regularly throughout this section. It is a quick and easy way to produce a graph. However, it is only as good as the person inputting the information. In order to become proficient with the graphing calculator it is important to use it regularly.

Review Exercises

In Exercises 1 through 12, find the indicated values of the given functions.

1. $f(x) = x + 3;$ $f(0), f(-1)$

2. $f(s) = 2s - 1;$ $f(-1), f(2)$

3. $f(x) = 2 - x;$ $f(-2), f(\frac{1}{3})$

4. $f(y) = 4y - 2;$ $f(-5), f(-2)$

5. $f(x) = 2x^2 - 1;$ $f(\sqrt{2}), f(-\frac{1}{2})$

6. $f(t) = -t^2;$ $f(3), f(-4)$

7. $f(z) = z^2 - 2z - 3;$ $f(-3), f(0.2)$

8. $f(x) = 7 - x - 4x^2;$ $f(-1), f(-v)$

9. $f(r) = -r^3;$ $f(0), f(-2)$

10. $f(n) = 12 - n^3;$ $f(-2), f(3)$

11. $f(x) = \sqrt{4x + 1};$ $f(0), f(6)$

12. $f(s) = \dfrac{2}{5 - \sqrt{s}};$ $f(4), f(16)$

In Exercises 13 through 20, use the grid in Figure 5.48 to graph the given ordered pairs and indicate the quadrant in which the point is located.

13. $(4, 2)$

14. $(-2, 3)$

15. $(3, -2)$

16. $(0, 5)$

17. $(2, 0)$

18. $(-5, -2)$

19. $(0, -6)$

20. $(-3, 0)$

In Exercises 21 through 32, graph the given functions. Use your graphing calculator to verify your graph.

21. $y = 5x - 1$

22. $y = 3 - 4x$

23. $y = 3 - \dfrac{1}{2}x$

24. $y = 4x - \dfrac{2}{3}$

25. $y = 3x^2 - 4$

26. $y = 1 - 3x - x^2$

27. $y = 2x^2 - x^3$

28. $y = 2^{x-1}$

29. $y = 4 - \dfrac{1}{x}$

30. $y = \dfrac{1}{x - 1}$

31. $y = 2\sqrt{x} - 3$

32. $y = \sqrt{x^2 + 9}$

In Exercises 33 through 38, find the slope of the line passing through the given pairs of points.

33. $(2, 6)$ and $(-5, 7)$

34. $(4, 10)$ and $(-3, -6)$

35. $(-2, 5)$ and $(5, -6)$

36. $(3, -3)$ and $(6, -3)$

37. $(5, -13)$ and $(0, 12)$

38. $(2, 8)$ and $(2, -8)$

In Exercises 39 through 43, find the x- and y-intercept of the line with the given equation. Sketch the graph using the x- and y-intercepts. Use your graphing calculator to check your graph.

39. $2x + y = 8$

40. $y = -6x + 9$

41. $8x - 6y = 36$

42. $-5y = 5 - x$

43. $y = 0.5x + 6.5$

In Exercises 44 through 47, find the slope and y-intercept of the line with the given equation. Sketch the graph using the

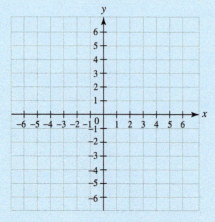

Figure 5.48

slope and y-intercept. Use your graphing calculator to check your graph.

44. $y = -x + 2$

45. $y = -2x$

46. $y = 2x - 14$

47. $10x - 10y = -40$

In Exercises 48 through 57, graph the given inequalities. Use your graphing calculator to check your graph.

48. $x \geq 5$

49. $y \leq -1$

50. $y > 2 + 3x$

51. $y < 2x - 4$

52. $8y \leq 4 - x$

53. $4y \geq -2x + 1$

54. $y \geq -x^2 + 2$

55. $y \leq 4 + 3x^2$

56. $y < 2x^2 - 3$

57. $y > x^3 - 1$

In Exercises 58 through 62, solve the given equations graphically. Use your graphing calculator to check your graph.

58. $7x - 9 = 0$

59. $6x + 11 = 0$

60. $3x^2 - x - 2 = 0$

61. $5 - 7x - x^2 = 0$

62. $x^3 - x^2 + 2x - 1 = 0$

In Exercises 63 through 66, set up the required functions.

63. The sales tax in a certain state is 6%. Express the tax T on a certain item as a function of the cost C of the item.

64. A salesperson earns $200 plus 3% commission on monthly sales. Express the salesperson's monthly income I as a function of the monthly dollar sales S.

65. The rate H at which heat is developed in a filament of an electric light bulb is proportional to the square of the electric current I. A current of 0.5 A produces heat at the rate of 60 W. Express H as a function of I.

66. A kitchen exhaust fan should remove each minute a volume of air that is proportional to the floor area. A properly operating fan removes 20 m³/min for a kitchen of 10 m² of floor area. Express the volume of air that is removed per minute for a properly operating fan as a function of the floor area A.

In Exercises 67 through 70, find the required values graphically.

67. We can determine the cost of using a computer owned by a certain company by the graph in Figure 5.49. How much does it cost to use the computer for 5 h in a given month? For 16 h? For 28 h?

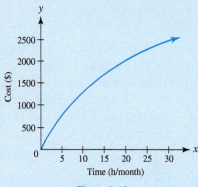

Figure 5.49

68. The velocity of a satellite is a function of its distance from the surface of the earth. The graph of the velocity and distance for a particular satellite is shown in Figure 5.50. The distance from the earth to the satellite varies from 200 to 1600 mi. What is the velocity of the satellite when it is 400 mi from the earth? When it is 750 mi? When it is 1400 mi?

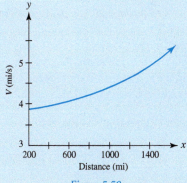

Figure 5.50

69. The length of a cable hanging between two equal supports 100 ft apart is given by

$$L = 100(1 + 0.0003y^2)$$

where y is the sag (vertical distance from top of support to bottom of cable) in the cable. Determine the length of a cable for which the sag is 5 ft; 10 ft; 20 ft.

70. Under certain conditions, the distance from a source of light where the illumination is least is found by solving the equation

$$x^3 + 8(x - 100)^3 = 0$$

Find the required distance (in meters).

In Exercises 71 and 72, find the required functions and graphs.

71. A temperature in degrees Fahrenheit (°F) equals 32 more than $\frac{9}{5}$ the number of degrees Celsius (°C). Express F as a function of C and plot the graph. From the graph determine the temperature at which $F = C$.

72. The path of a projectile is approximately parabolic. Given that y is the vertical distance from the ground to the projectile and x is the horizontal distance traveled, the equation

$$y = x - 0.0004x^2$$

is the equation of the path of a given projectile, where distances are measured in feet. Plot the graph, using units of 50 ft for x, and determine how far the projectile travels. (Assume level ground.)

Chapter Test

1. Determine the value of each of the following functions at the point indicated:
   a. $f(x) = 3x - 7$ find $f(0)$, $f(3)$, and $f(-1)$
   b. $f(x) = 2x^2 - 7x + 11$ find $f(-2)$ and $f(5)$
   c. $f(x) = 7\sqrt{x + 1} - 3$ find $f(8)$

2. Graph the following functions:
   a. $y = -4x - 6$
   b. $f(x) = x^2 - 20x + 10$

3. Find the slope of the line joining the points $(3, -4)$ and $(-7, 1)$.

4. Find the x- and y-intercept for the equation $y = 2x - 4$.

5. Find the slope and y-intercept of $4x - 3y = -6$.

6. Graph the following inequalities:
   a. $y < 3x + 9$
   b. $4y \geq -8x + 2$

7. Solve the following equations graphically:
   a. $y = 8x + 12$
   b. $f(x) = x^2 + 4x - 6$

8. The height H (in feet) above the ground of a submarine-launched missile is given by the function $H(x) = -16x^2 + 96x - 80$.
   a. Graph this function using an appropriate domain.
   b. Determine the times when the missile leaves and returns to the water.
   c. What is the greatest height reached by the missile?

9. For a small manufacturing company the profit P is a function of the number of items produced, x. The function that describes this relationship is $P(x) = -2x^2 + 210x - 1000$.
   a. Determine the profit when $x = 10$, $x = 45$, and $x = 80$.
   b. How many items must be produced to reach a maximum profit?

6

Introduction to Geometry

The word geometry comes from the Greek words *geo*, meaning *earth*, and *metro* meaning *measure*. Geometry is that branch of mathematics that deals with spatial relationships. It is the mathematics of points, lines, curves, and surfaces. We see evidence of basic geometric shapes in nature, and in the man-made structures around us. All architecture begins with a good understanding of geometry because this is crucial for understanding structural concepts and calculations.

The photo shows a sketch of the Pantheon that was originally built as a Roman temple and later served as a Catholic Church. The original design of the United States Capital in Washington, D.C. was based on the Roman Pantheon, although the original plans did go through several transformations before the final building was constructed.

As you progress through this chapter return to the Pantheon and try to identify the various geometric shapes found in the architecture of this building.

6.1   Basic Geometric Figures

Section Objectives
- Introduce basic concepts of points, lines, and angles
- Introduce the basic geometric shapes (triangles, quadrilaterals, circles)

Geometry deals with the properties and measurement of angles, lines, **surfaces,** and volumes and with the basic figures that are formed. In establishing the properties of the basic figures it is not possible to define every word **and prove** every statement. Certain words and concepts must be used without **definition.** In general, *the words **point, line,** and **plane** are accepted in geometry without being defined.* This gives us a starting point for defining other terms.

In Figure 6.1a we show a line passing through points *A* and *B*. The line extends indefinitely far in both directions. *A **line segment** is a portion of a line and is bounded by two fixed points.* In Figure 6.1a, the line segment bounded by the points A and B is denoted by *AB*. Although no line has a fixed length, **every line** segment does have a length that can be measured. *A **ray** consists of a fixed boundary point and the portion of a line to one of its sides* as in Figure 6.1b.

*The amount of rotation of a ray about its endpoint is called an **angle.*** (See Figure 6.1c.) Notation and terminology associated with angles are illustrated in **Example 1.**

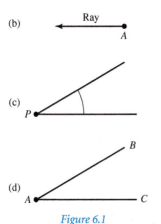

Figure 6.1

Example 1

In Figure 6.1d, the angle formed by *AB* and *AC* is denoted by ∠*BAC*. The symbol ∠ means angle, so ∠*BAC* means angle BAC. The **vertex** of the **angle is the** point *A,* and the **sides** of the angle are *AB* and *AC*. An angle is often **named by its** vertices, so ∠*BAC* can also be called ∠*A*.

One complete rotation (see Figure 6.2a) *of a ray about a point is defined to be an angle of* 360 **degrees,** *written as* 360°. A **straight angle** *contains* 180° (see

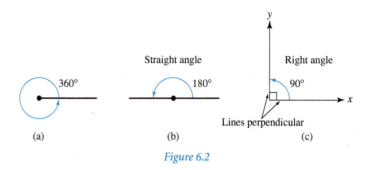

Figure 6.2

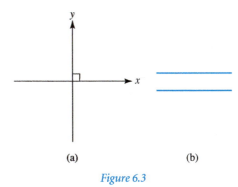

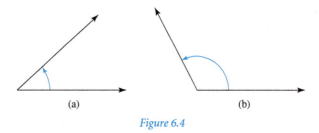

Figure 6.2b), *and a* **right angle** (*denoted, as in* Figure 6.2c, *by* ⌐) *contains* 90°. *If two lines meet so that the angle between them is a right angle, they are said to be* **perpendicular** (see Figure 6.3a). Lines in the same plane that do not intersect are said to be **parallel.** (see Figure 6.3b.)

A common method of stating the magnitude of an angle is to give the **measure** of the angle. That is, the measure of a straight angle is 180°, and the measure of a right angle is 90°. However, in our discussions we shall refer to the angle rather than the measure of the angle. Also, we shall use the same symbol for the angle and the measure of the angle.

An **acute angle** measures between 0° and 90° (see Figure 6.4a), and an **obtuse angle** measures between 90° and 180° (see Figure 6.4b).

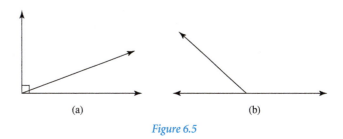

(a)   (b)

Figure 6.4

If the sum of the measure of two angles is 90°, the angles are called **complementary angles** (see Figure 6.5a). If the sum of the measure of two angles is 180°, the angles are called **supplementary angles** (see Figure 6.5b).

(a)   (b)

Figure 6.5

A device that can be used to measure angles approximately is a **protractor.** Figure 6.6 shows the protractor positioned to measure an angle of 50°.

When it is necessary to measure angles to an accuracy of less than 1°, **decimal parts of a degree** can be used. The use of decimal parts of a degree has become common with the extensive use of calculators. Previously, the common way of expressing angles to such accuracy was to divide each degree into 60 equal parts, called **minutes,** and to divide each minute into 60 equal

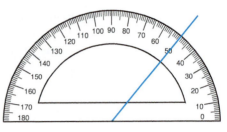

Figure 6.6

parts called **seconds.** Surveyors and astronomers frequently measure angles to the nearest second.

The notation for minutes is ′ and for seconds is ″. Thus an angle of 32 degrees, 15 minutes, and 38 seconds is denoted as $32°15'38''$. Because $1° = 60'$, we find a decimal part of a degree in minutes by multiplying the decimal by $60'$. Also, because $1' = \left(\frac{1}{60}\right)°$, we express minutes in terms of a decimal part of a degree by dividing by 60.

Example 2

1. $0.2° = (0.2)(60') = 12'$. Thus $17.2° = 17°12'$.

2. $36' = \left(\frac{36}{60}\right)° = 0.6°$. Thus $58°36' = 58.6°$.

3. $6.35° = 6°21'$, since $(0.35)(60') = 21'$.

Example 3

For a person standing on the surface of the earth and looking at the moon, the angle made by the "top edge" of the moon, the viewer, and the center of the moon is about $\left(\frac{1}{4}\right)°$. Express this angle in minutes.

Since $\left(\frac{1}{4}\right)° = 0.25°$, we get $0.25° = (0.25)(60') = 15'$.

Example 4

A surveyor measures the angle between two land boundaries and gets $72°11'$. Express this angle in degrees and decimal parts of a degree.

Since $11' = \left(\frac{11}{60}\right)° = 0.18°$ (to the nearest hundredth), we get $72°11' = 72.18°$.

A **triangle** *is a closed plane figure that has three sides,* and these three sides form three interior angles. *In an* **equilateral triangle,** *the three sides are equal in length and each angle is a 60° angle* (see Figure 6.7a). *In an* **isosceles triangle,** *two*

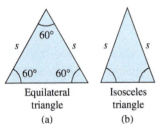

Equilateral triangle
(a)

Isosceles triangle
(b)

Scalene triangle
(c)

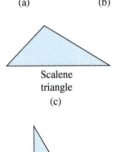

Leg   Hypotenuse

Leg

Right triangle
(d)

Figure 6.7

of the sides are equal in length and the **base angles** (*the angles opposite the equal sides*) *are also equal* (see Figure 6.7b). *In a* **scalene triangle,** *no two sides are equal in length and no two angles are equal* (see Figure 6.7c).

One of the most important triangles in scientific and technical applications is the **right triangle.** *In a right triangle, one of the angles is a right angle. The side opposite the right angle is called the* **hypotenuse,** *and the other two sides are called the* **legs** (see Figure 6.7d).

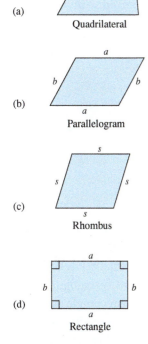

(a) Quadrilateral

(b) Parallelogram

(c) Rhombus

(d) Rectangle

(e) Square

(f) Trapezoid

Figure 6.9

Example 5

1. The triangle in Figure 6.8a, denoted as $\triangle ABC$, is equilateral. Since $AC = 5$, we also know that $AB = 5$ and $BC = 5$. All of the angles are 60° angles, and we note this by writing $\angle A = 60°$, $\angle B = 60°$, and $\angle C = 60°$.

2. The triangle in Figure 6.8b, $\triangle DEF$, is isosceles since $DF = 7$ and $EF = 7$. The base angles are $\angle D$ and $\angle E$. Therefore, since $\angle D = 70°$, we know that $\angle E = 70°$.

3. The triangle in Figure 6.8c, $\triangle PQR$, is scalene because none of the sides is equal in length to another side.

4. The triangle in Figure 6.8d, $\triangle LMN$, is a right triangle because $\angle M = 90°$. The hypotenuse is the side LN, and the legs are MN and LM.

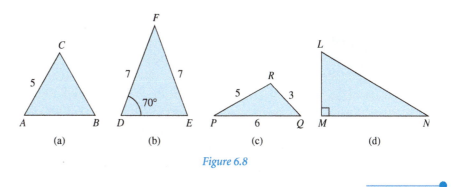

(a) (b) (c) (d)

Figure 6.8

A **quadrilateral** *is a closed plane figure that has four sides,* and these four sides form four interior angles. Figure 6.9a is a general quadrilateral.

A **parallelogram** *is a quadrilateral in which both pairs of opposite sides are parallel. This means that extensions of these sides will not intersect.* In a parallelogram, opposite sides are equal and opposite angles are equal (see Figure 6.9b). *A* **rhombus** *is a parallelogram with four equal sides* (see Figure 6.9c).

A **rectangle** *is a parallelogram in which intersecting sides are perpendicular, which means that all four interior angles are right angles.* Also, opposite sides of a rectangle are equal in length and parallel (see Figure 6.9d). In a rectangle the longer side is usually called the **length** and the shorter side the **width.** *A* **square** *is a rectangle with four equal sides* (see Figure 6.9e).

A **trapezoid** *is a quadrilateral in which exactly two sides are parallel. These parallel sides are called the* **bases** *of the trapezoid (see Figure 6.9f).*

Example 6

1. In parallelogram *ABCD*, shown in Figure 6.10a, opposite sides are parallel. This may be denoted as *AB*‖*DC* and *AD*‖*BC*, where ‖ means "parallel to." Also, opposite sides are equal, and opposite angles are equal. We write these as *AB* = *DC*, and *AD* = *BC*, and ∠*A* = ∠*C* and ∠*B* = ∠*D*.

2. In rectangle *EFGH*, shown in Figure 6.10b, the opposite sides are equal so that *EF* = *HG* and *EH* = *FG*. All interior angles are right angles. This means that ∠*E* = ∠*F* = ∠*G* = ∠*H* = 90°.

3. In quadrilateral *PQRS*, shown in Figure 6.10c, side *PQ* is parallel to side *SR*, which may be stated as *PQ*‖*SR*. It is a trapezoid with bases *PQ* and *SR*. The base angles of *PQ* are ∠*P* and ∠*Q*, and the base angles of *SR* are ∠*S* and ∠*R*.

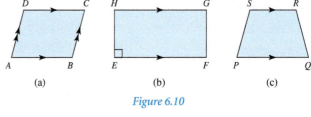

(a) (b) (c)

Figure 6.10

When we are identifying quadrilaterals, more than one designation may be technically correct. For example, a square can also be considered a rectangle (four right angles) or a rhombus (four equal sides). Only a square has both properties, however, and the word "square" should be used to identify the figure. It must also be kept in mind that a square has all the properties of a rectangle and a rhombus.

The last basic geometric figure we shall discuss in this section is the **circle.** *All the points on a circle are the same distance from a fixed point in the plane (see Figure 6.11a). This point O is the* **center** *of the circle. The distance ON (or OM) from the center to a point on the circle is the* **radius** *of the circle. The distance MN between two points on the circle and on a line passing through the center of the circle is the* **diameter** *of the circle.* Thus the diameter is twice the radius, which we may write as *d* = 2*r*.

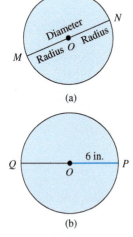

(a)

(b)

Figure 6.11

Example 7

For the circle shown in Figure 6.11b, the radius *OP* is 6 in. long. This means that the diameter *QP* is 12 in. long. We may also show these as *r* = 6 in. and *d* = 12 in.

6.1 Exercises

In Exercises 1 through 4, first estimate the measure of the angles given in Figure 6.12, and then use a protractor to measure them to the nearest degree. (It may be necessary to extend the lines.)

1.

2.

3.

4.

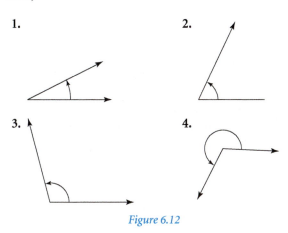

Figure 6.12

In Exercises 5 through 8, express each angle in degrees and minutes.

5. 56.4°

6. 18.9°

7. 136.45°

8. 79.05°

In Exercises 9 through 12, express each angle in degrees and decimal parts of a degree.

9. 156°15′

10. 33°48′

11. 67°6′

12. 16°57′

In Exercises 13 through 19, use Figure 6.13 to identify the indicated angles. Note that EB is drawn perpendicular to AC.

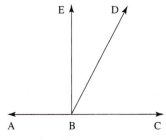

Figure 6.13

13. Two acute angles.

14. The straight angle.

15. An obtuse angle.

16. The acute angle adjacent to ∠*DBC*.

17. Two right angles.

18. If ∠*CBD* = 65°, find its complement.

19. If ∠*CBD* = 65°, find its supplement.

In Exercises 20 through 33, answer the given questions about the figures in Figure 6.14.

20. In Figure 6.14a, identify the isosceles triangle.

21. In Figure 6.14a, identify the scalene triangle.

22. In Figure 6.14a, identify the right triangle.

23. In Figure 6.14a, if ∠*DBC* = 54° (the angle with sides *DB* and *BC*, with vertex at *B*), how many degrees are there in ∠*C*?

24. In Figure 6.14b, identify and determine the lengths of the legs of the right triangle.

25. In Figure 6.14b, identify and determine the length of the hypotenuse.

26. In the parallelogram in Figure 6.14c, identify the pairs of parallel sides.

27. In the parallelogram in Figure 6.14c, identify the pairs of opposite angles.

28. In the parallelogram in Figure 6.14c, if ∠*ABC* = 135°, what is ∠*CDA*?

29. In the parallelogram in Figure 6.14c, what is the length of side *DC*?

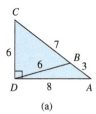
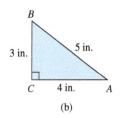

(a)   (b)

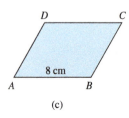
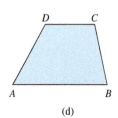

(c)   (d)

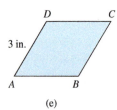
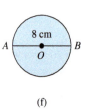

(e)   (f)

Figure 6.14

30. What are the bases of the trapezoid in Figure 6.14d?

31. What is the length of the side *DC* of the rhombus in Figure 6.14e?

32. For the circle in Figure 6.14f, identify a radius.

33. What is the radius of the circle in Figure 6.14f?

In Exercises 34 through 41, use a protractor and ruler to construct the figures required.

34. An equilateral triangle with sides of 2 in.

35. An isosceles triangle with equal sides of 3 cm and base angles of 75°.

36. A right triangle with a hypotenuse of 4 cm and one of the other sides of 2 cm.

37. A parallelogram with sides of 4 in. and 2 in. and one interior angle of 60°.

38. A rectangle with sides of 4 in. and 2 in.

39. A trapezoid with bases of 4 cm and 3 cm.

40. A rhombus with sides of 3 cm and an interior angle of 30°.

41. A quadrilateral with interior angles of 35°, 122°, and 95°.

In Exercises 42 through 52, answer the questions.

42. a. Is a square always a parallelogram?
  b. Is a parallelogram always a square?

43. a. Is an equilateral triangle always an isosceles triangle?
  b. Is an isosceles triangle always an equilateral triangle?

44. Suppose that the two opposite vertices of a rhombus are joined by a straight line. What are the figures into which it is divided?

45. Suppose that the opposite vertices of a figure are joined by a straight line and the figure is divided into two equilateral triangles. What is the figure?

46. If one leg of an isosceles right triangle is 10 cm in length, what is the length of the other leg?

47. The angle between a ray of light and a line perpendicular to a mirror is 36°. What is the angle between the ray of light and the plane of the mirror?

48. An airplane and a helicopter take off from the same spot. The path of the airplane makes an angle of 40° with the ground, whereas the helicopter goes straight up. What is the angle between their flight paths?

49. If a crater on Mercury is circular with a radius of 47 km, what is its diameter?

50. A circular access road is to be constructed around an office complex. If the distance between the center of the office complex and the road is 400 ft, what is the diameter of the circle formed by the road?

51. If a pilot is flying in a northerly direction and makes a 90° turn to the left, what is her new direction?

52. If a submarine captain is headed in an easterly direction and makes a 90° turn to the right, what is his new direction?

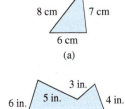

6.2 Perimeter

Section Objectives
- Understand the concept of perimeter
- Find the perimeter of any triangle or quadrilateral
- Find the circumference of a circle

Two of the most basic measures of plane geometric figures are perimeter and area. In this section we consider perimeter. Area will be presented in the following section.

In Chapter 1 we noted that *the* **perimeter** *of a plane figure is the distance around it*. In Example 1, perimeters are found directly from this definition.

Example 1

1. To find the perimeter of the triangular cooling fin shown in Figure 6.15a, we add the lengths of the sides. We get the perimeter p as follows.

$$p = 6 \text{ cm} + 7 \text{ cm} + 8 \text{ cm} = 21 \text{ cm}$$

2. To find the perimeter of the figure shown in Figure 6.15b, we simply add the lengths of the individual sides, even though the figure may appear to be somewhat complicated. In this case we find the perimeter to be

$$p = 10 \text{ in.} + 6 \text{ in.} + 5 \text{ in.} + 3 \text{ in.} + 4 \text{ in.} + 2 \text{ in.} = 30 \text{ in.}$$

Figure 6.15

We can use the definition of perimeter to derive formulas for many specific plane figures. However, it isn't necessary to memorize most of these formulas because we can easily apply the basic and general definition of perimeter. Consider the formulas in the following example.

For the figures shown in Figure 6.16, we have the following **perimeter formulas:**

Figure	*Perimeter*
(a) Triangle with sides a, b, c	$p = a + b + c$
(b) Equilateral triangle with side s	$p = 3s$
(c) Quadrilateral with sides a, b, c, d	$p = a + b + c + d$
(d) Rectangle of length l and width w	$p = 2l + 2w$
(e) Parallelogram with sides a and b	$p = 2a + 2b$
(f) Square of side s	$p = 4s$

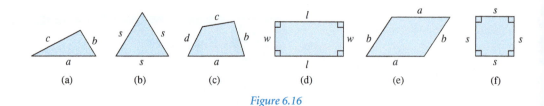

Figure 6.16

Example 2

1. The perimeter of an equilateral triangle with a side of 18 ft is

   $$p = 3(18 \text{ ft}) = 54 \text{ ft}$$

2. The perimeter of a parallelogram with sides of 21 cm and 15 cm is

   $$p = 2(21 \text{ cm}) + 2(15 \text{ cm})$$
   $$= 42 \text{ cm} + 30 \text{ cm} = 72 \text{ cm}$$

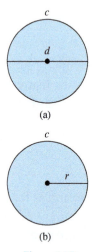

Figure 6.17

The perimeter of a circle is called the **circumference.** It cannot be found directly from the definition of perimeter. However, we shall use a basic geometric fact, without attempting to develop it, in order to arrive at the necessary formulas: *The ratio of the circumference to the diameter is the same for all circles.* This ratio is represented by π, a number equal to approximately 3.14. To eight significant digits, $\pi = 3.1415927$. We shall use the decimal form rather than the mixed-number form $3\frac{1}{7}$ (which is also only approximate) for π, because the decimal form is generally more convenient for calculations. (Your graphing calculator has a key whose secondary purpose is to provide the value of π expressed with as many significant digits as the calculator allows.)

Because the ratio of the circumference c to the diameter d equals π, we have $c/d = \pi$. This leads to the formula

$$\boxed{c = \pi d} \qquad \text{circumference of circle}$$

for the circumference. See Figure 6.17a. Also, because the diameter equals twice the radius, we have

$$\boxed{c = 2\pi r} \qquad \text{Note that } c = \pi d = \pi(2r) = 2\pi r$$

as a formula for the circumference in terms of the radius. See Figure 6.17b.

Although it isn't necessary to memorize most of the formulas for the perimeter, it is necessary to memorize at least one of the above two formulas because they don't follow directly from the general perimeter definition.

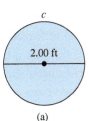

(a)

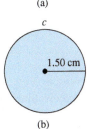

(b)

Figure 6.18

Example 3

1. The circumference of a wheel with a diameter of 2.00 ft is

$$c = \pi(2.00 \text{ ft}) = 3.14(2.00 \text{ ft})$$
$$= 6.28 \text{ ft}$$

See Figure 6.18a.

2. The circumference of a washer with a radius of 1.50 cm is

$$c = 2\pi(1.50 \text{ cm}) = 2(3.14)(1.50 \text{ cm})$$
$$= 9.42 \text{ cm}$$

See Figure 6.18b.

Example 4

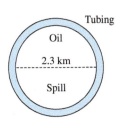

Figure 6.19

A circular oil spill has a diameter of 2.3 km. If this oil spill is to be enclosed within a length of specially designed flexible tubing, how long must this tubing be?

The length of the tubing must be as long as the circumference of the circle. Knowing the diameter, we can find the circumference of the circle as follows:

$$c = \pi(2.3 \text{ km}) = 3.14(2.3 \text{ km})$$
$$= 7.2 \text{ km}$$

The length of the tubing required is therefore 7.2 km. See Figure 6.19.

By using the definition of perimeter, we can find the perimeters of geometric figures that are combinations of basic figures. Consider the following examples.

Example 5

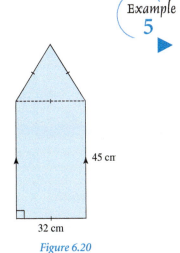

Figure 6.20

Figure 6.20 is a combination of a rectangle and an equilateral triangle. (The dashed line shows where the figures are joined, but ***it should not be counted as part of the perimeter because it is not on the outside of the figure.***) Because the lower part of the figure is a rectangle, we see that the dashed line is 32 cm long. This tells us that each side of the triangle is 32 cm long, because it is equilateral. The perimeter can now be found by adding the 32 cm along the bottom to the two 32-cm lengths at the top to the two 45-cm lengths at the sides. We get

$$p = \overset{\text{bottom}}{32 \text{ cm}} + \overset{\text{top}}{2(32 \text{ cm})} + \overset{\text{sides}}{2(45 \text{ cm})}$$
$$= 32 \text{ cm} + 64 \text{ cm} + 90 \text{ cm}$$
$$= 186 \text{ cm}$$

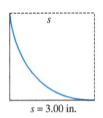

$s = 3.00$ in.

Figure 6.21

Example
6

A certain machine part is a square with a quarter circle removed (see Figure 6.21). The side of the square is 3.00 in. The perimeter of the part is to be coated with a special metal costing 25¢ per inch. What is the cost of coating this part?

Setting up a formula for the perimeter, we add the two sides of length s to **one-fourth of the circumference of a circle with radius s** to get

bottom   circular
and left   section

$$p = 2s + \frac{2\pi s}{4} = 2s + \frac{\pi s}{2}$$

where s is the side of the square as well as the radius of the circular part. Letting $s = 3.00$ in., we have

$$p = 2(3.00 \text{ in.}) + \frac{(3.14)(3.00 \text{ in.})}{2}$$

$$= 6.00 \text{ in.} + 4.71 \text{ in.} = 10.71 \text{ in.}$$

Because the coating costs 25¢ per inch, the cost is

$$c = 25p = (25¢/\text{in.})(10.71 \text{ in.}) = 268¢ = \$2.68$$

Example
7

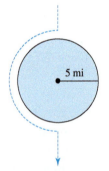

5 mi

Figure 6.22

A pilot must fly around a circular restricted airport zone that has a radius of 5 mi. (The 5 is exact.) See Figure 6.22. How much longer is this route than the route that passes directly through the restricted zone?

If the pilot could fly directly through the restricted zone, the distance traveled would be 10 mi (the diameter of the circle). Because the pilot must fly *around* the restricted zone, the distance traveled is one-half the circumference. Since

$$c = 2\pi r = 2\pi(5 \text{ mi})$$

$$= 10\pi \text{ mi} = 31.4 \text{ mi}$$

we see that the pilot must travel $\frac{1}{2}(31.4 \text{ mi}) = 15.7$ mi. This is 5.7 mi longer than the 10-mi route.

6.2   Exercises

In Exercises 1 through 24, find the perimeters (circumferences for circles) of the indicated figures for the given values.

1. The triangle shown in Figure 6.23a.

2. The triangle shown in Figure 6.23b.

3. The quadrilateral shown in Figure 6.23c.

4. The quadrilateral shown in Figure 6.23d.

5. Triangle: $a = 320$ m, $b = 278$ m, $c = 298$ m

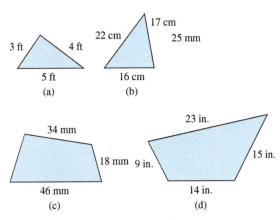

Figure 6.23

6. Triangle: $a = 52$ yd, $b = 49$ yd, $c = 64$ yd

7. Quadrilateral: $a = 16.5$ in., $b = 17.3$ in., $c = 21.8$ in., $d = 29.2$ in.

8. Quadrilateral: $a = 6.92$ cm, $b = 8.26$ cm, $c = 9.93$ cm, $d = 8.07$ cm

9. Equilateral triangle: $s = 64.6$ mm

10. Equilateral triangle: $s = 128$ ft

11. Isosceles triangle: equal sides of 15.3 in., third side of 26.5 in.

12. Isosceles triangle: equal sides of 36.2 cm, third side of 12.5 cm

13. Square: $s = 0.65$ m

14. Square: $s = 2.36$ yd

15. Rhombus: side of 0.15 mi

16. Rhombus: side of 178 mm

17. Parallelogram: $a = 47.2$ cm, $b = 36.8$ cm

18. Parallelogram: $a = 1.69$ in. $b = 1.46$ in.

19. Rectangle: $l = 68.7$ ft, $w = 46.6$ ft

20. Rectangle: $l = 4.57$ m, $w = 0.97$ m

21. Circle: $r = 15.1$ cm

22. Circle: $r = 10.6$ ft

23. Circle: $d = 6.74$ in.

24. Circle: $d = 42.0$ mm

In Exercises 25 through 32, find the perimeters of the indicated geometric figures shown in Figure 6.24. Dashed lines are used to identify figures but are not part of the figure where the perimeter is to be found.

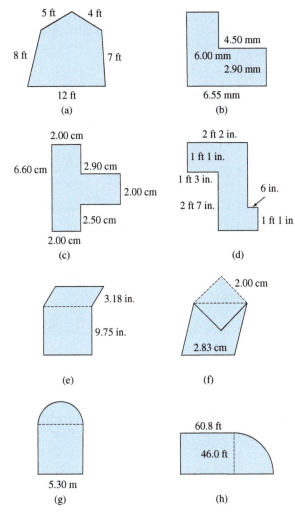

Figure 6.24

25. The figure in Figure 6.24a.

26. The figure in Figure 6.24b. All angles are right angles.

27. The figure in Figure 6.24c. All angles are right angles.

28. The figure in Figure 6.24d. All angles are right angles.

29. The figure in Figure 6.24e. A square is surmounted by a parallelogram.

30. The figure in Figure 6.24f. Half a square has been removed from a rhombus.

31. The figure in Figure 6.24g. A square is surmounted by a semicircle.

32. The figure in Figure 6.24h. A quarter circle is attached to a rectangle.

In Exercises 33 through 40, find a formula for the perimeter of the given figures.

33. An isosceles triangle with equal sides *s* and a third side *a*.

34. A rhombus with side *s*.

35. The semicircular figure shown in Figure 6.25a.

36. The quarter-circular figure shown in Figure 6.25b.

37. The figure shown in Figure 6.25c. A square is surmounted by an equilateral triangle.

38. The figure shown in Figure 6.25d. A quarter circle is attached to a triangle.

39. The isosceles trapezoid shown in Figure 6.25e.

40. The figure shown in Figure 6.25f. A semicircle and a quarter circle are attached to a rectangle.

In Exercises 41 through 58, solve the given problem.

41. Carpet binding is an edging tape that is normally sewn around area rugs in order to keep the edges of the rug from fraying. The price for such a binding often depends on who is doing the binding. By going directly to a binding service you are quoted a price of $1.65 per foot of binding. How much will it cost to bind a rectangular area run that is 9 ft wide by 12 ft long?

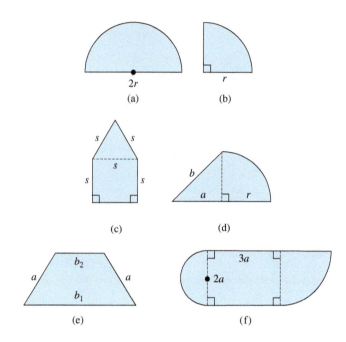

Figure 6.25

42. A rectangular bedroom floor measures 12 ft by 15 ft. The room has a large closet with a sliding door that is 6 ft wide. It also has a door leading from the bedroom that is 3 ft wide. What is the total length of floor molding needed for this room?

43. The radius of the earth's equator is 3960 mi. What is the circumference of the earth at the equator?

44. A CD-RW has a diameter of 5 in. The CD is kept in a plastic case. What is the perimeter of the smallest case that can be used to hold the CD?

45. A park ranger measures the distance around a circular lake to be 2800 ft. Find the distance across the center of the lake.

46. What is the total cost (at $12 per meter) of fencing needed for a plot of land that makes a right triangle? The lengths of the legs of the triangular plot are each 48 m and the hypotenuse is 68 m.

47. A flower bed is made up of a square plot with semicircles attached to each side of the square (see Figure 6.26). If one side of the square measures 4 ft, find the perimeter of the flower bed.

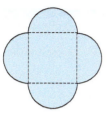

Figure 6.26

48. State law requires that all pools be enclosed by a fence in order to prevent accidents. The Monroe family pool has a radius of 25 ft and is surrounded by a deck that is 4 ft wide. The deck is to be surrounded by the fence as well. How many feet of fencing are needed?

49. Find the length of a belt needed to go around two pulleys, each with a radius of 4 in. whose centers are 15 in. apart.

50. As a ball bearing rolls along a straight track it makes 11 revolutions while traveling a distance of 109 mm. Find the radius of the ball bearing.

51. To prevent a basement from being flooded, drainage pipe is to be put around the outside of a building whose outline is shown in Figure 6.27a, consisting of three rectangles and a semicircle. How much drainage pipe is required to go around the building? (Ignore the fact that the pipe is slightly away from the walls.)

52. The area within a racetrack is a rectangle with semicircles at each end (see Figure 6.27b.) The radius of the circular parts is 30 yd, and the perimeter of the area within the track is 400 yd. How long is each straight section of the track?

53. As a wheel rolls along level ground, it makes one complete revolution as it travels a distance of 81.68 in. Find the diameter of the wheel.

54. If a car were to travel exactly one mile, one of its tires would make 830 revolutions. Find the radius of the tire in inches.

55. A point on a rotating pulley travels 20,109 cm while the pulley makes 360 revolutions. Find the distance between that point and the center of the pulley.

56. The cross-sectional drawing of a machine part consists of a semicircle surmounted on an equilateral triangle with sides of 5.28 cm. See Figure 6.28. Find the perimeter.

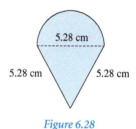

Figure 6.28

57. Find the perimeter of the figure eight that is formed when a circle with a diameter of 0.558 cm is surmounted with a circle of diameter 0.902 cm.

58. The centers of two pulleys are 88 cm apart and they both have diameters of 22 cm. How long is the belt that goes around the two pulleys? See Figure 6.29.

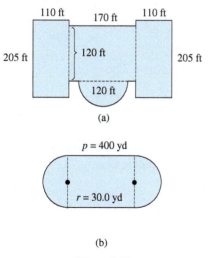

(a)

p = 400 yd

r = 30.0 yd

(b)

Figure 6.27

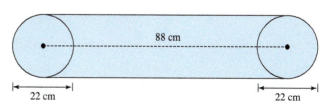

Figure 6.29

6.3 Area

Section Objectives
- Understand the concept of area
- Find the area of a triangle
- Find the area of select quadrilaterals
- Find the area of a circle

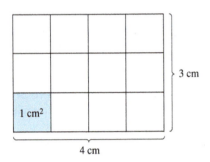

3 cm

1 cm²

4 cm

Figure 6.30

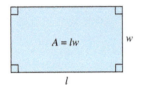

$A = lw$ w

l

Figure 6.31

$A = s^2$ s

Figure 6.32

In addition to perimeter, another basic measure associated with a plane geometric figure is its **area. Area** *gives us a measure of the surface of a geometric figure, just as perimeter gives us the measure of the distance around it.* There are many important real applications of area.

In finding the area of a geometric figure, we are finding the number of squares, one unit on a side, required to cover the surface of the figure. In the rectangle in Figure 6.30, we see that 12 squares, each 1 cm on a side, are needed to cover the surface. We therefore say that the area of the rectangle is 12 square centimeters, which we write as 12 cm².

If we note the rectangle in Figure 6.30, we see that its area can be determined by multiplying its length by its width. Because the area of any rectangle can be found in this way, we define the area of a rectangle to be

$$A = lw \quad \text{area of rectangle}$$

where l *and* **w** *must be measured in the same unit of length and* **A** *is in square units of length* (see Figure 6.31).

Because a square is a rectangle with all sides equal in length, its area is

$$A = s^2 \quad \text{area of square}$$

where **s** *is the length of one side* (see Figure 6.32).

Example 1

1. A given rectangle has a length of 9.0 ft and a width of 5.0 ft. With $l = 9.0$ ft and $w = 5.0$ ft the area of the rectangle is

   $A = (9.0 \text{ ft})(5.0 \text{ ft}) = 45 \text{ ft}^2$

2. A certain square has a side of 5.0 cm. With $s = 5.0$ cm the area of the square is

   $A = (5.0 \text{ cm})^2 = 25 \text{ cm}^2$

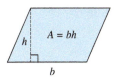

Figure 6.33

We can use the definition of the area of a rectangle to determine the area of a parallelogram. Examine the parallelogram in Figure 6.33 and note that a triangular area is formed to the left of the dashed line labeled h. This dashed line is perpendicular to the **base** b of the parallelogram. If we move the triangular area to the right of the parallelogram, a rectangle of length b and width h is formed. Because the area of the rectangle would be bh, the area of the parallelogram is

$$\boxed{A = bh} \quad \text{area of parallelogram}$$

Here **h** is called the **height,** or **altitude,** of the parallelogram. Remember: It is the perpendicular distance from one base **b** to the other. **The value of h is not the length of the side of the parallelogram** (unless it is also a rectangle).

Example 2

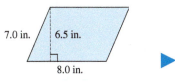

Figure 6.34

1. The area of a parallelogram for which $h = 9.0$ m and $b = 11$ m is

$$A = (11 \text{ m})(9.0 \text{ m}) = 99 \text{ m}^2$$

2. To find the area of the parallelogram shown in Figure 6.34, we need only the height (6.5 in., and the base (8.0 in.). **The length of the other side is not used** (although it would be used in finding the perimeter). Thus

$$A = (8.0 \text{ in.})(6.5 \text{ in.}) = 52 \text{ in.}^2$$

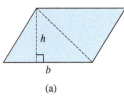

(a)

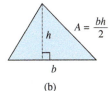

$$A = \frac{bh}{2}$$

(b)

Figure 6.35

If we draw a **diagonal** (*line joining opposite vertices*) in a parallelogram, we divide it into two triangles, with equal areas as shown in Figure 6.35a. Because the area of the triangle below the diagonal is one-half the area of the parallelogram, we have

$$\boxed{A = \frac{1}{2}bh} \quad \text{area of triangle}$$

as the area of the triangle (see Figure 6.35b). Here **h** is the **altitude,** or **height,** of the triangle. It is the length of the perpendicular line from a vertex to the opposite base of the triangle.

Example 3

1. The area of a triangle for which $b = 16$ in. and $h = 12$ in. is

$$A = \frac{1}{2}(16 \text{ in.})(12 \text{ in.}) = 96 \text{ in.}^2$$

It might be noted here that $\frac{1}{2} bh$ and $\frac{bh}{2}$ are the same.

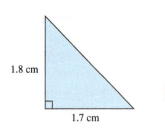

1.8 cm

1.7 cm

Figure 6.36

2. The area of the right triangle shown in Figure 6.36 is

$$A = \frac{1}{2}(1.7 \text{ cm})(1.8 \text{ cm}) = 1.5 \text{ cm}^2$$

▶ Because the legs of a right triangle are perpendicular, ***either leg can be considered as the base and the other as the altitude.***

Another useful formula for finding areas of triangles is Hero's formula. This formula is especially useful when we have *a triangle with three known sides and no right angle.*

Figure 6.37

$$A = \sqrt{s(s - a)(s - b)(s - c)}$$

$$\text{where } s = \frac{1}{2}(a + b + c)$$

See Figure 6.37.

 Example 4

A triangular fin has sides $a = 12$ cm, $b = 15$ cm, and $c = 23$ cm. Its area is found by first determining the value of s. See Figure 6.38.

$$s = \frac{1}{2}(12 \text{ cm} + 15 \text{ cm} + 23 \text{ cm}) = \frac{1}{2}(50 \text{ cm}) = 25 \text{ cm}$$

12 cm   15 cm

23 cm

Figure 6.38

We can now proceed as follows:

$$A = \sqrt{(25 \text{ cm})(25 \text{ cm} - 12 \text{ cm})(25 \text{ cm} - 15 \text{ cm})(25 \text{ cm} - 23 \text{ cm})}$$

$$= \sqrt{(25 \text{ cm})(13 \text{ cm})(10 \text{ cm})(2 \text{ cm})} = \sqrt{6500 \text{ cm}^4}$$

$$= 81 \text{ cm}^2$$

Note that in evaluating the square root, $\sqrt{\text{cm}^4} = \text{cm}^2$.

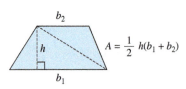

b_2

h $A = \frac{1}{2} h(b_1 + b_2)$

b_1

Figure 6.39

When we join opposite vertices of a trapezoid with a diagonal, two triangles are formed, as in Figure 6.39. The area of the lower triangle is $\frac{1}{2}b_1h$ and the area of the upper triangle is $\frac{1}{2}b_2h$. The sum of the areas is the area of the **trapezoid**, and it may be expressed as

$$A = \frac{1}{2}h(b_1 + b_2)$$ area of trapezoid

Example
5

The floor area of a trapezoidal room for which $h = 6.0$ yd, $b_1 = 4.0$ yd, and $b_2 = 3.0$ yd is

$$A = \frac{1}{2}(6.0 \text{ yd})(4.0 \text{ yd} + 3.0 \text{ yd})$$

$$= 21 \text{ yd}^2$$

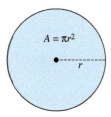

$A = \pi r^2$

r

Figure 6.40

The area of a circle cannot be found from the area of any quadrilateral. As in the case of the circumference, the area of a circle is expressed in terms of π. The area of a circle is given by

$$\boxed{A = \pi r^2} \qquad \text{area of circle}$$

See Figure 6.40.

Example
6

1. The area of a circle of radius 2.73 cm is

$$A = \pi(2.73 \text{ cm})^2 = (3.14)(7.4529 \text{ cm}^2)$$

$$= 23.4 \text{ cm}^2$$

2. To find the area of a circle of a given diameter, we *first divide the length of the diameter by 2 in order to obtain the length of the radius.* Thus if $d = 48.2$ in., then $r = 24.1$ in., and we get

$$A = \pi(24.1 \text{ in.})^2 = 3.14(581 \text{ in.}^2)$$

$$= 1820 \text{ in.}^2$$

So far we have included the units of measurement in the equations and algebraic work. Technically this should always be done, but it can be cumbersome (see Example 4). Recognizing that all lengths are in a specific type of unit, we know that the result is in terms of that unit. Therefore, from this point on, when the units of measurement of the given parts and the result are specifically known, we shall not include them in the equation or algebraic work.

The determination of area is important in many applications involving geometric figures. Consider the following example.

Example
7

A painter charges $1.25 per square foot for painting house exteriors. One side of a house is a rectangle surmounted by a triangle. The base of the rectangle is 30.0 ft, the height of the rectangle is 9.00 ft, and the height of the triangle is 7.00 ft. What

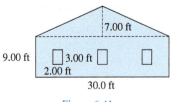

Figure 6.41

would his charge for this part of the job be? There are three rectangular windows, each 2.00 ft by 3.00 ft, in the side of the house (see Figure 6.41).

The area of the side of the house to be painted is the area of the rectangle plus that of the triangle minus the area of the windows. Thus

rectangle   triangle   windows

$$A = (30.0)(9.00) + \frac{1}{2}(30.0)(7.00) - 3(2.00)(3.00)$$

$$= 270 + 105 - 18.0 = 357 \text{ ft}^2$$

The charge is

$$c = \$1.25(357) = \$446.25$$

6.3   Exercises

In Exercises 1 through 40, find the areas of the indicated geometric figures.

1. Rectangle: $l = 60.0$ cm, $w = 45.0$ cm

2. Rectangle: $l = 152$ ft, $w = 85.0$ ft

3. A rectangle with length 24 in. and width 7.0 in.

4. A rectangle with length and width of 2.34 m and 5.68 m, respectively

5. Square: $s = 7.60$ in.

6. Square: $s = 0.160$ km

7. A square with a side of 2.54 in.

8. A square with perimeter 8.0 cm

9. Parallelogram: $b = 72.0$ mm, $h = 34.0$ mm

10. Parallelogram: $b = 1.50$ yd, $h = 1.20$ yd

11. A parallelogram with a base of 5.30 cm and an altitude of 4.60 cm

12. A parallelogram with a base of and altitude of 12.4 ft and 20.3 ft, respectively

13. Rhombus: side $= 16.5$ in., altitude $= 6.40$ in.

14. Rhombus: side $= 240$ cm, altitude $= 150$ cm

15. A rhombus with a side of 8 ft and an altitude of 5 ft

16. A rhombus with a side and altitude of 12.4 m and 6.02 m, respectively

17. Triangle: $b = 0.750$ m, $h = 0.640$ m

18. Triangle: $b = 64.0$ in., $h = 14.5$ in.

19. A triangle with a base of 42 cm and an altitude of 16 cm

20. A triangle with a base of 14.0 ft and an altitude of 8.25 ft

21. Right triangle: legs $= 16.5$ ft and 28.8 ft

22. Right triangle: legs $= 396$ mm and 250 mm

23. A right triangle with legs 153 cm and 205 cm

24. A right triangle with legs 80.1 in. and 20.5 in.

25. Triangle: $a = 21.2$ cm, $b = 12.3$ cm, $c = 25.4$ cm

26. Triangle: $a = 2.45$ m, $b = 3.62$ m, $c = 3.97$ m

27. A triangle with sides 5.2 ft, 6.4 ft, and 4.7 ft

28. A triangle with sides 10.2 in., 12.8 in., and 14.9 in.

29. Trapezoid: $h = 0.0120$ km, $b_1 = 0.0250$ km, $b_2 = 0.0180$ km

30. Trapezoid: $h = 1.23$ yd, $b_1 = 3.74$ yd, $b_2 = 2.36$ yd

31. A trapezoid with an altitude of 6.33 m and bases of 4.55 m and 5.25 m

32. A trapezoid with an altitude of 1.05 ft and bases of 2.37 ft and 3.85 ft

33. Circle: $r = 0.478$ ft   **34.** Circle: $d = 5.38$ cm

35. Circle: $r = 23.21$ ft   **36.** Circle: $d = 93.7$ m

37. A circle with a radius of 2.625 in.

38. A circle with a diameter of 203 mm

39. A circle with a circumference of 4.00 ft

40. A circle with a circumference of 12.28 m

In Exercises 41 through 48, determine (a) the perimeter and (b) the area of the given figures.

41. Figure 6.42a   **42.** Figure 6.42b

43. Figure 6.42c   **44.** Figure 6.42d

45. Figure 6.42e   **46.** Figure 6.42f

47. Figure 6.42g   **48.** Figure 6.42h

In Exercises 49 through 62, solve the given problems.

49. A rectangular swimming pool is 16.0 ft wide and 32.0 ft long. What is the cost of a solar cover that is sold for 35¢ per square foot?

50. A panel of sheetrock is rectangular with a length of 4.0 ft and a width of 8.0 ft. Two 1.0-ft diameter holes are cut out for heating ducts. What is the area of the remaining piece?

51. A gallon of paint will cover 300 ft^2. How much paint is required to paint the walls and ceiling of a room 12 ft by 16 ft by 8.0 ft, given that the room has three windows 2.0 ft by 3.0 ft and two doors 3.0 ft by 6.5 ft?

52. A metal plate is in the shape of a right triangle with sides of 3.56 cm, 4.08 cm, and 5.41 cm. Determine the area of the plate.

53. Find the area of the triangular tract of land shown in Figure 6.43.

54. A fence is to be made of trapezoidal slats as shown in Figure 6.44. What is the area of the fence if it is placed along a distance of 13.34 m?

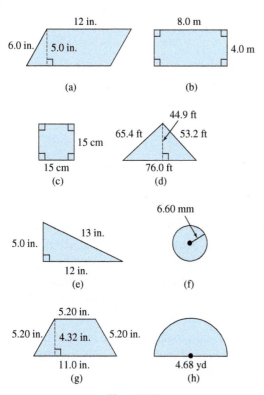

Figure 6.42

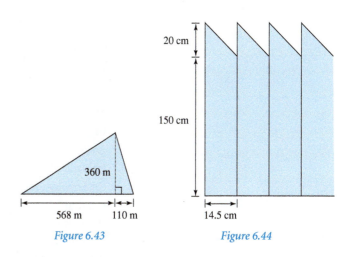

Figure 6.43   Figure 6.44

55. A flat steel ring 38.0 cm in diameter has a hole 8.90 cm in diameter in the center. What is the area of one face of the ring?

56. What is the area of the largest circle that can be cut from a rectangular plate 21.2 cm long and 15.8 cm wide?

57. An oil spill forms a circle with a diameter of 2.34 km. If cleanup costs $3000/km^2, find the total cost of cleaning up this oil spill.

58. If the lengths of the sides of a rectangle are doubled, how does the area change?

59. If you triple the lengths of the sides of a rectangle, how does the area change?

60. A triangular plot of land has the dimensions 95 ft, 109 ft and 107 ft.
a. How much fencing is needed to enclose this plot of land?
b. What is the area of this plot of land?

61. Suppose you want to lay linoleum flooring on your kitchen floor. The floor is divided into two parts, as shown in Figure 6.45. The rectangular section is 12 ft by

14 ft and the second, smaller section is a breakfast area that is triangular in shape. This section extends 5 ft beyond from the 12-ft side. The cost of linoleum is $21 per square yard.
a. How much linoleum do you need?
b. What is the cost for new flooring?

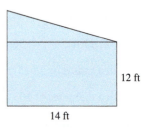

12 ft

14 ft

Figure 6.45

62. The Pentagon is one of the world's largest office buildings. It is a regular pentagon (five equal sides and all internal angles are equal in measure). Each side measures 921 ft with a diagonal of length 1490 ft. Find the area of the Pentagon.

6.4   Volume

Section Objectives
- Understand the concept of volume
- Find the volume of a rectangular solid
- Find the volume of a cylinder
- Find the volume of a pyramid
- Find the volume of a cone
- Find the volume of a sphere

So far we have examined perimeter and area as measures of plane geometric figures. An important measure associated with a solid figure is its **Volume.** *Volume gives us a measure of the amount of space occupied by a solid figure.* In finding the volume of a solid geometric figure we are finding the number of cubes, one unit on an edge, required to fill the figure. In the **rectangular solid** in Figure 6.46, we see that 24 cubes, each 1 cm on an edge, are required to fill the solid. We say that the volume is 24 cubic centimeters, or 24 cm^3.

If we note the rectangular solid in Figure 6.46 we see that the volume can be determined by finding the product of its length, width, and height. In general, the volume of a rectangular solid is

$$V = lwh$$ volume of rectangular solid

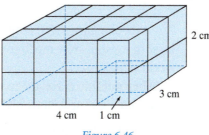

Figure 6.46   Figure 6.47

where **l, w,** *and* **h** *are in the same unit of length and* **V** *is in cubic units of length (see Figure 6.47).*

Example 1

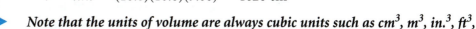

A computer peripheral device is in the shape of a rectangular solid 18.0 cm long, 10.0 cm wide, and 9.00 cm high. Find its volume.

The volume is computed as

$$V = lwh = (18.0)(10.0)(9.00) = 1620 \text{ cm}^3$$

▶ *Note that the units of volume are always cubic units such as* ***cm³, m³, in.³, ft³,*** *and so on.*

The **cube,** which we have already used in naming units of volume, *is a rectangular solid in which all edges are equal in length.* Because the length, width, and height equal the length of an edge *e*, the volume of a cube is

$$\boxed{V = e^3}$$ volume of cube

Figure 6.48

See Figure 6.48.

Example 2

A farmer stores grain in a bin that is in the shape of a cube. If each edge of this cube is 8.0 ft, find the volume of the grain that can be stored.

With *e* = 8.0 ft, we find the volume as follows:

$$V = e^3 = 8.0^3 = 510 \text{ ft}^3$$

Example 3

How many cubic feet are in 1 cubic yard?

A cubic yard is a cube in which each edge has a length of 1 yd. Since 1 yd = 3 ft, that same volume can be represented by a cube in which each edge has a length of 3 ft, and the volume can now be computed as

$$V = e^3 = 3^3 = 27 \text{ ft}^3 \text{ Therefore, 1 yd}^3 = 27 \text{ ft}^3.$$

Example 4

A rectangular lawn 150 ft long and 36 ft wide is to be covered with a 6 in or 0.50 ft layer of topsoil. How much topsoil must be ordered? (Topsoil is usually ordered by the cubic yard.)

When put into place, the topsoil will form a rectangular solid 150 ft long, 36 ft wide, and 6 in. (or 0.50 ft) high. Its volume is found by

$$V = (150)(36)(0.50) = 2700 \text{ ft}^3$$

We should now reduce the result to cubic yards. From Example 3 we see that $1 \text{ yd}^3 = 27 \text{ ft}^3$ so that

$$2700 \text{ ft}^3 = \overset{100}{\cancel{2700 \text{ ft}^3}} \cdot \frac{1 \text{yd}^3}{\cancel{27 \text{ft}^3}} = 100 \text{yd}^3$$

Example 5

A chemical waste holding tank is in the shape of a rectangular solid that is 3.00 ft long, 3.00 ft wide, and 2.00 ft high. How many gallons of liquid waste can be stored in this tank? One cubic foot can contain 7.48 gal.

The volume of the tank is

$$V = (3.00)(3.00)(2.00) = 18.0 \text{ ft}^3$$

Since each cubic foot holds 7.48 gal, we find the total capacity c as follows:

$$c = 18.0 \times 7.48 = 135 \text{ gal}$$

The tank can hold 135 gal of chemical waste.

A **right circular cylinder** is formed by rotating a rectangle on one of its sides. Each base is a circle, and the cylindrical surface is perpendicular to each of the bases (see Figure 6.49). *The volume of a right circular cylinder is the product of the area of the base of the cylinder and the height of the cylinder.* Because the base is a circle, the area of the base is πr^2 and we have the formula

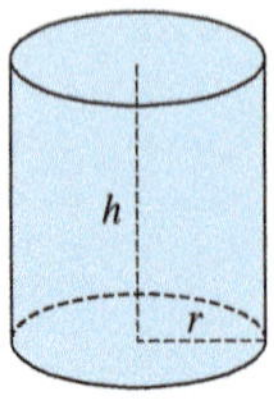

$$\boxed{V = \pi r^2 h} \qquad \text{volume of a cylinder}$$

where r is the radius of each base and h is the height.

Figure 6.49

Example 6

Find the volume of a container in the shape of a right circular cylinder for which $d = 16.5$ in. and $h = 6.75$ in.

In this case we are given the diameter which means the radius, r, is 8.25 inches. The volume is as follows:

$$V = \pi(8.25)^2(6.75)$$
$$V = (3.14)(68.1)(6.75)$$
$$V = 1440 \text{ in.}^3$$

Example 7

A storage tank for a kerosene heater is in the shape of a right circular cylinder with a diameter of 1.50 ft and a height of 4.16 ft. How many gallons can be stored in this tank? ($1.00 \text{ ft}^3 = 7.48$ gal).

Since the diameter is 1.50 ft we know that the radius, r, is 0.75 ft. The volume of the storage tank can be found as follows:

$$V = \pi r^2 h$$
$$V = (3.14)(0.75)^2(4.16)$$
$$V = 7.35 \text{ ft}^3$$

Since each cubic foot can hold 7.48 gal, we get the total storage capacity, C, of the tank by

$$C = 7.48(7.35)$$
$$= 55.0 \text{ gal}$$

A **pyramid** is a three-dimensional polygon whose sides are triangles that meet at a common point called the vertex (see Figure 6.50). The formula for the volume of a pyramid is given by

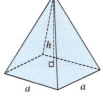

$$\boxed{V = \frac{1}{3}Bh}$$ volume of pyramid

Figure 6.50

where B represents the area of the base and h represents the height.

Example 8

The base of a pyramid is a square 12.0 cm on each side. The height of the pyramid is 18.5 cm.

To find the volume we note that the base area B is a square so

$$V = \frac{1}{3}(12.0)^2(18.5) = 888 \text{ cm}^3$$

A **right circular cone** is generated by rotating a right triangle around one of its legs. Therefore, the base of the cone is a circle, the height of the cone is the length

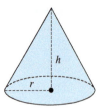

Figure 6.51

of one leg of the right triangle, and the radius of the base is the length of the other leg of the right triangle (see Figure 6.51). The volume of a right circular cone is

$$V = \frac{1}{3}\pi r^2 h$$

volume of a cone

Example 9

Find the volume of a cone with $r = 5.50$ in. and $h = 3.75$ in.

$$V = \frac{1}{3}(3.14)(5.50)^2(3.75) = 119 \text{ in.}^3$$

Example 10

A high-pressure atomizing oil burner propels a spray of oil in the shape of a cone a radius of 2.02 in. and a height of 1.80 in. What is the volume of this cone?

$$V = \frac{1}{3}\pi(2.02)^2(1.80)$$

$$V = \frac{1}{3}(3.14)(4.08)(1.80)$$

$$V = 7.69 \text{ in.}^3$$

Figure 6.52

The **sphere** is the last solid figure we will consider in this chapter, but others will be included in Chapter 15. A sphere can be formed by rotating a circle about its diameter. The **radius** of the sphere is the length of the line segment connecting the center to a point on the surface. Like a circle, the **diameter** of a sphere is twice its radius (see Figure 6.52). The volume of the sphere is computed by using the following equation:

$$V = \frac{4}{3}\pi r^3$$

volume of sphere

Here r is the radius of the sphere.

Example 11

The radius of a tennis ball is 1.25 in. Its volume is

$$V = \frac{4}{3}\pi(1.25)^3 = \frac{4}{3}(3.14)(1.95)$$

$$= 8.16 \text{ in.}^3$$

Example 12

The radius of the earth (which is approximately a sphere) is about 3960 mi. The volume of the earth is

$$V = \frac{4}{3}\pi(3960)^3$$

$$= 260{,}000{,}000{,}000 \text{ mi}^3$$

(In Chapter 10 we will see that such numbers can be concisely expressed in a form called *scientific notation*. This answer could be written as 2.60×10^{11} mi^3.)

Example 13

A grain storage container is in the shape of a cylinder surmounted by a hemisphere (half a sphere) on the top (see Figure 6.53). Given that the radius of the cylinder is 40 ft and the height is 120 ft, find the volume of the container.

The volume of the container is the volume of the cylinder plus the volume of the hemisphere. The volume of the cylinder is $\pi r^2 h$ and the volume of the hemisphere is one-half the volume of a sphere or $\frac{1}{2}\left(\frac{4}{3}\pi r^3\right)$. Therefore,

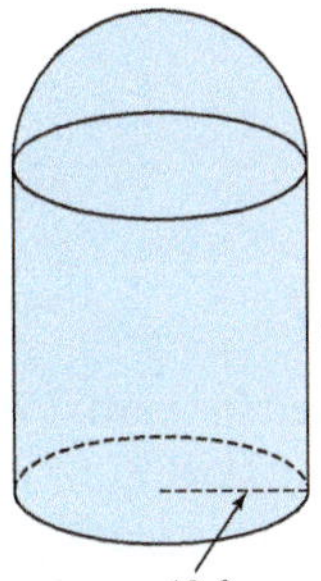

$$V = \pi r^2 h + \frac{1}{2}\left(\frac{4}{3}\pi r^3\right)$$

$$V = (3.14)(40)^2(120) + \frac{1}{2}\left(\frac{4}{3}\right)(3.14)(40)^3$$

$$V = 602{,}880 + 133{,}973$$

$$V = 736{,}853 \text{ ft}^3 \approx 737{,}000 \text{ ft}^3$$

$h = 120$ ft

$r = 40$ ft

Figure 6.53

6.4 Exercises

In Exercises 1 through 8, find the volumes of the rectangular solids with the given dimensions.

1. $l = 73.0$ mm, $w = 17.2$ mm, $h = 16.0$ mm

2. $l = 3.21$ ft, $w = 5.12$ ft, $h = 6.34$ ft

3. $l = 0.87$ cm, $w = 0.61$ cm, $h = 0.15$ cm

4. $l = 1.2$ cm, $w = 3.9$ cm, $h = 9.7$ cm

5. $l = 16$ m, $w = 16$ m, $h = 18$ m

6. $l = 9.0$ in., $w = 12$ in., $h = 15$ in.

7. $l = 120$ cm, $w = 85$ cm, $h = 150$ cm

8. $l = 342$ ft, $w = 20.5$ ft, $h = 1.50$ ft

In Exercises 9 through 16, find the volumes of the cubes with the given edges.

9. $e = 15$ cm

10. $e = 12$ ft

11. $e = 0.800$ in.

12. $e = 9.36$ m

13. $e = 0.22$ m

14. $e = 17.3$ in.

15. $e = 19.27$ cm

16. $e = 17.39$ cm

In Exercises 17 through 22, determine the volumes of the cylinders.

17. $r = 20.0$ cm; $h = 15.0$ cm

18. $r = 7.0$ in; $h = 4.0$ in.

19. $r = 15.0$ ft; $h = 3.60$ ft

20. $r = 1.59$ m; $h = 8.48$ m

21. $d = 366$ mm; $h = 140$ mm

22. $d = 0.634$ yd; $h = 0.156$ yd

In Exercises 23 through 26, determine the volumes of the pyramids.

23. Base is a polygon with $B = 3600$ ft^2; $h = 45.0$ ft.

24. Base is a polygon with $B = 7850$ cm^2; $h = 38.4$ cm.

25. Base is a square of side 25.0 mm; $h = 4.60$ mm.

26. Base is a square of side 8.50 ft; $h = 4.85$ ft.

In Exercises 27 through 30, determine the volumes of the cones.

27. $r = 10.0$ ft; $h = 14.0$ ft

28. $r = 25.0$ cm; $h = 40.0$ cm

29. $d = 62.8$ cm; $h = 26.3$ cm

30. $d = 17.8$ ft; $h = 22.3$ ft

In Exercises 31 through 38, find the volumes of the given spheres.

31. $r = 27.3$ cm

32. $r = 78$ ft

33. $r = 15.2$ m

34. $r = 55.3$ mm

35. $d = 34$ in.

36. $d = 764$ cm

37. $d = 1.12$ mm

38. $d = 8.2$ km

In Exercises 39 through 58, answer the given questions.

39. How many cubic inches are in 1 ft^3?

40. How many cubic inches are in 1 yd^3?

41. How many cubic centimeters are in 1 m^3?

42. How many cubic millimeters are in 1 m^3?

43. In studying the effects of a poisonous gas, a fire science specialist must calculate the volume of a room 12 ft wide, 14 ft long, and 8 ft high. Find that volume.

44. A heating duct is 2 ft wide, 1.5 ft high, and 30 ft long. Find the volume of air contained in the duct. See Figure 6.54.

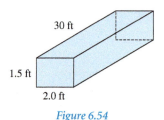

30 ft

1.5 ft

2.0 ft

Figure 6.54

45. In road construction, fill is measured in "yards" where a "yard" is understood to be one cubic yard. How many of these "yards" of fill can be carried by a truck whose trailer is 2.32 yd wide, 1.28 yd high, and 5.34 yd long?

46. A lawn has the shape shown in Figure 6.55. How many gallons of water fall on the lawn in a 0.75-in. rainfall? Assume that 1.00 ft^3 contains 7.48 gal.

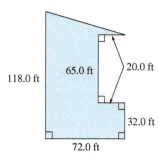

118.0 ft

65.0 ft

20.0 ft

32.0 ft

72.0 ft

Figure 6.55

47. A book is 4 in. wide, 7 in. long, and 1.8 in. thick. Can 30,000 copies of this book be stored in a room that is 9 ft long, 12 ft wide, and 8 ft high?

48. In constructing the foundation for a house, a hole 6 ft deep, 52 ft long, and 25 ft wide must be dug. How many cubic *yards* of earth will be removed?

49. A cylindrical grain storage container 82.0 ft high has a radius of 24.3 ft. One bushel of grain occupies about 1.24 ft^3. How many bushels can be stored in the container?

50. A lawn roller is a cylinder 96.0 cm long and 30.0 cm in radius. Suppose the cylinder of a lawn roller is filled with water for weight. How many liters of water does the cylinder hold? (1000 cm^3 = 1 L)

51. The Great Pyramid of Egypt has a square base that measures approximately 250 yd on each side. The height of the Pyramid is about 160 yd. What is its volume?

52. A fire sprinkler sprays water in the shape of a cone with a radius of 12 ft and height of 8 ft. Find the volume of this cone.

53. A cistern is a container designed to hold water or other liquids, and is most often associated with catching and storing rainwater. A conical cistern 10.0 ft high has a radius at the top of 6.0 ft. Water weighs 62.4 lb/ft^3. How many pounds of water does the cistern hold?

54. An oil storage tank is in the shape of a cylinder with a hemisphere on each end. The length of the cylinder is 45 ft and the diameter is 12.5 ft. What is the volume, in gallons, of the tank? (1.00 ft^3 = 7.48 gal)

55. A weather balloon with a diameter of 2.08 m is filled with helium. Find its volume.

56. The diameter of a basketball is about 9.5 in. What is its volume?

57. What is the difference in volume between a baseball with a diameter of 2.75 in. and a soccer ball with a diameter of 8.50 in.?

58. A sample of an alloy is in the shape of a solid sphere with a radius of 6.40 cm. If this sphere is melted down and then cast into a cube, what is the length of an edge of the cube?

Chapter Summary

Key Terms

point	right angle	quadrilateral	area
line	perpendicular lines	parallelogram	volume
plane	parallel lines	rhombus	right circular cylinder
line segment	triangle	rectangle	pyramid
ray	equilateral triangle	trapezoid	cone
angle	base angles	circle	right-circular
vertex sides of angle	scalene triangle	radius	sphere
degree	right triangle	diameter	
straight angle	hypotenuse	perimeter	

Figure	*Perimeter*
(a) Triangle with sides a, b, c	$p = a + b + c$
(b) Equilateral triangle with side s	$p = 3s$
(c) Quadrilateral with sides a, b, c, d	$p = a + b + c + d$
(d) Rectangle of length l and width w	$p = 2l + 2w$
(e) Parallelogram with sides a and b	$p = 2a + 2b$
(f) Square of side s	$p = 4s$

Formulas

$c = \pi d$	circumference of circle
$c = 2\pi r$	circumference of circle
$A = lw$	area of rectangle
$A = s^2$	area of square
$A = bh$	area of parallelogram
$A = \dfrac{1}{2}bh$	area of triangle
$A = \sqrt{s(s-a)(s-b)(s-c)}$ where $s = \frac{1}{2}(a+b+c)$	area of triangle
$A = \dfrac{1}{2}h(b_1 + b_2)$	area of trapezoid
$A = \pi r^2$	area of circle
$A = lwh$	volume of rectangular solid
$V = e^3$	volume of cube
$V = \pi r^2 h$	volume of cylinder

$V = \dfrac{1}{3}bh$	volume of pyramid
$V = \dfrac{1}{3}\pi r^2 h$	volume of cone
$V = \dfrac{4}{3}\pi r^3$	volume of sphere

Review Exercises

In Exercises 1 through 8, change the angles given in degrees and decimal parts of a degrees to degrees and minutes. Change the angles given in degrees and minutes to degrees and decimal parts of a degree.

1. 37.5°

2. 43.95°

3. 12.55°

4. 45.27°

5. 63°30′

6. 82°20′

7. 105°54′

8. 215°45′

In Exercises 9 through 24, find the perimeters of the indicated geometric figures.

9. Triangle: $a = 17.5$ in., $b = 13.8$ in., $c = 8.9$ in.

10. Quadrilateral: $a = 22.4$ cm, $b = 68.5$ cm, $c = 37.3$ cm, $d = 29.9$ cm

11. Equilateral triangle: $s = 8.5$ mm

12. Isosceles triangle: equal sides of 0.38 yd, third side of 0.53 yd

13. Square: $s = 6.8$ m

14. Rhombus: side of 15.2 in.

15. Parallelogram: $a = 692$ ft, $b = 207$ ft

16. Parallelogram: $a = 7.8$ m, $b = 6.2$ m

17. Rectangle: $l = 96$ cm, $w = 43$ cm

18. Rectangle: $l = 108$ in., $w = 92$ in.

19. Circle: $r = 4.25$ ft

20. Circle: $d = 38.0$ cm

21. The figure in Figure 6.56 (equilateral triangle attached to rectangle).

22. The figure in Figure 6.57 (square and rhombus attached).

23. The figure in Figure 6.58 (rectangle with a half circle removed).

24. The figure in Figure 6.59 (rectangle with a triangle and half-circle affixed).

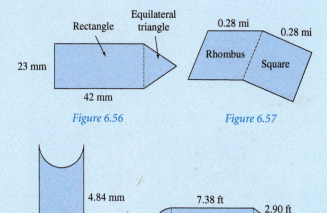

Figure 6.56

Figure 6.57

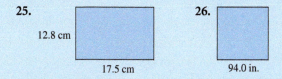

Figure 6.58

Figure 6.59

In Exercises 25 through 40, find the area of the given figures.

25.
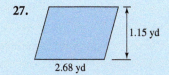
12.8 cm

17.5 cm

26.
94.0 in.

27.
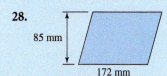
1.15 yd

2.68 yd

28.
85 mm

172 mm

29.
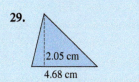
2.05 cm

4.68 cm

30.
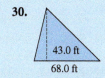
43.0 ft

68.0 ft

31.

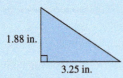

1.88 in.

3.25 in.

32.

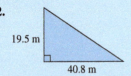

19.5 m

40.8 m

33.   0.067 km

0.016 km

0.118 km

34.   11.1 in.

36.8 in.

12.4 in.

35.

3.36 in.

36.

81.2 cm

37.

420 mm

38.

0.608 ft

39.

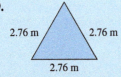

2.76 m   2.76 m

2.76 m

40.

3.27 ft   5.86 ft

7.13 ft

In Exercises 41 through 48, find the volumes of the indicated figures.

41. Rectangular solid: $l = 2.00$ m, $w = 1.50$ m, $h = 1.20$ m

42. Rectangular solid: $l = 25.0$ in., $w = 10.0$ in., $h = 15.0$ in.

43. Cube: $e = 3.50$ yd

44. Cube: $e = 220$ mm

45. Sphere: $r = 5.45$ cm

46. Sphere: $d = 6.20$ ft

47. Sphere with diameter 3.74 yd

48. Sphere with radius 0.9874 m

In Exercises 49 through 60, solve the given problems.

49. A rectangular swimming pool is 48 ft long and 22 ft wide. It is bordered by a walkway that is 8.5 ft. wide. What is the perimeter around the pool? What is the perimeter around the outside edge of the walkway? See Figure 6.60.

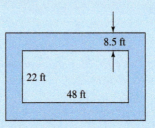

8.5 ft

22 ft

48 ft

Figure 6.60

50. A retaining wall 1.50 m high is built around a rectangular piece of land that is 36.0 m long and 32.0 m wide. What is the area of the wall?

51. Calculate the total weight of a brass sphere with a radius of 2.432 ft. Brass weighs 510 lb/ft^3.

52. An isosceles triangle has a perimeter of 75 cm. Each of the two equal sides is twice as long as the base. Find the length of the base.

53. A circle is drawn on a blueprint so that its circumference is 1088 mm longer than its diameter. Find the diameter of this circle.

54. If the diameter of the earth is 7920 mi and a satellite is in orbit at an altitude of 212 mi above the earth's surface, how far does the satellite travel in one rotation around the earth?

55. If water weighs 62.5 lb/ft^3, find the total weight of water that fills a rectangular pool that is 32 ft long, 16 ft wide, and 4 ft deep.

56. Find the area of lots *A* and *B* shown in Figure 6.61. Lot *A* has a frontage on Main Street of 140 ft. Lot *B* has a frontage on Main Street of 84 ft. The boundary

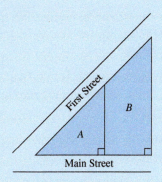

Figure 6.61

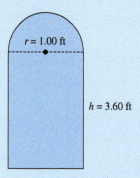

Figure 6.62

between the lots is 120 ft, and the right boundary line for lot *B* is 192 ft.

57. Find the area of the Norman window shown in Figure 6.62. It is a rectangle surmounted by a semicircle.

58. A hollow metal sphere has an outside radius of 4.55 cm and an inside radius of 4.05 cm. What is the volume of the metal in the sphere?

59. A brick is placed in the water-holding tank of a toilet in order to conserve water. The brick is $2\frac{1}{4}$ in. high, $3\frac{3}{4}$ in. wide, and 8 in. long. How much water is saved in 500 flushes? Assume that 1.0 in.^{3} can hold 0.00433 gal.

60. How many acres of land are covered by a JFK Airport runway that is 14,572 ft long and 150 ft wide? (1 acre = 43,560 ft^2.)

Chapter Test

1. Change 45.7° to degrees and minutes.

2. Change 62° 35′ to degrees and decimal parts of a degree.

3. Find the perimeter of each of the following geometric figures:
   a. a triangle with sides 35 in., 27.6 in., and 17.8 in.
   b. a quadrilateral with sides 11.2 cm, 34.25 cm, 18.7 cm, and 15.5 cm
   c. a circle with a radius of 7.75 in.

4. Find the area of each of the following geometric figures:
   a. a triangle with a base of 3.25 ft and a height of 1.88 ft
   b. a circle with a diameter of 98.4 mm
   c. a parallelogram with a height of 2.5 yd and a base of 5.5 yd

5. Find the volume of each of the following geometric figures:
   a. a cylinder with a base radius of 36.0 in. and a height of 2.40 in.

   b. a pyramid with a base area of 3850 ft^2 and a height of 125 ft
   c. a sphere with a diameter of 22.1 mm

6. A machine part is in the shape of a square with equilateral triangles attached to two sides, as shown in Figure 6.63. The length of one side of the square is 2 cm. Find the perimeter of the machine part.

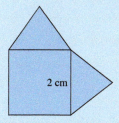

Figure 6.63

7. The diameter of the earth is 7920 mi and a satellite is in orbit at an altitude of 210 mi above the earth. How far does the satellite travel in one rotation about the earth?

8. A rectangular piece of wallboard with two holes cut out for heating ducts is shown in Figure 6.64. The

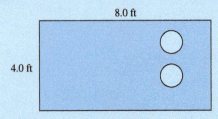

8.0 ft

4.0 ft

Figure 6.64

diameter of each of the heating duct holes is 1 ft. Find the area of the wallboard with the holes cut out (the shaded portion of the figure).

9. The Alaskan oil pipeline is 750 miles long and has a diameter of 4 feet. What is the maximum volume of the pipeline?

10. Determine the volume of a sphere which has a diameter of 165 feet.

7 Simultaneous Linear Equations

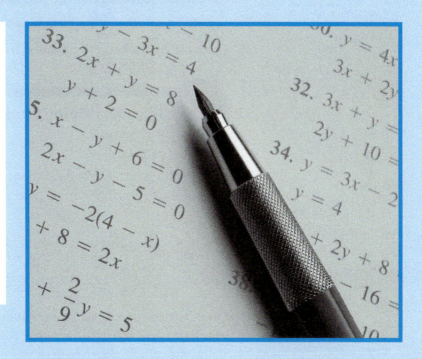

Two laws, considered fundamental in the study of electrical engineering, were developed by Gustav Kirchhoff, a German physicist who was born on March 12, 1824. These two laws are Kirchhoff's Current Law and Kirchhoff's Voltage Law. Kirchhoff's Current Law states that the sum of the current entering a point in a circuit is equal to the sum of the currents leaving the point. Kirchhoff's Voltage Law tells us that the total voltage around a closed loop must be zero.

To work successfully with Kirchhoff's laws requires that a system of linear equations be created that can be used to analyze the series and parallel circuits. Kirchoff's Voltage and Current laws apply to all electric circuits and a firm understanding of these laws is central to the understanding of how an electronic circuit works. Consequently, a strong algebraic background and the ability to solve a system of linear equations is critical for anyone entering the electronics field.

In the preceding chapters we solved equations for one unknown quantity. In this chapter we turn our attention to solving systems of linear equations.

A **system of linear equations** is two or more linear equations that are being solved simultaneously. We will focus our attention on those systems that contain two linear equations and two unknowns. It is important to remember that, in general, a **solution of a system** with two unknowns will require a value for each variable that makes *both* equations true.

7.1　Graphical Solution of Two Simultaneous Equations

Section Objectives
- Graph a system of linear equations
- Solve a system of linear equations graphically

We begin this section by considering a situation that leads to the construction of a system of two simultaneous equations with two unknowns.

Example 1

Two batteries are connected so that the sum of their voltages is 7.5 V. If one of the batteries is reversed, the difference between their voltages is 1.5 V. To determine the voltage of each battery, we could use the equations

$$x + y = 7.5$$
$$x - y = 1.5$$

Here x and y are the voltage levels of the two batteries.

Two planes leave the same airport at the same time and fly in opposite directions. One plane travels 200 km/h faster than the other, and at the end of 4 h they are 4400 km apart. The speeds of the planes can be found by solving the simultaneous equations

$$v_1 - v_2 = 200$$
$$4v_1 + 4v_2 = 4400$$

Here v_1 and v_2 are the speeds of the planes.

The ordered pair that identifies the **point of intersection** of the two linear functions being graphed is the **solution to the system** of the linear equations.

We now consider the problem of solving for the variables when we have two simultaneous equations with two variables. In this section we shall see how the solution may be found graphically.

Previously we graphed several different kinds of functions, including the linear function. We saw that *the graph of $y = mx + b$ is always a straight line.* The number of points on a line is infinite, but if a second line were to cross a given line, there would be only one point in common. The coordinates of this point satisfy the equations of both lines simultaneously.

Example 2

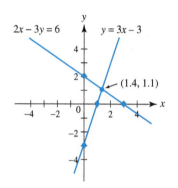

Figure 7.1

Using intercepts or other points, we graph the line $y = 3x - 3$ and $2x + 3y = 6$ as shown in Figure 7.1. The exact coordinates of the point of intersection are $(\frac{15}{11}, \frac{12}{11})$. These exact coordinates were obtained by an algebraic method we will discuss in Section 7.2. This result can be shown to be correct, since these coordinates satisfy the equation of each line. For the first line, we have

$$y = 3x - 3$$

substituting

$$\frac{12}{11} = 3\left(\frac{15}{11}\right) - 3 = \frac{45}{11} - \frac{33}{11} = \frac{12}{11}$$

and for the second line, we have

$$2x + 3y = 6$$

substituting

$$2\left(\frac{15}{11}\right) + 3\left(\frac{12}{11}\right) = \frac{30}{11} + \frac{36}{11} = \frac{66}{11} = 6$$

Even though we cannot obtain this accuracy for the coordinates of the point of intersection from the graph, *we can see from the graph that the point of intersection is about* (1.4, 1.1). This is the type of solution we shall attempt to obtain in this section. *The answers we get using this graphic approach might not be exact,* but they should be reasonably close to the exact solutions. If the values we obtain are not exact, they will not satisfy the equations exactly. The estimated values are acceptable if, upon substitution, both sides of each equation become *approximately* equal.

To solve a pair of simultaneous linear equations, we graph the two equations on the same set of axes and determine the point of intersection. The x coordinate of this point is the desired value of x; the y coordinate of this point is the desired value of y. Together they are the solution to the system of two equations in two variables. Because the coordinates of no other point satisfy both equations, the solution is unique.

Example
3

Graphically solve the system of equations

$$x - y = 6$$
$$2x + y = 3$$

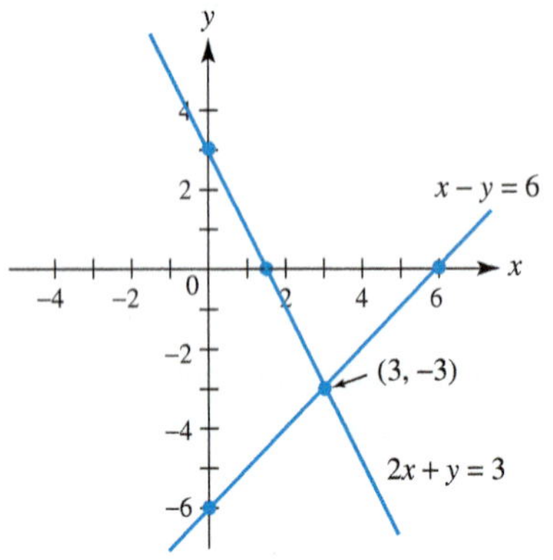

We determine the intercepts and one checkpoint for each equation. Then we graph each equation as shown in Figure 7.2.

Figure 7.2

$x - y = 6$			$2x + y = 3$	
x	*y*		*x*	*y*
0	−6		0	3
6	0		$\frac{3}{2}$	0
1	−5		1	1

We see from the graph that the lines cross at about $(3, -3)$ *so that the solution to this set of equations is* $x = 3$, $y = -3$. (Actually this solution is exact, although this cannot be shown without substitution of the values into the given equations.)

Example
4

A navigational technique involves two synchronized radio signals from two different sources. A position is determined from the sum and difference of the two signals by solving the equations

$$x + y = 9$$
$$x - y = 2$$

Graphically solve this system of equations.

We begin by graphing the two lines from the intercepts and checkpoints indicated in the following tables:

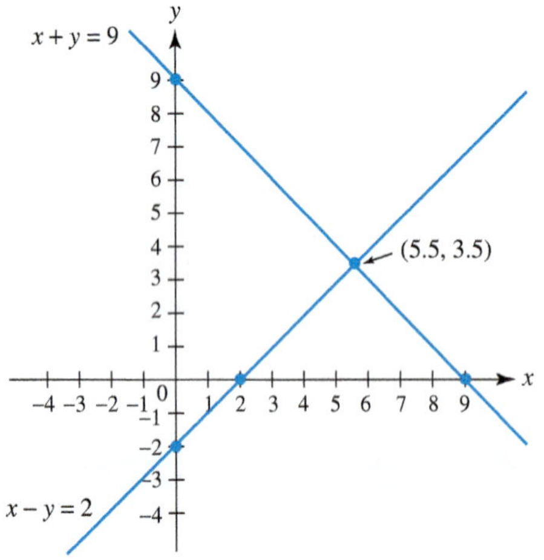

Figure 7.3

$x + y = 9$			$x - y = 2$	
x	*y*		*x*	*y*
0	9		0	−2
9	0		2	0
2	7		3	1

The resulting graph is shown in Figure 7.3. From the graph we estimate the coordinates of the point of intersection as $(5.5, 3.5)$ so that $x = 5.5$ and $y = 3.5$.

Systems of equations are also commonly referred to as **simultaneous equations.**

When graphing two linear equations on the same set of axes we can expect to see one of three possible scenarios: the graphs have one point in common, no points in common, or infinitely many points in common.

If the system of two linear equations has **one solution,** it is an ordered pair that is a solution to both equations. One ordered pair makes both equations true. We call this system **independent.**

If the system of two linear equations has **no solutions,** the two lines are parallel to each other and will never intersect. Therefore, they do not have any points in common. We call this system **inconsistent.**

If the system of two linear equations has **infinitely many solutions,** the two lines lie on top of each other. In this case, all the points of one line are the same as all the points of the other line so any solution that would work in one equation will work in the other. We call this system **dependent.**

Example 5 illustrates an inconsistent system of equations, and Example 6 illustrates a dependent system.

Example 5

Graphically solve the system of equations

$$x + y = 3$$
$$2x + 2y = 9$$

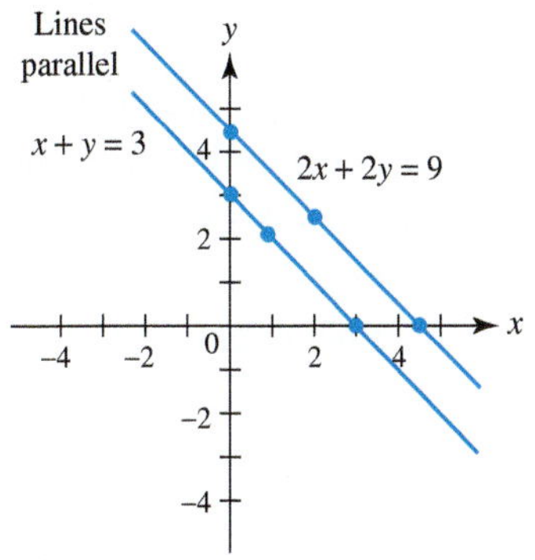

We set up the following tables to indicate the intercepts and checkpoint for each line. The lines are then graphed as in Figure 7.4.

Lines parallel

Figure 7.4

$x + y = 3$	
x	y
0	3
3	0
1	2

$2x + 2y = 9$	
x	y
0	$\frac{9}{2}$
$\frac{9}{2}$	0
2	$\frac{5}{2}$

We observe that within the limits of accuracy of the graphing, the lines appear to be parallel. (They are, in fact, parallel. This is always the case when one equation of the system can be multiplied through by a constant so that the coefficients of the variables are the same, respectively, as those in the other equation. In this case, if the first equation is multiplied by 2, we have $2x + 2y = 6$, which is the same as the second equation except for the constant.) *Because the lines are parallel, the system is inconsistent and there is **no solution.***

Example 6

Graphically solve the system of equations

$$x + 2y = 6$$
$$3x + 6y = 18$$

We set up the following tables to indicate the intercepts and checkpoint for each line. The lines are then graphed as in Figure 7.5.

Figure 7.5

x + 2y = 6	
x	y
0	3
6	0
2	2

3x + 6y = 18	
x	y
0	3
6	0
2	2

We note that *both intercepts are the same.* The two lines are coincident. *Hence every point on the two lines is a solution,* because the coordinates satisfy both equations. This, in turn, means that there is an infinite number of solutions. ▶ ***Because no unique solution may be determined, we call the system* dependent.**

Using Technology

A graphing calculator can display the graphs of several equations at the same time in the graph window. This allows us to use the graphing calculator to solve systems of linear equations.

We will use the linear equations from Example 3 to explore how we find the solution to a system of equations. Although we will focus on a system of linear equations, this method can be used for any system of equations.

Given $x - y = 6$ and $2x + y = 3$, we first rewrite each equation so that it is in $y = mx + b$ format and follow the steps as outlined below:

1. Enter the functions into Y_1 and Y_2 in the Y = screen.

2. Press ZOOM, then #6 to go to the standard window.

3. Press 2nd TRACE to access the "Calc" menu. (This is the same menu we used to find the zeros of a function.)

4. Press option #5 to select the "intersect" feature. You should see a screen similar to the one shown in Figure 7.6. You will see a blinking cursor, the coordinates of the point where the cursor is located, and the question "First curve?" on the screen.

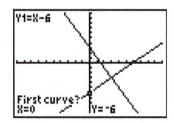

Figure 7.6

5. Use your right arrow key to move the cursor closer to the point of intersection and press ENTER. You will now see the question "Second Curve?" on the screen as shown in Figure 7.7.

6. Press ENTER. You will now see the question "Guess?" on your screen.

7. Press ENTER again.

8. You can now see that the two lines intersect at the point $(3, -3)$, as shown in Figure 7.8.

Try graphing the system of linear equations presented in Examples 5 and 6 in order to see what an inconsistent system of linear equations (Example 5) and a dependent system of linear equations (Example 6) look like on the graphing calculator.

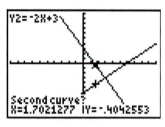

Figure 7.7

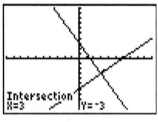

Figure 7.8

7.1 Exercises

In Exercises 1 through 24, solve the given systems of equations graphically. Where possible, estimate the coordinates of the point of intersection to the nearest 0.1 of a unit. If variables other than x and y are used, plot the first along the x axis and the second along the y axis. Verify your results by graphing these systems on your graphing calculator, and then compare your results.

1. $x + y = 3$
   $x - y = 3$

2. $x + y = 5$
   $2x + y = 7$

3. $x + 2y = 4$
   $x - 2y = 0$

4. $3x - y = 1$
   $2x + y = 4$

5. $r - x = 7$
   $2r + x = 5$

6. $m + n = 6$
   $3m + n = 2$

7. $x + 3y = 8$
   $x - y = 0$

8. $x + y = 1$
   $2x - 8y = 1$

9. $3y + 2x = 4$
   $4y + 5x = 3$

10. $2R - T = 7$
    $4R + 3T = 9$

11. $x + 4y = 4$
    $-2x + 2y = 7$

12. $3x - 2y = 4$
    $-9x - 4y = 13$

13. $a + 2b = -3$
    $3a - 5b = -3$

14. $2u - 3v = 8$
    $u + 4v = 6$

15. $s + t = 7$
    $3s + 3t = 5$

16. $2x - 5y = 10$
    $-6x + 15y = -30$

17. $3p - q = 6$
    $2p + 2q = 7$

18. $2x - 4y = 7$
    $3x + 2y = 3$

19. $3x - y = 8$
    $2x - 5y = 15$

20. $7r - 2s = 14$
    $2r + 3s = 9$

21. $0.5x - 1.6y = 3.2$
    $1.2x + 3.3y = 6.6$

22. $0.3x - 0.2y = 1.2$
    $1.3x + 2.5y = 5.0$

23. $\frac{1}{2}x + y = 4$
    $x + \frac{1}{3}y = 2$

24. $\frac{1}{3}x - 2y = 8$
    $3x - \frac{1}{5}y = 2$

In Exercises 25 through 32, each system of equations is either inconsistent or dependent. Identify those systems that are inconsistent and those that are dependent.

25. $x + y = 4$
    $x + y = 5$

26. $x + 2y = 6$
    $-3x - 6y = -18$

27. $2x - 3y = 6$
$12x - 18y = 36$

28. $-3x + 7y = 8$
$6x - 14y = 16$

29. $0.2x - 0.3y = 2.4$
$x - 1.5y = 12$

30. $1.1x + 2.0y = 3.0$
$-2.2x - 4.0y = 6.0$

31. $ax + by = 4$
$ax + by = 2$

32. $ax + by = c$
$3ax + 3by = 3c$

In Exercises 33 through 40, solve the given problems graphically.

33. Two hanger rods are connected to a box beam that supports a concrete deck. Forces acting on the rods sometimes push in the same direction and sometimes push in opposite directions. The forces x and y can be found by solving the equations

$$x + y = 120 \qquad x - y = 70$$

34. One alloy is 70% lead and 30% zinc whereas another alloy is 40% lead and 60% zinc. We can find the amount (in pounds) of each alloy needed to make 100 lb of another alloy 50% lead and 50% zinc by solving the equations

$$x + y = 100 \qquad 0.7x + 0.4y = 50$$

35. An electrician rents a generator and a heavy-duty drill for 5 h at a total cost of $50. On another job, it costs $56 to rent the generator for 4 h and the drill for 8 h. The hourly rates g and d can be found by solving the equations

$$5g + 5d = 50 \qquad 4g + 8d = 56$$

36. A current of 2 A passes through a resistor R_1 and a current of 3 A passes through a resistor R_2, the total voltage across the resistors is 8 V. Then the current in the first resistor is changed to 4 A and that in the second resistor is changed to 1 A; the total voltage is 11 V. The resistances (in ohms) can be found by solving the equations

$$2R_1 + 3R_2 = 8 \qquad 4R_1 + R_2 = 11$$

37. If two ropes with tensions T_1 and T_2 support a 20-lb sign, the tensions can be found by solving the equations

$$0.7T_1 - 0.6T_2 = 0 \qquad 0.7T_1 + 0.8T_2 = 20$$

38. The equation relating the currents for the circuit shown in Figure 7.9 are given by the equations $2i_1 + 6(i_1 + i_2) = 12$ and $4i_2 + 6(i_1 + i_2) = 12$. Find the currents i_1, i_2 to the nearest 0.1A.

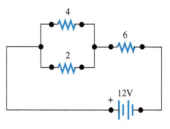

Figure 7.9

39. An architect designing a parking lot has a row 202 feet wide to divide into spaces for compact cars and full-size cars. The architect determines that 16 compact car spaces and 6 full-size car spaces can fill the row, or that 12 compact car spaces and 9 full-size car spaces use all but 1 foot of the width of this row. Determine the width (to the nearest 0.1 ft) of the spaces needed for each type of car.

40. The costs for redesigning a new household appliance for the first year include fixed costs of $35,000 and $275 to produce each appliance. The revenue from the new product is $325.
  a. Find the total cost for producing x redesigned appliances.
  b. Find the total revenue from the sale of x redesigned appliances.
  c. The breakeven point for any company occurs when the total cost and the total revenue are equal. Find the breakeven point graphically.

7.2 Substitution Method

Section Objectives
• Solve a system of equations using the substitution method

Now Try It! 1

Solve for the indicated variables:

a. $x = 7y$ for y
b. $2x + y = 80$ for y
c. $-6y = 8 - x$ for x
d. $2y + x = 5$ for y
e. $x + y = -10$ for x
f. $4y - 3x = 20$ for x

We have just seen how a system of two linear equations can be solved graphically. This technique is good for obtaining a "picture" of the solution of two equations. One of the difficulties of graphical solutions is that the solutions are not exact. If exact solutions are required, we must use algebraic methods. This section presents one basic algebraic method of solution, and we shall discuss another one in Section 7.3. Before moving on, it will be beneficial to work through the problems on the left in order to practice solving an equation for one variable in terms of another.

The method of this section is called **algebraic substitution.** *We first solve one of the equations for one of the two variables and then substitute the result into the other equation for that variable.* The result is a simple equation in one variable. We solve that equation to determine the value of one of the variables. We can then substitute that value into either equation so that the value of the other variable can be found. The following examples illustrate the method of algebraic substitution.

Procedure for Solving Simultaneous Equations Using the Substitution Method

1. Solve one of the equations for one of the variables.
2. **Substitute** the expression from (1) into the second equation.
3. Solve the second equation for the remaining variable.
4. Substitute the value found in step (3) into the first equation to determine the value of the other variable.
5. Check the solutions in both of the original equations.

Example 1

Use substitution to solve the system of equations

$$2x + y = 4$$
$$3x - y = 1$$

The first step is to solve one of the equations for one of the variables. We might quickly examine both equations to determine which equation and which variable would involve the easiest steps. In this system it is slightly easier to solve for y than for x because fractions can be avoided. Inspection shows that one algebraic step is all that is required to solve the first equation for y. Consequently, the first equation would be a good choice. It should be emphasized that *either equation can be solved for either variable,* and the final result will be the same.

▶ Solving the first equation for y, we have $y = 4 - 2x$. **This is now substituted into the second equation,** giving

$$3x - (4 - 2x) = 1$$

——————— in second equation, y is replaced by $4 - 2x$

We now have an equation with only one variable, and we proceed to solve this equation for x:

$$3x - 4 + 2x = 1$$
$$5x = 5$$
$$x = 1$$

▶ Now **the value of y that corresponds to $x = 1$ is found by substituting $x = 1$ into either equation.** Because the first has already been solved for y, we have

$$y = 4 - 2(1) = 2$$

We now know that the solution is $x = 1$ and $y = 2$. We *check this by substituting both values into the other equation.* We get

$$3(1) - 2 = 1$$

so that the solution checks.

Example
2

Use substitution to solve the system of equations

$$2x - 4y = 23$$
$$3x + 5y = -4$$

With these two equations it doesn't make any difference which variable or equation we select. Simply choosing to solve the first equation for x, we have $2x = 23 + 4y$, or

$$x = \frac{23 + 4y}{2}$$

Substituting this expression into the second equation and then solving for y, we get

$$3\left(\frac{23 + 4y}{2}\right) + 5y = -4$$

——————— in second equation, x is replaced by $\dfrac{23 + 4y}{2}$

We now multiply both sides by the lowest common denominator of 2 to get

$$3(23 + 4y) + 10y = -8$$
$$69 + 12y + 10y = -8$$
$$22y = -77$$
$$y = -\frac{7}{2}$$

Substituting this value into the solution for x in the first equation, we get

$$x = \frac{23 + 4\left(-\dfrac{7}{2}\right)}{2} = \frac{23 - 14}{2} = \frac{9}{2}$$

The solution is $x = \frac{9}{2}$ and $y = -\frac{7}{2}$. Checking this solution *in the second equation,* we have

$$3\left(\frac{9}{2}\right) + 5\left(-\frac{7}{2}\right) = \frac{27}{2} - \frac{35}{2} = -\frac{8}{2} = -4$$

This means that the solution checks.

Example 3

Two electric currents I_1 and I_2 (in amperes) can be found by solving the equations

$$-5I_1 + I_2 = 6 \qquad 10I_1 + 3I_2 = -2$$

Solve by substitution.

Solving for I_2 in the first equation we get

$$I_2 = 6 + 5I_1$$

We now substitute $6 + 5I_1$ for I_2 in the second equation to get

$$10I_1 + 3(6 + 5I_1) = -2$$
$$10I_1 + 18 + 15I_1 = -2$$
$$25I_1 = -20 \qquad I_1 = -\frac{20}{25} = -\frac{4}{5}$$

Substituting this value into the solution for I_2 we get

$$I_2 = 6 + 5\left(-\frac{4}{5}\right) = 6 - 4 = 2$$

We obtain the solution $I_1 = -\frac{4}{5}$ A and $I_2 = 2$ A. We check this solution in the second equation. Since

$$10\left(-\frac{4}{5}\right) + 3(2) = -8 + 6 = -2$$

we conclude that the solution is correct.

In Section 7.1 we noted that when two lines are parallel to each other, they have no points in common, or no solution. We called this system of equations inconsistent. *If a system of linear equations is inconsistent, the equation after substitution can be reduced to 0 = a, where a is some number other than zero. If the system of linear equations has infinitely many solutions, we call this system dependent. When we try to solve a system of linear equations that turns out to be dependent using substitution, we can reduce the equation to 0 = 0.*

Example 4

Use substitution to solve the system of equations

$$2x - y = 3$$
$$4x - 2y = 6$$

Solving the first equation for y and substituting this expression into the second equation, we get

$$y = 2x - 3$$
$$4x - 2(2x - 3) = 6$$
$$4x - 4x + 6 = 6$$
$$6 = 6$$

▶ By subtracting 6 from each side, we have $0 = 0$. **The system is dependent,** and has an infinite number of solutions (see Example 6 of Section 7.1).

Example 5

Use substitution to solve the system of equations

$$3x - 2y = 4$$
$$9x - 6y = 2$$

Solving the first equation for x, we get

$$x = \frac{2y + 4}{3}$$

Substituting this expression into the second equation, we obtain

$$9\left(\frac{2y + 4}{3}\right) - 6y = 2$$
$$3(2y + 4) - 6y = 2$$
$$6y + 12 - 6y = 2$$
$$12 = 2$$
$$10 = 0$$

▶ **Because this cannot be true, the system is inconsistent** and has no solution (see Example 5 of Section 7.1).

Now Try It! 2

Solve the following systems of equations by substitution

a. $y = x - 8$
   $3x + y = 4$
b. $5x + 5 = y$
   $y - 3x = 9$
c. $2x + 3y = 0$
   $3x + y = 7$

7.2 Exercises

In Exercises 1 through 24, solve the given systems of equations by substitution.

1. $x = y - 3$
   $x + y = 13$

2. $y = x + 2$
   $x + 2y = 7$

3. $x - y = 2$
   $2x - y = 8$

4. $-x + y = -6$
   $-x - 2y = 6$

5. $x + y = 1$
   $5x + 10y = 8$

6. $y + x = 1$
   $2y - 8x = 1$

7. $2x + y = 0$
   $3x + 2y = 1$

8. $2x = 3y$
   $x - 4y = 0$

9. $x - 2y = 0$
   $2x - 3y = 1$

10. $a - 2b = 4$
    $2a + b = 6$

11. $3x = 2y$
    $-x + 4y = 0$

12. $2x + 3y = 6$
    $x - y = 4$

13. $x - 3y = 6$
    $3x = 6y + 18$

14. $x = y + 1$
    $x - y = 5$

15. $k - 8u = 4$
    $2k + u = 2$

16. $z + 2u = 3$
    $3z - u = 1$

17. $\frac{1}{3}x + y = 9$
    $x - \frac{1}{3}y = 6$

18. $\frac{1}{2}x + y = 8$
    $x + 4y = 7$

19. $2x - 5y = 1$
    $3x - 2y = -4$

20. $2u + 3w = 6$
    $4u + 2w = 8$

21. $2r + 3s = 2$
    $3r - 2s = 16$

22. $3m + 2n = 7$
    $5m + 3n = 10$

23. $3x - 4k = 12$
    $4x + 5k = -7$

24. $3y - 2z = 6$
    $2y + 9z = 8$

In Exercises 25 through 32, solve the given systems of equations by substitution.

25. A space shuttle is used to launch a communications satellite. The satellite is launched either in the same direction

or the opposite direction of the shuttle, and the speed, in kilometers per hour, of both objects can be found by solving the equations

$$x + y = 1200 \qquad x - y = 800$$

26. A manager allocates x hours to manufacturing and y hours to machine maintenance. The values of x and y can be found by solving the equations

$$x + y = 480 \qquad 3x - 5y = 0$$

27. Two furnaces consume fuel oil at the rates of r_1 gal/h, and r_2 gal/h, respectively. The total consumption is 2.00 gal/h, and one furnace burns oil at a rate that is 0.50 gal/h more than the other. The two rates can be found by solving the equations

$$r_1 + r_2 = 2.00 \qquad r_1 = r_2 + 0.50$$

28. To determine how many milliliters of a 30% solution of hydrochloric acid should be drawn off from 100 mL and replaced by a 10% solution to give an 18% solution, we use the equations

$$x + y = 100 \qquad 0.30x + 0.10y = 18$$

Here we must determine the value of the variable y.

29. The voltage between two points in an electric circuit is 60 V. The contact point of a voltage divider is placed between these two points so that the voltage is divided into two parts; one of these parts is twice the other. These two voltages V_1 and V_2 can be found by solving the equations

$$V_1 = 2V_2 \qquad V_1 + V_2 = 60$$

30. The purchasing agent for a company acquires hardware and software for a computer system with a total selling price of $5000. The hardware has a 5% sales tax whereas the software has a 4% sales tax, and the total tax is $230. The prices x and y of the hardware and software, respectively, can be found by solving the equations

$$x + y = 5000 \qquad 0.05x + 0.04y = 230$$

31. A roof truss is in the shape of an isosceles triangle. The perimeter of the truss is 50 ft, and the base doubled is 9 ft more than three times the length of a rafter (neglecting overhang). See Figure 7.10. The length of the base b and a rafter r can be found by solving the equations

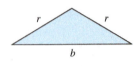

Figure 7.10

$$b + 2r = 50$$
$$2b = 3r + 9$$

32. Under certain conditions, the tensions T_1 and T_2 (in newtons) supporting a derrick are found by solving the equations

$$0.68T_1 + 0.57T_2 = 750$$
$$0.73T_1 - 0.82T_2 = 0$$

Answers

Now Try It! 1

a. $y = \dfrac{x}{7}$ **b.** $y = -2x + 80$ **c.** $x = 6y + 8$ **d.** $y = \dfrac{5 - x}{2}$ **e.** $x = -y - 10$ **f.** $x = \dfrac{20 - 4y}{-3}$

Now Try It! 2

a. $x = 3, y = -5$ **b.** $x = 2, y = 15$ **c.** $x = 3, y = -2$

7.3 Addition-Subtraction Method for Solving Simultaneous Equations

Section Objectives

• Solve a system of equations using the addition-subtraction method

So far we have discussed the graphical method and substitution method for solving a system of two simultaneous linear equations. In this section we introduce a second algebraic method known as the **addition-subtraction method.** *The basis of this method is that if one of the variables appears on the same side of each equation, and if the coefficients of this variable are numerically the same, it is possible to add (or subtract) the left-hand sides (along with the right-hand sides) in order to obtain an equation with only one of the variables remaining.* The resulting equation with only one variable can then be solved by the methods introduced in Chapter 5. Examples 1 and 2 illustrate this addition-subtraction method.

Procedure for Solving Simultaneous Equations Using the Addition-Subtraction Method

1. Rewrite each equation in the form $ax + by = c$.

2. If necessary, multiply each equation by a constant so that the coefficients of one of the variables will add up to zero.

3. Add the two equations together to eliminate a variable.

4. Solve the resulting equation for the remaining variable.

5. Substitute this value into one of the original equations and solve for the remaining variable.

6. Check by substituting both values into both original equations.

Example 1

Use the addition-subtraction method to solve the system of equations

$$2x - y = 1$$
$$3x + y = 9$$

We note that if the left-hand sides of the two equations are added, the terms $+y$ and $-y$ will be combined to produce a zero so that the result will not contain a term that includes y. The two sides of each equation are equal. If the left-hand sides are added, the sum will equal the sum of the right-hand sides. Because y is not present, the resulting equation will contain only x and can then be solved for x. We can then find y by substituting the value of x into either of the original equations and then solving for y. By adding the left-hand sides and equating this sum to the sum of the right-hand sides, we get

$$
\begin{array}{rl}
2x - y = & 1 \\
\underline{3x + y = } & \underline{9} \quad \text{add} \\
2x + 3x \longrightarrow 5x \qquad = & 10 \longleftarrow 1 + 9 \\
-y + y \qquad \qquad x = & 2
\end{array}
$$

Substituting $x = 2$ into the second equation, we have

$$3(2) + y = 9$$
$$6 + y = 9$$
$$y = 3$$

The solution is $x = 2$ and $y = 3$. Because we used the second equation to solve for y, we check this solution in the first equation and get

$$2(2) - 3 = 1$$

which means that the solution is correct.

Example 2

Use the addition-subtraction method to solve the system of equations

$$2x + 3y = 7$$
$$2x - y = -5$$

If we subtract the left-hand side of the second equation from the left-hand side of the first equation, x will not appear in the result. Subtracting the **respective** sides of the two equations we get

$$2x + 3y = 7$$
$$2x - y = -5$$

$2x - 2x = 0$ $4y = 12$ subtract $7 - (-5)$

$3y - (-y)$ $y = 3$

Substituting $y = 3$ into the first equation, we obtain

$$2x + 3(3) = 7$$
$$2x + 9 = 7$$
$$2x = -2$$
$$x = -1$$

The solution is $x = -1$, $y = 3$. Checking this solution in the second **equation**, we have $2(-1) - (3) = -2 - 3 = -5$.

In many systems it is necessary to multiply one (or both) of the **equations by** constants so that the resulting coefficients of one unknown are numerically the same. Examples 3 and 4 illustrate this procedure.

Example 3

Use the addition-subtraction method to solve the system of equations

$$2x + 3y = -5$$
$$4x - y = 4$$

If each term of the second equation is multiplied by 3, it will be **possible to** eliminate y by addition and continue to the solution. Multiplying **each term,** ▶ *including the constant on the right,* of the second equation by 3, we **have**

$$2x + 3y = -5$$ first equation
$$12x - 3y = 12$$ each term of second equation multiplied by 3

Adding the equations, we get

$3y + (-3y) = 3y - 3y = 0$ $14x = 7$

$$x = \frac{1}{2}$$

Substituting $x = \frac{1}{2}$ into the first equation, we get

$$2\left(\frac{1}{2}\right) + 3y = -5$$
$$1 + 3y = -5$$
$$3y = -6$$
$$y = -2$$

The solution is $x = \frac{1}{2}$, $y = -2$. Checking this solution by substituting in the second equation, we have

$$4\left(\frac{1}{2}\right) - (-2) = 2 + 2 = 4$$

so that the solution is verified.

Example 4

A power supply for a space shuttle requires x inductors and y capacitors, where x and y are found by solving the equations

$$2x + 5y = 46$$
$$3x - 2y = 12$$

It is not possible to multiply just one of these equations by an integer to make the coefficients of one of the variables numerically the same. We must multiply both equations by appropriate integers. We can multiply the first equation by 3 and the second equation by 2 in order to make the coefficients of x equal. Or we may multiply the first equation by 2 and the second one by 5 in order to eliminate y. That is, we must find the least common multiple of the coefficients of one variable. Choosing the least common multiple of 5 and -2, we obtain a new set of equations by making both coefficients of y equal to 10. We get

$$
\begin{array}{ll}
4x + 10y = 92 & \text{multiply first equation by 2} \\
\underline{15x - 10y = 60} & \text{multiply second equation by 5} \\
19x = 152 & \text{add}
\end{array}
$$

$$10y + (-10y) = 10y - 10y = 0$$

$$x = \frac{152}{19} = 8$$

$$2(8) + 5y = 46 \qquad \text{substitute } x = 8 \text{ in first equation}$$
$$5y = 46 - 16$$
$$5y = 30$$
$$y = 6$$

Checking, we get

$$3(8) - 2(6) = 24 - 12 = 12$$

The solution of $x = 8$ and $y = 6$ has therefore been verified. That is, the power supply requires 8 inductors and 6 capacitors.

Now Try It!

Solve the following system of equations using the addition-subtraction method.

a. $2a + b = -5$
   $a + 3b = 35$
b. $5p - 2q = -6$
   $2p + 3q = 9$
c. $4x - 3y = -8$
   $3x + 5y = -6$

If the system is dependent or inconsistent, we have the same type of result as that mentioned in Section 7.2. That is, if the system is dependent, the addition-subtraction method will result in $0 = 0$ after the equations are combined, or the result will be $0 = a$ (a not zero) if the system is inconsistent.

> Example
> 5
>
> Use the addition-subtraction method to solve the system of equations
>
> $$3x - 6y = 8$$
> $$-x + 2y = 3$$
>
> Multiplying the second equation by 3 and adding the equations, we get
>
> $$3x - 6y = 8$$
> $$\underline{-3x + 6y = 9}$$
> $$0 = 17 \qquad \text{add}$$
>
> Because this result is not possible, the system is *inconsistent*.

7.3 *Exercises*

In Exercises 1 through 24, solve the given systems of equations by the addition-subtraction method.

1. $x + y = 7$
     $x - y = 3$

2. $x + y = 5$
     $x - 2y = 1$

3. $2x + y = 5$
     $x - y = 1$

4. $x + y = 11$
     $2x - y = 1$

5. $m + n = 12$
     $2m - n = 0$

6. $2x + y = 9$
     $2x - y = -1$

7. $d + t = 3$
     $2d + 3t = 10$

8. $2r + s = 1$
     $r + 2s = -1$

9. $x + 2n = 11$
     $3x - 5n = -22$

10. $2x + y = 2$
      $5x + y = 8$

11. $x - 2y = 5$
      $2x + y = 20$

12. $4t - x = 0$
      $3t - x = -2$

13. $7x - y = 5$
      $7x - y = 4$

14. $3x - y = 4$
      $-6x + 2y = -8$

15. $a + 7b = 15$
      $3a + 2b = 7$

16. $-8x + 7y = 2$
      $3x - 5y = 9$

17. $13p - 18q = 21$
      $4p + 12q = 5$

18. $11k + 15t = 1$
      $2k + 3t = 1$

19. $\frac{2}{7}x - \frac{1}{3}y = -3$
      $\frac{5}{14}x + \frac{5}{3}y = -10$

20. $\frac{7}{8}x + \frac{5}{12}y = -4$
      $x + y = 8$

21. $2m = 10 - 3n$
      $3m = 12 - 4n$

22. $62x = 4y + 43$
      $15x + 12y = 17$

23. $\dfrac{1}{x} + \dfrac{2}{y} = 3$
      $\dfrac{1}{x} - \dfrac{2}{y} = -1$

24. $\dfrac{4}{x} - \dfrac{1}{y} = 1$
      $\dfrac{6}{x} + \dfrac{2}{y} = 5$

In Exercises 25 through 27, use the addition-subtraction method to solve the stated problems.

25. Two batteries produce a total voltage of 7.5 V. The difference in their voltages is 4.5 V. (See Figure 7.11.) We can determine the two voltage levels by solving the equations

$$V_1 + V_2 = 7.5 \qquad V_1 - V_2 = 4.5$$

26. The perimeter of a rectangular sign is 26 ft and the length is 3 ft longer than the width. See Figure 7.12. The dimensions can be found by solving the equations

$$2x + 2y = 26 \qquad x - y = 3$$

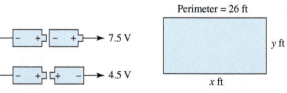

Figure 7.11

Perimeter = 26 ft

y ft

x ft

Figure 7.12

27. In determining the optimal oil-fuel mixture for a chain saw, the amount of oil x and gasoline y can be found by solving the equations

$$16x + y = 2 \qquad 32x - y = 1$$

In Exercises 28 through 32, set up appropriate systems of linear equations and solve the system algebraically.

28. One line printer can produce x lines per minute; a second line printer can produce y lines per minute. They print 7500 lines if the first prints for 2 min and the second prints for 1 min. They can print 9000 lines if the second prints for 2 min and the first for 1 min. Determine the printing rates for each printer.

29. A prospective employee must take a test of verbal skills and a test of mathematical skills. The total score is 1150, and the math score exceeds the verbal score by 150. Determine the score received on each test.

30. If x dollars are invested at an annual rate of 5% and y dollars are invested at 7%, their total annual interest amounts to $405. The amount y exceeds x by $1500. Determine the amount invested at each rate.

31. A college bookstore receives two shipments of calculators. The first shipment has 30 graphing calculators and 50 scientific calculators for a total cost of $4850; the second shipment has 75 graphing calculators and 25 scientific calculators for a total cost of $9625. Determine the price of each type of calculator.

32. An office building with 54 offices has two types of offices. One type of office rents for $900/month; the other type rents for $1250/month. If all of the offices are occupied and the total rental income is $55,600/month, determine how many of each office type of office there are.

Now Try It! Answers

a. $a = -10, b = 15$ **b.** $p = 0, q = 3$ **c.** $x = -2, y = 0$

7.4 Determinants in Two Equations

Section Objectives
• Solve a system of linear equations using determinants

In many systems of linear equations, the methods we have discussed are difficult because of the numbers involved or because the systems are simply too large. In this section we describe a method that is more systematic. We first describe the procedure for two equations with two unknowns, and then we explain why it works.

Consider a general system of equations with two variables that are written in the format

$$a_1x + b_1y = c_1$$
$$a_2x + b_2y = c_2$$

where a_1 and a_2 are the coefficients of the variable x, and b_1 and b_2 are the coefficients of the variable y.

The method we use is based on **determinants.** For systems of two linear equations with two unknowns we use **second-order determinants** defined as follows.

$$\begin{vmatrix} a_1 & b_1 \\ a_2 & b_2 \end{vmatrix} = a_1 b_2 - a_2 b_1$$

A **determinant** is a number associated with a square array of numbers. The order of a determinant refers to the number of rows (or columns) in the array. The value of a determinant is found by taking the product of the two numbers along the downward diagonal, minus the product of the two numbers along the upward diagonal. See Figure 7.13.

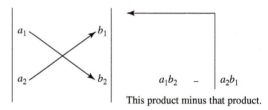

Figure 7.13

Example 1

Find the values of the given determinants.

$$\begin{vmatrix} 1 & 2 \\ 3 & 4 \end{vmatrix} = 1(4) - 3(2) = 4 - 6 = -2$$

$$\begin{vmatrix} 2 & 5 \\ 3 & -9 \end{vmatrix} = 2(-9) - 3(5) = -18 - 15 = -33$$

$$\begin{vmatrix} 2 & 1 \\ -3 & 7 \end{vmatrix} = 2(7) - (-3)(1) = 14 + 3 = 17$$

Given a general system of two equations with two unknowns,

$$a_1 x + b_1 y = c_1$$
$$a_2 x + b_2 y = c_2$$

the solutions for x and y can be expressed using determinants as follows.

$$x = \frac{\begin{vmatrix} c_1 & b_1 \\ c_2 & b_2 \end{vmatrix}}{\begin{vmatrix} a_1 & b_1 \\ a_2 & b_2 \end{vmatrix}} \quad \text{and} \quad y = \frac{\begin{vmatrix} a_1 & c_1 \\ a_2 & c_2 \end{vmatrix}}{\begin{vmatrix} a_1 & b_1 \\ a_2 & b_2 \end{vmatrix}}$$

We should examine the patterns in these two equations. Note that the denominators are the same. Both denominators use the coefficients of x and y in the original system of equations. In the numerators, the first variable x has the first column made up of the column of **c**'s; the second variable y has the second column consisting of **c**'s. If an unknown is missing from either equation, its coefficient is zero, and we should enter zero in the determinant.

This result of using determinants to solve systems of equations is referred to as **Cramer's rule.** Example 2 illustrates this procedure.

Example **2**

Use determinants to solve the given system of equations.

$$2x + \ y = 0$$
$$3x - 2y = 7$$

Use the equations for x and y given above. Noting that the solutions for both x and y involve the same denominator, it would be wise to first evaluate that particular determinant which uses the coefficients of the variables.

$$\begin{vmatrix} 2 & 1 \\ 3 & -2 \end{vmatrix} = 2(-2) - 3(1) = -4 - 3 = -7$$

To find the value of the determinant in the numerator for x, we start with the same coefficients but replace the *first* column with the constants on the right-hand side of the equations. We get

$$\begin{vmatrix} 0 & 1 \\ 7 & -2 \end{vmatrix} = 0(-2) - 7(1) = 0 - 7 = -7$$

To find the value of the determinant in the numerator for y, we start with the coefficients and replace the *second* column with the constants 0 and 7 to get

$$\begin{vmatrix} 2 & 0 \\ 3 & 7 \end{vmatrix} = 2(7) - 3(0) = 14 - 0 = 14$$

Finally, we express the solutions as shown below.

$$x = \frac{\begin{vmatrix} 0 & 1 \\ 7 & -2 \end{vmatrix}}{\begin{vmatrix} 2 & 1 \\ 3 & -2 \end{vmatrix}} = \frac{-7}{-7} = 1$$

$$y = \frac{\begin{vmatrix} 2 & 0 \\ 3 & 7 \end{vmatrix}}{\begin{vmatrix} 2 & 1 \\ 3 & -2 \end{vmatrix}} = \frac{14}{-7} = -2$$

We now know that $x = 1$ and $y = -2$. We should again check the solutions by substituting them in the original equations.

Be careful

Make sure that the equations are written in the form

$$a_1x + b_1y = c_1$$
$$a_2x + b_2y = c_2$$

before you set up your determinants.

Example 3

Use determinants to solve the given system of equations.

$$3x + 4x = -6$$
$$5x - 2y = 16$$

Be careful

Be sure that you multiply on the diagonal rather than across.

Find the solution for x and y as follows.

$$x = \frac{\begin{vmatrix} -6 & 4 \\ 16 & -2 \end{vmatrix}}{\begin{vmatrix} 3 & 4 \\ 5 & -2 \end{vmatrix}} = \frac{-6(-2) - 16(4)}{3(-2) - 5(4)} = \frac{12 - 64}{-6 - 20} = \frac{-52}{-26} = 2$$

$$y = \frac{\begin{vmatrix} 3 & -6 \\ 5 & 16 \end{vmatrix}}{\begin{vmatrix} 3 & 4 \\ 5 & -2 \end{vmatrix}} = \frac{3(16) - 5(-6)}{-26} = \frac{48 + 30}{-26} = \frac{78}{-26} = -3$$

same

We conclude that $x = 2$ and $y = -3$. Checking, we substitute those values in the original equations and simplify to get $-6 = -6$ and $16 = 16$, so the solution is verified.

Example 4

The currents (in amperes) of two circuits can be found from the following equations. Use determinants to solve the given system of equations.

$$1.34x - 2.73y = 9.44$$
$$-8.35x + 7.22y = 5.36$$

Using determinants we get

$$x = \frac{\begin{vmatrix} 9.44 & -2.73 \\ 5.36 & 7.22 \end{vmatrix}}{\begin{vmatrix} 1.34 & -2.73 \\ -8.35 & 7.22 \end{vmatrix}} = \frac{9.44(7.22) - 5.36(-2.73)}{1.34(7.22) - (-8.35)(-2.73)} = \frac{82.7896}{-13.1207} = -6.31$$

$$y = \frac{\begin{vmatrix} 1.34 & 9.44 \\ -8.35 & 5.36 \end{vmatrix}}{\begin{vmatrix} 1.34 & -2.73 \\ -8.35 & 7.22 \end{vmatrix}} = \frac{1.34(5.36) - (-8.35)(9.44)}{-13.1207} = \frac{86.0064}{-13.1207} = -6.56$$

The solutions $x = -6.31$ A and $y = -6.56$ A were rounded to three significant digits. When checking these solutions by substituting them in the original equations, we cannot expect to get identities, but both sides of each equation should be approximately the same.

> ### Now Try It!
>
> *Use determinants to solve the following systems of equations.*
>
> **1.** $2a + b = -5$
> $a + 3b = 35$
> **2.** $5p - 2q = -6$
> $2p + 3q = 9$
> **3.** $4x - 3y = -8$
> $3x + 5y = -6$

In some systems of equations, the determinant used in both denominators may be zero. Because division by zero is undefined, no solution will be found.

This indicates that the system is either inconsistent or dependent. With a zero denominator determinant, the system is inconsistent if the numerators are nonzero; it is dependent if the numerators are also zero.

We have described the procedure for using determinants to solve systems of two linear equations with two unknowns. We now explain why this procedure works. Again consider

$$a_1 x + b_1 y = c_1$$
$$a_2 x + b_2 y = c_2$$

If we multiply the first equation by b_2 and the second by b_1, we get

$$a_1 b_2 x + b_1 b_2 y = c_1 b_2$$
$$a_2 b_1 x + b_2 b_1 y = c_2 b_1$$

Subtracting, we obtain

$$a_1 b_2 x - a_2 b_1 x = c_1 b_2 - c_2 b_1$$

Factoring out x, we get

$$(a_1 b_2 - a_2 b_1)x = c_1 b_2 - c_2 b_1$$

Solving for x we find that

$$x = \frac{c_1 b_2 - c_2 b_1}{a_1 b_2 - a_2 b_1}$$

which corresponds exactly with

$$x = \frac{\begin{vmatrix} c_1 & b_1 \\ c_2 & b_2 \end{vmatrix}}{\begin{vmatrix} a_1 & b_1 \\ a_2 & b_2 \end{vmatrix}} = \frac{c_1 b_2 - c_2 b_1}{a_1 b_2 - a_2 b_1}$$

By similar reasoning, we can show that

$$y = \frac{a_1 c_2 - a_2 c_1}{a_1 b_2 - a_2 b_1}$$

which corresponds exactly with the expression for y.

 *Using Technology*

A graphing calculator can be used to evaluate determinants. This can be done using the *matrix* feature. A *matrix* is a tabular display of values. In this graphing calculator tutorial we will look at (a) how to enter values into a matrix, and (b) how to find the determinant of a matrix using the following system of equations (from Example 3):

$$3x + 4y = -6$$
$$5x - 2y = 16$$

Entering Data into a Matrix

1. Locate the MATRIX functions just over the x^{-1} key on your calculator.

2. Press 2nd x^{-1} to access the MATRIX menu. Then use the right arrow key to move the cursor to EDIT.

3. Select the matrix you wish to create, in this case Matrix A. Press #1. You should see a screen similar to the one shown in Figure 7.14. You must now indicate the number of rows and columns in your matrix (in this case we have 2 rows and 2 columns).

4. Type in the appropriate signed values followed by ENTER. After you have entered all of the values, your screen should look like the shown in Figure 7.15.

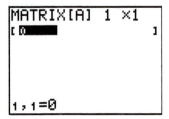
Figure 7.14

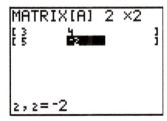
Figure 7.15

5. Repeat this process to create MATRIX [B] and MATRIX [C], which will be used to evaluate the determinant for the numerator of x and y, respectively, as shown in Figures 7.16a and 7.16b.

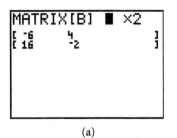
(a)

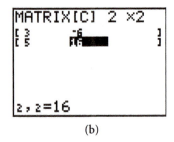
(b)

Figure 7.16

Finding the Determinant of this Matrix

To find the determinant of a single matrix:

1. Once the data has been entered into the matrix, press MATRIX, then 2nd QUIT to leave the MATRIX screen.

2. Access the MATRIX menu and use the right arrow key to highlight MATH. Press #1. You should see "det (" on your home screen.

3. Access the MATRIX menu again and press #1.

4. Press ENTER. If all went well, you should see the value of the determinant as shown in Figure 7.17.

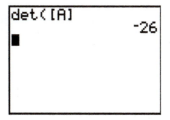

Figure 7.17

Using Determinants to Solve for *x* and *y*

The solution for *x* and *y* is a fraction whose denominator is a determinant made up of the coefficients of *x* and *y* in the original equations. In the numerators, we replace the column of the variable we are trying to find with the values of the constant term.

$$x = \frac{\begin{vmatrix} c_1 & b_1 \\ c_2 & b_2 \end{vmatrix}}{\begin{vmatrix} a_1 & b_1 \\ a_2 & b_2 \end{vmatrix}} \quad \text{and} \quad y = \frac{\begin{vmatrix} a_1 & c_1 \\ a_2 & c_2 \end{vmatrix}}{\begin{vmatrix} a_1 & b_1 \\ a_2 & b_2 \end{vmatrix}}$$

Once all three matrices have been created using the instructions under "Entering data into a matrix," it is possible to find the values for *x* and *y* as shown in Figures 7.18a and 7.18b.

To find the value of *x*

Figure 7.18(a)

To find the value of *y*

Figure 7.18(b)

Compare these results with those found in Example 3.

7.4   *Exercises*

In Exercises 1 through 12, evaluate the given determinants.

1. $\begin{vmatrix} 4 & 3 \\ 2 & 1 \end{vmatrix}$

2. $\begin{vmatrix} 4 & -3 \\ 2 & 1 \end{vmatrix}$

3. $\begin{vmatrix} 2 & 3 \\ -5 & 8 \end{vmatrix}$

4. $\begin{vmatrix} 2 & -4 \\ -5 & 1 \end{vmatrix}$

5. $\begin{vmatrix} 8 & 12 \\ 4 & 6 \end{vmatrix}$

6. $\begin{vmatrix} -9 & 5 \\ 8 & -6 \end{vmatrix}$

7. $\begin{vmatrix} 2 & -5 \\ -4 & 9 \end{vmatrix}$

8. $\begin{vmatrix} -4 & -2 \\ -9 & 8 \end{vmatrix}$

9. $\begin{vmatrix} -5 & -8 \\ -3 & -6 \end{vmatrix}$

10. $\begin{vmatrix} -4 & -6 \\ -2 & -9 \end{vmatrix}$

11. $\begin{vmatrix} 10 & -4 \\ -2 & 6 \end{vmatrix}$

12. $\begin{vmatrix} -5 & -9 \\ 12 & 14 \end{vmatrix}$

In Exercises 13 through 32, use determinants to solve the given systems of equations.

13. $x + y = 6$
$x - y = -4$

14. $x + 2y = 7$
$x - y = -2$

15. $2x - y = 1$
$x + y = 5$

16. $2x + 2y = -2$
$2x - y = 4$

17. $2s = 8$
$s - t = 1$

18. $2R_2 = 16$
$2R_1 - R_2 = -2$

19. $2v_1 - v_2 = 7$
$3v_1 - 2v_2 = 9$

20. $m + n = 0$
$3m + 2n = -20$

21. $x + 2y = 8$
$4x - 5y = -59$

22. $2x - y = -17$
$x + 5y = 41$

23. $x + 2y = 30$
$2x + y = 33$

24. $4x - 3y = 14$
$3x + 2y = 36$

25. $8x - 9y = -6$
$-3x + 4y = -4$

26. $7x + 12y = -9$
$4x + 8y = 4$

27. $3x + 4y = 11$
$5x - 2y = 14$

28. $5x + y = 3$
$10x - 3y = -4$

29. $5x + 10y = 4$
$x + y = 0.5$

30. $3x - 2y = 2.1$
$4x + y = 1.7$

31. $\frac{1}{2}x + \frac{1}{2}y = \frac{3}{8}$
$x - y = \frac{1}{4}$

32. $x + y = \frac{5}{6}$
$x - y = \frac{1}{6}$

In Exercises 33 through 36, use determinants to solve the stated problems.

33. Two different thermistors have resistances R_1 and R_2 (in ohms) that can be found from the following equations.

$R_1 + R_2 = 32{,}000$
$R_1 - R_2 = 16{,}000$

34. One machine produces x parts per hour whereas a second machine produces y parts per hour. Those values can be found from the following equations.

$x + y = 410$
$8x + 4y = 2440$

35. One solution of coolant consists of 30% water and 70% ethyl alcohol. A second solution is 60% water and 40% ethyl alcohol. A more desirable mixture is obtained by mixing x liters of the first solution with y liters of the second solution. Those amounts can be found by solving the following equations.

$0.3x + 0.6y = 8$
$0.7x + 0.4y = 8$

36. A company pays 4% sales tax on a microcomputer costing x dollars, and pays 5% sales tax on a printer costing y dollars. Those costs can be found by solving the following equations.

$0.04x + 0.05y = 340$
$0.04x - 0.05y = 140$

In Exercises 37 through 40, set up appropriate systems of linear equations and solve the system using determinants.

37. A receipt shows that it cost a total of $29.25 to purchase 6 files and 3 blades. Another receipt shows that 2 files and 4 blades cost $12.00. Determine the cost for 1 file and 1 blade.

38. Four pine seedlings and 3 fir seedlings cost $29.00. Two pine seedlings and 7 fir seedlings cost $42.00. Determine the cost of each type of seedling.

39. The sum of two resistances is 55.7 Ω and their difference is 24.7 Ω. Determine the value of each resister (in ohms).

40. A machinery sales representative receives a fixed salary plus a sales commission each month. If $6200 is earned on sales of $70,000 in one month and $4700 is earned on sales of $45,000 in the following month, find the fixed salary and commission percent for the machinery sales representative.

In Exercises 41 through 44, use determinants and a calculator to solve the given systems of equations.

41. $2.56x - 3.47y = 5.92$
$3.76x + 1.93y = 4.11$

42. $345x + 237y = -412$
$-207x + 805y = 623$

43. $3725x - 4290y = 16.25$
$4193x + 2558y = 29.36$

44. $23.07x - 19.13y = 5736$
$18.12x + 12.14y = 2053$

Now Try It! Answers

1. $a = -10, b = 15$ **2.** $p = 0, q = 3$ **3.** $x = -2, y = 0$

7.5 Problem Solving Using Systems of Linear Equations

Section Objectives
• Set up systems of equations from application problems
• Use appropriate methods to solve systems of linear equations

In the first section of this chapter we mentioned that many types of stated problems lead to systems of simultaneous equations. This has also been illustrated with examples and exercises in each of the previous sections. Now that we have seen how such systems are solved, we shall set up equations from stated problems and then solve these equations. Examples 1–6 illustrate this procedure.

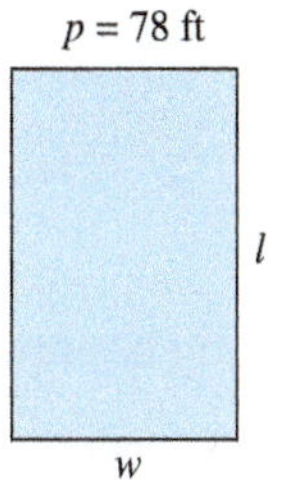

Example 1

In a certain house, the living room (rectangular) has a perimeter of 78 ft. The length is 3 ft less than twice the width. Find the dimensions of the room. See Figure 7.19.

First we let l = length of the room and w = width of the room. Because *the perimeter is 78 ft*, we know that

$$2l + 2w = 78 \qquad p = 2l + 2w$$

$p = 78$ ft

l

w

Figure 7.19

Because *the length is 3 ft less than twice the width*, we have

$$l = 2w - 3$$

Dividing through the first equation by 2 and rearranging the second equation, we have

$$l + \ w = 39$$
$$l - 2w = -3$$

Subtracting the sides of the second equation from the corresponding sides of the first equation, we get

$$3w = 42$$
$$w = 14 \text{ ft}$$

Having found the value of *w*, we now proceed to determine the value of *l* as follows.

$$l = 2(14) - 3 = 25 \text{ ft}$$

Thus the length is 25 ft and the width is 14 ft. This checks with the *statement of the problem.*

Example
2

A manager has 10 employees working 35 h each in one week. The weekly employee time must be allocated so that the machine maintenance time is $\frac{1}{5}$ of the manufacturing time. Find the number of hours spent on maintenance and on manufacturing.

Let x = manufacturing time and y = maintenance time. Because *there are 10 × 35 = 350 h to be allocated,* we get

$$x + y = 350$$

Also, because *the maintenance time is $\frac{1}{5}$ of the manufacturing time,* we get

$$y = \frac{1}{5}x$$

Choosing the method of substitution, we get

$$x + \left(\frac{1}{5}x\right) = 350 \qquad \text{in first equation, } y \text{ is replaced by } \frac{1}{5}x$$
$$5x + x = 1750 \qquad \text{multiply all terms by 5}$$
$$6x = 1750$$
$$x = 291.7$$

Because $x + y = 350$ and $x = 291.7$, we get the following equation for *y*.

$$291.7 + y = 350$$
$$y = 58.3$$

The sum of x and y is 350 h and 58.3 is $\frac{1}{5}$ of 291.7, so the solution checks. The manager should allocate 291.7 h to manufacturing and 58.3 h to maintenance.

Example **3**

Two types of electromechanical carburetors are being assembled. Each type A carburetor requires 15 min of assembly time, whereas type B carburetors require 12 min each. Type A carburetors require 2 min of testing whereas type B carburetors require 3 min each. There are 222 min of assembly time and 45 min of testing time available. How many of each type should be assembled and tested if all of the available time is to be used?

▶ We let x = number of type A carburetors and y = number of type B carburetors. *All of the type* **A** *carburetors require a total of* **15x** *min for assembly whereas all of the type* **B** *carburetors require a total of* **12y** *min for assembly.* Because *the assembly time available is 222 min*, we get $15x + 12y = 222$. By similar reasoning we express the testing time as $2x + 3y = 45$. We now rewrite the first equation and multiply the second equation by 4 to get

$$15x + 12y = 222 \qquad \text{\textcolor{blue}{assembly time}}$$
$$8x + 12y = 180 \qquad \text{\textcolor{blue}{testing time}}$$

Subtracting the second equation from the first we get

$$7x = 42$$
$$x = 6$$

Also,

$$15(6) + 12y = 222 \qquad \text{\textcolor{blue}{first equation with $x = 6$}}$$
$$90 + 12y = 222$$
$$12y = 132$$
$$y = 11$$

With the time available, 6 type A carburetors and 11 type B carburetors should be processed. Checking, we see that these results agree with the *statement of the problem.*

Example **4**

Two jet planes start at cities A and B, 6400 km apart, traveling the same route toward each other. They pass each other 2 h later. The jet that starts from city A travels 200 km/h faster than the other. How far are they from city A when they pass each other? See Figure 7.20.

Because we wish to determine the distance from city A to the place at which they pass each other, we let x = distance from city A to the point at which they pass and y = distance from city B to the point at which they pass.

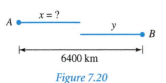

Figure 7.20

Because *the cities are* 6400 *km apart,* we have

$$x + y = 6400$$

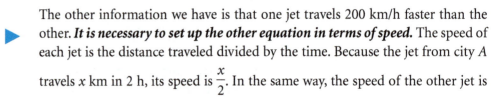

The other information we have is that one jet travels 200 km/h faster than the other. ***It is necessary to set up the other equation in terms of speed.*** The speed of each jet is the distance traveled divided by the time. Because the jet from city *A* travels *x* km in 2 h, its speed is $\dfrac{x}{2}$. In the same way, the speed of the other jet is $\dfrac{y}{2}$. Because *the jet from city A travels* 200 *km/h faster,* we have

$$\frac{x}{2} = \frac{y}{2} + 200$$

We now have the necessary two equations.

Proceeding with the solution, we repeat the first equation and multiply both sides of the second equation by 2. This gives

$$x + y = 6400$$
$$x = y + 400$$

We can easily use either the substitution method or the addition-subtraction method of solution. Using the addition-subtraction method, we have

$x + y = 6400$	first equation
$\underline{x - y = 400}$	from second equation
$2x = 6800$	add
$x = 3400 \text{ km}$	

This means that $y = 3000$ km. The faster jet traveled at 1700 km/h, and the slower one at 1500 km/h, so the solution checks with the statement of the problem.

An alternative method would be to let $u =$ speed of the jet from city *A* and $v =$ speed of the jet from city *B*. Even though the required distance from city *A* is not one of the unknowns, if we find *u* we can find the distance by multiplying *u* by 2, the time that elapses before the planes meet. Using these unknowns, we have the first equation $u = v + 200$ from the fact that the jet from city *A* travels 200 km/h faster. Also, because distance equals speed times time, the distance from city *A* to the meeting place is $2u$ and that from city *B* is $2v$. The total of these distances is 6400 km. Therefore $2u + 2v = 6400$ is the other equation.

Example

5

How many pounds of sand must be added to a mixture that is 50% cement and 50% sand in order to get 200 lb of a mixture that is 80% sand and 20% cement?

Because the unknown quantities are the amounts of sand and original mixture, we let $x =$ number of pounds of sand to be added and $y =$ number of pounds of original mixture required.

We know that the final mixture will have a total weight of 200 lb. Therefore,

$$x + y = 200 \qquad \text{total weight}$$

Next, the amount of sand, x, and the amount of sand present in the original mixture, $0.5y$, equals the amount of sand in the final mixture, $0.8(200)$. This gives us

$$x + 0.5y = 0.8(200) \qquad \text{weight of sand}$$

where the terms are labeled: x = added sand, $0.5y$ = 50% orig. mix., $0.8(200)$ = 80% final sand, wt. final mix.

Rewriting the first equation and multiplying the second equation by 2, we get

$$x + y = 200$$
$$2x + y = 320$$

Subtracting, we obtain

$$x = 120 \text{ lb}$$

which also gives us

$$y = 80 \text{ lb}$$

Because half the original mixture is sand, the final mixture has 160 lb of sand, and this is 80% of the 200 lb. The solution checks, and the required answer is 120 lb of sand.

Example

6

An airplane begins a flight with a total of 36.0 gal of fuel stored in two separate wing tanks. During a flight, $\frac{1}{4}$ of the fuel in one tank is consumed while $\frac{3}{8}$ of the fuel in the other tank is consumed; the total amount of fuel consumed is 11.25 gal. How much fuel is left in each tank?

We let x = original amount of fuel in one tank and y = original amount of fuel in the other tank. Because the plane began with 36.0 gal, we know that $x + y = 36.0$. We get

$$x + y = 36.0 \qquad \text{The two tanks began with a total of 36.0 gal.}$$
$$\tfrac{1}{4}x + \tfrac{3}{8}y = 11.25 \qquad \text{fuel consumed}$$

Using determinants we get

$$x = \frac{\begin{vmatrix} 36.0 & 1 \\ 11.25 & \frac{3}{8} \end{vmatrix}}{\begin{vmatrix} 1 & 1 \\ \frac{1}{4} & \frac{3}{8} \end{vmatrix}} = \frac{36.0(\frac{3}{8}) - 11.25(1)}{1(\frac{3}{8}) - \frac{1}{4}(1)} = \frac{2.25}{0.125} = 18.0$$

Although we could now solve for y using determinants, we might also note that since $x + y = 36.0$ and $x = 18.0$, it is obvious that $y = 18.0$. Each tank began with 18.0 gal. One tank has $\frac{3}{4}$ of the 18.0 gal remaining, so it contains 13.5 gal. The other tank has $\frac{5}{8}$ of the 18.0 gal remaining, so it contains 11.25 gal.

7.5 Exercises

In Exercises 1 through 20, solve the problems by designating the unknown quantities by appropriate letters and then solving the system of equations that is found.

1. At one location in an electric circuit, the sum of two currents is 5 A whereas at another location their difference is 11 A. Find the two currents.

2. The sum of two voltages is 100 V. If the higher voltage is doubled and the other halved, the sum becomes 155 V. What are the voltages?

3. In a certain process, five drills of one type will last as long as two drills of another type. The sum of the average number of hours of operation for the two drills is 105 h. What is the average number of hours each can be used?

4. Two wind turbines are used to supply a total of 50 kW of electricity. If the larger of the two wind turbines supplies twice as much electrical power as the smaller one, find the amount supplied by each.

5. The side of one square metal plate is 5 mm longer than twice the side of another square plate. The perimeter of the larger plate is 52 mm more than that of the smaller. Find the sides of the two plates. See Figure 7.21.

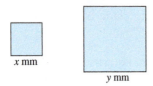

x mm

y mm

Figure 7.21

6. A company maintains two separate accounts for the investment of $45,000. Part of this amount is invested at 6% and the remainder at 8%. If the total annual income is $3400, how much is invested at each rate?

7. A purchase order includes items that cost a total of $8200. Some of those items require an additional sales tax of 4% whereas the remaining items require a 5% sales tax. If the total sales tax is $330, how much is taxed at each rate?

8. A rectangular field that is 100 ft longer than it is wide is divided into three smaller fields by placing two dividing fences parallel to those along the width. A total of 2600 ft of fencing is used. What are the dimensions of the field? See Figure 7.22.

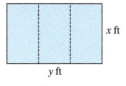

x ft

y ft

Figure 7.22

9. A fixed amount of a certain fuel contains 150,000 Btu of potential heat. Part is burned at 80% efficiency and part is burned at 70% efficiency so that the total amount of heat actually delivered is 114,000 Btu. Find the number of Btu burned at 80% and at 70%.

10. It takes a machine 2 h to process six items of one type and two items of a second type. It takes the same machine 4 h to process five items of each type. Find the time required to process each of the two types of items.

11. Based upon estimates of air pollution, fuel consumption for transportation contributes a percentage of pollution that is 16% less than all other sources combined. What percentage of air pollution is a result of this one source?

12. The voltage across an electric resistor equals the current times the resistance. The sum of two resistances is 14 Ω. When a current of 3 A passes through the smaller resistor and 5 A through the larger resistor, the sum of the voltages is 60 V. What is the value of each resistor?

13. Two gears together have 64 teeth. The larger gear has three times as many teeth as the smaller one. Find the number of teeth in each gear.

14. Suppose that 120 children and 140 adults attend a show for which the total receipts are $530. If a parent and two young children pay $5.50, how much are the adult tickets and how much are the children's tickets?

15. By mass, one alloy consists of 60% copper and 40% zinc. Another is 30% copper and 70% zinc. How many grams of each alloy are needed to make 120 g of an alloy that consists of 50% of each metal?

16. How many liters of a mixture containing 70% alcohol should be added to a mixture containing 20% alcohol to give 16L of a mixture containing 50% alcohol?

17. A shipment of 320 cell phones and radar detectors was destroyed due to a truck accident. On the insurance claim, the shipper stated that each phone was worth $110 and each radar detector was worth $160. The shipper also indicated that the total value of this shipment was $40,700. Determine the number of cell phones and the number of radar detectors included in this shipment.

18. A mechanic makes $220 for tuning an 8-cylinder engine and $150 for tuning a 4-cylinder engine. If she does 26 tune-ups in a month and makes $4600, how many 8-cylinder engines and how many 4-cylinder engines did she service?

19. A manufacturing company produces breadboards and testers. It takes 7 minutes to produce a breadboard and 5 minutes to produce a tester. The quality control department takes 4 minutes to inspect a breadboard and 2.5 minutes to inspect a tester. The assembly line works for 1000 minutes each day and the quality control department works 550 minutes each day. If each item is inspected on the same day it is assembled, determine the number of breadboards and testers produced daily.

20. A company can hire a total of 18 telecommunication specialists and technicians for a special project. The specialists earn $55/hour whereas the technicians earn $37.50/hour. How many specialists and technicians can the company hire if they have budgeted $797.50 per day for salaries for this project?

Chapter Summary

Key Terms

system of linear equations

simultaneous equations

independent system

inconsistent system

dependent system

algebraic substitution

addition-subtraction method

determinants

Cramer's rule

Key Concepts

Procedure for Solving Simultaneous Equations Using the Substitution Method

1. Solve one of the equations for one of the variables.
2. *Substitute* the expression from (1) into the second equation.
3. Solve the second equation for the remaining variable.
4. Substitute the value found in step (3) into the first equation to determine the value of the other variable.
5. Check the solutions in both of the original equations.

Procedure for Solving Simultaneous Equations Using the Addition-Subtraction Method

1. Rewrite each equation in the form $ax + by = c$.
2. If necessary, multiply each equation by a constant so that the coefficients of one of the variables will add up to zero.
3. Add the two equations together to eliminate a variable.
4. Solve the resulting equation for the remaining variable.
5. Substitute this value into one of the original equations and

solve for the remaining variable.

6. Check by substituting both values into both original equations.

Working with Determinants

$$\begin{vmatrix} a_1 & b_1 \\ a_2 & b_2 \end{vmatrix} = a_1b_2 - a_2b_1 \quad \text{determinant}$$

$$a_1x + b_1y = c_1$$
$$a_2x + b_2y = c_2$$

$$x = \frac{\begin{vmatrix} c_1 & b_1 \\ c_2 & b_2 \end{vmatrix}}{\begin{vmatrix} a_1 & b_1 \\ a_2 & b_2 \end{vmatrix}} \quad \text{and} \quad y = \frac{\begin{vmatrix} a_1 & c_1 \\ a_2 & c_2 \end{vmatrix}}{\begin{vmatrix} a_1 & b_1 \\ a_2 & b_2 \end{vmatrix}}$$

Review Exercises

In Exercises 1 through 32, solve the given systems algebraically.

1. $2x + y = 6$
$3x - y = 6$

2. $3x + y = 12$
$3x - y = 24$

3. $x + 2y = 12$
$x + 3y = 16$

4. $x + 2y = 12$
$2x - y = 4$

5. $2x - y = 4$
$x + 3y = 16$

6. $y + 2x = -3$
$3y - 5x = -3$

7. $2p + 7q = 10$
$3p - 2q = 10$

8. $m = n + 3$
$m = 5 - 3n$

9. $2u - 3v = 5$
$-u + 4v = -5$

10. $4x + 3y = -1$
$5x + 2y = 4$

11. $3a + 7b = 15$
$2a - 5b = 39$

12. $2p = 4x + 1$
$2p = 2x - 3$

13. $3x = -16 - y$
$7x = -8 - 5y$

14. $3y = 11 - 5h$
$15y - 15h = 7$

15. $-6y + 4z = -8$
$5y + 6z = 2$

16. $7x + 9y = 3$
$5x + 4y = 1$

17. $3n + d = 20$
$6n + 2d = 40$

18. $x - 2y = 6$
$-2x + 4y = 5$

19. $3x + y = -5$
$6x + 8y = 2$

20. $4x - y = 7$
$8x + 3y = -26$

21. $2x - 3y = 8$
$5x + 2y = 9$

22. $8x - 9y = 3$
$3x + 4y = 2$

23. $8s + 5t = 6$
$7s - 9t = 8$

24. $17x - 9y = 10$
$8x + 7y = -13$

25. $0.03x + 2y = 1$
$0.02x - 3y = 5$

26. $0.10x - 0.05y = 3$
$0.03x + 0.20y = 2$

27. $2x - 13y = 5$
$15x + 3y = 8$

28. $8r - 9t = 9$
$6r + 12t = 13$

29. $\dfrac{2}{3}x + \dfrac{3}{4}y = 26$

$\dfrac{1}{3}x - \dfrac{3}{4}y = -14$

30. $\dfrac{4x + 5y}{20} = 3$

$\dfrac{1}{2}x - \dfrac{1}{3}y = 2$

31. $\dfrac{9}{r} - \dfrac{5}{s} = 2$

$\dfrac{7}{r} - \dfrac{3}{s} = 6$

32. $\dfrac{2}{x} - \dfrac{2}{y} = 4$

$\dfrac{1}{x} - \dfrac{3}{y} = -2$

In Exercises 33 through 40, solve the given systems of equations graphically.

33. $x + 3y = 2$
$6x - 3y = 5$

34. $2x + 3y = 0$
$x + y = 2$

35. $3x - y = 6$
$x - 2y = 4$

36. $4x - y = 9$
$4x + 3y = 0$

37. $4u + v = 2$
$3u - 4v = 30$

38. $r + 4s = 6$
$r + 6s = 7$

39. $3x - y = 5$
$-6x + 2y = 3$

40. $m + 3n = 7$
$3m - 2n = 5$

In Exercises 41 through 44, evaluate the given determinants.

41. $\begin{vmatrix} 2 & 3 \\ 4 & 5 \end{vmatrix}$

42. $\begin{vmatrix} -4 & 2 \\ 6 & 4 \end{vmatrix}$

43. $\begin{vmatrix} 3 & -4 \\ 2 & -5 \end{vmatrix}$

44. $\begin{vmatrix} -12 & 9 \\ 7 & 8 \end{vmatrix}$

In Exercises 45 through 48, use determinants to solve the given systems of equations.

45. $x + 3y = -1$
$2x + 8y = 2$

46. $5x + y = -3$
$8x + 2y = -8$

47. $2x + 3y = 13$
$6x - 2y = -5$

48. $0.2x + 0.4y = -1.6$
$0.7x - 0.3y = 4.6$

In Exercises 49 through 52, solve the applied problems by any appropriate method.

49. When analyzing a certain electric circuit, the equations

$$4i_1 - 10i_2 = 3$$
$$-10i_2 - 5(i_1 + i_2) = 6$$

are obtained. Solve for the indicated electric currents i_1 and i_2 (in amperes).

50. The equation $s = v_0 t + \frac{1}{2}at^2$ is used to relate the distance s traveled by an object in time t, given that the initial velocity is v_0 and the acceleration is a. For an object moving down an inclined plane, it is noted that $s = 32$ ft when $t = 2$ s and $s = 63$ ft when $t = 3$ s. Find v_0 (in feet per second) and a (in feet per second squared).

51. Nitric acid is produced from air and nitrogen compounds. To determine the size of the equipment required, chemists often use a relationship between the air flow rate m (in moles per hour) and exhaust nitrogen rate n. For a certain operation, the following equations

$$1.58m + 41.5 = 38.0 + 2.00n$$
$$0.424m + 36.4 = 189 + 0.0728n$$

are obtained. Solve for m and n.

52. Two ropes support a weight. The equations used to find the tensions in the ropes are

$$0.866T_2 - 0.500T_3 = 0$$
$$0.500T_2 + 0.866T_3 = 50$$

Find T_2 and T_3 (in pounds).

In Exercises 53 through 60, set up the appropriate systems of equations and solve.

53. A welder works for 3 h on a particular job and his apprentice works on it for 4 h. Together they weld 255 spots. If the hours worked are reversed, they would have welded 270 spots together. How many spots can each one weld in an hour?

54. Two men together make 120 castings in one day. If one of them turns out half again as many as the other, how many does each make in a day?

55. The billing department of a large manufacturer reviews 884 invoices and finds that 700 require no changes. If the number of invoices requiring additional payments is 32 less than twice the number of invoices requiring refunds, how many invoices require additional payment and how many require refunds?

56. The cost of booklets at a printing firm consists of a fixed charge and a charge for each booklet. The total cost of 1000 booklets is $550; the total cost of 2000 booklets is $800. What is the fixed charge and the cost of each booklet?

57. A car travels 355 mi by going at one speed for 3 h and at another speed for 4 h. If it travels 3 h at the second

speed and 4 h at the first speed, it would have gone 345 mi. Find the two speeds.

58. An airplane travels 1200 km in 6 h with the wind. The trip takes 8 h against the wind. Determine the speed of the plane relative to the air and the speed of the wind.

59. For proper dosage a certain drug must be a 10% solution. How many milliliters each of a 5% solution and a 25% solution should be mixed to obtain 1000 mL of the required solution?

60. How many milliliters each of a 10% solution and a 25% solution of nitric acid must be used to make 95 mL of a 20% solution?

Chapter Test

1. Solve the following system of equations graphically.

$$5x = 15 - 3y$$
$$y = 6x - 12$$

2. Solve the following system of equations using an appropriate algebraic method.

$$x + y = 5$$
$$4y = 6 - 2x$$

3. Solve the following system of equations using an appropriate algebraic method.

$$7x - 2y = 22$$
$$4x + y = 4$$

4. Evaluate the given determinant: $\begin{vmatrix} 2 & -5 \\ 1 & 7 \end{vmatrix}$

5. Solve the following system of equations using determinants.

$$3x - 2y = 4$$
$$2x + 5y = 9$$

6. In applying Kirchhoff's law to an electric circuit the following equations are found:

$$22I_1 + 27I_2 = 220$$
$$-27I_1 + 76I_2 = 410$$

Find the indicated currents I_1 and I_2.

7. A contractor has been hired to build a deck for a new home owner. He pays a total of $1323 for 6 sets of stainless steel brackets and screws used for installing 2-inch deck boards and 13 sets of stainless steel brackets and screws used for installing $\frac{5}{4}$-inch deck boards. On a return trip to the building supply store he spends $324 for 3 sets of brackets and screws for the 2-inch deck boards and 2 sets of brackets and screws for the $\frac{5}{4}$-inch deck boards. Find the cost of each of the two different types of sets used.

8 Factoring

The history of algebra began in ancient Egypt and Babylon. It is often claimed that the Babylonians were the first to solve quadratic equations. This is probably an oversimplification of history because the Babylonians really had no notion of the concept of an equation as we know it. However, the Babylonians did develop a systematic (or algorithmic) approach to solving problems. Over time this approach to solving equations found its way to the Islamic world, where it became known as the "science of restoration and balancing." In fact, the Arabic word *al-jabr* is the root of the word *algebra* and means *connection* or *completion*. It is interesting to know that ancient civilizations always wrote out algebraic expressions using only a few short-hand notations or abbreviations. By medieval times Islamic mathematicians were tackling problems that involved arbitrarily high powers of the unknown value, and worked out the basic algebra of polynomials all without the use of symbols. The introduction of symbols to represent an unknown quantity and for algebraic powers and operations took place in the sixteenth century. Prior to this time, most problems and their solutions were stated in words. The use of symbols in algebra made it more useful in all fields of mathematics as well as other disciplines, most notably science and technology. Historically, writing polynomials in factored form was important for finding solutions to equations. Until the sixteenth century, the solution of polynomial equations involved systematic guessing-and-testing for possible linear factors involving integers.

8.1 The Distributive Property and Common Factors

Section Objectives
- Recognize common factors in a polynomial expression
- Factor an algebraic expression containing a common monomial factor
- Factor by grouping

We begin this chapter with a review of some basic terminology introduced in Chapter 3.

An **algebraic expression** is used to describe a combination of constants and variables joined by arithmetic operations (addition, subtraction, multiplication, division, roots, and powers). A polynomial is an algebraic expression having a single term or the sum of terms, each of which is one of the following: a constant, a variable with an exponent that is a positive integer, a product of constants, and variables with exponents that are positive integers. A **monomial** is a polynomial with one term. A **binomial** is a polynomial with two terms. A **trinomial** is a polynomial with three terms.

The **distributive law** was also introduced in Chapter 3. This basic law of algebra is useful when multiplying algebraic expressions.

> The **distributive law** for numbers a, x, and y states that
> $$a(x + y) = ax + ay$$

Finally, **factors** are those quantities that, when multiplied together, give us the original product.

Example 1 illustrates the process of **factoring.**

Example
1

Because $3x + 6 = 3(x + 2)$, the quantities 3 and $x + 2$ are factors of $3x + 6$. In starting with the expression $3x + 6$ and determining that the factors are 3 and $x + 2$, we have factored $3x + 6$. Also, we refer to $3(x + 2)$ as the factored from of $3x + 6$.

Because $2ax - ax^2 = ax(2 - x)$, the quantities ax and $2 - x$ are factors of $2ax - ax^2$. In fact, ax itself has factors of a and x, which means that we actually have three factors: a, x, and $2 - x$.

In our work with factoring we shall consider only the factoring of polynomials, all terms of which will have integers as coefficients. *We call a polynomial* **prime** *if it contains no factors other than $+1$ or -1, or plus or minus itself. We then say that an expression is* **factored completely** *if it is expressed as a product of its* **prime** *factors.*

> An expression is **factored completely** when it cannot be factored any further.

Any set of factors can be checked by multiplication. That is, we can check that the factors have been correctly determined by multiplying them to verify that the product is the original expression.

Example
2

The expression $x + 2$ is prime because it cannot be factored. The expression $3x + 6$ is not prime because it can be factored as $3x + 6 = 3(x + 2)$.

The expression $2ax - ax^2$ is not prime because it can be factored. If we write $2ax - ax^2 = a(2x - x^2)$, we have factored the expression, but have not factored it completely because $2x - x^2 = x(2 - x)$. Factoring $2ax - ax^2$ completely, we write $2ax - ax^2 = ax(2 - x)$. Here each factor is prime.

Methods of determining factors such as those in Example 2 will now be considered. Methods of factoring other types of expressions are considered in the ensuing sections. However, we must be able to recognize prime factors when they occur.

▶ Often an algebraic expression contains a factor or factors that are common to each term of the expression. ***The first step in factoring any expression is to determine whether there is such a*** **common monomial factor.** To do this we note the common factor by inspection and then use the reverse of the distributive law to show the factored form.

Example
3

Factor $ax - 5a$.

We note that each term of the expression $ax - 5a$ contains a factor of a. Using the reverse of the distributive law, we have

$$ax - 5a = a(x - 5)$$

Here a is the common monomial factor and $a(x - 5)$ is the required factored form of $ax - 5a$. We **check this result by multiplying** and find that

$$a(x - 5) = ax - 5a$$

which is the original expression.

Example 4

Factor $6ax^2 - 18x^3$.

The numerical factor 6 and the literal factor x^2 are common to each term of the given polynomial so that 6 and x^2 are **common monomial factors.** Technically, 6 and x^2 can themselves be factored as 3×2 and $x \times x$, but the usual practice is not to reduce *monomial* factors to prime factors. We therefore say that $6x^2$ is the common monomial factor (instead of $3 \times 2 \times x \times x$). We now express the factoring of $6ax^2 - 18x^3$ as

$$6ax^2 - 18x^3 = 6x^2(a) + 6x^2(-3x) = 6x^2(a - 3x)$$

Multiplying these factors, we obtain the original expression to verify that the factoring is correct.

When we factor by reversing the distributive law, the expression is written as the product of the **common monomial factor** and the quotient obtained by dividing the expression by the monomial factor. This is illustrated in Example 5.

Example 5

Factor $3ax^2 + 3x^3y - 6x^2z$.

The numerical factor 3 and the literal factor x^2 are common to each term of the polynomial so that $3x^2$ is the **common monomial factor.** Dividing the polynomial by $3x^2$, we have

$$\frac{3ax^2 + 3x^3y - 6x^2z}{3x^2} = \frac{3ax^2}{3x^2} + \frac{3x^3y}{3x^2} - \frac{6x^2z}{3x^2} = a + xy - 2z$$

Therefore, the factoring of $3ax^2 + 3x^3y - 6x^2z$ is expressed as

$$3ax^2 + 3x^3y - 6x^2z = 3x^2(a + xy - 2z)$$

Example 6

Factor $4c^4x^2 - 8c^3x + 2c^3$.

The numerical factor 2 and the literal factor c^3 are common to each term, which means that $2c^3$ is the **common monomial factor.** Dividing the expression by $2c^3$, we have

$$\frac{4c^4x^2 - 8c^3x + 2c^3}{2c^3} = \frac{4c^4x^2}{2c^3} - \frac{8c^3x}{2c^3} + \frac{2c^3}{2c^3} = 2cx^2 - 4x + 1$$

Therefore,

$$4c^4x^2 - 8c^3x + 2c^3 = 2c^3(2cx^2 - 4x + 1)$$

Note the presence of the 1 in the second factor. It is the result of dividing the term $2c^3$ by the factor $2c^3$. Because the term and the common factor are the same, it is a common error to omit the 1 in the second factor. ***It is incorrect to omit the 1,*** which can be seen through checking by multiplication.

As we have seen, we generally determine the common monomial factor of a polynomial by inspection. That is, we survey each term and identify any common numerical and literal factor. *When each term of a polynomial is expressed as the product of its prime factors, the* **greatest common factor** *is the product of the factors common to all terms.*

Now Try It!

Find the greatest common factor for each of the following:

1. $2x - 2y$
2. $4a^2b + a^2$
3. $3ax + 6a^2$
4. $4x^2 - 8xy - 20xy^2$
5. $6a^2b^3c - 9a^3bc^2 + 3a^2bc$

Example 7

We can identify the greatest common factor in each of the examples presented in this section.

In Example 3 the greatest common factor is a.

In Example 4 the greatest common factor is $6x^2$.

In Example 5 the greatest common factor is $3x^2$.

In Example 6 the greatest common factor is $2c^3$.

Example 8

When two resistors R_1 and R_2, are connected in parallel, they produce a total resistance of R_t. An equation describing the relationship among those resistance is

$$R_2 = \frac{R_1R_2}{R_t} - R_1$$

Factor the right side of this equation.

Examination of the right side of the equation shows that R_1 is the only common factor. We get

$$R_2 = R_1\left(\frac{R_2}{R_t} - 1\right)$$

As in Example 6, we must again include the factor of 1.

Factoring by grouping is a technique that uses two properties introduced in Chapter 3: the associative property, which allows us to regroup terms in a polynomial; and the distributive property, which is highlighted at the start of this section. Factoring by grouping is not used all that often, but when it is used it can be helpful.

Example 9

a. When factoring by grouping, you start by looking for common factors. The binomials

$2x - 8, 3x - 12,$ and $5x - 20$ all have $x - 4$ as a common factor.

$2x - 8$ can be rewritten as $2(x - 4)$

$3x - 12$ can be rewritten as $3(x - 4)$

$5x - 20$ can be rewritten as $5(x - 4)$

b. When we examine terms such as

$2x^4 - 6x^3, 3x^5 - 9x^4,$ and $x^2 - 3x$ we see that they all have $x - 3$ as a common factor.

$2x^4 - 6x^3$ can be rewritten as $2x^3(x - 3)$

$3x^5 - 9x^4$ can be rewritten as $3x^4(x - 3)$

$x^2 - 3x$ can be rewritten as $x(x - 3)$

We can use the following guidelines when factoring a polynomial using factoring by grouping.

1. Use parentheses to group terms into binomials with common factors.

2. Factor out the common factor in each binomial.

3. Factor out the common binomial. If no common binomial exists, try different groupings.

Example 10

a. Factor $x^3 - 2x^2 + 3x - 6$

$(x^3 - 2x^2) + (3x - 6)$ group terms into binomials with common factors

$x^2(x - 2) + 3(x - 2)$ factor out the common factor in each binomial

$(x^2 + 3)(x - 2)$ factor out the common binomial

b. Factor $12x^3 - 9x^2 - 8x + 6$

$(12x^3 - 9x^2) + (-8x + 6)$ group terms into binomials with common factors

$3x^2(4x - 3) - 2(4x - 3)$ factor out the common factor in each binomial

$(3x^2 - 2)(4x - 3)$ factor out the common binomial

8.1 Exercises

In Exercises 1 through 36, factor the given expressions by determining any common monomial factors that may exist.

1. $5x + 5y$

2. $3x^2 - 3y$

3. $7a^2 - 14bc$

4. $3s - 12t$

5. $a^2 + 2a$

6. $3p + pq$

7. $2x^2 - 4x$

8. $5h + 10h^2$

9. $3ab - 3c$

10. $4x^2 - 4x$

11. $4p - 6pq$

12. $6s^3 - 15st$

13. $3y^2 - 9y^2z$

14. $12x - 48xy$

15. $abx - abx^2y$

16. $2xy - 8x^2y^2$

17. $6x - 18xy$

18. $12xy - 8axy$

19. $3a^2b + 9ab$

20. $6xy^3 - 9xy^2$

21. $a^2bc^2f - 4acf$

22. $2rs^2t - 8r^2st^2$

23. $ax^3y^2 + ax^2y^3$

24. $2a^2x^3y - 6a^2x^3$

25. $2x + 2y - 2z$

26. $3r - 3s + 3t$

27. $5x^2 + 15xy - 20y^3$

28. $4rs - 14s^2 - 16r^2$

29. $6x^2 + 4xy - 8x$

30. $5s^3 + 10s^2 - 20s$

31. $12pq^2 - 8pq - 28pq^3$

32. $18x^2y^2 - 24x^2y^3 + 54x^3y$

33. $35a^3b^4c^2 + 14a^2b^5c^3 - 21a^3b^2$

34. $15x^2yz^3 - 45x^3y^2z^2 + 16x^2y^2z$

35. $6a^2b - 3a + 9ab^2 - 12a^2b^2$

36. $4r^2s - 8r^3s^2 + 16r^4s - 4r^2s^3$

In Exercises 37 through 42, factor the given expressions by grouping.

37. $x^3 + 3x^2 + 2x + 6$

38. $4x^3 + 3x^2 + 8x + 6$

39. $3x^3 - 15x^2 + 5x - 25$

40. $2x^4 - x^3 + 4x - 2$

41. $xy + x + 3y + 3$

42. $ab - 3a + 2b - 6$

In Exercises 43 through 48, solve the given problems.

43. For a change in temperature for a gas we get the expression

$$\frac{nRT_2}{T_1} - nR$$

Factor this expression.

44. A transmitter used in a security system is to be contained in a cannister with a surface area given by

$$A = 2\pi r^2 + 2\pi rh$$

Factor the right side of this expression.

45. An electric car has four batteries whose total volume is expressed as

$$V = 6wh + 8wh + 4w^2h$$

Factor the right side of this formula.

46. An air gun can shoot BBs upward so that the distance s (in feet) above the muzzle is given by the formula

$$s = 450t - 16t^2$$

where t is the time in seconds. Factor the right side of this formula.

47. A company found that its profit P for selling x items was given by

$$P = 2x^3 - 6x^2 + 10x$$

Factor the right side of this equation.

48. In computing the value of a 6-month loan on which payments of R dollars per month are made, we use the equation

$$P = Rv + Rv^2 + Rv^3 + Rv^4 + Rv^5 + Rv^6$$

where v is a factor involving the interest that is paid. Factor the right side of this equation.

In Exercises 49 through 52, determine whether or not the expressions have been factored completely.

49. a. $5x - 25x^2 = 5x(1 - 5x)$

    b. $2x^2 - 8x = 2(x^2 - 4x)$

50. a. $6ax^2 - 12a = 3a(2x^2 - 4)$
b. $4c^2y^2 - 2cy^2 = 2cy^2(2c - 1)$

51. a. $3x^2y^3 + 6x^3y^2 = 3x^2y^2(y + 2x)$
b. $12axy^2 - 36ay = 12y(axy - 3a)$

52. a. $6st - 24st^2 = 6st(1 - 4t)$
b. $16 + 36x^2 = 4(4 + 9x^2)$

In Exercises 53 through 56, determine whether or not the expressions are prime.

53. a. $3x - 8$ **b.** $4x - 8$ **c.** $3x - 9$

54. a. $x^2 + 2$ **b.** $x^2 + 2x$ **c.** $x^2 + x$

55. a. $xy + y$ **b.** $x^2y^2 + 9$ **c.** $3xy + 9y$

56. a. $3ax^2 + ay$ **b.** $3ax^2 + y$ **c.** $3ax^2y + y$

Now Try It! Answers

1. 2   **2.** a^2 **3.** $3a$ **4.** $4x$ **5.** $3a^2bc$

8.2   Factoring Trinomials

Section Objectives
- Factor trinomials of the form $ax^2 + bx + c$ where $a = 1$

In this section we discuss the factoring of trinomials of the form $ax^2 + bx + c$, where a, b, and c are integers and $a = 1$. In Section 8.3 we will consider the factoring of trinomials of this form where a does not equal 1.

In Chapter 3 we discussed multiplication of binomials such as $(x + 8)(x + 2)$. We used the distributive property to multiply this as follows:

$$(x + 8)(x + 2) = x^2 + 2x + 8x + 8(2)$$
$$= x^2 + (2 + 8)x + 8(2)$$
$$= x^2 + 10x + 16$$

Let's look at this problem and consider what it would look like in factored form. If the factors of trinomials are binomials, then:

$$x^2 + 10x + 16 = (x + \mathbf{p})(x + \mathbf{q})$$

When factoring trinomials of the form $ax^2 + bx + c$, where $a = 1$, we need to find values for **p** and **q** that will satisfy the following:

Multiplying $(x + \mathbf{p})(x + \mathbf{q})$ will give us

$$(x + \mathbf{p})(x + \mathbf{q}) \qquad = x^2 + \mathbf{p}x + \mathbf{q}x + \mathbf{pq}$$
$$= x^2 + (\mathbf{p} + \mathbf{q})x + \mathbf{pq}$$

Therefore, we need to find values for **p** and **q** whose sum will be the value of the coefficient of the x term and whose product will be the value of the constant term, that is,

$$\mathbf{\textit{p} + \textit{q} = \textit{b}} \text{ and } \mathbf{\textit{pq} = \textit{c}.}$$

> ### General Guidelines for Factoring Trinomials of the Form $ax^2 + bx + c, a = 1$
>
> Find two integers p and q such that $p + q = b$ and $pq = c$
>
> 1. Both integers must be positive if both b and c are positive.
>
> 2. Both integers must be negative if b is negative and c is positive.
>
> 3. One integer must be positive and one must be negative if c is negative.

Example
1

In order to factor the trinomial $x^2 + 5x + 4$, we set up the factoring as

integers to be placed here

$$x^2 + 5x + 4 = (x \quad)(x \quad)$$

sum product

and we are to find two integers such that their product is $+4$ and their sum is $+5$. The possible factors of $+4$ are

$$+2 \text{ and } +2, \quad -2 \text{ and } -2, \quad +4 \text{ and } +1, \quad -4 \text{ and } -1$$

Of these possibilities

▶ *only $+4$ and $+1$ have a sum of $+5$.*

Hence the required factors are $(x + 4)$ and $(x + 1)$. We conclude that

$$x^2 + 5x + 4 = (x + 4)(x + 1)$$

When we factor $x^2 - 5x + 4$, an examination of the possible factors of $+4$ reveals that

▶ *the only factors whose sum is -5 are -4 and -1*

Therefore, the factors are $(x - 4)$ and $(x - 1)$. Hence,

$$x^2 - 5x + 4 = (x - 4)(x - 1)$$

When we try to factor $x^2 + 3x + 4$, an examination of the possible factors of $+4$, which are indicated in the first part of this example, reveals that none of the combinations has a sum of $+3$. Therefore,

▶ $x^2 + 3x + 4$ *is prime and cannot be factored.*

Example
2

In order to factor $x^2 + x - 6$ we must find two integers whose product is -6 and whose sum is $+1$. The possible factors of -6 are -1 and $+6$, $+1$ and -6,

Be sure to check your answers by multiplying the binomials to insure that you get back the problem you started with.

Now Try It!

Factor the following

1. $x^2 + 3x + 2$
2. $x^2 - 2x - 15$
3. $x^2 - 13x + 42$

Caution

This technique can be used only for trinomials of the form $ax^2 + bx + c$, when $a = 1$.

-2 and $+3$, and $+2$ and -3. Of these factors, only -2 and $+3$ have a sum of $+1$ so that

$$x^2 + x - 6 = (x - 2)(x + 3)$$

When we factor $x^2 - 5x - 6$, we see that the necessary factors of -6 are $+1$ and -6 because *their sum is* -5. Hence,

$$x^2 - 5x - 6 = (x + 1)(x - 6)$$

When we try to factor $x^2 + 4x - 6$, we see that this expression is prime, because *none of the pairs of factors of* -6 *has a sum of* $+4$.

Example 3

In order to factor $x^2 - 14x + 24$ we must find two integer factors whose product is 24 and whose sum is -14.

Consider the following factors of 24:

24 and 1 -24 and -1 2 and 12 -2 and -12
3 and 8 -3 and -8 4 and 6 -4 and -6.

Because the middle term of the polynomial is negative, we need to consider the negative factors of 24. Checking the product and sum of each pair of factors we find that -2 and -12 meet our criteria because $-2(-12) = 24$ and $-2 + -12 = -14$; therefore,

$$x^2 - 14x + 24 = (x - 2)(x - 12)$$

8.2 Exercises

In Exercises 1 through 26, factor the given trinomials when possible.

1. $x^2 + 3x + 2$
2. $x^2 - x - 2$
3. $x^2 + x - 12$
4. $s^2 + 7s + 12$
5. $y^2 - 4y - 5$
6. $x^2 + 6x - 7$
7. $x^2 + 10x + 25$
8. $t^2 + 3t - 10$
9. $x^2 + 9x + 14$
10. $x^2 - 2x - 15$
11. $x^2 - x - 42$
12. $x^2 + 5x + 4$
13. $x^2 + 12x + 32$
14. $x^2 + 5x - 24$
15. $x^2 + 2x + 1$
16. $x^2 - 4x + 4$

17. $x^2 - 8x + 15$
18. $x^2 - 20x + 100$
19. $x^2 + 5x - 14$
20. $x^2 - 8x - 20$
21. $x^2 + 8x + 12$
22. $x^2 + 4x - 21$
23. $x^2 - 6x - 40$
24. $x^2 - 18x + 45$
25. $x^2 - 3x - 108$
26. $x^2 - 2x - 48$

In Exercises 27 through 30, each of the given trinomials is a perfect square in the sense that they each have two identical factors. Factor each of these trinomials.

27. $x^2 + 2x + 1$
28. $x^2 - 6x + 9$
29. $x^2 - 8x + 16$
30. $x^2 + 12x + 36$

In Exercises 31 through 34, factor the indicated expression.

31. In analyzing projected population growth, a city planner uses the expression

$$N(r^2 + 2r + 1)$$

which is not completely factored. Factor this expression.

32. When determining the deflection of a certain beam of length L at a distance x from one end, we encounter the expression

$$x^2 - 3Lx + 2L^2$$

Factor this expression.

33. In calculating interest earned in an escrow account, the bank uses the expression

$$PR^2 + 2PR + P$$

Factor this expression.

34. The area of a rectangle can be described by the expression

$$x^2 - 3x - 180$$

Factor this expression.

Now Try It! Answers

1. $(x + 2)(x + 1)$ **2.** $(x - 5)(x + 3)$ **3.** $(x - 6)(x - 7)$

8.3 Factoring General Trinomials

Section Objectives

- Factor trinomials of the form $ax^2 + bx + c$ where $a \neq 1$

We now consider factoring the trinomial $ax^2 + bx + c$, where a does not equal 1. The binomial factors of this trinomial will be of the form $(rx + p)$ and $(sx + q)$. That is, we shall determine whether factors exist such that

$$ax^2 + bx + c = (rx + p)(sx + q)$$

Multiplying out the right-hand side of this equation, we obtain

$$(rx + p)(sx + q) = rsx^2 + rqx + spx + pq$$
$$= rsx^2 + (rq + sp)x + pq$$

We now see that $a = rs, b = rq + sp$, and $c = pq$. It is therefore necessary to find four such integers, r, s, p, and q.

You probably realize that finding these integers is relatively easy so far as the ax^2 term and the c term are concerned. We simply need to find integers r and s whose product is a and integers p and q whose product is c. However, we must find the right integers so that the bx term is correct. This is usually the most troublesome part of factoring trinomials. We must remember that **the bx term will result from the sum of two terms** when the factors are multiplied together.

Summarizing this information on the factors of $ax^2 + bx + c$, where a does not equal 1, we have

$$a = rs \qquad c = pq$$
$$ax^2 + bx + c = (rx + p)(sx + q)$$
$$b = rq + ps$$

We see that *the "outer" product rqx and the "inner" product psx are the two terms that together form the bx term.* In finding the factors, we shall always take *r* and *s* to be positive if *a* is positive.

Example
1

When we factor $2x^2 - 11x + 5$, we take the factors of 2 to be $+2$ and $+1$ (we use only positive factors of *a* because it is $+2$). We now set up the factoring as

$(2x \quad)(x \quad)$

Because the product of the integers to be found is positive ($+5$), only integers of the same sign need be considered. Also, because **the sum of the outer and inner products is negative (-11),** the integers are negative. The factors of $+5$ are $+1$ and $+5$, and -1 and -5, which means that -1 and -5 is the only possible pair. Now trying the factors

$$(2x - 5)(x - 1)$$

$-5x$

$-2x \qquad\qquad -2x - 5x = -7x$

we see that $-7x$ is not the correct sum for the middle term. Next, trying

$$(2x - 1)(x - 5)$$

$-x$

$-10x \qquad\qquad -10x - x = -11x$

we have the correct sum of $-11x$. Therefore,

$$2x^2 - 11x + 5 = (2x - 1)(x - 5)$$

Example
2

Factor $4y^2 + 7y + 3$.

The positive factors of $+4$ are $+1$ and $+4$, and $+2$ and $+2$. Because $c = +3$, its factors must have the same sign, and because $b = +7$ *the sign is positive.* This means we need consider only positive factors of $+3$, and these factors are $+1$ and $+3$. Trying the factors

$$(y + 1)(4y + 3)$$

$4y$

$3y \qquad\qquad 3y + 4y = 7y$

we see that *we have the correct sum of $+7y$ for the outer and inner products.* This means

$$4y^2 + 7y + 3 = (y + 1)(4y + 3)$$

It is not necessary to try the other pair of factors $+2$ and $+2$.

Example 3

Factor $6x^2 + 5x - 4$.

First we find all positive factors of 6 (6 and 1, 2 and 3); then we find all factors of 4 (4 and 1, 2 and 2). We must choose a factor from each group so that the resulting pair can be used to form a 5. Because b is positive ($+5$), we know that the outer or inner product of the larger absolute value will be the product of positive numbers. This eliminates several possible combinations, such as

$$(6x + 1)(x - 4)$$

$$x$$

$$-24x \qquad\qquad -24x + x = -23x$$

We find that the factors $+6$ and $+1$ do not work with any of the factors of -4. Trying the factors $+3$ and $+2$, we find the combination

$$(3x + 4)(2x - 1)$$

$$8x$$

$$-3x \qquad\qquad -3x + 8x = +5x$$

has a sum of outer and inner products of $+5x$, the required value. Therefore,

$$6x^2 + 5x - 4 = (3x + 4)(2x - 1)$$

As you work through more and more of these types of problems, it becomes easier to recognize the correct combination of factors to use. The more you practice, the easier these become.

▶ As mentioned in previous sections, *we must always* **be on the alert for a common monomial factor.** A check should be made for such a factor first. Consider the following example.

Example 4

The height (in feet) of a projectile is given by $16t^2 - 48t - 64$, where t is the time (in seconds) of travel. Factor this expression.

At first it might appear that we should find factors of $+16$ and -64, but note

▶ that *there is a common factor of 16 in each term* so that we can write

$$16t^2 - 48t - 64 = 16(t^2 - 3t - 4)$$

It is now necessary to factor only $t^2 - 3t - 4$. The factors of this trinomial are $(t + 1)$ and $(t - 4)$ so that

$$16t^2 - 48t - 64 = 16(t + 1)(t - 4)$$

Many trinomials that are factorable by the methods of this section contain a literal factor in each term of the binomial factors. Example 5 illustrates this type of problem.

Now Try It!

Factor the following polynomials:

1. $2x^2 - 7x + 3$
2. $3a^2 + 20a + 12$
3. $6x^2 + 7x - 20$
4. $2y^2 - y - 6$

Example 5

Factor $4x^2 + 22xy - 12y^2$.

We first note that there is a common factor of 2 in each term. This means that $4x^2 + 22xy - 12y^2 = 2(2x^2 + 11xy - 6y^2)$. The problem now is to factor $2x^2 + 11xy - 6y^2$. Because the product of

$$(rx + py)(sx + qy) = rsx^2 + (rq + sp)xy + pqy^2$$

we see that this is essentially the same as the previous problems. The only difference is the presence of xy in the middle term and y^2 in the last term. However, the process of finding the values of r, s, p, and q is the same. We find that the arrangement

$$(2x - y)(x + 6y)$$

$$-xy$$

$$12xy \qquad\qquad 12xy - xy = 11xy$$

works for this combination. Therefore,

$$4x^2 + 22xy - 12y^2 = 2(2x - y)(x + 6y)$$

In this example we began by factoring out the common monomial factor of 2. A common error is to forget such a factor when expressing the final result. Be sure to include any common monomial factors in the final result. Checking the result through multiplication will help eliminate many errors.

Caution

Check your answers by multiplying the binomials to insure that you get back to the problem you started with.

8.3 Exercises

In Exercises 1 through 36, factor the given trinomials when possible.

1. $2q^2 + 11q + 5$

2. $2a^2 - a - 3$

3. $3x^2 - 8x - 3$

4. $2x^2 - 7x + 5$

5. $5c^2 + 34c - 7$

6. $3x^2 + 4x - 7$

7. $3x^2 - x - 5$

8. $7n^2 - 16n + 2$

9. $2s^2 - 13st + 15t^2$

10. $3x^2 - 14x + 8$

11. $5x^2 + 17x + 6$

12. $3x^2 - 17x - 6$

13. $4x^2 - 8x + 3$

14. $6x^2 + 19x - 7$

15. $12q^2 + 20q + 3$

16. $8y^2 + 5y - 3$

17. $6t^2 + 7tu - 10u^2$

18. $4x^2 + 33xy + 8y^2$

19. $8x^2 + 6x - 9$

20. $12x^2 - 7xy - 12y^2$

21. $4x^2 + 21x - 18$

22. $6s^2 - 7s - 20$

23. $8n^2 + 2n - 15$

24. $9y^2 + 28y - 32$

25. $2x^2 - 22x + 48$

26. $3x^2 - 12x - 15$

27. $4x^2 + 2xz - 12z^2$

28. $5x^2 + 15x + 25$

29. $2x^3 + 6x^2 + 4x$

30. $2x^4 + x^3 - 10x^2$

31. $10ax^2 + 23axy - 5ay^2$

32. $54a^2 - 45ab - 156b^2$

33. $3ax^2 + 6ax - 45a$

34. $2r^2x^2 - 7r^2x - 15r^2$

35. $14a^3x^2 - 7a^3x - 7a^3$

36. $5cx^2 + 5cx + 15c$

In Exercises 37 through 40, each of the given trinomials is a perfect square in the sense that they each have two identical factors. Factor each of these trinomials.

37. $4x^2 + 4x + 1$ **38.** $4x^2 + 12x + 9$

39. $9x^2 - 6x + 1$ **40.** $16x^2 - 24x + 9$

In Exercises 44 through 50, factor the indicated expression.

41. To find the total profit of an article selling for p dollars, it is sometimes necessary to factor the expression

$$2p^2 - 108p + 400$$

Factor this expression.

42. In calculating interest earned in an escrow account, the bank uses the expression

$$PR^2 + 2PR + P$$

Factor this expression.

43. In designing a parabolic mirror, an engineer uses the expression

$$y = 2x^2 - 24x + 64$$

Factor the right-hand side of this equation.

44. The resistance of a certain electric resistor varies with the temperature according to the equation

$$R = 10{,}000 + 600T + 5T^2$$

Factor the right-hand side of this equation.

45. An open box is made from a sheet of cardboard 12 in. square by cutting equal squares of side x from the corners and bending up the sides. Show that the volume of the box is given by the formula

$$V = 4x^3 - 48x^2 + 144x$$

Factor the right-hand side of this formula. See Figure 8.1.

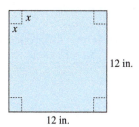

12 in.

12 in.

Figure 8.1

46. The total production cost for x core support plates used in nuclear reactors is expressed as

$$C = 6kx^2 + 11kx + 3k$$

Factor the right-hand side of this equation.

47. A manufacturer determines that the optimal size of a container yields the volume that is expressed as

$$V = 6x^3 - 15x^2 + 6x$$

Factor the right-hand side of this equation.

48. The change in surface area of a heated rectangular plate gives rise to the expression

$$Ak^2D^2 + 2AkD + A$$

Factor this expression.

49. In a study of pollution, an object is shot upward. The time this object is in the air can be found from the expression

$$16t^2 - 124t - 32$$

Factor this expression.

50. A rectangular panel is to be constructed so that wind pressures can be studied. The frame requires two wood sides of width x, but the bottom length will be along the ground so that no support is required there. With an enclosed area of 4000 ft^2 and wood sides totaling 180 ft, the dimensions can be found from the expression

$$2x^2 - 180x + 4000 = 0$$

Factor the left-hand side of this equation. See Figure 8.2.

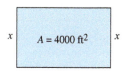

x $A = 4000$ ft^2 x

Figure 8.2

Now Try It! Answers

1. $(2x - 1)(x - 3)$ **2.** $(3a + 2)(a + 6)$ **3.** $(2x + 5)(3x - 4)$ **4.** $(2y + 3)(y - 2)$

8.4   The Difference Between Two Squares

Section Objectives
- Factoring binomials that are the difference of two squares

In Section 8.1 we discussed the factoring out of any common monomial factors. This should normally be the first step in any factoring process. In addition to factoring out common monomial factors, we must also do any factoring that involves more than one term. In this section we consider a very common and standard factoring procedure involving expressions that include a difference between two square quantities.

If we multiply $(x + y)$ by $(x - y)$, the product is $x^2 - y^2$. This product contains two perfect squares, x^2 and y^2. In general, *a* **perfect square** *is any quantity that is an exact square of a rational quantity.*

Having noted that

$$(x + y)(x - y) = x^2 - y^2$$

we see that the multiplication results in the *difference* between the *squares* of the two terms that appear in the binomials. Therefore, we can easily recognize the factors of the binomial $x^2 - y^2$, which are $(x + y)$ and $(x - y)$. It will be necessary to recognize the perfect squares represented by x^2 and y^2 when we factor expressions of this type. We can now apply the difference between any two squares.

Example 1

$$x^2 - 9 = x^2 - 3^2 = (x + 3)(x - 3)$$

Once we determine that an expression is the difference between two perfect squares, the factoring process becomes easy. First take the square root of each term:

$$\sqrt{x^2} = x \quad \text{and} \quad \sqrt{9} = 3$$

Now write the sum of these square roots $(x + 3)$ *and their difference* $(x - 3)$. We then indicate their product as $(x + 3)(x - 3)$. Additional illustrations of this procedure are given in Example 2.

Example
2

$$9 - 4y^2 = 3^2 - (2y)^2 = (3 + 2y)(3 - 2y)$$
$$36p^2 - 49q^2 = (6p)^2 - (7q)^2 = (6p + 7q)(6p - 7q)$$
$$4s^4 - 25t^4 = (2s^2)^2 - (5t^2)^2 = (2s^2 + 5t^2)(2s^2 - 5t^2)$$
$$4x^2 - 1 = (2x)^2 - 1^2 = (2x + 1)(2x - 1)$$
$$16y^4x^2 - p^6 = (4y^2x)^2 - (p^3)^2 = (4y^2x + p^3)(4y^2x - p^3)$$

In actually writing down the result of the factoring, we do not usually need to write the middle step shown in this example. The first illustration would be written simply as $x^2 - 9 = (x + 3)(x - 3)$. The middle steps are given here to indicate the perfect squares that are used.

As indicated previously, if it is possible to factor out a common monomial factor, this should be done first. If the resulting factor is the difference between squares, the factoring is not complete until this difference is itself factored. This illustrates that **complete factoring** *often requires more than one step.* Also, when you are writing the result in complete factoring, be sure to include *all* factors.

The procedure applicable to Examples 3–7 can be summarized in these steps:

1. Factor out any common monomial factors.

2. To factor the difference between two squares, first find the square root of each of the two terms. Then one factor is the sum of the resulting square roots; the other factor is the difference of the resulting square roots.

3. Check the resulting factors to see if any additional factoring is possible.

Example
3

In factoring $3x^2 - 12$, we note that there is a common factor of 3 in each term. Therefore, $3x^2 - 12 = 3(x^2 - 4)$. However, *the factoring is not yet complete* because $x^2 - 4$ is itself the difference between two perfect squares. Therefore, $3x^2 - 12$ is completely factored as

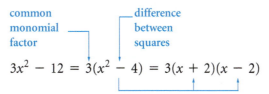

common monomial factor

difference between squares

$$3x^2 - 12 = 3(x^2 - 4) = 3(x + 2)(x - 2)$$

Example
4

In factoring $x^4 - 16$, we note that we have the difference between two squares. Therefore, $x^4 - 16 = (x^2 + 4)(x^2 - 4)$. However, the factor $x^2 - 4$ is again the difference between squares. Therefore

$$x^2 - 16 = (x^2 + 4)(x^2 - 4) = (x^2 + 4)(x - 2)(x + 2)$$

▶ **The factor $x^2 + 4$ is prime.** It is not equal to $(x + 2)^2$ because

$$(x + 2)^2 = (x + 2)(x + 2) = x^2 + 2x + 2x + 4 = x^2 + 4x + 4$$

Example
5

In factoring $2a^4b - 2b$, we first factor out the common monomial of $2b$ to get $2b(a^4 - 1)$. However, the factor $a^4 - 1$ is the difference between two squares. Therefore,

$$2a^4b - 2b = 2b(a^4 - 1) = 2b(a^2 + 1)(a^2 - 1)$$
$$= 2b(a^2 + 1)(a + 1)(a - 1)$$

Note that the factor $a^2 - 1$ is also the difference between two perfect squares. After factoring an expression, we should check the result to see if more factoring is necessary.

Example
6

An aircraft uses Venturi tubes for measuring airspeed, which is displayed on the airspeed indicator. Analysis of the Venturi tubes leads to the expression

$$\frac{1}{2}krv_1^2 - \frac{1}{2}krv_2^2$$

which is to be factored.

We first note that $\frac{1}{2}kr$ is a common factor of both terms. Factoring out $\frac{1}{2}kr$, we get

$$\frac{1}{2}kr(v_1^2 - v_2^2)$$

We now proceed to factor $v_1^2 - v_2^2$ as the difference between two squares. We get

$$\frac{1}{2}krv_1^2 - \frac{1}{2}krv_2^2 = \frac{1}{2}kr(v_1^2 - v_2^2) = \frac{1}{2}kr(v_1 + v_2)(v_1 - v_2)$$

The factoring is now complete.

Example
7

In the expression $(x + y)^2 - 25$, we have the difference between two squares. Treating $(x + y)$ as the first term, we get

$$(x + y)^2 - 25 = [(x + y) + 5][(x + y) - 5]$$
$$= (x + y + 5)(x + y - 5)$$

We expressed the final result by eliminating the unnecessary innermost sets of parentheses. Because only one type of grouping symbols was necessary, we use parentheses instead of brackets.

Factoring the difference between two squares, as discussed in this section, is a very common method of factoring and it is used frequently in algebra.

8.4 Exercises

In Exercises 1 through 30, factor the given expressions completely.

1. $a^2 - 1$
2. $x^2 - 4$
3. $t^2 - 9$
4. $E^2 - 16$
5. $16 - x^2$
6. $25 - y^2$
7. $4x^4 - y^2$
8. $25s^6 - 1$
9. $100 - a^4b^2$
10. $64 - x^2y^4$
11. $a^2b^2 - y^2$
12. $4q^2 - (rs)^2$
13. $81x^4 - 4y^6$
14. $49b^6 - 25c^4$
15. $5x^4 - 45$
16. $24x^2 - 54a^2$
17. $4x^2 - 100y^2$
18. $9x^2 - 81$
19. $x^4 - 1$
20. $a^4 - 81b^4$
21. $4x^2 + 36y^2$
22. $75a^2x^2 + 27b^4y^4$
23. $3s^2t^2 + 12s^2$
24. $9a^2b^6 + 81a^2b^2$
25. $25 - (x + y)^2$
26. $36 - (s + t)^2$
27. $(x + y)^2 - (x - y)^2$
28. $(x + 1)^2 - (x - 1)^2$
29. $a(x + y)^2 - a$
30. $ax^2(x + y)^2 - 9ax^2$

8.5 The Sum and Difference of Cubes

Section Objectives
• Factoring binomials that are the sum or difference of two perfect cubes

In this chapter we have already discussed common monomial factors, the difference between two squares, and factoring trinomials. In this section we consider algebraic expressions that are either the sum of two perfect cubes or the

Common Cubes	
$2^3 = 8$	$5^3 = 125$
$3^3 = 27$	$6^3 = 216$
$4^3 = 64$	$10^3 = 1000$

difference between two perfect cubes. In general, a **perfect cube** is any quantity that is an exact cube of a rational quantity.

We will base our factoring procedures on the equations

$$x^3 + y^3 = (x + y)(x^2 - xy + y^2)$$

$$x^3 - y^3 = (x - y)(x^2 + xy + y^2)$$

We can verify these equations by multiplying out their right sides. After combining like terms and simplifying, both sides of each equation will agree. Consider the following example.

Example 1

$x^3 - 1 = x^3 - 1^3$

$\qquad = (x - 1)[(x)^2 + (1)(x) + (1)^2]$

$\qquad = (x - 1)(x^2 + x + 1)$

$27x^3 - 8 = (3x)^3 - 2^3$

$\qquad = (3x - 2)[(3x)^2 + (3x)(2) + (2)^2]$

$\qquad = (3x - 2)(9x^2 + 6x + 4)$

$x^3 + 8 = x^3 + 2^3$

$\qquad = (x + 2)[(x)^2 - 2x + (2)^2]$

$\qquad = (x + 2)(x^2 - 2x + 4)$

$8x^6 + 27 = (2x^2)^3 + 3^3$

$\qquad = (2x^2 + 3)[(2x^2)^2 - (2x^2)(3) + (3)^2]$

$\qquad = (2x^2 + 3)(4x^4 - 6x^2 + 9)$

As in the preceding sections of this chapter, we should again begin any factoring process by first checking for the presence of a monomial factor that is common to each term. If there is a common monomial factor, it should be factored out first.

Although none of the illustrations in Example 1 included a common monomial factor, Examples 2–6 do include such a factor.

Example 2

In factoring $cx^3 + c$, we first note that c is a factor common to both terms and we factor it out to get $c(x^3 + 1)$. The factor $x^3 + 1$ is itself the sum of two perfect cubes and it is therefore factorable. We get

$$cx^3 + c = c(x^3 + 1) = c(x + 1)(x^2 - x + 1).$$

Example
3

In factoring $ax^5 - ax^2$, we begin by factoring out the common monomial of ax^2. We get the factored form $ax^2(x^3 - 1)$. However, this expression has not been completely factored because $x^3 - 1$ is the difference between two cubes and is therefore factorable. Continuing, we factor $x^3 - 1$ into $(x - 1)(x^2 + x + 1)$. We now know that

$$ax^5 - ax^2 = ax^2(x^3 - 1)$$
$$= ax^2(x - 1)(x^2 + x + 1)$$

Example
4

In factoring $a^3x^6 + a^3x^3y^3$, we first note that a^3x^3 is a monomial common to both terms. Factoring it out, we are left with $x^3 + y^3$, which can be factored. We get

$$a^3x^6 + a^3x^3y^3 = a^3x^3(x^3 + y^3)$$
$$= a^3x^3(x + y)(x^2 - xy + y^2)$$

Example
5

In factoring $x^6 - y^6$, we could actually consider this expression to be the difference between two squares or two cubes. Although either approach will work, it is simpler to consider the given expression as the difference between two squares because that type of factoring is easier. We get

$$x^6 - y^6 = (x^3)^2 - (y^3)^2$$
$$= (x^3 + y^3)(x^3 - y^3)$$

Again, this expression has not been completely factored because $x^3 + y^3$ and $x^3 - y^3$ are both factorable. We get

$$x^6 - y^6 = (x^3)^2 - (y^3)^2$$
$$= (x^3 + y^3)(x^3 - y^3)$$
$$= (x + y)(x^2 - xy + y^2)(x - y)(x^2 + xy + y^2)$$

Example
6

The volume of material used to construct a steel bearing with a hollow core is given by the expression

$$\frac{4}{3}\pi x^3 - \frac{4}{3}\pi y^3$$

Factor this expression.

We should begin by factoring out the common monomial of $\frac{4}{3}\pi$. We get

$$\frac{4}{3}\pi x^3 - \frac{4}{3}\pi y^3 = \frac{4}{3}\pi(x^3 - y^3)$$

It is now necessary to factor only $x^3 - y^3$, which is the difference between two perfect cubes. We get

$$\frac{4}{3}\pi x^3 - \frac{4}{3}\pi y^3 = \frac{4}{3}\pi(x - y)(x^2 + xy + y^2)$$

In general, the procedures of this section closely parallel the procedures of Section 8.4, where we considered the difference between two squares. Whether we have the difference between two squares, the difference between two cubes, or the sum of two cubes, we must always begin by factoring out any common monomial factors. We must then recognize the remaining form as a special case that is factorable, and be able to properly execute the complete factoring.

8.5   *Exercises*

In Exercises 1 through 28, completely factor the given expressions.

1. $a^3 - 1$

2. $a^3 + 1$

3. $t^3 + 8$

4. $t^3 - 27$

5. $1 - x^3$

6. $8 + s^3$

7. $8x^3 + 27a^3$

8. $64s^3 - 27t^3$

9. $8x^6 - y^3$

10. $x^3 + 27y^9$

11. $a^3x^3 - y^6$

12. $a^3x^6 + b^6y^3$

13. $8x^4 + 8x$

14. $12xy - 12xy^4$

15. $ax^2 - ax^2y^3$

16. $4x^6 + 4$

17. $2kR_1^3 + 16kR_2^3$

18. $I_1^3R_1^3 + I_2^3R_2^3$

19. $54t_1^6 - 2t_2^3$

20. $3L_1^3 - 81L_2^{12}$

21. $125s^6 - 64t^9$

22. $250ax^4 - 128ax$

23. $a^3b^3 + c^{15}$

24. $a^9 + b^3c^{12}$

25. $1 - a^6x^6$

26. $x^{12} - y^6$

27. $1 - (x + y)^3$

28. $(x + y)^3 + 1$

In Exercises 29 through 34, solve the given problems.

29. In experimenting with different container sizes, a manufacturer reduces a cube from an edge of length x to an edge of length y. The total change in volume for N such containers is expressed as

$$D = Nx^3 - Ny^3$$

Factor the right-hand side of this equation.

30. Two different spheres are submerged in a fluid. The total volume displaced is expressed as

$$V = \frac{4}{3}\pi r_1^3 + \frac{4}{3}\pi r_2^3$$

Factor the right-hand side of this equation.

31. In programming a video game, an image is made to follow a path described by the equation

$$y = 8x^3 + 27$$

Factor the right-hand side of this equation.

32. A glass flask is filled with mercury and the temperature is then raised. The amount of mercury that will overflow is found from the expression

$$c_1^3 V(T_1 - T_2) - c_2^3 V(T_1 - T_2)$$

Factor this expression.

33. In a certain electric circuit, the current leaving a junction must be subtracted from the current entering that junction. Kirchhoff's rule states that the difference must be zero. If those currents change with time and the entering current (in amperes) is at^6 and the exiting current (in amperes) is at^3, factor the expression representing the difference.

34. A machine part requires that cubes of edge y are removed from each corner of a cube with edge x. Factor the expression that represents the remaining volume.

Chapter Summary

Key Terms

factoring

factored completely

prime

common monomial factor

greatest common factor

factoring by grouping

Key Concepts

• Factors are those quantities which, when multiplied together, give us the original product.

• An expression is said to be *factored completely* when it cannot be factored any further.

• *Guidelines for factoring by grouping*

1. Use parentheses to group terms into binomials with common factors.

2. Factor out the common factor in each binomial.

3. Factor out the common binomial. If no common binomial exists, try a different grouping.

• *General guidelines for factoring trinomials of the form* $ax^2 + bx + c$, $a = 1$

• Find two integers p and q such that $p + q = b$ and $pq = c$

1. Both integers must be positive if both b and c are positive.

2. Both integers must be negative is b is negative and c is positive.

3. One integer must be positive and one must be negative if c is negative.

Key Formulas

$a(x + y) = ax + ay$ distributive property

$(x + y)(x - y) = x^2 - y^2$ difference between two squares

$x^2 + bx + c = (x + p)(x + q)$ factoring $ax^2 + bx + c$, $a = 1$

$ax^2 + bx + c = (rx + p)(sx + q)$ factoring a general trinomial

$x^3 + y^3 = (x + y)(x^2 - xy + y^2)$ factoring the sum of cubes

$x^3 - y^3 = (x - y)(x^2 + xy + y^2)$ factoring the difference of cubes

Review Exercises

In Exercises 1 through 62, factor the given expression if possible.

1. $5a - 5c$

2. $4r + 8s$

3. $3a^2 + 6a$

4. $6t^3 - 8t^2$

5. $12a^2b + 4ab$

6. $15t^3y - 10ty^2$

7. $8stu^2 - 24s^3tu^2$

8. $16x^2yz^4 - 4xyz^4$

9. $4x^2 - y^2$

10. $p^2 - 9u^2v^2$

11. $16y^4 - x^2$

12. $r^2s^2t^2 - 4x^2$

13. $x^2 + 2x + 1$

14. $x^2 - 2x - 3$

15. $x^2 - 7x + 6$

16. $x^2 + 2x - 63$

17. $ax^2 + 3a^2x - a^3$

18. $18r^2t - 9r^3t^2 - 6r^2t^3$

19. $2nm^3 - 4n^2m^2 + 6n^3m$

20. $8y^2 + 24y^3z - 32y^2z^4$

21. $4p^3t^2 - 12t^4 - 4t^2 + 4at^2$

22. $22r^2s^2t^2 - 121rst^2 - 22r^3st^2 + 33r^4s^2t^2$

23. $2xy^3 - 14x^2y^3 + 16xy^4 - 6x^3y^5$

24. $3st^2 - 6s^3t^3u - 12st^2u + 9st^3$

25. $(4rs)^2 - 9y^2$ **26.** $49r^4t^4 - y^6$

27. $36w^2x^2 + y^4$ **28.** $(a + b)^2 - c^2$

29. $2x^2 + 9x + 7$ **30.** $3y^2 + y - 10$

31. $5s^2 - 3s - 2$ **32.** $3a^2 + 20a - 7$

33. $14t^2 - 19t - 3$ **34.** $5x^2 + x - 4$

35. $9x^2 + 6x + 1$ **36.** $8r^2 + 2r - 15$

37. $x^2 + 3xy + 2y^2$ **38.** $6a^2 - 17ab + 12b^2$

39. $10c^2 + 23cd - 5d^2$ **40.** $4p^2 - 12pq + 9q^2$

41. $88x^2 - 19x - 84$ **42.** $16y^2 + 56y + 49$

43. $2x^2 - 18y^2$ **44.** $4r^2t^2 - 36p^2q^2$

45. $8y^4x^6 - 32y^2x^4$ **46.** $3m^5n - 27mn^3$

47. $3ax^2 + 3ax - 36a$ **48.** $36c^2x - 34cx - 30x$

49. $54r^3 - 45r^2s - 156rs^2$ **50.** $8c^2x^2 + 52c^2x + 72c^2$

51. $48y^3 - 64y^4 + 16y^5$ **52.** $18a^2u^2 + 23a^2u - 6a^2$

53. $5x^4 - 125$ **54.** $4a^8 - 64$

55. $16x^4 - 1$ **56.** $x^8 - 1$

57. $x^3 + 27$ **58.** $t^3 - 64$

59. $8x^3 + 1$ **60.** $8x^6 - 8y^3$

61. $ax^4y - axy^4$ **62.** $ab^4 + a^4b$

In Exercises 63 through 70, factor the indicated expressions.

63. The total voltage from three circuits is given as

$$iR_1 + iR_2 + iR_3$$

Factor this expression.

64. An industrial safety specialist analyzes the upward motion of a projectile ejected from a machine. The distance s (in feet) above the machine and the time t (in seconds) are related by the formula

$$s = 64t - 16t^2$$

Factor the right-hand side of this equation.

65. In designing a gear for an industrial robot, an engineer determines the number of teeth by the expression

$$PN + 2P$$

Factor this expression.

66. In determining the flow velocity of a viscous fluid, we use the formula

$$v = k(R^2 - r^2)$$

Factor the right-hand side of this equation.

67. When determining the velocity of a fluid flowing through a pipe, we may encounter the expression

$$kD^2 - 4kr^2$$

Factor this expression.

68. When studying the pressure P and the volume V of a certain gas, we find the expression

$$CP(V_2 - V_1) + PR(V_2 - V_1)$$

where C and R are constants. Factor this expression.

69. In the theory of magnetism, the expression

$$b(x^2 + y^2) - 2by^2$$

may be used. Factor this expression.

70. At an altitude of h ft above sea level, the boiling point of water is lower by a certain number of degrees than the boiling point at sea level, which is 212°F. The difference is given by the approximate equation

$$T^2 + 520T - h = 0$$

Assuming that $h = 5300$ ft, factor the left-hand side of this equation.

Chapter Test

Factor each of the following expressions completely.

1. $6x - 24y$

2. $12xy + 3xy^2 + 6x^2y^2$

3. $xy - x + 3y - 3$

4. $x^2 + 11x + 24$

5. $7x^2 - 9x + 2$

6. $4x^2 - 25$

7. $8x^3 - 64y^3$

8. $x^4 + 27x$

9. In order to find two resistors that will give an equivalent resistance of 400 Ω when wired in series and 75 Ω when wired in parallel, we must be able to solve the equation $R^2 - 400R + 30,000 = 0$. Factor the left-hand side of this equation.

10. An object is tossed into the air with an initial velocity of 32 ft/sec from a height of 6 feet above the ground. The following expression represents the height of the object over the time it is in the air: $16x^2 - 112x + 96$. Factor this expression.

9 Algebraic Fractions

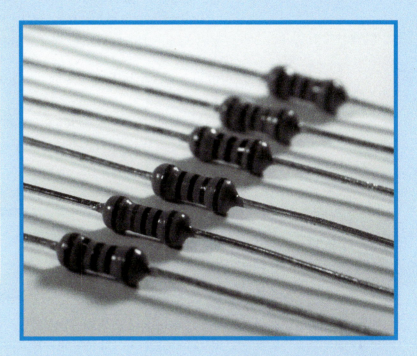

In electronics, a resistor is a two-terminal component that resists electrical current. It is used to regulate both the voltage and the current in a circuit. Simply put, a resistor resists the flow of electricity.

Resistors can be connected in one of three ways: in series, in parallel, or in some combination of both series and parallel. Resistors that are connected in series are connected in a line. The same current flows through all of them, one resistor after another. The total resistance of the circuit is found by simply adding up the resistance values of the individual resistors. Resistors that are connected in parallel are connected so that they branch out from a single point, called a node, and meet again somewhere else in the circuit. The current in a parallel circuit breaks up, with some flowing along each parallel branch and recombining when the branches meet again.

Therefore, the same current does not flow through all of the resistors. The total resistance of a set of resistors in parallel is found by adding up the reciprocals of the resistance values, and then taking the reciprocal of the following total, that is, $\dfrac{1}{R_1} + \dfrac{1}{R_2} + \dfrac{1}{R_3} + \cdots$.

In the diagram in Figure 9.1 we have combined a resistor in series with a parallel circuit. This is known as a **combination circuit.** The total resistance R for this circuit can be found using the formula $\dfrac{R_1 R_2}{R_1 + R_2} + R_3$.

Algebraic fractions are important in many areas of mathematics, science, and technology. We must be able to work with these algebraic fractions effectively.

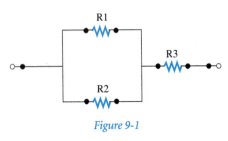

Figure 9-1

9.1 Equivalent Algebraic Fractions

Section Objectives
- Simplify or reduce rational expressions

Because algebraic expressions are representations of numbers, the basic operations on fractions from arithmetic form the basis of the algebraic operations. This means that the same basic rules of Chapter 1 will also apply to algebraic fractions.

One very important principle underlies the operations with fractions. It is the **fundamental principle of fractions:**

> *If the numerator and denominator of any fraction are multiplied or divided by the same quantity, not zero, the value of the fraction will remain unchanged.*

This principle can be expressed in symbols as

$$\frac{a}{b} = \frac{ac}{bc}$$

where a, b, and c can be numbers or algebraic expressions as long as c is not zero and b is not zero.

Caution	
$\dfrac{a}{b}$ ←	numerator
←	denominator

The use of the fundamental principle of fractions on any given fraction will yield an **equivalent fraction.** Note that this fundamental principle of fractions refers only to *multiplication* or *division of both* the numerator and denominator.

Example 1

From our work in Chapter 1, we know that if we multiply the numerator and denominator of the fraction $\frac{3}{4}$ by 2, we obtain the equivalent fraction $\frac{6}{8}$. If we divide the numerator and denominator of the fraction $\frac{18}{24}$ by 6, we obtain the equivalent fraction $\frac{3}{4}$. Therefore, $\frac{3}{4}$, $\frac{6}{8}$, and $\frac{18}{24}$ are equivalent. We can express this equivalence as follows:

$$\frac{3}{4} = \frac{6}{8} = \frac{18}{24}$$

Example 1 illustrates the use of the basic arithmetic operations on a fraction with whole numbers. Example 2 shows how those basic operations can be extended to algebraic fractions.

Example 2

If we multiply the numerator and denominator of the algebraic fraction

$$\frac{3x^2}{5x^3}$$

by x^2, we obtain the equivalent fraction

$$\frac{3x^4}{5x^5}$$

This can be seen by

$$\frac{(3x^2)(x^2)}{(5x^3)(x^2)} = \frac{3x^4}{5x^5}$$

If we divide the numerator and denominator of the fraction

$$\frac{3x^2}{5x^3}$$

by x^2, we obtain the equivalent fraction

$$\frac{3}{5x}$$

This can be shown as follows.

$$\frac{\dfrac{3x^2}{x^2}}{\dfrac{5x^3}{x^2}} = \frac{3}{5x}$$

We can now conclude that the algebraic fractions

$$\frac{3x^2}{5x^3}, \quad \frac{3x^4}{5x^5}, \quad \text{and} \quad \frac{3}{5x}$$

are equivalent, and we can write

$$\frac{3x^2}{5x^3} = \frac{3x^4}{5x^5} = \frac{3}{5x}$$

With an algebraic fraction, as with a fraction in arithmetic, *one of the most important operations is* **reducing** *it to its* **lowest terms,** *or* **simplest form.** As in the case of an arithmetic fraction, *we remove the factors that are common to both the numerator and denominator by dividing both the numerator and the denominator by the common factors.* By finding the greatest common factor of the numerator and denominator, and dividing each by this factor, we obtain the simplest form, and we say that the fraction has been reduced to its lowest terms.

Example 3

In simplifying the fraction $\frac{21}{28}$, we note that both the numerator and denominator have a factor of 7. Therefore, dividing each by 7, we obtain $\frac{3}{4}$, the simplest form of the fraction $\frac{21}{28}$.

In order to simplify the fraction

$$\frac{15x^2y^3z}{20xy^4z}$$

we note that the greatest common factor of the numerator and denominator is $5xy^3z$. Dividing both the numerator and denominator by this greatest common factor, we get

$$\frac{(15x^2y^3z) \div (5xy^3z)}{(20xy^4z) \div (5xy^3z)} = \frac{3x}{4y}$$

Example 4

When we are simplifying the fraction

$$\frac{x^2 - x - 2}{x^2 + 3x + 2}$$

▶ **we must first factor both the numerator and denominator** in order to determine whether there is a common factor. We obtain

$$\frac{x^2 - x - 2}{x^2 + 3x + 2} = \frac{(x - 2)(x + 1)}{(x + 2)(x + 1)} = \frac{x - 2}{x + 2}$$

We can now see that the greatest common factor is $(x + 1)$. We divide both the numerator and denominator by this factor so that the indicated result is obtained.

Example 5

When simplifying the fraction

$$\frac{2x^2 + 7x - 4}{4x^2 - 1}$$

we *first factor the numerator and denominator* to obtain

$$\frac{2x^2 + 7x - 4}{4x^2 - 1} = \frac{(2x - 1)(x + 4)}{(2x - 1)(2x + 1)}$$

Now we note the common factor of $2x - 1$, which is to be divided out (cancelled). Proceeding, we get

$$\frac{2x^2 + 7x - 4}{4x^2 - 1} = \frac{\overset{1}{\cancel{(2x - 1)}}(x + 4)}{\underset{1}{\cancel{(2x - 1)}}(2x + 1)} = \frac{x + 4}{2x + 1}$$

Special note: *Remember that in simplifying fractions we must* **divide** *both the numerator and denominator by the common* **factor.** *As in arithmetic, this process is sometimes called* **cancellation.** Many students are tempted to remove **terms** that appear in both the numerator and denominator. This is an **incorrect** application of cancellation. Examples 6–9 illustrate common errors in the simplification of fractions.

Example 6

In attempting to simplify the expression

$$\frac{x^2 + 4}{x^2} \qquad \text{— a term, but not a factor, of numerator}$$

a common error is to "cancel" the x^2 in both the numerator and denominator to get a result of 4, but the given fraction is not equivalent to 4. *The original fraction cannot be further simplified* because the numerator and denominator have no common *factor* that can be canceled. *The x^2 in the numerator is not a common factor of the entire numerator.*

Example 7

When simplifying the expression

$$\frac{x^2 - x - 2}{x^2 + 3x + 2}$$ terms, not factors, of numerator

terms, not factors, of denominator

many students would "cancel" the terms x^2 and 2, because they appear in both the numerator and denominator. **This is incorrect.** Removing them in this way is equivalent to *subtracting $x^2 + 2$ from the numerator and from the denominator*, which violates the fundamental principle of fractions. The proper reduction of this fraction is shown in Example 4.

When simplifying the expression

factor of $2x$, not of numerator

term of numerator

$$\frac{2x + 3}{2(x - 1)(x + 3)}$$

term of $x + 3$

many students would "cancel" the 3's and 2's. Again, this is *incorrect*. Actually, *this expression cannot be further simplified,* because there is no common *factor* in the numerator and the denominator. **The 2 in the numerator is not a factor of the entire numerator, and the 3 is not a factor of either the numerator or the denominator.**

There are cases, especially in the addition and subtraction of fractions, when a fraction in its simplest form must be changed to an *equivalent* form. This is accomplished by the *multiplication* of both the numerator and denominator by the same quantity.

When simplifying fractions, we often find that the numerator and denominator have factors that differ only in *sign*. We can show that

$$(a - b) = -(b - a)$$

by noting that $-(b - a) = -b + a = a - b$. This equation shows us that we *can change the signs of all the terms of a factor so long as we also introduce a factor of -1.* Example 8 illustrates the simplification of fractions where a change of signs is necessary.

Example 8

We can simplify the fraction

$$\frac{x - 3}{3 - x}$$

▶ **by expressing the denominator as $-(x - 3)$.** After rewriting the denominator, we divide out the common factor of $x - 3$. This leads to

$$\frac{x - 3}{3 - x} = \frac{x - 3}{-(x - 3)} = \frac{1}{-1} = -1$$

Example
9

Simplify the following fraction, which represents a mathematical model of the price-to-earnings ratio for a certain stock issue.

$$\frac{4 - x^2}{x^2 - 6x + 8}$$

The fraction

$$\frac{4 - x^2}{x^2 - 6x + 8} = \frac{(2 - x)(2 + x)}{(x - 2)(x - 4)}$$

has factors in the numerator and denominator that differ only in sign. Changing $(2 - x)$ to $-(x - 2)$, we get

▶ $$\frac{(2 - x)(2 + x)}{(x - 2)(x - 4)} = \frac{-(x - 2)(2 + x)}{(x - 2)(x - 4)} = \frac{-(2 + x)}{x - 4}$$

This result can also be expressed as

$$\frac{2 + x}{4 - x}$$

by replacing $x - 4$ by $-(4 - x)$.
   Acceptable forms of the simplest result are

$$\frac{-(2 + x)}{x - 4} = -\frac{2 + x}{x - 4} = \frac{2 + x}{-(x - 4)} = \frac{2 + x}{4 - x}$$

Changing signs of factors of fractions often causes students difficulty. This need not be the case if the equation $(a - b) = -(b - a)$ is understood and used correctly.

9.1 Exercises

In Exercises 1 through 12, divide the numerator and denominator of each given fraction by the given factor and obtain an equivalent fraction.

1. $\dfrac{16}{28}$; 4

2. $\dfrac{27}{39}$; 3

3. $\dfrac{6a^2x}{9a^3x}$; $3ax$

4. $\dfrac{14r^2st}{28rst^2}$; $7rt$

5. $\dfrac{2(x - 1)}{(x - 1)(x + 1)}$; $x - 1$

6. $\dfrac{(x + 5)(x - 3)}{3(x + 5)}$; $x + 5$

7. $\dfrac{2x^2 + 5x - 3}{x^2 + 4x + 3}$; $\ x + 3$ **8.** $\dfrac{6x^2 - 11x + 3}{2x^2 - 3x}$; $\ 2x - 3$

9. $\dfrac{3x^2 - 12}{x^2 + 4x + 4}$; $\ x + 2$ **10.** $\dfrac{x^2 - 6x + 9}{2x^2 - 2x - 12}$; $\ x - 3$

11. $\dfrac{4 - x^2}{x^2 - 5x + 6}$; $\ x - 2$ **12.** $\dfrac{2x^2 - 12x + 18}{9 - x^2}$; $\ x - 3$

In Exercises 13 through 36, determine the simplest form of each given fraction.

13. $\dfrac{2x}{6x}$ **14.** $\dfrac{5a}{30a}$

15. $\dfrac{5a^2b}{20a}$ **16.** $\dfrac{12xyz^2}{21xyz^3}$

17. $\dfrac{16a^3}{18a^3}$ **18.** $\dfrac{12x^6}{15x^2}$

19. $\dfrac{(2x - 1)(x + 4)}{(x + 4)(x - 2)}$ **20.** $\dfrac{(5x - 3)(2x - 1)}{(2x + 1)(5x - 3)}$

21. $\dfrac{(x - 1)(x + 1)(x + 2)}{2(x - 1)(x + 3)}$

22. $\dfrac{3(x + 5)(2x - 1)(3x - 2)}{9(5x + 1)(3x - 2)(2x - 1)}$

23. $\dfrac{x^2 - 1}{x^2 - 2x + 1}$ **24.** $\dfrac{x^2 + 5x + 4}{x^2 + 3x - 4}$

25. $\dfrac{3x^2 - x}{3x^2 + 5x - 2}$ **26.** $\dfrac{4x^2 + 9x - 9}{4x^2 - 8x}$

27. $\dfrac{6x^2 - 19x + 10}{8x^2 - 14x - 15}$ **28.** $\dfrac{4x^2 + 10x + 6}{12x^2 + 8x - 4}$

29. $\dfrac{x^2 - 9y^2}{3xy - 9y^2}$ **30.** $\dfrac{5a^2 + 39ab - 8b^2}{5a^2 + 4ab - b^2}$

31. $\dfrac{20 - 9x + x^2}{8 + 2x - x^2}$ **32.** $\dfrac{4ab - b^2}{b^2 - 16a^2}$

33. $\dfrac{3x - 9x^2}{3x - 1}$ **34.** $\dfrac{5a^2 - 10a}{4 - a^2}$

35. $\dfrac{x^2 - 5x + 6}{9 - x^2}$ **36.** $\dfrac{x^2 - 6x + 5}{20 - 4x}$

In Exercises 37 and 38, solve the given problems.

37. Evaluate the fraction

$$\dfrac{4a - 8}{4(a + 2)}$$

for the value $a = 5$ before and after reducing it to its simplest form.

38. Evaluate the fraction

$$\dfrac{2x^2 + 5xy - 3y^2}{2x^2 + 3xy - 2y^2}$$

for the values $x = 2$ and $y = 3$ before and after reducing it to its simplest form.

Now Try It! Answers

1. $\dfrac{3x}{4}$ **2.** $\dfrac{3a - 2b}{2a - b}$ **3.** $3x$ **4.** $\dfrac{x - 4}{x + 4}$ **5.** $\dfrac{s - 5}{2s - 1}$

9.2   **Multiplication and Division of Algebraic Fractions**

Section Objectives
• Multiply and divide algebraic fractions

Recall from Chapter 1 that *the product of two fractions is a fraction whose numerator is the product of the numerators and whose denominator is the product of the denominators of the given fractions.* Because algebra is a generalization of arithmetic, the same definition holds in algebra. Symbolically, we write this as

$$\frac{a}{b} \times \frac{c}{d} = \frac{ac}{bd} \quad (b \text{ and } d \text{ not zero})$$

Example 1

$$\frac{5}{6} \times \frac{7a^2}{9x} = \frac{(5)(7a^2)}{(6)(9x)} = \frac{35a^2}{54x}$$

$$\frac{3a}{b} \times \frac{x^2}{5b} = \frac{(3a)(x^2)}{(b)(5b)} = \frac{3ax^2}{5b^2}$$

$$\frac{2x}{y} \times \frac{x^2}{y^3} = \frac{2x^3}{y^4}$$

Closely associated with the multiplication of fractions is the **power** of a fraction. To find $\left(\dfrac{a}{b}\right)^n$ we would have n factors of $\left(\dfrac{a}{b}\right)$, which would result in a numerator of a^n and a denominator of b^n. Therefore,

$$\left(\frac{a}{b}\right)^n = \frac{a^n}{b^n} \quad (b \text{ not zero})$$

Example 2

$$\left(\frac{2}{3}\right)^3 = \frac{2^3}{3^3} = \frac{8}{27}$$

$$\left(\frac{2a^2}{x}\right)^4 = \frac{(2a^2)^4}{x^4} = \frac{(2^4)(a^2)^4}{x^4} = \frac{16a^8}{x^4}$$

To divide one fraction by another, *we invert the divisor (the fraction in the denominator) and multiply the dividend (the fraction in the numerator) by the inverted fraction.* If we want to divide $\dfrac{a}{b}$ by $\dfrac{c}{d}$, we multiply the numerator and denominator by $\dfrac{d}{c}$. This can be written as

$$\frac{\dfrac{a}{b}}{\dfrac{c}{d}} = \frac{\dfrac{a}{b} \times \dfrac{d}{c}}{\dfrac{c}{d} \times \dfrac{d}{c}} = \frac{\dfrac{ad}{bc}}{\dfrac{cd}{dc}} = \frac{\dfrac{ad}{bc}}{1} = \frac{ad}{bc}$$

Showing this division of fractions symbolically, we get

$$\frac{\dfrac{a}{b}}{\dfrac{c}{d}} = \frac{a}{b} \div \frac{c}{d} = \frac{a}{b} \times \frac{d}{c} = \frac{ad}{bc} \qquad \text{where } b, c, \text{ and } d \text{ are not zero}$$

Example 3

multiply

$$\frac{2x}{5} \div \frac{3}{7c} = \frac{2x}{5} \times \boxed{\frac{7c}{3}} = \frac{(2x)(7c)}{(5)(3)} = \frac{14cx}{15}$$

invert divisor

$$\frac{\dfrac{4xy}{3b}}{\dfrac{7b}{xyz}} = \frac{4xy}{3b} \times \frac{xyz}{7b} = \frac{(4xy)(xyz)}{(3b)(7b)} = \frac{4x^2y^2z}{21b^2}$$

Another descriptive word that we met in connection with the division of fractions is **reciprocal.** Recalling that *the reciprocal of a number is 1 divided by that number,* we see that the reciprocal of a is $1/a$ (a not zero). Also, because

$$\frac{1}{\dfrac{a}{b}} = 1 \times \frac{b}{a} = \frac{b}{a}$$

we recall that the reciprocal of the fraction a/b is the inverted fraction b/a.

Example 4

1. The reciprocal of $2a$ is $\dfrac{1}{2a}$.

2. The reciprocal of $\dfrac{1}{b}$ is $\dfrac{b}{1}$, or b.

3. The reciprocal of $-\dfrac{xy}{a}$ is $-\dfrac{a}{xy}$. When we are finding the reciprocal of a fraction, the sign of the fraction is unaltered.

One immediate use of the term "reciprocal" is in an alternative definition of the division of one fraction by another. We can state that *when we divide one fraction by another we multiple the dividend by the reciprocal of the divisor.*

As we stated in Section 9.1, when working with algebraic fractions, results are usually expressed in simplified form. Remember that in simplifying a fraction we must divide the numerator and denominator by factors that are common to both. Generally it is easier to *factor the numerators and denominators of the fractions being multiplied or divided before performing the indicated operation.* If this is not done, it is very possible that the result obtained by multiplying out the fractions will be a fraction containing expressions that are extremely difficult to factor. This, in turn, will make the simplification difficult.

In general, when multiplying algebraic fractions, we should:

1. *Indicate* the product of both the numerators and denominators in factored form.

2. Divide out any common factors.

3. Leave the result in factored form.

Example 5–8 illustrate the multiplication of algebraic fractions.

Example 5

Multiply

$$\frac{8x^2y^3}{21abc} \quad \text{by} \quad \frac{15ax^2}{16b^2y^2}$$

$$\frac{8x^2y^3}{21abc} \times \frac{15ax^2}{16b^2y^2} = \frac{(8)(15)ax^4y^3}{(21)(16)ab^3cy^2}$$

Now we see that the greatest common factor of the numerator and denominator is $(8)(3)ay^2$. Dividing the numerator and denominator by this factor, we have $5x^4y/14b^3c$. Therefore,

$$\frac{8x^2y^3}{21abc} \times \frac{15ax^2}{16b^2y^2} = \frac{5x^4y}{14b^3c}$$

Example 6

The voltage (in volts) in an electric circuit is found by multiplying the current and the resistance. Under certain conditions, the voltage in a circuit is found by performing the multiplication shown below.

$$\frac{x^2 - y^2}{x + 2y} \times \frac{3x + 6y}{x - y}$$

If we perform the multiplication directly, by multiplying numerators together and denominators together, we obtain

$$\frac{3x^3 + 6x^2y - 3xy^2 - 6y^3}{x^2 + xy - 2y^2}$$

Although the numerator is factorable, it is not of a simple form for factoring. The numerator is not factorable by methods we have developed. However, if we only *indicate the multiplication* (not actually perform it) as

$$\frac{x^2 - y^2}{x + 2y} \times \frac{3x + 6y}{x - y} = \frac{(x^2 - y^2)(3x + 6y)}{(x + 2y)(x - y)}$$

the numerator in this result is already partially factored and the denominator is completely factored. The additional work now required is considerably less than that required after direct multiplication. Completing the factoring of the numerator, we note that $x^2 - y^2 = (x + y)(x - y)$ and $3x + 6y = 3(x + 2y)$, and we get

$$\frac{x^2 - y^2}{x + 2y} \times \frac{3x + 6y}{x - y} = \frac{(x^2 - y^2)(3x + 6y)}{(x + 2y)(x - y)}$$

$$= \frac{3(x - y)(x + y)(x + 2y)}{(x + 2y)(x - y)} \quad \text{factors of } 3x + 6y$$

Here we see that the greatest common factor of the numerator and denominator is $(x - y)(x + 2y)$. Dividing both the numerator and denominator by this greatest common factor, we get the final result of

$$\frac{3(x + y)}{1} \quad \text{or} \quad 3(x + y)$$

Now Try It!

Perform the indicated operations.

1. $\dfrac{9x^2}{4x} \times \dfrac{4x + 4}{81}$

2. $\dfrac{x + 2}{x + 5} \times \dfrac{x^2 - 5x - 6}{x^2 + 3x + 2}$

3. $\dfrac{4 - 8x}{x + 1} \div \dfrac{1 - 2x}{x^2 - 1}$

4. $\dfrac{x + 2}{x^2 - 4x + 4} \div \dfrac{x^2 + x}{x - 2}$

Example 7

Divide

$$\frac{7x^2y}{3ad} \div \frac{2xy^2}{3a^2d}$$

We first invert the divisor and change the operation to indicate multiplication.

$$\frac{7x^2y}{3ad} \times \frac{3a^2d}{2xy^2} = \frac{(7)(3)a^2dx^2y}{(3)(2)adxy^2}$$

Now we see that the greatest common factor of the numerator and the denominator is $3adxy$. Divide the numerator and the denominator by this common factor and we have

$$\frac{7x^2y}{3ad} \times \frac{3a^2d}{2xy^2} = \frac{7ax}{2y}$$

$\widehat{Example}$
8

Perform the division

$$\frac{b^2 - 2b + 1}{9b^2 - 1} \div \frac{5b - 5}{12b - 4}$$

We first indicate the product of the dividend and the reciprocal of the divisor. We then factor both the numerator and denominator. Next we divide out the greatest common factor. These steps lead to

$$\frac{b^2 - 2b + 1}{9b^2 - 1} \div \frac{5b - 5}{12b - 4} = \frac{(b^2 - 2b + 1)(12b - 4)}{(9b^2 - 1)(5b - 5)} \quad \text{indicated multiplication}$$

> **Important**
>
> Do not cancel common factors until after you invert the divisor and change the operation to multiplication.

$$= \frac{(b - 1)(b - 1)(4)(3b - 1)}{(3b + 1)(3b - 1)(5)(b - 1)} \quad \text{factor and show cancellation}$$

$$= \frac{4(b - 1)}{5(3b + 1)} \quad \text{simplify}$$

The greatest common factor that was divided out was $(b - 1)(3b - 1)$. The result has been left in factored form. Although it would be correct to multiply out each of the numerators and denominators, the factored form is often more convenient and usually the preferred form of the result.

$\widehat{Example}$
9

Divide

$$\frac{a^2 - a - 2}{5a - 1} \div \frac{a - 2}{10a^2 + 13a - 3}$$

We first invert the divisor and change the operation to indicate multiplication.

$$\frac{a^2 - a - 2}{5a - 1} \times \frac{10a^2 + 13a - 3}{a - 2}$$

We now factor the numerator and denominator before canceling out common factors

$$\frac{(a - 2)(a + 1)}{5a - 1} \times \frac{(5a - 1)(2a + 3)}{a - 2} = (a + 1)(2a + 3)$$

$$= 2a^2 + 5a + 3$$

9.2 Exercises

In Exercises 1 through 8, find the reciprocal of each expression.

1. $8n;\quad \dfrac{1}{13s}$

2. $a^2b;\quad \dfrac{a^2}{b}$

3. $\dfrac{a}{3b};\quad \dfrac{3b}{a}$

4. $-\dfrac{2x^2}{y};\quad -\dfrac{5cd}{3ax}$

5. $\dfrac{x+y}{x-y};\quad \dfrac{x^2+y^2}{x^2}$

6. $-\dfrac{s^2t^2}{s-t};\quad \dfrac{x^2-9}{16-y^2}$

7. $\dfrac{a}{a+b};\quad \dfrac{-IR}{V}$

8. $\dfrac{4\pi r^2}{3};\quad \dfrac{2L+2H}{W}$

In Exercises 9 through 42, perform the indicated multiplications and divisions, expressing all answers in simplest form.

9. $\dfrac{12}{18}\times\dfrac{6}{9t}$

10. $\dfrac{15R}{20}\times\dfrac{40}{45R}$

11. $\dfrac{2}{9}\times\dfrac{3a}{5}$

12. $\dfrac{6x}{13}\times\dfrac{7}{a}$

13. $\dfrac{17rs}{12t}\times\dfrac{3t^2}{51s}$

14. $\dfrac{24xy^2}{5z}\times\dfrac{125z^2}{8y^3}$

15. $\left(\dfrac{3}{4}\right)^4$

16. $\left(\dfrac{7}{5}\right)^3$

17. $\left(\dfrac{a^2}{2x}\right)^5$

18. $\left(\dfrac{4xy^4}{8z^2}\right)^5$

19. $\left(\dfrac{ax^3}{b^2}\right)^3$

20. $\left(\dfrac{3x^2y^3}{6x^3}\right)^6$

21. $\dfrac{2}{5x}\div\dfrac{7c}{13}$

22. $\dfrac{6a}{17}\div\dfrac{5}{11c}$

23. $\dfrac{6x}{17}\div\dfrac{7}{68m}$

24. $\dfrac{15}{8z}\div\dfrac{25}{18y}$

25. $\dfrac{3x}{25y}\div\dfrac{27x^2}{5y^2}$

26. $\dfrac{3a^2x}{4ay^2}\div\dfrac{6ax}{5a^3x^2}$

27. $\dfrac{9a^2b^2}{10ab^3}\div\dfrac{72a^3b^4}{40}$

28. $\dfrac{4x}{3x^2y^2}\div\dfrac{20xy}{9y^3}$

29. $\dfrac{a-5b}{a+b}\times\dfrac{a+3b}{a-5b}$

30. $\dfrac{5n+10}{3n-9}\times\dfrac{n-3}{15}$

31. $\dfrac{x^2-y^2}{14x}\times\dfrac{35x^2}{3x+3y}$

32. $\dfrac{a+b}{4a}\times\dfrac{3a^2}{(a+b)^2}$

33. $\dfrac{x}{x-1}\times\dfrac{x^2-1}{x^2+2x}$

34. $\dfrac{a+2}{3}\times\dfrac{6a-3}{a^2+2a}$

35. $\dfrac{x^2+2x-3}{x^2-4}\times\dfrac{x^2-x-6}{x^2-5x+4}$

36. $\dfrac{3x^2+10x-8}{36x^2-16}\times\dfrac{9x^2+15x+6}{x^2+3x-4}$

37. $\dfrac{2b+3}{5}\div\dfrac{4b+6}{5b-2}$

38. $\dfrac{6}{3x+4}\div\dfrac{2x+2}{9x^2+12x}$

39. $\dfrac{3a^2-3b^2}{a^2-4b^2}\div\dfrac{a^2+2ab+b^2}{a+2b}$

40. $\dfrac{20x}{x^2-8x+15}\div\dfrac{10x}{x^2-2x-15}$

41. $\dfrac{s^2-5s-14}{s^2-9s-36}\div\dfrac{s^2+4s-77}{s^2+10s+21}$

42. $\dfrac{p^2+5pq+6q^2}{2p^2+pq-q^2}\div\dfrac{p^2-9q^2}{(p+q)^2}$

In Exercises 43 through 46, solve the given problems.

43. In attempting to compute the minimum cost of a parts container, the product of the factors 3.2, x, and $x+2$ must be divided by x^2. Perform the indicated operations and express the answer in its simplest form.

44. Under certain conditions, the product of the price and the demand for a certain commodity is constant. If the price of each of x units is

$$\dfrac{2x+6}{3}$$

and the demand for these is

$$\dfrac{6000}{x^2+3x}$$

determine the expression for the product of the price and demand.

45. The optical reflection coefficient R in a semiconductor is given by

$$R = \left(\frac{1-n}{1+n}\right)^2$$

Perform the operation indicated on the right-hand side of this equation and express the result in its simplest form.

46. In a certain electric circuit, the voltage and resistance vary with time according to the equations

$$V = \frac{t^2 - 1}{6t^2 + 3t}; \qquad R = \frac{t-1}{2t^2}$$

The current is found by dividing the voltage by the resistance. Find a formula for the current I in terms of t. Simplify your results.

In the development of the theory of relativity, we can find the expression

$$\left[\frac{\dfrac{c^2 - v^2}{c^2(c^2 - v^2)^2 - (c^2v^2)^2v^2}}{(c^2 + v^2)^2}\right]\left(\frac{c^2 - v^2}{c^2 + v^2}\right)^2$$

Show that this expression equals 1.

Now Try It! Answers

1. $\dfrac{x(x+1)}{9}$ **2.** $\dfrac{x-6}{x+6}$ **3.** $4(x-1)$ **4.** $\dfrac{1}{x(x-2)}$

9.3 The Lowest Common Denominator

Section Objectives
• Correctly determine the lowest common denominator

When adding fractions that have the same denominator, we add the numerators and keep the same denominator. We can express this in symbolic form as

$$\frac{a}{b} + \frac{c}{b} = \frac{a+c}{b} \qquad \text{where } b \neq 0$$

Example 1

$$\frac{3a}{x^2} + \frac{4a}{x^2} = \frac{3a + 4a}{x^2} = \frac{7a}{x^2}$$

We know that finding the sum of a set of fractions having the same denominator is a relatively easy task. However, if the fractions do not all have the same denominator, we also know that we must first change each to an equivalent fraction such that each resulting fraction has the same denominator. Because this step is of utmost importance in the addition and subtraction of fractions, and because it is also the step that causes the most difficulty with algebraic fractions, we shall devote this entire section to finding the common denominator of a given set of fractions. In Section 9.4, we discuss the complete method for the addition and subtraction of algebraic fractions.

From our previous work with fractions, we know that the most convenient and useful denominator for a set of fractions is the **lowest common denominator.** We know *that this is the denominator that contains the smallest number of prime factors of the given denominators and is exactly divisible by each denominator. For algebraic fractions, it is the simplest expression into which all given denominators will divide evenly.*

Example 2

When we are finding the sum

$$\frac{3}{8a} + \frac{5}{16a}$$

we note that both $8a$ and $16a$ will divide evenly into $16a$, which is the lowest common denominator. If we use $16a^2$ as the common denominator, we do not have the simplest possible expression for the common denominator.

With algebraic fractions, to determine the lowest common denominator of a set of fractions *we first find all the prime factors of each denominator. We then form the product of these prime factors, giving each factor the largest exponent it has in any of the given denominators.*

> *This might help . . .*
>
> If denominators have no common factors, the lowest common denominator is the product of all of the denominators.
>
> If denominators have common factors, list all of the different factors raised to the highest power that there is for each factor. The lowest common denominator is the product of the listed factors.

The process for finding the lowest common denominator can be summarized by the steps:

1. For each denominator, list all of the prime factors.

2. Identify each of the different factors that appear in the denominators.

3. ***Raise each factor to the largest exponent it has in any of the denominators.***

4. The lowest common denominator (l.c.d.) is the product of all of the results from step 3.

Example 3

1. When we are finding the lowest common denominator of $\frac{5}{24}$ and $\frac{7}{36}$, we first factor 24 and 36 into their prime factors. This gives

$$24 = 2 \times 2 \times 2 \times 3 = 2^3 \times 3 \quad \text{and} \quad 36 = 2 \times 2 \times 3 \times 3 = 2^2 3^2$$

The prime factors to be considered are 2 and 3. The largest exponent to which 2 appears is 3. The largest exponent to which 3 appears is 2. Therefore the lowest common denominator is $2^3 3^2 = 72$. No number smaller than 72 is divisible evenly by 24 and 36.

2. When we are finding the lowest common denominator of

$$\frac{1}{4x^2 y} \quad \text{and} \quad \frac{7a}{6xy^4}$$

we factor the denominators into their prime factors. This gives

largest exponent

of 2   of x of 3   of y

$$4x^2 y = 2^2 x^2 y \quad \text{and} \quad 6xy^4 = 2 \times 3 \times xy^4$$

The prime factors to be considered are 2, 3, x, and y. The largest exponent of 2 is 2; the largest exponent of 3 is 1; the largest exponent of x is 2; and the largest exponent of y is 4. Therefore, the lowest common denominator is $2^2 \times 3 \times x^2 y^4 = 12x^2 y^4$. This is the simplest expression into which $4x^2 y$ and $6xy^4$ both divide evenly.

When we are finding the lowest common denominator of a set of algebraic fractions, and when some of the factors are binomials, *we* **indicate** *the multiplication of the necessary factors but do not multiply out the product.* This prevents us from losing the identity of the several factors and therefore makes the process of finding the appropriate equivalent fractions much easier. The reasoning for this will be shown in Section 9.4 when we actually add and subtract algebraic fractions.

Examples 4–6 illustrate the procedure for finding the lowest common denominator for given sets of algebraic fractions.

Example 4

Find the lowest common denominator of the fractions

$$\frac{x - 9}{x^2 + x - 6} \quad \text{and} \quad \frac{x + 2}{x^2 - 5x + 6}$$

Factoring the denominators, we get

$x - 2$ appears in each, but is raised to first power only in each

$$x^2 + x - 6 = (x + 3)(x - 2) \qquad x^2 - 5x + 6 = (x - 2)(x - 3)$$

The necessary factors are $(x - 2), (x - 3)$, and $(x + 3)$. Because the highest power to which each appears is 1, the lowest common denominator is

$$(x - 2)(x - 3)(x + 3)$$

▶ Here we note that $(x - 2)$ appears as a factor in both factorizations. **Although it appears twice, its largest exponent is 1. Thus *we use only one factor of $(x - 2)$* in the lowest common denominator.**

The product of these three factors is $x^3 - 2x^2 - 9x + 18$. It is obvious that we would lose the identity of the factors in the multiplication. In forming the equivalent fractions, we would have to divide this expression by each of the individual denominators. This would create much extra work, because in the factored form this division can be done by inspection. Therefore, the factored form of $(x - 2)(x - 3)(x + 3)$ is preferable to the form $x^3 - 2x^2 - 9x + 18$.

Example 5

Find the lowest common denominator of the fractions

$$\frac{2}{5x^2} \qquad \frac{3x}{10x^2 - 10} \qquad \frac{5}{x^2 - x}$$

Factoring the denominators, we get

$$5x^2 = 5x^2 \quad \longleftarrow \text{largest exponent of } x$$

$$10x^2 - 10 = 10(x^2 - 1) = 2 \cdot 5(x - 1)(x + 1) \quad \boxed{\begin{array}{l}\text{appears in each;}\\ \text{largest exponent is 1}\end{array}}$$

$$x^2 - x = x(x - 1) \qquad \underline{}2^1 \qquad \underline{}(x + 1)^1$$

The prime factors of the lowest common denominator are 2, 5, x, $x - 1$, and $x + 1$. Each appears only to the first power except x, which appears to the

▶ second power. ***Although 5 and $(x - 1)$ appear in two factorizations, the largest exponent of each is 1.*** Therefore, the lowest common denominator is $2 \cdot 5x^2(x - 1)(x + 1) = 10x^2(x - 1)(x + 1)$. (Here we have multiplied the numerical factors in the coefficient. However, the others have been left in factored form.)

Example 6

Three resistors are connected in series so that the total resistance can be found by adding the individual values. Under certain conditions, the resistances are described by the fractions given below. In order to add these fractions (as discussed in Section 9.4), we must first find the lowest common denominator.

$$\frac{3x}{x^2 + 4x + 4} \qquad \frac{5}{x^2 - 4} \qquad \frac{x}{6x + 12}$$

Factoring the denominators, we get

$$x^2 + 4x + 4 = (x + 2)(x + 2) = (x + 2)^2$$

$$x^2 - 4 = (x - 2)(x + 2) \qquad \begin{array}{l}\text{appears in each;}\\ \text{largest exponent is 2}\end{array}$$

$$6x + 12 = 6(x + 2) = 2 \times 3(x + 2)$$

The prime factors of the lowest common denominator are 2, 3, $x + 2$, and $x - 2$. All appear only to the first power except $x + 2$, for which **the largest exponent is 2**. Therefore, the lowest common denominator is

$$2 \times 3(x + 2)^2(x - 2) = 6(x + 2)^2(x - 2)$$

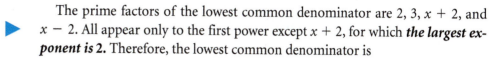

9.3 Exercises

In the following exercises, find the lowest common denominator of each given set of fractions.

1. $\dfrac{a}{3}$; $\dfrac{a}{18}$

2. $\dfrac{x}{2}$; $\dfrac{3x}{7}$

3. $\dfrac{s}{12}$; $\dfrac{5s}{18}$

4. $\dfrac{r}{6}$; $\dfrac{7r}{40}$

5. $\dfrac{1}{4a}$; $\dfrac{1}{6a}$

6. $\dfrac{7}{15x}$; $\dfrac{8}{25x}$

7. $\dfrac{3a}{8t}$; $\dfrac{7b}{20t}$

8. $\dfrac{3V}{100R}$; $\dfrac{9V}{40R}$

9. $\dfrac{5}{18}$; $\dfrac{11}{45y}$

10. $\dfrac{7}{40}$; $\dfrac{7}{72n}$

11. $\dfrac{2}{3x}$; $\dfrac{5}{9x^2}$

12. $\dfrac{5}{56t^3}$; $\dfrac{17}{196t^2}$

13. $\dfrac{1}{4x^2}$; $\dfrac{3}{8x}$

14. $\dfrac{2}{15a^3}$; $\dfrac{3}{5a^3}$

15. $\dfrac{9}{60ax}$; $\dfrac{9}{28a}$

16. $\dfrac{3a}{8rst}$; $\dfrac{9b}{20st}$

17. $\dfrac{36}{125ax^2}$; $\dfrac{7}{15ax}$

18. $\dfrac{9}{16a^2b^2}$; $\dfrac{1}{12a^2b}$

19. $\dfrac{1}{25a^2}$; $\dfrac{7}{3a^3}$

20. $\dfrac{5}{4x^5}$; $\dfrac{9}{8x^2}$

21. $\dfrac{15}{32ab^3}$; $\dfrac{11}{12a^3b^2}$

22. $\dfrac{8}{27abc^5}$; $\dfrac{7x}{3a^2bc^4}$

23. $\dfrac{3}{5a^2}$; $\dfrac{2}{15a}$; $\dfrac{4}{3a^2}$

24. $\dfrac{5}{2x^3}$; $\dfrac{3}{8x}$; $\dfrac{1}{4x^2}$

25. $\dfrac{27}{4acx^3}$; $\dfrac{5}{12a^2cx}$; $\dfrac{13b}{20acx}$

26. $\dfrac{4}{25p^2q}$; $\dfrac{8}{15q^2r}$; $\dfrac{16}{27prs^2}$

27. $\dfrac{5}{4x - 4}$; $\dfrac{3}{8x}$

28. $\dfrac{3}{2x + 8}$; $\dfrac{5}{4x}$

29. $\dfrac{4}{3a + 9}$; $\dfrac{5}{a^2 + 3a}$

30. $\dfrac{5}{3a^2x - 9ax}$; $\dfrac{7x}{6a^2 - 18a}$

31. $\dfrac{3x}{2x - 2y}$; $\dfrac{5}{x^2 - xy}$; $\dfrac{7x}{6x^2 - 6y^2}$

32. $\dfrac{a + 3b}{a^2 - ab - 2b^2}$; $\dfrac{a + b}{a^2 - 4b^2}$

33. $\dfrac{x - 5}{x^2 - 3x + 2}$; $\dfrac{x}{2x^2 - 4x + 2}$

34. $\dfrac{x - 1}{4x^2 - 36}$; $\dfrac{x + 7}{3x^2 + 18x + 27}$

35. $\dfrac{7}{2t^2 - 5t - 12}$; $\dfrac{5t}{2t^2 + 10t + 6}$

36. $\dfrac{3}{x^2 - 8x + 16}$; $\dfrac{2}{4 - x}$

37. $\dfrac{x}{x^2 - x - 6}$; $\dfrac{2x}{x^2 + 6x + 9}$; $\dfrac{x}{x^2 - 9}$

38. $\dfrac{x - 7y}{3x^2 - 7xy - 6y^2}$; $\dfrac{x + y}{x^2 - 9y^2}$; $\dfrac{7}{2x - 6y}$

39. $\dfrac{5}{2x^3 - 3x^2 + x}$; $\dfrac{x}{x^4 - x^2}$; $\dfrac{2 - x}{2x^2 + x - 1}$

40. $\dfrac{5}{6x^4 - 6y^4}$; $\dfrac{x + y}{4x^2 - 4y^2}$; $\dfrac{2x - 5y}{3x^2 - 6xy + 3y^2}$

9.4 Addition and Subtraction of Algebraic Fractions

Section Objectives
- Write equivalent algebraic fractions
- Add and subtract algebraic fractions

Having discussed the basic method of finding the lowest common denominator, we can perform the operations of addition and subtraction of algebraic fractions. As we have pointed out, if we want to add or subtract fractions whose denominators are the same, we place the sum of the numerators over the denominator. *If the denominators differ, we must first change the fractions to equivalent ones with a common denominator equal to the lowest common denominator. We then combine the numerators of the equivalent fractions algebraically, placing this result over the common denominator.* Examples 1–5 illustrate this method.

Recall that in order to create an equivalent fraction you must multiply the numerator and denominator by the same common factor.

$$\frac{a}{b} = \frac{a}{b} \times \frac{c}{c} = \frac{ac}{bc}$$

equivalent fractions

Example 1

Combine:

$$\frac{7}{5ax} + \frac{3b}{5ax} - \frac{4b}{5ax}$$

Because the denominators are the same for each fraction, we need only to combine the numerators over the common denominator. We therefore have

$$\frac{7}{5ax} + \frac{3b}{5ax} - \frac{4b}{5ax} = \frac{7 + 3b - 4b}{5ax} = \frac{7 - b}{5ax}$$

Note that the signs of the terms of the numerators are the same as those of the fractions from which they were obtained.

Example 2

Combine:

$$\frac{3}{ax} + \frac{x}{2a^2} - \frac{7}{2x}$$

By examining the denominators, we see that the factors necessary in the lowest common denominator are 2, a, and x. Both 2 and x appear only to the first power and a appears squared. Thus the lowest common denominator is $2a^2x$. We now wish to write each fraction in an equivalent form with $2a^2x$ as a denominator. We have

$$\frac{3}{ax} = \frac{3}{ax} \times \frac{2a}{2a} = \frac{6a}{2a^2x}$$

$$\frac{x}{2a^2} = \frac{x}{2a^2} \times \frac{x}{x} = \frac{x^2}{2a^2x}$$

$$\frac{7}{2x} = \frac{7}{2x} \times \frac{a^2}{a^2} = \frac{7a^2}{2a^2x}$$

This leads to

$$\frac{3}{ax} + \frac{x}{2a^2} - \frac{7}{2x} = \frac{3(2a)}{(ax)(2a)} + \frac{x(x)}{(2a^2)(x)} - \frac{7(a^2)}{(2x)(a^2)} \qquad \text{change to equivalent fractions}$$

factors needed

$$= \frac{6a}{2a^2x} + \frac{x^2}{2a^2x} - \frac{7a^2}{2a^2x}$$

$$= \frac{6a + x^2 - 7a^2}{2a^2x} \qquad \text{combine numerators}$$

Example 3

Two batteries are connected in series so that their combined voltage (in volts) can be found by adding the fractions given below. Perform the indicated addition to find an expression representing the total voltage.

$$\frac{4}{x + 3} + \frac{3}{x + 2}$$

Because there is only one factor in each denominator and they are different, the lowest common denominator of these fractions is the product $(x + 3)(x + 2)$.

▶ This in turn means that we must ***multiply the numerator and denominator of the first fraction by x + 2 and the numerator and denominator of the second fraction by x + 3*** in order to make equivalent fractions with $(x + 3)(x + 2)$ as the denominator. Therefore, performing the addition, we get

$$\frac{4}{x + 3} + \frac{3}{x + 2} = \frac{4(x + 2)}{(x + 3)(x + 2)} + \frac{3(x + 3)}{(x + 2)(x + 3)}$$

change to equivalent fractions

factors needed

$$= \frac{4(x + 2) + 3(x + 3)}{(x + 3)(x + 2)}$$ combine numerators

$$= \frac{4x + 8 + 3x + 9}{(x + 3)(x + 2)}$$ simplify

$$= \frac{7x + 17}{(x + 3)(x + 2)}$$

Guidelines for Adding and Subtracting Algebraic Expressions

1. Find the lowest common denominator, if necessary.

2. Write equivalent fractions by multiplying both the numerator and denominator by any factor the algebraic fraction is missing from the common denominator.

3. Add or subtract the numerators and place over the common denominator.

4. Simplify the numerator by combining like terms.

5. Simplify the algebraic fraction to its lowest terms.

Example 4

Combine:

$$\frac{x}{x + y} - \frac{2y^2}{x^2 - y^2}$$

Factoring the denominator of the second fraction, we have $x^2 - y = (x + y)(x - y)$. Because the factor $x + y$ appears in the first fraction and there is no other (third) factor, the lowest common denominator is $(x + y)(x - y)$. Therefore, we must ▶ ***multiply the numerator and denominator of the first fraction by x − y, whereas the second fraction remains the same.*** This leads to

$$\frac{x}{x+y} - \frac{2y^2}{x^2-y^2} = \frac{x}{x+y} - \frac{2y^2}{(x+y)(x-y)}$$

change to equivalent fractions each with l.c.d.

$$= \frac{x(x-y)}{(x+y)(x-y)} - \frac{2y^2}{(x+y)(x-y)}$$

factor needed

$$= \frac{x^2 - xy - 2y^2}{(x+y)(x-y)}$$

combine numerators over l.c.d.

$$= \frac{(x+y)(x-2y)}{(x+y)(x-y)}$$

factor numerator

$$= \frac{x-2y}{x-y}$$

simplify

We note here that when the numerators are combined, the result is factorable. One of the factors also appears in the denominator, which means it is possible to reduce the resulting fraction. Remember: ***We must express the result in its simplest form.***

<table><tr><td>

Now Try It!

Perform the indicated operations.

1. $\dfrac{3}{ab^2} + \dfrac{5}{abc}$

2. $\dfrac{8}{x+1} - \dfrac{3}{x}$

3. $\dfrac{3}{x^2-x} + \dfrac{x}{x^2-3x+2}$

4. $\dfrac{-5}{x^2-x-6} - \dfrac{1}{x+2}$

</td></tr></table>

Example 5

Combine:

$$\frac{5}{a^2-a-6} + \frac{1}{a^2-5a+6} - \frac{2}{a^2-4a+4}$$

Factoring the denominators, we get

$$a^2 - a - 6 = (a-3)(a+2) \quad \text{\color{blue}largest exponent of } a-3 \text{ and } a+2 \text{ is } 1$$
$$a^2 - 5a + 6 = (a-3)(a-2)$$

$$a^2 - 4a + 4 = (a-2)(a-2) = (a-2)^2 \quad \text{\color{blue}largest exponent of } a-2$$

The prime factors of the lowest common denominator are $a-3$, $a+2$, and $a-2$, where $a-2$ appears to the second power. The lowest common denominator is therefore $(a-3)(a+2)(a-2)^2$. To have equivalent fractions with the lowest common denominator, it is necessary to ***multiply the numerator and the denominator of the first fraction by*** $(a-2)^2$***, those of the second fraction by*** $(a-2)(a+2)$***, and those of the third fraction by*** $(a+2)(a-3)$***.*** This leads to

$$\frac{5}{a^2-a-6} + \frac{1}{a^2-5a+6} - \frac{2}{a^2-4a+4}$$
$$= \frac{5}{(a-3)(a+2)} + \frac{1}{(a-3)(a-2)} - \frac{2}{(a-2)^2}$$

denominators factored

$$= \frac{5(a-2)^2}{(a-3)(a+2)(a-2)^2} + \frac{1(a-2)(a+2)}{(a-3)(a-2)(a-2)(a+2)} - \frac{2(a+2)(a-3)}{(a-2)^2(a+2)(a-3)}$$

change to equivalent fractions with l.c.d.

$$= \frac{5(a-2)^2 + (a-2)(a+2) - 2(a+2)(a-3)}{(a-3)(a+2)(a-2)^2}$$ combine numerators over l.c.d.

$$= \frac{5(a^2 - 4a + 4) + a^2 - 4 - 2(a^2 - a - 6)}{(a-3)(a+2)(a-2)^2}$$ simplify numerator

$$= \frac{5a^2 - 20a + 20 + a^2 - 4 - 2a^2 + 2a + 12}{(a-3)(a+2)(a-2)^2}$$

$$= \frac{4a^2 - 18a + 28}{(a-3)(a+2)(a-2)^2}$$

$$= \frac{2(2a^2 - 9a + 14)}{(a-3)(a+2)(a-2)^2}$$

We should now check $2a^2 - 9a + 14$ to determine whether it is factorable. Because it is not factorable, we cannot further simplify the form of the result.

9.4 Exercises

In Exercises 1 through 8, change the indicated sum of the fractions to an indicated sum of the equivalent fractions with the proper lowest common denominator. Do not combine.

1. $\dfrac{5}{9a} - \dfrac{7}{12a}$

2. $\dfrac{1}{40ax} + \dfrac{5}{84a}$

3. $\dfrac{5}{ax} + \dfrac{1}{bx} - \dfrac{4}{a}$

4. $\dfrac{6}{a^3b} - \dfrac{5}{4ab} - \dfrac{1}{6a^2}$

5. $\dfrac{4}{x^2 - x} - \dfrac{3}{2x^3 - 2x^2}$

6. $\dfrac{5b}{3ab - 6ac} - \dfrac{7c}{3b^2 - 12c^2}$

7. $\dfrac{x}{2x - 4} + \dfrac{5}{x^2 - 4} - \dfrac{3x}{x^2 + 4x + 4}$

8. $\dfrac{a - 1}{3a - 3} - \dfrac{8}{a^2 - 5a + 4} - \dfrac{1}{9}$

In Exercises 9 through 34, combine the given fractions, expressing all results in simplest form.

9. $\dfrac{2}{5x} + \dfrac{3}{2x}$

10. $\dfrac{3}{8s} - \dfrac{5}{12s}$

11. $\dfrac{2}{3a} - \dfrac{5}{3b}$

12. $\dfrac{7}{8s} + \dfrac{5}{20t}$

13. $\dfrac{6}{x} + \dfrac{3}{x^2}$

14. $\dfrac{4}{3a} - \dfrac{5}{2a^2}$

15. $\dfrac{1}{2} - \dfrac{5}{8b} + \dfrac{3}{20}$

16. $\dfrac{11}{12} + \dfrac{13}{40} - \dfrac{4}{15s}$

17. $\dfrac{8}{by} - \dfrac{1}{y^2}$

18. $\dfrac{1}{ax} + \dfrac{2}{a^2x}$

19. $\dfrac{3}{x^3y} + \dfrac{2}{3xy}$

20. $\dfrac{5}{p^2q} - \dfrac{1}{6pq}$

21. $\dfrac{2}{x} + \dfrac{5}{y} - \dfrac{3}{x^2 y}$

22. $\dfrac{2}{a^2} - \dfrac{5}{6b} + \dfrac{3}{8ab}$

23. $\dfrac{7}{2x} - \dfrac{5}{4y} + \dfrac{1}{6z}$

24. $\dfrac{1}{6pq} + \dfrac{7}{3p} - \dfrac{9}{2p}$

25. $\dfrac{3}{a-2} + \dfrac{5}{a+2}$

26. $\dfrac{2}{2x-1} + \dfrac{3}{2x+3}$

27. $\dfrac{x}{2x-6} - \dfrac{3}{4x+12}$

28. $\dfrac{x-1}{4x+6} + \dfrac{2x}{6x+9} - \dfrac{5}{6}$

29. $\dfrac{4}{4-9x^2} - \dfrac{x-5}{2+3x}$

30. $\dfrac{q-3}{q^2-q-12} + \dfrac{q+1}{q^2-4q}$

31. $\dfrac{3x}{x^2-4} + \dfrac{2}{x-2} - \dfrac{5x}{x+2}$

32. $\dfrac{3}{2x-2} + \dfrac{5}{x+1} - \dfrac{x}{x^2-1}$

33. $\dfrac{2}{3} + \dfrac{3}{x+5} - \dfrac{2}{x-5}$

34. $\dfrac{4}{x} - \dfrac{1}{3x-1} + \dfrac{x+1}{3x+1}$

In Exercises 35 through 42, solve the given problems.

35. In attempting to determine the minimum cost of producing a component for a helium-neon laser, a scientist used the expression

$$26.500 + 5.080x + \dfrac{0.004}{x^2}$$

Combine and simplify.

36. In analyzing the heat transfer involved in a thermograph, we encounter the expression

$$\dfrac{R_2}{R_1} - 1$$

Combine this expression into a single fraction.

37. In the theory dealing with the motion of the planets, we find the expression

$$\dfrac{p^2}{2mr^2} - \dfrac{gmM}{r}$$

Combine and simplify.

38. The total resistance for the circuit shown at the start of this chapter can be found using the algebraic expression

$$\dfrac{R_1 R_2}{R_1 + R_2} + R_3$$

Combine this expression so that it can be written as a single fraction.

39. In a parallel circuit, the total resistance is described by

$$\dfrac{1}{R} = \dfrac{1}{R_1} + \dfrac{1}{R_2}$$

Combine the two fractions on the right side of the equation into one fraction.

40. Airplanes must be equipped with emergency locator transmitters that give a radio signal in the event of a crash. A value describing a characteristic of the transmittal signal is found from the expression

$$1 - \dfrac{x^2}{2} + \dfrac{x^4}{24} - \dfrac{x^6}{120}$$

Combine these fractions.

41. In the study of heat transfer, the thermal resistance of deposits is described by the formula

$$R_d = \dfrac{1}{u_d} - \dfrac{1}{u}$$

Combine the fractions on the right side of the equation.

42. In determining the characteristics of a specific optical lens, we use the expression

$$\dfrac{2n^2 - n - 4}{2n^2 + 2n - 4} + \dfrac{1}{n-1}$$

Combine and simplify.

Now Try It! Answers

1. $\dfrac{3c + 5b}{ab^2 c}$

2. $\dfrac{5x - 3}{x(x+1)}$

3. $\dfrac{x^2 + 3x - 6}{x(x-1)(x-2)}$

4. $-\dfrac{1}{x-3}$

9.5   Solving Fractional Equations

Section Objectives
- Solve equations involving fractions
- Solve real world application problems involving fractional equations

In science and technology, many equations involve fractions. One way to solve equations that contain algebraic fractions is to *multiply each term on both sides of the equation* by the least common denominator. This method eliminates all of the fractions contained in the given equation. This procedure is fundamentally different from the procedure used to combine fractions. When combining fractions we often multiply both numerator and denominator by some quantity, and the result is an equivalent fraction. With an equation involving fractions we "clear" the fractions by multiplying only the numerators of all the terms. The following examples illustrate how to solve equations with fractions.

Example 1

Solve for x:

$$\frac{x+1}{6} + \frac{3}{2} = x$$

We first note that the lowest common denominator of the terms of the equation is 6. We *multiply each term on both sides of the equation by* 6 to get

$$\frac{6(x+1)}{6} + \frac{6(3)}{2} = 6x$$

We now reduce each term to its lowest terms and solve the resulting equation:

$$(x+1) + 9 = 6x \quad \text{simplify}$$
$$x + 1 + 9 = 6x \quad \text{remove parentheses}$$
$$10 = 5x \quad \text{subtract } x \text{ from each side}$$
$$x = 2 \quad \text{divide each side by 5 and reverse sides}$$

When we check this solution *in the original equation,* we obtain 2 on each side of the equation. Therefore, the solution checks.

It is very important to *check the solution* because even when correct procedures are followed, it sometimes happens that what appears to be a solution is actually discarded because it does not satisfy the original equation.

Example 2

Solve for x:

$$\frac{x}{a} - \frac{3}{4a} = \frac{1}{2}$$

We first determine that the lowest common denominator of the terms of the equation is $4a$. We then ***multiply each term on both sides of the equation by $4a$*** and continue with the solution:

$$\frac{4a(x)}{a} - \frac{4a(3)}{4a} = \frac{4a(1)}{2} \qquad \text{multiply each term by } 4a$$

$$4x - 3 = 2a \qquad \text{simplify}$$

$$4x = 2a + 3 \qquad \text{add 3 to each side}$$

$$x = \frac{2a + 3}{4} \qquad \text{divide each side by 4}$$

Note that ***we cannot cancel a factor of*** **2,** because 2 is not a factor of the numerator. Checking shows that upon substitution of $\dfrac{(2a + 3)}{4}$ for x, the left-hand side of the equation is $\frac{1}{2}$, which means that the solution checks.

Example 3

Solve for x:

$$\frac{x}{a} - \frac{1}{2b} = \frac{3x}{ab}$$

The lowest common denominator of the terms of the equation is $2ab$. We therefore multiply each term on both sides of the equation by $2ab$ and proceed with the solution:

$$\frac{2ab(x)}{a} - \frac{2ab(1)}{2b} = \frac{2ab(3x)}{ab} \qquad \text{multiply each term by } 2ab$$

$$2bx - a = 6x \qquad \text{simplify}$$

$$2bx - 6x = a \qquad \text{add } a, \text{ subtract } 6x \text{ (each side)}$$

To isolate x, we must factor x from each term on the left. This leads to

$$x(2b - 6) = a \qquad \text{factor out } x \text{ on left}$$

$$x = \frac{a}{2b - 6} \qquad \text{divide each side by } 2b - 6$$

Checking will lead to the expression

$$\frac{3}{2b(b - 3)}$$

on each side of the original equation so that the solution is correct.

Example
4

Solve for t:

$$\frac{2}{t} = \frac{3}{t + 2}$$

After determining that the lowest common denominator of the terms of the equation is $t(t + 2)$, we *multiply each term on both sides of the equation by this expression* and proceed with the solution:

$$\frac{t(t + 2)(2)}{t} = \frac{t(t + 2)(3)}{t + 2}$$ multiply each term by $t(t + 2)$

$$2(t + 2) = 3t$$ simplify

$$2t + 4 = 3t$$ remove parentheses

$$4 = t \quad \text{or} \quad t = 4$$ subtract $2t$ from each side

Checking shows that with $t = 4$, each side of the original equation is $\frac{1}{2}$.

Example
5

Solve for x:

$$\frac{3}{x} + \frac{2}{x - 2} = \frac{4}{x^2 - 2x}$$

Factoring the denominator of the term on the right, $x^2 - 2x = x(x - 2)$, we find that the lowest common denominator of the terms of the equation is $x(x - 2)$. We get

$$\frac{3x(x - 2)}{x} + \frac{2x(x - 2)}{x - 2} = \frac{4x(x - 2)}{x(x - 2)}$$ multiply each term by $x(x - 2)$

$$3(x - 2) + 2x = 4$$ simplify

$$3x - 6 + 2x = 4$$

$$5x = 10$$

$$x = 2$$

Substituting this value in the original equation, we obtain

$$\frac{3}{2} + \frac{2}{0} = \frac{4}{0}$$

▶ Because division by zero is an undefined operation, the value $x = 2$ cannot be a solution. *We conclude that **there is no solution** to this equation.*

Now Try It!

Solve for x

1. $\dfrac{x}{14} + \dfrac{x}{8} = 11$

2. $\dfrac{1}{x - 1} = \dfrac{2}{x + 3}$

3. $\dfrac{2}{x + 1} - \dfrac{1}{x} = -\dfrac{2}{x^2 + x}$

Example 5 shows us that *whenever we multiply through by a lowest common denominator that contains the unknown, it is possible that a solution will be introduced into the resulting equation that is not a solution of the original equation. Such a solution is called an* **extraneous solution.** Only certain equations will lead to extraneous solutions, but we must be careful to identify them when they occur.

Many stated problems lead to equations that involve fractions. Example 6 illustrates such a problem.

Example 6

A plane can fly with a ground speed of 160 mi/h if there is no wind. It travels 350 mi with a headwind in the same time it takes to go 450 mi with a tailwind. Find the speed of the wind.

Let x = the speed of the wind. The ground speed of the plane as it flies into the headwind then becomes $160 - x$ mi/h. (The speed of 160 mi/h is reduced by the wind of x mi/h.) When flying with the tailwind, the speed becomes $160 + x$ mi/h.

The solution is based on the fundamental and important formula $d = rt$ (distance = rate × time). The key is to recognize that the upwind *time* is equal to the downwind *time*. With $d = rt$, we know that $t = d/r$, and we now express the equality of the upwind and downwind times as follows:

$$\text{upwind } \frac{d}{r} = \text{downwind } \frac{d}{r}$$

or

$$\frac{350}{160 - x} = \frac{450}{160 + x}$$

We now proceed to solve for x. We first clear the fractions by multiplying both sides of the equation by the lowest common denominator of $(160 - x)(160 + x)$ and get this result.

$$350(160 + x) = 450(160 - x)$$

We can now eliminate the parentheses and proceed with the usual method of solution.

$$56{,}000 + 350x = 72{,}000 - 450x$$
$$800\,x = 16{,}000$$
$$x = 20 \text{ mi/h}$$

Checking the solution with the original statement, we see that the upwind and downwind speeds are 140 mi/h and 180 mi/h, respectively. Dividing 350 mi by 140 mi/h yields a time of 2.5 h. Dividing 450 mi by 180 mi/h also yields a time of 2.5 h so that the solution checks.

Example 6 involved distance, rate, and time so that it is necessary to know the formula $d = rt$. However, the solution also required the key observation that the two values of time (or d/r) are equal.

Example 7 involves a situation that often occurs in real applications. The general key question is this: If one element can do a job in time t_1 while another element can do a job in time t_2, how long will the job take if both elements work together? This same general problem can be applied to two pumps emptying a tank, or two painters painting a house, and so on.

Example
7

One company makes a special boat that can remove the algae in a certain lake in 120 h. Another boat can remove the algae in 180 h. If both boats work together, how long will it take to remove the algae from this lake?

▶   Relative to the lake under consideration, the first boat works at the **rate of** $\frac{1}{120}$ **lake per hour** while the rate of the second boat is $\frac{1}{180}$ **lake per hour.** When the boats work together, the algae is being removed at the combined rate of

$$\frac{1}{120} + \frac{1}{180}$$

lake per hour. The key to this problem is to identify the two *rates,* which are then combined through addition. Letting t represent the time required for both boats working together to clear this lake, we note that if we multiply t by the combined total *rate,* the result should be 1 (lake).

$$t\left(\frac{1}{120} + \frac{1}{180}\right) = 1 \qquad \text{hours}\left(\frac{\text{lake}}{\text{hour}}\right) = \text{lake}$$

We must now solve for t. We first multiply by t on the left side and then multiply both sides of the equation by the lowest common denominator of 360 to get

$$\frac{t}{120} + \frac{t}{180} = 1$$

$$3t + 2t = 360 \qquad \text{multiply each term by 360}$$

$$5t = 360 \qquad \text{simplify}$$

$$t = 72$$

Both boats working together can clear this lake in 72 h.

In general, if one element can do a job in time t_1 and a second element can do a job in time t_2, both elements working together can do the job in time t, where t is found by solving the equation

$$t\left(\frac{1}{t_1} + \frac{1}{t_2}\right) = 1 \quad \text{or} \quad t = \frac{1}{\dfrac{1}{t_1} + \dfrac{1}{t_2}}$$

This generalization can be easily extended to include cases involving more than two elements.

In Chapter 7 we looked at how to solve systems of linear equations using the graphing calculator and its "intersect" feature. We can apply some of those same ideas here to solve equations with algebraic fractions in a similar way.

Solve the equation $\dfrac{2x - 1}{9} - \dfrac{1}{3} = 2x$ graphically.

1. Start by entering $Y1 = \dfrac{2x - 1}{9} - \dfrac{1}{3}$ and $Y2 = 2x$ as seen in Figure 9.2. Notice the parentheses in the numerator of Y1. They are essential in this problem because they help to identify the entire numerator.

   We know that for $Y1 = Y2$ we must find the point they have in common, their point of intersection.

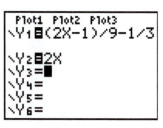

Figure 9.2

2. Graph the two equations on the same set of axes. Using a window setting of $[-3, 3, 1]$ for both X and Y, we get a graph that looks like the picture in Figure 9.3.

   Using the same steps that were used to solve a system of linear equations graphically

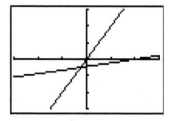

Figure 9.3

3. Press 2nd TRACE to access the "Calc" menu.

4. Press option #5 to select the "intersect" feature.

5. Press ENTER in response to the question "First Curve?"

6. Press ENTER in response to the question "Second Curve?"

7. Press ENTER in response to the question "Guess?"

8. You will see that Y1 and Y2 intersect at the point $(-.25, -.5)$, as shown in Figure 9.4. Because our original problem was written only in terms of the variable x, the solution to the equation

$$\frac{2x - 1}{9} - \frac{1}{3} = 2x \text{ is } x = -0.25.$$

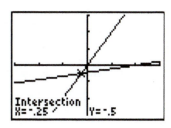

Figure 9.4

9.5   Exercises

In Exercises 1 through 24, solve for s, t, v, x, or y.

1. $\dfrac{x}{8} = \dfrac{1}{2}$

2. $\dfrac{y}{6} = \dfrac{1}{3}$

3. $\dfrac{2s}{4} - 1 = s$

4. $\dfrac{x}{4} + 6 = x$

5. $\dfrac{2x}{5} + 3 = \dfrac{7x}{10}$

6. $\dfrac{3y}{8} - 2 = \dfrac{y}{2}$

7. $\dfrac{t+1}{2} + \dfrac{1}{4} = \dfrac{3}{8}$

8. $\dfrac{2x-1}{9} - \dfrac{1}{3} = 2x$

9. $\dfrac{x}{b} + 3 = \dfrac{1}{b}$

10. $\dfrac{y}{c} - 2 = \dfrac{3}{c}$

11. $\dfrac{y}{2a} - \dfrac{1}{a} = 2$

12. $\dfrac{x}{b} + \dfrac{1}{3b} = 1$

13. $\dfrac{1}{2a} - \dfrac{3s}{4} = \dfrac{1}{a^2}$

14. $\dfrac{2}{5c} + \dfrac{2t}{5} = \dfrac{1}{c^3}$

15. $\dfrac{2x}{b} + \dfrac{4}{b^2} = \dfrac{x}{4b}$

16. $\dfrac{3s}{4a} - \dfrac{s}{a^2} = \dfrac{1}{8}$

17. $\dfrac{1}{x+2} = \dfrac{1}{2x}$

18. $\dfrac{1}{2v-1} = \dfrac{1}{3}$

19. $\dfrac{2}{3s+1} = \dfrac{1}{4}$

20. $\dfrac{3}{3x+2} = \dfrac{2}{x}$

21. $\dfrac{5}{2x+4} + \dfrac{3}{x+2} = 2$

22. $\dfrac{2}{s-3} - \dfrac{3}{2s-6} = 4$

23. $\dfrac{1}{y^2-y} - \dfrac{1}{y} = \dfrac{1}{y-1}$

24. $\dfrac{2}{x^2-1} - \dfrac{2}{x+1} = \dfrac{1}{x-1}$

In Exercises 25 through 30, solve for the indicated letter.

25. An equation obtained in analyzing a certain electric circuit is

$$\frac{V-6}{5} + \frac{V-8}{15} + \frac{V}{10} = 0$$

Solve for *V*.

26. In biology, an equation used when analyzing muscle reactions is

$$P + \frac{a}{V^2} = \frac{k}{V}$$

Solve for *a*.

27. In optics, the thin lens equation is given as

$$\frac{1}{p} + \frac{1}{q} = \frac{1}{f}$$

where *p* is the distance of the object from the lens, *q* is the distance of the image from the lens, and *f* is the focal length of the lens. Solve for *q*.

28. A formula relating the number of teeth *N* of a gear, the outside diameter D_0 of the gear, and the pitch diameter D_p is

$$D_p = \frac{D_0 N}{N+2}$$

Solve for *N*.

29. An equation found in the thermodynamics of refrigeration is

$$\frac{W}{Q_1} = \frac{T_2}{T_1} - 1$$

where *W* is the work input, Q_1 is the heat absorbed, T_1 is the temperature within the refrigerator, and T_2 is the external temperature. Solve for T_2.

30. An equation which arises in the study of the motion of a projectile is

$$m^2 - \frac{2v^2 m}{gx} + \frac{2v^2 y}{gx^2} + 1 = 0$$

In Exercises 31 through 36, solve the given problems.

31. An investor has $6000 in one account and $8000 in another account. Each account is increased by the same amount so that they have a ratio of $\frac{4}{5}$. Find that amount.

32. An advertiser wants a rectangular ad with the length and width in the ratio of $\frac{3}{2}$. If the ad has a border with a total length of 25 in., find the dimensions.

33. Water is pumped through a section of pipe so that its speed is 3.0 mi/h. If the speed is increased to 4.0 mi/h, it would take the same time for the water to travel an additional 0.60 mi. How long is the original section of pipe?

34. A firefighter has two pumps available for draining a flooded basement. One pump can do the job in 4.0 h, whereas the other pump would require only 3.0 h. If both pumps are used, how long does it take to drain the basement?

35. One painting team can paint a bridge in 600 h while a different team can do the same job in 400 h. If both teams work together, how long will it take them to paint the bridge?

36. A car passes a second car on the threeway traveling 5 mi/h faster than the car it passed. If the faster car can travel 340 miles in the same time at takes the slower car to travel 315 miles, find the speed of each car.

Now Try It! Answers

1. $x = 12$ **2.** $x = 5$ **3.** $x = -1$

Chapter Summary

Key Terms

fundamental principle of fractions

equivalent fractions

lowest common denominator

Key Concepts

• To *simplify* an algebraic fraction, factor the numerator and denominator.

• When *multiplying* algebraic fractions, multiply numerators together and multiply denominators together.

• When *dividing* algebraic fractions we multiply by the reciprocal of the

algebraic fraction that represents the divisor.

• To *add* or *subtract* algebraic fractions begin by determining the lowest common denominator. Rewrite each fraction as an equivalent fraction with the lowest common denominator. Add or subtract the numerators keeping the same denominator.

• To *solve an algebraic equation* begin by multiplying each term on each side of the equation by the lowest common denominator in order to clear the fractions from the equation. Then solve the remaining equation and check all answers.

Key Formulas

$$(a - b) = -(b - a)$$

$$\left(\frac{a}{b}\right)^n = \frac{a^n}{b^n} \quad (b \text{ not zero})$$

$$\frac{a}{b} \times \frac{c}{d} = \frac{ac}{bd} \quad (b \text{ and } d \text{ not zero})$$

$$\frac{\dfrac{a}{b}}{\dfrac{c}{d}} = \frac{ad}{bc} \quad (b, c, \text{ and } d \text{ not zero})$$

Review Exercises

In Exercises 1 through 12, reduce the given fractions to simplest form.

1. $\dfrac{9rst^6}{3s^4t^2}$

2. $\dfrac{-14y^2z^3}{84y^5z^2}$

3. $\dfrac{2a^2bc}{6ab^2c^3}$

4. $\dfrac{76x^3yz^3}{19xz^5}$

5. $\dfrac{4x + 8y}{x^2 - 4y^2}$

6. $\dfrac{a^2x - a^2y}{ax^2 - ay^2}$

7. $\dfrac{p^2 + pq}{3p + 2p^3}$

8. $\dfrac{4a - 12ab}{5b - 15b^2}$

9. $\dfrac{2a^2 + 2ab - 2ac}{4ab + 4b^2 - 4bc}$

10. $\dfrac{p^2 - 3p - 4}{p^2 - p - 12}$

11. $\dfrac{6x^2 - 7xy - 3y^2}{4x^2 - 8xy + 3y^2}$

12. $\dfrac{2y^2 - 14y + 20}{7y - 2y^2 - 6}$

In Exercises 13 through 20, find the lowest common denominator of each of the given sets of fractions.

13. $\dfrac{x}{6}; \dfrac{x}{9}$

14. $\dfrac{a}{2x^2}; \dfrac{b}{5x^2}$

15. $\dfrac{4}{3t^2}; \dfrac{3r}{4t}$

16. $\dfrac{c}{bt^3}; \dfrac{c}{at^2}$

17. $\dfrac{7}{12bt}; \dfrac{9}{16b^2t}$

18. $\dfrac{cx}{ay^2z^3}; \dfrac{cy}{byz^4}$

19. $\dfrac{2}{4x^3 - 8x^2}; \dfrac{3x}{5x - 10}$

20. $\dfrac{x-1}{x^2+2x-15}$; $\dfrac{x+4}{x^2+8x+15}$

In Exercises 21 through 56, perform the indicated operations.

21. $\dfrac{2x}{3a}\times\dfrac{5a^2}{x^3}$ **22.** $\dfrac{5b^3}{6c}\times\dfrac{3c^5}{10b}$ **23.** $\dfrac{3x^2}{4y^3}\times\dfrac{5y^4}{x^3}$

24. $\dfrac{7p}{12q^3}\times\dfrac{20q^4}{35p^3}$ **25.** $\dfrac{3a}{4}\div\dfrac{a^2}{8}$ **26.** $\dfrac{ab^3}{bc}\div a^2c$

27. $\dfrac{au^2}{4bv^2}\div\dfrac{a^2u}{8b^2v}$ **28.** $\dfrac{20m^3n^2}{12mn^4}\div\dfrac{27mn}{8m^2n^3}$

29. $\dfrac{2}{a^2}-\dfrac{3}{5ab}$ **30.** $\dfrac{3x}{4y}-\dfrac{5y}{6x}$ **31.** $\dfrac{5}{c}-\dfrac{3}{c^2d}+\dfrac{1}{2cd}$

32. $\dfrac{3}{2x^2}+\dfrac{1}{4x}-\dfrac{a}{6x^3}$ **33.** $\left(\dfrac{2}{3}\right)^3$ **34.** $\left(\dfrac{I}{Rt}\right)^3$

35. $\left(\dfrac{ax}{3y^2}\right)^4$ **36.** $\left(\dfrac{2x^2y}{xyz^3}\right)^5$

37. $\dfrac{2x}{x^2-1}\times\dfrac{x-1}{x^2}$ **38.** $\dfrac{5a}{a^2-a-6}\times\dfrac{a+2}{10a^2}$

39. $\dfrac{x^2-2x-15}{x^2-9}\times\dfrac{x^2-6x+9}{4x-12}$

40. $\dfrac{2x^2-14x+24}{x^2-4x+3}\times\dfrac{3x^2-27}{4x-20}$

41. $\dfrac{a^2-1}{4a}\div\dfrac{2a+2}{8a^2}$ **42.** $\dfrac{6x}{x^2+x-2}\div\dfrac{3}{2x+4}$

43. $(9x^2-4y^2)\div\dfrac{3x-2y}{y-2x}$

44. $\dfrac{6r^2-rs-s^2}{4r^2-16s^2}\div\dfrac{2r^2+rs-s^2}{r^2+3rs+2s^2}$

45. $\dfrac{3}{x^2}+\dfrac{2}{x^2+3x}$ **46.** $\dfrac{1}{a^2-2a}-\dfrac{3}{2a}$

47. $\dfrac{2x}{x-2}-\dfrac{x^2-3}{x^2-4x+4}$

48. $\dfrac{3}{3x+y}-\dfrac{7}{3x^2-5xy-2y^2}$

49. $\dfrac{2x-3}{2x^2-x-15}-\dfrac{3x}{x^2-9}$

50. $\dfrac{2x}{x^2+2x-8}+\dfrac{3x}{2x+8}$

51. $\dfrac{3}{x^2+5x}-\dfrac{x}{x^2-25}+\dfrac{2}{2x^2+9x-5}$

52. $\dfrac{3}{8x+16}-\dfrac{1-x}{4x^2-16}+\dfrac{5}{2x-4}$

53. $\left(\dfrac{2}{a-b}\times\dfrac{a+b}{5}\right)\div\left(\dfrac{a^2+2ab+b^2}{a^2-b^2}\right)$

54. $\left(\dfrac{x^4-1}{(x-1)^2}\div(x^2+1)\right)\times\left(\dfrac{x-1}{x+1}\right)$

55. $\left(\dfrac{1}{x-2}\div\dfrac{3+x}{x+1}\right)\div\left(\dfrac{2+x}{x^2-x-6}\right)$

56. $\left(\dfrac{2s}{s-1}+\dfrac{s^2}{s^2-1}\right)\div\left(\dfrac{s^3}{s-1}\right)$

In Exercises 57 through 64, solve the given equations.

57. $\dfrac{1}{2}-\dfrac{x}{6}+3=\dfrac{2x}{9}$ **58.** $\dfrac{6}{7}-\dfrac{x}{4}=2x-\dfrac{5}{14}$

59. $\dfrac{bx}{a}-\dfrac{1}{4}=x+\dfrac{3}{2a}$, for x

60. $\dfrac{4(x-y)}{5b}-\dfrac{x}{b^2}=\dfrac{y}{5b}$, for x

61. $\dfrac{5}{x+1}-\dfrac{3}{x}=\dfrac{-5}{x(x+1)}$

62. $\dfrac{20}{y^2-25}-\dfrac{1}{y+5}=\dfrac{2}{y-5}$

63. $\dfrac{5x}{x^2+2x}+\dfrac{1}{x}=\dfrac{3}{4x+8}$

64. $\dfrac{2}{2t+1}-\dfrac{t-1}{4t^2-4t-3}=\dfrac{1}{2t-3}$

In Exercises 65 through 74, solve the given problems.

65. A sphere of radius r floats in a liquid. When determining the height h that the sphere protrudes above the surface, the expression

$$\dfrac{1}{4r^2}-\dfrac{h}{12r^3}$$

is found. Combine and simplify.

66. In biology, when discussing the theory of x-rays, the expression

$$\frac{ds}{L} - \frac{ds^3}{24L^3}$$

arises. Combine the terms of this expression.

67. When considering the impedance of an electronic amplifier, the expression

$$1 - \frac{2Z_1}{Z_1 + Z_2}$$

is encountered. Combine and simplify.

68. In economics, when determining discounts the equation

$$r = \frac{A}{Pt} - \frac{1}{t}$$

arises. Solve for A.

69. In the study of the convection of heat, the equation

$$\frac{1}{U} = \frac{x}{k} + \frac{1}{h}$$

is used. Solve for x.

70. In the study of the motion of the planets, the equation

$$\frac{mv^2}{2} - \frac{kmM}{r} = -\frac{kmM}{2a}$$

is found. Solve for r.

71. A company allocates $60,000 more for research than for manufacturing. If there is a $\frac{5}{3}$ ratio of research to manufacturing, find the amount spent on research.

72. One automatic packaging machine can package 100 boxes of machine parts in 12 minutes and a second machine can do it in 10 minutes. A newer model of the same machine can do it in 8 minutes. How long will it take the three machines to pack 100 boxes of machines parts if they are all working at the same time?

73. The estate of a recently deceased relative was divided among three cousins as follows: 1/5 went to one cousin, 2/3 went to a second cousin, and the remainder of the estate went to the remaining cousin. This person received $25,000. How much did the other two cousins inherit?

74. A marathoner completes a training run at a rate of 6.5 mi/h. Two hours after she begins her run a friend starts out along the same trail at a pace of 8.5 mi/h. How many hours will it take for the second runner to overtake the first runner?

Chapter Test

In Exercises 1–3, reduce the given fractions to simplest form.

1. $\dfrac{12m^2n}{15mn^2}$ 2. $\dfrac{abx - bx^2}{acx - cx^2}$ 3. $\dfrac{3a^2 + 6a}{a^2 + 4a + 4}$

In Exercises 2–6, perform the indicated operations.

4. $\dfrac{15a^2}{7b^2} \times \dfrac{28ab}{9a^3c}$

5. $\dfrac{5x - 1}{x^2 - x - 2} \times \dfrac{x - 2}{10x^2 + 13x - 3}$

6. $\dfrac{5xy}{a - x} \div \dfrac{10xy}{a^2 - x^2}$ 7. $\dfrac{3}{4x^2} - \dfrac{2}{x^2 - x} - \dfrac{x}{2x - 2}$

8. $\dfrac{4x}{x^2 + 3x + 2} + \dfrac{7x - 1}{x^2 + 4x - 5}$

9. Solve the following fractional equation.

$$\frac{3}{2x^2 - 3x} + \frac{3}{x} = \frac{3}{2x - 3}$$

10. One company determines that it will take its crew 450 hours to clean up a chemical dump site, and a second company determines that it will take its crew 600 hours to clean up the site. How long will it take the two crews working together the clean up the chemical dump site?

10 Exponents, Roots, and Radicals

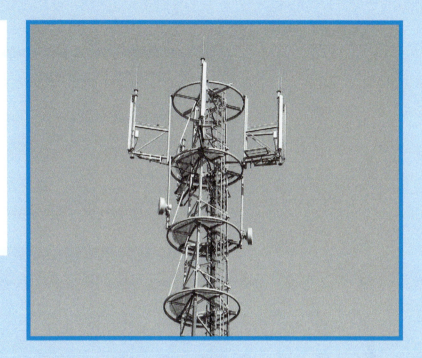

A cellular network is composed of cellular towers. A call is placed from your cell phone and travels by radio waves to a cellular tower that picks up the signal from your phone. The signal is then sent to a **M**obile **T**elephone **S**witching **O**ffice (MTSO) that handles calls from several towers called a cluster. The MTSO transfers the signal to a local phone line that sends the signal to the final destination via landline, microwave signals, or satellite. If you are driving while making your hands-free call, the MTSO links your call to neighboring cells. It senses when the signal is becoming weak and hands your call off to the next tower or the MTSO in the next cluster. It controls the quality of your radio link by keeping you connected to the best possible signal as you move from cell to cell. The process is seamless as far as the cell phone user is concerned. Cell tower construction is costly. Therefore, the placement of the cell towers is important.

The cell tower transmission area is generally represented as hexagonal in shape. The area of the hexagon is given by the formula $\dfrac{3\sqrt{3}}{2}L^2$, where L represents the length of the side of the hexagon. In this chapter we will evaluate such expressions as we explore the rules for exponents and work with radical expressions.

10.1   Integral Exponents

Section Objectives
- Review the laws of exponents
- Use the laws of exponents to simplify expressions with integral exponents

The laws of **exponents** were introduced in Chapter 3. We restate them here:

Summary of the Laws of Exponents

$$(a^m)(a^n) = a^{m+n}$$
$$(a^m)^n = a^{mn}$$
$$(ab)^n = a^n b^n$$
$$\frac{a^m}{a^n} = a^{m-n} \qquad a \neq 0$$
$$a^{-n} = \frac{1}{a^n} \qquad a \neq 0$$

These rules for working with exponents were originally defined for positive integers; however, they are valid for all integral exponents. Later in this chapter we will show how fractions can also be used as exponents.

The following examples illustrates some of these laws of exponents.

Example 1

$$(a^m)(a^n) = a^{m+n}$$
$$(a^m)^n = a^{mn}$$
$$(ab)^n = a^n b^n$$

$$(2^3)(2^4) = 2^7$$
$$(2^3)^5 = 2^{15}$$
$$(4x)^2 = 16x^2$$

$$\frac{a^m}{a^n} = a^{m-n} \quad \text{if} \quad m > n \quad \text{and} \quad a \neq 0 \qquad \frac{x^8}{x^3} = x^5$$

$$\frac{a^m}{a^n} = \frac{1}{a^{n-m}} \quad \text{if} \quad m < n \quad \text{and} \quad a \neq 0 \qquad \frac{x^3}{x^8} = \frac{1}{x^5}$$

$$\frac{a^m}{a^n} = 1 \quad \text{if} \quad m = n \quad \text{and} \quad a \neq 0 \qquad \frac{a^6}{a^6} = 1$$

Now Try It!

1. $x^5 x^2$

2. $(a^{-3})^3$

3. $(3x)^3$

4. $\dfrac{z^7}{z^3}$

5. $\dfrac{-18x^2yz^4}{9xy^3}$

6. $\dfrac{20x^3y^3z^3}{4x^3yz^5}$

We now need to further develop those concepts so that we will be prepared to consider some of the topics that will arise in later sections.

Although the above rules involve only positive integer exponents, this section shows how negative integers and zero can also be used as exponents. Chapter 12 also involves the use of negative integer and zero exponents.

Consider the case in which the exponent in the denominator is larger than the numerator so that $m - n$ is a negative integer. This is illustrated in Example 2.

Example **2**

$$\frac{a^3}{a^5} = a^{3-5} = a^{-2}$$

Using $a^{-n} = \dfrac{1}{a^n}$ for $a \neq 0$

we see that

$$a^{-2} = \frac{1}{a^2}$$

This shows us that we can use negative integers as exponents. Also, we see that *we can move a **factor** from the numerator to the denominator or from the denominator to the numerator by changing the sign of its exponent.* Note that only the sign of the exponent changes with such a move, but the sign of the fraction remains the same. Consider Example 3.

Example **3**

In the division $b^4 \div b^7$, we can write

$$\frac{b^4}{b^7} = b^{4-7} = b^{-3} = \frac{1}{b^3}$$

As with positive exponents, we must be careful to note just what number or expression has the negative exponent. Consider the steps in Example 4.

Example 4

1. $4x^{-2} = \dfrac{4}{x^2}$ since **only x has the exponent of -2.**

2. $(4x)^{-2} = \dfrac{1}{(4x)^2} = \dfrac{1}{16x^2}$ since the quantity $4x$ has the exponent of -2, as indicated by the parentheses.

3. $\dfrac{1}{3a^{-1}} = \dfrac{1}{3\left(\dfrac{1}{a}\right)} = \dfrac{1}{\dfrac{3}{a}} = 1 \times \dfrac{a}{3} = \dfrac{a}{3}$

In general, with certain specific exceptions, *negative exponents are not used in the expression of a final result.* They are, however, used in intermediate steps in many operations.

Example 5

In reducing the fraction

$$\frac{x^2yz^3}{xy^4z^2}$$

to its simplest form, we can place all the factors in the numerator by the use of negative exponents and combine the exponents of the same base to determine the result. This leads to

$$\frac{x^2yz^3}{xy^4z^2} = x^2x^{-1}yy^{-4}z^3z^{-2} = x^{2-1}y^{1-4}z^{3-2} = xy^{-3}z = \frac{xz}{y^3}$$

In dividing a^m by a^n, where $m = n$, we note that

$$\frac{a^m}{a^m} = 1 \quad \text{a quantity divided by itself equals 1} \qquad \frac{a^m}{a^m} = a^{m-m} = a^0$$

Therefore

$$\boxed{a^0 = 1 \quad \text{if} \quad a \neq 0}$$

This equation states that *any algebraic expression that is not zero and that is raised to the zero power is 1.*

Example 6

$$5^0 = 1 \qquad\qquad x^0 = 1$$
$$(rs)^0 = 1 \qquad\qquad (ax - p)^0 = 1$$
$$5x^0 = 5(1) = 5 \qquad\qquad (5x)^0 = 1$$

▶ Note that in the expression $5x^0$ *only x has the exponent zero.*

Zero and negative exponents can be used in the other laws of exponents just as any positive exponent is used. For reference, we now expand the list the laws of exponents we have encountered:

$$(a^m)(a^n) = a^{m+n}$$
$$(a^m)^n = a^{mn}$$
$$(ab)^n = a^n b^n$$
$$\frac{a^m}{a^n} = a^{m-n} \qquad (a \neq 0)$$
$$\left(\frac{a}{b}\right)^n = \frac{a^n}{b^n} \qquad (b \neq 0)$$
$$a^{-n} = \frac{1}{a^n} \qquad (a \neq 0)$$
$$a^0 = 1 \qquad (a \neq 0)$$

Example 7

Consider the expression $(x + 2)^{-2}$.

$$(x + 2)^{-2} = \frac{1}{(x + 2)^2} = \frac{1}{(x + 2)(x + 2)} = \frac{1}{x^2 + 4x + 4}$$

Caution

Negative exponents are generally not used when we write our final results. There are exceptions, such as some numbers expressed in scientific notation.

Note that $(x + 2)^{-2} \neq x^{-2} + 2^{-2}$. This is a mistake that is commonly made when simplifying this type of expression.

We see that $x^{-2} + 2^{-2} = \frac{1}{x^2} + \frac{1}{2^2} = \frac{1}{x^2} + \frac{1}{4}$

Clearly $(x + 2)^{-2} \neq x^{-2} + 2^{-2}$ since $\frac{1}{x^2 + 4x + 4} \neq \frac{1}{x^2} + \frac{1}{4}$

Example
8

In developing a computer program to search for data, we find that we must solve the equation

$$\frac{1}{2^n} = 8$$

for *n*. We proceed as follows

equal $\begin{array}{c} \dfrac{1}{2^n} = 2^{-n} \\[2mm] 8 = 2^3 \end{array}$ equal

We conclude that $-n = 3$, or $n = -3$.

Common Errors

Parentheses are important when working with exponents:

$(7x)^0 = 1$ while $7x^0 = 7(1) = 7$

$(7x)^{-1} = \dfrac{1}{7x}$ while $7x^{-1} = \dfrac{7}{x}$

$(x + y)^{-1} \neq x^{-1} + y^{-1}$

$\dfrac{1}{x + y} \neq \dfrac{1}{x} + \dfrac{1}{y}$

10.1 Exercises

In Exercises 1 through 44, express each of the given expressions in the simplest form. Write all answers with only positive exponents.

1. t^{-5}

2. R^{-3}

3. x^{-4}

4. s^{-8}

5. $\dfrac{1}{x^{-3}}$

6. $\dfrac{1}{a^{-7}}$

7. $\dfrac{1}{R_1^{-3}}$

8. $\dfrac{1}{t^{-8}}$

9. $3c^{-2}$

10. $(3c)^{-1}$

11. $\dfrac{1}{3c^{-1}}$

12. $\dfrac{1}{(3c)^{-1}}$

13. 5^0

14. $\dfrac{1}{5^0}$

15. $(c - 5)^0$

16. $\dfrac{1}{(c - 5)^0}$

17. $5c^0$

18. $3x^0$

19. $\dfrac{9x^0}{y^{-2}}$

20. $\dfrac{2x^{-1}}{3y^0}$

21. $\dfrac{3}{3^{-5}}$

22. $\dfrac{7^{-2}}{7^6}$

23. $\dfrac{6^{-4}}{6^{-5}}$

24. $\dfrac{9^{-5}}{9^{-4}}$

25. $a^2 x^{-1} a^{-3}$

26. $b^{-2} c^{-3} b^5$

27. $2(c^{-2})^4$

28. $6(x^{-1})^{-1}$

29. $\dfrac{x^{-5} y^{-1}}{xy^{-3}}$

30. $\dfrac{3^{-1} a}{3a^{-1}}$

31. $\dfrac{5^{-2} x}{5x^{-1}}$

32. $\dfrac{2a^{-3}}{2^{-1} a}$

33. $\dfrac{st^{-2}}{(2t)^0}$

34. $\dfrac{(m^{-2} n)^0}{m^3 n^{-4}}$

35. $\dfrac{2^{-2} x^{-2} y^{-4}}{2x^{-6} y^0}$

36. $\dfrac{6^{-1} a^{-2} b}{6a^{-3} b^{-2}}$

37. $\dfrac{(3a)^{-1} b^2}{3a^0 b^{-5}}$

38. $\dfrac{(3p^0)^{-2}}{3p^2 q^{-8}}$

39. $\left(\dfrac{2a^{-1}}{5b} \right)^2$

40. $\left(\dfrac{x^2}{y^{-1} z} \right)^{-2}$

41. $\dfrac{(xy^{-1})^{-2}}{x^2 y^{-3}}$

42. $\dfrac{r^0 st^{-2}}{(r^{-1} s^2 t)^{-3}}$

43. $\dfrac{(3a^{-2} bc^{-1})^{-1}}{6a^{-3}(bc)^{-1}}$

44. $\dfrac{(axy^{-1})^{-2}}{(a^{-2} x^{-1} y)^{-3}}$

In Exercises 45 through 50, solve the given problems.

45. The resistance in a certain parallel electric circuit is described as

$$R_1^{-1} + R_2^{-1}$$

Rewrite this expression so that it contains only positive exponents.

46. In certain economic circumstances, the price p for each of x commodities is given by the formula

$$p = 10(3^{-x})$$

Express this formula without the use of negative exponents.

47. Negative exponents are often used to express the units of a quantity. For example, velocity can be expressed in units of ft/s of ft $\times$ s^{-1}. In this same manner, express the units of density in g/cm^3 and the units of acceleration m/s^2 by the use of negative exponents.

48. Hall's constant for certain semiconductors is given as

$$2 \times 10^{-3} m^3 c^{-1}$$

Express this constant using only positive exponents.

49. In determining heating requirements, a solar engineer uses the expression

$$362 \text{ Btu} \times h^{-1} \times ft^{-2}$$

Write this expression using only positive exponents.

50. In determining the number of seedlings that must be planted now in order to produce a certain number of trees later, a forester encounters the equation

$$\left(\frac{1}{2}\right)^{n} = 16$$

Find a value for n that makes this equation true.

Now Try It! Answers

1. x^7 **2.** a^{-9} **3.** $27x^3$ **4.** z^4 **5.** $\dfrac{-2xz^4}{y^2}$ **6.** $\dfrac{5y^2}{z^2}$

10.2 **Fractional Exponents**

Section Objectives
- Understand the relationship between radicals and fractional exponents

So far, the only numbers used for exponents have been positive integers, negative integers, and zero. In this section we will show how fractions can also be used as exponents. We will see that these fractional exponents are very convenient in certain topics of mathematics.

In Chapter 1 we introduced the concept of both the *power of a number* and the *root of a number*. We stated that finding the root of a number was the reverse of raising a number to a power. Therefore, the square root of 25 is 5 because 5^2 is 25.

In general we define $a^{1/n}$ to be the **principal root** of a or

$$a^{1/n} = \sqrt[n]{a} \text{ and}$$
$$a^{m/n} = (\sqrt[n]{a})^m$$
$$a^{m/n} = \sqrt[n]{a^m}$$

This tells us that any value or expression raised to a fractional exponent can also be written as a radical.

For fractional exponents to be meaningful, they must satisfy the basic laws of exponents already established.

Example 1

$$4^{1/2} = \sqrt{4} = 2 \qquad\qquad (27)^{1/3} = \sqrt[3]{27} = 3$$

$$(-32)^{1/5} = \sqrt[5]{-32} = -2 \qquad\qquad (x^4)^{1/2} = \sqrt{x^4} = \sqrt{(x^2)^2} = x^2$$

Normally it is not necessary, and in fact it is often cumbersome, to include the radical interpretation when evaluating expressions with fractional exponents. Thus the preceding results would usually be written as

$$4^{1/2} = 2 \qquad (27)^{1/3} = 3 \qquad (-32)^{1/5} = -2 \qquad (x^4)^{1/2} = x^2$$

Example 2

$(a^{1/3})^2 = a^{2/3}$, which can be written as $(a^2)^{1/3}$. Thus $(a^2)^{1/3} = (a^2)^{1/3}$, or $(\sqrt[3]{a})^2 = \sqrt[3]{a^2}$.

Example 3

$$8^{2/3} = (8^{1/3})^2 = 2^2 = 4 \qquad\qquad 4^{7/2} = (4^{1/2})^7 = 2^7 = 128$$

$$(27)^{4/3} = [(27)^{1/3}]^4 = 3^4 = 81 \qquad (x^4)^{3/2} = [(x^4)^{1/2}]^3 = (x^2)^3 = x^6$$

In Example 3, although $(8^{1/3})^2 = (8^2)^{1/3}$, it is usually easier to take the root first and then raise the result to the power.

Example 4

$8^{2/3}$ written in radical form is $\sqrt[3]{8^2}$ or $(\sqrt[3]{8})^2$. Since $\sqrt[3]{8^2} = \sqrt[3]{64} = 4$ and $(\sqrt[3]{8})^2 = 2^2 = 4$, we see that $8^{2/3} = \sqrt[3]{8^2} = (\sqrt[3]{8})^2$.

Also, $4^{7/2} = \sqrt{4^7} = (\sqrt{4})^7$.

Examples 5–7 illustrate the use of the basic laws of exponents with expressions that have fractional exponents.

Example 5

$$8^{-2/3} = \frac{1}{8^{2/3}} = \frac{1}{(8^{1/3})^2} = \frac{1}{2^2} = \frac{1}{4}$$

$$(8^{4/5})^0 = 1$$

$$2^{1/2}2^{1/3} = 2^{1/2 + 1/3} = 2^{5/6}$$

$$\frac{2^{1/2}}{2^{1/3}} = 2^{1/2 - 1/3} = 2^{1/6}$$

$$\frac{4^{-1/2}}{2^{-5}} = \frac{2^5}{4^{1/2}} = \frac{32}{2} = 16$$

Example 6

$$x^{1/2}x^{1/4} = x^{1/2 + 1/4} = x^{3/4}$$

$$\frac{a}{a^{1/2}b^{-2}} = a^{1 - 1/2}b^2 = a^{1/2}b^2$$

$$\frac{x^{-1/3}y^{2/3}}{x^{2/3}y^{-2/3}} = \frac{y^{2/3 + 2/3}}{x^{2/3 + 1/3}} = \frac{y^{4/3}}{x}$$

Example 7

In order for an aircraft wing to produce the required lift of 1000 N/m², the velocity of air over the upper wing surface is found by applying Bernoulli's equation and evaluating

$$V_1 = (2L + RV_2^2)^{1/2}R^{-1/2}$$

Find V_1 (in meters per second) if the velocity on the lower wing surface is $V_2 = 108$ m/s, the air density is $R = 1.29$ kg/m³, and the lift is $L = 1000$ N/m².

First, evaluating $2L + RV_2^2$, we get $2(1000) + (1.29)(108)^2 = 17050$. We now evaluate

$$V_1 = (2L + RV_2^2)^{1/2}R^{-1/2}$$

$$= (17,050)^{1/2}(1.29)^{-1/2} = \frac{17,050^{1/2}}{1.29^{1/2}}$$

$$= \frac{\sqrt{17,050}}{\sqrt{1.29}} = 115 \text{ m/s}$$

The velocity of air over the upper wing surface must be 115 m/s.

10.2 Exercises

In Exercises 1 through 8, change the given expressions to radical form.

1. $5^{1/2}$ **2.** $7^{1/3}$ **3.** $a^{1/4}$ **4.** $b^{1/5}$

5. $x^{3/5}$ **6.** $x^{3/4}$ **7.** $R^{7/3}$ **8.** $s^{9/7}$

In Exercises 9 through 16, use fractional exponents to express the following:

9. $\sqrt[3]{a}$ **10.** $\sqrt[7]{b}$ **11.** $\sqrt{x}$ **12.** $\sqrt[5]{ax}$

13. $\sqrt[3]{x^2}$ **14.** $\sqrt[4]{a^3}$ **15.** $\sqrt[5]{b^8}$ **16.** $\sqrt[10]{s^5}$

In Exercises 17 through 38, evaluate the given expressions.

17. $9^{1/2}$ **18.** $64^{1/3}$ **19.** $16^{1/4}$ **20.** $(-8)^{1/3}$

21. $4^{3/2}$ **22.** $9^{3/2}$ **23.** $8^{4/3}$ **24.** $27^{2/3}$

25. $81^{3/4}$ **26.** $16^{3/4}$ **27.** $(-8)^{2/3}$ **28.** $(-27)^{5/3}$

29. $36^{-1/2}$ **30.** $100^{-1/2}$ **31.** $16^{-1/4}$ **32.** $32^{-2/5}$

33. $-(-32)^{-1/5}$ **34.** $-64^{-2/3}$ **35.** $(4^{1/2})(27^{2/3})$

36. $\dfrac{64^{1/2}}{64^{1/3}}$ **37.** $2^{-1/2} \times 2^{3/2}$ **38.** $\dfrac{25^{-3/2}}{81^{3/4}}$

In Exercises 39 through 46, use the laws of exponents to simplify the given expressions. Express all answers with positive exponents.

39. $a \times a^{1/2}$ **40.** $\dfrac{b}{b^{1/3}}$ **41.** $\dfrac{ab}{a^{1/4}}$ **42.** $\dfrac{xyz^2}{z^{-3/2}}$

43. $\dfrac{x(x^{1/3})^4}{x^{2/5}}$ **44.** $\dfrac{xy^{1/2}}{(y^{2/3})^{-2}}$ **45.** $\dfrac{x^{-1/2}x^2}{x^{2/3}}$ **46.** $\dfrac{xy^{-1/2}z}{z^{2/3}x^{-1/2}}$

In Exercises 47 through 52, solve the given problems.

47. A restaurant finds that it can make better use of its floor space by using tables with shapes described by the equation

$$\sqrt{x^5} + \sqrt{y^5} = k$$

Express this equation with fractional exponents instead of radicals.

48. An expression which arises in Einstein's theory of relativity is

$$\sqrt{\frac{c^2 - v^2}{c^2}}$$

Express this with the use of fractional exponents and then simplify.

49. In analyzing curvature in the deflection of a beam, the equation

$$R = \frac{(1 + D_1^2)^{3/2}}{D_2}$$

is found. Express this equation in radical form.

50. Considering only the volume of gasoline vapor and air mixture in the cylinder, the efficiency of an internal combustion engine is approximately

$$E = 100\left(1 - \frac{1}{R^{2/5}}\right)$$

where E is in percent and R is the compression ratio of the engine. Rewrite this equation in radical form.

51. In determining the number of electrons involved in a certain calculation with semiconductors, the expression

$$9.60 \times 10^{18} T^{3/2}$$

is used. Evaluate this expression for $T = 289\ K$.

52. In studying the magnetism of a circular loop, the equation

$$B = \frac{2\pi k I R^2}{(x^2 + R^2)^{3/2}}$$

is found. Simplify this expression for the center of the loop, where $x = 0$.

10.3 Imaginary Roots

Section Objectives
- Introduce the concept of an imaginary number
- Simplify radicals having negative values under the radical sign

Earlier we defined *a square root of a given number to be that number which, when squared, equals the given number.* The square of a positive or negative number is always positive, yet it is not possible to square any real number and get a negative result. Intuitively we know that there are actually two numbers that could be considered to be the square root of a given positive number.

Because $(+3)^2 = 9$ and $(-3)^2 = 9$, we see that, by the definition we have used up to this point, the square root of 9 can be either $+3$ or -3. Also, because the square of $+3$ and the square of -3 are both 9, we have no apparent result for the square root of -9.

Because there are two numbers, one positive and the other negative, whose squares equal a given number, we define *the **principal square root** of a positive number to be positive.* That is:

$$\boxed{\sqrt{N} = x}$$

*where $\sqrt{N}$ is the principal square root of N and **equals (positive) x.***

By defining the principal square root of a number in this way, we do not mean that problems involving square roots cannot have negative results. By $-\sqrt{N}$ we indicate the negative of the principal value, which means the negative of a positive signed number, and this is a negative number.

Example **1**

$$-\sqrt{16} = -(+4) = -4 \qquad\qquad -\sqrt{25} = -5$$
$$-\sqrt{81} = -9 \qquad\qquad\qquad\quad -\sqrt{0.09} = -0.3$$

Returning to the problem of the square root of a negative number, we find it necessary to define a new kind of number to provide the required result. We define

$$\boxed{j^2 = -1 \quad \text{or} \quad j = \sqrt{-1}}$$

*The number j is called the **imaginary unit.*** (It is sometimes referred to as the *j-operator.*) Some fields of science, technology, and mathematics normally represent

Imaginary numbers do exist and, in spite of their name, they are not really imaginary at all. The name dates back to when they were first introduced. At that time, people tried to imagine what it would be like to have a number system that contained square roots of negative numbers, hence the name "imaginary." Eventually it was realized that such a number system does in fact exist, but by then the name had stuck.

the imaginary unit by i instead of j, but we will use j because electric current is also represented by i.

There are occasions where it will be necessary to evaluate powers of the imaginary unit, particularly j^2.

$$j = \sqrt{-1}$$
$$j^2 = \sqrt{-1}\sqrt{-1} = -1$$
$$j^3 = j^2 \times j = -1j = -j$$
$$j^4 = j^2 \times j^2 = (-1)(-1) = 1$$
$$j^5 = j^4 \times j = 1 \times j = j$$
$$j^6 = j^4 \times j^2 = 1 \times j^2 = 1(-1) = -1$$

Notice that $j^5 = j$ and $j^6 = j^2$. In fact, the powers of j go through a cycle with the first four powers repeating themselves. In a similar manner you can easily show that $j^7 = -j$ and $j^8 = 1$.

Since j^2 is defined to be -1, it follows that

$$\sqrt{-a} = \sqrt{a(-1)} = \sqrt{aj^2} = \sqrt{a}\sqrt{j^2} = \sqrt{a}\,j = j\sqrt{a}$$

where a is a positive number. That is,

$$\boxed{\begin{aligned} \sqrt{-a} &= \sqrt{a}\,j \quad \text{(where } a > 0\text{)} \\ &= j\sqrt{a} \end{aligned}}$$

Example
2

$$\sqrt{-16} = 4j \qquad\qquad \sqrt{-25} = 5j$$

$$-\sqrt{-9} = -3j \qquad\qquad -\sqrt{-100} = -10j$$

Also,

$$(6j)^2 = 6^2j^2 = 36(-1) = -36$$

$$(2j)^2 = 2^2j^2 = 4(-1) = -4$$

$$(-3j)^2 = (-3)^2j^2 = 9(-1) = -9$$

Most of our work with roots will deal with square roots, but we shall have some occasion to consider other roots of a number. In general: *The* **principal nth root** *of a number N is designated as*

$$\boxed{\sqrt[n]{N} = x}$$

where $x^n = N$. The number n is called the **index** and N is called the **radicand.** (If $n = 2$, it is usually not written, as we have already seen). *If N is positive, x is positive. If N is negative and n is odd, x is negative.* That is, odd roots of negative numbers are negative. We shall not consider the case of N being negative when n is even and larger than 2, as in expressions such as $\sqrt[4]{-16}$ or $\sqrt[6]{-64}$.

Example 3

$$\sqrt[3]{64} = 4 \quad \text{since} \quad 4^3 = 64$$
$$\sqrt[3]{-64} = -4 \quad \text{since} \quad (-4)^3 = -64$$
$$\sqrt[4]{81} = 3 \quad \text{since} \quad 3^4 = 81$$
$$\sqrt[5]{32} = 2 \quad \text{since} \quad 2^5 = 32$$
$$\sqrt[5]{-32} = -2 \quad \text{since} \quad (-2)^5 = -32$$

Caution

When using your graphing calculator be aware that the TI-84 uses the imaginary unit i instead of j.

If a negative sign precedes the radical sign, we express the result as the negative of the principal value. That is, we first determine the value of the nth root and then consider the sign preceding the radical. This is illustrated in Example 4.

Example 4

$$-\sqrt[3]{64} = -4 \qquad -\sqrt[3]{-64} = -(-4) = 4$$

$$-\sqrt[4]{16} = -2 \qquad -\sqrt[5]{-32} = -(-2) = 2$$

$$-\sqrt[7]{-128} = -(-2) = 2$$

Example 5

Find the surface area of a cube with a volume of 8.00 ft³.
We must use the formula for the volume of a cube and solve for e. We get

$$e^3 = 8.00 \quad \text{or} \quad e = \sqrt[3]{8.00} = 2.00 \text{ ft}$$

Since each edge is 2.00 ft, each of the six square sides must have an area of 4.00 ft² so that the total surface area is 24.0 ft².

We conclude this section with a brief discussion of the type of number that the root of a number represents. Recall that a *rational number is a number that can be expressed as the division of an integer by another integer.* If the root of a number can be expressed as the ratio of one integer to another, then it is **rational.** Otherwise, it is **irrational.** All of the rational and irrational numbers combined constitute the **real numbers.** The **imaginary numbers** introduced in this section constitute another type of number.

Example
6

$\sqrt{4} = 2$, and since 2 is an integer, it is rational.

$\sqrt[3]{0.008} = 0.2 = \frac{1}{5}$, and since $\frac{1}{5}$ is the ratio of one integer to another, it is rational.

$\sqrt{2}$ is approximately 1.414, but it is not possible to find two integers whose ratio is exactly $\sqrt{2}$. Therefore $\sqrt{2}$ is irrational.

$\sqrt[3]{7}$ and $\sqrt[5]{-39}$ are irrational since it is impossible to express either number as the ratio of one integer to another.

$\sqrt{-4} = 2j$, which we have already found to be an imaginary number.

In discussing roots of numbers we are considering numbers of which many—in fact most—are irrational.

10.3 *Exercises*

In Exercises 1 through 32, find the indicated principal roots.

1. $\sqrt{49}$

2. $\sqrt{36}$

3. $-\sqrt{144}$

4. $-\sqrt{81}$

5. $\sqrt{0.16}$

6. $\sqrt{0.01}$

7. $-\sqrt{0.04}$

8. $-\sqrt{0.36}$

9. $\sqrt{400}$

10. $\sqrt{900}$

11. $-\sqrt{1600}$

12. $-\sqrt{3600}$

13. $\sqrt[3]{8}$

14. $\sqrt[3]{27}$

15. $\sqrt[3]{-8}$

16. $\sqrt[3]{-27}$

17. $-\sqrt[3]{125}$

18. $-\sqrt[3]{-125}$

19. $\sqrt[3]{0.125}$

20. $\sqrt[3]{-0.001}$

21. $\sqrt[4]{16}$

22. $-\sqrt[4]{625}$

23. $\sqrt[5]{243}$

24. $-\sqrt[5]{-243}$

25. $\sqrt[6]{64}$

26. $-\sqrt[8]{256}$

27. $\sqrt{-4}$

28. $\sqrt{-49}$

29. $-\sqrt{-400}$

30. $-\sqrt{-900}$

31. $\sqrt{-0.49}$

32. $-\sqrt{-0.0001}$

In Exercises 33 through 40, perform the indicated operations.

33. $(5j)^2$

34. $5j^2$

35. $(5j)(4j)$

36. $(-7j)^2$

37. $(-3j)(-4j)$

38. $(-8j)^2$

39. $-(8j)^2$

40. $-8j^2$

In Exercises 41 through 46, solve the given problems.

41. A water tank is in the shape of a cube with a volume of 216 ft³. What is the length of an edge of this tank? See Figure 10.1.

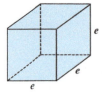

$V = 216$ ft³

Figure 10.1

42. The period T, in seconds, of a pendulum of length l, in feet, is given by the equation

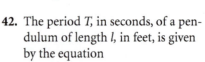

$$T = \frac{\pi}{2}\sqrt{\frac{l}{2}}$$

What is the period of a pendulum with a length of 18.0 ft?

43. The ratio of the rates of diffusion of two gases is given by

$$\frac{r_1}{r_2} = \frac{\sqrt{m_2}}{\sqrt{m_1}}$$

where m_1 and m_2 are the masses of the molecules of the gases. Find the ratio r_1/r_2 if $m_1 = 25$ units and $m_2 = 81$ units.

44. In a study of the population growth of a bacteria culture, the number of bacteria currently present is found to be 1000 times the fourth root of the number present in the original culture. If the original culture had 810,000 bacteria, find the number currently present.

45. The annual yield (in decimal form) of an investment is found by evaluating

$$\sqrt[3]{\frac{V}{I}} - 1$$

where V is the value of the investment after three years and I is the original amount invested. Evaluate this expression for $V = \$532.40$ and $I = \$400$.

46. The diameter (in millimeters) of a water droplet can be found by using the formula

$$d = \sqrt[3]{0.95493V}$$

where V is the volume. Find d if $V = 28.275$ mm^3.

10.4 Simplifying Radicals

Section Objectives
- Introduce the root of a product rule
- Introduce the root of a quotient rule
- Write a radical expression in its simplest form
- Rationalize a denominator

In the last section we showed how a quantity raised to a fractional exponent can be written as a **radical**. In general we stated this relationship as follows:

> For each of the following assume that a is a positive real number.
>
> $$a^{1/n} = \sqrt[n]{a}$$
>
> $$a^{m/n} = (\sqrt[n]{a})^m$$
>
> $$\sqrt[n]{a^n} = a^{n/n} = a$$

We now introduce several additional rules for radicals that will be useful to us as we simplify radical quantities. To avoid difficulties with imaginary numbers, we will assume that all letters represent positive real numbers.

> $$\sqrt[n]{ab} = \sqrt[n]{a} \times \sqrt[n]{b}$$
>
> $$\sqrt[n]{\frac{a}{b}} = \frac{\sqrt[n]{a}}{\sqrt[n]{b}}$$
>
> where a and b are positive real numbers.

In each of the rules listed above, the number under the **radical sign** is called the **radicand,** and the number indicating the root being taken is called the **index.** We will use these as we work to simplify a radical to its simplest form.

Example 1

We know that $\sqrt{36} = 6$. If we now consider

$$\sqrt{36} = \sqrt{4 \times 9} = \sqrt{4}\sqrt{9} = 2 \times 3$$

we see that the same result of 6 is obtained.

If the root of either a or b can be found exactly, we can simplify the radical. Consider Example 2.

Example 2

By writing $\sqrt{48} = \sqrt{16 \times 3}$, we get the following result:

$$\sqrt{48} = \sqrt{16 \times 3} = \sqrt{16}\sqrt{3} = 4\sqrt{3}$$

Thus $\sqrt{48}$ can be expressed in terms of $\sqrt{3}$. Also, in the same way we have

$$\sqrt{175} = \sqrt{25 \times 7} = \sqrt{25}\sqrt{7} = 5\sqrt{7}$$

To Reduce a Radical to its Simplest Form

1. Factor the expression under the radical sign so that one or more of the factors is a perfect n^{th} power (where n is the index of the radical).

2. Write each factor in the product under its own radical sign.

3. Reduce the perfect n^{th} power root you created.

A radical is not simplified completely until the number or expression under the radical sign contains no perfect n^{th} power factors.

Example 3

In Example 2 we could have written $\sqrt{48} = \sqrt{6 \times 8}$. However, with these factors no simplification can be done because neither 6 nor 8 is a perfect square. If we noted that $\sqrt{48} = \sqrt{4 \times 12} = \sqrt{4}\sqrt{12} = 2\sqrt{12}$, the radical has been simplified, *but not completely.* We must then note that $\sqrt{12} = \sqrt{4 \times 3} = \sqrt{4}\sqrt{3} = 2\sqrt{3}$, which means that $\sqrt{48} = 2\sqrt{12} = 2(2\sqrt{3}) = 4\sqrt{3}$. Radicals should always be expressed in simplest form.

Other illustrations of simplifying radicals follow:

> To simplify means to find another expression with the same value. It does not mean to find a decimal approximation.

perfect square

$$\sqrt{72} = \sqrt{36 \times 2} = \sqrt{36}\sqrt{2} = 6\sqrt{2}$$

$$\sqrt{54} = \sqrt{9 \times 6} = \sqrt{9}\sqrt{6} = 3\sqrt{6}$$

$$\sqrt{126} = \sqrt{9 \times 14} = \sqrt{9}\sqrt{14} = 3\sqrt{14}$$

$$\sqrt{-25} = \sqrt{25 \times (-1)} = \sqrt{25}\sqrt{-1} = 5j$$

With a number such as 126, the perfect square factor of 9 might not be immediately obvious, and we may have to experiment with different combinations of factors until the right combination is identified.

An algebraic product can be expressed as the product of square roots. Then, for any of these products the expression can be simplified. Example 4 gives three illustrations.

Example 4

$$\sqrt{a^2 b^2 c} = \sqrt{a^2}\sqrt{b^2}\sqrt{c} = ab\sqrt{c}$$

$$\sqrt{25a^4} = \sqrt{25}\sqrt{a^4} = 5\sqrt{(a^2)^2} = 5a^2$$

$$\sqrt{6a^2 b^6 x} = \sqrt{6}\sqrt{a^2}\sqrt{b^6}\sqrt{x} = (\sqrt{6})(a)(\sqrt{(b^3)^2})(\sqrt{x})$$

$$= (\sqrt{6})(a)(b^3)(\sqrt{x})$$

not perfect squares

$$= ab^3\sqrt{6}\sqrt{x} = ab^3\sqrt{6x}$$

Another operation frequently performed on fractions with radicals is to write the fraction so that no radicals appear in the denominator. Before the extensive use of calculators, one reason for this was ease of calculation, and therefore this reason is no longer important. However, the procedure of writing a fraction with radicals in a form in which no radicals appear in the denominator, called **rationalizing the denominator**, is also useful for other purposes.

To Simplify a Radical Expression by Rationalizing the Denominator

1. Multiply the numerator and the denominator by a factor that will eliminate the radical sign in the denominator. Choose the smallest possible factor that will yield a perfect n^{th} root, where n is the index.

2. Make sure all of the remaining radicals are simplified.

3. Simplify the fraction if possible.

Examples 5–8 illustrate the rationalization of denominators.

Example 5

$$\frac{3}{\sqrt{2}} = \frac{3}{\sqrt{2}}\frac{\sqrt{2}}{\sqrt{2}} = \frac{3\sqrt{2}}{2}$$

choose a multiplying factor that eliminates the radical in the denominator

$$\frac{a}{\sqrt{5}} = \frac{a}{\sqrt{5}}\frac{\sqrt{5}}{\sqrt{5}} = \frac{a\sqrt{5}}{5}$$

choose the smallest factor that yields a perfect square

$$\frac{a}{\sqrt{b}} = \frac{a}{\sqrt{b}}\frac{\sqrt{b}}{\sqrt{b}} = \frac{a\sqrt{b}}{b}$$

Example 6

$$\sqrt{\frac{1}{3}} = \sqrt{\frac{1 \times 3}{3 \times 3}} = \frac{\sqrt{3}}{\sqrt{3^2}} = \frac{\sqrt{3}}{3}$$

$$\sqrt{\frac{12}{5}} = \sqrt{\frac{4 \times 3}{5}} = \frac{\sqrt{4 \times 3 \times 5}}{5 \times 5} = \frac{2\sqrt{3 \times 5}}{\sqrt{5^2}} = \frac{2\sqrt{15}}{5}$$

Example 7

$$\sqrt{\frac{a^2}{b}} = a\sqrt{\frac{1}{b}} = a\sqrt{\frac{b}{b^2}} = \frac{a\sqrt{b}}{b}$$

$$\sqrt{\frac{a^4 b}{c^3}} = \sqrt{\frac{a^4 b}{c^2 c}} = \frac{a^2}{c}\sqrt{\frac{b}{c}} = \frac{a^2}{c}\sqrt{\frac{bc}{c^2}} = \frac{a^2\sqrt{bc}}{c^2}$$

$$\sqrt{\frac{4a^3}{7b}} = 2a\sqrt{\frac{a}{7b}} = 2a\sqrt{\frac{a \times 7b}{(7b)^2}} = \frac{2a\sqrt{7ab}}{7b}$$

Example 8

The time (in seconds) it takes an object to fall 1000 ft can be expressed as

$$\sqrt{\frac{1000}{16}}$$

We can simplify this expression as follows.

$$\sqrt{\frac{1000}{16}} = \sqrt{\frac{100 \times 10}{16}} = \sqrt{\frac{100}{16}} \times \sqrt{10} = \frac{10\sqrt{10}}{4} = \frac{5\sqrt{10}}{2}$$

Example 9

$$\sqrt[3]{16} = \sqrt[3]{8}\sqrt[3]{2} = 2\sqrt[3]{2}$$

$\underline{\quad}$ perfect cube

$$\sqrt[4]{3a^7} = \sqrt[4]{3}\sqrt[4]{a^4}\sqrt[4]{a^3} = (\sqrt[4]{3})(a)(\sqrt[4]{a^3}) = a\sqrt[4]{3a^3}$$

$\underline{\quad}$ perfect fourth power

$$\sqrt[5]{32x^8} = (\sqrt[5]{32})(\sqrt[5]{x^5})(\sqrt[5]{x^3}) = (2)(x)(\sqrt[5]{x^3}) = 2x\sqrt[5]{x^3}$$

$\underline{\quad}$ perfect fifth powers

In these three illustrations, we have taken out the factors that are a perfect cube, a perfect fourth power, and a perfect fifth power, respectively. Note that the powers must be taken into account. We cannot simplify $\sqrt[3]{25}$ because no perfect cube factor is present. It is incorrect to conclude that $\sqrt[3]{25}$ simplifies to 5.

10.4 Exercises

In Exercises 1 through 12, rationalize the denominators and simplify.

1. $\dfrac{1}{\sqrt{2}}$

2. $\dfrac{1}{\sqrt{3}}$

3. $\dfrac{2}{\sqrt{5}}$

4. $\dfrac{3}{\sqrt{6}}$

5. $\dfrac{1}{\sqrt{a}}$

6. $\dfrac{a}{\sqrt{b}}$

7. $\dfrac{\sqrt{a}}{\sqrt{b}}$

8. $\dfrac{\sqrt{r}}{\sqrt{2s}}$

9. $\sqrt{\dfrac{3}{5}}$

10. $\sqrt{\dfrac{2}{7}}$

11. $\sqrt{\dfrac{a^3}{3}}$

12. $\sqrt{\dfrac{a^4}{bc}}$

In Exercises 13 through 36, express the given radicals in simplest form. If a radical appears in the denominator, rationalize the denominator.

13. $\sqrt{12}$

14. $\sqrt{27}$

15. $\sqrt{28}$

16. $\sqrt{44}$

17. $\sqrt{45}$

18. $\sqrt{99}$

19. $\sqrt{150}$

20. $\sqrt{98}$

21. $\sqrt{147}$

22. $\sqrt{162}$

23. $\sqrt{243}$

24. $\sqrt{640}$

25. $\sqrt{ac^2}$

26. $\sqrt{3a^4}$

27. $\sqrt{a^3b^2}$

28. $\sqrt{12a^5}$

29. $\sqrt{4a^2bc^3}$

30. $\sqrt{9a^3b}$

31. $\sqrt{80x^4yz^5}$

32. $\sqrt{240xy^7z^6}$

33. $\sqrt{\dfrac{ab^2}{12}}$

34. $\sqrt{\dfrac{c^3e^5}{44}}$

35. $\sqrt{\dfrac{2x^2y}{5a^8}}$

36. $\sqrt{\dfrac{13xy^5}{40a^3}}$

In Exercises 37 through 48, simplify the given radicals.

37. $\sqrt[3]{54}$

38. $\sqrt[3]{24}$

39. $\sqrt[3]{8a^4}$

40. $\sqrt[3]{6a^{10}}$

41. $\sqrt[4]{16a^9}$

42. $\sqrt[4]{81a^4b^5}$

43. $\sqrt[4]{243a^{11}}$

44. $\sqrt[5]{64x^7}$

45. $\sqrt[4]{162a^{10}x^{12}}$

46. $\sqrt[5]{243r^6s^{10}t^{12}}$

47. $\sqrt[7]{256r^7s^{14}t^{16}}$

48. $\sqrt[8]{a^9b^{16}c^5}$

In Exercises 49 through 56, solve the given problems.

49. A radar antenna is mounted on a circular plate with area A. The diameter of the plate is given by

$$d = 2\sqrt{\dfrac{A}{3.14}}$$

Express this equation in rationalized form.

50. The expression for the resonant frequency of a computer's electric circuit is given by

$$f = \dfrac{1}{2\pi}\sqrt{\dfrac{1}{LC}}$$

Express this equation in rationalized form.

51. A parking lot is planned as a rectangle with a length twice as long as its width. If its area is 12,800 ft^2, find its dimensions.

52. A beam is in the shape of a rectangular solid with a square top and bottom. The sides are rectangles with the length equal to 36 times the width. If the surface area of a side is 576 in.^{2}, find the volume of the beam.

53. The open-sea speed of an oil tanker is determined by

$$V = k\sqrt[3]{\dfrac{P}{W}}$$

Express this equation in rationalized form.

54. The radius of a spherical droplet of jet fuel can be found by using the formula

$$r = \sqrt[3]{\frac{3V}{4\pi}}$$

Express this equation in rationalized form.

55. The diameter of a crankshaft for a diesel engine is determined by

$$d = k\sqrt[3]{\frac{16J}{C}}$$

Express this equation in rationalized form.

56. In analyzing the rate of fluid flow through a pipe, the expression

$$r = \sqrt[4]{\frac{8nLR}{\pi D}}$$

is encountered. Express this equation in rationalized form.

10.5 Operations with Radicals

Section Objectives
- Add or subtract radical expressions
- Multiply or divide radical expressions

Now that we have seen how radicals can be simplified, we can show how the basic operations of addition, subtraction, multiplication, and division are performed on them. These operations follow the basic algebraic operations, the only difference being that radicals are involved. Also, these radicals must be expressed in simplest form so that the final result is in simplest form.

In adding and subtracting radicals, we add and subtract like radicals, just as we add and subtract like terms in the algebraic expressions we have already encountered. Radicals are considered to be like radicals if they have the same index. Here, like radicals are those that have the same simplest radical form. Consider Example 1.

Example 1

similar radicals

$$\sqrt{3} + 4\sqrt{5} - 3\sqrt{3} + \sqrt{5} = (\sqrt{3} - 3\sqrt{3}) + (4\sqrt{5} + \sqrt{5})$$

similar radicals

$$= -2\sqrt{3} + 5\sqrt{5}$$

Be Careful

Do not try to combine radicals that are not like radicals.

$$\sqrt{3} + \sqrt{5} \neq \sqrt{8}$$

However, when working with radicals, we must be certain that all radicals are in simplest form. This is necessary because many radicals that do not appear to be similar in form actually *are* similar, as shown in Example 2.

Example 2

$\sqrt{3} + \sqrt{80} - \sqrt{27} + \sqrt{5}$ appears to have four different types of terms, none of them similar. However, noting that

$$\sqrt{80} = \sqrt{16 \times 5} = 4\sqrt{5}$$

and

$$\sqrt{27} = \sqrt{9 \times 3} = 3\sqrt{3}$$

we have the equivalent expression $\sqrt{3} + 4\sqrt{5} - 3\sqrt{3} + \sqrt{5}$, which is the same expression found in Example 1. Now we can see that the third term and the first term are similar, as are the second and fourth terms. We get

$$\sqrt{3} + \sqrt{80} - \sqrt{27} + \sqrt{5} = \sqrt{3} + 4\sqrt{5} - 3\sqrt{3} + \sqrt{5}$$
$$= -2\sqrt{3} + 5\sqrt{5}$$

Examples 3 and 4 further illustrate the method of addition and subtraction of radicals.

Example 3

$$\sqrt{8} + \sqrt{18} - 6\sqrt{2} + 3\sqrt{32} = \sqrt{4 \times 2} + \sqrt{9 \times 2} - 6\sqrt{2} + 3\sqrt{16 \times 2}$$
$$= 2\sqrt{2} + 3\sqrt{2} - 6\sqrt{2} + 12\sqrt{2} \longleftarrow \text{all similar radicals}$$
$$= 11\sqrt{2}$$
$$\sqrt{20} + 3\sqrt{5} - 7\sqrt{12} + 2\sqrt{45} = \sqrt{4 \times 5} + 3\sqrt{5} - 7\sqrt{4 \times 3} + 2\sqrt{9 \times 5}$$
$$= 2\sqrt{5} + 3\sqrt{5} - 14\sqrt{3} + 6\sqrt{5}$$
$$= 11\sqrt{5} - 14\sqrt{3}$$

not similar to other radicals

Example 4

$$\sqrt{a^2 b} + \sqrt{9b} - 2\sqrt{16c^2 b} = a\sqrt{b} + 3\sqrt{b} - 2(4c)\sqrt{b}$$
$$= (a + 3 - 8c)\sqrt{b}$$
$$\sqrt{3a^3 c^2} - \sqrt{12ab^2} + \sqrt{24ac^3} = \sqrt{(3a)(a^2 c^2)} - \sqrt{(3a)(4b^2)} + \sqrt{(6ac)(4c^2)}$$
$$= ac\sqrt{3a} - 2b\sqrt{3a} + 2c\sqrt{6ac}$$

not similar to others
due to factor c

When multiplying expressions containing radicals, we proceed just as in any algebraic multiplication. When simplifying the result, we use the relation

$$\sqrt{a}\sqrt{b} = \sqrt{ab}$$

which was introduced in the last section. Examples 5–9 illustrate this principle.

Example
5

$$\sqrt{3}\sqrt{7} = \sqrt{21}$$
$$\sqrt{2}\sqrt{t} = \sqrt{2t}$$

$$\sqrt{2}(\sqrt{6} + 3\sqrt{5}) = \sqrt{2}\sqrt{6} + 3\sqrt{2}\sqrt{5}$$
$$= \sqrt{12} + 3\sqrt{10}$$
$$= 2\sqrt{3} + 3\sqrt{10}$$

In Example 6 we illustrate the procedure for multiplying binomial expressions that contain radical terms. We must be careful to use the distributive law correctly so that we include all terms of the result.

Example
6

$$5\sqrt{3}\sqrt{2} - 6\sqrt{3}\sqrt{2}$$

$$(\sqrt{3} - 2\sqrt{2})(3\sqrt{3} + 5\sqrt{2}) = 3\sqrt{3}\sqrt{3} - \sqrt{3}\sqrt{2} - 10\sqrt{2}\sqrt{2}$$
$$= 3\sqrt{9} - \sqrt{6} - 10\sqrt{4}$$
$$= 3(3) - \sqrt{6} - 10(2)$$
$$= 9 - \sqrt{6} - 20$$
$$= -11 - \sqrt{6}$$

$$\sqrt{a}\sqrt{b} - 6\sqrt{a}\sqrt{b}$$

$$(\sqrt{a} - 3\sqrt{b})(2\sqrt{a} + \sqrt{b}) = 2\sqrt{a}\sqrt{a} - 5\sqrt{a}\sqrt{b} - 3\sqrt{b}\sqrt{b}$$
$$= 2\sqrt{a^2} - 5\sqrt{ab} - 3\sqrt{b^2}$$
$$= 2a - 5\sqrt{ab} - 3b$$

$$(\sqrt{2} - 3\sqrt{x})(\sqrt{2} + \sqrt{3}) = \sqrt{2}\sqrt{2} + \sqrt{2}\sqrt{3} - 3\sqrt{x}\sqrt{2} - 3\sqrt{x}\sqrt{3}$$
$$= 2 + \sqrt{6} - 3\sqrt{2x} - 3\sqrt{3x}$$

not similar

In Section 10.4 we learned how to rationalize a fraction in which a radical appears in the denominator. However, we restricted our attention to the simpler cases of rationalization. Example 7 illustrates rationalizing the denominator of a fraction in which the numerator is a sum of terms.

Example
7

$$\frac{\sqrt{3} + 5}{\sqrt{2}} = \frac{(\sqrt{3} + 5)}{\sqrt{2}} \times \frac{\sqrt{2}}{\sqrt{2}} = \frac{\sqrt{3}\sqrt{2} + 5\sqrt{2}}{\sqrt{4}}$$

$$= \frac{\sqrt{6} + 5\sqrt{2}}{2}$$

If the denominator is the sum of two terms, one or both of which are radicals, the fraction can be rationalized by multiplying both the numerator and denominator by the difference of the same two terms.

This is so because

$$(\sqrt{a} + \sqrt{b})(\sqrt{a} - \sqrt{b}) = a - b$$

which yields an expression without radicals. *The* **conjugate** *of* $a + b$ *is* $a - b$, *and the conjugate of* $a - b$ *is* $a + b$. In rationalizing a fraction with a denominator having a sum or difference that involves radicals, we can multiply the numerator and denominator by the conjugate of the denominator. We illustrate this in Example 8.

Example 8

1. Rationalize $\dfrac{\sqrt{3}}{\sqrt{2} - \sqrt{5}}$.

In rationalizing this expression we multiply the numerator and the denominator by $\sqrt{2} + \sqrt{5}$, which is found from the denominator by changing the sign between the terms.

$$\frac{\sqrt{3}}{\sqrt{2} - \sqrt{5}} = \frac{\sqrt{3}}{(\sqrt{2} - \sqrt{5})} \times \frac{(\sqrt{2} + \sqrt{5})}{(\sqrt{2} + \sqrt{5})} = \frac{\sqrt{6} + \sqrt{15}}{2 - 5} = \frac{\sqrt{6} + \sqrt{15}}{3}$$

conjugate

2. $\dfrac{\sqrt{a}}{2\sqrt{a} + 3} = \dfrac{\sqrt{a}}{(2\sqrt{a} + 3)} \times \dfrac{(2\sqrt{a} - 3)}{(2\sqrt{a} - 3)} = \dfrac{2\sqrt{a}\sqrt{a} - 3\sqrt{a}}{2 \times 2 \times \sqrt{a} \times \sqrt{a} - 3 \times 3}$

conjugate

$$= \frac{2a - 3\sqrt{a}}{4a - 9}$$

Example 9

The resistance (in ohms) in an electric circuit can be expressed as

$$\frac{\sqrt{60}}{\sqrt{6} + \sqrt{10}}$$

We can rationalize this expression by using the conjugate as shown.

$$\frac{\sqrt{60}}{(\sqrt{6} + \sqrt{10})} \times \frac{(\sqrt{6} - \sqrt{10})}{(\sqrt{6} - \sqrt{10})} = \frac{\sqrt{360} - \sqrt{600}}{6 - 10} = \frac{6\sqrt{10} - 10\sqrt{6}}{-4}$$

conjugate

$$= \frac{3\sqrt{10} - 5\sqrt{6}}{-2}$$

10.5 Exercises

In Exercises 1 through 48, perform the indicated operations, expressing each answer in simplest form. In Exercises 39 through 48, rationalize the denominators.

1. $\sqrt{7} + 3\sqrt{7} - 2\sqrt{7}$

2. $\sqrt{3} - 2\sqrt{3} + 3\sqrt{3}$

3. $\sqrt{7} - \sqrt{5} + 3\sqrt{7} + 2\sqrt{5}$

4. $\sqrt{11} + 8\sqrt{11} - \sqrt{17} + 3\sqrt{11}$

5. $2\sqrt{3} + \sqrt{27}$

6. $3\sqrt{5} + \sqrt{75}$

7. $2\sqrt{40} + 5\sqrt{10}$

8. $3\sqrt{44} - 2\sqrt{11}$

9. $\sqrt{2}\sqrt{50}$

10. $\sqrt{3}\sqrt{9}$

11. $\sqrt{5}\ \sqrt{15}$

12. $\sqrt{2}\sqrt{24}$

13. $\sqrt{8} + \sqrt{18} + \sqrt{32}$

14. $\sqrt{12} + \sqrt{27} + \sqrt{48}$

15. $\sqrt{28} - 2\sqrt{63} + 5\sqrt{7}$

16. $2\sqrt{24} + \sqrt{6} - 2\sqrt{54}$

17. $2\sqrt{8} - 2\sqrt{12} - \sqrt{50}$

18. $2\sqrt{20} - \sqrt{44} + 3\sqrt{11}$

19. $\sqrt{a} + \sqrt{9a}$

20. $\sqrt{5x} + \sqrt{20x}$

21. $\sqrt{2a} + \sqrt{8a} + \sqrt{32a^3}$

22. $\sqrt{ac} + \sqrt{4ac} + \sqrt{16a^3c}$

23. $a\sqrt{2} + \sqrt{72a^2} - \sqrt{12a}$

24. $\sqrt{x^2yz} - \sqrt{y^3z} + 2x\sqrt{y^2z}$

25. $\sqrt{3}(\sqrt{7} - 3\sqrt{6})$

26. $\sqrt{5}(\sqrt{20} - 6\sqrt{3})$

27. $\sqrt{2}(\sqrt{8} - \sqrt{32} + 5\sqrt{18})$

28. $\sqrt{7}(\sqrt{14} + 2\sqrt{6} - \sqrt{56})$

29. $\sqrt{a}(\sqrt{ab} + 3\sqrt{ac})$

30. $\sqrt{2a}(\sqrt{8} + \sqrt{6a^3})$

31. $(\sqrt{2} + \sqrt{3})(2\sqrt{2} - \sqrt{3})$

32. $(\sqrt{7} - 2\sqrt{5})(3\sqrt{7} + \sqrt{5})$

33. $(\sqrt{5} - 3\sqrt{3})(2\sqrt{5} + \sqrt{27})$

34. $(2\sqrt{11} - \sqrt{8})(3\sqrt{2} + \sqrt{22})$

35. $(\sqrt{a} - 3\sqrt{c})(2\sqrt{a} + 5\sqrt{c})$

36. $(\sqrt{2b} + 3)(\sqrt{2} - \sqrt{b})$

37. $(\sqrt{3} + 2)^2$

38. $(2\sqrt{3} - 1)^2$

39. $\dfrac{\sqrt{3} + \sqrt{2}}{\sqrt{2}}$

40. $\dfrac{\sqrt{3} + \sqrt{5}}{\sqrt{3}}$

41. $\dfrac{\sqrt{8} + \sqrt{50}}{\sqrt{2}}$

42. $\dfrac{\sqrt{45} - \sqrt{12}}{\sqrt{3}}$

43. $\dfrac{\sqrt{7}}{\sqrt{3} + \sqrt{7}}$

44. $\dfrac{2\sqrt{5}}{\sqrt{5} - \sqrt{11}}$

45. $\dfrac{\sqrt{3} + \sqrt{6}}{2\sqrt{3} - \sqrt{6}}$

46. $\dfrac{\sqrt{2} - 3\sqrt{7}}{2\sqrt{2} + \sqrt{7}}$

47. $\dfrac{\sqrt{a}}{\sqrt{a} + 2\sqrt{b}}$

48. $\dfrac{\sqrt{x}}{2\sqrt{x} - 3\sqrt{y}}$

In Exercises 49 through 52, solve the given problems.

49. The resistance in an electric circuit is described by

$$\dfrac{\sqrt{R_1 R_2}}{\sqrt{R_1} + \sqrt{R_2}}$$

Rationalize this expression.

50. A surveyor determines the area of a certain parcel of land by evaluating

$$\dfrac{1}{2}\sqrt{a}(\sqrt{a} + 3\sqrt{a})$$

Find the simplest form of this expression.

51. In the theory of waves in wires, we may encounter the expression

$$\dfrac{\sqrt{d_1} - \sqrt{d_2}}{\sqrt{d_1} + \sqrt{d_2}}$$

Evaluate this expression for $d_1 = 10$ and $d_2 = 3$.

52. Three square pieces of land are along a straight road, as shown in Figure 10.2. Find the distance x along the road in simplest radical form.

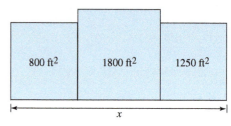

Figure 10.2

10.6 Working with Radical Equations

Section Objectives
- Solve equations containing radical expressions

Equations often contain radical expressions. An equation in which the variable is under a radical sign is called a **radical equation.** For example, $\sqrt{x} + 3 = 7$ is an example of a radical equation whereas $x + \sqrt{3} = 7$ is not a radical equation. The radical in the radical equation can be any root (square root, cube root, fourth root . . .).

To Solve Radical Equations

1. Isolate the radical term on one side of the equation.
2. Raise both sides of the equation to whatever power is necessary to eliminate the radical sign.
3. If there is still a radical in the equation, repeat steps 1 and 2.
4. Solve the equation.
5. Check your result by substituting your result into the original equation.

Example
1

Solve the equation $\sqrt{x - 4} = 2$.

Because the radical term is already on one side of the equation by itself, we begin by squaring each side of the equation.

$$(\sqrt{x - 4})^2 = 2^2$$
$$x - 4 = 4$$
$$x = 8$$

To check our work we substitute $x = 8$ back into the original equation:

$$\sqrt{8 - 4} = 2$$
$$\sqrt{4} = 2$$
$$2 = 2$$

Example
2

Solve $\sqrt{9 - x} + 3 = 5$.

First we must isolate the radical to one side of the equation. This is done by subtracting 3 from both sides of the equation. This gives us

$$\sqrt{9 - x} = 2$$

Next we need to square both sides of the equation in order to eliminate the radical sign.

$$(\sqrt{9 - x})^2 = 2^2$$
$$9 - x = 4$$
$$9 - 4 = x$$
$$5 = x$$

The solution $x = 5$ checks when we place it back into the original equation.

Example
3

Solve $\sqrt{x^2 - 6x} = 9 - x$.

Squaring both sides we get

$$(\sqrt{x^2 - 6x})^2 = (9 - x)^2$$
$$x^2 - 6x = x^2 - 18x + 81$$
$$-6x = -18x + 81$$
$$x = \frac{81}{12} = \frac{27}{4}$$

Caution

To square the entire right-hand side of the equation, we do not square term by term.

Example 4

Solve $\sqrt{x+1} + \sqrt{x-4} = 5$.

In order to solve an equation that contains two radical expressions we begin by first isolating one of the radicals. This is done by moving the other radical to the right-hand side of the equation. We will then square both sides of the resulting equation.

$$\sqrt{x+1} = 5 - \sqrt{x-4}$$
$$(\sqrt{x+1})^2 = (5 - \sqrt{x-4})^2$$
$$x+1 = 25 - 10\sqrt{x-4} + (\sqrt{x-4})^2$$
$$x+1 = 25 - 10\sqrt{x-4} + x - 4$$

Because we still have a radical remaining in the equation we need to combine like terms and then isolate the remaining radical on one side of the equation. We will again square both sides of the equation to try to eliminate the radical sign.

$$x+1 = 25 - 10\sqrt{x-4} + x - 4$$
$$x+1 = x + 21 - 10\sqrt{x-4}$$
$$10\sqrt{x-4} = 20$$
$$\sqrt{x-4} = 2$$
$$(\sqrt{x-4})^2 = 2^2$$
$$x - 4 = 4$$
$$x = 8$$

This solution checks.

Finally, we will look at an example that contains a radical with an index other than 2. The process that we will follow will not change. However, because the index of the radical is 3, we cube each side of the equation.

Example 5

Solve $\sqrt[3]{2x-16} = 2$.

Cubing both sides of the equation we get $(\sqrt[3]{2x-16})^3 = 2^3$.

This gives us $2x - 16 = 8$.

Solving for this equation we find that $x = 12$.

This solution checks.

10.6 Exercises

In Exercises 1 through 16, solve the given equation.

1. $\sqrt{x-8} = 2$

2. $\sqrt{x+4} = 3$

3. $\sqrt{8-2x} = 4$

4. $2\sqrt{2x+5} = 8$

5. $2\sqrt{3-x} = 6$

6. $\sqrt{x+1} = \sqrt{2x-7}$

7. $-3\sqrt{2x+1} = -9$

8. $\sqrt[3]{y-5} = 3$

9. $\sqrt[4]{5-x} = 2$

10. $\sqrt{x^2-9} = 4$

11. $\dfrac{6}{\sqrt{3+x}} = \sqrt{3+x}$

12. $x + 2 = \sqrt{x^2 + 6}$

13. $\sqrt{2x+1} = \dfrac{9}{3\sqrt{2x+1}}$

14. $\sqrt{x^2 - 2} + 3 = x$

15. $\sqrt{x+1} - \sqrt{x+4} = 1$

16. $\sqrt{12 + x} = 2 + \sqrt{x}$

In Exercises 17 through 26, solve the given problem.

17. The resonant frequency f in an electric circuit with an inductance L and a capacitance C is given by the formula

$$f = \dfrac{1}{2\pi\sqrt{LC}}.$$

Solve for L.

18. A formula used in analyzing a certain type of concrete beam is

$$k = \sqrt{2np + (np)^2} - np.$$

Solve for p.

19. The formula $r = \sqrt{\dfrac{A}{\pi}}$ is one that should be familiar to all of us. It is written in a form that we do not normally see. Solve this equation for A and identify this common equation.

20. The speed, V, of a vehicle involved in a car crash can often be determined by using the formula

$$V = v\sqrt{\dfrac{D}{d}},$$

where D represents the length of the skid marks at the scene of the accident, and d are the skid marks left by a test vehicle traveling at v miles per hour. Solve this equation for d.

21. The velocity, v, of an object falling to the ground is determined by the formula

$$v = \sqrt{v_0^2 - 2gh}$$

where v_0 is the initial velocity, g is the gravitational force, and h represents the height from which the object is dropped. Solve this equation for h.

22. The velocity of a proton can be described in terms of its kinetic energy, KE, and the mass of the proton, m, using the formula

$$v = \sqrt{\dfrac{2KE}{m}}.$$

Solve this equation for m.

23. The maximum speed at which a car can negotiate a curve without skidding out of control can be found using the formula

$$v = \sqrt{usgR}.$$

Solve this equation for R.

24. The natural frequency, f, of an object under simple harmonic motion can be found using the formula

$$f = \dfrac{1}{2\pi}\sqrt{\dfrac{kg}{W}}$$

where g is the gravitational force, k is a constant, and W represents the weight of the object. Solve this equation for W.

25. The lateral surface area of a cone is found using the formula

$$S = \pi r\sqrt{r^2 + h^2}$$

where r is the radius of the base of the cone and h is the height of the cone. Solve this equation for h.

26. A right triangle has one side with a length of 12 in. and a perimeter of 30 in. The perimeter of the right triangle can be represented by the formula

$$x + 12 + \sqrt{x^2 + (12)^2} = 30.$$

Solve this equation for x.

Chapter Summary

Key Terms

exponents

principal root

imaginary number

index

radicand

rational number

irrational number

real numbers

radical

radical sign

rationalizing the denominator

conjugate

radical equation

Key Concepts

• Finding the root of a number is the inverse of raising a number to a power. Any value or expression raised to a fractional exponent can also be written as a radical.

• Any value or expression raised to the *zero power* is, by definition, equal to 1.

• We defined an *imaginary number*, $j = \sqrt{-1}$, in order to provide a mechanism for working with problems that contain a negative value under a square root sign.

To Reduce a Radical to its Simplest Form

1. Factor the expression under the radical sign so that one or more of the factors is a perfect n^{th} power (where n is the index of the radical).

2. Write each factor in the product under its own radical sign.

3. Reduce the perfect n^{th} power root you created.

To Simplify a Radical Expression by Rationalizing the Denominator

1. Multiply the numerator and the denominator by a factor that will eliminate the radical sign in the denominator. Choose the smallest possible factor that will yield a perfect n^{th} root.

2. Make sure all of the remaining radicals are simplified.

3. Simplify the fraction if possible.

• To *add or subtract* radical expressions, we add or subtract like radicals.

• To *multiply* radical expressions, we proceed as in algebraic multiplication. In cases where we are multiplying binomial expressions containing radicals, we apply the distributive property.

• To *divide* radical expressions, where the denominator is the sum or difference of square roots simply rationalize the expression by multiplying both the numerator and the denominator by the conjugate of the denominator.

To Solve Radical Equations

1. Isolate the radical term on one side of the equation.

2. Raise both sides of the equation to whatever power is necessary to eliminate the radical sign.

3. If there is still a radical in the equation, repeat steps 1 and 2.

4. Solve the equation.

5. Check your result by substituting your result into the original equation.

Key Formulas

$(a^m)(a^n) = a^{m+n}$

$(a^m)^n = a^{mn}$

$(ab)^n = a^n b^n$

$\dfrac{a^m}{a^n} = a^{m-n} \qquad (a \neq 0)$

$\left(\dfrac{a}{b}\right)^n = \dfrac{a^n}{b^n} \qquad (b \neq 0)$

$a^{-n} = \dfrac{1}{a^n} \qquad (a \neq 0)$

$a^0 = 1 \qquad (a \neq 0)$

$j^2 = -1 \text{ or } j = \sqrt{-1}$

$\sqrt{-a} = \sqrt{a}j = j\sqrt{a}$

$\sqrt[n]{x^n} = x$

$\sqrt{ab} = \sqrt{a}\sqrt{b} \qquad (a \text{ and } b \text{ are positive})$

$\sqrt{a^2} = a \qquad (a > 0)$

$\sqrt{\dfrac{a}{b}} = \sqrt{\dfrac{a \times b}{b \times b}} = \dfrac{\sqrt{ab}}{\sqrt{b^2}} = \dfrac{\sqrt{ab}}{b}$

$(\sqrt{a} + \sqrt{b})(\sqrt{a} - \sqrt{b}) = a - b$

$a^{1/n} = \sqrt[n]{a}$

$a^{m/n} = (\sqrt[n]{a})^m = \sqrt[n]{a^m}$

Review Exercises

In Exercises 1 through 25, evaluate the given expressions.

1. 10^{-1}

2. 2^{-4}

3. $\dfrac{1}{3^{-2}}$

4. $\dfrac{1}{4^{-1}}$

5. $3^0 6^{-1}$

6. $\dfrac{9^0}{4^{-3}}$

7. $\sqrt{169}$

8. $-\sqrt{900}$

9. $\sqrt[3]{125}$

10. $-\sqrt[3]{-125}$

11. $\sqrt{\dfrac{1}{16}}$

12. $\sqrt{\dfrac{4}{25}}$

13. $-\sqrt{\dfrac{9}{121}}$

14. $-\sqrt{\dfrac{144}{169}}$

15. $100^{1/2}$

16. $1000^{1/3}$

17. $49^{3/2}$

18. $121^{3/2}$

19. $8^{7/3}$

20. $\dfrac{16^{3/4}}{27^{2/3}}$

21. $\dfrac{25^{3/2}}{5^{-1}}$

22. $\sqrt{-81}$

23. $\sqrt{-144}$

24. $-\sqrt{-0.64}$

25. $-\sqrt{-0.01}$

In Exercises 26 through 39, write the given expressions in simplest form, expressing all results with positive exponents.

26. $3a^{-2}b$

27. $2xy^{-1}$

28. $\dfrac{mn^{-2}}{m^{-3}}$

29. $\dfrac{2rs^{-1}}{t^{-5}}$

30. $\dfrac{2x^3 y^{-1}}{3x^{-2} y^2}$

31. $\dfrac{(2a)^0 (b^{-1}c)}{4a^2 bc^{-3}}$

32. $a^{1/4} a^{1/3}$

33. $x^{1/5} x^{2/3}$

34. $\dfrac{a^{2/3}}{a^{-1/2}}$

35. $\dfrac{b^{-1}}{b^{-1/2}}$

36. $(xy^{-2})^{1/2}$

37. $(8x^{-3} y^{3/2})^{1/3}$

38. $\dfrac{(st^{1/2})^{2/3}}{t^{-2}}$ **39.** $\dfrac{(16c^2)^{3/4}}{ac^{-1/5}}$

In Exercises 40 through 59, express the given radicals in simplest form. Rationalize the denominator if a radical appears in the denominator.

40. $\sqrt{44}$ **41.** $\sqrt{27}$

42. $\sqrt{72}$ **43.** $\sqrt{54}$

44. $\sqrt{128}$ **45.** $\sqrt{124}$

46. $\sqrt[3]{40}$ **47.** $\sqrt[3]{108}$

48. $\sqrt{\dfrac{1}{11}}$ **49.** $\sqrt{\dfrac{4}{7}}$

50. $\sqrt{4a^2}$ **51.** $\sqrt{28a}$

52. $\sqrt{125b^2c}$ **53.** $\sqrt{90b^3}$

54. $\sqrt{\dfrac{6}{a}}$ **55.** $\sqrt{\dfrac{3}{a^2}}$

56. $\sqrt{\dfrac{28}{3a}}$ **57.** $\sqrt{\dfrac{400{,}000}{ab}}$

58. $\sqrt[3]{16a^3}$ **59.** $\sqrt[3]{81x^2y^4}$

In Exercises 60 through 74, perform the indicated operations and simplify. Rationalize the denominator if a radical appears in the denominator.

60. $\sqrt{63} - 2\sqrt{28}$

61. $2\sqrt{45} - 3\sqrt{80}$

62. $3\sqrt{7} - 2\sqrt{6} + \sqrt{28}$

63. $\sqrt{2a^2} + 3\sqrt{8} - \sqrt{32a^2}$

64. $\sqrt{20a} + 4\sqrt{5a^3} - 3\sqrt{45a}$

65. $\sqrt{2}(\sqrt{6} - 2\sqrt{24})$

66. $\sqrt{3}(3\sqrt{6} + \sqrt{54})$

67. $\sqrt{a}(\sqrt{ab} - 3\sqrt{5b})$

68. $\sqrt{ab}(\sqrt{b} - 3\sqrt{a})$

69. $(\sqrt{6} - \sqrt{5})(2\sqrt{6} - 3\sqrt{5})$

70. $(\sqrt{a} - 3\sqrt{b})(2\sqrt{a} + \sqrt{b})$

71. $(\sqrt{ab} - \sqrt{c})(2\sqrt{ab} + \sqrt{c})$

72. $\dfrac{\sqrt{2}}{\sqrt{5} - \sqrt{2}}$

73. $\dfrac{\sqrt{2} - 1}{2\sqrt{2} + 3}$

74. $\dfrac{\sqrt{11} - \sqrt{5}}{2\sqrt{11} + \sqrt{5}}$

In Exercises 75 through 81, solve the given problems.

75. The center of gravity of a half-ring is found by using the expression

$$(r^2 + 4R^2)(2\pi R)^{-1}$$

where r is the radius of the cross section and R is the radius of the ring. Write this expression without negative exponents.

76. The speed of sound through a medium is given by

$$v = \sqrt{\dfrac{E}{d}}$$

where E and d are constants depending on the medium. Rationalize this expression.

77. In analyzing electrode properties in a cathode ray tube, the expression

$$v = \sqrt{\dfrac{2eV}{m}}$$

is found. Express this equation so that the right-hand side is in rationalized form.

78. The radius (in inches) of a floppy disk is expressed in terms of its top surface area as follows.

$$r = \dfrac{\sqrt{441\pi}}{\sqrt{65\pi}}$$

Rationalize and simplify the right-hand side of this equation.

79. In analyzing the plastic deformation within single crystals, the expression

$$\left(\frac{a}{\sqrt{2}}\right) \div \left(\frac{a\sqrt{6}}{2}\right)$$

is derived. Simplify and express in rationalized form.

80. The density of an object equals its weight divided by its volume. A certain metal has a density of 1331 lb/ft³. What is the edge of a cube of this metal that weighs 8.00 lb?

81. The thermodynamic temperature of the filament of a light bulb equals approximately 1000 times the fourth root of the wattage. Find the temperature in degrees Celsius (273° less than the thermodynamic temperature) of the filament of a 25-W bulb.

In Exercises 82 through 86, solve the given equation algebraically.

82. $\sqrt{x + 2} = 6$

83. $\sqrt{3x^2 - 8} = x$

84. $\sqrt{x + 2} = \sqrt{2x + 9}$

85. $\sqrt{2x - 1} = \sqrt{x + 3}$

86. $\sqrt[3]{3x} = 9$

Chapter Test

For those problems with exponents, express your results with only positive exponents.

1. Simplify $2\sqrt{20} - \sqrt{125}$

2. Simplify $\dfrac{100^{\frac{3}{2}}}{8^{-\frac{2}{3}}}$

3. Simplify $\sqrt{27a^4b^3}$

4. Simplify $\left(\dfrac{4a^{-\frac{1}{2}}b^{\frac{3}{4}}}{b^{-2}}\right)\left(\dfrac{b^{-1}}{2a}\right)$

5. Simplify $(\sqrt{2x} - 3\sqrt{y})^2$

6. Rationalize the denominator $\dfrac{3 - 2\sqrt{x}}{2\sqrt{x}}$

7. Rationalize the denominator $\dfrac{2\sqrt{15x} + \sqrt{3}}{\sqrt{15x} - 2\sqrt{3}}$

8. Solve the following equation $\sqrt{x^2 - 3x} = x - 7$

9. The speed v of a ship of weight W whose engines produce power P is

$$v = k\sqrt[3]{\frac{P}{W}}.$$

Express this equation with
a. fractional exponents
b. as a radical with the denominator rationalized

10. In a study of the biological effects of sound, the expression

$$\left(\frac{2n}{\omega r}\right)^{-\frac{1}{2}}$$

is found. Express this expression in simplest rationalized radical from.

11 Quadratic Equations

Suppose a baseball is tossed up into the air. Now suppose that we are able to record the height of the baseball at regular time intervals. We record the height of the baseball every second it is in flight from the moment it is tossed into the air.

Table 11.1 shows a chart of the height of the ball at each second.

Table 11.1 **Height of Tossed Baseball**

Time (sec)	0	1	2	3	4	5	6	7
Height (ft)	6	90	142	162	150	106	30	–

As we read the data presented in Table 11.1, we notice that the height of the ball increases up to a certain point and then begins to decrease as the ball falls back to earth. The **quadratic equation** $h = -16t^2 + 100t + 6$ is a good mathematical model for the height of the baseball from the time it is tossed into the air until the time it hits the ground.

Each term in this quadratic equation has a physical explanation. From physics we know that the $-16t^2$ represents the effect of the acceleration of the baseball due to gravity. We will see later in this chapter that the negative sign tells us that the baseball reaches a maximum height before changing its direction and falling back to earth. The 6 is simply the initial height of the ball when it is tossed in the air.

Using this equation we can determine a number of things about the flight of the baseball. We will be able to determine its maximum height above the ground, the time at which it achieved this height, as well as the time at which the ball actually hit the ground.

Quadratic equations are used in many applications. In this chapter we will discuss the nature of a quadratic equation and methods of solving these equations. We will also look at the graph of a quadratic equation and at the useful information we can take from the graph.

11.1   The Quadratic Equation

Section Objectives
- Write a quadratic equation in standard form
- Identify information about a quadratic equation from this form

Given that a, b, and c are constants, the equation

$$ax^2 + bx + c = 0 \quad \text{where } a \neq 0$$

is called **the general quadratic equation in x.**

It is the x^2 term that distinguishes the quadratic equation from other types of equations, *if a = 0 the equation is not considered to be quadratic.* However, either b or c, or both, can be zero and the equation is still a quadratic. Therefore, $a \neq 0$, but b and c can be any numbers.

Examples 1 and 2 show examples of quadratic equations.

Example 1

The equation $2x^2 + 3x + 7 = 0$ is a quadratic equation, where $a = 2$, $b = 3$, and $c = 7$.

$$\begin{array}{ccc} a & b & c \\ \downarrow & \downarrow & \downarrow \\ 2x^2 + & 3x + & 7 = 0 \\ \uparrow & \uparrow & \uparrow \end{array}$$

second   first   constant   zero
power   power   quadratic equation form
term   term

The equation $4x^2 - 7x + 9 = 0$ is also a quadratic equation in **standard form,** despite the presence of the minus sign, because we can identify $a = 4$, $b = -7$, and $c = 9$. That is,

$$4x^2 - 7x + 9 = 4x^2 + (-7)x + 9.$$

The equation $x^3 - 9x + 8 = 0$ is *not* a quadratic equation in standard form because of the term x^3.

Example 2

The equation $3x^2 - 19 = 0$ is a quadratic equation, where $a = 3$, $b = 0$, and $c = -19$.

The equation $x^2 - 8x = 0$ is a quadratic equation, where $a = 1$, $b = -8$, and $c = 0$.

The equation $x^2 = 0$ is a quadratic equation, where $a = 1$, $b = 0$, and $c = 0$.

Consider the illustrations of Example 3 in which the quadratic equation is not in standard form. Using basic operations we can rewrite the equation so that it is in the form $ax^2 + bx + c = 0$. Once this is done we can find the values of a, b, and c.

Example 3

The equation $3x^2 - 6 = 7x$ is not in standard form, but it can easily be put in this form by subtracting $7x$ from each side of the equation. Performing this operation, we get

$$\begin{array}{ccc} 3x^2 - & 7x - & 6 = 0 \\ \uparrow & \uparrow & \uparrow \\ a & b & c \end{array}$$

which means $a = 3$, $b = -7$, and $c = -6$.

The equation $2x^2 = (x - 8)^2$ is not in standard form. To determine whether or not it may be put in this form, we must square the right-hand side as indicated. By then collecting terms on the left, we can establish the form of the equation. This leads to

$$2x^2 = x^2 - 16x + 64$$

or

$$x^2 + 16x - 64 = 0 \quad \text{standard form}$$

$$a = 1 \quad b \quad c$$

In this equation we can see that $a = 1$, $b = 16$, and $c = -64$.

The equation $3x^2 = (3x - 1)(x + 2)$ becomes

$$3x^2 = 3x^2 + 5x - 2 \quad \text{or} \quad -5x + 2 = 0$$

We recognize that $-5x + 2 = 0$ is not a quadratic equation.

Recall that we defined the solution of an equation as the value of the variable that when substituted into the equation upholds the equality. That means that the two sides of the equation will be equal.

As we shall see in Section 11.2, *the solution of a quadratic equation is generally a pair of numbers, although occasionally only one number satisfies the equation.* In any case, there cannot be more than two numbers that satisfy a quadratic equation. The Example 4 illustrates checking possible values of x as solutions of a quadratic equation. Note that we are not yet finding solutions to quadratic equations. Example 4 simply illustrates how to *check* potential solutions after they have been identified.

Example 4

The shape of a parabolic antenna is related to the equation $2x^2 + 5x - 3 = 0$. Determine which, if any, of the given values $x = 1$, $x = -3$, $x = \frac{1}{2}$, $x = 2$ are solutions of that equation.

Testing $x = 1$, we get

$$2(1)^2 + 5(1) - 3 = 2 + 5 - 3 = 4$$

Because the resulting value is not zero, substitution of 1 for x does *not* make the equation true and $x = 1$ is not a solution.

Testing $x = -3$, we get

$$2(-3)^2 + 5(-3) - 3 = 2(9) - 15 - 3 = 18 - 15 - 3 = 0$$

Because the value is zero, substitution of -3 for x does make the equation true and $x = -3$ is a solution.

Testing $x = \frac{1}{2}$, we get

$$2\left(\frac{1}{2}\right)^2 + 5\left(\frac{1}{2}\right) - 3 = 2\left(\frac{1}{4}\right) + \frac{5}{2} - 3 = \frac{1}{2} + \frac{5}{2} - 3 = 3 - 3 = 0$$

Because the value is zero, substitution of $\frac{1}{2}$ for x does make the equation true and $x = \frac{1}{2}$ is also a solution. Because we have now found two such values, we know that the solutions to this equation are $x = -3$ and $x = \frac{1}{2}$. Any other value of x cannot be a solution because there can be at most two different solutions. The value of $x = 2$, for example, cannot be a solution because we already have two solutions.

Our primary concern is with quadratic equations that have real roots. However, it is possible that the solution of a quadratic equation can contain numbers with $j = \sqrt{-1}$. It is also possible that the real solutions of a quadratic equation are equal, so that only one value satisfies the equation.

Example

5

The equation $x^2 + 4 = 0$ has the solutions of $2j$ and $-2j$. This can be verified by substitution:

$$(2j)^2 + 4 = 4j^2 + 4 = 4(-1) + 4 = -4 + 4 = 0$$
$$(-2j)^2 + 4 = (-2)^2 j^2 + 4 = 4(-1) + 4 = -4 + 4 = 0$$

The equation $x^2 + 4x + 4 = 0$ has the solution $x = -2$ only. The two solutions here are equal. The reason for this will be seen in Section 11.2.

11.1 Exercises

In Exercises 1 through 12, determine whether or not the given equations are quadratic by performing algebraic operations that could put each in the standard form $ax^2 + bx + c = 0$. If the standard quadratic form of the equation is obtained, identify a, b, and c.

1. $x^2 - 7x = 4$

2. $3x^2 = 5 - 9x$

3. $x^2 = (x - 1)^2$

4. $2x^2 - x = 2x(x + 8)$

5. $(x + 2)^2 = 0$

6. $x(x^2 - 1) = x^3 + x^2$

7. $x^2 = x(1 - 6x)$

8. $x(1 + 2x) = 3x - 2x^2$

9. $x^2(1 - x) = 4$

10. $x(x^2 + 2x - 1) = 0$

11. $3x(x^2 - 2x + 1) = 1 + 3x^2$

12. $x(x^3 + 6) - 1 = x^2(x^2 + 1)$

In Exercises 13 through 23, test the given values to determine which, if any, are solutions of the equation.

13. $x^2 - 5x + 6 = 0$; $x = 1, x = 2, x = 3$

14. $x^2 + x - 6 = 0$; $x = 0, x = 2, x = 3$

15. $x^2 - 4x + 4 = 0$; $x = 2, x = -2, x = 4$

16. $x^2 + 1 = 0$; $x = 0, x = -1, x = 1$

17. $x^2 - x - 2 = 0$; $x = 1, x = -1, x = 2$

18. $x^2 + 4x + 3 = 0$; $x = -1, x = 1, x = -3$

19. $2x^2 - 3x + 1 = 0$; $x = -1, x = 0, x = \dfrac{1}{2}, x = 1$

20. $3x^2 - x - 2 = 0$; $x = 1, x = 2, x = -\dfrac{2}{3}, x = -1$

21. $V^2 + V = 12$; $V = -1, V = -3, V = 3, V = -4$

22. $y^2 = 3y + 10$; $y = 2, y = -1, y = 5$

23. $x^2 + 16 = 0$; $x = -4, x = 2, x = 4$

In Exercises 24 through 28, solve the given problems.

24. A computer simulation program involves a particle whose path is described by the equation

$$y = -x^2 + 6x - 8$$

Given that $y = -3$, find the resulting quadratic equation. Show that $x = 5$ is a solution of the quadratic equation.

25. A manufacturer approximates the cost of producing x items by the equation

$$c = -5x^2 + 8x + 600$$

Given that $c = 468$, find the resulting quadratic equation. Show that $x = 6$ is a solution of the quadratic equation.

26. An object is shot upward so that its distance above the ground is given by

$$s = 96t - 16t^2$$

where t is the time in seconds. Given that $s = 128$ ft, find the resulting quadratic equation. Show that $t = 2.00$ s is a solution of the equation.

27. The value of an account after two years is described by

$$V = P(1 + r)^2$$

where P is the initial amount invested and r is the interest rate in decimal form. If $P = 500$ and $V = 605$, show that the resulting equation is quadratic and that $r = 0.1$ is a solution.

28. Conforming to a zoning ordinance, a store owner wants to make a rectangular sign with an area of 48.0 ft^2. He wants the length l to exceed the width w by 2.0 ft. Show that the resulting equation involving the width is quadratic. See Figure 11.1.

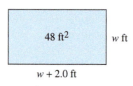

48 ft^2 w ft

w + 2.0 ft

Figure 11.1

11.2   Solving Quadratic Equations by Factoring

Section Objectives
- Introduce the zero product rule
- Solve a quadratic equation by factoring

In solving a quadratic equation by the method of factoring, we should begin by putting the quadratic equation in the standard form $ax^2 + bx + c = 0$. If the left-hand side is factorable, the solution is easily obtained. Solving a quadratic equation by factoring it uses a principle known as the *zero product rule,* which states that if the product of two factors is zero then one of the factors must be zero.

Zero Product Rule

If $pq = 0$ then either $p = 0$ or $q = 0$ (or both).

This is true because the only way we can get zero as an answer when we multiply is if one of the numbers used in the multiplication is zero. Therefore, if you can factor an expression *that is equal to zero,* you can set each factor equal to zero and solve it for the unknown.

The procedure for solving a quadratic equation by factoring is summarized in these steps:

1. Arrange the quadratic equation so that it is in the standard form of $ax^2 + bx + c = 0$.

2. Factor the left-hand side.

▶ 3. Individually **set each of the two factors equal to zero and solve.**

4. Check both solutions by substituting them in the original equation.

We use the zero product rule in Examples 1–3.

Example 1

Solve the equation $x^2 - x - 2 = 0$.

We first write the left-hand side in factored form as follows.

$$(x - 2)(x + 1) = 0$$

The product of these two factors is zero if at least one of them is zero. Setting each factor equal to zero and solving the resulting equations, we get

$$x - 2 = 0 \qquad\qquad x + 1 = 0$$
$$x = 2 \qquad\qquad x = -1$$

The solutions are $x = 2$ and $x = -1$. Checking each of these values, we get

$$2^2 - 2 - 2 = 4 - 2 - 2 = 0$$

and

$$(-1)^2 - (-1) - 2 = 1 + 1 - 2 = 0$$

Checking the solution is strongly recommended.

Be Careful

The *zero product rule* can be used only when the product of two factors is equal to zero.

If $(x - 2)(x + 1) = 6$, it would be *incorrect* to say $x - 2 = 6$ or $x + 1 = 6$.

In Example 1, if $x - 2 = 0$, the left-hand side can be written $0 \times (x + 1)$, and this product is zero regardless of the value of $x + 1$. Therefore, we see that setting this factor equal to zero should give us a solution to the original equation. The same is true if $x + 1 = 0$. We must keep in mind, however, that **the equation must be written in the standard form before we factor the left-hand side. This is necessary because we must have zero on the right-hand side of the equation in order to use the zero product rule.** If any number other than zero appears on the right, this method doesn't work.

Example 2

Solve the equation $x^2 - 10x = -21$.

The correct procedure is to first get this quadratic equation into standard form

$$x^2 - 10x + 21 = 0$$

which factors into $(x - 7)(x - 3) = 0$. In this way, we get the correct solutions $x = 7$ and $x = 3$.

Many students would be tempted to factor the left-hand side into $x(x - 10)$ and then set $x = -21$ and $x - 10 = -21$. The "solutions" obtained, $x = -21$ and $x = -11$, are not correct, as can be verified by substitution.

Example 3

Solve the equation $6x^2 = 5 - 7x$.

Following the procedure described above, we rewrite the equation in standard form

$$6x^2 + 7x - 5 = 0$$

$$(2x - 1)(3x + 5) = 0 \qquad \text{factor}$$

$$2x - 1 = 0 \qquad 3x + 5 = 0 \qquad \text{set each factor} = 0$$

$$2x = 1 \qquad 3x = -5 \qquad \text{solve}$$

$$x = \frac{1}{2} \qquad x = -\frac{5}{3}$$

To be certain that no improper algebraic steps have been taken, it is best to check each solution in the original equation, even though it is not in quadratic form. Checking, we obtain the following:

$$6\left(\frac{1}{2}\right)^2 \overset{?}{=} 5 - 7\left(\frac{1}{2}\right) \qquad 6\left(-\frac{5}{3}\right)^2 \overset{?}{=} 5 - 7\left(-\frac{5}{3}\right)$$

$$6\left(\frac{1}{4}\right) \overset{?}{=} 5 - \frac{7}{2} \qquad 6\left(\frac{25}{9}\right) \overset{?}{=} 5 + \frac{35}{3}$$

$$\frac{3}{2} = \frac{3}{2} \qquad \frac{50}{3} = \frac{50}{3}$$

Because the values on each side are equal, the solutions check. That is, substitution of the solutions will make the original equation true so that the solutions are verified.

In Section 11.1 we mentioned that *the two solutions, or* **roots** *of the equation, can be equal. This is true when the two factors are the same.* Example 4 illustrates this type of quadratic equation.

Example **4**

Solve the equation $4x^2 - 12x + 9 = 0$.
Factoring this equation, we obtain

$$(2x - 3)(2x - 3) = 0$$

Setting each factor equal to zero gives

$$2x - 3 = 0$$

$$x = \frac{3}{2}$$

Because both factors are the same, the equation has a **double root** of $\frac{3}{2}$. Checking verifies the solution.

Many students improperly solve a quadratic equation in which $c = 0$. Example 5 illustrates this type and the error often made in its solution.

Example **5**

Solve the equation $x^2 - 6x = 0$.
Noting that the two terms of this equation contain x, *a common error is to divide through by x.* This results in the equation $x - 6 = 0$ and the solution $x = 6$. However, *the solution $x = 0$ was lost through the division by x.*
Instead of dividing through by x, we should factor the equation into

$$x(x - 6) = 0$$

By setting each factor equal to zero, we obtain

$$x = 0 \qquad x - 6 = 0$$
$$x = 6$$

There are *two solutions: $x = 0$ and $x = 6$.* Checking verifies these roots.

A number of different verbally stated problems may lead to quadratic equations. Example 6 illustrates setting up and solving such problems.

Now Try It!

Solve these quadratic equations by factoring.

1. $x^2 - 2x - 35 = 0$
2. $4x^2 + 12x = 16$
3. $-18x + 20 = 2x^2$
4. $3x^2 + 4 = 8x$

Example
6

Above sea level, the boiling point of water is a lower temperature than at sea level where it is 212°F. The difference (in Fahrenheit degrees) is given by the approximate equation

$$T^2 + 520T - h = 0$$

where h is the altitude in feet. Find the approximate boiling point of water in a light airplane flying at 5300 ft.

With $h = 5300$ ft, the quadratic equation becomes

$$T^2 + 520T - 5300 = 0$$

Factoring is not easy with numbers that are this large, but the above expression does factor as follows

$$(T - 10)(T + 530) = 0 \qquad \text{factor}$$

Proceeding, we set each factor equal to zero and solve.

$$T - 10 = 0 \qquad T + 530 = 0 \qquad \text{set each factor} = 0$$
$$T = 10 \qquad T = -530 \qquad \text{solve}$$

We are finding the amount by which the temperature is *lowered* so that only the positive solution of $T = 10$ is reasonable. Because the 212°F boiling point is lowered by 10°F, we have a boiling point of 202°F at 5300 ft. (Some pilots have been injured by exploding thermos containers when their hot contents began to boil at higher altitudes.)

11.2 Exercises

In Exercises 1 through 32, solve the given quadratic equations by factoring.

1. $x^2 - 9 = 0$

2. $4x^2 - 25 = 0$

3. $x^2 + x - 2 = 0$

4. $x^2 - 5x + 6 = 0$

5. $t^2 - 3t = 10$

6. $s^2 = 6s + 7$

7. $3x^2 + 5x - 2 = 0$

8. $4x^2 - 7x + 3 = 0$

9. $2x^2 + 3x = 2$

10. $3y^2 = 8 - 10y$

11. $6x^2 + 13x - 5 = 0$

12. $6x^2 - 7x - 20 = 0$

13. $6n^2 - n = 2$

14. $2v^2 + 30 = 17v$

15. $5x^2 + 4 = 21x$

16. $9p^2 + 20 = -27p$

17. $2x^2 = 9 + 7x$

18. $6x^2 + 11x = 10$

19. $x^2 + 4x + 4 = 0$

20. $x^2 + 9 = 6x$

21. $R^2 - 8R = 0$

22. $3t^2 + 5t = 0$

23. $8m^2 + 24m = 14$

24. $18x^2 - 3x = 6$

25. $4x^2 + 49 = 28x$

26. $9x^2 + 30x + 25 = 0$

27. $R^2 = 7R$

28. $4r^2 = 48r$

29. $8m + 3 = 3m^2$

30. $3 - x = 4x^2$

31. $x^2 - 4a^2 = 0$ (*a* is constant)

32. $2x^2 - 8ax + 8a^2 = 0$ (*a* is constant)

In Exercises 33 through 42, solve any resulting quadratic equations by factoring.

33. The number *x* of items produced by a certain company and the corresponding profit are related by the equation

$$P = -x^2 + 19x - 34$$

Find the "break-even" point. That is, find the number of items for which the profit *P* is zero.

34. Under certain conditions, the motion of an object suspended by a helical spring requires the solution of the equation

$$D^2 + 8D + 12 = 0$$

Solve for *D*.

35. The perimeter of a rectangular garden is 70 m and the area is 250 m². Find the dimensions of the garden. See Figure 11.2.

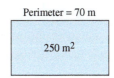

Perimeter = 70 m

250 m²

Figure 11.2

36. At an altitude of *h* ft above sea level, the boiling point of water is lower by a certain number of degrees than the boiling point at sea level, which is 212°F. The *difference* is given by the approximate equation

$$T^2 + 520T - h = 0$$

Compute the approximate boiling point of water on the top of Mt. Baker in Washington (altitude about 10,800 ft). See Example 6.

37. The distance an object falls due to gravity is given by $s = 16t^2$, where *s* is the distance in feet and *t* is the time in seconds. How long does it take an object to fall 100 ft?

38. Under specified conditions, the deflection of a beam requires the solution of the equation

$$4Lx - x^2 - 4L^2 = 0$$

where *x* is the distance from one end and *2L* is the length of the beam. Solve for *x* in terms of *L*.

39. A nuclear power plant supplies a fixed power level at a constant voltage, and the current is determined from the equation

$$100I^2 - 1700I + 1600 = 0$$

Solve for *I*.

40. A box for shipping sheet metal screws must be designed so that it contains a volume of 84 ft³. Various considerations require that the box be 3 ft wide, and the height must be 3 ft less than the length. Find the dimensions of the box. See Figure 11.3.

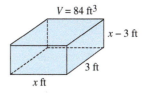

$V = 84$ ft³

$x - 3$ ft

3 ft

x ft

Figure 11.3

41. The mass *m* (in Mg) of the fuel supply in the first stage booster of a rocket is given by the formula $m = 135 - 6t - t^2$ where *t* is the time (in *s*) after launch. After how many seconds does the booster run out of fuel?

42. A large computer company has a profit function that can be described by the formula $p = -2x^2 + 520x - 5000$ where *p* represents the profit (in $) and *x* represents the number of computers sold (in thousands). Find the number of computers sold at the breakeven point for this company (the point at which profit is zero).

Now Try It! Answers

1. $x = 7, -5$ **2.** $x = -4, 1$ **3.** $x = -10, 1$ **4.** $x = \frac{2}{3}, 2$

11.3　Completing the Square

Section Objectives
• Solve a quadratic equation by completing the square

Although factoring is generally the easiest way to solve a quadratic equation, most equations of this type are not easily factorable. In this section we present a method called ***completing the square*** which can be used to solve a quadratic equation. Let's begin by looking at what happens when you square a binomial like $(x + a)^2 = (x + a)(x + a) = x^2 + 2ax + a^2$. Notice that the coefficient of the middle term is twice the square root of the constant term (or last term). Notice also that the first and last terms are perfect squares.

The process of completing the square provides us with a way to take a trinomial that is not a perfect square, and make it into one by inserting the correct constant term (which is the square of half the coefficient of x). When adding in the new constant term it is important that we make sure to follow a proper algebraic procedure and add that new constant term to both sides of the equation.

Use the following guidelines as we work through the next few examples.

Solving a Quadratic Equation by Completing the Square

1. Rewrite the equation with the constant on one side of the equation. If the coefficient of the x^2 term is not 1, divide each side by the coefficient of this term.

2. Take one-half of the coefficient of the x term and then square that value.

3. Add this value to both sides of the equation created in step 1.

4. Write the left-hand side as a square and simplify the right-hand side.

5. Take the square root of both sides of the equation.

6. Solve each resulting equation.

Example
1

Solve the quadratic equation $x^2 - 8x + 6 = 0$

We notice that this equation is not factorable. We now work through the process of completing the square.

1. Begin by subtracting 6 from both sides of the equation to get

$$x^2 - 8x = -6$$

2. Take one-half of the coefficient of the x term and square it.

$$\left[\frac{1}{2}(-8)\right]^2 = (-4)^2 = 16$$

3. Add 16 to both sides of the equation from step 1.

$$x^2 - 8x + 16 = -6 + 16$$

4. Write the left-hand side as a square and simplify the right-hand side to get

$$(x - 4)^2 = 10$$

5. Take the square root of both sides of the equation.

$$x - 4 = \pm\sqrt{10}$$

6. Solve $x - 4 = +\sqrt{10}$ and $x - 4 = -\sqrt{10}$

$$x = 7.16 \quad \text{and} \quad x = 0.838$$

In Examples 2 and 3 we will look at how we solve a quadratic equation using completing the square when the coefficient of the x^2 term is not 1.

Example
2

Solve $2x^2 + 16x - 9 = 0$ by completing the square.

1. Since the coefficient of the x^2 term is not 1, we begin by dividing each side by 2 to get

$$x^2 + 8x - \frac{9}{2} = 0$$

2. Rewrite the equation with the constant on one side.

$$x^2 + 8x = \frac{9}{2}$$

3. Take one-half of the coefficient of the x term and then square that value.

$$\left[\frac{1}{2}(8)\right]^2 = 4^2 = 16$$

> ### Remember
> When you add the value needed to complete the square to the left-hand side of the equation, you must also add it to the right-hand side as well.

4. Add this value to both sides of the equation created in step 2.

$$x^2 + 8x + 16 = \frac{9}{2} + 16$$

5. Write the left-hand side as a square and simplify the right-hand side.

$$(x + 4)^2 = \frac{9}{2} + 16 = \frac{41}{2} = 20.5$$

6. Take the square root of both sides of the equation.

$$x + 4 = \pm\sqrt{20.5}$$

7. Solve $x + 4 = +\sqrt{20.5}$ and $x + 4 = -\sqrt{20.5}$

Using a calculator we approximate the solution to this quadratic equation to be

$$x = 0.5277 \quad \text{and} \quad x = -8.528$$

Example 3

Solve $5x^2 + 6x - 3 = 0$ by completing the square.

1. Because the coefficient of the x^2 term is not 1, we begin by dividing each side by 5 to get

$$x^2 + 1.2x - 0.6 = 0$$

> **Remember**
>
> When the coefficient of the x^2 term is not 1, the first thing you must do is divide each side of the equation by the coefficient of the x^2 term.

2. Rewrite the equation with the constant on one side.

$$x^2 + 1.2x = 0.6$$

3. Take one-half of the coefficient of the x term and then square that value.

$$\left[\frac{1}{2}(1.2)\right]^2 = 0.6^2 = .36$$

4. Add this value to both sides of the equation created in step 2.

$$x^2 + 1.2x + .36 = 0.6 + .36$$

5. Write the left-hand side as a square and simplify the right-hand side.

$$(x + 0.6)^2 = .96$$

6. Take the square root of both sides of the equation.

$$x + 0.6 = \pm\sqrt{.96}$$

7. Solve $x + 0.6 = +\sqrt{.96}$ and $x + 0.6 = -\sqrt{.96}$

Using a calculator we approximate the solution to this quadratic equation to be

$$x = .38 \quad \text{and} \quad x = -1.58$$

The technique of completing the square is used to develop the quadratic formula, which will be presented in Section 11.4.

11.3 Exercises

In Exercises 1 through 20, solve the given quadratic equations by completing the square.

1. $x^2 + 4x - 9 = 0$

2. $x^2 + 7x - 3 = 0$

3. $x^2 - 3x + 6 = 0$

4. $x^2 - x - 6 = 0$

5. $x^2 + 3x + 2 = 0$

6. $a^2 = 4a - 2$

7. $s^2 + 6s = 4$

8. $R^2 + 10R + 9 = 13$

9. $v^2 + 2v = 15$

10. $x^2 + 12 = 8x$

11. $3x^2 + 6x - 4 = 0$

12. $4x^2 + 7x - 5 = 0$

13. $8x^2 - 4x - 3 = 0$

14. $5x^2 - 2x - 1 = 0$

15. $2x^2 + 5x = 3$

16. $4x^2 + x = 3$

17. $3y^2 = 3y + 2$

18. $3x^2 = 3 - 4x$

19. $9x^2 + 1 = -6x$

20. $2x^2 + 2 = 3x$

In Exercises 21 and 22, use completing the square to solve the given problems.

21. The voltage V across a certain electronic device is related to the temperature T (in °C) by $V = 4.0T - 0.2T^2$. For what temperature(s) is $V = 15V$?

22. A rectangular storage area is 8.0 m longer than it is wide. If the area is 28 m^2, what are the dimensions of the storage area?

11.4 Solving Quadratic Equations Using the Quadratic Formula

Section Objectives

• Solve a quadratic equation using the quadratic formula

Because many quadratic equations cannot be solved by factoring, solution can be found directly from a special formula called the **quadratic formula.**

$$x = \frac{-b \pm \sqrt{b^2 - 4ac}}{2a} \qquad \text{quadratic formula}$$

If an equation is in the form $ax^2 + bx + c = 0$, then direct substitution of the appropriate values a, b, and c into the quadratic formula gives the solutions. Notice that the symbols $\pm$ precede the radical. This indicates that *there are two solutions, one for the + sign and one for the − sign of the $\pm$ that precedes the radical.* Examples 1–4 illustrate the use of the quadratic formula.

Example
1

Solve the equation $x^2 - 5x + 6 = 0$ using the quadratic formula.

$$a = 1 \quad b = -5 \quad c = 6$$

Because the equation is in the proper form, we recognize that $a = 1, b = -5$, and $c = 6$. Therefore,

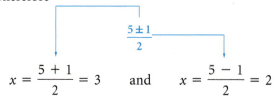

$$x = \frac{-(-5) \pm \sqrt{(-5)^2 - 4(1)(6)}}{2(1)} = \frac{5 \pm \sqrt{25 - 24}}{2} = \frac{5 \pm 1}{2}$$

Therefore

$$x = \frac{5 + 1}{2} = 3 \quad \text{and} \quad x = \frac{5 - 1}{2} = 2$$

These roots could have been found by factoring, but we see that the same results are obtained by using the formula. The roots are easily checked through substitution in the original equation.

A very common error in applying the quadratic formula is to place the denominator of $2a$ under the radical, but not under the $-b$ term in the numerator.

It is important to always divide the denominator of 2a into* both *terms of the numerator, not just the radical term.

Example 2

Solve $2x^2 + 7x = 3$ by the quadratic formula.

Before the formula can be used, the equation must be put in the form of $ax^2 + bx + c = 0$. This gives

$$2x^2 + 7x - 3 = 0$$

Always make sure that your quadratic equation is in standard form before trying to use the quadratic formula.

so that $a = 2, b = 7$, and $c = -3$. The solutions are

$$x = \frac{-7 \pm \sqrt{49 - 4(2)(-3)}}{2(2)} = \frac{-7 \pm \sqrt{49 + 24}}{4} = \frac{-7 \pm \sqrt{73}}{4}$$

This form of the result is generally acceptable. If decimal approximations are required, then, using $\sqrt{73} = 8.544$, we get

$$x = \frac{-7 + 8.544}{4} = 0.386 \quad \text{and} \quad x = \frac{-7 - 8.544}{4} = -3.886$$

Example
3

Solve the equation $3x^2 = 2x - 5$ using the quadratic formula.

Putting the equation in the proper form as $3x^2 - 2x + 5 = 0$, we have $a = 3, b = -2$, and $c = 5$. Therefore,

$$x = \frac{-(-2) \pm \sqrt{(-2)^2 - 4(3)(5)}}{2(3)} = \frac{2 \pm \sqrt{4 - 60}}{6}$$

$$= \frac{2 \pm \sqrt{-56}}{6} = \frac{2 \pm 2\sqrt{-14}}{6} = \frac{1 \pm \sqrt{-14}}{3}$$

Since $\sqrt{-14} = j\sqrt{14}$, we see that the result contains imaginary numbers. *There are no real roots.* The two solutions can be expressed as

$$\frac{1 + j\sqrt{14}}{3} \quad \text{and} \quad \frac{1 - j\sqrt{14}}{3}$$

or as

$$\frac{1}{3} + j\frac{\sqrt{14}}{3} \quad \text{and} \quad \frac{1}{3} - j\frac{\sqrt{14}}{3}$$

When the solution to a quadratic equation involves imaginary numbers, the two solutions will always be identical except for the sign change that precedes the imaginary part. Such complex numbers that differ only in the sign that precedes the imaginary part are called conjugates. The complex numbers $a + bj$ and $a - bj$ are conjugates.

There are many real problems that require quadratic equations. In Example 4 we illustrate one such problem.

Example
4

A nuclear waste holding facility is situated on a rectangular parcel of land that is 100 m long and 80 m wide. A safety zone in the form of a uniform strip constitutes the outer limits of this parcel as shown in Figure 11.4. The facility itself has an interior area of 6000 m². Find the width of the safety zone. (Assume an accuracy of two significant digits.)

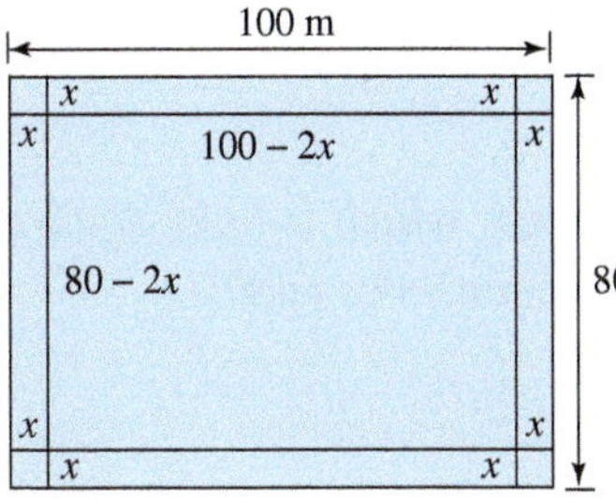

In Figure 11.4 we let $x =$ the width of the safety zone. The interior area of the facility itself is rectangular with length $= 100 - 2x$, width $= 80 - 2x$, and area $= 6000$ m². Using the area formula for a rectangle we get

length × width   = area
$(100 - 2x)(80 - 2x) = 6000$

Figure 11.4

Now Try It!

Write each of the following quadratic equations in standard form (if needed) and identify a, b, and c.

1. $x^2 + 4x - 9 = 0$
2. $x^2 - x - 6 = 0$
3. $4x^2 + 7x - 5 = 0$
4. $4x^2 + x = 3$
5. $a^2 = 4a - 2$
6. $9x^2 + 1 = -6x$

Simplifying, we get

$$8000 - 200x - 160x + 4x^2 = 6000$$
$$4x^2 - 360x + 2000 = 0$$
$$x^2 - 90x + 500 = 0$$

Solving this last equation by the quadratic formula, we have (with $a = 1$, $b = -90$ and $c = 500$):

$$x = \frac{-(-90) \pm \sqrt{(-90)^2 - 4(1)(500)}}{2(1)}$$

$$= \frac{90 \pm \sqrt{8100 - 2000}}{2}$$

$$= \frac{90 \pm \sqrt{6100}}{2} = \frac{90 \pm 10\sqrt{61}}{2} = 45 \pm 5\sqrt{61}$$

Because $\sqrt{61} = 7.810$, it follows that $5\sqrt{61} = 39.05$. Thus $x = 6.0$ and $x = 84$. However, x cannot be 84 because that would make the safety zone wider than the whole parcel, which means that this answer cannot be true. If the safety zone is 6.0 m wide, the interior area is 88 m by 68 m, and this area is 6000 m^2. This means that the safety zone is approximately 6.0 m wide. Although the mathematical solution yields two answers, only one of them is valid in this situation.

In Section 11.3 we introduced **completing the square.** We shall now use this method to derive the quadratic formula.

First we start with the general quadratic equation

$$ax^2 + bx + c = 0$$

Next we divide through by a, obtaining

$$x^2 + \frac{b}{a}x + \frac{c}{a} = 0$$

Let us subtract c/a from each side, which gives

$$x^2 + \frac{b}{a}x = -\frac{c}{a}$$

Now *to complete the square we take one-half of b/a, which is b/2a, square it, obtaining b²/4a², and add this to each side of the equation* to get

$$x^2 + \frac{b}{a}x + \frac{b^2}{4a^2} = \frac{b^2}{4a^2} - \frac{c}{a}$$

The left-hand side is now the perfect square of $(x + b/2a)$. Indicating this and combining fractions on the right-hand side, we have

$$\left(x + \frac{b}{2a}\right)^2 = \frac{b^2 - 4ac}{4a^2}$$

Taking the square root of each side, we get

$$x + \frac{b}{2a} = \frac{\sqrt{b^2 - 4ac}}{2a} \quad \text{and} \quad x + \frac{b}{2a} = \frac{-\sqrt{b^2 - 4ac}}{2a}$$

Subtracting $b/2a$ from each side and combining fractions we get

$$x = \frac{-b + \sqrt{b^2 - 4ac}}{2a} \quad \text{and} \quad x = \frac{-b - \sqrt{b^2 - 4ac}}{2a}$$

These two solutions are combined into one expression which is the quadratic formula. When we apply the quadratic formula, we eliminate the intermediate steps that are included in completing the square, but the results will be the same.

Up to this point we have presented three methods for solving quadratic equations:

1. Factoring

2. Quadratic formula

3. Completing the square

As a practical consideration, the method of completing the square is rarely used because the quadratic equation is more efficient and yields the same results. In choosing between factoring and the quadratic formula, it is generally a good strategy to use factoring if the quadratic expression is factorable. If the quadratic equation cannot be solved by factoring, then the quadratic formula should be used. Here is a quick test that reveals if the quadratic equation can be solved by factoring: First put the equation in standard form and calculate the value of $b^2 - 4ac$. *If that value is zero or a positive number that is a perfect square, then factoring can be used; otherwise, the quadratic formula should be used.* Figure 11.5 summarizes the strategy for solving quadratic equations. This strategy helps to prevent wasted time and effort with attempts to factor expressions that cannot be factored.

Example

5

Following the strategy of Figure 11.5 we can solve the quadratic equation $3x^2 - 5x + 2 = 0$ by factoring because $b^2 - 4ac = 25 - 4(3)(2) = 1$ is a perfect square. However, the quadratic equation $5x^2 - 3x - 4 = 0$ can be solved with the quadratic formula because $b^2 - 4ac = 9 - 4(5)(-4) = 9 + 80 = 89$ is not a perfect square so that *factoring will not be possible.*

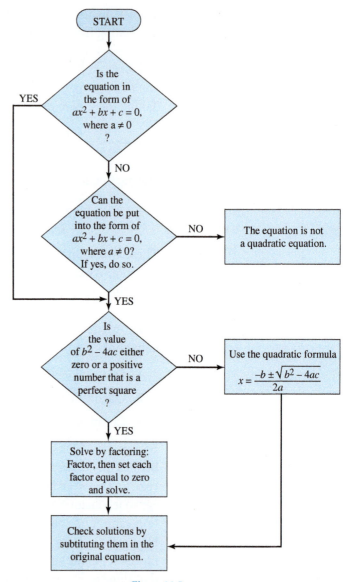

START

Is the equation in the form of $ax^2 + bx + c = 0$, where $a \neq 0$?

YES

NO

Can the equation be put into the form of $ax^2 + bx + c = 0$, where $a \neq 0$? If yes, do so.

NO → The equation is not a quadratic equation.

YES

Is the value of $b^2 - 4ac$ either zero or a positive number that is a perfect square ?

NO → Use the quadratic formula $x = \dfrac{-b \pm \sqrt{b^2 - 4ac}}{2a}$

YES

Solve by factoring: Factor, then set each factor equal to zero and solve.

Check solutions by substituting them in the original equation.

Figure 11.5

11.4 Exercises

In Exercises 1 through 28, solve the quadratic equations by using the quadratic formula.

1. $x^2 - 2x - 3 = 0$

2. $x^2 + 3x - 10 = 0$

3. $2x^2 + 7x + 3 = 0$

4. $3x^2 - 5x - 2 = 0$

5. $x^2 + 5x + 3 = 0$

6. $x^2 - 3x - 1 = 0$

7. $s^2 = 4s + 2$

8. $n^2 - 2n = 6$

9. $4t^2 = 8t - 3$

10. $3x^2 = x + 10$

11. $9x^2 - 16 = 0$

12. $3R^2 = 75$

13. $4x^2 - 12x = 7$

14. $2x^2 + 7x + 2 = 0$

15. $R^2 + 2R = -5$ **16.** $t^2 - t = -2$

17. $t^2 - t = 8$ **18.** $x^2 - 8 = 5x$

19. $4x^2 = 8 - 2x$ **20.** $3u^2 = 18 - 6u$

21. $I^2 - 7I = 0$ **22.** $5s^2 = 7s$

23. $2t^2 - 3t = -8$ **24.** $4x^2 = 9x - 6$

25. $10t + 8 = 3t^2$ **26.** $7 - 15x = -2x^2$

27. $6a^2x^2 + 11ax + 3 = 0$ (*a* is constant)

28. $2x^2 + (a + 2)x + a = 0$ (*a* is constant)

In Exercises 29 through 38, solve by means of the quadratic formula any quadratic equations that may arise.

29. Find two consecutive positive integers whose product is 272.

30. When a number is added to its square, the result is 240. Find the number.

31. The perimeter of a rectangular sign is 58 cm and its area is 204 cm^2. Find its dimensions.

32. Under certain conditions, the distance *s* that an explosion propels an object above the ground is given by

$$s = 143t - 16t^2$$

where *t* is the time in seconds. After what amount of time is the object 60.0 ft above the ground?

33. In studying the effects of gasoline vapors, a firefighter finds that under certain conditions, the partial pressure *P* of a certain gas (in atmospheres) is found by solving the equation

$$P^2 - 3P + 1 = 0$$

Solve and then choose the value of *P* which is less than one atmosphere.

34. The mechanical power developed in an electric motor is given by

$$P = T - I^2R$$

Find the current *I* (in amperes) if the power *P* is 448 W, the total power *T* is 480 W, and the resistance *R* is 2.00 Ω.

35. A farmer has a rectangular field 400 m by 500 m. On this field, a uniform strip on the outside will be left unplanted so that half of the total area will be planted. How wide is the strip? See Figure 11.6.

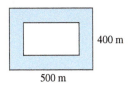

500 m

400 m

Figure 11.6

36. A rectangular piece of sheet metal has a length that is twice the width. In each corner, a 3-in. square is cut out, and the outer strips are then bent up to form an open box with a volume of 168 in^3. Find the dimensions of the original sheet. See Figure 11.7.

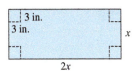

3 in.

3 in.

x

2*x*

Figure 11.7

37. A computer analysis shows that the cost *C* (in dollars) for a company to make *x* units of a certain product is given by $C = 0.1x^2 + 0.8x + 7$. How many units can be made for $50?

38. At an altitude *h* (in ft) above sea level, the boiling point of water is lower by $T°$F than the boiling point at sea level, which is 212° F. The difference can be approximated by solving the equation $T^2 + 520T - h = 0$. What is the boiling point in Boulder, Colorado, which has an altitude of 5300 ft?

Now Try It! Answers

1. $a = 1, b = 4, c = -9$ **2.** $a = 1, b = -1, c = -6$ **3.** $a = 4, b = 7, c = -5$

4. $a = 4, b = 1, c = -3$ **5.** $a = 1, b = -4, c = 2$ **6.** $a = 9, b = 6, c = 1$

11.5   Graphing the Quadratic Function

Section Objectives
- Recognize the graph of a quadratic equation
- Find the vertex of the graph of a quadratic equation
- Find x and y intercepts of the graph of a quadratic equation
- Determine the axis of symmetry
- Solve a quadratic equation graphically

In this section we will discuss the graph of the quadratic polynomial $ax^2 + bx + c$. We will explore various features of a quadratic graph including the shape, vertex, x and y intercepts, axis of symmetry, and maximum and minimum points.

Recall the problem introduced at the start of this chapter. The height of a baseball tossed in the air was recorded over a period of time. We presented the **quadratic equation** $h = -16t^2 + 100t + 6$ as a good mathematical model for the height of the baseball from the time it is tossed into the air until the time it hits the ground. To graph this equation we would create the following table of values (Figure 11.8) and use these values to create a graph of the function (Figure 11.9).

X	Y₁
0	6
1	90
2	142
3	162
4	150
5	106
6	30

Figure 11.8

Notice that the graph presented is shown only in the first quadrant of the rectangular coordinate system. This is because both our time t and height h are positive numbers.

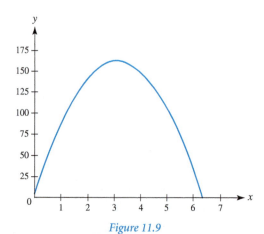

Figure 11.9

The graph of any quadratic function $y = ax^2 + bx + c$ is called a **parabola.** The graph of a parabola is recognizable by its classic U-shaped curve. A parabola can either *open up* or *open down.* This is determined by the sign of the ax^2 term. If $a > 0$, the parabola will open up. If $a < 0$, the parabola will open down. All parabolas have a **vertex,** or turning point. If the graph of a quadratic equation opens up ($a > 0$), then the vertex will be a **minimum point.** If the graph of a quadratic equation opens down ($a < 0$), then the vertex will be a **maximum point.**

Example
1

The graph of the quadratic equation $y = 2x^2 - 8x + 6$ is shown in Figure 11.10a and the graph of $y = -2x^2 - 8x + 6$ is shown in Figure 11.10b.

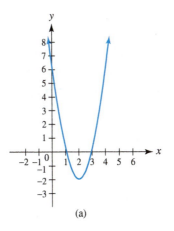

(a)

For this graph $a = 2$. Because $a > 0$ we see that the graph of the parabola opens up. The vertex is a minimum point whose coordinates are $(2, -2)$. We can read this value directly from our graph.

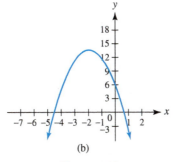

(b)

Figure 11.10

For this graph $a = -2$. Because $a < 0$ we see that the graph of the parabola opens down. The vertex is a maximum point whose coordinates are $(-2, 14)$.

Recall that the y-intercept is the point at which the graph crosses the y axis. We know that $x = 0$ at the y-intercept.

We can sketch the graph of a parabola by using its basic shape and knowing the location of two or three points, including the vertex. Another point that is easy to identify is the *y-intercept.* For $y = ax^2 + bx + c$, if $x = 0$, then $y = c$. This means that the point $(0, c)$ is the **y-intercept.**

Example
2

We can graph the quadratic equation $y = 2x^2 + 8x + 3$ by creating a table of values and then plotting the points. We sketch a smooth U shaped curve as seen in Figure 11.11.

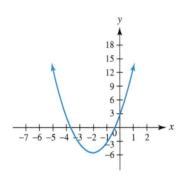

For this graph $a = 2$. Because $a > 0$ we see that the graph of the parabola opens up. The vertex is a minimum point whose coordinates are $(-2, -5)$. The y-intercept is $(0, 3)$.

Figure 11.11

If we look carefully at each of the graphs presented so far it appears that if we were to fold our graph along a line that passes through the vertex, the left- and right-hand side of the graphs would match up. This line is called the **axis of symmetry**. It is found using the formula $x = \dfrac{-b}{2a}$ which gives us the x coordinate of the vertex. The y coordinate is found by substituting this x value back into the original equation.

Example
3

In this example we will use the axis of symmetry to find the coordinates of the vertices for the graphs presented in Example 1.

For $y = 2x^2 - 8x + 6$ we identify $a = 2$ and $b = -8$.

Using the formula for the axis of symmetry to find the x coordinate gives us

$$x = \frac{-b}{2a} = \frac{-(-8)}{2(2)} = \frac{8}{4} = 2.$$

and the y coordinate is $y = 2(2)^2 - 8(2) + 6 = -2$.

Therefore the vertex is $(2, -2)$.

For $y = -2x^2 - 8x + 6$ we identify $a = -2$ and $b = -8$.

Using the formula for the axis of symmetry to find the x coordinate gives us

$$x = \frac{-b}{2a} = \frac{-(-8)}{2(-2)} = \frac{8}{-4} = -2.$$

and the y coordinate is $y = -2(-2)^2 - 8(-2) + 6 = 14$.

Therefore the vertex is $(-2, 14)$.

We can pull this information together to help us to sketch a fairly accurate graph of a given quadratic equation. This can be seen in Example 4.

Example
4

Given the quadratic equation $y = -8x^2 + 24x - 3$, find the vertex and the y-intercept. Use this information to sketch the graph.

First we identify $a = -8$ and $b = 24$. Using the formula for the axis of symmetry to find the x coordinate gives us

$$x = \frac{-b}{2a} = \frac{-24}{2(-8)} = \frac{-24}{-16} = 1.5.$$

The y coordinate is $y = -8(1.5)^2 + 24(1.5) - 3 = 15$.

Therefore the vertex is $(1.5, 15)$. Because $a < 0$ we know that the parabola will open down and that the vertex is a maximum point.

To find the y-intercept we let $x = 0$ in the equation $y = -8x^2 + 24x - 3$. This gives us an intercept at $(0, -3)$.

We use the maximum point $(1.5, 15)$ and the y-intercept $(0, -3)$, along with the fact that the graph opens down, to get an approximate sketch of the graph. We are careful to sketch our graph so that it is symmetric to a vertical line through the vertex. Our sketch should look similar to the graph shown in Figure 11.12.

In the previous sections of this chapter we looked at algebraic methods for solving a quadratic equation. We will now consider a graphical solution. To solve the equation $ax^2 + bx + c = 0$, let $y = ax^2 + bx + c$ and graph the equation. The **zeros** or **roots** of the equation are the **x-intercepts.** These are the points for which $y = 0$. Example 5 illustrates how we might solve a quadratic equation graphically.

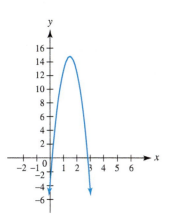

Figure 11.12

Example
5

Graphically find the roots for the equation $y = x^2 - x - 6$.

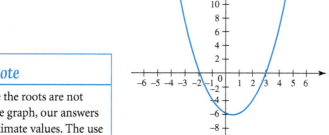

Figure 11.13

We graph this function to get the graph shown in Figure 11.13. The x-intercepts are the roots or solutions to this equation. In this example they are easily identified as $x = -2$ and $x = 3$. Generally the roots are not easily read from the graph.

Note

In those cases where the roots are not easy to read from the graph, our answers are generally approximate values. The use of the graphing calculator can help us in these cases, as will be see in "Using Technology" found at the end of this section.

Summary — The Graph of a Quadratic Equation:

1. The graph of $y = ax^2 + bx + c$ is called a **parabola,** where $a \neq 0$.

2. If $a > 0$, the parabola will open up. If $a < 0$, the parabola will open down.

3. The turning point of a parabola is referred to as its **vertex.**

4. For $a > 0$ the vertex is a **minimum point.** For $a < 0$ the vertex is a **maximum point.**

5. The y-intercept occurs at $(0, c)$

6. The **axis of symmetry** is found by $x = \dfrac{-b}{2a}$

7. The graphical solution for a quadratic equation occurs at the point(s) where the graph crosses the x axis.

 *Using Technology*

In this section we will focus on how to solve a quadratic equation using the **zero** feature of the calculator, previously introduced in Section 5.6.

Consider the problem of the ball whose height was recorded as it was tossed into the air. The quadratic equation $h = -16t^2 + 100t + 6$ represents the height of the baseball from the time it is tossed into the air until it hits the ground. Graph this function and use the graph to determine when the ball hits the ground. The height, h, of the ball when it is on the ground is equal to zero and the solution to the equation is its x-intercepts. The points are also commonly referred to as the **zeros** of a function.

We begin by entering the function into the graphing calculator.

- Press Y = to enter $Y_1 = -16x^2 + 100x + 6$.

- Press WINDOW and set Xmin = 0, Xmax = 7, Ymin = 0, Ymax = 165, and Yscl = 15.

- Press GRAPH to view a graph similar to Figure 11.14.

> **Note**
>
> If an equation is not in the form $y = ax^2 + bx + c$, it should be rewritten so that it is in standard form before it is entered into your graphing calculator.

- While in the graph screen, press 2nd TRACE to access the CALC menu (Figure 11.15). Then press #2 to access the **zero** feature.

- Notice that a blinking cursor appears on the graph and that "Left Bound?" is displayed on the bottom of the graph. Because we want to determine when the ball struck the ground, you are interested only in the right intercept.

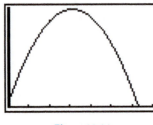

Figure 11.14

- Move the blinking cursor so that it is on the graph *to the left of the x-intercept.* Then press ENTER.

- You are now asked for the "Right Bound?". Move your cursor so that it is on the graph *to the right of the same x-intercept.* Press ENTER.

- At this point you will see "Guess?" on the bottom of the screen. Just press ENTER once again.

- You should see that the ball struck the ground at approximately 6.3 seconds after it was tossed in the air (Figure 11.16).

Figure 11.15

Figure 11.16

We can also use the graphing calculator to determine the coordinates of the vertex of our parabola. This will tell us the maximum height the ball reached, as well as the time at which it reached that height.

- While in the graph screen, press 2nd TRACE to access the CALC menu Figure 11.17. Then press #4 to access the **maximum** feature. If the graph opened up, we would use option #3 to find the coordinates of the minimum point.

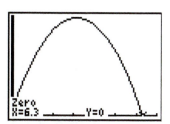

Figure 11.17

- Notice that a blinking cursor appears on the graph and that "Left Bound?" is displayed on the bottom of the graph. You need to move the blinking cursor so that it is on the graph *to the left of the vertex.* Then press ENTER.

- You are now asked for the "Right Bound?". Move your cursor so that it is on the graph *to the right of the vertex.* Press ENTER.

- At this point you will see "Guess?" on the bottom of the screen. Just press ENTER once again.

- You should now see the screen shown in Figure 11.18. Notice that your cursor is at the vertex of the parabola. The bottom of the screen displays the coordinates for the vertex. This tells us that the axis of symmetry is 3.1.

- We also know that the ball reached a maximum height of 162.2 feet 3.1 seconds after it was tossed into the air.

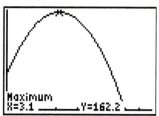

Figure 11.18

11.5 Exercises

In Exercises 1 through 6, determine if the graph in each of the following parabolas will open up or down. Indicate whether the vertex of the graph will be a maximum point or a minimum point.

1. $y = x^2 - 6x + 5$

2. $y = x^2 - 4x$

3. $y = -3x^3 + 10x - 4$

4. $y = -x^2 - 4x - 3$

5. $y = 2x^2 + 8x - 5$

6. $y = -2x^2 - 5x$

In Exercises 7 through 12, graph each of the following parabolas by first finding the vertex and the y-intercept. If possible, check your solutions with a graphing calculator.

7. $y = x^2 - 6x + 5$

8. $y = x^2 - 4x$

9. $y = -3x^2 + 10x - 4$

10. $y = -x^2 - 4x - 3$

11. $y = 2x^2 + 8x - 5$

12. $y = -2x^2 - 5x$

In Exercises 13 through 16, sketch a graph of each parabola by using the vertex and y- and x-intercepts. If possible, check your solutions with a graphing calculator.

13. $y = x^2 - 4$

14. $y = x^2 + 3x$

15. $y = -2x^2 - 6x + 8$

16. $y = -3x^2 + 12x - 5$

In Exercises 17 through 22, sketch a graph of each **parabola by** using the axis of symmetry and the y-intercept.

17. $y = 2x^2 + 3$

18. $y = x^2 + 2x + 2$

19. $2x^2 = 7x + 4$

20. $y = -2x^2 - 2x - 6$

21. $y = -3x^2 - x$

22. $3x^2 - 25 = 20x$

In Exercises 23 through 26, solve the given applied **problems.**

23. The vertical distance d (in cm) of the end of **a robotic** arm above a conveyor belt in its 8-s cycle is **given by** $d = 2t^2 - 16t + 47$. Sketch a graph of this **function.**

24. The height h (in ft) above the ground of a submarine-launched missile is given by the function $h = -16t^2 + 96t - 80$ where t represents the time in seconds. Sketch a graph of this function. Determine the time at **which the** missile leaves and returns to the water.

25. Using the equation in Exercise 24, determine **the maxi-**mum height reached by the missile.

26. The recommended dosage for most medicines is deter-mined by patient weight. The formula $D = 0.1w^2 + 5w$, wherein D represents the dosage in mg and w **represents** patient body weight in lbs, is used to determine **dosages for** a particular medicine. If 1000 mg are prescribed, **find the** appropriate body weight for a person receiving **this dosage.**

Chapter Summary

Key Terms

quadratic equation

standard form of a quadratic equation

quadratic formula

parabola

vertex

maximum point

minimum point

roots

axis of symmetry

Key Concepts

- **Zero Product Rule:** If $pq = 0$, then either $p = 0$ or $q = 0$ (or both)

Solving a Quadratic Equation by Factoring

- Collect all terms to one side and simplify.
- Factor the quadratic expression.
- Set each factor equal to zero.
- Solve for the variable.

Solving a Quadratic Equation by Completing the Square

- Rewrite the equation with the constant on one side of the equation. If the coefficient of the x^2 term is not 1, divide each side by the coefficient of this term.

- Take one-half of the coefficient of the x term and then square that value.
- Add this value to both sides of the equation created in Step 1.
- Write the left-hand side as a square and simplify the right-hand side.
- Take the square root of both sides of the equation.
- Solve each equation.

The Graph of a Quadratic Equation

- The graph of a quadratic equation is called a parabola.
- If $a > 0$, the parabola will open up. If $a < 0$, the parabola will open down.

- The vertex is the turning point of a parabola.
- For $a > 0$ the vertex is a minimum point.
- For $a < 0$ the vertex is a maximum point.
- The y-intercept occurs at $(0, c)$.
- The axis of symmetry is found by $x = \dfrac{-b}{2a}$.
- The graphical solution for a quadratic equation occurs at the point(s) where the graph crosses the x axis.

Key Formulas

$ax^2 + bx + c = 0 \quad a \neq 0$ standard form of a quadratic equation

$x = \dfrac{-b \pm \sqrt{b^2 - 4ac}}{2a}$ quadratic formula

$x = \dfrac{-b}{2a}$ axis of symmetry

Review Exercises

In Exercises 1 through 12, solve the given quadratic equations by factoring.

1. $x^2 + 7x + 12 = 0$
2. $x^2 - 7x - 8 = 0$
3. $2x^2 - 11x - 6 = 0$
4. $4x^2 = 4x + 3$
5. $6n^2 - 35n - 6 = 0$
6. $8s^2 - 9s = 0$

7. $16x^2 - 24x + 9 = 0$

8. $4p^2 + 49 = 28p$

9. $15R^2 + 21R = 0$

10. $9r^2 - 55r + 6 = 0$

11. $t^2 + t = 110$

12. $2x^2 + 39x = 20$

In Exercises 13 through 24, solve the given quadratic equations by (a) factoring and (b) the quadratic formula.

13. $x^2 + 5x - 24 = 0$

14. $x^2 - 4x - 21 = 0$

15. $x^2 + 9 = -6x$

16. $2x^2 + 3 = 5x$

17. $6m^2 + 5 = 11m$

18. $7x^2 + 24x = 16$

19. $3x^2 - 6x + 3 = 0$

20. $5x^2 + 40x = 0$

21. $x^2 + 4x + 2 = 0$

22. $2x^2 + 5x - 1 = 0$

23. $y^2 - 6y = 6$

24. $5t^2 = 5t + 3$

In Exercises 25 through 28, solve the given equation by completing the square.

25. $x^2 - x - 30 = 0$

26. $2t^2 = t + 4$

27. $x^2 = 2x + 5$

28. $4x^2 - 8x = 3$

In Exercises 29 through 36, solve the given quadratic equation using an appropriate algebraic method.

29. $x^2 + 8x - 9 = 0$

30. $x^2 + 3x = 5$

31. $2x^2 - 5 = 2x$

32. $4x^2 + 16 = 16x$

33. $x^2 = 7 + 2x$

34. $2x^2 = 3x + 5$

35. $2x + 8 = 5x^2$

36. $3x^2 + 4x = 4$

In Exercises 37 through 40, sketch the graphs of the given functions by using the vertex, the y-intercept, and one or two other points. If possible, use a graphing calculator to verify your graphs.

37. $y = 2x^2 - x - 1$

38. $y = -4x^2 - 1$

39. $y = x - 3x^2$

40. $y = 2x^2 + 8x - 10$

In Exercises 41 through 44, solve the given quadratic equations graphically. If there are no real roots, state this as the answer. If possible, use a graphing calculator to verify your graphs.

41. $2x^2 + x - 4 = 0$

42. $-4x^2 - x - 1 = 0$

43. $3x^2 = -x - 2$

44. $x(15x - 12) = 8$

In Exercises 45 through 54, solve the given or resulting quadratic equations by any appropriate method unless otherwise specified.

45. Two positive numbers differ by 14 and their product is 351. Find the larger number.

46. Two consecutive positive integers have a product of 420. Find these integers.

47. The length of a rectangle is 1 in. more than the width. See Figure 11.19. If 3 in. is added to the length and 1 in. to the width, the new area is twice the original area. Find the dimensions of the original rectangle.

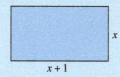

Figure 11.19

48. If the radius of a circle is doubled, its area increases by 147π m². Find the radius of the circle.

49. The relation of cost C (in dollars) and units x (in hundreds) produced by a certain company is

$$C = 3x^2 - 12x + 13$$

For what value of $x(x < 2)$ is $C = \$3$?

50. A projectile is fired vertically into the air. The distance (in feet) above the ground, in terms of the time in seconds, is given by the formula

$$s = 96t - 16t^2$$

How long will it take to hit the ground?

51. A company finds that its profit for producing x units weekly is

$$p = 25x - x^2 - 100$$

For which values of x is the profit zero?

52. A machine designer finds that if she changes a square part to a rectangular part by increasing one dimension of the square by 4 mm, the area becomes 77 mm². Determine the side of the square part. See Figure 11.20.

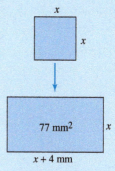

Figure 11.20

53. It is important to maintain proper air pressure in your car tires in order to maintain the life of your tire. The quadratic equation $L = -1106 + 69.7x - 1.06x^2$ can be used to predict the life L (in miles) of a particular tire given different amounts of air pressure x (in psi). Determine the air pressure that will guarantee the longest life of the tire.

54. A manufacturer of a common seasonal item has determined that the revenue function most effective for predicting his income is given by $R = -2x^2 + 200x - 50$ where R represents income (in \$) and x represents the number of items sold. Use this function to determine the number of items needed to maximize the manufacturer's income and what that maximum income will be.

Chapter Test

Solve problems 1 through 4 using the appropriate algebraic method.

1. $x^2 + 4x + 2 = 0$

2. $2x^2 = 9x - 4$

3. $4x^2 - 8x = 3$

4. $2x^2 - x = 6 - 2x(3 - x)$

5. Solve by completing the square: $x^2 - 6x - 9 = 0$

6. Sketch an accurate graph of $y = 2x^2 + 8x + 5$. Determine whether the graph opens up or down. Identify the coordinates of the vertex and the y-intercept.

7. Solve $y = -3x^2 + 17x - 12$ graphically. Include a labeled sketch of your graph with your solution.

8. If a ball is tossed into the air, its height h (in ft) is recorded and represented by the quadratic equation $h = -16t^2 + 22t + 100$ where t represents the time in seconds. Determine when the ball will reach a height of 80 feet.

9. A rectangular plot of land has an area of 520 square feet. The length of this plot of land is 6 feet longer than its width. Determine the dimensions of the plot of land.

12 Exponential and Logarithmic Functions

The Internet has revolutionized our world in ways that could not have been imagined twenty years ago. It is used to disseminate information. It allows us to interact with other individuals regardless of geographic location. The Internet represents one of the most successful examples of a commitment to research involving a partnership between the government, industry and academia.

The Internet had its roots in the 1960s and was originally designed as a project of the United States government's Department of Defense. The idea was to create a noncentralized network of computers designed to survive cataclysmic events (e.g., nuclear war) and still function when parts of the network were down or destroyed. Researchers and academics in other fields began to tap into and make use of the network. In 1985, the National Science Foundation began a program to establish a distributed network of networks (the Internet) access

397

across the United States capable of handling far greater traffic. By the 1990s the Internet had experienced explosive growth. According to Internet World Statistics, there were 16 million Internet users worldwide in December 1995, 451 million in December 2000, 817 million in December 2004, and 957 million in September 2005.

The function that would serve as a model for this data is known as an **exponential function,** so called because the independent variable appears as an exponent. Simply stated, an exponential function demonstrates extremely rapid growth (or decay). It models growth in which the amount being added to the system is proportional to the amount already present; the bigger the system, the greater the increase. We will present a more formal definition of the exponential function in Section 12.1. Exponential functions are used in a broad spectrum of disciplines. They can be used to study the spread of AIDS, the projected growth or decay of the national debt, population growth and decline, compound interest, nuclear physics, mechanical systems, and much more. Logarithms were used extensively before calculators became common. In this chapter we will introduce the fundamental concepts of logarithms. Applications of logarithmic functions include the decibel sound scale, radioactive decay, and the Richter scale, which is used for measuring the intensity of earthquakes.

12.1 Exponential Functions

Section Objectives:
- Recognize an exponential equation
- Identify the base in an exponential equation
- Evaluate exponential functions
- Graph exponential functions

In Chapter 3 we introduced the concept of an exponent. In earlier chapters we have introduced several different types of functions, most notably linear equations and quadratic equations. In a linear function we saw that the rate of change (slope) of the line was constant. In this section we turn our attention to a new type of relationship, one that changes at a constant *percent rate.* We define this new equation as follows:

A function $y = ab^x$ where $b > 0$, $b \neq 1$, and x is any real number is called an **exponential function**. The number b is called the *base*.

Notice that the variable x is in the exponent of this equation.

(Example 1

In each of the following determine whether or not the given function is an exponential function.

a. $y = 3^x$ **b.** $y = x^3$

c. $f(x) = 2(7)^x$ **d.** $f(x) = 2(-7)^x$

Both (b) and (d) **do not represent** exponential functions. In (b) the variable is not in the exponent. In (d) the base is -7. According to the definition of an exponential function, the base b must be greater than zero.

Unlike other functions we have looked at so far, *exponential functions* require that the independent variable of the equation be in the exponent of the function.

(Example 2

Identify the base in each of the following.

a. $y = 3^x$ **b.** $f(x) = 2(7)^x$

c. $y = 4^{x-3}$ **d.** $y = 3a^x$

In (a) the base is 3.
In (b) the base is 7. Notice that only the value of 7 is raised to the x power.
In (c) the base is 4. In this case the exponent is the expression $x - 3$.
In (d) the base is a.

(Example 3

Evaluate the following exponential functions for the given value of x.

a. $y = 3^x$ for $x = 4$ becomes $y = 3^4 = 81$

b. $f(x) = 2(7)^x$ for $x = 3$ becomes $f(3) = 2(7)^3 = 2(343) = 686$

c. $y = -2(4^x)$ for $x = 3/2$ becomes $y = -2(4^{\frac{3}{2}}) = -2(8) = -16$

Graphing an Exponential Function

The graph of an exponential function is useful to show the properties of an exponential equation. We graph this type of function in the same way that we graphed other functions throughout this book. Create a set of ordered pairs by first choosing values for the independent variable x, and then using those values to find corresponding values of the dependent variable y. Plot the ordered pairs and connect the points with a smooth curve.

Example
4

Graph the exponential function $y = 2^x$ as x goes from -3 to $+3$.

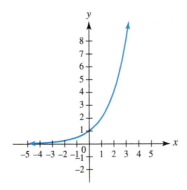

Figure 12.1

Table 12.1

x	-3	-2	-1	0	1	2	3
y	$\dfrac{1}{8}$	$\dfrac{1}{4}$	$\dfrac{1}{2}$	1	2	4	8

Note that the graph of the exponential function in Figure 12.1 is very different from the graphs of other functions that we have looked at. As x increases, we notice that y increases rather quickly and that the graph of the function gets steeper quickly. This is characteristic of an *exponential growth* function. Notice that the *y-intercept* is the point $(0, 1)$ and it appears that there is no *x-intercept*. In fact, this is true. We can see that as x gets smaller and smaller, y gets closer and closer to zero. The graph gets closer and closer to the x axis but never actually touches or crosses it.

In Example 4, the value for b was greater than 1. We now consider the characteristics of an exponential function in which the value for b is less than 1 (but greater than 0).

Example
5

Graph the function $y = \left(\dfrac{1}{2}\right)^x$ as x goes from -3 to 3

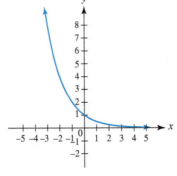

Figure 12.2

Table 12.2

x	-3	-2	-1	0	1	2	3
y	8	4	2	1	$\dfrac{1}{2}$	$\dfrac{1}{4}$	$\dfrac{1}{8}$

Notice that this graph of an exponential function shows that as the value of x decreases, y decreases at a rapid rate and that the graph of the function declines. Because the base b is less than 1, the graph falls. This is characteristic of an *exponential decay* function which occurs when $0 < b < 1$. We note that the *y-intercept* is once again the point $(0, 1)$ and that there is no *x-intercept*.

From these two graphs we can see that exponential functions have the following characteristics:

Basic Characteristics of Exponential Functions

For $y = ab^x$

1. The domain is the set of all values of x, the range is the set of all $y > 0$.

2. The *y-intercept* is $(0, 1)$.

3. There are no *x-intercepts*.

4. For $b > 1$, as x increases, y increases.

5. For $0 < b < 1$ as x increases, y decreases.

 *Using Technology*

Given the rapid growth (or decay) of an exponential function, it if often easier to create such graphs using a graphing calculator as compared to the more traditional methods using graph paper.

To graph the exponential function $y = 2(7)^x$, we proceed as usual.

1. Press Y= and type the function $2(7)^x$. To raise 7 to the x power you will need to find and use the ∧ symbol. You will type in $2(7)\wedge x$ into your calculator as in Figure 12.3.

2. Create a graph in the standard window by pressing ZOOM 6.

Figure 12.3

3. This graph appears to be increasing rapidly. In order to get a better look at the function we need adjust our viewing window. The graph appears to begin around $x = -2$. The values for y become too large after $x = 1$. We also note that $y > 0$. We adjust our viewing window accordingly and extend Ymax in order to get a better look at the graph. We use the window settings shown in Figure 12.4 in order to get a better look at the graph of this exponential function.

4. This produces the following graph of $y = 2(7)^x$, as shown in Figure 12.5.

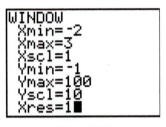

Figure 12.4   *Figure 12.5*

12.1 Exercises

In Exercises 1 through 8, determine whether the given functions are exponential functions.

1. $y = 5^x$

2. $y = 5^{-x}$

3. $y = -3^x$

4. $y = (-3)^{-x}$

5. $y = x^3$

6. $y = x^{\frac{1}{2}}$

7. $y = -7(-4)^x$

8. $y = 3a^x$

In Exercises 9 through 12, identify the base of each exponential function.

9. $y = 3^x$

10. $y = \left(\dfrac{2}{3}\right)^x$

11. $y = 0.5\pi^x$

12. $y = 2(7)^x$

In Exercises 13 through 16, determine whether the exponential function represents an increasing function or a decreasing function.

13. $y = 3^x$

14. $y = \left(\dfrac{2}{3}\right)^x$

15. $y = 0.5\pi^x$

16. $y = 2(7)^x$

In Exercises 17 through 20, evaluate the exponential function for the given value of x.

17. $f(x) = 4^x$ for $x = -2$

18. $f(x) = 0.5(3)^x$ for $x = 3$

19. $f(x) = 3(1.5)^{3x}$ for $x = 1.5$

20. $f(x) = 1.2(6^{-x})$ for $x = 2$

In Exercises 21 through 26, graph the following exponential functions. If possible, check your results with your graphing calculator.

21. $y = 4^x$

22. $y = 2\left(\dfrac{1}{2}\right)^x$

23. $y = 0.5(3)^x$

24. $f(x) = 3(1.5)^{3x}$

25. $y = 0.2(10^{-x})$

26. $f(x) = 0.25^x$

In Exercises 27 through 32, solve the given problems.

27. The value V of a bank account in which \$250 is invested at 5% interest compounded annually is given by the formula

$$V = 250(1.05)^x$$

where x represents the time in years. Find the value of the account after 4 years.

28. The intensity I of an earthquake is given by

$$I = I_0(10)^R$$

where I_0 is a minimum intensity for comparison and R is the Richter scale magnitude of the earthquake. Find I in terms of I_0 when $R = 5.5$.

29. The amount (in milligrams) of a drug in the body t hours after taking a pill is given by the function

$$A(t) = 25(0.85)^t.$$

Determine the amount of drug left in the body after 10 hours.

30. Let the formula $P(t) = 32(1.047)^t$ represent the total population P (in thousands) of a small town in New York State t years after the last census. What is the population of this town 5 years after the last census?

31. In 1980 there were about 203,000 cell phone subscribers in the United States. Since then, there has been an exponential growth in the number of cell phone subscribers.

This growth can be modeled using the function $C(t) = 0.0272(1.495)^{t-1980}$ where C represents the number of cell subscribers (in millions) and t represents the year. Determine the number of cell phone subscribers in 2005.

32. The electric current i (in mA) in a given circuit is $i = 2.5(1 - e^{-0.10t})$ where t is the time (in s). Determine i for $t = 0.005$ s ($e = 2.718$).

12.2 Logarithms

Section Objectives
• Convert expressions between exponential and logarithmic form
• Evaluate logarithms using a calculator
• Graph logarithmic equations

The concept of logarithm is closely linked to the concept of exponents. In fact, *a logarithm is actually an exponent.* Recall that in an expression such as b^x, *the number b is called the* **base,** and *the number x is called the* **exponent.** We now define the **logarithm** of a number as follows.

If $b^x = N$, then $\log_b N = x$. (where $b > 0$ and $b \neq 1$)

From this definition we see that $\log_b N$ is the exponent to which b must be raised to get N. There are two different ways to express the same relationship, as illustrated in Examples 1 through 3.

Example 1

The equation $2^5 = 32$ can be expressed in

logarithmic form as $\log_2 32 = 5$. $2^5 = 32$

The equation $10^3 = 1000$ can be expressed in

logarithmic form as $\log_{10} 1000 = 3$.

The equation $10^{-1} = 0.1$ can be expressed in

logarithmic form as $\log_{10} 0.1 = -1$.

The equation $\log_5 25 = 2$ can be expressed in

exponential form as $5^2 = 25$.

The equation $\log_{10}100 = 2$ can be expressed in exponential form as $10^2 = 100$.

The equation $\log_{10}0.001 = -3$ can be expressed in exponential form as $10^{-3} = 0.001$.

Now Try It!

Convert each of the following to the indicated form:

1. $\log_4 \frac{1}{64} = -3$ to exponential form
2. $2^4 = 16$ to logarithmic form
3. $\log_7 \sqrt{7} = \frac{1}{2}$ to exponential form
4. $7^0 = 1$ to logarithmic form
5. $\log_{27} 9 = \frac{2}{3}$ to exponential form
6. $6^{\frac{1}{4}} = \sqrt[4]{6}$ to logarithmic form

In Example 2 we can see that the logarithm of a number yields a result that is actually an exponent. It would be helpful to stop reading at this point and practice converting back and forth between the two forms given in the above definition. It is essential that we understand that an expression such as $\log_2 32 = 5$ is equivalent to $2^5 = 32$. We must be able to easily convert from one form to the other.

Although *the base b can be any* **positive** *number except 1*, there are special values which are often used. The base 2 has a special importance for computer applications. We now consider the base 10, which is so common that it is referred to as the **common logarithm.** Because the base 10 is used so often, we have a *special notation that consists of omitting the subscript.* See Example 2.

Example 2

$\log 10000 = 4$ since $10^4 = 10000$.   (log 10000 is the same as $\log_{10} 10000$)

$\log 10 = 1$ since $10^1 = 10$.

$\log 0.01 = -2$ since $10^{-2} = 0.01$.

exponent

$$M = \log_b N \qquad b^M = N$$

base

Keep in mind that

$M = b^N$ is in exponential form
$\log_b M = N$ is in logarithmic form

In the remainder of this section we consider only common logarithms. So far, all of the logarithms involved bases and results that could be determined by inspection. We can evaluate log 100 because we know that $10^2 = 100$; we get $\log 100 = 2$. However, we can't use inspection to evaluate a number such as log 23 because we don't know a number x that will satisfy the equation $10^x = 23$. In evaluating such numbers, we can use calculators or computers.

The key point to recognize is that we can raise 10 to powers that are fractions or decimals.

We can use a scientific or graphing calculator to find the values of a common log. No matter what calculator you are using you should be able to locate a key marked LOG. In some calculators we press the LOG key first and then enter the value, followed by the ENTER key. In other calculators we enter the value first, followed by the LOG key.

Example 3

Use your calculator to verify the following:

a. $\log(58.24) = 1.7652$

b. $\log(0.01279) = -1.8931$

Example 4

Find the value of x in the logarithmic expression $\log x = 0.75$.

We know that $\log x = 0.75$ can be rewritten in the exponential form $10^{0.75} = x$. We can evaluate the expression $10^{0.75}$ in our calculator to get 5.623. Therefore, $\log(5.623) = 0.75$.

In a similar way we can find the value of x given that $\log x = -1.735$. We begin by rewriting $\log x = -1.735$ as the exponential form $10^{-1.735} = x$. Using our calculator we determine that the value of $x = 0.01841$.

Therefore, $\log(0.01842) = -1.735$.

Graphing a Logarithmic Function

Just as we explored the characteristics of an exponential function through its graph, we can also explore the characteristics of the **logarithmic function** by graphing this function.

> **A logarithmic function** is given by the equation
> $y = \log_b(x)$ with $b > 0$ and $b \neq 1$.

We graph this function by first choosing values for the independent variable x, and using those values to find corresponding values of the dependent variable y. Plot the ordered pairs and connect the points with a smooth curve.

Example 5

Graph the logarithmic function $y = \log(x)$ as x goes from 1 to 6.

Table 12.3

x	1	2	3	4	5	6	7	8	9	10
y	0	0.30	0.48	0.60	.70	0.78	0.85	0.90	0.95	1

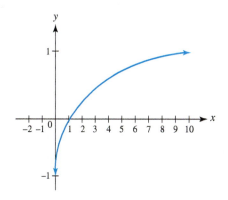

Figure 12.6

Note that the graph of the logarithmic function in Figure 12.6 is very different from the graph of the exponential function. We see that the domain of x is all positive numbers and that the range of y is all real numbers. We also can see that as x increases, y increases. Notice that the *x-intercept* is the point $(1, 0)$ and it appears that there is no *y-intercept*. In fact, this is true. We can see that as x gets smaller and smaller, *y* approaches negative infinity. The graph gets closer and closer to the *y axis* but never actually touches or crosses it so that *x* is never zero. If you were to enter log(0) in your calculator you would get an ERROR message.

Basic Characteristics of Logarithmic Functions

For $y = \log_b x$

1. The domain is the set of all $x > 0$, the range is the set of all values of *y*.
2. The *x-intercept* is $(1, 0)$.
3. There are no *y-intercepts*.
4. If $0 < x < 1, \log_b x < 0$. If $x = 1, \log_b x = 0$. If $x > 1, \log_b x > 0$.

12.2 Exercises

In Exercises 1 through 12, write the logarithmic form of the given exponential equations.

1. $10^2 = 100$

2. $10^3 = 1000$

3. $10^{-2} = 0.01$

4. $10^{-4} = 0.0001$

5. $10^{3.4600} = 2884$

6. $10^{-1.57} = 0.02692$

7. $10^{-3.4444} = 0.0003594$

8. $10^{0.6667} = 4.6419$

9. $2^{10} = 1024$

10. $3^4 = 81$

11. $6^3 = 216$

12. $5^4 = 625$

In Exercises 13 through 24, write the exponential form of the given logarithmic equations.

13. $\log 10 = 1$

14. $\log 100 = 2$

15. $\log 1000 = 3$

16. $\log 1,000,000 = 6$

17. $\log 0.01 = -2$

18. $\log 0.0001 = -4$

19. $\log 567 = 2.7536$

20. $\log 0.00344 = -2.4634$

21. $\log_2 8 = 3$

22. $\log_2 64 = 6$

23. $\log_3 243 = 5$

24. $\log_5 125 = 3$

In Exercises 25 through 36, find the value of the given logarithms without using a calculator.

25. $\log 10,000$

26. $\log 100,000$

27. $\log 0.001$

28. $\log 0.00001$

29. $\log 10^6$

30. $\log 10^{12}$

31. $\log 10^{-3}$

32. $\log 10^{-8}$

33. $\log_2 16$

34. $\log_3 9$

35. $\log_4 64$

36. $\log_6 1296$

In Exercises 37 through 44, determine the logarithm of each given number using a calculator.

37. 7.26

38. 8.15

39. 2600

40. 83,000

41. 208   **42.** 0.0533

43. 0.00358   **44.** 0.0305

In Exercises 45 through 56, find the value of x.

45. $\log_4 16 = x$ **46.** $\log_5 125 = x$ **47.** $\log_{16} \frac{1}{4} = x$

48. $\log_2 32 = x$ **49.** $\log_7 x = 3$ **50.** $\log_5 x = 4$

51. $\log_2 x = 8$ **52.** $\log x = 5$ **53.** $\log_b 81 = 2$

54. $\log_b 625 = 4$ **55.** $\log_b 4 = -\frac{1}{3}$ **56.** $\log_b 4 = \frac{2}{3}$

In Exercises 57 through 60, plot the graph for the following functions. Check your results with a graphing calculator when possible.

57. $y = 3\log(x)$ **58.** $y = \log(5x)$

59. $y = -\log(-x)$ **60.** $y = -2\log(-2x)$

In Exercises 61 through 64, solve the following problems as indicated.

61. Using the Richter scale to measure the magnitude of an earthquake of intensity I is done using the formula

$$R = \log\left(\frac{I}{I_0}\right)$$

where I_0 is the minimum base level used for comparison. Find the Richter scale reading of the 1906 San Francisco earthquake where $I = (2.5 \times 10^8)I_0$.

62. The loudness of a sound, measured in decibels (dB), is given by

$$10\log\left(\frac{I}{I_0}\right)$$

where I is the intensity of the sound and I_0 is the intensity of the faintest sound that can be heard. A busy city street has a noise level with an intensity level of $I = 10^7 I_0$. Evaluate this logarithm to find the decibel level.

63. Using the information presented in Exercise 62, find the decibel level of a sound whose physical intensity I is 100 times that of I_0.

64. A certain strain of bacteria doubles every three hours. The time t (in hours) needed to grow the bacteria to N bacteria is given by the formula

$$t = 3\left(\frac{\log\left(\frac{N}{50}\right)}{\log 2}\right).$$

Find the time required for the colony to grow to a million bacteria.

Now Try It! Answers

1. $4^{-3} = \frac{1}{64}$ **2.** $\log_2 16 = 4$ **3.** $7^{\frac{1}{2}} = \sqrt{7}$ **4.** $\log_7 1 = 0$ **5.** $27^{\frac{2}{3}} = 9$ **6.** $\log_6 \sqrt[4]{6} = \frac{1}{4}$

12.3   Properties of Logarithms

Section Objectives
- Simplify logarithmic expressions using the properties of logarithms
- Evaluate logarithmic expressions using the properties of logarithms

Logarithms were once extremely important for scientific and technical calculations, but the widespread availability of calculators and computers has reduced this use considerably. As we have already indicated, a logarithm is just

Recall

$b^m b^n = b^{m+n}$

$\dfrac{b^m}{b^n} = b^{m-n}$

$(b^m)^n = b^{mn}$

an exponent. Therefore, it is natural that we would expect logarithms to follow the laws of exponents.

In this section we illustrate the use of logarithms for finding products, quotients, powers, and roots of numbers, and thereby determine these **important** properties of logarithms.

Example 1

We know that $2^3 \times 2^4 = 2^7$. We also know that $2^3 = 8$, $2^4 = 16$, and $2^7 = 128$. In the display below we show those values and their logarithmic forms to illustrate a basic property of logarithms.

$$8 \times 16 = 128$$

$$2^3 \times 2^4 = 2^{3+4} \qquad 3 = \log_2 8 \qquad 4 = \log_2 16 \qquad 3 + 4 = \log_2 128$$

$$\log_2 128 = \log_2 8 + \log_2 16$$

We see that $128 = 8 \times 16$ and that $\log_2 128 = \log_2 8 + \log_2 16$.

Following the method of Example 1, we shall now find a general **expression** for the logarithm of the product of two numbers. Considering the **numbers** P and Q, which can be expressed as 10^x and 10^y, respectively, we have

$$PQ = (10^x)(10^y) = 10^{x+y}$$

where

$$x = \log P \qquad y = \log Q \qquad x + y = \log PQ$$

or

$$\boxed{\log PQ = \log P + \log Q}$$

In other words

The logarithm of the product of two numbers is equal to the sum of the logarithms of the numbers.

Example 2

Evaluate $\log(1.30 \times 120)$ using the product property of logarithms.

$$\log(1.30 \times 120) = \log 1.30 + \log 120$$

$$\log 1.30 = 0.1139$$

$$\log 120 = \underline{2.0792} \quad \text{add}$$

$$2.1931 \quad = \text{log of the product}$$

Be Careful

Do not make the following mistake

$$\log(P + Q) \neq \log P + \log Q$$

Following the same line of reasoning used to develop the method for multiplication, we can use the following property to find the logarithm of the quotient P/Q:

$$\log\left(\frac{P}{Q}\right) = \log P - \log Q$$

In other words

The logarithm of the quotient of two numbers is equal to the logarithm of the numerator minus the logarithm of the denominator.

In order to divide one number by another using logarithms, we subtract the logarithm of the denominator from the logarithm of the numerator. The following example illustrates this procedure.

Example **3**

Evaluate $\log\left(\dfrac{629}{27.0}\right)$ using the quotient property of logarithms

Be Careful

$\log(P - Q) \neq \log P - \log Q$

and

$\dfrac{\log P}{\log Q} \neq \log P - \log Q$

$$\log\left(\frac{629}{27.0}\right) = \log 629 \overset{\text{quotient property of logs}}{-} \log 27.0$$

$$\begin{aligned}\log 629 &= 2.7987\\ \log 27.0 &= \underline{1.4314} \qquad \text{subtract}\\ &\quad\ \ 1.3673\end{aligned}$$

To find a power of a number P, which may be expressed as 10^x, we have

$$P = 10^x \qquad P^n = (10^x)^n = 10^{nx}$$

where $x = \log P$ and $nx = \log P^n$. This means that

$$\log P^n = n \log P$$

In other words

The logarithm of the nth power of a number is equal to n times the logarithm of the number.

▶ The exponent can be integral or fractional. ***If we wish to find the root of a number, we interpret n as a fractional exponent.***

Example **4**

Find $\log 2^4$.

Because $\log P^n = n \log P$, we get $\log 2^4 = 4 \log 2$ ($\log 2 = 0.3010$ so that $4 \log 2 = 4(0.3010) = 1.2040$).

$\log 2^4 = 4 \log 2$

Example 5

Evaluate $\log 3.85^{1.4}$.

Because $\log P^n = n \log P$, we know that $\log 3.85^{1.4} = 1.4 \log 3.85$

$$= 1.4(0.5855) = 0.8196.$$

Example 6

Evaluate $\log \sqrt[3]{0.916}$.

We know that $\sqrt[3]{0.916}$ can be rewritten as $(0.916)^{\frac{1}{3}}$ so $\log \sqrt[3]{0.916}$ is equivalent to $\log (0.916)^{\frac{1}{3}}$. Using $\log P^n = n \log P$ we have

$$\log (0.916)^{\frac{1}{3}} = \frac{1}{3} \log (0.916) = \frac{1}{3}(-0.0381) = -0.0127.$$

Example 7

Evaluate $\log \left(\dfrac{(65.1)\sqrt{804}}{6.82} \right)$.

Using all the properties of logarithms introduced in this section we can rewrite this logarithm as follows

$$\log\left(\frac{(65.1)\sqrt{804}}{6.82} \right) = \log(65.1) + \frac{1}{2}\log (804) - \log 6.82$$

$$= 1.8136 + 1.4526 - 0.8338$$

$$= 2.4324$$

Example 8

Express $\log(x^2 y^3)$ as a sum of simpler logarithms.

Using the product property we rewrite $\log(x^2 y^3)$ as $\log x^2 + \log y^3$.

Using the power property we rewrite $\log x^2 + \log y^3$ as $2 \log x + 3 \log y$.

Therefore, $\log(x^2 y^3) = 2 \log x + 3 \log y$.

Example 9

Express $\log \left(\dfrac{\sqrt{x}}{y^4} \right)$ as the difference of logarithms.

$$\log \left(\frac{\sqrt{x}}{y^4} \right) = \log \sqrt{x} - \log y^4 = \frac{1}{2}\log x - 4 \log y$$

> **Example 10**
>
> Express $\log 3 + 2 \log x - \log y$ as the logarithm of a single quantity.
>
> $$\log 3 + 2 \log x - \log y = (\log 3 + \log x^2) - \log y$$
> $$= \log 3x^2 - \log y = \log\left(\frac{3x^2}{y}\right)$$

> **Example 11**
>
> Given $\log 2 = 0.3010$ and $\log 3 = 0.4771$, find (a) log 6 and (b) log 81 by using the properties of logarithms presented in this section.
>
> **a.** $\log 6 = \log(2 \times 3) = \log 2 + \log 3 = 0.3010 + 0.4771 = 0.7781$
>
> **b.** $\log 81 = \log 3^4 = 4 \log 3 = 4(0.4771) = 1.9084$

In this section we have developed some very important properties of logarithms. Also, it can be seen that powers and roots of numbers can be easily determined by use of logarithms. We should also note that calculators use logarithms to perform many of these types of calculations quickly and easily.

12.3 *Exercises*

In Exercises 1 through 12, write each expression as the sum, difference, or multiple of logarithms.

1. $\log(ab)$

2. $\log(4x)$

3. $\log(xyz)$

4. $\log\left(\frac{5}{2}\right)$

5. $\log\left(\frac{x}{3}\right)$

6. $\log(x^4)$

7. $\log a^5$

8. $\log\left(\frac{abc}{d}\right)$

9. $\log\left(\frac{2ac}{3b^2}\right)$

10. $\log \sqrt[3]{xy}$

11. $\log\left(\frac{\sqrt{x}}{a^2}\right)$

12. $\log\left(\frac{\sqrt[3]{y}}{7}\right)$

In Exercises 13 through 20, express each expression as a single logarithm.

13. $\log a + \log c$

14. $\log 3 + \log x$

15. $\log 9 - \log 3$

16. $\log 7 - \log 5$

17. $\log \sqrt{x} + 2 \log x$

18. $\log 3^3 + \log 9$

19. $2 \log 2 + 3 \log n$

20. $\frac{1}{2} \log a - 2 \log 5$

In Exercises 21 through 28, write each expression as the sum, difference, or multiple of logarithms and then evaluate.

21. $\log\left(\frac{1}{32}\right)$

22. $\log\left(\frac{4}{7}\right)$

23. $\log (3)^{2.4}$

24. $\log (5)^{0.15}$

25. $\log \sqrt[4]{27}$

26. $\log \sqrt[3]{6}$

27. $\log (75)$

28. $\log (0.05)$

In Exercises 29 through 36, perform the indicated calculations for each problem.

29. The pH of a solution measures the acidity of that solution. Acidic solutions have a pH < 7; solutions with pH > 7 are called basic. The formula for finding the pH of a solution is given by the formula

$$\text{pH} = -\log[\text{H}_3\text{O}^+]$$

where H_3O^+ is the hydrogen ion concentration in moles per liter. The hydrogen concentration of milk is 3.97×10^{-7}. Find the pH of milk.

30. In astronomy, the equation for the distance modulus of an object that describes the difference between the apparent and absolute magnitude of the object based on the distance to the object is given by the formula

$$m - M = 5 \log\left(\frac{d}{10}\right)$$

where m represents the apparent magnitude of the object, M represents the absolute magnitude of the object, and d represents the distance to the object. Suppose we have discovered a star that is 250 parsecs from us has an apparent magnitude of 6. Determine the absolute magnitude for this star using this given formula.

31. Fechner's law states that within limits, the intensity of a sensation increases as the logarithm of the stimulus. This is represented by the equation

$$S = k \log\left(\frac{I}{I_0}\right),$$

where S is the subjective intensity of the stimulus, I is the physical intensity of the stimulus, I_0 stands for the threshold physical intensity, and k is a constant that is different for each sensory stimulus. Determine the intensity of sound whose physical intensity is 100 times the threshold physical intensity given $k = 10$.

32. The formula

$$N = \frac{\log\left(\frac{A}{P}\right)}{\log(1 + i)}$$

is used to determine the number of payments N needed to repay a loan of amount A given the amount of each payment P and an interest rate of i%. How many payments are necessary to repay a loan of $3500 at 6% interest if you make a payment of $100 each month?

33. The power gain or loss of an amplifier can be found using the equation

$$n = 10 \log\left(\frac{P_0}{P_1}\right)$$

where P_0 is the power output, P_1 is the power input and n represents the gain or loss (in decibels). Find the power gain when $P_0 = 12$ W and $P_1 = .25$ W.

34. The loudness of sound, measured in decibels is given by the formula

$$10 \log\left(\frac{I}{I_0}\right)$$

where I is the intensity of the sound and I_0 is the intensity of the faintest sound that can be heard. A busy street has a noise level with intensity $I = 10^7 I_0$. Determine the decibel level.

35. The energy of an expanding gas at a constant temperature can be found using the equation

$$E = P_0 V_0 \log\left(\frac{V_1}{V_0}\right)$$

where V_0 is the initial volume of the gas, V_1 is the new volume of the gas and P_0 is a constant. Find E if $V_1 = 518$, $V_0 = 265$ and $P_0 = 85$.

36. In computer programming, the job size of the quickest algorithm is given by

$$N \frac{\log N}{\log 2}$$

where N is the size of an array. Find the job size for an array of size 256.

In Exercises 37 through 44, use a scientific or graphing calculator and logarithms to perform the indicated calculations.

37. 16^{160}

38. 64^{1024}

39. 2^{500}

40. 8^{128}

41. $\sqrt{16^{256}}$

42. $\sqrt{3^{213}}$

42. $(16^{256})(455^{40})$

44. $\sqrt[5]{1753^{300}}$

12.4 **Natural Logarithms**

Section Objectives
- Introduce the natural logarithm
- Evaluate expressions involving natural logarithms
- Simplify natural logarithmic expressions using the properties of logarithms

We have seen that in the basic definition of the logarithm of a number, the base b of the logarithm can be any positive number except 1. Most of our work has been with common logs (base 10), but there is another number that is very useful as a base of logarithms. This number is e, an irrational number equal to approximately 2.718. *Logarithms to the base e are called* **natural logarithms.** We denote the natural logarithm of x by writing $\ln x$.

> If $e^x = N$, then $\ln N = x$.

Whenever we see the "ln" notation, we must remember that the base is the number e (about 2.718). Many formulas in science and technology incorporate these natural logarithms.

Example 1

Since $e^0 = 1, \ln 1 = 0$. Since $e^1 = e, \ln e = 1$.

In the same way, $\ln e^{\frac{1}{2}} = \frac{1}{2}$, or $\ln e^{0.5} = 0.5$, and $\ln e^2 = 2$. Using the approximate value of $e = 2.718$, we have

$$\ln e = \ln 2.718 = 1$$
$$\ln e^{0.5} = \ln 2.718^{0.5} = \ln 1.649 = 0.5$$
$$\ln e^2 = \ln 2.718^2 = \ln 7.389 = 2$$

common logarithm
  $\log_{10} x$ is the same as $\log x$
natural logarithm
  $\log_e x$ is the same as $\ln x$

In Example 1 we evaluated natural logarithms by knowing the correspondence between the logarithmic and exponential forms of a relationship. We can also obtain values of natural logarithms by using a scientific calculator or a graphing calculator. No matter what calculator you are using you should be able to locate a key marked LN. On some calculators we press the LN key first and then enter the value, followed by the ENTER key (this is the sequence we follow if we are using a TI graphing calculator). On other calculators we enter the value first, followed by the LN key.

It is sometimes necessary to find antilogarithms of natural logarithms. That is, we seek a value of N where $\ln N$ is equal to some given value. Example 2 shows how a calculator can be used for problems of this type.

In developing common logarithms for calculation, we obtained certain important properties of logarithms. *These properties are not dependent on that is used and are therefore valid for any base* so that the following properties apply to natural logarithms:

$$\ln PQ = \ln P + \ln Q$$

$$\ln \frac{P}{Q} = \ln P - \ln Q$$

$$\ln P^n = n \ln P$$

These properties are similar to those introduced in the previous section. In fact, these properties hold for all logarithms, regardless of the base of the logarithm. We illustrate these properties for natural logarithms in Examples 2 through 4. Use your scientific or graphing calculator to find the necessary values.

Example 2

Using the fact that $600 = 60 \times 10$, find $\ln 600$.

$$\ln 600 = \ln (60 \times 10) = \ln 60 + \ln 10$$

product property of logs

Thus

$$\ln 600 = 4.0943 + 2.3026$$
$$= 6.3969$$

Example 3

Evaluate $\ln \left(\dfrac{7}{4} \right)$.

quotient property of logs

$$\ln \frac{7}{4} = \ln 7 - \ln 4$$
$$= 1.9459 - 1.3863$$
$$= 0.5596$$

Thus

$$\ln \frac{7}{4} = \ln 1.75 = 0.5596.$$

Example 4

Evaluate $\ln (1.5)^2$.

$$\ln (1.5)^2 = 2 \ln 1.5$$
$$= 2(0.4055)$$
$$= 0.8110$$

Thus

$$\ln (1.5)^2 = \ln 2.25 = 0.8110$$

Examples 5 through 8 illustrate the use of natural logarithms in applied problems.

Example
5

If Q_0 mg of radium decays through radioactivity, an equation relating the amount Q that remains after t years is

$$t = \frac{\ln Q_0 - \ln Q}{4.27 \times 10^{-4}}$$

Find the "half-life" of radium by finding the number of years for 500 mg of radium to decay such that 250 mg remains.

From the given information $Q_0 = 500$ mg and $Q = 250$ mg. Thus,

$$t = \frac{\ln 500 - \ln 250}{4.27 \times 10^{-4}}$$

We can evaluate ln 500 and ln 250.

$$t = \frac{\ln\dfrac{500}{250}}{4.27 \times 10^{-4}} = \frac{\ln 2.00}{4.27 \times 10^{-4}}$$

$$= \frac{0.6931}{4.27 \times 10^{-4}}$$

$$= 1.62 \times 10^3 \text{ years}$$

It therefore takes 1620 years for 500 mg of radium to decay such that 250 mg remains. That is, the "half-life" of radium is 1620 years.

Example
6

A certain electric circuit is used in an airplane's navigational system. The time (in seconds) it takes the current in the circuit to reach 2.0 A is given by

$$t = -\frac{25\left(\ln\dfrac{2.0}{15}\right)}{16}$$

Find that time.

We note that $\ln\frac{2.0}{15} = \ln 2.0 - \ln 15$. Since $\ln 2.0 = 0.6931$ and $\ln 15 = 2.7081$, we get

$$t = -\frac{25(0.6931 - 2.7081)}{16}$$

$$= -\frac{25(-2.0150)}{16} = \frac{50.3750}{16}$$

$$= 3.1 \text{ s}$$

12.4 Exercises

In Exercises 1 through 8, use your calculator to find the indicated natural logarithms.

1. $\ln 8.4$

2. $\ln 18$

3. $\ln 0.9$

4. $\ln 0.2$

5. $\ln 200$

6. $\ln 640$

7. $\ln 0.001$

8. $\ln 0.012$

In Exercises 9 through 16, use natural logarithms to perform the indicated calculations.

9. $(6.92)(1.43)$

10. $(0.082)(7.4)$

11. $\dfrac{81{,}350}{267.4}$

12. $\dfrac{0.0208}{0.00345}$

13. $\sqrt{787}$

14. $\sqrt[3]{28.36}$

15. 2.37^{50}

16. 0.901^{80}

In Exercises 17 through 20, use $\ln 5 = 1.6094$, $\ln 3 = 1.0986$, and the properties of logarithms to determine the value of each of the following:

17. Given that $15 = 3 \times 5$, find $\ln 15$.

18. Given that $0.6 = \frac{3}{5}$, find $\ln 0.6$.

19. Given that $5^3 = 125$, find $\ln 125$.

20. Given that $3^5 = 243$, find $\ln 243$.

In Exercises 21 through 30, solve the applied problems.

21. In studying the population of birds in a certain forest, the equation

$$t = \frac{-\ln \dfrac{P}{P_0}}{0.05}$$

is used. Solve for t if $P = 400$ and $P_0 = 500$.

22. Assuming daily compounding, the approximate time (in years) it takes for a deposit of P dollars to grow to a value of V dollars is given by

$$t = \frac{\ln V - \ln P}{r}$$

where r is the interest rate in decimal form. How long does it take for a \$2500 deposit to grow to \$3000 if the interest rate is 6.00%?

23. If interest is compounded continuously (daily compounded interest closely approximates this), a bank account can double in t years according to the equation

$$i = \frac{\ln 2}{t}$$

where i is the interest rate expressed in decimal form. What interest rate is required for an account to double in 7 years?

24. Under certain conditions, the electric current i in a circuit containing a resistance R and an inductance L is given by

$$\ln \frac{i}{I} = -\frac{Rt}{L}$$

where I is the current at $t = 0$. Calculate how long, in seconds, it takes i to reach 0.430 A if $I = 0.750$ A, $R = 7.50\ \Omega$, and $L = 1.25$ H.

25. A savings account is set up so that compounding is done continuously. If P dollars are deposited for time t (in years), the interest rate r (in decimal form) is found from

$$r = \frac{\ln \dfrac{V}{P}}{t}$$

where V is the latest value of the account. Find r if a \$500 deposit results in a value of \$645.23 after exactly 3 years.

26. Insulation resistance R is described by

$$R = \frac{10^6 t}{C(\ln V_0 - \ln V)}$$

Find the resistance if $t = 120$, $C = 0.080$, $V_0 = 140$, and $V = 110$.

27. At the start of an experiment a bacteria colony has a mass of 0.000000002 grams. After 2 hours the mass of the colony is 0.00000015 grams. Find the growth rate r of the colony given the formula

$$r = \frac{\ln\left(\dfrac{M}{M_0}\right)}{t}$$

where M_0 represents the initial mass of the colony and M represents the mass after t hours.

28. The distance x traveled by a motorboat in t seconds after the engine is cut off is given by the formula

$$x = \frac{\ln(kv_0t + 1)}{k},$$

where v_0 is the velocity of the boat at the time the engine is cut and k is a constant. Find out how far a boat travels if $t = 21.7$ s, $v_0 = 12.0$ m/s, and $k = 6.80 \times 10^{-3}$ m.

29. A formula used by technicians for analyzing connection heat transfer is

$$C = \frac{4.92}{(\ln R)^{0.584}}$$

Find $\ln R$ if $R = 22$.

30. The work (in joules) done by a gas as it expands from volume V_0 to volume V_1 is found by using the formula

$$W = nRT \ln\left(\frac{V_1}{V_0}\right)$$

Find the work W if $n = 1$ mole, $R = 8.31$ J/mole, $T = 295$ K, $V_0 = 5.00$ L and $V_1 = 12.0$ L.

12.5 Exponential and Logarithm Equations

Section Objectives
- Solve exponential equations
- Solve logarithmic equations

Exponential Equations

In this section we will focus on techniques solving exponential equations algebraically. Recall that an exponential equation is one in which the variable occurs in the exponent. Understanding the relationship that exists between an exponential equation and a logarithmic equation makes it easier to see that in order to solve exponential equations we use logarithms. Before we begin to solve exponential equations consider that if $b^m = b^n$ we can conclude that $m = n$. Because logarithms are exponents we can see that

If m and n are positive and $m = n$, then $\log m = \log n$.

In other words,

We can take the log of both sides of an equation in order to solve that equation.

Although some exponential equations can be solved by changing to the logarithmic form, they are generally more easily solved by taking the log of both sides of the equation and applying the basic properties of logarithms.

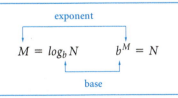

$$M = \log_b N \qquad b^M = N$$

exponent / base

$$\log(P \times Q) = \log P + \log Q$$
$$\log\left(\frac{P}{Q}\right) = \log P - \log Q$$
$$\log P^n = n \log P$$

Example 1

Solve the equation $3^{2x} = 5$.

Taking the log of both sides of the equation gives us

$$\log 3^{2x} = \log 5.$$

We can now apply the power property on the left-hand side.

$$2x \log 3 = \log 5$$

We can solve for x as follows

$$2x \log 3 = \log 5$$
$$2x = \frac{\log 5}{\log 3}$$
$$2x = 1.465$$
$$x = 0.733$$

This solution means that $3^{2x} = 3^{2(0.732)} \approx 5$.

Example 2

Solve the equation $2(4^{x-1}) = 17^x$.

Taking the log of both sides of the equation gives us

$$\log 2(4^{x-1}) = \log 17^x$$
$$\log 2 + \log 4^{x-1} = \log 17^x$$

We can now apply the power property where appropriate on each side of the equation

$$\log 2 + (x - 1)\log 4 = x \log 17$$

We can solve for x as follows

$$\log 2 + (x - 1)\log 4 = x \log 17$$
$$\log 2 + x \log 4 - \log 4 = x \log 17$$
$$\log 2 - \log 4 = x \log 17 - x \log 4$$
$$\log 2 - \log 4 = x(\log 17 - \log 4)$$
$$\frac{\log 2 - \log 4}{\log 17 - \log 4} = x$$
$$-0.479 = x$$

Example 3

Population growth is often described as being exponential in nature. The formula that models this type of growth is $P = P_0 e^{rt}$ where P_0 is the initial population, r is the rate of growth, and P represents the population after t years. The population of the world was estimated at 5.5 billion in 1992. If the world's population grows at an annual rate of 2% per year, determine how long it will take for the world population to double (i.e., when will $P = 11$ billion).

We start with $11 = 5.5e^{0.02t}$. Because this equation is defined in terms of e, we can solve it more easily by the natural logarithm of each side of the equation.

Taking the natural log of both sides of the equation gives us

$$\ln 11 = \ln 5.5e^{0.02t}$$
$$\ln 11 = \ln 5.5 + \ln e^{0.02t}$$

We can now apply the power property where appropriate on the right-hand side of the equation

$$\ln 11 - \ln 5.5 = 0.02t \ln e$$

We can solve for t as follows

$$\ln 11 - \ln 5.5 = 0.02t \ln e$$
$$\frac{\ln 11 - \ln 5.5}{\ln e} = 0.02t$$
$$0.693 = 0.02t$$
$$34.65 = t$$

Therefore, it will take almost 35 years for the world's population to double, or sometime in the year 2026.

> **Remember**
>
> $\log 10 = 1$
> $\ln e = 1$

Logarithmic Equations

In this section we will focus on techniques solving logarithmic equations algebraically. Some important formulas in the sciences and in many technical areas are written in terms of logarithms. Logarithmic equations are those equations in which we are taking the logarithm of an expression that contains a variable. In general we will want to use the basic properties of logarithms to simplify the equation as much as possible. We then rewrite the logarithmic equation to an equivalent exponential equation in order to solve it.

Example 4

Solve the equation $\log_4 x - \log_4 7 = 2$.

We use the quotient property of logarithms to rewrite the left-hand side of the equation.

$$\log_4\left(\frac{x}{7}\right) = 2$$

Write this new equation in exponential form.

$$4^2 = \frac{x}{7}$$

$$16 = \frac{x}{7}$$

$$112 = x$$

Example 5

Solve the equation $\log 2x + \log 5 = 3$.

We use the product property of logarithms to rewrite the left-hand side of the equation.

$$\log(5(2x)) = 3$$
$$\log 10x = 3$$

Write this new equation in exponential form.

$$10x = 10^3$$
$$10x = 1000$$
$$x = 100$$

Earlier in this section we noted that if $m = n$, then $\log m = \log n$. The converse of this statement is also true.

If $\log m = \log n$, then $m = n$

We use this fact in Example 6.

Example 6

Solve the equation $\log (x + 2) + \log(x - 2) = \log 5$.

We use the product property on the left-hand side of the equation.

$$\log((x + 2)(x - 2)) = \log 5$$

Applying the property presented before the start of this example we get

$$(x + 2)(x - 2) = 5$$
$$x^2 - 4 = 5$$
$$x^2 = 9$$
$$x = \pm 3$$

We need to check each answer to determine which value of x is the solution to this equation. Logarithms of negative numbers are not defined and we see that $x = -3$ cannot be used in either term on the left-hand side of the original equation.

Therefore, the solution is $x = 3$.

Finally, we will work through one example involving the natural logarithm.

Solve the equation $3 \ln 2 + \ln (x - 1) = \ln 24$.

We apply both the power property and the product property to this problem.

$$\ln 2^3 + \ln (x - 1) = \ln 24$$
$$\ln 8(x - 1) = \ln 24$$
$$8x - 8 = 24$$
$$8x = 32$$
$$x = 4$$

12.5 Exercises

In Exercises 1 through 24, solve the given equation.

1. $2^x = 16$

2. $3^x = \dfrac{1}{81}$

3. $5^x = 0.3$

4. $3^{-x} = 0.525$

5. $e^{-x} = 17.54$

6. $6^{x+1} = 10$

7. $5^{x-1} = 2$

8. $3(14^x) = 40$

9. $2(10^{x+2}) = 35$

10. $3.1^{2x} - 4 = 16$

11. $e^x - 1 = 6$

12. $2e^{4x} = 15$

13. $3 \log 8 = x$

14. $2 \ln x = 1$

15. $5 \log_{32} x = -3$

16. $2 \log(3 - x) = 1$

17. $3 \log(2x - 1) = 1$

18. $\log_2 x + \log_2 7 = \log_2 21$

19. $\log 12x^2 - \log 3x = 3$

20. $\ln x - \ln \left(\dfrac{1}{3}\right) = 1$

21. $\ln (2x - 1) - 2 \ln 4 = 3 \ln 2$

22. $\dfrac{1}{2}\log(x + 2) + \log 5 = 1$

23. $\log(2x - 1) + \log(x + 4) = 1$

24. $\log(x - 3) + \log x = \log 4$

In Exercises 25 through 30, find the indicated value.

25. The temperature T (in °C) of a cooling gold ingot is given by the formula

$$T = 22 + 98(0.30)^t$$

where t is the time in minutes. How long will it take a gold ingot to cool to 35°C?

26. The pH of a solution measures the acidity of that solution. Acidic solutions have a pH < 7; solutions with pH > 7 are called basic. The formula for finding the pH of a solution is given by the formula

$$\text{pH} = -\log [H_3O^+],$$

where H_3O^+ is the hydronium ion concentration in moles per liter. If the pH of a sample of rain is 4.764, find the hydronium concentration.

27. The approximate density d (in lb/ft^3) of seawater at a depth of h miles is given by the formula

$$d = 64.0e^{0.00676h}.$$

Find the depth at which the density will be 130 lb/ft^3.

28. The number of decibels gained or lost in a device with power input P_{in} and power output P_{out} is given by the formula

$$G = 10 \log \left(\frac{P_{out}}{P_{in}} \right).$$

If a transmission line has a loss of 3.25 dB, find the power transmitted out if the input is 2750 kW.

29. In a certain environment, a colony of bacteria grows according to the formula

$$P = P_0 e^{rt}$$

where P_0 is the initial population, r is the rate of growth, and P represents the population after t hours. Suppose a

colony of bacteria grows from 400 to 2600 in 6 hours. Determine the growth rate and express your answer as a percent.

30. The Richter scale is used to measure the intensity of an earthquake. On the Richter scale, the magnitude of the earthquake R is found using the formula

$$R = \log \left(\frac{I}{I_0} \right),$$

where I is the intensity of the earthquake and I_0 is a standard intensity. How many times I_0 was the 1906 San Francisco earthquake, whose magnitude was 8.3 on the Richter scale?

12.6 Log-Log and Semi-Log Graphs

Section Objectives
• Create graphs on logarithmic and semi-logarithmic paper

When constructing graphs of some functions we may see that one variable grows much more rapidly than the other variable. This is very common when graphing exponential and logarithmic functions. Up until this point all of our graphs have been done using the rectangular coordinate system. In this system the lines are all equally spaced. For graphs with large changes in values, we can create a more accurate graph by using a scale that is marked off in distances proportional to the logarithms of the values being represented. Such a scale is called a **logarithmic scale.** On logarithmic scales the distances between integers are not equal. However, this scale does allow for a much greater range of values. It also provides us with an opportunity to view a complex graph in a much simpler form. This makes the analysis of the curve much easier.

If we wish to use a large range of values for only one variable, we use what is known as **semilogarithmic,** or **semi-log** graph paper. In this case only one axis uses a logarithmic scale. Exponential functions take on a linear appearance when graphed on log-log paper. If we wish to use a large range of values for both variables, we use **logarithmic** or **log-log** graph paper. In this case both axes use a logarithmic scale. Power functions take on a linear appearance when graphed on log-log paper.

Figures 12.7 and 12.8 show an example of a **log-log scale** and a **semi-log scale.**

We use *log-log* and *semi-log* graph paper when the range of values is too large for ordinary graph paper.

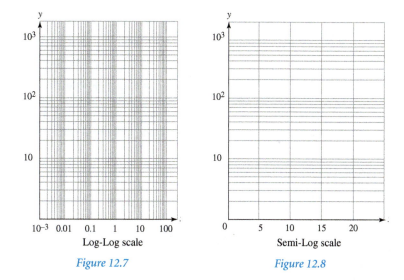

Log-Log scale

Figure 12.7

Semi-Log scale

Figure 12.8

Examples 1 and 2 illustrate the use of log-log and semi-log graph paper.

 Example 1

Construct the graph of $y = 4(3^x)$ using semi-log graph paper.

When we create a table of values for this function we see that

Table 12.4

x	−1	0	1	2	3	4	5	6
y	1.3	4	12	36	108	324	972	2916

One of the first things we notice is that the values for y become large rather quickly. When we plot this on a rectangular coordinate system we need to use very large values along the y axis. This makes the first few values appear to be practically the same. However, when we use semi-log graph paper we can create a graph so that all y values are accurately plotted along with the x values. Notice in Figure 12.9 that the vertical scale is cyclical and labeled with powers of 10. The horizontal axis is labeled in increments of 1. We also notice that an exponential function takes on the appearance of a straight line when graphed on semi-log axes.

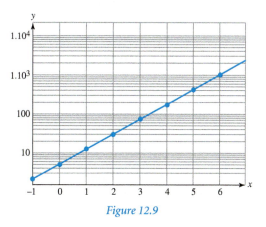

Figure 12.9

Algebraically we start with $y = 4(3^x)$. Taking the log of both **sides of the** equation gives us

$$\log y = \log 4(3^x) = \log 4 + \log 3^x$$
$$\log y = \log 4 + x \log 3$$

However, because log y is graphed automatically we focus on **graphing** $u = \log 4 + x \log 3$ where $u = \log y$.

Example 2

Construct the graph of $y = \dfrac{1}{x^2}$ using log-log graph paper.

When we create a table of values for this function we see that

Table 12.5

x	0.5	1	2	4	8	12	16	20
y	4	1	0.25	0.0625	0.0156	0.0069	0.0039	0.0025

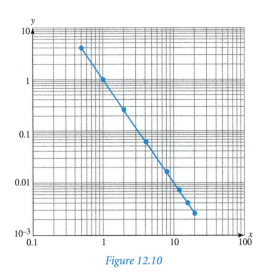

Figure 12.10

One of the first things we notice is that the values for y become smaller and smaller. It would be difficult to plot these y values using the rectangular coordinate system. We plot these points on log-log paper in which the scale for both the x and y axes is logarithmic. Notice in Figure 12.10 that both scales are cyclical and labeled with powers of 10. We also notice that this power function takes on the appearance of a straight line when graphed on log-log axes.

Algebraically we start with $y = \dfrac{1}{x^2}$. Take the log of both sides of **the equation** gives us

$$\log y = \log \frac{1}{x^2} = \log 1 - \log x^2$$
$$\log y = \log 1 - 2 \log x$$

However, because $\log y$ is graphed automatically we focus on graphing $u = \log 1 - 2 \log x$ where $u = \log y$.

Logarithmic and semi-logarithmic paper can be used for plotting data derived from experimentation. Often real world data gathered covers too large a range of values to be plotted on regular graph paper. We choose log-log or semi-log graph paper when this occurs.

12.6 Exercises

In Exercises 1 through 12, plot the graphs of the given functions on semi-logarithmic graph paper.

1. $y = 2^x$ **2.** $y = 5^x$

3. $y = 2(4^x)$ **4.** $y = 2^{-x}$

5. $y = x^3$ **6.** $y = 2x^4$

7. $y = 4^{x/2}$ **8.** $y = 3^{x/4}$

9. $y = e^x$ **10.** $y = e^{-2x}$

11. $y = 3e^{2x/3}$ **12.** $y = 5e^{-x}$

In Exercises 13 through 24, plot the graphs of the given functions on log-log paper.

13. $y = 0.01x^4$ **14.** $y = \sqrt{x}$

15. $y = x^{\frac{2}{3}}$ **16.** $y = \dfrac{4}{x^2}$

17. $y = 3x^6$ **18.** $y = 0.2x^3$

19. $y = 2\sqrt[3]{3}$ **20.** $y = 10x^{-3}$

21. $y = \left(\dfrac{1}{4}\right)^{-x}$ **22.** $y = \sqrt[3]{\dfrac{10}{x}}$

23. $y = x^4 + x$ **24.** $y = \dfrac{1}{\sqrt[3]{x}}$

In Exercises 25 through 28, plot the indicated graphs.

25. On the moon the distance s (in ft) a rock will fall due to gravity is given by the formula

$$s = 2.66t^2$$

where t is the time (in s) of the fall. Plot the graph of this function on graph paper.

26. By pumping, the air pressure in a tank is reduced by 18% each second. Therefore, the pressure p (in kPa) in the tank is given by the formula

$$p = 101(0.82)^x$$

where x is the time (in s). Plot the graph on semi-log graph paper.

27. Graph the exponential radioactive decay modeled by the formula

$$A = 100e^{-0.04t}$$

using both log-log and semi-log graph paper.

28. Suppose the population of a certain city is distributed in suburbs around the city's center. Let P be the number of people (in millions) living within r miles of the city's center. Then P can be approximated by the formula

$$P = \frac{4}{1 + 500e^{-2r}}.$$

Create a graph of this function using both log-log and semi-log paper.

Chapter Summary

Key Terms

base

exponent

logarithm

common logarithm

natural logarithm

log-log graph

semi-log graph

Key Concepts

• A function $y = b^x$ where $b > 0, b \neq 1$, and x is any real number is called an **exponential function.** The number b is called the *base*.

Basic Characteristics of Exponential Functions

For $y = b^x$

1. The domain is all values of x, the range is all $y > 0$.
2. The *y-intercept* is $(0, 1)$
3. There are no *x-intercepts*.
4. For $b > 1$, as x increases, y increases.

5. For $0 < b < 1$, as x increases, y decreases.

• A **logarithmic function** is given by the equation $y = \log_b(x)$ with $b > 0$ and $b \neq 1$.

Basic Characteristics of Logarithmic Functions

For $y = \log_b x$

1. The domain is all $x > 0$, the range is all values of y.
2. The *x-intercept* is $(1, 0)$
3. There are no *y-intercepts*.

4. If $0 < x < 1, \log_b x < 0$. If $x = 1, \log_b x = 0$. If $x > 1, \log_b x > 0$

• To convert from the exponential form of an equation to the logarithmic form, and vice versa, remember that

$$M = \log_b N \qquad b^M = N$$

exponent

base

Formulas

$$\log PQ = \log P + \log Q \qquad \log\left(\frac{P}{Q}\right) = \log P - \log Q \qquad \log P^n = n \log P$$

$$\ln PQ = \ln P + \ln Q \qquad \ln \frac{P}{Q} = \ln P - \ln Q \qquad \ln P^n = n \ln P$$

Review Exercises

In Exercises 1 through 8, write the logarithmic form of the given exponential equations.

1. $10^4 = 10000$
2. $10^{-3} = 0.001$
3. $6^4 = 1296$
4. $5^{-3} = 0.008$
5. $10^1 = 10$
6. $10^{1.23} = 17.0$
7. $10^{-2.54} = 0.00288$
8. $10^{0.4} = 2.5$

In Exercises 9 through 16, write the exponential form of the given logarithmic equations.

9. $\log_2 128 = 7$
10. $\log_3 81 = 4$
11. $\log 100000 = 5$
12. $\log 0.1 = -1$
13. $\ln 20 = 3.0$
14. $\ln 0.25 = -1.4$

15. $\log_4 1024 = 5$

16. $\log_5 0.04 = -2$

In Exercises 17 through 36, write each of the following using the product, quotient, or power properties of logarithms. Evaluate when possible.

17. $\log(2.07)(3.45)$

18. $\log(8.75)(0.205)$

19. $\log(xy)$

20. $\log(3z)$

21. $\log(2abc)$

22. $\log\left(\dfrac{7.14}{9310}\right)$

23. $\log\left(\dfrac{89.15}{9.176}\right)$

24. $\log\left(\dfrac{xz}{3}\right)$

25. $\log\left(\dfrac{ab}{cd}\right)$

26. $\log(6.184)^3$

27. $\log(1.034)^{0.3}$

28. $\log(2x^4)$

29. $\log a^5 b^2$

30. $\log\sqrt[3]{1.17}$

31. $\log\sqrt[3]{0.9006}$

32. $\log(6.12)(\sqrt{128})$

33. $\log\left(\dfrac{\sqrt{86000}}{45.8}\right)$

34. $\log\sqrt[5]{xy}$

35. $\log\left(\dfrac{\sqrt{2x}}{a^2}\right)$

36. $\log\left(\dfrac{\sqrt[3]{5y}}{7}\right)$

In Exercises 37 through 40, find the indicated natural logarithms.

37. $\ln 4.1$

38. $\ln 12$

39. $\ln 0.6$

40. $\ln 0.1$

In Exercises 41 through 44, use $\ln 2 = 0.6931$, $\ln 3 = 1.0986$, and the properties of natural logarithms to find the given values.

41. Given that $6 = 2 \times 3$, find $\ln 6$.

42. Given that $8 = 2^3$, find $\ln 8$.

43. Given that $1.5 = 3 \div 2$, find $\ln 1.5$.

44. Given that $32 = 2^5$, find $\ln 32$.

In Exercises 45 through 56, write each expression as a single logarithm.

45. $\log 4 + \log c$

46. $\log 3x - \log 7$

47. $\dfrac{1}{2}\log 7 - \dfrac{1}{2}\log x$

48. $\log 7x + \log 5y$

49. $-\log\sqrt{2x} + 2\log y$

50. $\log x^3 + \log y$

51. $2\log 2x + 3\log y$

52. $\dfrac{1}{2}\log a - 2\log 5$

53. $2(\log y + 2\log x)$

54. $\dfrac{\log x}{\log 4} - \log y$

55. $2(\log y - 3\log x)$

56. $\dfrac{2\log x}{\log y} - \log 7$

In Exercises 57 through 64, solve the given equations.

57. $e^{2x} = 5$

58. $2(5^x) = 15$

59. $3^{x+2} = 5^x$

60. $6^{x+2} = 12^{x-1}$

61. $\log(x) + \log(6) = \log(12)$

62. $2\log(2) - \log(x + 1) = \log(5)$

63. $\log(x + 2) = 2 - \log(2)$

64. $\log(x + 2) + \log(x) = 0.4771$

In Exercises 65 through 68, plot the graphs of the following functions on semi-log or log-log graph paper as appropriate.

65. $y = 6^x$

66. $y = 5x^3$

67. $y = \sqrt[3]{x}$

68. $y = \sqrt[4]{\dfrac{16}{x}}$

In Exercises 69 through 78, solve the given problems. Perform all calculations by means of logarithms.

69. The magnitudes of tsunamis (popularly known as tidal waves) are measured by

$$m = 3.32 \log_{10} h$$

where h is the maximum wave height (in meters) measured at the coast. Find the magnitude of a tsunami that has a 25-m height. (Such a tsunami hit Sumatra on December 26, 2004.)

70. An earthquake is measured at 5.2 on the Richter scale so that

$$5.2 = \log \frac{I}{I_0}$$

where I is the earthquake's intensity and I_0 is the reference level used for comparison. Find the value of the ratio I/I_0.

71. The loudness of a sound, measured in decibels (dB), is given by

$$10 \log \frac{I}{I_0}$$

where I is the intensity of the sound and I_0 is the intensity of the faintest sound that can be heard. At a distance of 100 ft, the noise of a jet has an intensity level of $I = 1 \times 10^{14} I_0$. Find the decibel level.

72. The number of bacteria in a certain culture after t hours is given by

$$N = (1000)10^{0.0451t}$$

How many are present after 3.00 h?

73. In chemistry, the pH value of a solution is a measure of its acidity. The pH value is defined by the relation

$$pH = -\log (H^+)$$

where H^+ is the hydrogen-ion concentration. If the pH of a certain wine is 3.4065, find the hydrogen-ion concentration. (If the pH value is less than 7, the solution is acid. If the pH value is above 7, the solution is basic.)

74. Find the value of the hydrogen-ion concentration in pure water. Pure water has a pH of 7. (See Exercise 73.)

75. Under certain conditions the temperature T (in degrees Celsius) of a cooling object is given by the equation

$$T = 50.0(10^{-0.1t})$$

where t is the time in minutes. Find the temperature T after 5.00 min.

76. The velocity v of a rocket is given by the formula

$$v = 2.30u \log \frac{m_0}{m}$$

where u is the exhaust velocity, m_0 is the initial mass of the rocket, and m is the final mass of the rocket. Calculate v, given that $u = 2.05$ km/s, $m_0 = 1250$ Mg, and $m = 6.35$ Mg.

77. Under specific conditions, an equation relating the pressure P and volume V of a gas is

$$\ln P = C - 1.50 \ln V$$

Find P (in atmospheres) if $C = 3.00$ and $V = 2.20$ ft^3.

78. If an electric capacitor C is discharging through a resistor R, the time t of discharge is given by

$$t = RC \ln \frac{I_0}{I}$$

where I is the current and I_0 is the current for $t = 0$. Find the time in seconds if $I_0 = 0.0528$ A, $I = 0.00714$ A, $R = 100\ \Omega$, and $C = 100\ \mu$F.

Chapter Test

In problems 1 through 6, solve the given equations for x.

1. Evaluate $\dfrac{2 \ln 0.9564}{\ln 6011}$

2. Evaluate $\dfrac{\log 732}{\log 5^2}$

3. $5^{x-1} = 2$

4. $15.6^{x+2} = 23^x$

5. $2 \log (3 - x) = 1$

6. $\ln x - \ln \dfrac{1}{3} = 1$

7. Graph the function $y = 2(3)^x$ on semi-log graph paper.

8. Express $\log \left(\dfrac{4a^5}{9} \right)$ as the sum, difference, and multiple of logarithms.

9. If A_0 dollars is invested at 8%, and compounded continuously for t years, the value A of the investment is given by the formula $A = A_0 e^{0.08t}$. Determine how long it will take for an investment to double in value.

Geometry and Right Triangle Trigonometry

Most students, when called upon to do so, can easily recite the Pythagorean theorem as $a^2 + b^2 = c^2$. This famous theorem concerning any triangle that contains one right angle was founded by Pythagoras approximately 2500 years ago in 532 BC (see Figure 13.1).

Very little is known about Pythagoras. We know that he was born sometime around 580 BC in the area of the world now known as Turkey. Although historians do not always agree as to the dates surrounding Pythagoras's life, it is believed that he settled in Croton, on the southeastern coast of Italy in about 520 BC. It is here that Pythagoras began his famed Pythagorean society. Life in the Pythagorean society was based on the

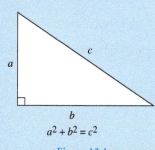

$$a^2 + b^2 = c^2$$

Figure 13.1

idea of true equality. Men and women were regarded equally and all property was communal. Even mathematical discoveries were communal and, as founder of the community, were always attributed to Pythagoras. Members of the Pythagorean society were sworn to secrecy and were not allowed to discuss information with anyone outside of the society. Information was passed along by word of mouth rather than writing things down in order to help ensure this secrecy. Consequently, it is difficult to know exactly what Pythagoras personally discovered. Pythagoras was interested in the principles of mathematics, the concept of numbers, the concept of a triangle or other mathematical figures, and the abstract idea of proof. Pythagoras saw numbers in everything. He was convinced that the divine principles of the universe could be expressed in terms of numbers. He associated numbers with form, relating arithmetic to geometry. The Pythagorean theorem grew from this line of thought.

In Chapter 6 we introduced some basic concepts of geometry. In this chapter we expand upon those topics in order to acquire a more comprehensive understanding of geometry. The content of this chapter will allow us to solve many more applied problems.

13.1 Angles and Their Measure

Section Objectives
- Introduce basic concepts of angles
- Introduce terminology associated with different types of angles

As stated in Chapter 6, geometry deals with the properties and measurement of angles, lines, surfaces, and volumes, as well as the basic figures that are formed. We noted that certain concepts, such as *a* **point,** *a* **line,** *and a* **plane,** *are accepted as being understood intuitively.* We shall make no attempt to define these terms, although we shall use them in defining and describing other terms. This in itself points out an important aspect in developing a topic: Not everything can be defined or proved; some concepts must be accepted and used as a basis for studying geometry.

In Chapter 6 we defined an **angle** and how it is measured in terms of degrees, minutes, seconds, and decimal parts of a degree. *We also defined a* **straight angle** *as an angle of* 180° *and a* **right angle** *as an angle of* 90°.

Two other basic types of angles are identified by whether or not they are less or greater than 90°. *An angle between* 0° *and* 90° *is an* **acute angle** (see Figure 13.2). *An angle greater than* 90° *but less than* 180° *is an* **obtuse angle** (see Figure 13.3).

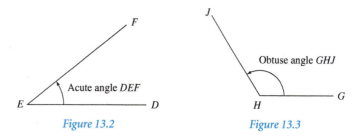

Figure 13.2   Figure 13.3

Example 1

An angle of 73° is an acute angle because 73° is between 0° and 90°.

An angle of 154° is an obtuse angle because 154° is between 90° and 180°.

In geometry, it is often necessary to refer to specific pairs of angles. *Two angles that have a common vertex and a side common between them are known as* **adjacent angles.** *If two lines cross to form equal angles on opposite sides of the point of intersection, which is the common vertex, these equal angles are called* **vertical angles.**

Example 2

In Figure 13.4, $\angle BAC$ and $\angle CAD$ have a common vertex at A and the common side AC between them so that $\angle BAC$ and $\angle CAD$ are adjacent angles.

In Figure 13.5 lines AB and CD intersect at O. Here $\angle AOC$ and $\angle BOD$ are vertical angles, and they are equal. Also, $\angle BOC$ and $\angle AOD$ are vertical angles and are equal.

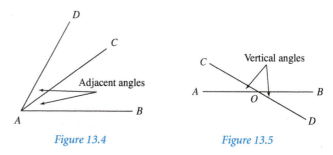

Figure 13.4   Figure 13.5

If the sum of the measures of two angles is 180°, the angles are called **supplementary angles.** *If the sum of the measures of two angles is 90°, the angles are called* **complementary angles.** These types of angles are illustrated in Example 3.

Example 3

In Figure 13.6a, $\angle BAC = 55°$, and in Figure 13.6b $\angle DEF = 125°$. Because $\angle BAC + \angle DEF = 55° + 125° = 180°$, $\angle BAC$ and $\angle DEF$ are supplementary angles.

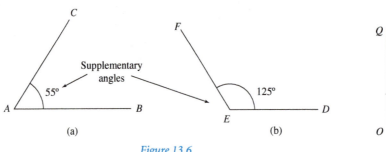

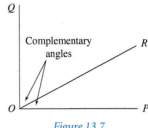

Figure 13.6   *Figure 13.7*

In Figure 13.7, $\angle POQ$ is a right angle, which means that $\angle POQ = 90°$. Since $\angle POR + \angle ROQ = \angle POQ = 90°$, $\angle POR$ and $\angle ROQ$ are complementary angles.

Because the sum of supplementary angles is 180°, if one angle is known, the other angle (its **supplement**) can be found by subtracting the known angle from 180°. Also, since the sum of complementary angles is 90°, one angle can be found by subtracting the other angle (its **complement**) from 90°.

Example 4

If one angle is 36°, then its supplement is 144°, since

$$180° - 36° = 144°$$

The complement of an angle of 36° is 54°, since

$$90° - 36° = 54°$$

We define parallel lines as lines whose extensions will not meet. In a plane, *if a line crosses two parallel or nonparallel lines, it is called a* **transversal.** In Figure 13.8, $AB \parallel CD$, which means that AB is parallel to CD, and the transversal of these parallel lines is the line *EF*.

When a transversal crosses a pair of parallel lines, certain pairs of equal angles result. In Figure 13.8, the **corresponding angles** are equal. (That is, $\angle 1 = \angle 5, \angle 2 = \angle 6, \angle 3 = \angle 7,$ and $\angle 4 = \angle 8$.) Also, the **alternate interior**

Figure 13.8

angles are equal ($\angle 3 = \angle 6$ and $\angle 4 = \angle 5$). The **alternate exterior** angles are also equal ($\angle 1 = \angle 8$ and $\angle 2 = \angle 7$).

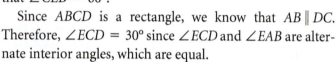

Example 5

A surveyor has established that the tract of land shown in Figure 13.9 has the angles shown. Also, it is known that the lot is rectangular. Find the measure of $\angle DEC$, $\angle CEB$, $\angle ECD$, and $\angle DAE$.

$\angle DEC = 120°$ since $\angle DEC$ and $\angle AEB$ are vertical angles, which are known to be equal. Since $\angle AEB = 120°$ (as shown), it follows that $\angle DEC = 120°$.

Since $\angle AEC$ is a straight angle of $180°$, it follows that $\angle AEB$ and $\angle CEB$ are supplementary and their sum is therefore $180°$. Since $\angle AEB$ is known to be $120°$, it follows that $\angle CEB = 60°$.

Since $ABCD$ is a rectangle, we know that $AB \parallel DC$. Therefore, $\angle ECD = 30°$ since $\angle ECD$ and $\angle EAB$ are alternate interior angles, which are equal.

$\angle DAE = 60°$ since $\angle DAE$ and $\angle EAB$ are complementary angles, and we already know that $\angle EAB = 30°$. We know that those angles are complementary because $\angle DAB$ is one of the right angles in the rectangle.

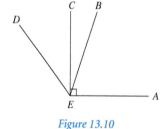

Figure 13.9

Now Try It!

Given $\triangle BAC$, identify

1. two acute angles
2. one right angle
3. supplementary angles
4. complementary angles

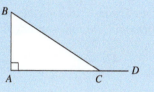

13.1 Exercises

In Exercises 1 through 4, determine the values of the described angles.

1. The complement of $37°$

2. The complement of $83°$

3. The supplement of $159°$

4. The supplement of $27°$

In Exercises 5 through 8, use Figure 13.10. Identify the indicated angles.

5. An acute angle with one side CE

6. The obtuse angle

7. Two pairs of adjacent angles

8. One pair of complementary angles

Figure 13.10

In Exercises 9 through 12, use Figure 13.11. Identify the indicated angles.

9. One pair of supplementary right angles

10. Two pairs of adjacent angles

11. The complement of $\angle DBE$

12. The supplement of $\angle CBD$

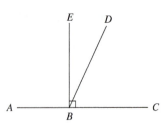

Figure 13.11

In Exercises 13 through 16, use Figure 13.12. Determine the measure of the indicated angles.

13. ∠DBE   **14.** ∠EBF   **15.** ∠DBA   **16.** ∠FBA

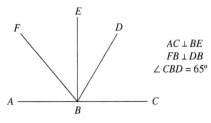

$AC \perp BE$
$FB \perp DB$
$\angle CBD = 65°$

Figure 13.12

In Exercises 17 and 18, use Figure 13.13. Determine the measure of the indicated angles.

17. ∠4   **18.** ∠3

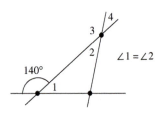

Figure 13.13

In Exercises 19 and 20, use Figure 13.14. Determine the measure of the indicated angles (DA ⊥ CF, ∠AOB = 28°).

19. ∠EOF   **20.** ∠EOC

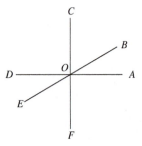

Figure 13.14

In Exercises 21 through 24, use Figure 13.15. Identify the indicated pairs of angles from those that are numbered.

21. Two pairs of vertical angles

22. One pair of alternate interior angles

23. One pair of corresponding angles

24. One pair of supplementary angles

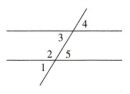

Figure 13.15

In Exercises 25 through 28, use Figure 13.16. Lines that appear to be parallel are parallel. Determine the indicated angles.

25. ∠1   **26.** ∠2   **27.** ∠4   **28.** ∠3

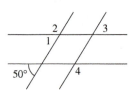

Figure 13.16

In Exercises 29 through 32, use Figure 13.17. Determine the measure of the indicated angles.

29. ∠FCE   **30.** ∠ECD   **31.** ∠BCE   **32.** ∠BFC

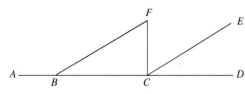

$BF \parallel CE$
$FC \perp AD$
$\angle ABF = 148°$

Figure 13.17

In Exercises 33 through 40, use Figure 13.18 and determine the measure of the indicated angles. In Figure13.18 the line through points A, B, C, and D is the deck of a bridge. Lines BF and CE represent pole supports and they form right angles with the bridge deck. All other lines are support cables. Lines that appear to be parallel are parallel.

33. ∠FBG   **34.** ∠FEG   **35.** ∠CDE   **36.** ∠CED

37. ∠BEC   **38.** ∠FGE   **39.** ∠BGF   **40.** ∠BED

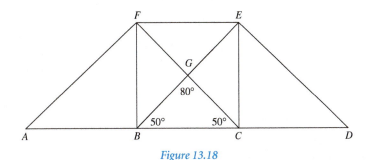

Figure 13.18

Now Try It! Answers

1. ∠ACB and ∠ABC   **2.** ∠BAC   **3.** ∠BCA and ∠BCD   **4.** ∠ACB and ∠ABC

13.2 Other Geometric Figures

Section Objectives
- Determine the measures of the angles of any triangle
- Determine the measures of the angles of any quadrilateral
- Introduce additional concepts related to the circle

When part of a plane is bounded and closed by straight line segments, it is called a **polygon.** In general, polygons are named according to the number of sides they have. *A* **triangle** *has three sides, a* **quadrilateral** *has four sides, a* **pentagon** *has five sides, a* **hexagon** *has six sides, and so on.* The polygons of greatest importance are the triangle and the quadrilateral. This section is devoted to properties of triangles and quadrilaterals, as well as the properties of the circle.

In a polygon, *a line segment that joins any two nonadjacent vertices is called a* **diagonal.** From this definition, we can see that a triangle cannot have diagonals whereas polygons of four or more sides do have diagonals. In Figure 13.19 the diagonals of a hexagon are shown as dashed lines.

In Chapter 6 we discussed certain types of triangles. These included the equilateral triangle, isosceles triangle, scalene triangle, and right triangle. We shall now develop certain additional important properties of triangles.

One extremely important property of a triangle is that *the sum of its interior angles is 180°.*

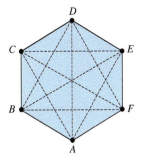

Figure 13.19

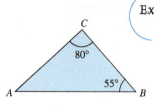

Figure 13.20

Example 1

In the triangle shown in Figure 13.20, we may find $\angle A$ as follows:

$$\angle B + \angle C = 55° + 80° = 135°$$
$$\angle A = 180° - 135°$$
$$= 45°$$

Example 2

In the triangle shown in Figure 13.21, side AB is parallel to DC. We may find $\angle ACB$ as follows:

$$\angle A = 38° \qquad \text{∠A and ∠DCA are alternate interior angles}$$
$$\angle B = 90° \qquad \text{it is a right angle}$$
$$\angle A + \angle B = 38° + 90° = 128°$$
$$\angle ACB = 180° - 128° = 52°$$

Figure 13.21

We now know the three angles in triangle ABC.

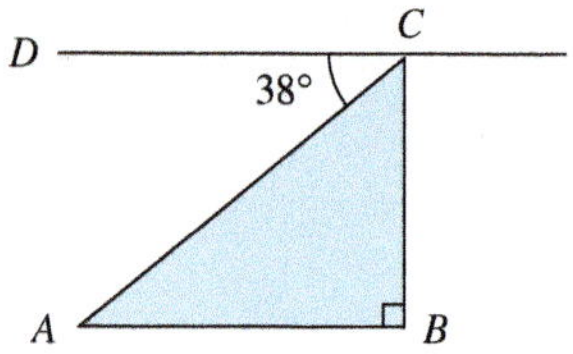

Example 3

A forest ranger uses orienteering to find the location of two known bird sanctuaries. The path followed is triangular as shown in Figure 13.22. Find the angle at point B.

Since all triangles have the property that the sum of the three angles is 180°, we get

$$\angle B + 37° + 93° = 180° \qquad \text{sum of angles of triangle}$$
$$\angle B + 130° = 180°$$
$$\angle B = 50°$$

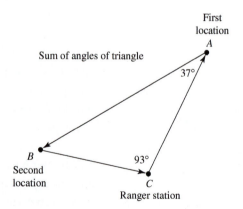

Figure 13.22

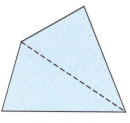

Figure 13.23

We will now consider quadrilaterals. In Chapter 6 we discussed the rectangle, square, parallelogram, rhombus, and trapezoid. Here we consider an important property of all quadrilaterals.

Consider a quadrilateral to be divided into two triangles as shown in Figure 13.23. The sum of the angles of each triangle is 180°. Because these angles give the total of the angles of the quadrilateral, we conclude that: *The sum of the angles of a quadrilateral is* 360°.

Example 4

In the parallelogram in Figure 13.24, we know that $\angle A = \angle C$ and $\angle D = \angle B$ since they are opposite angles. Since $\angle A = 68°$, we get $\angle C = 68°$. The four angles total 360° and

$$\angle A + \angle C = 68° + 68° = 136°$$

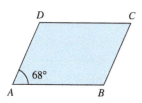

Figure 13.24

so that $\angle D + \angle B = 360° - 136° = 224°$. Since $\angle D = \angle B$, we know that $\angle D = 112°$ and $\angle B = 112°$.

We conclude this section with an extension of the discussion of the circle. In Chapter 6 we defined the **radius** and **diameter** of the circle and gave formulas for finding the area and circumference. We now note that *two circles with the same center are* **concentric,** as shown in Figure 13.25.

A line segment having its endpoints on a circle is a **chord.** *A* **tangent** *is a line that touches a circle at one point (**does not pass through**). A* **secant** *is a line that passes through two points of a circle.* These are illustrated in Figure 13.26.

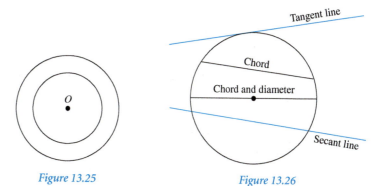

Figure 13.25   *Figure 13.26*

A **central angle** *of a circle is an angle with its vertex at the center of the circle. An arc of a circle consists of that part of the circle between and containing two specific points on the circle. An arc is measured by its central angle.* If an arc has a central angle of 50°, then the measure of the arc is also 50°. These arcs are illustrated in Example 5.

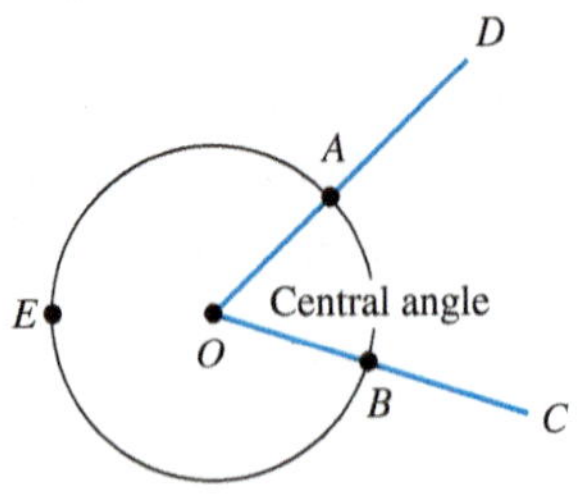

Example
5

For the circle in Figure 13.27, ∠*DOC* is a central angle. (We could also have designated it as ∠*AOB*.) Also, that part of the circle between and including *A* and *B* is the arc *AB*. If ∠*DOC* = 50°, then *AB* = 50° and *AEB* = 310°.

Figure 13.27

An angle is **inscribed** in an arc (see Figure 13.28) if the sides of the angle contain the endpoints of the arc and the vertex of the angle is a point (not an endpoint) of the arc. An important property associated with inscribed angles is this: ***An inscribed angle has a measure that is one-half the measure of its intercepted arc.*** This is illustrated in Example 6.

Example
6

In the circle shown in Figure 13.28, ∠*ABC* is inscribed in *ABC*, and it intercepts *AC*. If *AC* = 60°, ∠*ABC* = 30°.

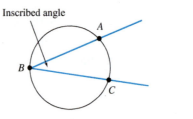

Inscribed angle

Figure 13.28

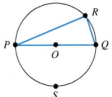

Figure 13.29

In the circle shown in Figure 13.29, *PQ* is a diameter and ∠*PRQ* is inscribed in semicircle *PRQ*. Since *PSQ* = 180°, ∠*PRQ* = 90°. We conclude that: ***An angle inscribed in a semicircle is a right angle.***

13.2 Exercises

In Exercises 1 through 12, find ∠A in the indicated figures.

1. Figure 13.30a

2. Figure 13.30b

3. Figure 13.30c

4. Figure 13.30d

5. Figure 13.31a

6. Figure 13.31b

7. Figure 13.31c

8. Figure 13.31d

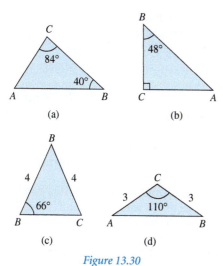

Figure 13.30

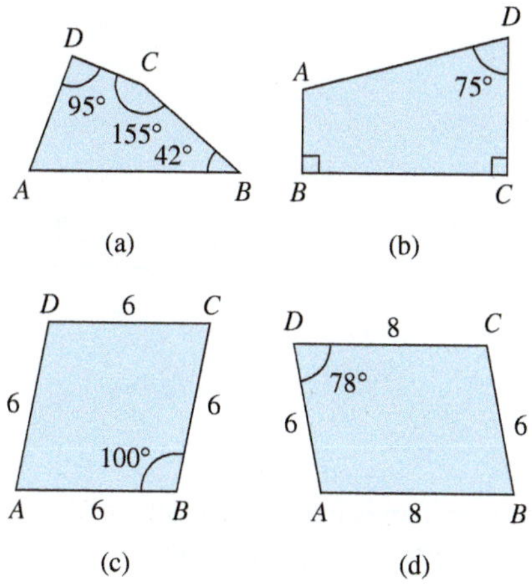

Figure 13.31

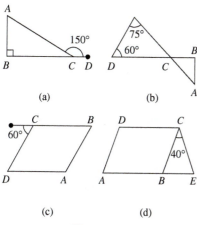

Figure 13.32

9. Figure 13.32a

10. Figure 13.32b

11. Figure 13.32c; *ABCD* is a parallelogram.

12. Figure 13.32d; *ABCD* is a parallelogram. $CE = CB$.

In Exercises 13 through 16, find the measure of the indicated angle or arc.

13. Find $\angle AOB$ in Figure 13.33a; $AB = 32°$.

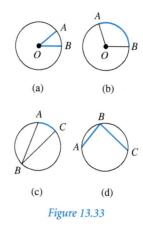

Figure 13.33

14. Find AB in Figure 13.33b; $\angle AOB = 118°$.

15. Find AC in Figure 13.33c; $\angle ABC = 38°$.

16. Find $\angle ABC$ in Figure 13.33d; $AC = 180°$.

In Exercises 17 through 24, use Figure 13.34. In the figure, O is the center of the circle. Identify the following:

17. Two secant lines

18. A tangent line

19. Two chords

20. An inscribed angle

21. An acute central angle

22. An obtuse central angle

23. Two minor arcs

24. Two major arcs

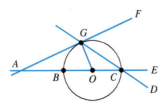

Figure 13.34

In Exercises 25 through 28, use Figure 13.34 ($\angle BOG = 60°$). Determine each of the following:

25. BG

26. CG

27. $\angle BCG$

28. $\angle GAO$

In Exercises 29 through 32, use Figure 13.35. In the figure, O is the center of the circle, line BT is tangent to the circle at B, and $\angle ABC = 55°$. Determine the indicated arcs and angles.

29. AC

30. CAB

31. $\angle CBT$

32. $\angle BTC$

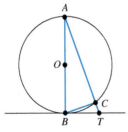

Figure 13.35

In Exercises 33 through 36, answer the given questions.

33. What is the maximum number of acute angles a triangle may contain?

34. What is the maximum number of acute angles a quadrilateral may contain?

35. How many diagonals does a pentagon have?

36. How many diagonals can be drawn from a single vertex of an octagon?

In Exercises 37 through 42, solve the given problems.

37. A transmitting tower is supported by a wire that makes an angle of 50° with the ground. What is the angle between the tower and the wire?

38. The surface of a road makes an angle of 2.4° with level ground (see Figure 13.36). If a survey stake is located at the high side, find the angle A between the stake and the road surface.

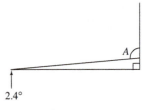

Figure 13.36

39. If a stop sign is made in the shape of a regular octagon, find its interior angles. See Figure 13.37.

Figure 13.37

40. If a building is in the shape of a regular pentagon, find its interior angles. See Figure 13.38.

Figure 13.38

41. The streets in a certain city meet at the angles shown in Figure 13.39. Find the angle x between the indicated streets.

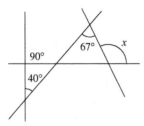

Figure 13.39

42. Metal braces support a beam as shown in Figure 13.40. Find the indicated angle.

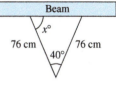

Figure 13.40

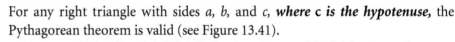

13.3 Right Triangles and Pythagorean Theorem

Section Objectives
- Introduce the Pythagorean theorem
- Find the missing sides and angles of a right triangle
- Solve application problems involving right triangles

One property of a right triangle is so important that we will devote this section to developing it and showing some of its applications. This property is stated in the **Pythagorean theorem:**

> *In a right triangle, the square of the length of the hypotenuse equals the sum of the squares of the lengths of the other two sides.*

Recall that in a right triangle, the hypotenuse is the side opposite the right angle.

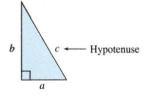

Figure 13.41

$$a^2 + b^2 = c^2 \qquad \text{Pythagorean theorem}$$

For any right triangle with sides a, b, and c, **where c is the hypotenuse,** the Pythagorean theorem is valid (see Figure 13.41).

We now present five examples illustrating the use of the Pythagorean theorem.

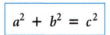

Example
1

For a right triangle ABC with the right angle at C, if $AC = 5$, and $BC = 12$, find AB (see Figure 13.42).

Letting AC, BC, and AB denote the sides, we find that the Pythagorean theorem, applied to this triangle, is

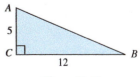

Figure 13.42

$$(AC)^2 + (BC)^2 = (AB)^2$$

legs — hypotenuse

Substituting the values for AC and BC, we get

$$5^2 + (12)^2 = (AB)^2$$
$$25 + 144 = (AB)^2$$
$$169 = (AB)^2$$
$$13 = AB$$

We find the final value $AB = 13$ by taking the square root of both sides. We use the positive square root because lengths are considered positive.

Example 2

A pole is on level ground and perpendicular to the ground. Guy wires, which brace the pole on either side, are attached at the top of the pole. Each guy wire is 25.0 ft long and the pole is 20.0 ft high. How far are the grounded ends of the guy wires from each other? (See Figure 13.43.)

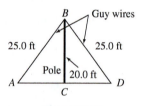

Figure 13.43

From the figure we see that we are to find AD. In order to find AD we shall find AC, which in turn equals CD, and therefore $AD = 2AC$. From the Pythagorean theorem, we have $(AC)^2 + (BC)^2 = (AB)^2$. Substituting the known values of AB and BC, we get

$$(AC)^2 + (20.0)^2 = (25.0)^2 \quad \text{hypotenuse}$$
$$(AC)^2 + 400 = 625$$
$$(AC)^2 = 225$$
$$(AC) = 15.0 \text{ ft}$$

Therefore, $AD = 30.0$ ft.

Example 3

A jet travels 7.50 km while gaining altitude at a constant rate. If it travels between horizontal points 5.80 km apart, what is its gain in altitude?

From the statement of the problem, we set up the diagram in Figure 13.44. We are to find the vertical distance x:

$$(5.80)^2 + x^2 = (7.50)^2$$
$$33.64 + x^2 = 56.25$$

Now Try It!

Given right triangle $\triangle ABC$ with the right angle at C, find the missing side of the triangle if:

1. $a = 4.0, b = 8.0$
2. $b = 21, c = 36$
3. $a = 14.2, c = 20.7$

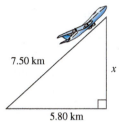

Figure 13.44

$$x^2 = 56.25 - 33.64$$
$$= 22.61$$
$$x = 4.75 \text{ km}$$

The jet gains 4.75 km in altitude.

Example
4

A surveyor wants to determine the distance between two points, but a large building is an obstruction. However, the distance can be found by selecting a convenient third point that can be used to form a right triangle. Find the distance between points *A* and *B* (see Figure 13.45).

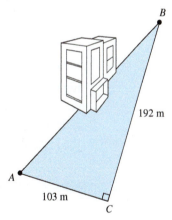

Figure 13.45

Since the right angle is at *C*, the hypotenuse is the required distance *AB*. Applying the Pythagorean theorem we get

$$(103)^2 + (192)^2 = (AB)^2$$
$$10609 + 36864 = (AB)^2$$
$$47473 = (AB)^2$$
$$AB = 218 \text{ m}$$

The distance between points *A* and *B* is 218 m.

Example
5

A cable is attached to the top of a 22.0-ft pole. A man holding the cable moves a certain distance from the pole and notes that there are 30.0 ft of cable from the ground to the top of the pole. He then moves another 10.0 ft from the pole. How long must the cable be to reach the ground at his feet? (See Figure 13.46.)

From the figure, we see that if we first find his original distance x from the pole, we can then proceed to calculate the required distance y. Applying the Pythagorean theorem to the triangle with sides x, 22.0, and 30.0, we get the following solution.

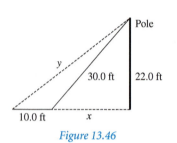

Figure 13.46

$$x^2 + (22.0)^2 = (30.0)^2$$
$$x^2 + 484 = 900$$
$$x^2 = 416$$
$$x = 20.4 \text{ ft}$$

Now we add 10.0 to 20.4 and apply the Pythagorean theorem to the triangle with sides 30.4, 22.0, and y. We get

$$(30.4)^2 + (22.0)^2 = y^2$$
$$924 + 484 = y^2$$
$$y^2 = 1408$$
$$y = 37.5 \text{ ft}$$

When applying the Pythagorean theorem, we must be sure that we are working with a *right* triangle. Also, we must take care to ensure that $a^2 + b^2 = c^2$ is arranged correctly with the hypotenuse and legs in their proper locations. In a right triangle, the hypotenuse is always opposite the right angle and it is always the longest side.

13.3 Exercises

In Exercises 1 through 20, find the indicated sides of the right triangle shown in Figure 13.47. Where necessary, round off results to three significant digits.

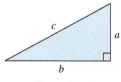

Figure 13.47

	a	b	c			a	b	c
1.	3	4	?		**2.**	9	12	?
3.	8	15	?		**4.**	24	10	?
5.	6	?	10		**6.**	2	?	4
7.	5	?	7		**8.**	3	?	9
9.	?	12	16		**10.**	?	10	18
11.	?	15	32		**12.**	?	5	36

	a	b	c		a	b	c
13.	56	?	82	**14.**	?	125	230
15.	5.62	40.5	?	**16.**	23.5	4.33	?
17.	0.709	?	2.76	**18.**	?	0.0863	0.145
19.	?	16.5	42.4	**20.**	73.7	?	86.1

In Exercises 21 through 40, set up the given problems and solve by use of the Pythagorean theorem.

21. What is the length of the diagonal of a square with 4.00 cm sides?

22. What is the length of the diagonal of a rectangle 8.50 in. long and 4.60 in. wide?

23. A square is inscribed in a circle. See Figure 13.48 (All four vertices of the square touch the circle.) Find the length of the sides of the square if the radius of the circle is 8.50 m.

Figure 13.48

24. An airplane is 3000 ft directly above one end of a 10,000-ft runway. What is the distance between the airplane and a glide-slope indicator located on the ground at the other end of the runway?

25. A ramp for the physically challenged must be designed so that it rises a total of 1.2 m over a flat distance of 7.8 m. How long is the third side of the ramp? See Figure 13.49.

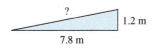

Figure 13.49

26. A 22.0-m-nigh tree casts a shadow 15.6 m long. How far is it from the top of the tree of the tip of the shadow? See Figure 13.50.

Figure 13.50

27. A searchlight is 520 ft from a wall, and its beam reaches a point 38.0 ft up the wall. What is the length of the beam?

28. The guy wires bracing a telephone pole on a level ground area and the line along the ground between the grounded ends of the wires form an equilateral triangle whose sides are 20.0 ft. Find the height at which the wires are attached to the pole.

29. Figure 13.51 shows a roof truss. The rafters are 21.0 ft long, including a 1.5-ft overhang, and the height of the truss is 6.50 ft. Determine the length of the base of the truss.

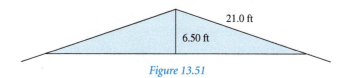

Figure 13.51

30. Figure 13.52 shows a metal plate with two small holes bored in it. What is the center-to-center distance between the holes? The given distances are measured center to center.

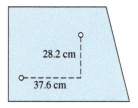

Figure 13.52

31. In an alternating-current circuit containing a resistor and a capacitor, the capacitor contributes an effective resistance to the current, called the *reactance*. The total effective resistance in the circuit, called the *impedance Z*, is related to the resistance R and reactance X in exactly the same way that the hypotenuse of a right triangle is related to the sides. Find Z for a circuit in which $R = 17.0 \, \Omega$ and $X = 8.25 \, \Omega$.

32. The electric intensity at point P due to an electric charge Q is in the direction from the charge shown in Figure 13.53. The electric intensity E at P is equivalent to the two intensities. E_h and E_v, which are horizontal and vertical, respectively. Given that $E = 35.0 \text{ kV/m}$ and $E_h = 17.3 \text{ kV/m}$, find E_v.

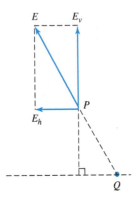

Figure 13.53

33. Two forces, F_1 and F_2, acting on an object are at right angles to each other. Their net resultant force F on the object is related to F_1 and F_2 in the same way that the hypotenuse of a right triangle is related to the sides. Given that $F_1 = 865 \text{ lb}$ and $F_2 = 225 \text{ lb}$, find F.

34. A source of light L, a mirror M, and a screen S are situated as shown in Figure 13.54. Find the distance a light ray travels in going from the source to the mirror and then to the screen.

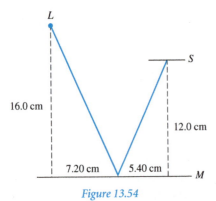

Figure 13.54

35. A rectangular dining room is 6.50 m long, 4.75 m wide, and 2.45 m high. What is the distance from a corner on the floor to the opposite corner at the ceiling? See Figure 13.55.

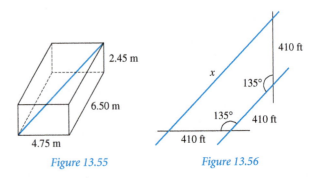

Figure 13.55   *Figure 13.56*

36. Figure 13.56 shows four streets of a city. What is the indicated distance?

37. Figure 13.57 shows a quadrilateral tract of land. What is the length of the indicated side?

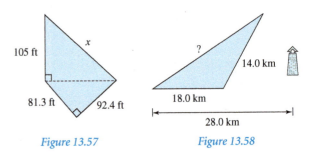

Figure 13.57   *Figure 13.58*

38. A ship 28.0 km due west of a lighthouse travels 18.0 km toward the lighthouse and then turns. After traveling another 14.0 km in a straight line it is due north of the lighthouse. How far, on a direct line, is the ship from its starting point? See Figure 13.58.

39. The hypotenuse of a right triangle is 24.0 cm, and one leg of the triangle is twice the other. Find the perimeter of the triangle.

40. One leg of a right triangle is 15 ft long and the area of the triangle is 180 ft^2. What is the length of the hypotenuse?

Now Try It! Answers

1. $c = 8.9$ **2.** $a = 29$ **3.** $b = 15.1$

13.4 Similar Triangles

Section Objectives
- Understand the concept of similar triangles
- Demonstrate that two triangles are similar
- Use the properties of similar triangles to solve problems

In this section, we consider the properties of triangles that have the same basic shape although not necessarily the same size.

Two triangles are **similar** *if they have the same shape (but not necessarily the same size).* There are two very important properties of similar triangles:

1. The corresponding angles of similar triangles are equal.

2. The corresponding sides of similar triangles are proportional.

Before considering some implications of these two properties, Examples 1 and 2 illustrate what is meant by corresponding angles and corresponding sides.

Example
1

In Figure 13.59 the triangles are similar and are lettered so that corresponding angles have the same letters. That is, angles A and A' are corresponding angles, B and B' are corresponding angles, and C and C' are corresponding angles. Also, the sides between these vertices are corresponding sides. That is, AB corresponds to side $A'B'$, and so forth.

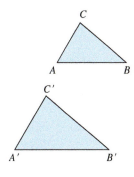

Figure 13.59

Example
2

In Figure 13.60 we show another pair of triangles that are also similar, even though these triangles are not drawn so that corresponding parts are in the same position relative to the page. The triangles are lettered so that the corresponding parts can be

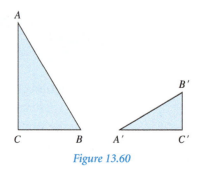

Figure 13.60

identified. Angles A and A' are corresponding angles, B and B' are corresponding angles, and C and C' are corresponding angles. Sides AB and $A'B'$ are corresponding sides, and so on.

If we wish to show that two triangles are similar, we must show that one of the two properties listed above is valid. It is a characteristic of triangles that *if one of these two conditions is valid, then the other condition must also be valid.*

That is, if we can show that the corresponding angles of the two triangles are equal, then we can conclude that the corresponding sides are proportional. If the corresponding sides are proportional, the corresponding angles are equal.

Example 3

The two triangles in Figure 13.59 have been drawn so that the corresponding angles are equal. This means that the ***corresponding sides are also proportional which is shown as***

$$\frac{AB}{A'B'} = \frac{BC}{B'C'} = \frac{CA}{C'A'} \begin{array}{l} \longleftarrow \text{ sides of } \triangle ABC \\ \longleftarrow \text{ sides of } \triangle A'B'C' \end{array}$$

If we know that two triangles are similar, we can use the two listed properties to determine the unknown parts of one triangle from the known parts of the other. Example 4 illustrates how this can be done.

Example 4

In Figure 13.61, given triangle ABC where DE is parallel to BC, show that triangle ADE is similar to triangle ABC.

We shall review certain symbols that are commonly used in geometry. The symbol $\triangle$ denotes "triangle," $\sim$ denotes "similar," and as we noted earlier, $\parallel$ denotes "parallel." The statement above can be stated as: Given $\triangle ABC$ where $DE \parallel BC$, show that $\triangle ADE \sim \triangle ABC$.

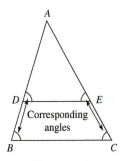

Figure 13.61

To show that the triangles are similar, we will show that the corresponding angles are equal. Recall that the corresponding angles of parallel lines cut by a transversal are equal. This implies that

$$\angle ADE = \angle ABC \quad \text{and} \quad \angle AED = \angle ACB$$

Since $\angle DAE$ is common to both triangles, we have one angle in each triangle equal to one angle in the other triangle. *Since the corresponding angles are equal,* $\triangle ADE \sim \triangle ABC$.

⟍ Example
5

In Figure 13.62, in right $\triangle ABC$ with the right angle at $C, CD \perp AB$. (As we noted earlier, $\perp$ means "perpendicular.") Show that $\triangle ADC \sim \triangle CDB$.

First, both triangles contain a right angle. That is, $\angle CDA = \angle CDB = 90°$. Since the sum of the angles in a triangle is $180°$, **the sum of the other two angles in $\triangle ADC$ is $90°$,** or $\angle CAD + \angle ACD = 90°$. Also, $\angle ACB$ is a right angle, which means that $\angle ACD + \angle DCB = 90°$. These two equations can be written as

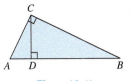

Figure 13.62

$$\angle CAD = 90° - \angle ACD \qquad \angle DCB = 90° - \angle ACD$$

Since the right-hand sides of these equations are equal, we conclude that $\angle CAD = \angle DCB$. We have now shown that *two angles are respectively equal in the two triangles.* Since, in any triangle, the sum of the angles is $180°$, *the remaining angles must also be equal.* Therefore, $\angle ACD = \angle CBD$. The triangles are therefore similar. We do note, however, that the corresponding sides appear in different positions with respect to the page. This is indicated by writing the ratio of the corresponding sides as

$$\frac{AD}{CD} = \frac{DC}{DB} = \frac{AC}{BC} \quad \longleftarrow \text{sides of } \triangle ADC$$
$$\qquad\qquad\qquad\qquad \longleftarrow \text{sides of } \triangle CDB$$

where side CD (or DC) is part of both triangles.

Example
6

In Figure 13.63, $\triangle ABC$ and $\triangle DEF$ were designed to be similar, and the known lengths of certain sides are as shown. Find the lengths of sides CB and AB.

Because the triangles are similar, the corresponding sides are proportional so that

$$\frac{AC}{DF} = \frac{CB}{FE} = \frac{BA}{ED}$$

Substituting the known values, we get

$$\frac{6}{4} = \frac{CB}{3} = \frac{BA}{2}$$

▶ Since *the middle and right ratios are both equal to* $\frac{6}{4}$, we can solve for the unknown in each case. We get

$$\frac{CB}{3} = \frac{6}{4} \quad \text{or} \quad CB = \frac{6(3)}{4} = \frac{9}{2} \text{ cm}$$

$$\frac{BA}{2} = \frac{6}{4} \quad \text{or} \quad BA = \frac{6(2)}{4} = 3 \text{ cm}$$

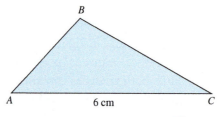

Figure 13.63

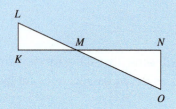
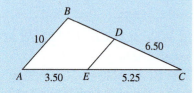

Now Try It!

1. Given

with $KN = 15$, $MN = 9$, and $MO = 12$, find LM.
2. In the following figure, $AB \parallel DE$. Find the length of
   a. DE
   b. BD
   c. BC

Example
7

On level ground a silo casts a shadow 24 ft long. At the same time, a pole 4.0 ft high casts a shadow 3.0 ft long. How tall is the silo? (See Figure 13.64.)

The rays of the sun are essentially parallel. The two triangles indicated in Figure 13.64 are similar, because *each has a right angle and the angles at the tops are equal.* The lengths of the hypotenuses are of no importance in this problem, so we use only the other sides in stating the ratios of the corresponding sides. Denoting the height of the silo by h, we get

$$\frac{h}{4.0} = \frac{24}{3.0} \quad h = 32 \text{ ft}$$

We conclude that the silo is 32 ft tall.

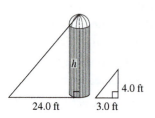

Figure 13.64

One of the most practical uses of similar figures is that of **scale drawings.** Maps, charts, blueprints, and most drawings that appear in books are familiar examples of scale drawings. Actually, there have been many scale drawings used in the previous sections of this book.

In any scale drawing, all distances are drawn a certain ratio of the distances they represent and all angles equal the angles they represent. Consider Examples 8 and 9.

Example 8

In drawing a map, a scale of 1 cm = 200 km is used. In measuring the distance between Chicago and Toronto on the map, we find it to be 3.5 cm. The actual distance x between Chicago and Toronto is found from the proportion

$$\text{actual distance} \longrightarrow \frac{x}{3.5 \text{ cm}} = \frac{200 \text{ km}}{1 \text{ cm}} \quad \text{or} \quad x = 700 \text{ km}$$
$$\text{distance on map} \longrightarrow$$

(scale points to 200 km / 1 cm)

If we did not have the scale but know that the distance between Chicago and Toronto is 700 km, then by measuring distances on the map between Chicago and Toronto (3.5 cm) and between Toronto and Philadelphia (2.7 cm), we could find the distance between Toronto and Philadelphia. It is found from the following proportion, determined by use of similar triangles:

$$\frac{700 \text{ km}}{3.5 \text{ cm}} = \frac{y}{2.7 \text{ cm}}$$

$$y = \frac{2.7(700)}{3.5} = 540 \text{ km}$$

Example 9

A satellite photograph reveals three missile silos an equal distance apart. The images of the silos on the photograph are 2.80 mm apart. Analysis of the photograph and other satellite data implies that the scale is 1.00 mm = 0.429 km. Find the actual distance between the silos.

The triangle formed by the silos in the photograph is similar to the triangle formed by the actual silos. Representing the actual distance between the silos as x, we get

$$\frac{\text{actual distance}}{\text{distance in photo}} = \frac{x}{2.80 \text{ mm}} = \frac{0.429 \text{ km}}{1.00 \text{ mm}} \quad \text{or} \quad x = 1.20 \text{ km}$$

(scale points to 0.429 km / 1.00 mm)

Similarity requires *equal* angles and *proportional* sides. *If the corresponding angles and the corresponding sides of two triangles are equal, then the two triangles are said to be* **congruent.** As a result of this definition, the areas and perimeters of congruent triangles are also equal. Informally, we can say that similar triangles have the same shape whereas congruent triangles have the same shape and same size.

Example
10

A right triangle with legs of 2 in. and 4 in. is congruent to any other right triangles with legs of 2 in. and 4 in. However, it is similar to any right triangle with legs of 5 in. and 10 in., since the corresponding sides are proportional. See Figure 13.65a.

One equilateral triangle with 6-cm sides, is congruent to any other equilateral triangle with 6-cm sides, and it is similar to any other equilateral triangle, regardless of the length of the sides. We know that corresponding angles are equal and that the ratios of the corresponding sides must be equal. See Figure 13.65b.

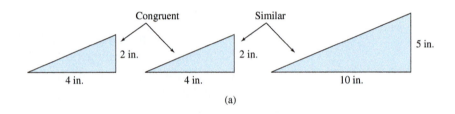

(a)

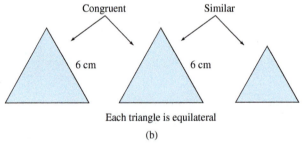

Each triangle is equilateral

(b)

Figure 13.65

13.4 Exercises

In Exercises 1 through 8, assume that $\triangle ABC \sim \triangle A'B'C'$ *with angles A and A' corresponding, angles B and B' corresponding, and angles C and C' corresponding. Find the indicated missing parts.*

1. If $\angle A = 20°$ and $\angle B = 100°$, find angles $\angle A'$, $\angle B'$, and $\angle C'$.

2. If $AB = 5$, $BC = 6$, and $A'B' = 15$, find $B'C'$.

3. If $AB = 10.0$, $BC = 11.0$, $AC = 13.0$, and the shortest side of $\triangle A'B'C'$ is 15.0, find the other two sides of $\triangle A'B'C'$.

4. If $\angle B = 30°$ and $\angle A' = 80°$, find $\angle C$.

5. If $\angle B = 65°$ and $\angle C = 75°$, find angles $\angle A'$, $\angle B'$, and $\angle C'$.

6. If $AB = 3.0, BC = 4.0, AC = 5.0$, and $A'B' = AC$, find $B'C'$ and $A'C'$.

7. If $AC = BC = A'C' = 12.7$ and $\angle C = 40°$, find $B'C'$ and angles $\angle A'$ and $\angle B'$.

8. If $A'B' = 20.2, B'C' = 15.3$, and $A'C' = AB = 10.9$, find AC and BC.

In Exercises 9 through 12, determine whether or not the triangles in the given figures are (a) similar and (b) congruent. Angles that are equal are marked in the same way, and so are equal sides.

9. Figure 13.66　　　**10.** Figure 13.67

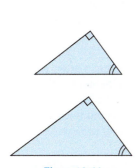

Figure 13.66

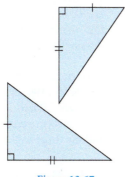

Figure 13.67

11. Figure 13.68　　　**12.** Figure 13.69

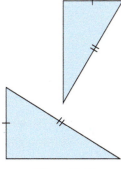

Figure 13.68

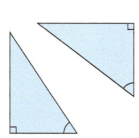

Figure 13.69

In Exercises 13 through 16, for the given pair of triangles, identify the corresponding angles and corresponding sides to the

given angles and sides. The triangles in each of the figures are similar.

13. In Figure 13.70, $\angle A$ corresponds to _____ and side AC corresponds to _____.

14. In Figure 13.70, $\angle F$ corresponds to _____ and side DF corresponds to _____.

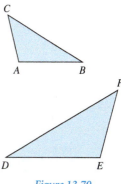

Figure 13.70

15. In Figure 13.71, $\angle Q$ corresponds to _____ and side RP corresponds to _____.

16. In Figure 13.71, $\angle S$ corresponds to _____ and side UT corresponds to _____.

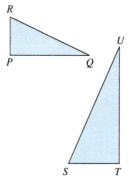

Figure 13.71

In Exercises 17 through 20, for the given sides of the triangles of Figures 13.70 and 13.71, find the indicated sides. In Figure 13.70, $AB = 5, BC = 7, AC = 4$, and $FE = 8$. In Figure 13.71: $RP = 6, PQ = 8, RQ = 10$, and $ST = 9$.

17. In Figure 13.70, find DE.

18. In Figure 13.70, find FD.

19. In Figure 13.71, find *TU*.

20. In Figure 13.71, find *SU*.

In Exercises 21 through 24, use Figure 13.72. In this figure BD ∥ AE, AE = 18, and DB = 6.

21. If *DC* = 7, find *CE*.   **22.** If *CA* = 27, find *CB*.

23. If *CB* = 10, find *BA*.   **24.** If *CE* = 24, find *DE*.

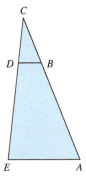

Figure 13.72

In Exercises 25 and 26, use △ABC in Figure 13.73. In this figure AC ⊥ BC and CD ⊥ AB.

25. Given that △*ABC* ~ △*ACD*, find *AB* if *AD* = 9 and *AC* = 12.

26. Given that △*ABC* ~ △*CBD*, find *BD* if *BC* = 6 and *AB* = 9.

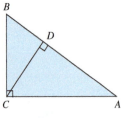

Figure 13.73

In Exercises 27 and 28, solve for the unknown side by use of an appropriate proportion. Parallel lines are marked with arrows.

27. In Figure 13.74, *KM* = 6, *MN* = 9, and *MO* = 12. Find *LM*.

28. In Figure 13.75, *BD* = 5, *BE* = 8, and *BA* = 10. Find *BC*.

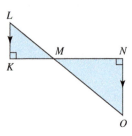
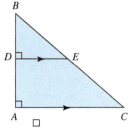

Figure 13.74   *Figure 13.75*

In Exercises 29 and 30, draw the appropriate figures.

29. Draw △*ABC* such that *AB* ⊥ *BC* and ∠*ACB* = ∠*CAB*.

30. Draw △*ABC* with *D* on *AB* and *E* on *AC* such that *DE* ∥ *BC*.

In Exercises 31 and 32, find the required values.

31. Two triangles are similar. The sides of the larger triangle are 3.0 cm, 5.0 cm, and 6.0 cm, and the shortest side of the other triangle is 2.0 cm. Find the remaining sides of the smaller triangle.

32. Two triangles are similar. The angles of the smaller triangle are 50°, 100°, and 30°, and the sides of the smaller triangle are 7.00 in., 9.00 in., and 4.57 in. The longest side of the larger triangle is 15.0 in. Find the other two sides and the three angles of the larger triangle.

In Exercises 33 and 34, show that the required triangles are similar.

33. In Figure 13.76, show that △*XYK* ~ △*NFK*. (*XY* ∥ *FN*)

34. In Figure 13.73, show that △*ACB* ~ △*ADC*.

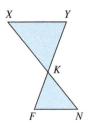

Figure 13.76

In Exercises 35 and 36, find the required values.

35. In Figure 13.77, △*ACB* ≅ △*EDC* (≅ means "congruent"). If *AD* = 16 in., how long is *AC*?

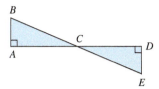

Figure 13.77

36. In Figure 13.78, $\triangle ABC \cong \triangle ADC$. If $\angle CAD = 40°$, how many degrees are there in $\angle CAB + \angle ABC$?

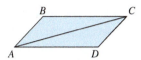

Figure 13.78

In Exercises 37 through 46, solve the given problems.

37. One stake used as a snow marker casts a shadow 5.0 ft long. A second stake casts a shadow 3.0 ft long. If the second stake is 4.0 ft tall, how tall is the first stake? (Both stakes are on level ground.) See Figure 13.79.

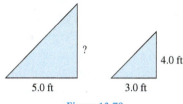

5.0 ft   3.0 ft

Figure 13.79

38. On level ground an oak tree casts a shadow 36 m long. At the same time, a pole 9 m high casts a shadow 12 m long. How high is the tree?

39. A 1-m stick is placed vertically in the shadow of a vertical pole such that the ends of their shadows are at the same point. If the shadow of the meter stick is 80 cm long and that of the pole is 280 cm long, how high is the pole? See Figure 13.80.

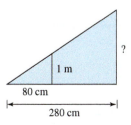

1 m

80 cm

280 cm

Figure 13.80

40. In constructing a metal support in the form of $\triangle ABC$ as shown in Figure 13.81, it is deemed necessary to strengthen the support with an added brace DE which is parallel to BC. How long must the brace be if $AB = 20$ in., $AD = 14$ in., and $BC = 25$ in.?

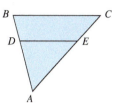

Figure 13.81

41. A certain house blueprint has a scale of $1\frac{1}{4}$ in. $= 10$ ft. The living room is 18 ft long. What distance on the blueprint represents this length?

42. On a map, 12 cm $= 100$ mi. What is the distance between two cities if the distance between them on the map is 7.5 cm?

43. A satellite photograph includes the images on three different military bases. In the photograph, the distances between the bases are 10.4 cm, 12.3 cm, and 5.2 cm. If the photograph has a scale of 1.000 cm $= 146.7$ km, find the actual distances between the military bases.

44. In a book on aerial photography, it is stated that the typical photograph scale is $\frac{1}{18,450}$. In an 8.0 in. by 10.0 in. photograph with this scale, what is the longest distance between two locations included in the photograph? Express the answer in miles and round to the nearest tenth.

45. A 4.0 ft wall stands 2.0 ft from a building. The ends of a straight pole touch the building and the ground 6.0 ft from the wall. A point on the pole touches the wall's top. How high on the building does the pole touch? See Figure 13.82.

46. To find the width ED of a river, a surveyor places markers at A, B, C, and D (see Figure 13.83). He places them so that $AB \parallel ED$, $BC = 50.0$ m, $DC = 300$ m, and $AB = 80.0$ m. How wide is the river?

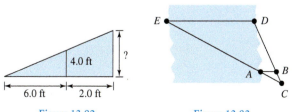

6.0 ft   2.0 ft

Figure 13.82   Figure 13.83

Now Try It! Answers

1. $LM = 8$ **2. a.** $DE = 6$ **b.** $BD = 4.33$ **c.** $BC = 10.83$

13.5 The Trigonometric Ratios

Section Objectives
- Define the trigonometric ratios of sine, cosine, and tangent
- Define the trigonometric ratios of cosecant, secant, and cotangent
- Define the reciprocal relationships between the six trigonometric ratios
- Find the value of the six trigonometric ratios of a right triangle given two sides of that triangle

Many applied problems in science and technology require the use of triangles, especially right triangles. Included among the applied problems are those involving forces acting on objects, air navigation, surveying, the motion of projectiles, and light refraction in optics. In **trigonometry** we develop methods for measuring the sides and angles of triangles, as well as solving related applied problems. Because of its many uses, trigonometry is generally recognized as one of the most practical and relevant branches of mathematics. In this section we will present the fundamental concept of a trigonometric ratio.

We know that a triangle is composed of three sides and three angles. If one side and any other two parts of the triangle are known we can find the other three parts. This is known as **solving a triangle.** We know that we can form a ratio between the lengths of any two sides of a right triangle. We know that if we are given two right triangles that are similar, the corresponding sides are proportional. The ratio between any two sides of a right triangle is called a **trigonometric ratio** and is defined as follows:

Given a right triangle *ABC* with the right angle at *C* (see Figure 13.84), we have the following ratios for angle *A* in terms of the sides of the triangle.

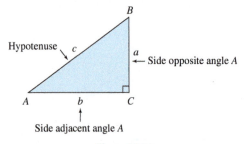

Figure 13.84

$$\text{Sine of angle } A \;=\; \frac{\text{side opposite angle } A}{\text{hypotenuse}}$$

$$\text{Cosine of angle } A \;=\; \frac{\text{side adjacent to angle } A}{\text{hypotenuse}}$$

$$\text{Tangent of angle } A \;=\; \frac{\text{side opposite angle } A}{\text{side adjacent to angle } A}$$

$$\text{Cotangent of angle } A \;=\; \frac{\text{side adjacent to angle } A}{\text{side opposite angle } A}$$

$$\text{Secant of angle } A \;=\; \frac{\text{hypotenuse}}{\text{side adjacent to angle } A}$$

$$\text{Cosecant of angle } A \;=\; \frac{\text{hypotenuse}}{\text{side opposite angle } A}$$

The names of these trigonometric ratios are usually abbreviated as **sin A, cos A, tan A, cot A, sec A,** and **csc A.** The definitions can be used to find the trigonometric ratios of any acute angle in any right triangle. Using the definitions and considering angle *B*, we can also state the following.

$$\sin B = \frac{\text{side opposite angle } B}{\text{hypotenuse}} \qquad\qquad \csc B = \frac{\text{hypotenuse}}{\text{side opposite angle } B}$$

$$\cos B = \frac{\text{side adjacent angle } B}{\text{hypotenuse}} \qquad\qquad \sec B = \frac{\text{hypotenuse}}{\text{side adjacent angle } B}$$

$$\tan B = \frac{\text{side opposite angle } B}{\text{side adjacent angle } B} \qquad\qquad \cot B = \frac{\text{side adjacent angle } B}{\text{side opposite angle } B}$$

Because these definitions are so important, they should be remembered. It is often helpful to use some memory device such as the word "SOH-CAH-TOA" which also summarizes the trigonometric ratios. The first three letters suggest that sine is related to the opposite and hypotenuse. "CAH" suggests that cosine is adjacent divided by hypotenuse, and "TOA" represents tangent as opposite divided by adjacent.

It would be helpful to memorize the definitions of sine, cosine, and tangent, then apply those definitions to the triangle given in Figure 13.85. You should be able to cover up the fractions given in Example 1 and obtain the same values by referring to Figure 13.85 and using the memorized definitions.

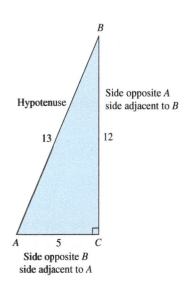

Figure 13.85

Example
1

In the triangle shown in Figure 13.85, we have

$$\sin A = \frac{12}{13} \qquad \sin B = \frac{5}{13} \qquad \text{\textcolor{blue}{side opposite}} \over \text{\textcolor{blue}{hypotenuse}}$$

$\sin A = \dfrac{12}{13}$	$\sin B = \dfrac{5}{13}$	side opposite / hypotenuse
$\cos A = \dfrac{5}{13}$	$\cos B = \dfrac{12}{13}$	side adjacent / hypotenuse
$\tan A = \dfrac{12}{5}$	$\tan B = \dfrac{5}{12}$	side opposite / side adjacent
$\cot A = \dfrac{5}{12}$	$\cot B = \dfrac{12}{5}$	side adjacent / side opposite
$\sec A = \dfrac{13}{5}$	$\sec B = \dfrac{13}{12}$	hypotenuse / side adjacent
$\csc A = \dfrac{13}{12}$	$\csc B = \dfrac{13}{5}$	hypotenuse / side opposite

In some cases we know only two sides of a right triangle. The third side can be found by using the Pythagorean theorem. The trigonometric ratios can then be found for the angles. This is illustrated in Example 2.

Example
2

In Figure 13.86 we know the two sides of the right triangle. We find the third side by using the Pythagorean theorem.

$$a^2 + b^2 = c^2$$
$$a^2 + (15)^2 = (17)^2$$
$$a^2 + 225 = 289$$
$$a^2 = 64$$
$$a = 8$$

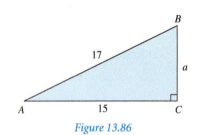

Figure 13.86

Knowing that $a = 8$ we can now find any of the trigonometric ratios. For example, $\sin A = \frac{8}{17}$ and $\tan B = \frac{15}{8}$.

If we compare the six trigonometric ratios we can see that cot A is the reciprocal of tan A, sec A is the reciprocal of cos A, and csc A is the reciprocal of sin A. These relationships are expressed below and they will be especially useful for many calculator computations, which will be discussed in the next section.

$$\cot A = \frac{1}{\tan A} \qquad \sec A = \frac{1}{\cos A} \qquad \csc A = \frac{1}{\sin A}$$

If we have a right triangle and know only one of the trigonometric ratios for a given angle, the values of the other five trigonometric ratios can be found. Example 3 illustrates how this can be done.

Example 3

Suppose we have a right triangle and know that $\sin A = \frac{9}{10}$. We know that the ratio of the side opposite angle A to the hypotenuse is 9 to 10. For the purposes of finding the other ratios, we may assume that the lengths of these two sides are 9 units and 10 units (see Figure 13.87). We find the third side by using the Pythagorean theorem and we get

$$x = \sqrt{10^2 - 9^2} = \sqrt{19}$$

We can now determine the other five trigonometric ratios of the angle whose sine is given as $\frac{9}{10}$.

$$\cos A = \frac{\sqrt{19}}{10} \qquad\qquad \cot A = \frac{\sqrt{19}}{9}$$

$$\tan A = \frac{9}{\sqrt{19}} = \frac{9\sqrt{19}}{19} \qquad \sec A = \frac{10}{\sqrt{19}} = \frac{10\sqrt{19}}{19}$$

$$\csc A = \frac{10}{9}$$

Using a calculator, we can easily obtain the decimal forms of the above values.

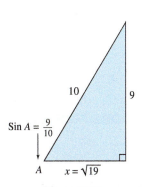

$\text{Sin } A = \frac{9}{10}$

Figure 13.87

Example 4

Given a right triangle in which $\tan A = 0.577$, we know that the ratio of the side opposite angle A to the side adjacent to it is 0.577. This means that **we can construct a triangle with sides of 0.577 opposite A and 1 adjacent to A,** as shown in Figure 13.88.

The length of the hypotenuse is found by using the Pythagorean theorem as follows:

$$a^2 + b^2 = c^2$$
$$(0.577)^2 + 1^2 = c^2$$
$$1.33 = c^2$$
$$c = 1.15$$

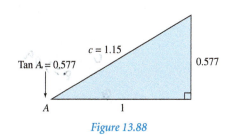

$\text{Tan } A = 0.577$

Figure 13.88

Now Try It!

Determine the six trigonometric functions in terms of angle A for the following triangles.

1.

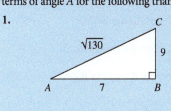

2.

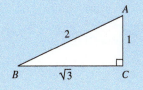

The other five trigonometric ratios for this angle can now be found, as shown below.

$$\sin A = \frac{0.577}{1.15} = 0.502 \qquad \cot A = \frac{1}{0.577} = 1.73$$

$$\cos A = \frac{1}{1.15} = 0.870 \qquad \sec A = \frac{1.15}{1} = 1.15$$

$$\csc A = \frac{1.15}{0.577} = 1.99$$

13.5 Exercises

In Exercises 1 through 4, find the indicated trigonometric ratios in fractional form from Figure 13.89.

1. $\sin A$, $\tan A$, $\cos B$

2. $\cos A$, $\sin B$, $\cot A$

3. $\cot B$, $\sec A$, $\tan B$

4. $\sec B$, $\csc A$, $\csc B$

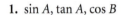

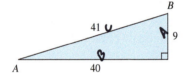

Figure 13.89

In Exercises 5 through 8, find the indicated trigonometric ratios in fractional form from Figure 13.90.

5. $\sin A$, $\sec B$, $\cot A$

6. $\csc A$, $\sin B$, $\cot B$

7. $\tan A$, $\cos B$, $\sec A$

8. $\tan B$, $\csc B$, $\cos A$

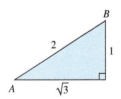

Figure 13.90

In Exercises 9 through 12, find the indicated trigonometric ratios in decimal form from Figure 13.91.

9. $\cos A$, $\tan A$, $\csc B$

10. $\sin B$, $\sec A$, $\cot B$

11. $\sin A$, $\cos B$, $\cot A$

12. $\csc A$, $\tan B$, $\sec B$

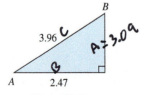

Figure 13.91

In Exercises 13 through 24, determine the indicated trigonometric ratios. The listed sides are those shown in Figure 13.92.

13. $a = 4$, $b = 3$. Find $\sin A$ and $\tan B$.

14. $a = 8$, $c = 17$. Find $\cos A$ and $\csc B$.

15. $b = 7$, $c = 25$. Find $\cot A$ and $\cos B$.

16. $a = 9$, $c = 25$. Find $\sin A$ and $\sec B$.

17. $a = 1$, $b = 1$. Find $\cos A$ and $\cot B$.

18. $a = 3$, $c = 4$. Find $\sec A$ and $\tan B$.

19. $b = 15$, $c = 22$. Find $\csc A$ and $\cos B$.

20. $a = 67$, $b = 119$. Find $\cos A$ and $\cot B$.

21. $a = 1.2$, $b = 1.5$. Find $\sin A$ and $\tan B$.

22. $a = 3.44$, $c = 6.82$. Find $\cot A$ and $\csc B$.

23. $b = 0.0446$, $c = 0.0608$. Find $\sin A$ and $\sec B$.

24. $a = 1673$, $c = 1944$. Find $\csc A$ and $\sin B$.

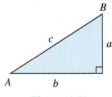

Figure 13.92

In Exercises 25 through 36, use the given trigonometric ratios to find the indicated trigonometric ratios.

25. If $\tan A = 1$, find $\sin A$.

26. If $\sin A = \frac{1}{2}$, find $\cos A$.

27. If $\cos A = 0.70$, find $\csc A$.

28. If $\sec A = 1.6$, find $\sin A$.

29. If $\cot A = 0.563$, find $\cos A$.

30. If $\csc A = 2.64$, find $\tan A$.

31. If $\sin A = 0.720$, find $\tan A$.

32. If $\tan A = 0.350$, find $\cos A$.

33. If $\cos A = 0.8660$, find $\sin A$.

34. If $\csc A = 5.55$, find $\sec A$.

35. If $\sec A = \sqrt{3.00}$, find $\tan A$.

36. If $\cot A = \sqrt{5.00}$, find $\sin A$.

In Exercises 37 through 44, solve the given problems.

37. From the definitions of the trigonometric ratios, it can be seen that $\sin A = \cos B$. What ratio associated with angle B equals $\tan A$? $\csc A$?

38. Construct three right triangles. The first triangle should have sides 3 in., 4 in., and 5 in. The second triangle should have sides 6 cm, 8 cm, and 10 cm. The third triangle should have sides 9 cm, 12 cm, and 15 cm. For each triangle determine the sine and tangent of the smallest angle. What is the relationship between the three triangles? What is true of the trigonometric ratios found?

39. State the definitions of all six trigonometric ratios of angle A in terms of the sides a, b, and c of the triangle shown in Figure 13.93.

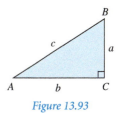

Figure 13.93

40. In Figure 13.93, if $a < b$, is $\sin A < \sin B$? Explain.

41. In Figure 13.93, if $a = 5$ and $b = 12$, what is the value of $(\sin A)^2 + (\cos A)^2$?

42. In Figure 13.93, if $a = 8$ and $b = 15$, what is the value of $(\sin A)^2 + (\cos A)^2$?

43. In Figure 13.93, given that $a = 5$ and $b = 12$, calculate the values of $\sin A$, $\cos A$, and $\tan A$. Then show that $(\sin A)/(\cos A) = \tan A$.

44. In Figure 13.93, given that $a = 8$ and $b = 15$, calculate the values of $\sin A$, $\cos A$, and $\tan A$. Then show that $(\sin A)/(\cos A) = \tan A$.

Now Try It! Answers

1. $\sin A = \dfrac{9}{\sqrt{130}}$ $\cos A = \dfrac{7}{\sqrt{130}}$ $\tan A = \dfrac{9}{7}$ $\csc A = \dfrac{\sqrt{130}}{9}$ $\sec A = \dfrac{\sqrt{130}}{7}$ $\cot A = \dfrac{7}{9}$

2. $\sin A = \dfrac{\sqrt{3}}{2}$ $\cos A = \dfrac{1}{2}$ $\tan A = \dfrac{\sqrt{3}}{1}$ $\csc A = \dfrac{2}{\sqrt{3}}$ $\sec A = \dfrac{2}{1}$ $\cot A = \dfrac{1}{\sqrt{3}}$

13.6 Values of the Trigonometric Ratios

Section Objectives
- Determine the values of the six trigonometric ratios for a 30°–60°–90° triangle
- Determine the values of the six trigonometric ratios for a 45°–45°–90° triangle
- Determine the values of the six trigonometric ratios using a calculator

In the previous section we defined the trigonometric ratios of an angle, but we did not mention the size of the angle. We did note, however, that the ratio of two sides of a right triangle is the same as the ratio of the two corresponding sides in any similar right triangle. Therefore, *for an angle of a particular size, a given trigonometric ratio has a particular value.* By using certain basic geometric properties we can establish the values of the trigonometric ratios for certain angles that are frequently used.

A basic geometric fact is that *in a 30°–60°–90° triangle, the length of the side opposite the 30° angle is one-half the length of the hypotenuse* (see Figure 13.94a). We can easily verify this statement by referring to the equilateral triangle shown in Figure 13.94b which has been divided into two smaller congruent triangles by the altitude. Each of the smaller triangles has angles of 30° (at top), 60°, and 90°.

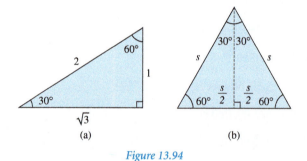

Figure 13.94

In Figure 13.94(a), the hypotenuse has been given the value 2, the side opposite the 30° angle has been given the value 1, and from the Pythagorean theorem we determine that the third side has the value $\sqrt{3}$. Using the known angles and sides in Figure 13.94(a), we can now establish all of the trigonometric ratios shown in the following table.

Angle	sin	cos	tan	cot	sec	csc
30°	$\dfrac{1}{2}$	$\dfrac{\sqrt{3}}{2}$	$\dfrac{\sqrt{3}}{3}$	$\sqrt{3}$	$\dfrac{2\sqrt{3}}{3}$	2
60°	$\dfrac{\sqrt{3}}{2}$	$\dfrac{1}{2}$	$\sqrt{3}$	$\dfrac{\sqrt{3}}{3}$	2	$\dfrac{2\sqrt{3}}{3}$

Example 1 illustrates the use of another geometric property to establish the values of the trigonometric ratios of 45°.

Example 1

To find the trigonometric ratios for 45°, we construct an isosceles right triangle, as shown in Figure 13.95. Because the triangle is isosceles, *both acute angles are 45° and the legs are equal.* Each of the equal sides can be given the value 1. From the Pythagorean theorem, we find that the hypotenuse will then be $\sqrt{2}$. Using

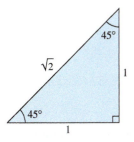

Figure 13.95

the values shown in Figure 13.95, we can apply the definitions of the trigonometric ratios to get the following.

$$\sin 45° = \frac{\sqrt{2}}{2} \qquad \cos 45° = \frac{\sqrt{2}}{2} \qquad \tan 45° = 1$$

$$\cot 45° = 1 \qquad \sec 45° = \sqrt{2} \qquad \csc 45° = \sqrt{2}$$

Combining the values from the preceding table for 30° and 60° and the values from Example 1 for 45°, we can set up the following short table of values of the trigonometric ratios in exact form.

Angle	sin	cos	tan	cot	sec	csc
30°	$\frac{1}{2}$	$\frac{\sqrt{3}}{2}$	$\frac{\sqrt{3}}{3}$	$\sqrt{3}$	$\frac{2\sqrt{3}}{3}$	2
45°	$\frac{\sqrt{2}}{2}$	$\frac{\sqrt{2}}{2}$	1	1	$\sqrt{2}$	$\sqrt{2}$
60°	$\frac{\sqrt{3}}{2}$	$\frac{1}{2}$	$\sqrt{3}$	$\frac{\sqrt{3}}{3}$	2	$\frac{2\sqrt{3}}{2}$

Figure 13.96

Values of other trigonometric ratios can be found by using a calculator.

When using your calculator to determine the value of the various trigonometric ratios, the keys SIN, COS, and TAN are used. It is important to make sure that your calculator is in **degree mode.**

To determine the value of a trigonometric function, press the appropriate trig function key and then enter the number of degrees in the angle you are working with (see Figure 13.96). This is illustrated in Example 2.

Example 2

Use a calculator to determine the value of sin 20°.

Verify that your calculator is in degree mode. Then use the key sequence SIN 20 ENTER. Your calculator screen will look like Figure 13.97.

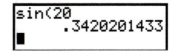

Figure 13.97

Most calculators do not have CSC, SEC, or COT keys. To determine the values of these trigonometric ratios we use the reciprocal relationships introduced in the last section along with the SIN, COS, and TAN keys.

Example

3

Use your calculator to find cot 65°, sec 65°, and csc 65°.

To find cot 65° use the key sequence

1 ÷ TAN 65 ENTER to get a result of 0.4663076582 (see Figure 13.98).

To find sec 65° use the key sequence

1 ÷ COS 65 ENTER to get a result of 2.366201583 (see Figure 13.98).

To find csc 65° use the key sequence

1 ÷ SIN 65 ENTER to get a result of 1.103377919 (see Figure 13.98).

Remember

$$\csc A = \frac{1}{\sin A}$$

$$\sec A = \frac{1}{\cos A}$$

$$\cot A = \frac{1}{\tan A}$$

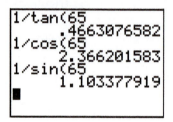

Figure 13.98

Sometimes we know the value of a trigonometric ratio and we want to find the angle involved. To find the angle when the ratio is known, use the 2nd function key to access $\sin^{-1}$, $\tan^{-1}$ or $\cos^{-1}$.

Example

4

Use your calculator to find angle A if sin A = 0.454.

To find angle A use the key sequence

2nd SIN 0.454 ENTER to get an angle measure of approximately 27° (see Figure 13.99).

Remember

Using the key sequence 2nd SIN accesses the $\sin^{-1}$ function.

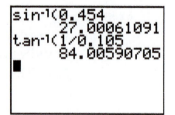

Figure 13.99

Example

5

Use your calculator to find angle A if cot A = 0.105.

To find angle A we use the key sequence

2nd TAN 1 ÷ 0.105 ENTER to get an angle measure of approximately 84° (see Figure 13.99).

Example 6

In the right triangle shown in Figure 13.100, the two acute angles add up to 90°. Since $\sin A = a/c$ and $\cos B = a/c$, we see that $\sin A = \cos B$. Also, since $\angle A + \angle B = 90°$,

$$\angle B = 90° - \angle A \quad \text{or} \quad \sin A = \cos(90° - \angle A)$$

If $\angle A = 40°$, for example, then

$$\sin 40° = \cos 50°$$

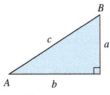

Figure 13.100

Now Try It!

Use your calculator to determine the value of the following trigonometric functions:

1. $\sin 26.5°$
2. $\tan 78.7°$
3. $\sec 68.0°$
4. $\cot 10.2°$
5. $\cos 59.6°$
6. $\csc 30°$

Use your calculator to determine the value of angle A to the nearest 0.1°:

7. $\sin A = 0.6371$
8. $\cos A = 0.707$
9. $\tan A = 1.3947$
10. $\csc A = 1.3972$

The following example illustrates the use of a trigonometric ratio in an applied problem.

Example 7

A surveyor needs to determine the angle α for the parcel of land shown in Figure 13.101. Because we already know the two legs of the triangle, we can directly obtain the value of the tangent of α to get

$$\tan \alpha = \frac{62.10 \text{ m}}{136.4 \text{ m}} \qquad \begin{array}{l} \text{side opposite } \alpha \\ \hline \text{side adjacent to } \alpha \end{array}$$

We use the following sequence on a calculator to find the angle α:

$$2^{\text{nd}} \text{ TAN } 62.1 \div 136.4 \text{ ENTER}$$

which tells us that $\alpha = 24.48°$ (rounded off to hundredths).

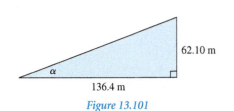

62.10 m

136.4 m

Figure 13.101

13.6 Exercises

In Exercises 1 through 16, use a calculator to determine the value of the indicated trigonometric ratio.

1. $\cos 32.0°$
2. $\sin 21.0°$
3. $\tan 24.5°$
4. $\cot 15.4°$
5. $\sec 48.2°$
6. $\csc 56.1°$
7. $\sin 66.6°$
8. $\cos 52.9°$
9. $\cot 76.6°$
10. $\tan 63.7°$
11. $\sin 44.8°$
12. $\cos 9.2°$

13. $\csc 13.7°$

14. $\sec 41.3°$

15. $\tan 68.0°$

16. $\cot 46.4°$

In Exercises 17 through 32, use a calculator to determine the value of the angle α to the nearest 0.1°.

17. $\sin \alpha = 0.5299$

18. $\tan \alpha = 0.2126$

19. $\sec \alpha = 1.057$

20. $\cos \alpha = 0.7944$

21. $\csc \alpha = 1.149$

22. $\cot \alpha = 0.8040$

23. $\cos \alpha = 0.4712$

24. $\sin \alpha = 0.9888$

25. $\cot \alpha = 0.7620$

26. $\sec \alpha = 1.666$

27. $\tan \alpha = 0.3250$

28. $\csc \alpha = 2.608$

29. $\cos \alpha = 0.09932$

30. $\sin \alpha = 0.9464$

31. $\cot \alpha = 0.6190$

32. $\tan \alpha = 1.900$

In Exercises 33 through 36, use a calculator and the given trigonometric ratio to find the indicated trigonometric ratio.

33. $\sin \alpha = 0.5592$; $\sec \alpha$

34. $\cos \alpha = 0.8290$; $\tan \alpha$

35. $\tan \alpha = 1.600$; $\csc \alpha$

36. $\cot \alpha = 0.1584$; $\sin \alpha$

In Exercises 37 through 44, find, to the nearest 0.1°, angle α in Figure 13.102 for the given sides of the triangle.

37. $a = 4, c = 5$

38. $a = 3, c = 7$

39. $b = 6.2, c = 8.2$

40. $a = 3.2, b = 2.0$

41. $a = 15.5, c = 27.3$

42. $b = 0.35, c = 0.84$

43. $a = 6580, b = 1230$

44. $a = 3.95, c = 45.2$

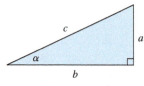

Figure 13.102

In Exercises 45 through 52, solve the given problems.

45. In an AC (alternating-current) electric circuit, the instantaneous voltage is found by using the formula

$$v_{inst} = V_p \sin \theta$$

Find the instantaneous voltage if $V_p = 15\ V$ and $\theta = 63°$.

46. Under certain conditions, the height of an overcast cloud layer is found by the formula

$$H = s \tan \theta$$

Find the height H when $s = 5750\ ft$ and $\theta = 27°$.

47. In surveying, the horizontal distance between two points is obtained by evaluating $H = s \cos \alpha$. Find H if

$$s = 132.8\ m \text{ and } \alpha = 17.7°.$$

48. In determining the angle a windshield makes with the dashboard of a car, it is necessary to find A given that $\tan A = 2.174$. Find that angle to the nearest 0.1°.

49. In the study of optics, it is known that a light ray entering glass is bent toward a line perpendicular to the surface, as shown in Figure 13.103. The *index of refraction* of the glass is defined to be

$$n = \frac{\sin i}{\sin r}$$

Find the index of refraction if $i = 72.0°$ and $r = 37.7°$.

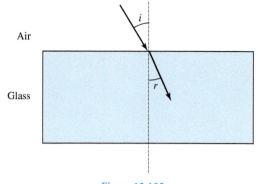

Figure 13.103

50. The work W done by a force F is defined as

$$W = Fd \cos \theta$$

where F is the magnitude of the force, d is the distance through which it acts, and θ is the angle between the direction of the force and the direction of motion. Given

that a 25.0-lb force acts through 20.0 ft and the angle between the force and motion is 32.0°, how much work is
done by the force?

51. One end of a 25.0-m metal rod lies on a flat surface; the rod
makes an angle of 17.0° with the surface. The length of the
shadow of the rod due to a light shining vertically down on
it is (25.0)(cos 17.0°). Find the length of the shadow.

52. The coefficient of friction for an object on an inclined
plane equals the tangent of the angle that the plane
makes with the horizontal if the object moves down the
plane with a constant speed. The coefficient of friction
between a wooden crate and a wooden plank is 0.340
when the create is moving with a constant speed. What
angle does the plank make with the horizontal?

Now Try It! Answers

1. 0.4462	**2.** 5.0045	**3.** 2.6700	**4.** 5.5578	**5.** 0.5060
6. 2	**7.** 39.6°	**8.** 45°	**9.** 54.4°	**10.** 45.7°

13.7 Right Triangle Applications

Section Objectives
- Find the missing sides and angles of a right triangle
- Outline a process for solving a right triangle
- Solve practical applications involving right triangles

In this section we use examples and exercises to investigate many of the applications
of the trigonometric ratios. First, we consider the general idea of **solving a triangle.**

 In every triangle there are three angles and three sides. *If three of these six
parts are known, the other three can be found provided that at least one of the
known parts is a side. By solving a triangle, we mean determining the values of the
six parts.* The strategy for solving **right triangles** will involve these four steps:

1. If two angles are known, the third angle can be found by using the property that the sum of the three angles is 180°.

2. If two sides are known, find the third side by using the Pythagorean theorem ($c^2 = a^2 + b^2$).

3. If only one side and one of the acute angles are known, find another side
by using a trigonometric ratio.

4. If neither of the acute angles is known but two sides are known, both acute
angles can be found by using trigonometric ratios.

When using these steps, we should try to avoid using derived values for finding other values. That is, when solving for a particular part of the triangle, try to
use only values that were given in the original statement of the problem.

Examples 1 and 2 illustrate solving right triangles.

Example 1

Given that the hypotenuse of a right triangle is 16.0 and that one of the acute angles is 35.0°, find the other acute angle and the two sides (see Figure 13.104).

Here we know one side ($c = 16.0$) and two angles ($\alpha = 35.0°$ and the right angle is 90°). Using step 1 we can determine that *the third angle is 55.0° since the three angles must add up to 180°.* Since only one side is known, we follow the suggestion of step 3 because we use the sine ratio to find the value of b. Since

$$\sin 35.0° = \frac{b}{16.0} \longleftarrow \text{hypotenuse given}$$

given angle

we have

$$b = 16.0 \sin 35.0° = 9.18$$

The calculator steps for this solution are

16 × SIN 35 ENTER gives us 9.1772230 for b

Side a can be found by using the cosine relation. Since we have $\cos 35.0° = a/16.0$,

$$a = 16.0 \cos 35.0° = 13.1$$

Although we could have found side a using the Pythagorean theorem, it is better to avoid using derived values (such as $b = 9.18$) if possible. We were able to find side a using the original values, not derived values. In addition to the given information, we now know that the other acute angle is 55.0° and the other sides are 9.18 and 13.1. *All six parts of the triangle are now known, and the triangle is solved.*

Example 2

In a right triangle the two legs are 5.00 and 8.00. Find the hypotenuse and the two acute angles (see Figure 13.105).

Because we know two sides, we can follow the suggestion of step 2 to use the Pythagorean theorem for finding the third side. We get

$$c = \sqrt{(5.00)^2 + (8.00)^2}$$

given sides

$$= 9.43$$

Because neither of the acute angles is known, we will *use a trigonometric ratio to find one of them.* From the figure we see that

$$\tan A = \frac{5.00}{8.00}$$

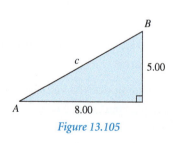

Figure 13.105

Figure 13.104 area:

16.0

35.0°

a

b

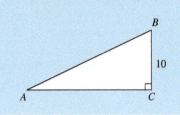

Now Try It!

Solve the following right triangle given $\angle A = 37°$, $\angle C = 90°$, and $a = 10$.

which means that $\tan A = 0.6250$. From the table or by using a calculator we can now establish that $\angle A = 32.0°$ (to the nearest 0.1°).

We can now find $\angle B$ by solving

$$\tan B = \frac{8.00}{5.00}$$

With $\tan B = 1.60$ we get $\angle B = 58.0°$. The triangle is now solved because we know all six parts (three angles and three sides). (The use of the tangent ratio is preferred here because both sine and cosine would have used the derived value of $c = 9.43$; if possible, we should not use derived values for other calculations.)

Using the trigonometric ratios in applied problems involves the same approach as solving triangles, although *it is usually one particular part of the triangle that we need to determine.* Examples 3–8 illustrate some of the basic applications.

Example 3

A section of a highway is 4.20 km long and rises along a uniform grade that makes an angle of 3.2° with the horizontal. What is the change in elevation? (See Figure 13.106, which is not drawn to scale.)

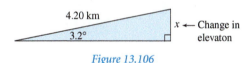

Figure 13.106

We seek the value of the side opposite the known angle and we know the hypotenuse. Because *we are involved with "opposite" and hypotenuse,* the solution is most directly obtained through use of the sine ratio.

$$\sin 3.2° = \frac{x}{4.20}$$

$$x = 4.20 \sin 3.2°$$
$$= 0.234 \text{ km}$$

The highway therefore changes in elevation by a height of 0.234 km.

Example 4

In a military training exercise, a rocket is aimed directly at an airplane as in Figure 13.107. If the rocket were to be fired, what angle would its path make with the ground?

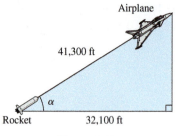

41,300 ft

α

Rocket   32,100 ft

Figure 13.107

Because we know the *hypotenuse and the adjacent side* of the desired angle, we can easily find that angle by using the cosine ratio.

required
angle

↓

$$\cos \alpha = \frac{32{,}100}{41{,}300} \xleftarrow{\text{given adjacent side}} \text{given adjacent side}$$

Using the 2nd and cos keys on a calculator, we find that the angle (to the nearest 0.1°) is 39.0°.

In Example 4 we expressed the angle by rounding it to the nearest 0.1°. If the sides of the triangle are given with three significant digits, the derived angles should be rounded to the nearest 0.1°. If the sides have four significant digits, the derived angles should be rounded to the nearest 0.01°, and so on.

Example 5

If a certain airplane loses its only engine, it can glide along a path that makes an angle of 18.0° with the level ground (see Figure 13.108). If the plane is 5500 ft above the ground, what is the maximum horizontal distance it can go?

In Figure 13.108 we seek the value of the distance x. Because the known side is *opposite* the known angle, and the side to be determined is *adjacent*, the solution can be completed by using the tangent.

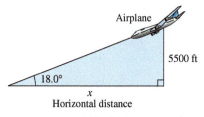

Airplane

5500 ft

18.0°

x
Horizontal distance

Figure 13.108

$$\tan 18.0° = \frac{5500}{x} \xleftarrow{} \text{given opposite side}$$
$$\xleftarrow{} \text{required adjacent side}$$

$$x = \frac{5500}{\tan 18.0°} = 17{,}000 \text{ ft}$$

The horizontal distance is 17,000 ft.

Example 7

While studying the behavior of wind, one weather observer flies a hot air balloon while another observer remains at the point of departure (see Figure 13.109). The **angle of depression** (the angle between the horizontal and the line of sight, *downward with respect to the balloon observer*) is measured by the pilot to be 71.2° as the balloon flies into clouds 8400 ft above ground. Find the horizontal distance between the ground observer and the point directly below the balloon when it entered the clouds.

In Figure 13.109 the angle of depression is equal to the **angle of elevation** (the angle between the horizontal and the line of sight, *upward with respect to the ground observer*) because the alternate interior angles are equal. We therefore know that the angle of elevation is also 71.2° as shown. We now have a right

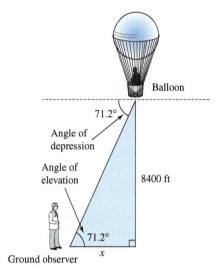

Figure 13.109

triangle with a known angle and *the opposite and adjacent sides involved.* Using the tangent we get

$$\tan 71.2° = \frac{8400}{x} \quad \begin{array}{l}\text{given opposite side}\\[4pt]\text{required adjacent side}\end{array}$$

$$x = \frac{8400}{\tan 71.2°} = 2900 \text{ ft}$$

Example
8

A television antenna is on the roof of a building. From a point on the ground 36.0 ft from the building, the angles of elevation of the top and the bottom of the antenna are 51.0° and 42.0°, respectively. How tall is the antenna?

In Figure 13.110 we let *x* represent the distance from the top of the building to the ground and *y* represent the distance from the top of the antenna to the ground. ***The solution will be the value of y − x.*** We proceed to find the values of *y* and *x*.

$$\frac{x}{36.0} = \tan 42.0°$$

$$x = 36.0 \tan 42.0°$$

$$= 32.4 \text{ ft}$$

and

$$\frac{y}{36.0} = \tan 51.0°$$

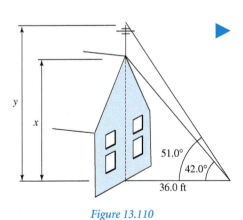

Figure 13.110

$$y = 36.0 \tan 51.0°$$
$$= 44.5 \text{ ft}$$

The length of the antenna is $y - x = 12.1$ ft. We were able to determine the length of the antenna without entering the building or climbing onto the roof.

13.7 Exercises

In Exercises 1 through 16, solve the triangles with the given parts. The parts are indicated in Figure 13.111. (Angles are indicated only by the appropriate capital letter.)

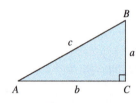

Figure 13.111

1. $A = 30.0°, a = 12.0$
2. $A = 45.0°, b = 16.0$

3. $B = 56.3°, c = 22.5$
4. $B = 17.1°, a = 15.7$

5. $A = 76.8°, c = 13.4$
6. $B = 35.7°, b = 1.45$

7. $a = 0.650, c = 1.35$
8. $a = 4.70, b = 7.40$

9. $b = 5.80, c = 45.0$
10. $a = 734, b = 129$

11. $a = 9.72, c = 10.8$
12. $b = 0.195, c = 0.321$

13. $A = 7.0°, b = 15.3$
14. $B = 84.5°, c = 1730$

15. $a = 65.1, b = 98.3$
16. $b = 1.89, c = 7.14$

In Exercises 17 through 36, solve the given problems by finding the appropriate part of a triangle. In each problem, draw a rough sketch.

17. The angle of elevation to the top of a statue is 10.0° from a point 165 ft from the base of the statue. Find the height of the statue. See Figure 13.112.

18. The angle of elevation of the sun is 51.3° at the time a tree casts a shadow 23.7 m long. Find the height of the tree.

19. A robot is on the surface of Mars. The angle of depression from a camera in the robot to a rock on the surface

of Mars is 23.7°. The camera is 122 cm above the surface. How far is the camera from the rock?

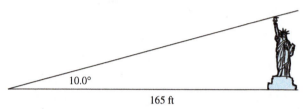

Figure 13.112

20. A guy wire whose grounded end is 16.0 ft from the pole it supports makes an angle of 56.0° with the ground. How long is the wire?

21. The bottom of a picture on a wall is on the same level as a person's eye. The picture is 135 cm high, and the angle of elevation to the top of the picture is 23.0°. How far is the person's eye from the bottom of the picture? See Figure 13.113.

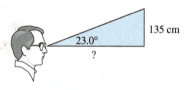

Figure 13.113

22. Along the shore of a river, from a rock at a height of 300 ft above the river, the angle of depression of the closest point of the opposite shore is 12.0°. What is the distance across the river from the base of the height to the closest point on the opposite shore?

23. A point near the top of the Leaning Tower of Pisa is about 50.5 m from a point at the base of the tower (measured along the tower). This top point is also directly above a point on the ground 4.25 m from the same base point.

What angle does the tower make with the ground? See
Figure 13.114.

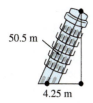

50.5 m

4.25 m

Figure 13.114

24. A roof rater is 5.25 m long (neglecting the overhang),
and its upper end is 1.70 m above the lower end. Find the
angle between the rafter and the horizontal.

25. Aviation weather reports include the *ceiling* which is the
distance between the ground and the bottom of overcast
clouds. A ground observer is 5000 ft from a spotlight
which is aimed vertically. The angle of elevation between
the ground and the spot of light on the clouds is measured by the observer to be 38.7°. What is the ceiling? See
Figure 13.115.

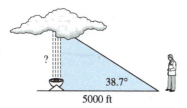

?

38.7°

5000 ft

Figure 13.115

26. A ship is traveling toward a port from the west when it
changes course by 18.0° to the north. It then travels
23.0 miles until it is due north of the port. How far was
it from port when it turned? See Figure 13.116.

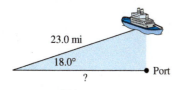

23.0 mi

18.0°

?   • Port

Figure 13.116

27. A person is building a swimming pool 12.5 m long. The
depth at one end is to be 1.00 m and the depth at the
other end is to be 2.50 m. Find, to the nearest 0.1°, the
angle between the bottom of the pool and the horizontal.
See Figure 13.117.

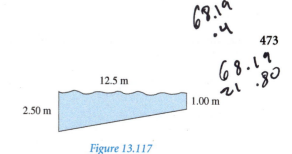

12.5 m

2.50 m   1.00 m

Figure 13.117

28. A shelf is supported by a straight 65.0-cm support attached to the wall at a point 53.5 cm below the bottom of
the shelf. What angle does the support make with the
wall? See Figure 13.118.

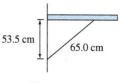

53.5 cm   65.0 cm

Figure 13.118

29. An observer in a helicopter 800 ft above the ground notes
that the angle of depression of an object is 26.0°. How far
from directly below the helicopter is this object? See
Figure 13.119.

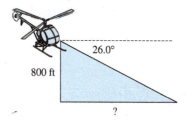

26.0°

800 ft

?

Figure 13.119

30. One way of finding the distance from the earth to the
moon is indicated in Figure 13.120. From point *P* the
moon is directly overhead, and from point *Q* the moon
is just visible. Both points are on the equator. The angle
at *E* about 89.0°. Given that the radius of the earth is
6360 km how far is it to the moon (*PM*)?

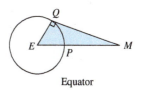

Q

E   *P*   *M*

Equator

Figure 13.120

31. An astronaut observes two cities that are 45.0 km apart;
the angle between the lines of sight is 10.6°. Given that he

is the same distance from each, how high above the surface of the earth is he? See Figure 13.121 and ignore the earth's curvature.

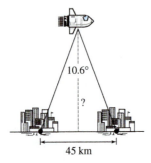

Figure 13.121

32. From an airplane 5000 ft above the surface of the water, a pilot observes two boats directly ahead. The angles of depression are 20.0° and 12.0°. How far apart are the boats?

33. One observer is directly below a jet flying at an altitude of 2575 ft. At the same time, a second observer measures the angle between the ground and the jet and finds that angle to be 28.68°. What is the distance between the two observers?

34. A surveyor wants to determine the height of a vertical cliff without actually climbing it. She walks 263 ft away from the bottom of the cliff and measures the angle between the ground and the top of the cliff. If this angle is 52.0°, how high is the cliff?

35. An aircraft encoding altimeter and a radar system indicate that a plane is 7350 ft high when it is 17,400 ft from the end of the active runway. If the plane flies directly toward the end of the runway, what angle does its path make with the ground? See Figure 13.122.

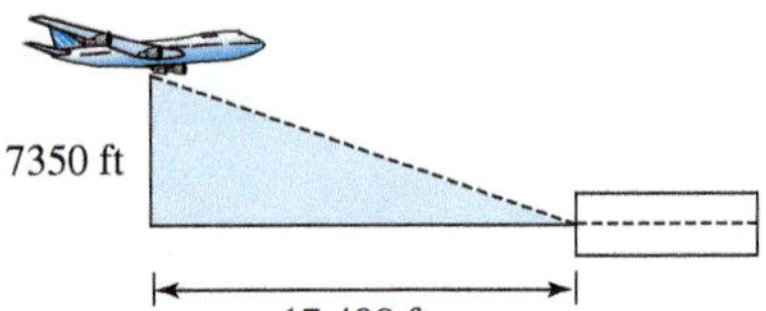

7350 ft

17,400 ft

Figure 13.122

36. An observer is a lighthouse uses an instrument to record the angle of depression of ships. If the angle of depression is denoted by A and the observer is 88 ft above sea level, find a formula for the distance d between the ship and the base of the lighthouse.

Now Try It! Answers

$\angle B = 53°$ $b = 13.3$ $c = 16.6$

Chapter Summary

Key Terms

angle	complementary angles	Pythagorean theorem
straight angle	polygon	similar triangle
right angle	diagonal	trigonometry
acute angle	chord	solving a triangle
obtuse angle	tangent line	angle of depression
adjacent angles	secant line	angle of elevation
vertical angles	central angle	
supplementary angles	inscribed angle	

Formulas

Pythagorean theorem $a^2 + b^2 = c^2$

Six trigonometric ratios

$$\sin A = \frac{\text{opposite}}{\text{hypotenuse}} \qquad \csc A = \frac{\text{hypotenuse}}{\text{opposite}}$$

$$\cos A = \frac{\text{adjacent}}{\text{hypotenuse}} \qquad \sec A = \frac{\text{hypotenuse}}{\text{adjacent}}$$

$$\tan A = \frac{\text{opposite}}{\text{adjacent}} \qquad \cot A = \frac{\text{adjacent}}{\text{opposite}}$$

Reciprocal relationship

$$\cot A = \frac{1}{\tan A} \qquad \sec A = \frac{1}{\cos A} \qquad \csc A = \frac{1}{\sin A}$$

Review Exercises

In Exercises 1 through 6, answer the given questions.

1. What is the complement of an angle of 29°?

2. What is the supplement of an angle of 29°?

3. What angle is formed when a tangent line touches a circle at a point that is connected to the center?

4. Find the measure of the fourth angle in a quadrilateral with angles of 40°, 120°, and 140°.

5. Find the smallest angle in a right triangle that contains an angle of 63°.

6. Triangles *ABC* and *DEF* are similar. If $\angle A = \angle E = 56°$ and $\angle B = \angle F = 83°$, find the missing angles.

In Exercises 7 through 10, use Figure 13.123 and identify the indicated angles. (Assume that BD ⊥ AC.)

7. An obtuse angle

8. Two acute angles

9. The complement of ∠ABE

10. The supplement of ∠ABE

Figure 13.123

In Exercises 11 through 14, use Figure 13.124 and identify or evaluate the following: (Assume that AB ∥ CD.)

11. A pair of alternate interior angles

12. A pair of corresponding angles

13. ∠2 + ∠3 = ?

14. ∠1 + ∠5 = ?

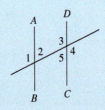

Figure 13.124

In Exercises 15 through 26, use the indicated figures. Determine the measure of the indicated angles.

15. ∠EBD (Figure 13.125)

16. ∠EBC (Figure 13.125)

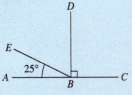

Figure 13.125

17. ∠CBD (Figure 13.126)

18. ∠ABC (Figure 13.126)

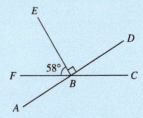

Figure 13.126

19. ∠BDC (Figure 13.127)

20. ∠CDE (Figure 13.127)

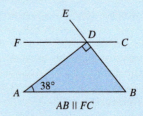

AB ∥ FC

Figure 13.127

21. ∠BFD (Figure 13.128)

22. ∠FDE (Figure 13.128)

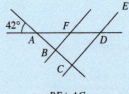

BF ⊥ AC
EC ⊥ AC

Figure 13.128

23. ∠C (Figure 13.129a)

24. ∠B (Figure 13.129b)

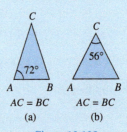

AC = BC   AC = BC
(a)   (b)

Figure 13.129

25. ∠*DCA* (Figure 13.130)

26. ∠*ADC* (Figure 13.130)

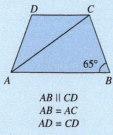

AB ∥ CD
AB = AC
AD = CD

Figure 13.130

In Exercises 27 through 30, use Figure 13.131. Determine the indicated arcs and angles.

27. *BC*   **28.** *AB*

29. ∠*ABC*   **30.** ∠*ACB*

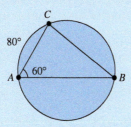

Figure 13.131

In Exercises 31 through 34, Use Figure 13.132. Line CT is tangent to the circle with the center at O. Determine the indicated angles.

31. ∠*BTA*   **32.** ∠*TAB*

33. ∠*BTC*   **34.** ∠*ABT*

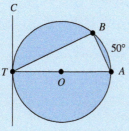

Figure 13.132

In Exercises 35 through 42, find the indicated sides of the right triangle shown in Figure 13.133. Where necessary, round off results to three significant digits.

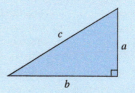

Figure 13.133

35. $a = 9, b = 40, c = ?$

36. $a = 14, b = 48, c = ?$

37. $a = 40, c = 58, b = ?$

38. $b = 56, c = 65, a = ?$

39. $a = 6.30, b = 3.80, c = ?$

40. $a = 126, b = 251, c = ?$

41. $b = 29.3, c = 36.1, a = ?$

42. $a = 0.782, c = 0.885, b = ?$

In Exercises 43 and 44, solve for the unknown side of the given triangle by use of an appropriate proportion.

43. In Figure 13.134, $AC = 12, BC = 8, BD = 5$. Find *AB*.

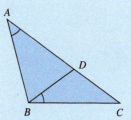

Figure 13.134

44. In Figure 13.135, $AB = 4, BC = 4, CD = 6$. Find *BE*.

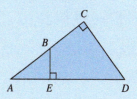

Figure 13.135

In Exercises 45 through 52, use a calculator to determine the values of the given trigonometric ratios. Round results to three significant digits.

45. $\sin 47.3°$ **46.** $\cos 61.2°$

47. $\tan 7.4°$ **48.** $\sin 52.6°$

49. $\cot 83.4°$ **50.** $\csc 29.8°$

51. $\sec 30.5°$ **52.** $\cot 75.9°$

In Exercises 53 through 60, use a calculator to determine the value of the angle A to the nearest 0.1°.

53. $\cos A = 0.8660$ **54.** $\tan A = 3.732$

55. $\sin A = 0.7071$ **56.** $\cos A = 0.0175$

57. $\cot A = 19.08$ **58.** $\sec A = 1.766$

59. $\csc A = 1.196$ **60.** $\csc A = 1.472$

In Exercises 61 through 64, find the indicated trigonometric ratios in fractional form for the angles of the triangle in Figure 13.136.

61. $\sin A, \cos B$ **62.** $\tan B, \cos A$

63. $\sec B, \tan A$ **64.** $\cot A, \csc B$

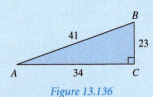

Figure 13.136

In Exercises 65 through 72, determine the indicated trigonometric ratios. The listed sides are those shown in Figure 13.137.

65. $a = 12.0, b = 5.00.$ Find $\cos A$ and $\tan B$.

66. $b = 9.00, c = 15.0.$ Find $\sec A$ and $\cot B$.

67. $a = 8.70, b = 2.30.$ Find $\tan A$ and $\cos B$.

68. $a = 125, c = 148.$ Find $\sin A$ and $\sec B$.

69. $a = 0.0890, c = 0.0980.$ Find $\cot A$ and $\sin B$.

70. $b = 1670, c = 4200.$ Find $\csc A$ and $\csc B$.

71. $b = 0.0943, c = 0.105.$ Find $\cos A$ and $\tan B$.

72. $a = 8.60, b = 7.90.$ Find $\sin A$ and $\cot B$.

In Exercises 73 through 80, find, to the nearest 0.1°, the indicated angle. The parts of the triangle are those shown in Figure 13.137

73. $a = 8.00, c = 13.0.$ Find A.

74. $b = 51.0, c = 58.0.$ Find A.

75. $a = 5.60, b = 1.30.$ Find B.

76. $a = 0.780, c = 2.40.$ Find B.

77. $b = 3420, c = 7200.$ Find B.

78. $a = 0.00670, b = 0.0156.$ Find A.

79. $a = 4910, b = 3650.$ Find A.

80. $b = 0.860, c = 0.915.$ Find B.

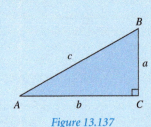

Figure 13.137

In Exercises 81 through 92, use a scientific calculator to solve the triangles with the given parts. The given parts are shown in Figure 13.137.

81. $A = 21.0°, c = 6.93$

82. $B = 18.0°, a = 0.360$

83. $B = 32.7°, b = 45.9$

84. $B = 7.4°, c = 1890$

85. $a = 8.70, b = 5.20$

86. $a = 0.00760, b = 0.0120$

87. $a = 97.0, c = 108$

88. $b = 17.4, c = 54.0$

89. $A = 37.25°, c = 13,872$

90. $B = 64.333°, b = 5280$

91. $b = 2.13520, c = 112.503$

92. $b = 0.0357, c = 112.03$

In Exercises 93 through 122, solve the given problems.

93. What is the diagonal distance along the floor between corners of a rectangular room 12.5 ft wide and 17.0 ft long?

94. An observer is 550 m from the launch pad of a rocket. After the rocket has ascended vertically to a point 750 m from the observer, what is its height above the launch pad? See Figure 13.138.

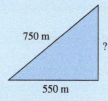

750 m

?

550 m

Figure 13.138

95. The hypotenuse of a right triangle is 24.0 m and one side is twice the other. Find the perimeter of the triangle.

96. The hypotenuse of a right triangle is 3.00 ft longer than one of the sides, which in turn is 5.00 ft longer than the other side. How long are the sides and the hypotenuse?

97. One of the acute angles of a right triangle is three times the other. How many degrees are there in each angle?

98. In a given triangle, the second angle is three times the first and the third angle is twice the first. How many degrees are there in each angle?

99. A tree casts a shadow of 12 ft and the distance from the end of the shadow to the top of the tree is 13 ft. How high is the tree?

100. A tree and a telephone pole cast shadows as shown in Figure 13.139. Find the height of the telephone pole.

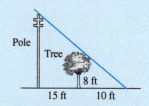

Pole

Tree

8 ft

15 ft   10 ft

Figure 13.139

101. Light is reflected from a mirror so that the angle of incidence i (see Figure 13.140) equals the angle of reflection r. Suppose that a light source is 6.38 centimeters from a mirror and a particular ray of light strikes the mirror at the point shown. How far is the screen S from the mirror?

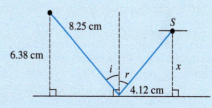

8.25 cm

S

6.38 cm

i r

x

4.12 cm

Figure 13.140

102. A good approximation of the height of a tree can be made by following the procedure suggested in Figure 13.141. By measuring DE, AE, and BC (use a ruler), the length of the tree $DE + EF$ can be found. Find the height of a tree if $AB = 50$ cm, $BC = 30$ cm, $AE = 2400$ cm, and $DE = 150$ cm.

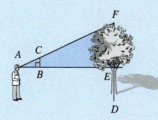

F

A C

B

E

D

Figure 13.141

103. What is the length of the steel support in the structure shown in Figure 13.142?

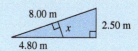

8.00 m

2.50 m

x

4.80 m

Figure 13.142

104. An approximate equation found in the diffraction of light through a narrow opening is

$$\sin \theta = \frac{\lambda}{d}$$

where λ (the Greek lambda) is the wavelength of light and d is the width of the opening. If $\theta = 1.1°$ and $d = 32.0 \, \mu m$, what is the wavelength of the light?

105. An equation used for the instantaneous value of electric current in an alternating-current circuit is

$$i = I_m \cos \theta$$

Calculate $I_m = 56.0 \, mA$ and $\theta = 10.5°$.

106. In determining the height h of a building that is 220 m away, a surveyor may use the equation

$$h = 220 \tan \theta$$

where θ is the angle of elevation to the top of the building. What should θ be if $h = 130 \, m$?

107. Find the amount of work W done by a force F if

$$W = Fd \cos \theta$$

and $F = 57.3 \, lb$, $d = 23.8 \, ft$, and $\theta = 27.3°$.

108. A blueprint drawing of a room is a rectangle with sides of 4.52 cm and 3.27 cm. What is the angle between the longer wall and a diagonal across the room?

109. When the angle of elevation of the sun is 40.0°, what is the length of the shadow of a tree 65.0 ft tall? See Figure 13.143.

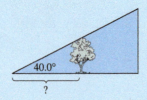

Figure 13.143

110. The distance from a point on the shore of a lake to an island is 3500 ft. From a point directly above the shore on a cliff the angle of depression to the island is 15.0°. How high is the cliff?

111. A standard sheet of plywood is rectangular with the dimensions 4.00 fit by 8.00 ft. If such a sheet is cut

in one straight line from one corner to the opposite corner, find the angles created.

112. A guy wire supporting a television tower is grounded 250 m from the base of the tower. Given that the wire makes an angle of 29.0° with the ground, how high up on the tower is it attached?

113. The span of a roof is 30.0 ft. Its rise is 8.00 ft at the center of the span. What is the angle the roof makes with the horizontal?

114. A roadway rises 85.0 ft for every 1000 ft along the road. Determine the angle of inclination of the roadway.

115. An observer 3000 ft from the launch pad of a rocket measures the angle of inclination to the rocket soon after liftoff to be 65.0°. How high is the rocket, assuming it has moved vertically?

116. A parachutist exits an airplane at a height of 4250 ft directly over his target. The path of his descent is a straight line that makes an angle of 82.3° with the ground. By what distance does the parachutist miss his target? See Figure 13.144.

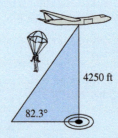

4250 ft

82.3°

Figure 13.144

117. A searchlight is 520 ft from a building. The beam lights up the building at a point 150 ft above the ground. What angle does the beam make with the ground?

118. Two laser beams from the same source are transmitted to two points 100 m apart on a flat surface. One beam is perpendicular to the surface and the other makes an angle of 87.5° with the surface. How far from the surface is the source? See Figure 13.145 on page 481.

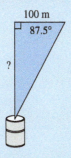

100 m

87.5°

?

Figure 13.145

119. While on a mountain top 2.0 mi high, the angle of depression to the horizon is measured as 1.8°. Use this information to estimate the radius of the earth.

120. The sides of an equilateral triangle are 24.0 in. How long is its altitude?

121. Given that A and B are acute angles, is cos A > cos B if A > B? Is tan A > tan B if A > B?

122. Given that A and B are acute angles, is sin A > sin B if A > B? Is cot A > cot B if A > B?

Chapter Test

Use Figure 13.146 to answer the following questions. ∠EBC *is a right angle.*

1. Identify two acute angles.

2. Identify an obtuse angle.

3. If ∠CBD = 65°, find its complement.

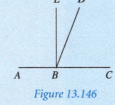

Figure 13.146

4. If ∠CBD = 65°, find its supplement.

5. Identify an acute angle adjacent to ∠DBC.

Use Figure 13.147 to find the measure of each of the following angles.

6. ∠1

7. ∠2

8. ∠3

9. ∠4

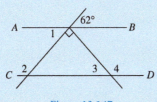

Figure 13.147

10. If $KN = 15$, $MN = 9$, and $MO = 12$, find LM in Figure 13.148.

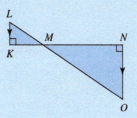

Figure 13.148

11. Use your calculator to determine
  a. sin 65.4°
  b. cos A = 0.6557

12. Find tan A if $\sin A = \dfrac{2}{3}$.

13. Find cot A if tan A = 1.294.

14. Solve the right triangle if ∠A = 37°, ∠C = 90°, and b = 52.8.

15. A water slide at an amusement park is 85 ft long and is inclined at an angle of 52° with the horizontal. How high is the top of the slide above the water level?

14 Oblique Triangles and Vectors

Physics is a mathematical science in that many of its concepts and principles have a mathematical basis. Motion is one such concept. The motion of an object can be described by words such as distance, speed, and acceleration. These quantities can be divided into two categories—vector quantities or scalar quantities.

In this chapter we will explore the concept of vectors which are quantities that are described by both a magnitude and a direction. Scalar quantities are described by their magnitude. We can describe a windy day in two different ways. We might say that the wind is blowing at 40 miles per hour. The local meteorologist might describe the wind as blowing out of the north at 40 miles per hour. In the first instance wind is described as a scalar quantity (40 miles per hour), whereas in the second instance the meteorologist describes wind as a vector quantity—using both wind speed (40 miles per hour) and direction (out of the north) in his description.

In addition to working with vectors in this chapter, we will also look at methods that can be used to solve oblique triangles (triangles that do not contain right angles). We begin by generalizing our definitions of the trigonometric ratios so that we are not restricted to working with only acute or right angles.

14.1   Trigonometric Functions of Any Angle

Section Objectives
- Define trigonometric ratios in terms of a point in the rectangular coordinate system
- Determine the sign of the trigonometric ratios in any quadrant
- Determine the reference angle for a given angle
- Determine the trigonometric values of angles in any quadrant
- Determine angle measure given the value of a trigonometric function

We begin our study of oblique triangles by positioning angles so that they can be discussed in terms of coordinate axes. *If the initial side of an angle extends from the origin to the right on the positive x axis in the rectangular coordinate system, the angle is said to be in* **standard position.** See Figure 14.1. The angle is then determined by the position of the terminal side. If the terminal side is in the first quadrant, the angle is called a "first-quadrant angle." Similar terms are used when the terminal side is in the other quadrants. *If the terminal side coincides with one of the axes, the angle is a* **quadrantal angle.**

The measure of an angle in standard position corresponds to the amount of rotation generated when the terminal side is rotated from the initial side. *If the terminal side is rotated in a counterclockwise direction, the angle is said to be* **positive.** *If the rotation is clockwise, the angle is* **negative,** *as in Figure 14.1.*

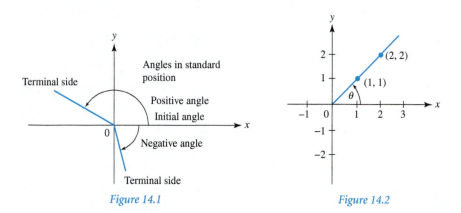

Figure 14.1   *Figure 14.2*

Example

1

An angle of 150° in standard position (see Figure 14.1) has its terminal side in the second quadrant so that 150° *is a second-quadrant angle.* The terminal side is rotated 150° in the counterclockwise direction as shown.

Example

2

See Figure 14.1. An angle of −60° in standard position has its terminal side in the fourth quadrant. The terminal side is rotated 60° in the *clockwise* direction as shown.

When an angle is in standard position, *the terminal side is determined if we know any point, other than the origin, on the terminal side.*

Example

3

See angle θ in Figure 14.2. The angle θ is in standard position and the terminal side passes through the points (1, 1) and (2, 2) as well as infinitely many other points. Knowing that the terminal side passes through any one of these points makes it possible to determine the terminal side.

If we were to stipulate for an angle β that the terminal side passes through (0, 3), we would know that β is one example of a quadrantal angle. Because the terminal side would coincide with one of the axes (the positive y axis in this case), β is a quadrantal angle.

Using angles that are in standard position, we can present more general definitions that are consistent with the definitions of the trigonometric ratios already presented in Chapter 13. Refer to Figure 14.3, where we show three important distances for a given point on the terminal side of the angle. *They are the* **abscissa** *(x coordinate), the* **ordinate** *(y coordinate), and the* **radius vector.** *The radius vector has a distance that extends from the origin to the point.* The radius vector is designated as r, as shown in Figure 14.3.

Using the abscissa (x coordinate), the ordinate (y coordinate), and radius vector $r = \sqrt{x^2 + y^2}$, we define the trigonometric functions as follows.

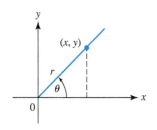

Figure 14.3

$$\text{sine } \theta = \frac{y}{r} \qquad \text{cosine } \theta = \frac{x}{r} \qquad \text{tangent } \theta = \frac{y}{x}$$

$$\text{cosecant } \theta = \frac{r}{y} \qquad \text{secant } \theta = \frac{r}{x} \qquad \text{cotangent } \theta = \frac{x}{y}$$

These relationships hold for any point on a given terminal side. The ratios will be different for a different angle having a different terminal side.

The trigonometric *ratios,* defined only in terms of the sides of a right triangle in Chapter 13 are special cases of these more general trigonometric *functions.* **These functions depend on the size of an angle in standard position and can be evaluated for angles of any magnitude** and with a terminal side in any quadrant.

As with the trigonometric ratios, the names of the various functions are abbreviated for convenience. These abbreviations are $\sin\theta$, $\cos\theta$, $\tan\theta$, $\cot\theta$, $\sec\theta$, and $\csc\theta$. Note that a given function is not defined if the denominator is zero. This occurs for $\tan\theta$ and $\sec\theta$ when $x = 0$ and for $\cot\theta$ and $\csc\theta$ when $y = 0$. In all cases **we assume that $r > 0$.** If $r = 0$, there would be no terminal side and therefore no angle.

Example 4

Determine the values of the trigonometric functions of an angle with a terminal side passing through the point $(3, 4)$.

By placing the angle in standard position, as shown in Figure 14.4, and drawing the terminal side through $(3, 4)$, we see that r is found by applying the Pythagorean theorem. Thus,

$$r = \sqrt{3^2 + 4^2} = 5$$

Using the values $x = 3, y = 4,$ and $r = 5$, we find the following values:

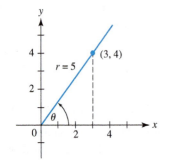

Figure 14.4

$$\sin\theta = \frac{\overset{y}{\downarrow}}{\underset{r}{\uparrow}5}\frac{4}{} \qquad \cos\theta = \frac{\overset{x}{\downarrow}}{\underset{r}{\uparrow}5}\frac{3}{} \qquad \tan\theta = \frac{\overset{y}{\downarrow}}{\underset{x}{\uparrow}3}\frac{4}{}$$

$$\csc\theta = \frac{5}{4} \qquad \sec\theta = \frac{5}{3} \qquad \cot\theta = \frac{3}{4}$$

Example 5

Determine the values of the trigonometric functions of the angle with a terminal side that passes through the point $(-1.50, 3.50)$. See Figure 14.5.

From the Pythagorean theorem, we find that

$$r = \sqrt{x^2 + y^2} = \sqrt{(-1.50)^2 + (3.50)^2} = \sqrt{2.25 + 12.25}$$
$$= \sqrt{14.50} = 3.81$$

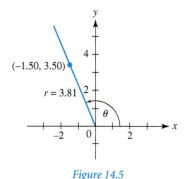

Figure 14.5

Using the values $x = -1.50$, $y = 3.50$, and $r = 3.81$, we have the following results for the trigonometric functions.

$$\sin\theta = \frac{3.50}{3.81} = 0.919 \qquad \cos\theta = \frac{-1.50}{3.81} = -0.394$$

$$\tan\theta = \frac{3.50}{-1.50} = -2.33 \qquad \cot\theta = \frac{-1.50}{3.50} = -0.429$$

$$\sec\theta = \frac{3.81}{-1.50} = -2.54 \qquad \csc\theta = \frac{3.81}{3.50} = 1.09$$

From Example 5 we see that **the value of a trigonometric function can sometimes be negative.** It is important that we determine the correct sign of each function, and that depends on the location of the terminal side of the angle. We will now determine the sign of each function in each quadrant.

The sign of $\sin\theta$ depends on the sign of the y coordinate of the point on the terminal side. This is because $\sin\theta = y/r$ and r is *positive*. We know that y is positive if the point defining the terminal side is above the x axis and that y is negative if this point is below the x axis. As a result, if the terminal side of the angle is in the first or second quadrant, the value of $\sin\theta$ will be positive, but if the terminal side is in the third or fourth quadrant, $\sin\theta$ is *negative*.

Now Try It! 1

If the terminal side of angle θ passes through the point $(8, -15)$, find the values of the six trigonometric ratios in fraction form.

	Quadrant			
	I	II	III	IV
$\sin\theta$	+	+	−	−

Example 6

The value of $\sin 20°$ is positive, because the terminal side of $20°$ is in the first quadrant. The value of $\sin(-200°)$ is positive, because the terminal side of $-200°$ is in the second quadrant. The values of $\sin 200°$ and $\sin 340°$ are negative, because the terminal sides of these angles are in the third and fourth quadrants, respectively.

	Quadrant			
	I	II	III	IV
$\cos\theta$	+	−	−	+

Because $\cos\theta = x/r$, the sign of $\cos\theta$ depends upon the sign of x. Because x is positive in the first and fourth quadrants, $\cos\theta$ is *positive* in these quadrants. In the same way, $\cos\theta$ is *negative* in the second and third quadrants.

Example 7

The values of $\cos 20°$ and $\cos(-30°)$ are positive, because these angles are first- and fourth-quadrant angles, respectively. The values of $\cos 160°$ and $\cos 200°$ are negative, because these angles are second- and third-quadrant angles, respectively.

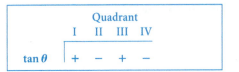

Quadrant			
I	II	III	IV
tan θ +	–	+	–

The sign of tan θ *depends upon the ratio of y to x*. In the first quadrant both *x* and *y* are positive, and therefore the ratio *y/x* is positive. In the third quadrant both *x* and *y* are negative, and therefore the ratio *y/x* is *positive*. In the second the fourth quadrants either *x* or *y* is positive and the other is negative, and so the ratio of *y/x* is *negative*.

Example 8

The values of tan 20° and tan 200° are positive, because the terminal sides of these angles are in the first and third quadrants, respectively. The values of tan 160° and tan (−20°) are negative, because the terminal sides of these angles are in the second and fourth quadrants, respectively.

Because csc θ is defined in terms of *r* and *y*, as is sin θ, the sign of csc θ is the same as that of sin θ. For the same reason, cot θ has the same sign as tan θ, and sec θ has the same sign as cos θ. A method for remembering the signs of the functions in the four quadrants is as follows:

Positive functions

Figure 14.6

All functions of first-quadrant angles are positive. The sin θ *and* csc θ *are positive for second-quadrant angles. The* tan θ *and* cot θ *are positive for third-quadrant angles. The* cos θ *and* sec θ *are positive for fourth-quadrant angles. All others are negative.* See Figure 14.6.

Example 9

sin 50°, sin 150°, sin (−200°), cos 300°, cos (−40°), tan 220°, tan (−100°), cot 260°, cot (−310°), sec 280°, sec (−37°), csc 140°, and csc (−190°) are all positive.

sin 190°, sin 325°, cos 100°, cos (−95°), tan 172°, tan 295°, cot 105°, cot (−6°), sec 135°, sec (−135°), csc 240°, and csc 355° are all negative.

This might help:

Here is a common mnemonic used to recall the sign of the trigonometric functions in each quadrant:

All Students Take Calculus

A tells us that **all** trig functions are positive in the first quadrant.
S tells us that **sine** is positive in the second quadrant.
T tells us that **tangent** is positive in the third quadrant.
C tells us that **cosine** is positive in the fourth quadrant.

Two angles in standard position are **coterminal** *if their terminal sides are the same.* It is a property of all angles that *any angle in standard position is coterminal with an angle between* 0° *and* 360°. Because the terminal sides of coterminal angles are the same, *the trigonometric functions of coterminal angles are the same.* The problem of finding the value of the trigonometric function of any angle can therefore be reduced to angles between 0° and 360°.

Example 10

The following pairs of angles are coterminal:

10°   and   370°   See Fig. 14.7a
90°   and   450°

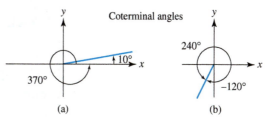

Figure 14.7

See Fig. 14.7b

Now Try It! 2

Determine the sign of the following trigonometric expressions:

1. tan 275°
2. sin 102°
3. cos 71°
4. csc 200°
5. sec 315°
6. cot 185°

$$350° \quad \text{and} \quad -10°$$
$$-120° \quad \text{and} \quad 240°$$

Because the coterminal angles share the same terminal side, we can conclude that for two coterminal angles, the values of the trigonometric functions are equal. For example, the following equations are valid:

$$\sin 10° = \sin 370°$$
$$\cos 90° = \cos 450°$$
$$\sec 350° = \sec(-10°)$$
$$\tan(-120°) = \tan 240°$$

When attempting to find trigonometric functions of an angle whose terminal side is not in the first quadrant, we will use a **reference angle** denoted by α. In general, *the reference angle α for the given angle θ is the positive acute angle formed by the x axis and the terminal side of θ.* From Figure 14.8 we see that if the given angle θ is in the second quadrant, then $\alpha = 180° - \theta$. Figure 14.9 suggests that for an angle in the third quadrant, the relationship between α and θ is $\alpha = \theta - 180°$. Figure 14.10 shows that when θ is in the fourth quadrant, the reference angle α can be found from $\alpha = 360° - \theta$.

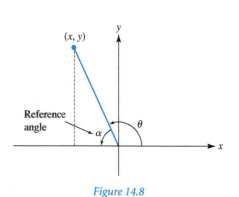

Figure 14.8

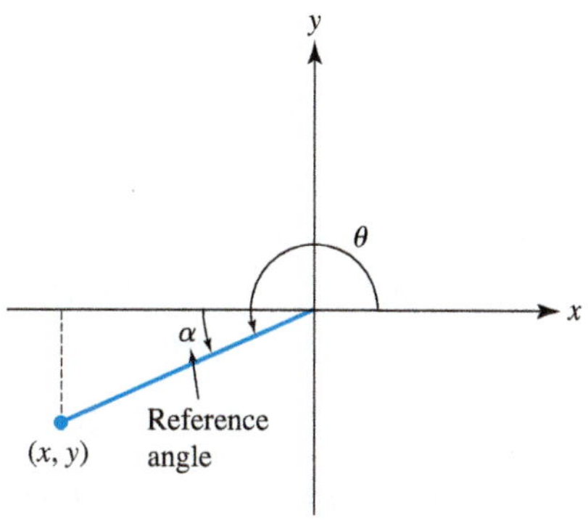

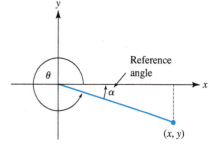

Figure 14.9

Figure 14.10

If θ is in the ...	then the reference angle α can be found from the equation ...
second quadrant third quadrant fourth quadrant	$\alpha = 180° - \theta$ $\alpha = \theta - 180°$ $\alpha = 360° - \theta$

The reason for using reference angles is that the trigonometric function of an angle θ and its reference angle α are *numerically* the same and might differ only in sign.

Example 11

The angle $\theta = 126.9°$ is in the second quadrant and the terminal side passes through the point $(-3, 4)$ as shown in Figure 14.11. The reference angle is $\alpha = 180° - 126.9° = 53.1°$ and we can see from the figure that the x, y, and r values of $(-3, 4)$ and $(3, 4)$ differ only in sign. We know that the trigonometric functions depend only on the values of x, y, and r. It therefore follows that ▶ *the trigonometric functions of 126.9° and 53.1° may differ only in sign.* In the second quadrant, the sine and cosecant are positive whereas the other four functions are negative. Therefore,

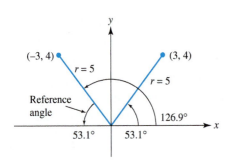

Figure 14.11

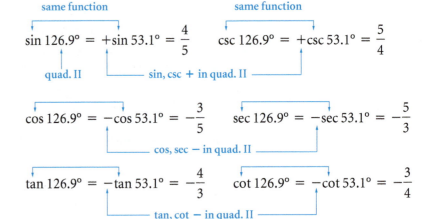

$$\sin 126.9° = +\sin 53.1° = \frac{4}{5} \qquad \csc 126.9° = +\csc 53.1° = \frac{5}{4}$$

quad. II — sin, csc + in quad. II

$$\cos 126.9° = -\cos 53.1° = -\frac{3}{5} \qquad \sec 126.9° = -\sec 53.1° = -\frac{5}{3}$$

cos, sec − in quad. II

$$\tan 126.9° = -\tan 53.1° = -\frac{4}{3} \qquad \cot 126.9° = -\cot 53.1° = -\frac{3}{4}$$

tan, cot − in quad. II

Example 12

An angle of $250.0°$ is a third-quadrant angle. We therefore can find the reference angle, as shown in Figure 14.12.

$$\alpha = 250.0° - 180° = 70.0°$$

We can now find the numerical values of the trigonometric functions at 250.0° by first finding the values of the same functions of 70.0°. We then adjust the signs to

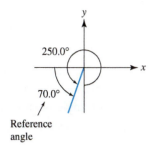

Figure 14.12

make use of the fact that in the third quadrant the tangent and cotangent are positive whereas the other four trigonometric functions are negative. We therefore get

same function   reference angle

$$\sin 250.0° = -\sin 70.0° = -0.940$$
$$\cos 250.0° = -\cos 70.0° = -0.342$$
$$\tan 250.0° = +\tan 70.0° = 2.75$$
$$\cot 250.0° = +\cot 70.0° = 0.364$$
$$\sec 250.0° = -\sec 70.0° = -2.92$$
$$\csc 250.0° = -\csc 70.0° = -1.06$$

quadrant III   proper sign for quadrant III

Example 13

An angle of 318.5° is a fourth-quadrant angle. We therefore can find the reference angle, as shown in Figure 14.13.

$$\alpha = 360° - 318.5° = 41.5°$$

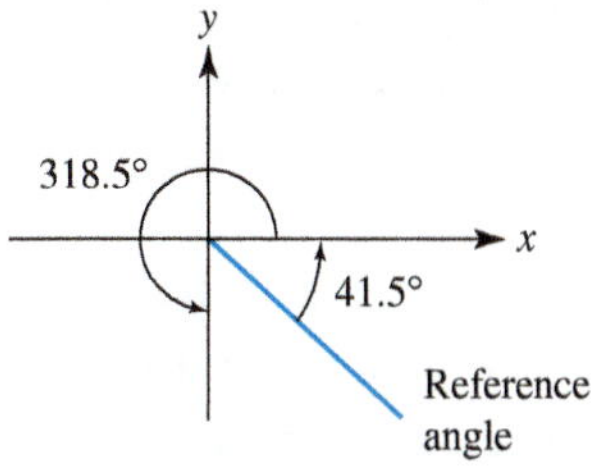

In the fourth quadrant, the cosine and secant are positive and the other functions are negative. Therefore,

$$\sin 318.5° = -\sin 41.5° = -0.663$$
$$\cos 318.5° = +\cos 41.5° = 0.749$$
$$\tan 318.5° = -\tan 41.5° = -0.885$$
$$\cot 318.5° = -\cot 41.5° = -1.13$$
$$\sec 318.5° = +\sec 41.5° = 1.34$$
$$\csc 318.5° = -\csc 41.5° = -1.51$$

quadrant IV   proper sign for quadrant IV

Figure 14.13

We now demonstrate how we find θ when the value of a trigonometric function of θ is known. Because the angles of primary importance are those from 0° to 360°, and because each function is positive in two quadrants and negative in the other two quadrants, *we must take care to obtain* both *values of θ for which $0° < \theta < 360°$.*

Example 14

Given that $\sin \theta = 0.342$, find θ for $0° < \theta < 360°$.

Here we are to find any angles between 0° and 360° for which $\sin \theta = 0.342$. Because $\sin \theta$ is positive for first- and second-quadrant angles, *there are two such angles.*

Using a calculator, we press the key sequence

2nd SIN 0.342 ENTER

and the display of 19.998772 is rounded to 20.0 so that one angle is 20.0°. To find the second-quadrant angle for which $\sin \theta = 0.342$, we note that $\sin (180° - 20.0°) = \sin 160°$. Thus the second-quadrant angle is 160.0°. The values of θ are therefore 20.0° and 160.0°. We note that *the calculator gives only one of the two answers.*

In finding an angle when the value of a trigonometric function is given, we may also know certain additional information. If this information is such that θ is restricted to a particular quadrant, we must be careful to obtain the proper value. This is illustrated in Example 15.

Example
15
Given that $\tan \theta = -1.505$ and $\cos \theta > 0$, find θ for $0° < \theta < 360°$.

Because $\tan \theta$ is negative in the second and fourth quadrants and $\cos \theta$ is positive in the first and fourth quadrants, θ *must be a fourth-quadrant angle.* (That is, where both conditions are met.) Using the calculator sequence

2nd TAN -1.505 ENTER

we get the negative result of $-56.4°$ (rounded). This is a fourth-quadrant angle expressed as a negative angle. The reference angle is therefore 56.4°, and we can express θ as a fourth-quadrant angle between 0° and 360° as $\theta = 360° - 56.4° = 303.6°$.

As we saw in Examples 14 and 15, if we are given the value of a trigonometric function and wish to find the corresponding angles between 0° and 360°, we must be careful when using a calculator. There may be *two solutions* (as in Example 14), or *the solution may not be in proper form* (as in Example 15). Therefore, we must recognize that **blind use of the calculator can lead to an incomplete or incorrect solution.**

Example
16
A pilot flies a triangular route connecting three airports. The distance d (in miles) between two of the airports can be found by solving the equation

$$\frac{134}{\sin 147.0°} = \frac{d}{\sin 12.2°}$$

Multiplying both sides of this equation by sin 12.2°, we get

$$d = \frac{134 \sin 12.2°}{\sin 147.0°} = \frac{134(0.2113)}{0.5446}$$
$$= 52.0$$

We now know that $d = 52.0$ mi.

14.1 Exercises

In Exercises 1 through 4, draw the given angles in standard position.

1. 30°, 135°

2. 100°, 240°

3. 200°, −60°

4. −30°, 400°

In Exercises 5 through 12, assume that the terminal side of θ passes through the given point and find the values of the given trigonometric ratios in fractional form.

5. (5, 12); $\sin \theta, \tan \theta$

6. $(\sqrt{3}, 1)$; $\cos \theta, \cot \theta$

7. (−3, 2); $\sec \theta, \sin \theta$

8. (−8, 15); $\tan \theta, \csc \theta$

9. (−1, −1); $\cot \theta, \cos \theta$

10. (−3, −4); $\sin \theta, \sec \theta$

11. (6, −8); $\cos \theta, \sin \theta$

12. (5, −6); $\csc \theta, \cos \theta$

In Exercises 13 through 20, assume that the terminal side of θ passes through the given point and find the values of the given trigonometric ratios in decimal form.

13. (1.00, 3.00); $\cos \theta, \cot \theta$

14. (2.50, 1.30); $\sin \theta, \tan \theta$

15. (−15.0, 6.20); $\tan \theta, \csc \theta$

16. (−2.30, 7.40); $\sec \theta, \sin \theta$

17. (−140, −170); $\sin \theta, \sec \theta$

18. (−0.175, −1.05); $\cot \theta, \csc \theta$

19. (27.3, −17.5); $\csc \theta, \cos \theta$

20. (1.75, −7.50); $\cos \theta, \sin \theta$

In Exercises 21 through 28, determine the algebraic sign of the given trigonometric functions.

21. sin 45°, cos 130°, tan 350°

22. sin (−25°), csc 120°, tan 200°

23. cos 250°, sec 160°, cot (−10°)

24. sin 182°, csc (−12°), cos 95°

25. tan 260°, csc 72°, cos 380°

26. cot (−93°), sec 295°, tan 110°

27. sin 718°, cot (−570°), sec 520°

28. cos 212°, tan 275°, sin (−380°)

In Exercises 29 through 36, determine the quadrant in which the terminal side of θ lies, subject to the given conditions.

29. sin θ is negative, cos θ is positive

30. sin θ is negative, tan θ is negative

31. cos θ is positive, cot θ is positive

32. cos θ is positive, csc θ is negative

33. csc θ is positive, cot θ is negative

34. csc θ is negative, tan θ is negative

35. cot θ is negative, sec θ is negative

36. cot θ is positive, csc θ is negative

In Exercises 37 through 44, use the SIN *and* COS *keys on a calculator to determine the quadrant in which the terminal side of the given angle θ lies. The quadrant can be determined by using the calculator to find the signs of* sin θ *and* cos θ.

37. $\theta = 852°$

38. $\theta = 963°$

39. $\theta = -986°$

40. $\theta = -1265°$

41. $\theta = 1537°$

42. $\theta = 1506°$

43. $\theta = -2070°$

44. $\theta = -2280°$

In Exercises 45 through 52, express the given trigonometric functions in terms of the same function of a positive acute angle.

45. sin 165°, cos 230°

46. tan 95°, sec 305°

47. cos 207°, csc 290°

48. sec 100°, cot 218°

49. tan 342°, sec $(-10°)$

50. cot 104°, sin $(-104°)$

51. cot 650°, tan $(-300°)$

52. sin 760°, csc $(-210°)$

In Exercises 53 through 72, determine the values of the given trigonometric functions.

53. sin 213.0°

54. cos 307.0°

55. tan 275.3°

56. cot 97.2°

57. sec 156.5°

58. csc 108.4°

59. cos 202.1°

60. sin 291.7°

61. cot 195.0°

62. tan 114.3°

63. csc 215.6°

64. sec 347.1°

65. sin 102.6°

66. cos 171.4°

67. tan 250.2°

68. cot 322.8°

69. sin $(-32.4°)$

70. cos $(-67.8°)$

71. tan $(-112.5°)$

72. sec $(-215.9°)$

In Exercises 73 through 88, find θ to the nearest 0.1° for $0° < \theta < 360°$. (In each case there are two values of θ.)

73. sin $\theta = -0.8480$

74. cot $\theta = -0.2126$

75. cos $\theta = 0.4003$

76. tan $\theta = -1.830$

77. cot $\theta = 0.5265$

78. sin $\theta = 0.6374$

79. tan $\theta = 0.2833$

80. cos $\theta = -0.9287$

81. sin $\theta = -0.7880$

82. csc $\theta = 1.580$

83. csc $\theta = -1.580$

84. sec $\theta = 6.188$

85. sec $\theta = -1.023$

86. cot $\theta = 0.3779$

87. cos $\theta = -0.9994$

88. tan $\theta = -20.4$

In Exercises 89 through 96, find θ to the nearest 0.1° for $0° < \theta < 360°$, subject to the given conditions.

89. sin $\theta = -0.4384$, cos $\theta > 0$

90. cos $\theta = 0.4083$, tan $\theta < 0$

91. tan $\theta = -1.200$, csc $\theta > 0$

92. sec $\theta = 1.526$, sin $\theta < 0$

93. csc $\theta = 1.150$, tan $\theta < 0$

94. cot $\theta = -1.540$, sec $\theta > 0$

95. cos $\theta = -0.9870$, csc $\theta < 0$

96. sin $\theta = 0.9860$, sec $\theta < 0$

In Exercises 97 through 104, solve the given problems.

97. When a weather balloon reaches an altitude of 1250 ft, it is 105 ft downwind from its departure point. The angle θ its path makes with the ground is found by solving the equation

$$\tan \theta = \frac{1250}{105}$$

Find the acute angle θ.

98. A certain alternating-current voltage can be found from the equation

$$V = 170 \cos 725.0°$$

Find the voltage V.

99. Under specified conditions, a force F (in pounds) is determined by solving the equation

$$\frac{F}{\sin 125.0°} = \frac{250}{\sin 35.4°}$$

Find the magnitude of the force.

100. A brace is used to support a frame so that cement can be poured. The angle θ between the brace and the form can be found by solving the equation

$$\frac{4.27}{\sin \theta} = \frac{6.35}{\sin 78.5°}$$

Solve for θ assuming that it is an acute angle.

101. A formula for finding the area of a triangle, knowing sides a and b and angle C, is

$$A = \frac{1}{2} ab \sin C$$

Find the area of a triangle for which $a = 27.3$, $b = 35.2$, and $C = 136.0°$. See Figure 14.14

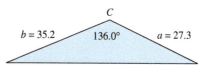

Figure 14.14

102. In calculating the area of a triangular tract of land, a surveyor uses the formula in Exercise 101. He inserts the values $a = 510$ m, $b = 345$ m, and $C = 125.0°$. Find the area of the tract of land.

103. A highway construction engineer uses the formula

$$\tan \theta = \frac{V^2}{k}$$

to calculate the angle at which a curved section of road should be banked. Find θ if the maximum speed is $V = 60$ mi/h and $k = 86,400$ mi^2/h^2.

104. A ray of light enters water and is refracted as shown in Figure 14.15. In this case, the angle of the refracted beam θ can be found by using the index of refraction

$$n = \frac{\sin 45°}{\sin \theta}$$

Find θ if n is 1.33 for water.

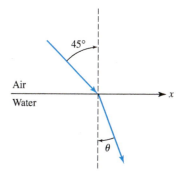

Figure 14.15

Answers

Now Try It! 1

1. $\sin \theta = \dfrac{-15}{17}$ **2.** $\cos \theta = \dfrac{8}{17}$ **3.** $\tan \theta = \dfrac{-15}{8}$ **4.** $\csc \theta = \dfrac{17}{-15}$ **5.** $\sec \theta = \dfrac{17}{8}$ **6.** $\cot \theta = \dfrac{8}{-15}$

Now Try It! 2

1. negative   **2.** positive   **3.** positive   **4.** negative   **5.** positive   **6.** positive

14.2 The Law of Sines

Section Objectives
- Find the missing parts of an oblique triangle using the Law of Sines
- Recognize the ambiguous case when working with the Law of Sines
- Solve application problems involving oblique triangles

Until now we have limited our study of triangle solutions to right triangles. However, many triangles that require solutions do not contain a right angle. As we noted earlier, such a triangle is termed an *oblique triangle*. In this and the following sections we learn methods for solving oblique triangles.

There are several ways in which oblique triangles can be solved, but we shall restrict our attention to the two most useful methods, the **Law of Sines** and the **Law of Cosines.** In this section we discuss the Law of Sines.

Let ABC be an oblique triangle with sides a, b, and c opposite angles A, B, and C, respectively. By drawing a perpendicular h from B to side b, we see from Figure 14.16 that $\sin A = h/c$ and $\sin C = h/a$. Solving for h, we have $h = c \sin A$ and $h = a \sin C$. Therefore, $c \sin A = a \sin C$, or

$$\frac{a}{\sin A} = \frac{c}{\sin C}$$

By drawing a perpendicular from A to a we also find that $c \sin B = b \sin C$, or

$$\frac{b}{\sin B} = \frac{c}{\sin C}$$

Combining these results, we have the **Law of Sines:**

$$\boxed{\frac{a}{\sin A} = \frac{b}{\sin B} = \frac{c}{\sin C}}$$

The Law of Sines is a statement of proportionality between the sides of a triangle and the sines of the angles opposite them. We should note that *there are three actual equations combined.* Also, we might note that the same result will be obtained if we choose another vertex in the derivation so that the Law of Sines is valid for all triangles.

Now we can see how the Law of Sines is applied to the solution of a triangle in which two angles and one side are known. If two angles are known, the third can be found from the fact that the sum of the angles in a triangle is 180°. At this point we must be able to *determine the ratio between the given side and the sine of the angle opposite it.* Then, by use of the Law of Sines, we can find the other sides.

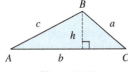

Figure 14.16

Example 1

Given that $c = 22.1$, $A = 37.4°$, and $B = 75.1°$, find a, b, and C (see Figure 14.17).

First, we can see that

$$C = 180° - (37.4° + 75.1°) = 180° - 112.5°$$
$$= 67.5°$$

We now know side c and angle C. ***Using the equation relating a, A, c, and C,*** we have

$$\frac{a}{\sin 37.4°} = \frac{22.1}{\sin 67.5°} \quad c$$
$$\quad\quad A \quad\quad C$$

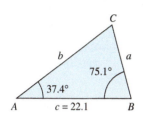

Figure 14.17

or

$$a = \frac{22.1 \sin 37.4°}{\sin 67.5°} = 14.5 \quad \text{round to three significant digits}$$

The calculator key sequence for this computation is as follows.

22.1 × SIN 37.4 ÷ SIN 67.5 ENTER gives us 14.528957

Now *using the equation relating b, B, c, and C,* we have

$$\frac{b}{\sin 75.1°} = \frac{22.1}{\sin 67.5°} \quad c$$
$$\quad\quad B \quad\quad C$$

or

$$b = \frac{22.1 \sin 75.1°}{\sin 67.5°} = 23.1$$

The calculator sequence is the same as that for a if 37.4 is replaced by 75.1. We now know that $a = 14.5$, $b = 23.1$, and $C = 67.5°$.

In Example 2 we illustrate the use of the Law of Sines in a situation where we know two sides and the angle opposite one of them.

Example 2

Electric supply lines must be built so that points A and B are connected, but a direct path cuts through a marsh. How much longer is the indirect path from point A to point C and then to point B, as shown in Figure 14.18?

From Figure 14.18 we can see that the indirect path from A to C to B is a distance of 197.0 m + 302.9 m = 499.9 m. We must now find the length of side c

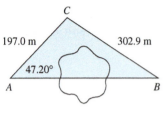

Figure 14.18

so that we can determine the amount by which 499.9 m exceeds the length of the direct path represented by side c.

We will first use the Law of Sines to find angle B, which in turn will allow us to find angle C. We will then use the Law of Sines a second time so that the length of side c can be determined. We begin by finding angle B and then angle C.

$$\frac{a}{\sin A} = \frac{b}{\sin B} \quad \text{becomes} \quad \frac{302.9}{\sin 47.20°} = \frac{197.0}{\sin B}$$

which leads to

$$\sin B = \frac{197.0 \sin 47.20°}{302.9} = 0.4772$$

With $\sin B = 0.4772$, we determine that $B = 28.50°$. (We might also get $B = 151.50°$, but we discard that possibility because the sum of angles A and B would then exceed 180°.) With $A = 47.20°$ and $B = 28.50°$, we conclude that $C = 104.30°$ because the sum of angles A, B, and C must be 180°.

We now use the Law of Sines a second time so that the length of side c can be determined.

$$\frac{a}{\sin A} = \frac{c}{\sin C} \quad \text{becomes} \quad \frac{302.9}{\sin 47.20°} = \frac{c}{\sin 140.30°}$$

which leads to

$$c = \frac{302.9 \sin 104.30°}{\sin 47.20°} = 400.0 \text{ m}$$

The indirect path is 499.9 m whereas the direct path is 400.0 m. The indirect path is therefore 99.9 m longer than the direct path.

In Example 2 we were able to solve the triangle by using the Law of Sines twice. Our ultimate goal was to find the length of side c, but it was necessary to first find angles B and C. Example 3 illustrates another use of the Law of Sines in solving an applied problem.

Example 3

A microwave relay tower 10.2 m tall is placed on the top of a building as shown in Figure 14.19. From point D, the angles of elevation to the top and bottom of the tower are measured as 50.0° and 45.0°, respectively. Find the height of the building.

Triangle BCD is a right triangle with angles of 45.0°, 45.0°, and 90.0°. We want to find the length of side BC and this can be easily computed if we could

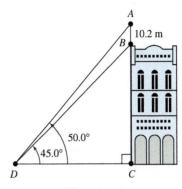

Figure 14.19

determine the length of the hypotenuse *BD*. To find the length of *BD* we will consider triangle *ABD* and use the Law of Sines. It is easy to see that $\angle ADB = 5.0°$. Also, $\angle DBC = 45.0°$ (since the angles of $\triangle BCD$ must yield a total of 180°) which implies that $\angle DBA = 135.0°$ (since $\angle DBA$ and $\angle DBC$ are supplementary). Thus, $\angle DAB = 180° - (135.0° + 5.0°) = 40.0°$ (since the angles of $\triangle ABD$ must yield a total of 180°). Using the Law of Sines in $\triangle ABD$ we get

side opposite

$$\frac{DB}{\sin \angle DAB} = \frac{AB}{\sin \angle ADB}$$

angle

or

$$\frac{DB}{\sin 40.0°} = \frac{10.2}{\sin 5.0°}$$

or

$$DB = \frac{10.2 \sin 40.0°}{\sin 5.0°}$$

so that

$$DB = 75.2 \text{ m}$$

Now that we know the length of side *DB*, we consider $\triangle BCD$, which is a right triangle, and we get

$$\sin 45.0° = \frac{BC}{75.2}$$

$$BC = (75.2)(\sin 45.0°) = 53.2 \text{ m}$$

The building is 53.2 m tall.

We now focus on the situation in which we know two sides and the angle opposite one of them. This is referred to as the **ambiguous case** because, depending on the particular values given, ***there may be one solution, two solutions, or no solution.*** Example 4 illustrates these possibilities.

Example
4

(a) In Figure 14.20a, we are given $a = 1$, $b = 4$, and $\angle A = 30°$. Using the Law of Sines we get

$$\frac{a}{\sin A} = \frac{b}{\sin B}$$

$$\frac{1}{\frac{1}{2}} = \frac{4}{\sin B}$$

$$\sin B = 2$$

However, this is impossible because $\sin B$ cannot exceed 1. We conclude that ***there is* no solution.**

(b) In Figure 14.20b, we are given $a = 2$, $b = 4$, and $\angle A = 30°$. Using the Law of Sines we get

$$\frac{a}{\sin A} = \frac{b}{\sin B}$$

$$\frac{2}{\frac{1}{2}} = \frac{4}{\sin B}$$

$$\sin B = 1$$

We conclude that $\angle B = 90°$ and there is **one solution.**

(c) In Figure 14.20c, we are given $a = 3$, $b = 4$, and $\angle A = 30°$. Using the Law of Sines we get

$$\frac{a}{\sin A} = \frac{b}{\sin B}$$

$$\frac{3}{\frac{1}{2}} = \frac{4}{\sin B}$$

$$\sin B = \frac{2}{3}$$

This result means that $\angle B = 41.8°$ or $138.2°$. In this case we have **two solutions.** They are represented by the two possible positions of side a in Figure 14.20c.

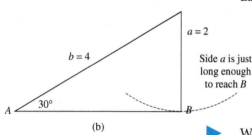

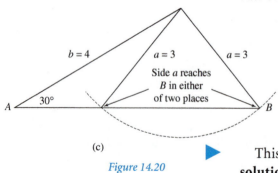

Figure 14.20

In Example 4 it would be easy to miss the second solution for the case that led to two solutions. Care must be taken to recognize this ambiguous case so that both solutions are detected. Also, it is possible to have one solution in the ambiguous case with a triangle that is not a right triangle as in Figure 14.20(b). In Example 4, for any value of a greater than 4, only one solution exists. In general, if the side opposite the given angle is longer than the other given side, then only one solution exists. In Figure 14.21 we illustrate the case for $a = 5$, $b = 4$, and $\angle A = 30°$.

Note that in this case, the side $a = 5$ opposite the given angle of $A = 30°$ is longer than the other given side of $b = 4$, so only one solution exists. That solution is the triangle shown in Figure 14.21. The dashed arc is intended to illustrate that no other angle B will produce a triangle.

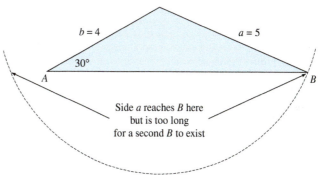

Figure 14.21

We have seen that solving triangles in the ambiguous case (case 2) can lead to two solutions, one solution, or no solution. Because it is so easy to miss a second solution when two solutions exist, we might consider the following. *When given two sides and the angle opposite one of them, there are two triangle solutions if the following conditions are all met.*

1. The given angle is acute.

2. The given angle's opposite side is shorter than the adjacent side.

3. The sine of the second angle is calculated to be a value less than one.

A scale drawing of the given data will also help to identify the correct number of solutions.

14.2 **Exercises**

In Exercises 1 through 24, solve the triangles with the given parts by use of the Law of Sines.

1. $a = 7.16, A = 70.8°, B = 36.6°$

2. $b = 36.8, B = 31.5°, C = 43.5°$

3. $c = 2190, A = 37.0°, B = 34.0°$

4. $a = 4.66, B = 17.9°, C = 82.6°$

5. $b = 155, c = 72.8, B = 20.7°$

6. $a = 45.0, c = 75.0, C = 76.4°$

7. $a = 0.926, b = 0.228, A = 143.2°$

8. $a = 232, b = 557, B = 112.8°$

9. $a = 63.8, B = 58.4°, C = 22.2°$

10. $a = 13.0, A = 55.2°, B = 67.5°$

11. $b = 438, B = 47.4°, C = 64.5°$

12. $b = 283, B = 13.7°, C = 76.3°$

13. $a = 26.2, b = 22.2, B = 48.1°$

14. $a = 89.4, c = 37.3, C = 15.6°$

15. $b = 576, c = 730, B = 31.4°$

16. $a = 0.841, b = 0.965, A = 57.1°$

17. $a = 94.2, c = 68.0, A = 69.1°$

18. $a = 630, b = 670, B = 102.0°$

19. $a = 1.43, b = 4.21, A = 30.4°$

20. $b = 15.0, c = 55.0, B = 75.0°$

21. $a = 100, c = 200, A = 30.0°$

22. $b = 17.2, c = 23.7, C = 125.2°$

23. $b = 2.14, c = 6.73, B = 85.2°$

24. $b = 16.35, c = 10.13, C = 74.9°$

In Exercises 25 through 32, use the Law of Sines to solve the given problems.

25. Find the lengths of the supports of the shelf shown in Figure 14.22.

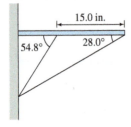

Figure 14.22

26. The front line of a triangular piece of land is 425 m long and makes angles of 75.0° and 62.5° with the other sides. What are the lengths of the other sides? See Figure 14.23.

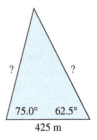

Figure 14.23

27. A rocket is fired at an angle of 42.0° with the ground from a point 2500 m directly behind an observer. Soon thereafter it is observed that the rocket is at an angle of elevation of 47.0° directly in front of the observer. How far on a direct line has the rocket moved?

28. The pilot of a light plane notes that the angles of depression of two objects directly ahead are 21.5° and 16.0°. The objects are known to be 1.25 mi apart. How far is the plane on a direct line from the nearer object?

29. A submarine leaves a port and travels due north. At a certain point it turns 45.0° to the left and travels an additional 75.0 km to a point 90.0 km from the port. How far from the port is the point where the submarine turned? See Figure 14.24.

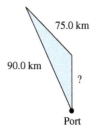

Figure 14.24

30. Two points A and B are located on opposite sides of a lake. A third point C is positioned so that $\angle B = 52.0°$. If BC is 162 ft and AC is 212 ft, what is the length of side AB? See Figure 14.25.

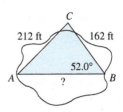

Figure 14.25

31. A communications satellite is directly above the extension of a line between receiving towers A and B. It is determined

from radio signals that the angle of elevation from tower A is 88.9° and the angle of elevation from tower B is 87.6°. If A and B are 658 km apart, how far is the satellite from A? (Ignore the curvature of the earth.) See Figure 14.26.

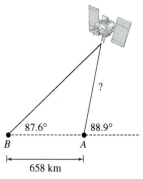

Figure 14.26

32. An airplane is sighted from points A and B on ground level. Points A and B are 357 ft apart. At point A, the angle of elevation to the airplane is 41.2°; for point B it is 46.9°. How high is the airplane? See Figure 14.27.

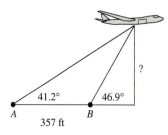

Figure 14.27

14.3 The Law of Cosines

Section Objectives
- Find the missing parts of an oblique triangle using the Law of Cosines
- Solve application problems involving oblique triangles

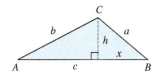

Figure 14.28

In this section we introduce the **Law of Cosines** as a means for solving the oblique triangles.

Consider any oblique triangle—for example, the one shown in Figure 14.28. In this triangle we see that $h/b = \sin A$ or that $h = b \sin A$. Also, $c - x = b \cos A$, or $x = c - b \cos A$. From the Pythagorean theorem, we see that $a^2 = h^2 + x^2$, or

$$a^2 = (b \sin A)^2 + (c - b \cos A)^2$$

Multiplying out, we get

$$a^2 = b^2 \sin^2 A + c^2 - 2bc \cos A + b^2 \cos^2 A$$
$$= b^2(\sin^2 A + \cos^2 A) + c^2 - 2bc \cos A$$

(Note that $\sin^2 A = (\sin A)^2$.) Recalling the definitions of the trigonometric functions, $\sin \theta = y/r$ and $\cos \theta = x/r$, which in turn means that $\sin^2 \theta + \cos^2 \theta = (x^2 + y^2)/r^2$. Since $x^2 + y^2 = r^2$, we then conclude that

$$\sin^2 \theta + \cos^2 \theta = 1$$

This equation holds for any angle θ, which means that it holds for angle A.

The above equation for a^2 can now be simplified to

$$a^2 = b^2 + c^2 - 2bc \cos A$$ Law of Cosines

This formula is known as the **Law of Cosines.**

Using the same method, we may also show that

$$b^2 = a^2 + c^2 - 2ac \cos B$$

and

$$c^2 = a^2 + b^2 - 2ab \cos C$$

These are simply alternative forms of the Law of Cosines. *All three forms express the same basic relationship among the sides of the triangle and the angle opposite one of the sides.*

If we know two sides and the included angle, we can directly solve **for the side** opposite the given angle. Then we normally would use the Law of Sines to complete the solution. If we are given all three sides, we can solve for the **angle opposite** one of them by the Law of Cosines. Again we usually use the **Law of Sines** to complete the solution.

Example 1

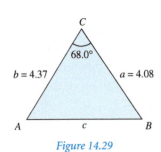

Figure 14.29

Solve the triangle with $a = 4.08$, $b = 4.37$, and $C = 68.0°$ (see Figure 14.29.)

Using $c^2 = a^2 + b^2 - 2ab \cos C$, with side c and angle C, we get

unknown side opposite known angle

$$c^2 = (4.08)^2 + (4.37)^2 - 2(4.08)(4.37) \cos 68.0°$$

known sides

$$c = \sqrt{(4.08)^2 + (4.37)^2 - 2(4.08)(4.37) \cos 68.0°} = 4.73$$

This type of computation can be easily performed on a calculator. We list below the key sequence that can be used to evaluate *c*. We should check **that our** calculator is in the degree mode before we begin.

2nd x^2 4.08 x^2 + 4.37 x^2 −2 × 4.08 × 4.37 × COS 68)) ENTER to get 4.7312925

From the Law of Sines, we now have

$$\frac{4.08}{\sin A} = \frac{4.37}{\sin B} = \frac{4.73}{\sin 68.0°}$$

sides opposite

angles

Solving for A, we obtain

$$\sin A = \frac{4.08 \sin 68.0°}{4.73}$$

from which we get

$$A = 53.1°$$

by using the calculator sequence

$$4.08 \times \text{SIN } 68 \div 4.73 \text{ ENTER to get } 53.108105$$

Now, since the sum of the three angles is 180°, we have $B = 58.9°$. We conclude that $A = 53.1°$, $B = 58.9°$, and $c = 4.73$.

Example 2

Solve the triangle with $b = 45.0$, $c = 62.5$, and $A = 126.3°$ (see Figure 14.30).

Using the form of the Law of Cosines with side a and angle A, we have the following. Since $A = 126.3°$, we note that $\cos A$ is negative.

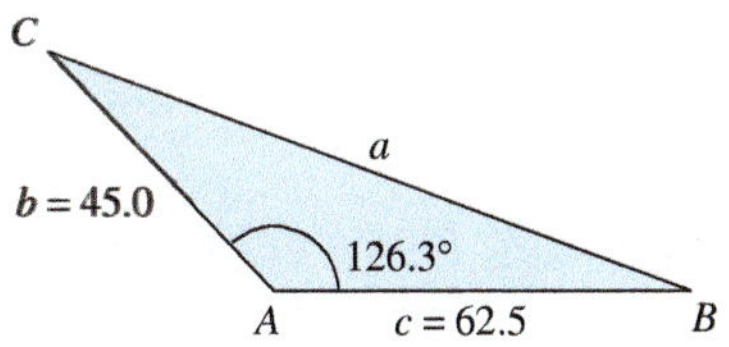

Figure 14.30

unknown side opposite known angle

$$a^2 = (45.0)^2 + (62.5)^2 - 2(45.0)(62.5) \cos 126.3°$$

$$a = \sqrt{(45.0)^2 + (62.5)^2 - 2(45.0)(62.5) \cos 126.3°} = 96.2$$

From the Law of Sines, we have

$$\frac{96.2}{\sin 126.3°} = \frac{45.0}{\sin B}$$

Solving for B we get

$$\sin B = \frac{45.0 \sin 126.3°}{96.2}$$

which gives us

$$B = 22.1°$$

Finally, $C = 180° - (126.3° + 22.1°) = 31.6°$. Therefore $a = 96.2$, $B = 22.1°$, and $C = 31.6°$.

Example 3

A survey team locates three benchmarks. They are equipped with a tape for measuring distances, but they don't have any instruments for measuring angles. Solve the triangle for which $a = 83.8$ ft, $b = 36.7$ ft, and $c = 72.4$ ft. See Figure 14.31.

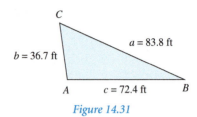

Figure 14.31

We can use any of the forms of the Law of Cosines to *find one of the angles.* Choosing the form with *a* and cos *A*, we solve for cos *A* to get

$$\cos A = \frac{b^2 + c^2 - a^2}{2bc} = \frac{(36.7)^2 + (72.4)^2 - (83.8)^2}{2(36.7)(72.4)} = -0.0816$$

which gives us

$$A = 94.7° \qquad \text{(rounded)}$$

▶ ***Because cos A is negative, we know that A is between 90° and 180°.***
Finally, *using the Law of Sines* we determine that

$$B = 25.9° \quad \text{and} \quad C = 59.4°$$

In Example 3 we used the Law of Cosines to *find the largest angle first.* That is always a good strategy. If the triangle contains an obtuse angle, that fact will then be discovered because the cosine of the obtuse angle will be negative. This will avoid problems when the Law of Sines is used, because the sine of an obtuse an-

▶ gle is positive and ***the proper result may not be given directly by the calculator.*** Example 4 is another illustration of the Law of Cosines for solving a triangle with two given sides and the included angle.

Example
4

In order to avoid a restricted area, an airplane flies from point *A* to point *C* to point *B* as shown in Figure 14.32. Following this path, the distance traveled is 18.0 km. If a military airplane is allowed to fly directly from *A* to *B*, how much shorter is the route?

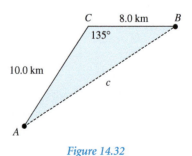

Figure 14.32

This problem involves finding the length of side c. Using the Law of Cosines we get

— unknown side opposite known angle —

$$c^2 = (8.00)^2 + (10.0)^2 - 2(8.00)(10.0)\cos 135.0°$$

$$c = \sqrt{(8.00)^2 + (10.0)^2 - 2(8.00)(10.0)\cos 135.0°}$$

$$= 16.6 \text{ km}$$

The direct path is therefore $18.0 - 16.6 = 1.4$ km shorter.

14.3 Exercises

In Exercises 1 through 24, use the Law of Cosines to solve the triangles with the given parts.

1. $a = 22.3, b = 16.4, C = 87.5°$

2. $b = 16.3, c = 30.5, A = 42.1°$

3. $a = 7720, c = 42,000, B = 58.9°$

4. $a = 7.00, b = 9.00, C = 133.9°$

5. $a = 1510, b = 308, C = 98.0°$

6. $a = 30.6, c = 15.2, B = 70.8°$

7. $a = 770, b = 934, c = 1600$

8. $a = 0.729, b = 0.789, c = 0.459$

9. $a = 51.2, b = 38.4, c = 72.6$

10. $a = 72.3, b = 65.9, c = 98.3$

11. $a = 7.20, b = 5.30, c = 9.19$

12. $a = 510, b = 650, c = 221$

13. $a = 238, b = 312, C = 24.7°$

14. $a = 7.92, b = 5.02, C = 100.6°$

15. $a = 5.12, c = 1.24, B = 37.3°$

16. $b = 102, c = 84.0, A = 56.9°$

17. $a = 637, b = 831, c = 987$

18. $a = 1.42, b = 1.95, c = 1.65$

19. $a = 3.47, b = 4.52, c = 2.25$

20. $a = 825, b = 770, c = 712$

21. $a = 0.1034, b = 0.1287, c = 0.1429$

22. $a = 2.160, b = 3.095, C = 109.10°$

23. $b = 10.21, c = 34.26, A = 9.65°$

24. $a = 2936, b = 1874, c = 3468$

In Exercises 25 through 32, use the Law of Cosines to solve the given problems.

25. To measure the distance AC, a person walks 620 m from A to B, then turns 35.0° to face C, and walks 730 m to C. What is the distance AC? See Figure 14.33.

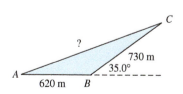

Figure 14.33

26. A forest ranger using orienteering walks directly north for a distance of 3.20 km. After making a turn, he then walks 3.82 km. From that location he returns to the starting point by walking 4.67 km. Find the angles of the triangle formed by this route. See Figure 14.34.

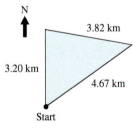

Figure 14.34

27. A triangular shelf is to be placed in a corner where the walls meet at an angle of 105.0°. If the edges of the shelf along the walls are 56.0 cm and 65.0 cm, how long is the outer edge of the shelf?

28. One ship 250 mi from a port is on a line 36.5° south of west of the port. A second ship is 315 mi from the port on a line 15.0° south of east of the port. How far apart are the ships?

29. One end of an 8.00-ft pole is 11.55 ft from an observer's eyes and the other end is 17.85 ft from his eyes. Through what angle does the observer see the pole? See Figure 14.35.

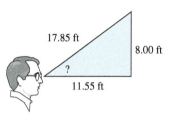

Figure 14.35

30. A triangular piece of land is bounded by 30.1 m of stone wall, 26.5 m of road frontage, and 50.5 m of fencing. What angle does the fence make with the road?

31. A plastic triangle used for drafting has sides of 4.58 in., 9.16 in., and 7.93 in. Find the angles formed by this triangle.

32. In order to find a single force equivalent to the net effect produced by two other forces acting on an object, it is necessary to find the length of the third side in a triangle. Find that length if the two known sides are 86.0 lb and 110 lb, and the angle between them is 107.0°. (See Figure 14.36.)

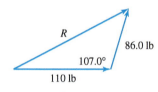

Figure 14.36

14.4 Introduction to Vectors

Section Objectives
- Understand the difference between a scalar and a vector quantity
- Determine the magnitude of a vector
- Determine the direction of a vector
- Determine the sum or difference of two or more vectors
- Determine the resultant of two or more vectors

At the beginning of the chapter we stated that a **vector** is a quantity having both magnitude and direction. We begin this section with examples of vectors, and we will then consider procedures for adding and subtracting them.

Example 1

An example of a vector is found in an aviation weather report describing the wind as coming from the north at 40 mi/h. This quantity has both a *direction* (from the north) and a *magnitude* (40 mi/h) and it is therefore a vector.

Example 2

If we are told that a missile travels 3270 km, we know only the distance traveled. This distance is a **scalar** quantity. If we later learn that the missile traveled 3270 km on a direct path from San Diego to Moscow, we would be specifying the direction as well as the distance traveled. *Because the direction and magnitude are both known, we now have a vector,* as well as a serious international incident.

Now Try It!

Determine whether the following situations describe a scalar or a vector quantity.

1. a north wind
2. a north wind blowing at 10 mph
3. a car driving at 30 mi/h
4. a car driving towards the center of town at 30 mi/h
5. applying 50 pounds of force while pulling a sled uphill
6. pulling a sled uphill

We know the **displacement** *of an object when we know its direction as well as the distance from the starting point.* Displacement is therefore a vector quantity.

To add scalar quantities we simply find the sum of the magnitudes. For example, 300 mi + 400 mi = 700 mi illustrates scalar addition. The addition of vector quantities is somewhat more complex, as illustrated by Example 3.

Example 3

An air traffic controller at a New York airport instructs the pilot of a jet to fly 300 mi due west, then turn and fly 400 mi due north. Following these instructions will clearly cause the jet to travel 700 mi, but it will not be 700 mi from New York. In fact, we see by using the Pythagorean theorem that the jet is 500 mi from New York (see Figure 14.37). The magnitude of the displacement is 500 mi, which means that the jet is actually 500 mi from New York. The direction of the displacement is indicated by the longest arrow shown in Figure 14.37. Measuring the angle, we find that it is directed at an angle of 53° north of west. Here we have added vectors with magnitudes of 300 and 400 and the result is a vector of magnitude 500. The "discrepancy" is due to the different directions of the two original vectors.

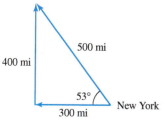

Figure 14.37

From Example 3 we see that *the addition of vectors must involve consideration of directions as well as magnitudes.* The method used to add the vectors in Example 3 can be generalized. We begin with some notation: Vector quantities will be represented by **boldface** type while the same letter in *italic* type will represent

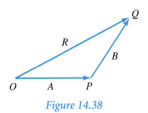

Figure 14.38

Notation for a vector **A**:

$\vec{A}$

magnitude only. Using this notation, **A** is a vector of magnitude *A*. Because handwriting does not lend itself easily to italic and boldface forms, the handwritten notation for a vector will be an arrow placed over the letter as in $\vec{A}$.

Let **A** and **B** represent vectors directed from *O* to *P* and *P* to *Q* respectively (see Figure 14.38). *The vector sum* **A** + **B** *is the vector* **R**, *from the* **initial point** *O to the* **terminal point** *Q. Here vector* **R** *is called the* **resultant.** *In general, a resultant is a single vector that can replace any number of other vectors and still produce the same physical effect.*

In this section we will look at adding vectors graphically. With the graphic approach, we add vectors by means of a diagram. There are two basic ways of adding vectors graphically: the **tail-to-head method** and the **parallelogram method**.

The tail-to-head method is illustrated in Figure 14.39. To add **B** to **A**, we shift **B** parallel to itself until its tail touches the head of **A**. In doing so we must *be careful to keep the magnitude and direction of* **B** *unchanged*. The vector sum **A** + **B** is the vector **R**, which is drawn from the tail of **A** to the head of **B**.

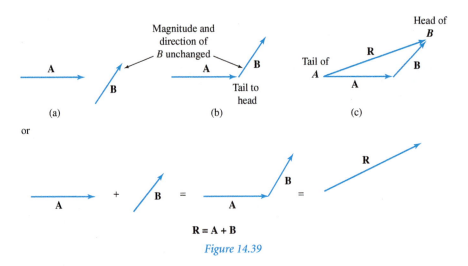

Figure 14.39

The tail-to-head method can also be used for adding more than two vectors. Starting with one of the vectors, we position the initial point of the second vector at the terminal point of the first. Then the third vector is located so that its initial point is at the terminal point of the second vector, and so on. *The resultant vector is determined by the initial point of the first vector and the terminal point of the last vector. The order of the vectors does not affect the result.*

Example 4

In Figure 14.40a we show the vectors **A**, **B**, and **C** which are to be added. In Figure 14.40b, the resultant **R** is found from one order, whereas Figure 14.40c shows that the same resultant is obtained with a different order of the same vectors. In addition to the orders of **A** + **B** + **C** and **B** + **A** + **C**, *other orders are also possible and they will produce the same resultant* **R**.

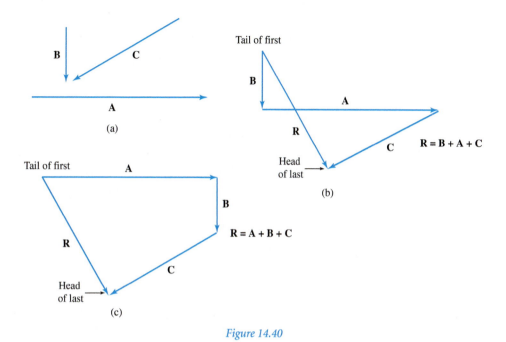

Figure 14.40

Example 4 illustrates the tail-to-head method of adding vectors. Another convenient method of adding two vectors is to *let the two vectors be the sides of a parallelogram. The resultant is then the diagonal of the parallelogram.* The initial point of the resultant is the common initial point of the vectors being added. In this method the vectors are first placed tail-to-tail. See the Example 5.

Example 5

Using the parallelogram method, add vectors **A** and **B** given in Figure 14.41a. Figure 14.41b shows the vectors placed tail-to-tail. Figure 14.41c shows the construction of the parallelogram and Figure 14.41d shows the resultant.

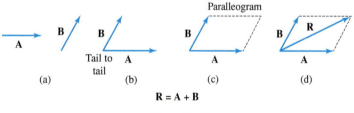

Figure 14.41

When we are considering applications of vectors, one method of addition may illustrate the situation better than the other method. For example, if two forces are acting on an object, the parallelogram method provides a good illustration of the actual situation. When we are adding displacements, the tail-to-head method is usually effective in showing the physical situation. Either method can be used to find the resultant for any set of vectors.

Example
6

A nuclear submarine travels 10 km east and then 5 km northeast. The two vectors can be shown as in Figure 14.42a, because the movement to the northeast started when the eastward movement ceased. Then it is natural to find the resultant displacement as in Figure 14.42b. By measurement in the figure, the magnitude of the displacement can be found to be 14 km.

A force of 10 lb acts on an object to the right. A second force of 5 lb acts on the same object upward to the right, at an angle of 45° with the horizontal. The two vectors can be shown as in Figure 14.42c, because the forces act on the same object. Then it is more natural to find the resultant force as in Figure 14.42d. The magnitude of the resultant force is about 14 lb.

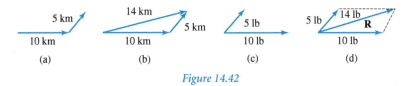

(a)   (b)   (c)   (d)

Figure 14.42

If vector **A** has the same direction as vector **B**, and **A** also has a magnitude n times that of **B**, we can state that **A** = n**B**. Thus 2**A** represents a vector twice as long as **A** and in the same direction as **A**.

Example
7

For the vectors **A** and **B** in Figure 14.41, find 3**A** + 2**B**. See Figure 14.43.

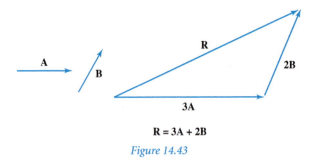

R = 3A + 2B

Figure 14.43

Vector **B** *can be subtracted from vector* **A** *by reversing the direction of* **B** *and proceeding as in vector addition.* Thus **A** − **B** = **A** + (−**B**), where the minus sign indicates that vector −**B** has the opposite direction of vector **B** (but the same magnitude). Vector subtraction is shown in Example 8.

For vectors **A** and **B** of Figure 14.39, find 2**A** − **B**. See Figure 14.44.

Although we can add, subtract, and find resultants of vectors using these techniques, we should remember that these graphic methods yield results that are approximate, not exact.

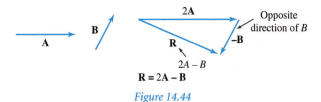

R = 2A − B

Figure 14.44

A pilot flies a small airplane with an air speed of 140 mi/h. The plane is pointed to the north, but is blown off course by a wind of 35 mi/h which is *from* a direction of 45° south of west (toward 45° north of east). See Figure 14.45. Find the true speed of the plane.

From Figure 14.45b, the resultant **R** has a magnitude of approximately 165 and we conclude that the true speed of the plane is 165 mi/h.

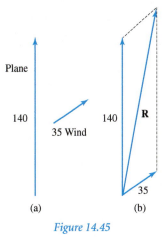

Figure 14.45

In Example 9 the solution of 165 mi/h is only an approximation because it depends upon a measurement. The true direction of the plane can be determined by measuring the angle with a protractor, but that is also an approximation. Applied uses of vectors often require more exact results. In the following sections we present methods that will yield more exact results.

14.4 Exercises

In Exercises 1 through 8, determine whether a scalar or a vector is described in part (a) and part (b). Explain your answers.

1. (a) A car travels at 50 mi/h. (b) A car travels at 50 mi/h due south.

2. (a) A person traveled 300 km to the southwest. (b) A person traveled 300 km.

3. (a) A vertical rope supports a 100-g object. (b) A force equivalent to 100 g is applied to an object.

4. (a) A small craft warning reports winds of 25 mi/h. (b) A small craft warning reports winds out of the north at 25 mi/h.

5. (a) An arm of an industrial robot pushes with a 10-lb force downward on a part. (b) A part is being pushed with a 10-lb force by an arm of an industrial robot.

6. (a) Spilled oil is being pushed by the 2-mi/h current in a river. (b) Spilled oil is being pushed south by the 2-mi/h current in a river.

7. (a) A ballistics test shows that a bullet hit a wall at a speed of 400 ft/s. (b) A ballistics test shows that a bullet hit a wall at a speed of 400 ft/s perpendicular to the wall.

8. (a) To denote the *speed* of an object we state how fast it is moving. Thus speed is a _____. (b) To denote the *velocity* of an object we state how fast and in what direction it is moving. Thus velocity is a _____.

*In Exercises 9 through 16, find the sum of the vectors **A** and **B** shown in the indicated figures.*

9. Figure 14.46a   **10.** Figure 14.46b

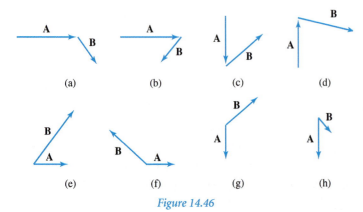

(a)   (b)   (c)   (d)

(e)   (f)   (g)   (h)

Figure 14.46

11. Figure 14.46c   **12.** Figure 14.46d

13. Figure 14.46e   **14.** Figure 14.46f

15. Figure 14.46g   **16.** Figure 14.46h

In Exercises 17 through 36, find the indicated vector sums and differences by means of diagrams, using the vectors in Figure 14.47.

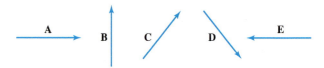

Figure 14.47

17. **A** + **B**   **18.** **A** + **C**

19. **A** + **A**   **20.** **B** + **B**

21. **A** + **D**   **22.** **B** + **C**

23. **A** + **E**   **24.** **C** + **D**

25. **A** + **B** + **C**   **26.** **B** + **C** + **A**

27. **A** + **D** + **E**   **28.** **A** + **B** + **E**

29. **A** + 3**B**   **30.** 2**B** + **C**

31. **A** − 3**B**   **32.** 2**B** − **C**

33. **A** + 2**B** + **C**   **34.** 2**A** + **B** + **C**

35. **A** − 2**B** + **E**   **36.** **A** − 2**B** + **C**

In Exercises 37 through 44, solve the given problems.

37. A surveyor sets a stake, then travels 150 ft north where a second stake is set. He then travels east 300 ft and sets a third stake. Use a diagram like Figure 14.48 to find the displacement between the first and third stakes.

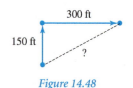

Figure 14.48

38. A motorist travels 20 mi due west and then 6 mi northeast. By means of a diagram find where the motorist is relative to the starting position.

39. A test of the behavior of a tire on ice involves two forces. One force of 500 lb is exerted to the side while another force of 1200 lb is exerted to the front. Use a diagram to find the resultant of these two forces.

40. The motor of a boat applies a force that would make the boat travel 10 mi/h in still water. The boat is pointed directly at the opposite shore on a river, and the current is downstream at 6 mi/h. Use a diagram to find the resultant of the two forces caused by the motor and the current.

41. In conducting a stress test on the end of a horizontal beam, two forces are applied. One force of 1800 lb is exerted vertically downward on the beam. Another force of 2700 lb is exerted horizontally at the same point on the

beam. By means of a diagram find the resultant of these two forces.

42. A car is stuck in snow. One pedestrian pushes the back of the car with a force having a magnitude of 50 lb while another pedestrian pushes on the driver's side with a force having a magnitude of 30 lb. By means of a diagram find the resultant of these two forces.

43. A balloon is rising vertically at the rate of 15 m/s and the wind is blowing from the northwest at 10 m/s. By means of a diagram find the resultant velocity of the balloon.

44. An accident investigator determines that when two cars crashed, one car was traveling on a northbound highway at a speed of 55 mi/h. The second car entered the highway at a speed of 30 mi/h from a side road, and its path was 45° north of west. Use a diagram to find the resultant of the two forces at the point of impact.

Now Try It! Answers

1. scalar **2.** vector **3.** scalar **4.** vector **5.** vector **6.** scalar

14.5 **Vector Components**

Section Objectives
• Resolve a vector into its components
• Solve application problems involving vectors

In the last section we used graphic techniques for adding and subtracting vectors and multiples of vectors. Beside being able to perform these vector operations, there is often a need to represent a given vector as the sum of two other vectors. These two vectors, when added, combine to produce the original vector. *The method is called* **resolving** *the given vector into its* **components.** This is usually done within the framework of the rectangular coordinate system. That is, the given vector is resolved into two components; one of the components has a direction along the *x* axis and the other component has a direction along the *y* axis. When the components have directions along the axes, they are called **rectangular component vectors.** Example 1 shows how to resolve a given vector into its rectangular components.

Example
1

Resolve the vector **A**, with $A = 20.0$, which makes an angle of 56.0° with the positive *x* axis, into two components, one directed along the *x* axis (the *x component*) and the other along the *y* axis (the *y component*). That is, resolve the given vector into its rectangular components. See Figure 14.49.

In the figure we see that two right triangles are formed. For the right triangle with sides **A** and **A**$_x$ we can use the cosine function to get

$$\cos 56.0° = \frac{A_x}{A}$$

(Remember that **A** represents a vector whereas A is a scalar.) Solving for A_x we get

$$A_x = A \cos 56.0°$$

Since $A = 20.0$ we have

$$A_x = 20.0 \cos 56.0° = 11.2 \qquad \text{(rounded)}$$

using the calculator sequence

20 × COS 56 ENTER to get 11.183858

Working with the same right triangle, it can be shown that the dashed side opposite the 56.0° angle is equal to the magnitude A_y. Using the sine function we get

$$\sin 56.0° = \frac{A_y}{A}$$

or

$$A_y = A \sin 56.0° = 16.6 \qquad \text{(rounded)}$$

using the calculator sequence

20 × SIN 56 ENTER to get 16.580751

We have now resolved vector **A** into two components. One component has a magnitude of 11.2 and is in the direction of the positive *x* axis. The second component has a magnitude of 16.6 and is in the direction of the positive *y* axis.

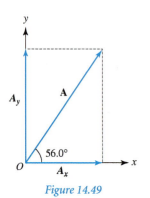

Figure 14.49

We can generalize the first method used in Example 1: When resolving a vector **A** into its rectangular components, *the magnitude of the horizontal component is $A \cos \theta$ and the magnitude of the vertical component is $A \sin \theta$, where θ is the angle of the vector in standard position.* Vector **A** has rectangular components A_x and A_y with magnitudes given by

$$\boxed{\begin{aligned} A_x &= A \cos \theta \\ A_y &= A \sin \theta \end{aligned}}$$

If the original vector is not in the first quadrant, the rectangular component vectors will be directed along their respective axes so that their sum is the original vector.

Rectangular components in the direction of the positive x axis or positive y axis are represented by positive values, whereas rectangular components in the direction of the negative x axis or negative y axis will be represented by negative values. This will make it easier to add vectors, as discussed in the following section.

Example 2

The velocity of a rocket is 350 km/s, directed to the right, at an angle of 23.5° below the horizontal. What are the horizontal (x component) and vertical (y component) components of the velocity?

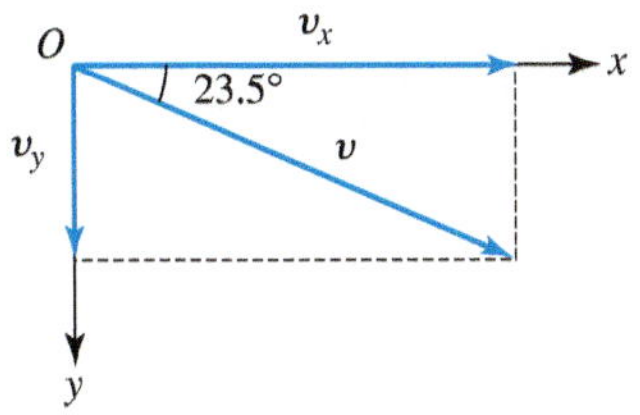

We can construct the figure as shown in Figure 14.50. Noting that the angle corresponds to a *clockwise* rotation, we represent it as $-23.5°$ to get

$$v_x = v \cos(-23.5°) = 350 \cos(-23.5°) = 321 \text{ km/s}$$

Also, since v_y is equal to the side opposite the 23.5° angle, we get

$$v_y = v \sin(-23.5°) = 350 \sin(-23.5°) = -140 \text{ km/s}$$

Remember, *we use negative values for rectangular components in the direction of the negative x axis or negative y axis.*

Figure 14.50

To Resolve a Vector into its Components

- $V_x = V \cos\theta$ gives the magnitude of the horizontal component of vector V
- $V_y = V \sin\theta$ gives the magnitude of the vertical component of vector V

Example 3

In analyzing the stress on a ceiling joist caused by snow on a roof, an architect finds it necessary to resolve the force vector in Figure 14.51a into its rectangular components. Find the rectangular component vectors.

From Figure 14.51b we can see that one rectangular component vector will be in the direction of the negative x axis whereas the other is in the direction of the negative y axis. Note that R_x and R_y are both negative because they are in the direction of the negative x axis and negative y axis.

Now Try It!

1. Resolve a vector with a magnitude of 8.6 and angle $\theta = 68.0°$ into its horizontal and vertical components.
2. Resolve a vector with a magnitude of 250 and angle $\theta = 320°$ into its horizontal and vertical components.

$R_x = R \cos 210.0° = 13.4 \cos 210.0° = -11.6$ directed along negative x axis

$R_y = R \sin 210.0° = 13.4 \sin 210.0° = -6.70$ directed along negative y axis

The horizontal component vector has a magnitude of 11.6 lb and is directed along the negative x axis, whereas the vertical component has a magnitude of 6.70 lb and is directed along the negative y axis.

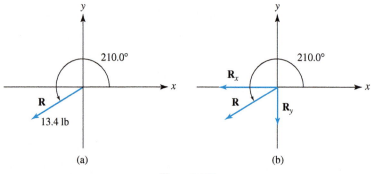

Figure 14.51

Example 4

A cable is anchored at a point as shown in Figure 14.52. The pull on the cable is equivalent to 1360 N and we want to provide weights X and Y so that the three forces at the anchor point offset each other. Note that the weight X actually produces a force in the direction of the negative x axis. Find the forces produced by X and Y.

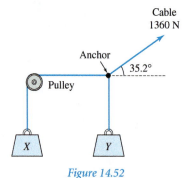

Figure 14.52

We will represent by **C** the force vector representing the cable. Resolving **C** into rectangular components we get

$C_x = C \cos 35.2°$

$\quad = 1360 \cos 35.2° = 1110 \text{ N}$

$$C_y = C \sin 35.2°$$
$$= 1360 \sin 35.2° = 784 \text{ N}$$

From these rectangular components we see that the force of the cable is equivalent to two other forces: one that pulls to the right with 1110 N and one that pulls vertically upward with 784 N. To offset the pull to the right of 1110 N, we should make $X = 1110$ N. To offset the pull upward, we should make $Y = 784$ N.

14.5 Exercises

In Exercises 1 through 4, use a diagram to resolve the vector given in the indicated figure into its x component and y component.

1. Figure 14.53a

2. Figure 14.53b

3. Figure 14.53c

4. Figure 14.53d

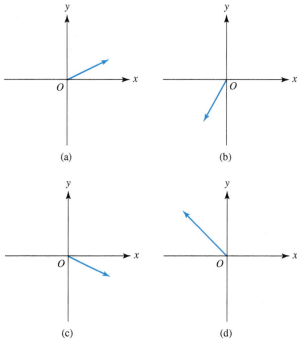

(a)

(b)

(c)

(d)

Figure 14.53

In Exercises 5 through 16, resolve the vector given in the indicated figure into its x component and y component.

5. Figure 14.54a

6. Figure 14.54b

7. Figure 14.54c

8. Figure 14.54d

9. Figure 14.54e

10. Figure 14.54f

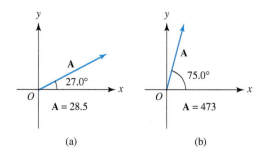

(a)

(b)

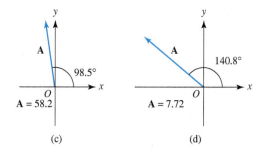

(c)

(d)

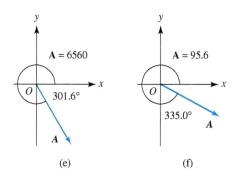

(e)

(f)

Figure 14.54

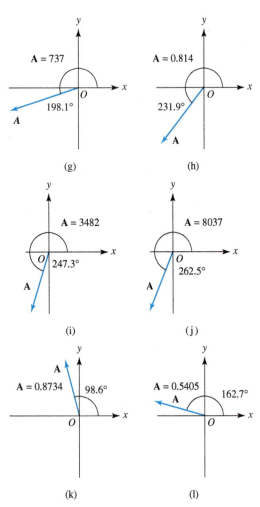

(g)   (h)

(i)   (j)

(k)   (l)

Figure 14.54 (continued)

11. Figure 14.54g

12. Figure 14.54h

13. Figure 14.54i

14. Figure 14.54j

15. Figure 14.54k

16. Figure 14.54l

In Exercises 17 through 32, resolve the given vector into its x component and y component. The given angle θ is measured counterclockwise from the positive x axis (in standard position).

17. Magnitude 25.8, $\theta = 45.0°$

18. Magnitude 465, $\theta = 30.0°$

19. Magnitude 6.34, $\theta = 57.0°$

20. Magnitude 136, $\theta = 18.3°$

21. Magnitude 219, $\theta = 263.0°$

22. Magnitude 34.5, $\theta = 201.9°$

23. Magnitude 1370, $\theta = 225.0°$

24. Magnitude 29.3, $\theta = 315.0°$

25. Magnitude 87.2, $\theta = 306.9°$

26. Magnitude 219, $\theta = 342.0°$

27. Magnitude 0.05436, $\theta = 233.5°$

28. Magnitude 0.0204, $\theta = 117.6°$

29. Magnitude 25872, $\theta = 356.20°$

30. Magnitude 13724, $\theta = 279.90°$

31. Magnitude 34.666, $\theta = 112.70°$

32. Magnitude 58.307, $\theta = 166.40°$

In Exercises 33 through 38, resolve the given problems.

33. A plane is 24.0° north of east from a city at a distance of 36.5 mi. Find the horizontal (east) and vertical (north) components of the displacement. See Figure 14.55.

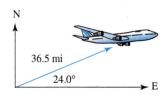

Figure 14.55

34. A bulldozer stuck in mud is being pulled by a cable connected to another bulldozer. The cable makes an angle of 18.0° with an imaginary line representing the forward path of the stuck bulldozer. The magnitude of the force is 8500 lb of pull. What is the magnitude of the component of the force that has the same direction as the stuck bulldozer? See Figure 14.56.

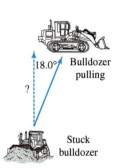

Figure 14.56

35. In studying the interference of harmonic sound waves, it becomes necessary to find the rectangular components of the vector that has a magnitude of 0.024 m. The vector makes an angle of 58.3° with the positive *x* axis. Find the rectangular component vectors.

36. A missile makes an angle of 84.8° with a level platform. What is the horizontal force applied by the rocket on its platform if the rocket exerts 155,000 lb of thrust? See Figure 14.57.

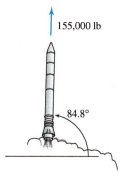

155,000 lb

84.8°

Figure 14.57

37. In a parallel resistance-capacitance (*RC*) electric circuit, the current I_C through the capacitance and the current I_R

through the resistance are out of phase by 90°, as depicted in Figure 14.58. If $I_C = 2.35$ A and $I_R = 2.89$ A, find the magnitude of the resultant and the phase angle θ.

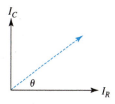

I_C

θ

I_R

Figure 14.58

38. A 6480-lb truck is parked on a ramp that is at a 10.3° angle with the horizontal. Find the component of the truck's weight that is parallel with the surface of the ramp. See Figure 14.59.

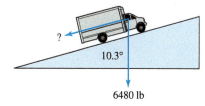

?

10.3°

6480 lb

Figure 14.59

Answers

1. $A_x = 3.2$ $A_y = 8.0$ **2.** $A_x = 192$ $A_y = -161$

14.6 Vector Addition

Section Objectives

- Add vectors using vector components
- Solve application problems involving vectors

In Section 14.4 we saw how to add vectors graphically, but the results were only approximate. Common problems found in science and technology require the addition of vectors with more exact results. We will use the Pythagorean theorem and the trigonometric ratios to develop a procedure giving the exactness required. This procedure is based on resolving vectors into components, and Section 14.5 provided the foundation for this technique.

We begin with a simple example in which the two given vectors are already at right angles so that, in this particular case, we need not resolve them into x component and y component vectors. The objective is to *add the two vectors to get a resultant vector with a known magnitude and a known direction.*

Example
1

Add vectors **A** and **B**, with $A = 86.4$ and $B = 67.4$. The vectors are at right angles as shown in Figure 14.60.

Because the vectors **A**, **B**, and their sum **R** form a right triangle, we can *find the magnitude R of the resultant* **R** by using the Pythagorean theorem as follows:

$$R = \sqrt{A^2 + B^2} = \sqrt{(86.4)^2 + (67.4)^2}$$

$$= 110 \quad \text{(rounded)}$$

Having found the magnitude of the resultant, we can *find its direction* angle θ by using an appropriate trigonometric ratio. We choose tangent because we know the sides opposite and adjacent to θ.

$$\tan \theta = \frac{B}{A} = \frac{67.4}{86.4}$$

so that

$$\theta = 38.0° \quad \text{(rounded)}$$

The sum of **A** and **B** is the resultant **R** which has a magnitude of 110 and direction $\theta = 38.0°$, where θ is located as in Figure 14.60. We must remember that **the solution is not complete without determining the direction.**

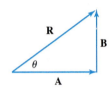

Figure 14.60

In Example 1 the given vectors were directed so that they were at right angles to each other. When adding vectors that are not at right angles, we follow these key steps:

1. Resolve each vector into its x and y components by using

$$\boxed{\begin{aligned} A_x &= A \cos \theta \\ A_y &= A \sin \theta \end{aligned}}$$

2. Combine the x components to determine the x component R_x of the resultant. Also combine the y components to determine the y component R_y of the resultant.

3. Find the magnitude of the resultant by using the Pythagorean theorem with the x and y components of the resultant (as found in step 2).

$$\boxed{R = \sqrt{R_x^2 + R_y^2}}$$

4. Find the direction of the resultant by using tangent with the x and y components of the resultant (as found in step 2). Use

$$\tan \theta = \frac{R_y}{R_x}$$

and adjust θ so that it corresponds to the correct quadrant.

It is important to remember that the resultant vector has not been determined unless both its magnitude **and direction** are known.

Examples 2 and 3 illustrate how to find the magnitude of the resultant of two vectors, and the angle the resultant makes with the x axis, by using steps 1 through 4.

Example
2

Add vectors **A** and **B** as shown in Figure 14.61a.

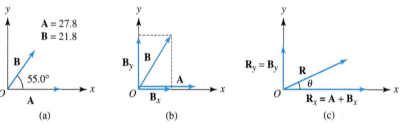

Figure 14.61

Step 1   Because vector **A** is directed along the x axis, we see that $A_x = A = 27.8$ and $A_y = 0$. It is necessary now to resolve vector **B** into its components:

$$B_x = B \cos 55.0° = 21.8 \cos 55.0° = 12.5$$
$$B_y = B \sin 55.0° = 21.8 \sin 55.0° = 17.9$$

These components are shown in Figure 14.61b.

Step 2   Now, because the only component in the y direction is $\mathbf{B}_y$, we have $R_y = B_y$. Also, because both **A** and $\mathbf{B}_x$ are directed along the positive x axis, we add their magnitudes together to get the magnitude of the x component of the resultant. We get

$$R_x = A + B_x = 27.8 + 12.5 = 40.3$$
$$R_y = B_y = 17.9$$

Step 3   We now find the magnitude of the resultant **R** from the Pythagorean theorem:

$$R = \sqrt{R_x^2 + R_y^2} = \sqrt{(40.3)^2 + (17.9)^2}$$
$$= 44.1$$

Step 4   The direction of **R** is specified by determining angle θ from

$$\tan \theta = \frac{R_y}{R_x} = \frac{17.9}{40.3} \quad \text{or} \quad \theta = 23.9° \longleftarrow \text{don't forget the direction}$$

Thus, the magnitude of the resultant is 44.1 and it is directed at an angle of 23.9° above the positive x axis (see Figure 14.61c).

After each vector has been resolved into its components, we must *be very careful to observe the direction of these components.* In Example 3, note that vector **B** has a positive vertical component and a negative horizontal component.

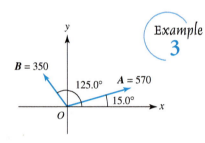

(a)

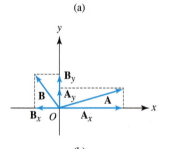

(b)

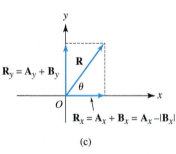

(c)

Figure 14.62

Example 3

Add vectors **A** and **B** as shown in Figure 14.62a.

Because neither vector is directed along an axis, it is necessary to resolve both vectors into their x and y components. Following the same procedure as in previous examples, we have (see Figure 14.62b):

$$A_x = A \cos 15.0° \quad \text{and} \quad A_y = A \sin 15.0°$$
$$B_x = B \cos 125.0° \quad \text{and} \quad B_y = B \sin 125.0°$$

Substituting the appropriate values, we get

$$A_x = 570 \cos 15.0° = 551$$
$$A_y = 570 \sin 15.0° = 148$$
$$B_x = 350 \cos 125.0° = -201 \longleftarrow \text{directed along negative } x \text{ axis}$$
$$B_y = 350 \sin 125.0° = 287$$

We can condense and summarize the above results by using the following table format.

	x components		*y components*	
A	570 cos 15.0° =	551	570 sin 15.0° = 148	
B	350 cos 125.0° = −201		350 sin 125.0° = 287	
R	Resultant	350 Add	435	Add

We now have

$$R = \sqrt{R_x^2 + R_y^2} = \sqrt{(350)^2 + (435)^2} = 558$$

$$\tan \theta = \frac{R_y}{R_x} = \frac{435}{350} \quad \text{so that} \quad \theta = 51.2°$$

We conclude that the magnitude of the resultant is 558 and its direction is 51.2° above the positive x axis.

Now Try It!

Add vectors **A** and **B**:

1. **A** = 18.0, 0.0°
   B = 12.0, 27.0°
2. **A** = 780, 28.0°
   B = 346, 320.0°

Examples 4 and 5 demonstrate two of the many applications of vectors.

Example
4

Two forces in the same vertical plane act on an object. One force of 6.85 lb acts to the right at an angle of 18.0° above the horizontal. The other force of 7.44 lb acts to the right at an angle of 51.5° below the horizontal. Find the resultant force.

From the statement of the problem, we show the forces as $\mathbf{F}_1$ and $\mathbf{F}_2$ in Figure 14.63a. We then resolve each force into its x and y components as follows. The result can be seen in Figure 14.63b.

	x components	*y components*
$\mathbf{F}_1$	$6.85 \cos 18.0° = 6.515$	$6.85 \sin 18.0° = 2.117$
	──────── clockwise angle ────────	
$\mathbf{F}_2$	$7.44 \cos(-51.5°) = \underline{4.632}$	$7.44 \sin(-51.5°) = \underline{-5.823}$
$\mathbf{R}$	11.147 Add	-3.706 Add

The resultant force and its components are shown in Figure 14.63c.

The magnitude of the resultant is found by the Pythagorean theorem:

$$R = \sqrt{R_x^2 + R_y^2} = \sqrt{(11.147)^2 + (-3.706)^2} = 11.7\,\text{lb}$$

The angle θ is found as follows:

$$\tan \theta = \frac{R_y}{R_x} = \frac{-3.706}{11.147}$$

$$\theta = -18.4°$$

The angle of $-18.4°$ could also be expressed as $341.6°$. The resultant force is 11.7 lb and is directed at an angle of 18.4° below the positive x axis.

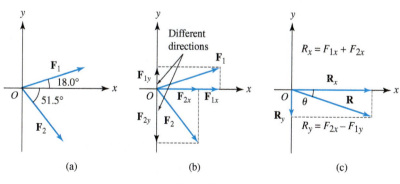

(a)   (b)   (c)

Figure 14.63

Example
5

A pilot flies her plane so that it is pointed in the direction shown in Figure 14.64. Her airspeed indicator shows 205 mi/h. At her altitude there is a 62.0 mi/h wind blowing from the southwest. Find the true speed of the plane relative to the ground, and find the direction the plane is actually headed. (Note that a wind *from* the southwest blows *to* the northeast.)

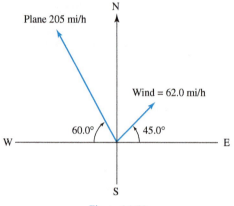

Figure 14.64

This problem involves finding the vector that is the sum of the two given vectors. The magnitude of the resultant is the actual speed of the plane; the direction of the resultant is the direction the plane will actually travel. We can treat Figure 14.64 as a regular *x-y* coordinate system with east corresponding to the positive *x* axis and north corresponding to the positive *y* axis. The following table shows the vectors resolved into rectangular components. Note that the angle of 60.0° becomes 120.0° when the plane vector is represented in standard position.

	x component		*y component*	
Wind	$62.0 \cos 45.0° =$	43.8	$62.0 \sin 45.0° =$	43.8
Plane	$205 \cos 120.0° = -102.5$		$205 \sin 120.0° = 177.5$	
R	Resultant	-58.7 Add		221.3 Add

standard position

The plane's horizontal component is negative because it corresponds to the negative *x* axis.

The magnitude of the resultant **R** can now be found by using the Pythagorean theorem.

$$R = \sqrt{R_x^2 + R_y^2} = \sqrt{(-58.7)^2 + (221.3)^2} = 229$$

We now proceed to find the direction as follows. With $R_x = -58.7$ and $R_y = 221.3$, we use the tangent function to get

$$\tan\theta = \frac{R_y}{R_x} = \frac{221.3}{-58.7} = -3.770$$

Thus $\theta = -75.1°$.

With a positive y rectangular component and a negative x rectangular component, we know that the resultant is in the second quadrant. In this case, the result of $-75.1°$ corresponds to $75.1°$ north of west.

In general, a negative angle indicates that the resultant vector is in the second or fourth quadrant. We can adjust the negative angle by adding 180° or 360° so that the result is a positive angle for a vector in standard position.

The plane travels at a true speed of 229 mi/h and flies in a direction of 75.1° north of west. (Air navigation headings are actually measured in degrees from 0° to 360° where 0° is north and the other angles are measured clockwise from north so that 90° is east, 180° is south, 270° is west, and so on.)

14.6 Exercises

*In Exercises 1 through 8, vectors **A** and **B** are at right angles. Determine the magnitude and the direction of the resultant relative to vector **A**.*

1. $A = 5.00, B = 2.00$
2. $A = 72.0, B = 15.0$
3. $A = 746, B = 1250$
4. $A = 0.962, B = 0.385$
5. $A = 34.7, B = 16.3$
6. $A = 126, B = 703$
7. $A = 12.05, B = 22.46$
8. $A = 0.0348, B = 0.0526$

In Exercises 9 through 16, use the Pythagorean theorem and trigonometric ratios to find the resultant of the vectors in the indicated figure. Determine the magnitude and direction of the resultant.

9. Figure 14.65a
10. Figure 14.65b
11. Figure 14.65c
12. Figure 14.65d
13. Figure 14.65e
14. Figure 14.65f
15. Figure 14.65g
16. Figure 14.65h

In Exercises 17 through 32, determine the resultant of the vectors with the given magnitudes and directions. Positive angles are measured counterclockwise from the positive x axis, and negative angles are measured clockwise from the positive x axis.

17. **A:** 8.0, 0°
    B: 6.0, 90°
18. **A:** 0.766, 44.0°
    B: 0.486, 90.0°
19. **A:** 47.2, 90.0°
    B: 28.5, 165.0°
20. **A:** 450, 20.0°
    B: 320, 65.0°
21. **A:** 566, 155.0°
    B: 1240, 221.0°
22. **A:** 20.6, 131.1°
    B: 45.1, 318.3°
23. **A:** 1324.5, 193.70°
    B: 6433.9, 171.30°
24. **A:** 0.03482, 26.5°
    B: 0.04463, 208.3°
25. **A:** 5.0, 0°
    B: 12.0, −90°
26. **A:** 137, 48.0°
    B: 256, −90.0°

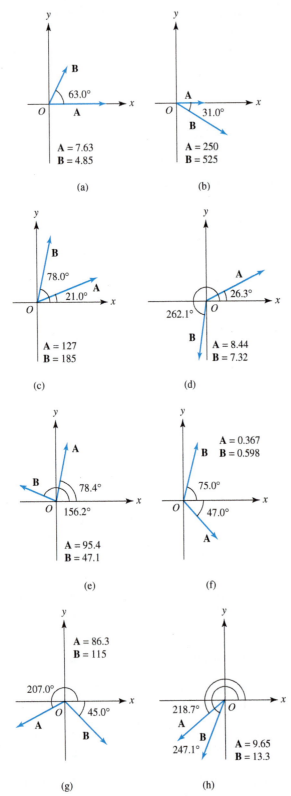

(a)

(b)

(c)

(d)

(e)

(f)

(g)

(h)

Figure 14.65

27. A: 16.4, −108.5°
    B: 29.6, 0.0°

28. A: 548, −21.4°
    B: 365, −67.2°

29. A: 2560, −32.6°
    B: 1830, −47.9°

30. A: 0.429, 124.6°
    B: 0.108, −118.3°

31. A: 0.859, 156.2°
    B: 0.635, −41.7°

32. A: 25.36, −16.4°
    B: 10.28, 212.8°

In Exercises 33 through 44, solve the given vector problems by using the Pythagorean theorem and the trigonometric ratios.

33. In an automobile, two forces act on the same object and at right angles to each other. One force is 212 lb, and the other is 158 lb. Find the resultant of these two forces.

34. Two forces in the same vertical plane act on an object. One force of 8.35 lb acts to the right at an angle of 16.0° above the horizontal. The other force of 9.83 lb acts to the right at an angle of 63.7° below the horizontal. Find the resultant force.

35. A ship travels 38.0 mi due east from a port and then turns south for 26.0 mi farther. What is the displacement of the ship from the port?

36. A motorboat that travels 7.00 km/h in still water heads directly across a stream that flows at 3.20 km/h. What is the resultant velocity of the boat? See Figure 14.66.

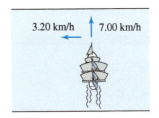

Figure 14.66

37. A plane is traveling horizontally at 1500 km/h. A rocket is fired horizontally ahead from the plane at an angle of 8.0° from the direction of the plane and with a velocity of 2500 km/h. What is the resultant velocity of the rocket?

38. Two forces in the same horizontal plane act on an object. One force of 58.0 lb acts to the right. The second force of 40.8 lb acts to the left at an angle of 23.5° from the line of action of the first force. Find the resultant force.

39. A 50.0-lb object is suspended from the ceiling by a rope. It is then pulled to the side by a horizontal force of 20.0 lb and held in place. What is the tension in the rope? (In order that the object be in equilibrium there should be no net force in any direction. This means that the tension in the rope must equal the resultant of the two forces acting on it. The forces in this case are the 20.0-lb force and the weight of the object.)

40. A plane is headed due north at a speed of 550 mi/h with respect to the air. There is a tail wind blowing from the southeast at 75.0 mi/h. What is the resultant velocity of the plane with respect to the ground? See Figure 14.67.

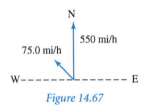

Figure 14.67

41. Two voltages V_1 and V_2 are out of phase so that they are represented by two vectors with the same initial point. Find the resultant vector if $V_1 = 120.0\,\text{V}$, $V_2 = 18.0\,\text{V}$, and the phase angle between the two vectors is 59.7°.

42. An object is affected by two different substances having magnetic intensities represented by vectors $\mathbf{H}_1$ and $\mathbf{H}_2$. If $H_1 = 106\,\text{A/m}$, $H_2 = 76.0\,\text{A/m}$, and the angle between the two vectors is 53.7°, find the resultant.

43. An object with a weight equivalent to 90.0 N is on the floor. Two forces are acting on the object. The first force, of 110 N, acts at an angle of 30.0° upward from the horizontal to the right. The second force, of 75.0 N, acts at an angle of 20.0° upward from the horizontal to the left. Will the object move? (Ignore frictional effects.) See Figure 14.68.

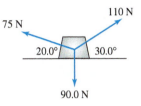

Figure 14.68

44. A wooden block is on an inclined plane, as shown in Figure 14.69. The block will slip down the plane if the component acting downward along the plane is greater than the frictional force acting upward along the plane. Given that the block weighs 85.0 lb and the plane is inclined at 13.0°, will the block slip if the frictional force is 18.0 lb?

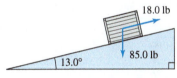

Figure 14.69

Answers

1. $R = 29.2$ $\theta = 10.8°$ **2.** $R = 965$ $\theta = 8.6°$

Chapter Summary

Key Terms

angle in standard position	coterminal angles	vector
quadrantal angle	reference angle	scalar
abscissa	Law of Sines	resultant vector
ordinate	Law of Cosines	resolving a vector
radius vector	ambiguous case	

Formulas

$$\text{sine } \theta = \frac{y}{r} \qquad \cos\text{ine } \theta = \frac{x}{r} \qquad \text{tangent } \theta = \frac{y}{x} \left.\vphantom{\begin{array}{c} \\ \\ \\ \end{array}}\right\}$$

$$\text{cotangent } \theta = \frac{x}{y} \qquad \text{secant } \theta = \frac{r}{x} \qquad \text{cosecant } \theta = \frac{r}{y}$$

If θ is in the . . .　then the reference angle α can be found from the equation . . .

second quadrant　　$\alpha = 180° - \theta$

third quadrant　　　$\alpha = \theta - 180°$

fourth quadrant　　$\alpha = 360° - \theta$

$$\frac{a}{\sin A} = \frac{b}{\sin B} = \frac{c}{\sin C} \qquad \text{Law of Sines}$$

$$\sin^2\theta + \cos^2\theta = 1$$

$$\left.\begin{array}{l} a^2 = b^2 + c^2 - 2bc\cos A \\ b^2 = a^2 + c^2 - 2ac\cos B \\ c^2 = a^2 + b^2 - 2ab\cos C \end{array}\right\} \quad \text{Law of Cosines}$$

$$A_x = A\cos\theta \qquad\qquad \text{horizontal component of A}$$

$$A_y = A\sin\theta \qquad\qquad \text{vertical component of A}$$

$$R = \sqrt{R_x^2 + R_y^2} \qquad\quad \text{magnitude of resultant of R}$$

$$\tan\theta = \frac{R_y}{R_x} \qquad\qquad\quad \text{direction of resultant of R}$$

Review Exercises

In Exercises 1 through 4, find the trigonometric functions of θ given that the terminal side of θ passes through the given point. Express results for Exercises 1 and 2 in fractional form; express results for Exercises 3 and 4 in decimal form.

1. $(4, 3)$

2. $(-5, 12)$

3. $(7, -2)$

4. $(-2, -3)$

In Exercises 5 through 8, determine the quadrant in which the terminal side of θ lies, subject to the given conditions.

5. sin θ is positive and tan θ is negative

6. cos θ is negative and sin θ is negative

7. sec θ is positive and csc θ is negative

8. cot θ is negative and sec θ is positive

In Exercises 9 through 12, express the given trigonometric functions in terms of the same function of a positive acute angle.

9. cos 132°, tan 194°   **10.** sin 243°, cot 318°

11. sin 289°, sec (−15°)   **12.** cos 103°, csc (−100°)

In Exercises 13 through 28, determine the values of the given trigonometric functions.

13. cos 243.0°   **14.** sin 139.0°

15. cot 287.0°   **16.** tan 188.0°

17. csc 247.5°   **18.** sec 96.3°

19. sin 215.2°   **20.** cos 337.2°

21. tan 291.4°   **22.** cot 109.3°

23. tan 256.7°   **24.** cos 172.5°

25. sin 763.0°   **26.** tan 482.0°

27. sec 715.4°   **28.** csc 992.0°

In Exercises 29 through 40, find θ to the nearest 0.1° for 0° < θ < 360°. Each case has two solutions.

29. tan θ = 0.7532   **30.** sin θ = −0.5293

31. cos θ = −0.4208   **32.** cot θ = −1.829

33. csc θ = 2.175   **34.** sec θ = 1.428

35. sin θ = −0.1639   **36.** cos θ = 0.2735

37. cot θ = −2.145   **38.** tan θ = 0.1459

39. sin θ = 0.8319   **40.** sec θ = −1.817

In Exercises 41 through 58, solve the triangles with the given parts.

41. $a = 100, A = 47.0°, B = 61.3°$

42. $a = 35.0, A = 48.2°, C = 73.3°$

43. $a = 89.7, B = 148.8°, C = 10.0°$

44. $A = 36.4°, B = 123.6°, c = 78.4$

45. $a = 3.20, b = 4.50, C = 91.7°$

46. $b = 3820, c = 2910, A = 84.8°$

47. $a = 283, c = 278, B = 62.0°$

48. $b = 913, c = 117, A = 57.6°$

49. $b = 16.3, c = 18.2, B = 54.3°$

50. $a = 902, b = 763, A = 148.9°$

51. $a = 65.5, b = 78.4, A = 51.0°$

52. $a = 840, b = 1390, A = 19.2°$

53. The triangle shown in Figure 14.70(a)

54. The triangle shown in Figure 14.70(b)

55. The triangle shown in Figure 14.70(c)

56. The triangle shown in Figure 14.70(d)

57. The triangle shown in Figure 14.70(e)

58. The triangle shown in Figure 14.70(f)

In Exercises 59 through 62, determine whether a scalar or a vector is described in part (a) and part (b). Explain your answers.

59. a. The airspeed indicator of an airplane shows a speed of 187 mi/h.

b. The vertical speed indicator of an airplane shows that the plane is descending at a rate of 500 ft/min.

60. a. A missile is launched straight up with a speed of 2560 mi/h.

b. A missile is cruising at the rate of 2560 mi/h.

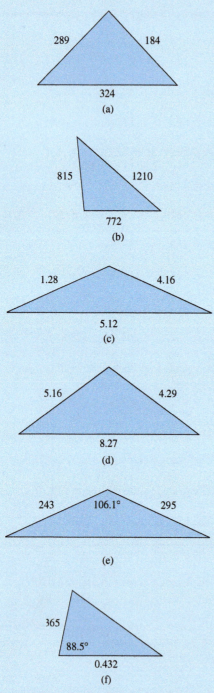

Figure 14.70

(a) 289, 184, 324

(b) 815, 1210, 772

(c) 1.28, 4.16, 5.12

(d) 5.16, 4.29, 8.27

(e) 243, 106.1°, 295

(f) 365, 88.5°, 0.432

61. a. A 257-lb force is applied to a beam at a particular point.

b. A force of 257 lb is applied to a beam in such a way that it is perpendicular down to the beam.

62. a. Two men are pulling on opposite ends of a rope in opposite directions with forces having magnitudes of 80 lb and 72 lb, respectively.

b. Two men are pulling on a rope and their efforts are measured as 80 lb and 72 lb, respectively.

*In Exercises 63 through 66, find the sum of vectors **A** and **B** shown in the indicated figures. Use diagrams.*

63. Figure 14.71a   **64.** Figure 14.71b

65. Figure 14.71c   **66.** Figure 14.71d

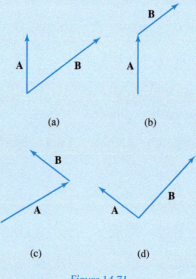

(a)   (b)

(c)   (d)

Figure 14.71

In Exercises 67 through 74, use appropriate diagrams to find the indicated vector sums and differences for the vectors shown in Figure 14.72

67. A + B

68. A − B

69. A + B + C

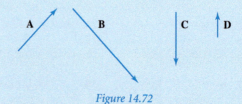

Figure 14.72

70. $\mathbf{A} + \mathbf{B} + \mathbf{C} + \mathbf{D}$

71. $2\mathbf{A} + \mathbf{B}$

72. $\mathbf{A} + \frac{1}{2}\mathbf{B}$

73. $2\mathbf{A} - \mathbf{B}$

74. $2\mathbf{D} - \mathbf{C}$

In Exercises 75 through 78, find the x and y components of the vectors shown in the indicated figures. Use the trigonometric ratios.

75. Figure 14.73a

76. Figure 14.73b

77. Figure 14.73c

78. Figure 14.73d

81. Vector $\mathbf{A}$, with $A = 16.48, \theta = -57.3°$

82. Vector $\mathbf{A}$, with $A = 0.8966, \theta = 262.9°$

In Exercises 83 through 86, resolve the described vector into appropriate vertical and horizontal components.

83. A wind of 36 mi/h is blowing in the direction of 15.0° north of west.

84. A Cessna Citation jet is flying at a speed of 270 mi/h in a direction of 20.0° south of west.

85. A cable is anchored on the ground and is connected to the top of a pole. The cable makes an angle of 39.0° with the pole and it pulls on the pole with a force having a magnitude of 555 lb.

86. A hammer strikes a board at a point. The direction of the hammer makes an angle of 85.0° with the board's surface, and the force of the strike is 27 lb.

In Exercises 87 through 90, add the vectors shown in the indicated figures by using the Pythagorean theorem and the trigonometric ratios.

87. Figure 14.74a   **88.** Figure 14.74b

89. Figure 14.74c   **90.** Figure 14.74d

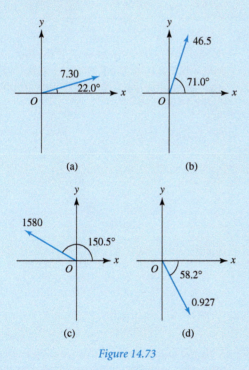

Figure 14.73

In Exercises 79 through 82, resolve the given vector into its x component and y component. Each angle is in standard position.

79. Vector $\mathbf{A}$, with $A = 17.2, \theta = 22.2°$

80. Vector $\mathbf{A}$, with $A = 127, \theta = 132.4°$

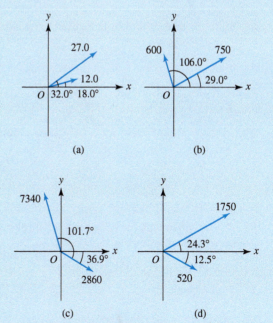

Figure 14.74

In Exercises 91 through 98, determine the resultant of the vectors with the given magnitudes and directions.

91. Vector **A**, with $A = 17.76, \theta = 90.00°$
Vector **B**, with $B = 21.20, \theta = 0.00°$

92. Vector **A**, with $A = 136.4, \theta = 50.8°$
Vector **B**, with $B = 635.0, \theta = 17.7°$

93. Vector **A**, with $A = 0.992, \theta = 18.4°$
Vector **B**, with $B = 0.545, \theta = 112.1°$

94. Vector **A**, with $A = 47.3, \theta = -36.2°$
Vector **B**, with $B = 98.1, \theta = 134.2°$

95. Vector **A**, with $A = 36, \theta = 25°$
Vector **B**, with $B = 49, \theta = -90°$

96. Vector **A**, with $A = 2.39, \theta = 51.2°$
Vector **B**, with $B = 5.08, \theta = 85.6°$

97. Vector **A**, with $A = 75.6, \theta = 29.4°$
Vector **B**, with $B = 48.3, \theta = 151.2°$

98. Vector **A**, with $A = 429, \theta = 248.1°$
Vector **B**, with $B = 301, \theta = 301.0°$

In Exercises 99 through 124, solve the given problems.

99. Solve the triangle shown in Figure 14.75a using the Law of Cosines.

100. Solve the triangle shown in Figure 14.76b using the Law of Cosines.

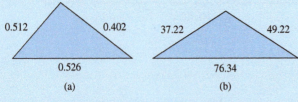

(a)   (b)

Figure 14.75

101. At a certain instant, the voltage in a certain alternating-current circuit is given by

$$v = 150 \cos 140.0°$$

Find the voltage.

102. At a given instant, the displacement of a particle moving with simple harmonic motion is given by

$$d = 16.0 \sin 250.0°$$

Find the displacement (in centimeters).

103. To find the distance between points *A* and *B* on opposite sides of a river, a distance *AC* is measured as 600 m, where *C* is on the same side of the river as *A*. Angle *BAC* is measured to be 110.0° and angle *ACB* is 34.5°. What is the distance between *A* and *B*? See Figure 14.76.

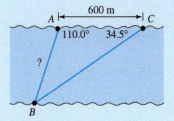

Figure 14.76

104. A 28.0-ft pole leans against a slanted wall, making an angle of 25.0° with the wall. If the foot of the pole is 12.5 ft from the foot of the wall, find the angle of inclination of the wall to the ground. See Figure 14.77.

Figure 14.77

105. A plumb line is dropped from the top peak of a tower that is 80.0 ft tall. If the plumb line reaches a point on the ground 3.10 ft from the center of the base of the tower, by how many degrees is the tower away from being vertical?

106. A triangular patio has sides of 40.0 ft, 32.0 ft, and 28.0 ft. What are the angles between the edges of the patio?

107. Two points on opposite sides of an obstruction are 156 m and 207 m, respectively, from a third point. The lines joining the first two points and the third point intersect at an angle of 100.5° at the third point. How far apart are the two original points? See Figure 14.78.

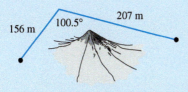

Figure 14.78

108. A crate is being held aloft by two ropes that are tied at the same point on the crate. They are 11.6 m and 6.69 m long, respectively, and the angle between them is 104.9°. Find the distance between the ends of the ropes. See Figure 14.79.

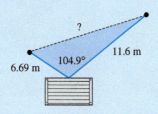

Figure 14.79

109. An observer measures the angle of elevation to the top of a mountain and obtains a value of 39.0°. After moving 100 ft farther away from the mountain, the angle of elevation is measured as 38.2°. How tall is the mountain? See Figure 14.80.

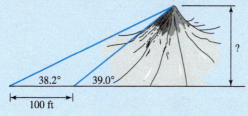

Figure 14.80

110. In a military training exercise, two bases are 380 km apart and they are trying to locate an unidentified airplane. The naval base is 40.0° south of west of the army base. Direction-finding instruments indicate that the plane is 74.0° south of east of the army base

and 8.3° north of east of the naval base. What is the distance between the plane and the nearest base? See Figure 14.81.

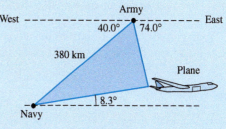

Figure 14.81

111. A tree surgeon attempts to drop a tree by making a cut and then pulling with a rope. The angle between the rope and the tree is 47.0° and the rope is pulled with a force that has a magnitude of 155 lb. Only the horizontal component of the force will cause the tree to fall. Find the horizontal component. See Figure 14.82.

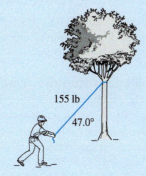

Figure 14.82

112. The force caused by a truck on an inclined loading ramp is represented by a vector that has a magnitude of 32,400 lb and a direction that is 82.0° below the negative x axis. Resolve this vector into its x and y components.

113. A missile is launched at an angle of 81.1° with respect to level ground. If it is moving at a speed of 2250 km/h, find the horizontal and vertical components of the velocity.

114. In conducting a ballistics test, it is determined that at one point, a bullet is traveling 800 ft/s along a path that makes an angle of 62.7° with level ground.

Find the horizontal and vertical components of the velocity vector.

115. A cable is connected to the front of a boat in a lake. The cable makes an angle of 40.5° with the water surface. If the cable is pulled with a force having a magnitude of 305 lb, find the force that pulls the boat forward along the surface of the lake. See Figure 14.83.

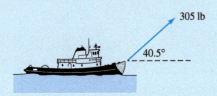

Figure 14.83

116. A jet climbs at a speed of 380 mi/h and its path makes an angle of 21.6° with level ground. Find the horizontal and vertical components of the velocity.

117. In conducting a stress test on an automobile part, two forces are applied at a point. A vertical force of 1200 lb pushes down on the part while a horizontal force of 850 lb pulls the part to the right. Find the resultant of these two forces.

118. A shearing pin is designed to break and disengage gears before damage is done to a machine. In conducting tests on such a pin, two forces are applied. One force pulls vertically upward with a magnitude of 9650 lb while another force pulls horizontally with a magnitude of 9370 lb. Find the resultant of these two forces. See Figure 14.84.

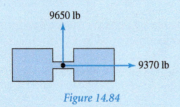

Figure 14.84

119. Two forces act on an object at right angles to each other. One force is 87.2 lb and the other is 13.7 lb. Find the resultant of these two forces.

120. Forces of 867 lb and 532 lb act on an object. The angle between these forces is 67.7°. Find the resultant of these two forces.

121. A machine pulls a cart with a rope in the same way that a child pulls on a wagon. The rope makes an angle of 26° with the horizontal, and the rope is pulled with a force of 36 lb. Find the horizontal component of the force. (This is the component that will cause the cart to move horizontally.) See Figure 14.85.

Figure 14.85

122. A 655-lb cart is on an incline that makes an angle of 34.8° with the horizontal. Find the force parallel to the surface of the incline. (This is the force that must be negated in order to prevent the cart from rolling down the incline.) See Figure 14.86.

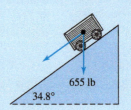

Figure 14.86

123. Two cables are connected to a weight and the angle between these cables is 52.7°. The cables are pulled with forces of 176 N and 221 N, respectively. Find the resultant of these two forces. See Figure 14.87.

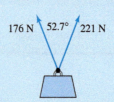

Figure 14.87

124. A plane travels 50.0 mi east from an airfield and then travels another 35.0 mi north. What is the plane's displacement from the airfield? See Figure 14.88.

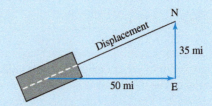

Figure 14.88

Chapter Test

1. Find the trigonometric functions of θ given that the terminal side of θ passes through $(8, 6)$.

2. Determine the quadrant in which the terminal side of θ lies given that $\sin \theta$ is positive and $\cos \theta$ is negative.

3. Solve the triangle with the given parts:

   $A = 48°, B = 68°$, and $a = 145$

4. Solve the triangle with the given parts:

   $a = 47.4, b = 40.0$, and $c = 45.5$

5. Find the x and y components of the vector shown below.

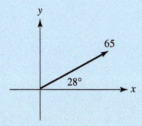

6. Determine the resultant of the vectors with the following magnitude and direction:

   $A = 449, \theta = 74.2°$
   $B = 285, \theta = 208.9°$

7. A communication satellite is directly above the extension of a line between receiving towers A and B. It is determined from radio signals that the angle of elevation of the satellite from tower A is 89.2° and the angle of elevation from tower B is 86.5°. If A and B are 1290 km apart, how far is the satellite from A? (Neglect the curvature of the earth.)

8. Two stakes, A and B, are 88.6 m apart. From a third stake C, an $\angle ACB$ is formed measuring 85.4°. At stake A, $\angle BAC$ measures 74.3°. Find the distance from stake C to each of the other stakes.

9. Two hockey players strike the puck at the same time, hitting it with horizontal forces of 5.75 lb and 3.25 lb that are perpendicular to each other. Find the resultant of these forces.

10. A surveyor locates a point at which an upgrade in a road is to begin. The point is 27.3 m northeast of the surveyor and 5.0 m north of a right-of-way marker. What is the surveyor's displacement from the right-of-way marker?

15 Graphs of Trigonometric Functions

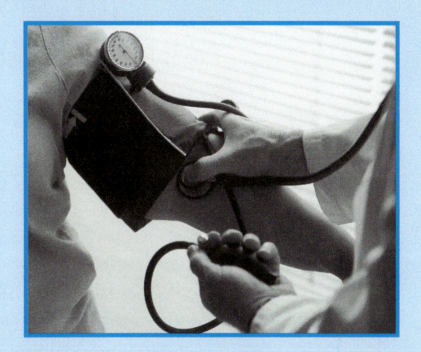

Periodic functions are functions that repeat a particular pattern, or cycle, over and over again. Tidal cycles and blood pressure are two examples of periodic functions that are familiar to many of us.

Have you ever been to the ocean and watched the tide come in and out? Have you ever checked a tide chart? If you have done either of these two things, you will have noticed that the tides often vary, but in predictable, repetitive ways. The range of the tides is affected by the relative positions of the sun and the moon. During the new moon and the full moon, the highest tides and lowest tides occur. During the first and third quarters of the lunar month, the lowest tides and the highest low tides occur. Throughout the month, the tidal range gradually increases and decreases between the minimum and the maximum values of the range. Tides are just one example of a

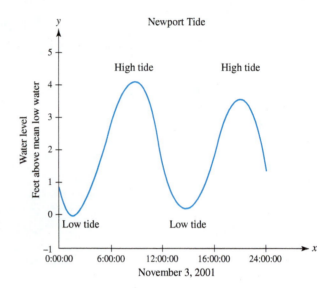

periodic function. When you go to the doctor's office for an annual checkup, one of the first things the nurse often does is to measure your blood pressure. The heart pumps blood by contracting periodically. Blood gushes through the body in waves and so blood pressure varies like a wave.

This is why blood pressure is always given as two numbers, say 120/60—the first is the peak blood pressure, and the second is the lowest blood pressure.

In this chapter we will introduce two trigonometric functions, the sine function and the cosine function, which can be used to model such repetitive behavior.

15.1   Radian Measure

Section Objectives
- Convert angle measures from degrees to radians
- Convert angle measures from radians to degrees

The **radian** is defined as the measure of the angle with these properties:

1. The angle has its vertex at the center of a circle.

2. The angle intercepts an arc on the circle, and the length of that arc equals the length of the radius of that circle.

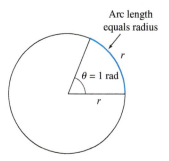

Figure 15.1

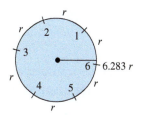

Figure 15.2

To understand this definition, refer to Figure 15.1. The indicated angle θ has a measure described as 1 radian (rad) because the arc length r equals the radius r.

Because the circumference of any circle is given by $c = 2\pi r$, it follows that the radius may be marked off 2π times (or about 6.283 times) along the circumference, as shown in Figure 15.2. *One complete rotation therefore corresponds to 2π rad.* Because one complete rotation also corresponds to 360°, we conclude that 2π rad = 360°. Dividing by 2 we get

$$\boxed{\pi \text{ rad } = 180°}$$

We can divide both sides of this equation by 180 to get

$$\boxed{1° = \frac{\pi}{180} \text{ rad } = 0.01745 \text{ rad}} \qquad \text{degrees to radians}$$

If we divide both sides of the original equation by π we get

$$\boxed{1 \text{ rad } = \frac{180°}{\pi} = 57.3°} \qquad \text{radians to degrees}$$

These equations express relationships between degrees and radians, and suggest ways of converting between radians and degrees. We summarize these procedures as follows:

1. To convert an angle measured in **degrees to radians** we multiply

$$degrees \cdot \frac{\pi}{180}$$

2. To convert an angle measured in **radians to degrees** we multiply

$$radians \cdot \frac{180}{\pi}$$

We will use these in Examples 1, 2, and 3, which follow.

Example 1

We convert 180° to radians as follows:

$$180° = (180)\left(\overbrace{\underbrace{\frac{\pi}{180}}_{180°}}^{\pi \text{ rad}}\right) = \pi = 3.14 \text{ rad}$$

We convert 240° to radians as follows:

$$240° = (240)\left(\frac{\pi}{180}\right) = \frac{4\pi}{3} = 4.19 \text{ rad}$$

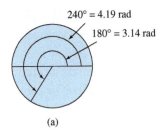

(a)

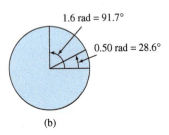

(b)

Figure 15.3

See Figure 15.3a. To convert 1.6 rad to degrees, we get

$$1.6 \text{ rad} = (1.6)\left(\frac{180}{\pi}\right) = \frac{288°}{\pi} = 91.7°$$

with labels: $180°$ and π rad

To convert 0.50 rad to degrees, we get

$$0.50 \text{ rad} = (0.50)\left(\frac{180}{\pi}\right) = \frac{90°}{\pi} = 28.6°$$

See Figure 15.3b.

One advantage of using radians is that it often simplifies formulas and calculations involving π. Because of this, radian measures are often expressed in terms of π. Example 2 illustrates this.

Example 2

Convert 30°, 45°, and 60° to radian measure and leave the answers in terms of π.

converting degrees to radians

$$30° = (30)\left(\frac{\pi}{180}\right) = \frac{\pi}{6} \text{ rad}$$

$$45° = (45)\left(\frac{\pi}{180}\right) = \frac{\pi}{4} \text{ rad}$$

$$60° = (60)\left(\frac{\pi}{180}\right) = \frac{\pi}{3} \text{ rad}$$

If we have a radian measure already expressed in terms of π, we can easily find the corresponding measure in degrees because the conversion will cause two values of π to cancel out through division. See Example 3.

Example 3

Convert $\pi/2$ and $2\pi/3$ to degree measure.

converting radians to degrees

$$\frac{\pi}{2} \text{ rad} = \left(\frac{\pi}{2}\right)\left(\frac{180}{\pi}\right) = 90°$$

$$\frac{2\pi}{3} \text{ rad} = \left(\frac{2\pi}{3}\right)\left(\frac{180}{\pi}\right) = 120°$$

Many modern uses of trigonometric ratios involve treating them as functions of real numbers instead of angle measurements. For example, we might consider $y = \sin x$ where x is a real number and not an angle measured in degrees or radians. This becomes important as we proceed to consider many applications in which angle measurements are inappropriate. In electronics, for example, we might describe the relationship between current i and time t by $i = 5 \sin 60\pi t$. Here, t is not an angle and it would be inappropriate to express t in degrees or radians. This brings us to a very important point. Because π is the *ratio* of the circumference of a circle to its diameter, it is one distance divided by another. As a result, radians really have no units and *radian measure amounts to measuring angles in terms of numbers*. As a result of this useful property of radians, **angle measurements lacking indicated units are understood to be in radian units**.

Example
4

Now Try It!

1. Express these angles in radians (leave your answer in terms of π).
   a. $30°$ **b.** $135°$
   c. $225°$ **d.** $305°$
2. Express these angles in degrees.
   a. $\dfrac{2\pi}{3}$ **b.** $\dfrac{7\pi}{12}$
   c. $\dfrac{8\pi}{9}$ **d.** $\dfrac{17\pi}{20}$

To convert 1.75 to degrees, we get

no units indicates radian measure

$$1.75 = (1.75)\left(\frac{180}{\pi}\right) = \frac{315°}{\pi} = 100°$$

To convert 270° to an equivalent number without units, we get

$$270° = (270)\left(\frac{\pi}{180}\right) = \frac{3\pi}{2} = 4.71 \quad \text{no units indicates radian measure}$$

In this example 1.75 and 4.71 are expressed without units, so they are understood to be radian measure.

Example
5

1. Evaluate $\cos \pi/3$. ⟵ no units indicates radian measure
   Since

   converting radians to degrees

   $$\frac{\pi}{3} = \left(\frac{\pi}{3}\right)\left(\frac{180}{\pi}\right) = 60° \quad \text{we get}$$

   $$\cos \frac{\pi}{3} = \cos 60° = 0.5000.$$

2. Evaluate $\tan 1.37$. ⟵ no units indicates radian measure
   Since

   $$1.37 = (1.37)\left(\frac{180}{\pi}\right) = 78.5° \quad \text{we get}$$

   $$\tan 1.37 = \tan 78.5° = 4.915.$$

Example
6

Find a number t (not in degrees) for which $\sin t = 0.9178$.

Since $\sin t = 0.9178$, we could get $t = 66.6°$, but we want t as a number not in degree units. We therefore convert $66.6°$ to radians to get

$$t = (66.6)\left(\frac{\pi}{180}\right) = 1.162$$

so that $\sin 1.162 \approx 0.9178$.

└── no units indicates radian measure

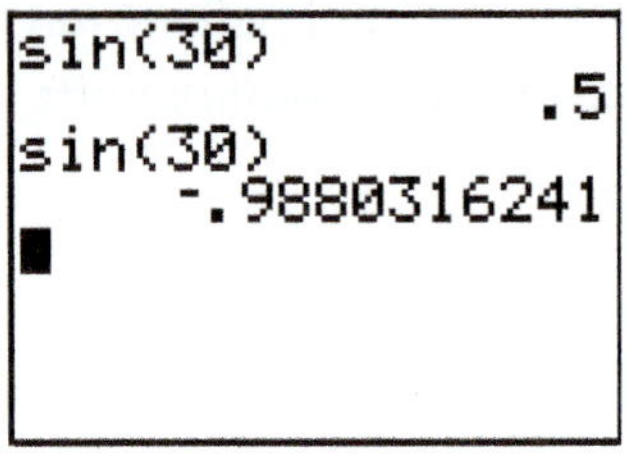

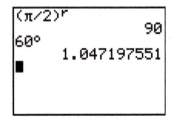

Figure 15.4

Most scientific and graphing calculators have a capability for selecting the degree mode or radian mode. If the radian mode is selected, then all entered angles are treated as radians.

An entry of sin 30 while in degree mode in your scientific calculator would yield a result of 0.5 (see Figure 15.4). The same entry would yield a result of -0.9880316241 if the calculator is in radian mode.

We can also use our calculator to change an angle measured in radians to an angle measured in degrees, or from an angle measured in degrees to an angle measured in radians. An angle of $\pi/2$ radians is easily converted to a $90°$ angle using this capability of the calculator (see Figure 15.5). Likewise, an angle measure of $60°$ can be converted to an angle of 1.047197551 radians with the calculator.

Figure 15.5

In any event, these conversions can be made by multiplying by the appropriate conversion factor as indicated in Examples 7 and 8.

Example
7

Convert $39.4°$ to radians by using a scientific calculator.

We have already converted degrees to radians by multiplying the given number of degrees by $\pi/180$. The correct key sequence is therefore

$$39.4 \times \pi \div 180 = 0.68765973$$

The key represented by π may involve a key whose secondary purpose is to give the value of π; if that is the case, precede the key by the key labeled 2nd; otherwise, use 3.1416 instead of π. This whole process can be streamlined by noting that $\pi/180 = 0.01745$ so that 39.4×0.01745 will yield the radian equivalent of $39.4°$. That is, to convert an angle to radians, we can simply multiply the angle by 0.01745.

Example
8

Convert 2.37 to degrees.

We can convert radians to degrees by multiplying by $180°/\pi$, or 57.296°. The entry of

$$2.37 \times 57.296 = 135.79152$$

shows that 2.37 is equivalent to 135.8° (rounded to tenths).

We should be sure that we understand the underlying theory so that errors will not arise from thoughtless use of calculators. Example 9 leads to two solutions even though the calculator might seem to suggest that there is only one solution.

Example
9

Find t in radians if $\sin t = 0.8415$ and $0 < t < 2\pi$.

Using a calculator *in the radian mode* we can easily determine that $\sin 1.0001 = 0.8415$ so that $t = 1.0001$. But in addition to this first-quadrant angle, **there is also a second-quadrant angle** for which $\sin t = 0.8415$. Using the idea of reference angles, we find the second-quadrant angle in radians as follows

For reference:
θ is standard position angle
α is reference angle

Quadrant
II $\alpha = 180° - \theta$

III $\alpha = \theta - 180°$

IV $\alpha = 360° - \theta$

┌── equivalent to 180°

$$\theta = \pi - 1.0001 = 2.1415$$

└── reference angle

Note that we solved for θ, with the angles expressed in radians instead of degrees. The two solutions are $t = 1.0001$ and $t = 2.1415$.

If we wish to find the value of a function of an angle greater than $\pi/2$, we must first determine which quadrant the angle is in and then find the reference angle. In this process we should note the radian measure equivalents of 90°, 180°, 270°, and 360°.

For 90° we have $\dfrac{\pi}{2} = 1.571$;

for 180° we have $\pi = 3.142$;

for 270° we have $\dfrac{3\pi}{2} = 4.712$;

for 360° we have $2\pi = 6.283$.

Example

10

Find sin 3.402 using the proper reference angle.

Because 3.402 is greater than 3.142 but less than 4.712, we know that this angle is in the third quadrant and has a reference angle of 3.402 − 3.142 = 0.260. Thus,

sin is − in third quadrant

$$\sin 3.402 = -\sin 0.260 = -0.2575$$

Now find cos 5.210. Because 5.210 is between 4.712 and 6.283, we know that this angle is in the fourth quadrant and its reference angle is 6.283 − 5.210 = 1.073. Thus,

cos is + in fourth quadrant

$$\cos 5.210 = \cos 1.073 = 0.4773$$

Example

11

The voltage (in volts) in a circuit is described by the equation

$$v = 112 \sin 0.540t$$

Find the voltage at $t = 1.35$ s.

With $t = 1.35$ s, we evaluate v as follows.

substitute for t

$$v = 112 \sin 0.540(1.35)$$

$$= 112 \sin 0.729 = 112(0.666)$$
$$= 74.6 \text{ V}$$

Using Technology

We can use the calculator to convert an angle measured in radians to an angle measured in degrees easily.

To convert $\pi/3$ to degrees:

1. Begin by checking to make sure that the calculator is in degree mode (see Figure 15.6).

2. Enter $(\pi/3)$.

3. Press 2nd APPS to access the ANGLE submenu.

4. Choose option #3 by scrolling down to #3 and pressing ENTER (see Figure 15.7).

Figure 15.6

5. Press ENTER again to see that $\pi/3$ is equal to an angle of 60 degrees.

To convert an angle measured in degrees to an angle measured in radians we follow a similar sequence of steps.

To convert 45° to radians:

1. Begin by checking to make sure that the calculator is in radian mode.

2. Enter (45).

3. Press 2nd APPS to access the ANGLE submenu.

4. Choose option #1 and press ENTER.

5. Press ENTER again to see that 45° is equivalent to 0.7853981634 radians (see Figure 15.8).

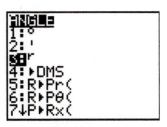

Figure 15.7

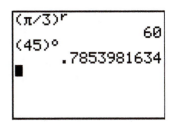

Figure 15.8

15.1 Exercises

In Exercises 1 through 8, express the given angle measurements in radians. (Leave answers in terms of π).

1. 40°, 16°

2. 36°, 240°

3. 55°, 330°

4. 20°, 335°

5. 30°, 135°

6. 60°, 300°

7. 175°, 210°

8. 52°, 280°

In Exercises 9 through 16, the given numbers express angle measure. Express the measure of each angle in terms of degrees.

9. $\dfrac{2\pi}{3}, \dfrac{\pi}{5}$

10. $\dfrac{\pi}{10}, \dfrac{4\pi}{5}$

11. $\dfrac{\pi}{12}, \dfrac{5\pi}{4}$

12. $\dfrac{2\pi}{9}, \dfrac{3\pi}{2}$

13. $\dfrac{14\pi}{15}, \dfrac{7\pi}{9}$

14. $\dfrac{11\pi}{12}, \dfrac{8\pi}{5}$

15. $\dfrac{7\pi}{36}, \dfrac{3\pi}{4}$

16. $\dfrac{51\pi}{30}, \dfrac{23\pi}{12}$

In Exercises 17 through 24, express the given angles in radian measure. (Round to three significant digits and use 3.14 as an approximation for π.)

17. 46.0°

18. 72.0°

19. 190.0°

20. 253.0°

21. 278.6°

22. 98.6°

23. 182.4°

24. 327.9°

In Exercises 25 through 32, the numbers express angle measure. Express the measure of each angle in terms of degrees.

25. 0.80

26. 0.25

27. 2.50

28. 1.75

29. 3.25

30. 5.00

31. 12.4

32. 75

In Exercises 33 through 44, evaluate the given trigonometric functions by first changing the radian measure to degree measure.

33. $\cos\dfrac{\pi}{3}$ **34.** $\sin\dfrac{\pi}{4}$ **35.** $\tan\dfrac{2\pi}{3}$

36. $\cos\dfrac{7\pi}{12}$ **37.** $\sin 1.05$ **38.** $\cos 0.785$

39. $\tan 0.875$ **40.** $\cot 1.43$ **41.** $\sec\dfrac{5\pi}{18}$

42. $\csc\dfrac{3\pi}{8}$ **43.** $\csc 3.00$ **44.** $\sec 5.66$

In Exercises 45 through 52, evaluate the given trigonometric functions directly, without first changing the radian measure to degree measure.

45. $\sin 1.0300$ **46.** $\tan 0.1450$

47. $\sec 3.650$ **48.** $\csc 4.766$

49. $\cot 6.180$ **50.** $\cos 5.050$

51. $\csc 2.875$ **52.** $\sec 3.135$

In Exercises 53 through 60, find the values of θ for $0 < \theta < 2\pi$. Each case has two solutions.

53. $\cos\theta = 0.5000$ **54.** $\sin\theta = 0.9975$

55. $\tan\theta = -2.8198$ **56.** $\cos\theta = -0.7900$

57. $\sin\theta = 0.4350$ **58.** $\cot\theta = 3.916$

59. $\csc\theta = -1.030$ **60.** $\sec\theta = 2.625$

In Exercises 61 through 66, solve the given problems.

61. Under certain conditions, the instantaneous voltage in a power line is given by

$$v = 170 \sin 377t$$

where t is time in seconds. Find v if $t = 0.0050$ s.

62. One pulley rotates through $340.2°$ while another pulley rotates through an angle of 6.00 radians. Which pulley rotates more?

63. In surveying, the horizontal distance between two points is calculated from

$$H = s \cos\alpha$$

Find H if $s = 37.4$ m, and $\alpha = 0.465$.

64. In analyzing the behavior of an object attached to a spring, the time t (in seconds) required for the object to move halfway in to the center from its initial position is found from

$$\cos kt = \frac{1}{2}$$

If $k = 10$, find the time t.

65. A flywheel rotates through 4.5 revolutions. What angle corresponds to this amount of rotation? Express the answer in degrees and in radians.

66. At cruising speed, the crankshaft of a light aircraft engine rotates 2300 r/min. Express 2300 complete revolutions as an angle in degree measure and in radian measure.

In Exercises 67 through 70, use a scientific calculator to convert the given angle measure from radians to degrees.

67. 2.55 **68.** 1.375

69. $\dfrac{\pi}{13}$ **70.** $\dfrac{7.3\pi}{9.2}$

In Exercises 71 through 74, use a scientific or graphing calculator to convert the given angle measure from degrees to radians.

71. $75°$ **72.** $135°$

73. $10°$ **74.** $170°$

In Exercises 75 through 78, use a scientific calculator to evaluate the given trigonometric functions.

75. $\sin 3.46227$ **76.** $\cos\dfrac{\pi}{4.9365}$

77. $\sin 0.000134\pi$ **78.** $\sin 0.000786$

Now Try It! Answers

1. a. $\dfrac{\pi}{6}$ **b.** $\dfrac{3\pi}{4}$ **c.** $\dfrac{5\pi}{4}$ **d.** $\dfrac{61\pi}{36}$

2. a. $120°$ **b.** $105°$ **c.** $160°$ **d.** $153°$

Section Objectives
- Determine arc length
- Determine the area of a circular sector
- Determine angular velocity
- Solve applied problems involving arc length or angular velocity

Radian measure has many applications in mathematics and technology. In this section we examine some of the common uses of radian measure in geometric and technical applications.

We know that a central angle of a circle is an angle with its vertex at the center of the circle. If the central angle is measured in radians, we can develop a simple and useful relationship between the *central angle θ*, the **length** *s of the corresponding* **arc,** and the radius *r* of the circle (see Figure 15.9). *The relationship between s, r, and θ is*

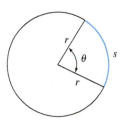

Figure 15.9

$$s = r\theta \qquad \text{(where } \theta \text{ is in radians)}$$

Example
1

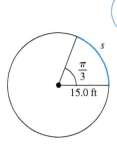

Figure 15.10

For a circle with a radius of *r* = 15.0 ft, find the length of the arc with a central angle of *θ* = *π*/3. See Figure 15.10.

Using *s* = *r θ*, we get

$$\theta \text{ in radians}$$

$$s = (15.0)\left(\frac{\pi}{3}\right)$$

$$= 15.7 \text{ ft}$$

Example
2

Given that a circle has an arc with a length of 4.60 cm and central angle of 120.0°, find the radius. See Figure 15.11.

With *s* = 4.60 cm and *θ* = (120.0)(*π*/180) so that **the central angle is in radian measure,** we setup the following equation.

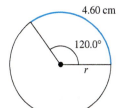

Figure 15.11

$$4.60 = r(120.0)\left(\frac{\pi}{180}\right) = \frac{120\pi r}{180}$$

$$\theta \text{ in radians}$$

or

$$r = \frac{(180)(4.60)}{120\pi} = 2.20 \text{ cm}$$

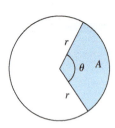

Figure 15.12

In Figure 15.12 we give an example of a **sector** of a circle. *A sector of a circle is the region enclosed by two radii and an arc of the circle.* Just as the arc length is proportional to the central angle, *the area of a sector is also proportional to the central angle* so that $A = k\theta$, where θ is in radian measure. Knowing that a complete circle has a central angle of $\theta = 2\pi$ and an area of $A = \pi r^2$, we can solve for the constant of proportionality k to get the following.

$$A = \frac{1}{2}\theta r^2 \qquad \text{(where } \theta \text{ is in radians)}$$

Here A is the area of the sector, r is the radius, and θ is the central angle in radians.

Example 3

A circle has a radius of 36.0 cm. Find the area of a sector with a central angle of 24.0°. See Figure 15.13.

▶ We must **be sure to express θ in radian measure.** Using the formula $A = \frac{1}{2}\theta r^2$ we get

Figure 15.13

$$A = \frac{1}{2}(24.0)\underbrace{\left(\frac{\pi}{180}\right)}_{\theta \text{ in radians}}(36.0)^2$$

$$= 271 \text{ cm}^2$$

Example 4

The area of a sector of a circle with a radius of 2.50 m is known to be 3.28 m^2. Find the central angle corresponding to this sector. See Figure 15.14.

With $A = 3.28$ m^2 and $r = 2.50$ m, we get

$$3.28 = \frac{1}{2}\theta(2.50)^2$$

Solving for θ we get

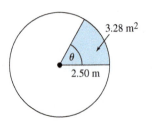

Figure 15.14

$$\theta = \frac{2(3.28)}{(2.50)^2} = 1.05 \qquad \text{no units indicates radian measure}$$

This means that the central angle is 1.05 rad, or 60.2°.

Our next application involves velocity. Recall that velocity is a vector with both magnitude and direction, whereas speed is a scalar that corresponds to the magnitude of velocity. When an object travels in a circular path at a constant speed, the distance traveled is the length of the arc that is its path. Using the formula for arc length, we divide both sides by t to get

$$\frac{s}{t} = \frac{r\theta}{t}$$

In this expression s/t is a speed because it represents distance per unit of time. Also, θ/t represents the change in radians per unit of time. We therefore express the above equation as follows.

$$\boxed{v = \omega r}$$ (where ω is in radians per unit time)

In this equation, v is the magnitude of the **linear velocity** and ω (the Greek omega) is the **angular velocity** *of an object moving around a circle of radius r*. See Figure 15.15. The units of ω are radians per unit of time. In practice, the angular velocity is often given in revolutions per unit of time, and for such cases it is necessary to convert ω to radians per unit of time before substitution in the formula for angular velocity. This conversion is not difficult if we remember that one revolution corresponds to 2π radians.

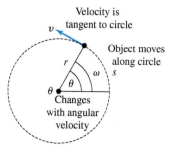

Figure 15.15

Example 5

An object is moving in a circular path with a radius of 10.0 cm, with an angular velocity of 3.80 rad/s. Find the linear velocity of this object. See Figure 15.16.

Using $v = \omega r$

$$v = (3.80)(10.0)$$
$$= 38.0 \text{ cm/s}$$

Therefore, the linear velocity of the object is 38.0 cm/s in the circular path.

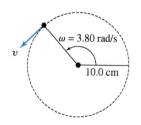

Figure 15.16

Example 6

A floppy disk commonly used in computer systems has a radius of 2.625 in. and it rotates with an angular velocity of 360 r/min. (We use r/min to designate revolutions per minute.) Find the linear velocity of a point on the outer edge. See Figure 15.17.

We can use the formula for angular velocity here, but we must be sure to express ω as radians per unit time, not revolutions per unit time. Because each complete revolution is equivalent to 2π rad, we have

$$\omega = (360)(2\pi) = 720\pi \text{ rad/min}$$

We now use the formula for angular velocity and get

$$v = (720\pi)(2.625)$$
$$= 5940 \text{ in./min}$$

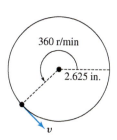

Figure 15.17

The linear velocity of a point on the outer edge is therefore 5940 in./min. We can see that a point nearer the middle has a lesser linear velocity because *r* is less for such a point.

Example 7

A connecting rod serves as a link between a piston and a rotating disk. **The disk rotates at 2800 r/min. If the rod is connected to the disk at a point 1.55 in. from the center, find the linear velocity of the point of connection. See Figure 15.18.**

We can again use the formula for angular velocity provided that **we express** the angular velocity ω in radians per unit time, not revolutions per **unit time.** Because 2800 r/min is equivalent to $(2800)(2\pi)$ rad/min, we get

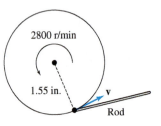

Figure 15.18

$$\underset{\omega \text{ in rad/min}}{v = (2800)(2\pi)(1.55)} = 27{,}300 \text{ in./min}$$

The linear velocity is 27,300 in./min.

Example 8

A light aircraft was damaged when the nosewheel collapsed. If the tip **of the pro**peller hit the runway and made an impression equivalent to an object **moving** 8640 ft/min, find the rate (in revolutions per minute) at which the **propeller was** turning. The propeller is 5.50 ft from tip to tip.

The velocity of 8640 ft/min is the linear velocity v. Noting that $v = 8640$ ft/min and $r = 5.50$ ft/2 = 2.75 ft, we use the formula for angular velocity

$$v = \omega r$$
$$8640 = \omega(2.75)$$

which means that $\omega = 8640/2.75 = 3140$ rad/min. We convert **the angular** velocity to revolutions per minute by dividing by 2π and we get $3140/2\pi = 500$ r/min. The propeller was turning at the rate of 500 r/min, indicating **that** the engine was idling.

15.2 Exercises

In Exercises 1 through 8, the radius of a circle and the central angle are given. (a) Find the length of the arc subtended by the central angle. (b) Find the area of the sector.

1. $r = 3.65$ cm, $\theta = 2.01$

2. $r = 15.78$ in., $\theta = \dfrac{\pi}{6}$

3. $r = 412$ mm, $\theta = 49.3°$

4. $r = 0.275$ m, $\theta = 98.6°$

5. $r = 2.37$ ft, $\theta = \dfrac{\pi}{12}$

6. $r = 9.27$ cm, $\theta = 120.0°$

7. $r = 6.55$ in., $\theta = 235.0°$

8. $r = 3960$ mi, $\theta = 2.38$

In Exercises 9 through 12, an object is moving in a circular path with the given radius and angular velocity. Find the linear velocity.

9. $r = 12.5$ cm, 5.65 rad/s

10. $r = 42.39$ mm, 362.0 rad/min

11. $r = 1.28$ ft, 861 r/min

12. $r = 3.75$ in., 453 r/s

In Exercises 13 through 30, solve the given problems.

13. In a circle with a radius of 36.0 cm, find the length of the arc subtended on the circumference by a central angle of 60.0°.

14. In a circle with a diameter of 12.4 m, find the length of the arc subtended on the circumference by a central angle of 36.0°.

15. Find the area of the circular sector indicated in Exercise 13.

16. Find the area of a sector of a circle given that the central angle is 150.0° and the diameter is 39.50 m.

17. Find the radian measure of an angle at the center of a circle with a radius of 73.0 cm that intercepts an arc length of 118 cm.

18. Find the central angle of a circle that intercepts an arc length of 928 mm when the radius of the circle is 985 mm.

19. Two concentric (same center) circles have radii of 12.5 ft and 18.0 ft. Find the portion of the area of the sector of the larger circle that is outside the smaller circle when the central angle is 60.0°. See Figure 15.19.

Figure 15.19

20. In a circle with a radius of 2.40 ft, the arc length of a sector is 4.00 ft. What is the area of the sector?

21. A pendulum 1.05 m long oscillates through an angle of 6.2°. Find the distance through which the end of the pendulum swings in going from one extreme position to the other.

22. The radius of the earth is about 3960 mi. What is the length, in miles, of an arc of the earth's equator for a central angle of 1.0°?

23. Two streets meet at an angle of 84.0°. What is the length of the piece of curved curbing at the intersection if it is constructed along the arc of a circle 18.0 ft in radius? See Figure 15.20.

Figure 15.20

24. In traveling three-fourths of the way along a traffic circle a car travels 0.203 mi. What is the radius of the traffic circle? See Figure 15.21.

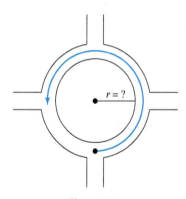

Figure 15.21

25. A truck whose tires are 33.0 in. in diameter is traveling at 55 mi/h. What is the angular velocity of the tires in radians per second?

26. A satellite circles the earth four times each 24.0 h. If its altitude is constant at 22,040 mi, what is its velocity? (The earth's radius is about 3960 mi.)

27. The armature of a dynamo is 54.0 cm in diameter and is rotating at 900 r/min. What is the linear velocity of a point on the rim of the dynamo?

28. A pulley belt 16.0 in. long takes 3.25 s to make one complete revolution. The diameter of the pulley is 4.57 in. What is the pulley's angular velocity in revolutions per minute?

29. A conical tent is made from a circular piece of canvas 16.0 ft in diameter, with a sector of central angle 120.0° removed. What is the surface area of the tent? See Figure 15.22.

120.0°

16.0 ft

Figure 15.22

30. A circular sector whose central angle is 210.0° is cut from a circular piece of sheet metal with a diameter of 18.0 cm. A cone is then formed by bringing the two radii of the sector together. What is the lateral surface area of the cone?

In Exercises 31 through 34, another use of radians is illustrated. Solve the given problems.

31. It can be shown through advanced mathematics that an excellent approximate method of evaluating $\sin \theta$ or $\tan \theta$ is given by

$$\sin \theta = \tan \theta = \theta$$

provided that the value of θ are small (the equivalent of a few degrees or less) and θ is expressed in radians. This equation is particularly useful for very small values of θ—even some scientific calculators cannot adequately handle angles of 1″ or 0.001° or less. Verify this equation by changing 1″ to radians and then evaluating $\sin 1''$ and $\tan 1''$.

32. Using $\sin \theta = \tan \theta = \theta$, evaluate $\tan 0.0005°$.

33. An astronomer observes that a star 14.5 light years away moves through an angle of 0.000054° in one year. Assuming it moves in a straight line perpendicular to the initial line of observation, how many miles does the star move? (One light year $= 5.88 \times 10^{12}$ mi)

34. In calculating the back line of a lot, a surveyor discovers an error of 0.05° in angle measurement. If the lot is 480.0 ft deep, by how much is the back line calculation in error? (See Figure 15.23.)

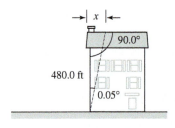

x

90.0°

480.0 ft

0.05°

Figure 15.23

15.3 Graphs of $y = a \sin x$ and $y = a \cos x$

Section Objectives
- Find the amplitude for a sine function and a cosine function
- Graph the sine function and cosine function

We have seen that the trigonometric functions can be used in many technical applications. Angles measured in degrees are not appropriate for many of those applications and radians are used instead. In this and the following two sections we develop procedures for graphing trigonometric functions using only radian measures. Also, because the sine and cosine functions are of the greatest importance, we shall restrict our attention to these two trigonometric functions.

We will construct our graphs on the rectangular coordinate system and *angles will be expressed as* **numbers,** and these numbers can have any appropriate unit of measurement. We begin with the function $y = \sin x$ and we construct a table of values of x and y for that function. Recall that x is the input value and y is the output value. In determining the values for the following table we use our calculator to determine the values for y. It is necessary for the calculator to be in radian mode. Then, if $x = \pi/6$ we see that $y = \sin(\pi/6) = 0.5$. We repeat this process to complete the following table.

x	0	$\dfrac{\pi}{6}$	$\dfrac{\pi}{3}$	$\dfrac{\pi}{2}$	$\dfrac{2\pi}{3}$	$\dfrac{5\pi}{6}$	π	$\dfrac{7\pi}{6}$	$\dfrac{4\pi}{3}$	$\dfrac{3\pi}{2}$	$\dfrac{5\pi}{3}$	$\dfrac{11\pi}{6}$	2π
y	0	0.5	0.87	1	0.87	0.5	0	-0.5	-0.87	-1	-0.87	-0.5	0

If we plot the points corresponding to the pairs of x and y values in the table, we obtain the graph shown in Figure 15.24.

Proceeding in the same way, we develop the table of values and graph that correspond to $y = \cos x$ (see Figure 15.20).

x	0	$\dfrac{\pi}{6}$	$\dfrac{\pi}{3}$	$\dfrac{\pi}{2}$	$\dfrac{2\pi}{3}$	$\dfrac{5\pi}{6}$	π	$\dfrac{7\pi}{6}$	$\dfrac{4\pi}{3}$	$\dfrac{3\pi}{2}$	$\dfrac{5\pi}{3}$	$\dfrac{11\pi}{6}$	2π
y	1	0.87	0.5	0	-0.5	-0.87	-1	-0.87	-0.5	0	0.5	0.87	1

The graphs shown in Figures 15.24 and 15.25 actually continue indefinitely to the right and to the left as they repeat the same basic pattern shown. This repetitive nature of both graphs is often referred to as the property of being **periodic.** It is this property that makes the trigonometric functions especially relevant in the study of phenomena that are cyclic in nature, such as electric current, sound waves, radio waves, orbital motion, fluid motion, vibration of a spring, and so on.

Comparison of the graphs in Figures 15.24 and 15.25 should reveal that the two graphs have the same shape. The cosine curve is actually the sine curve moved $\pi/2$ units to the left.

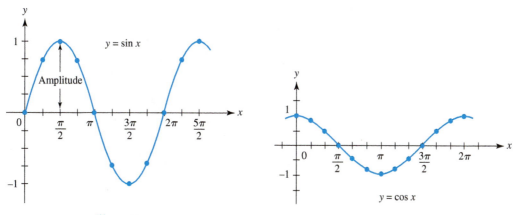

Figure 15.24   Figure 15.25

Having graphed $y = \sin x$ and $y = \cos x$, we will now consider a variation of these two basic equations. If we want to graph $y = a \sin x$, we simply note that each height y obtained for the graph of $y = \sin x$ is multiplied by the number a. Instead of reaching up to 1 and down to -1, the curve will now rise and fall between a and $-a$. The largest value of y is therefore $|a|$ instead of 1. We use the absolute value because a can be a negative number. *The number $|a|$ is called the* **amplitude** *of the curve and it represents the maximum y value of the curve.* These statements also apply to $y = a \cos x$. This can be seen in Examples 1 and 2 as follows.

Example
1

Plot the curve of $y = 4 \sin x$.

We can develop the following table of values. Note that each value of y is four times the corresponding value of y in the table for $y = \sin x$. The graph is shown in Figure 15.26.

x	0	$\dfrac{\pi}{6}$	$\dfrac{\pi}{3}$	$\dfrac{\pi}{2}$	$\dfrac{2\pi}{3}$	$\dfrac{5\pi}{6}$	π	$\dfrac{7\pi}{6}$	$\dfrac{4\pi}{3}$	$\dfrac{3\pi}{2}$	$\dfrac{5\pi}{3}$	$\dfrac{11\pi}{6}$	2π
y	0	2	3.5	4	3.5	2	0	-2	-3.5	-4	-3.5	-2	0

From the equation

amplitude

$y = 4 \sin x$

we determine that the amplitude is 4, as indicated. By comparing Figure 15.26 to Figure 15.24, we see that the shapes are the same except that the graph for $y = 4 \sin x$ reaches high points and low points that are four times as large as the graph for $y = \sin x$.

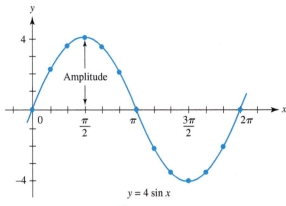

$y = 4 \sin x$

Figure 15.26

Example 2

Plot the curve of $y = -2 \cos x$.

The following table of values can be used for constructing the graph shown in Figure 15.27.

x	0	$\dfrac{\pi}{6}$	$\dfrac{\pi}{3}$	$\dfrac{\pi}{2}$	$\dfrac{2\pi}{3}$	$\dfrac{5\pi}{6}$	π	$\dfrac{7\pi}{6}$	$\dfrac{4\pi}{3}$	$\dfrac{3\pi}{2}$	$\dfrac{5\pi}{3}$	$\dfrac{11\pi}{6}$	2π
y	-2	-1.7	-1	0	1	1.7	2	1.7	1	0	-1	-1.7	-2

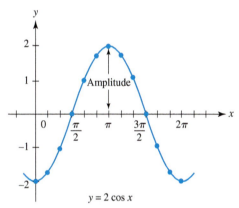

Figure 15.27

The amplitude of

$$\text{amplitude} = |-2| = 2$$

$$y = -2 \cos x$$

► is 2, as shown. ***The negative sign has the effect of inverting the curve.***

It is not always necessary to develop a table of values like the tables included in Examples 1 and 2. Given any equation of the form $y = a \sin x$ or $y = a \cos x$, we can quickly construct a sketch of the curve that includes the intercepts and amplitude. In Example 3 we will sketch the curve instead of plotting the curve from a table of values. We will use only a few important values along with our knowledge of the basic shape of the curve and the effect of the amplitude.

For graphs of the form $y = a \sin x$ or $y = a \cos x$, the values of x for which the curve either crosses the x axis or reaches its lowest or highest points are $x = 0$, $\pi/2$, π, $3\pi/2$, and 2π. It is these particular values of x that we should definitely include in our abbreviated table. We can begin by setting up the table as follows.

x	0	$\dfrac{\pi}{2}$	π	$\dfrac{3\pi}{2}$	2π
y					

The corresponding values of y are then found by substituting each value for x in the original equation. The pairs of values included in the completed table are then plotted and connected by a smooth curve.

Example
3

Sketch the graph of $y = \frac{1}{2}\cos x$.

Instead of developing a large table of values, we set up a table of values corresponding to points where the curve either crosses the x axis or where the curve achieves its highest and lowest heights. Starting with those values of x shown in the above incomplete table, we proceed to find the corresponding values of y as follows:

$$y = \frac{1}{2}\cos 0 = \frac{1}{2}(1) = \frac{1}{2} \qquad \text{one of highest points}$$

$$y = \frac{1}{2}\cos \frac{\pi}{2} = \frac{1}{2}(0) = 0 \qquad \text{crosses axis}$$

$$y = \frac{1}{2}\cos \pi = \frac{1}{2}(-1) = -\frac{1}{2} \qquad \text{one of lowest points}$$

$$y = \frac{1}{2}\cos \frac{3\pi}{2} = \frac{1}{2}(0) = 0 \qquad \text{crosses axis}$$

$$y = \frac{1}{2}\cos 2\pi = \frac{1}{2}(1) = \frac{1}{2} \qquad \text{one of highest points}$$

The table can now be completed as

x	0	$\dfrac{\pi}{2}$	π	$\dfrac{3\pi}{2}$	2π
y	$\dfrac{1}{2}$	0	$-\dfrac{1}{2}$	0	$\dfrac{1}{2}$

We now plot the points $\left(0, \frac{1}{2}\right)$, $\left(\frac{\pi}{2}, 0\right)$, $\left(\pi, -\frac{1}{2}\right)$, $\left(\frac{3\pi}{2}, 0\right)$, and $\left(2\pi, \frac{1}{2}\right)$ and connect them as shown in Figure 15.28.

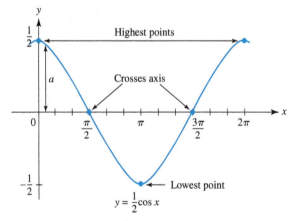

Figure 15.28

Example
4

In analyzing the current in an alternating-current circuit, a technician uses the equation $y = -110 \sin x$, where x represents time (in seconds) and y represents current (in amperes). Sketch the graph of $y = -110 \sin x$.

In the accompanying table we list the important values associated with this curve.

	crosses axis			
x	0	$\dfrac{\pi}{2}$	π	$\dfrac{3\pi}{2}$	2π
y	0	-110	0	110	0
	lowest		highest	

We now plot these five points and connect them by following the general shape of the sine curve. The sketch is shown in Figure 15.29.

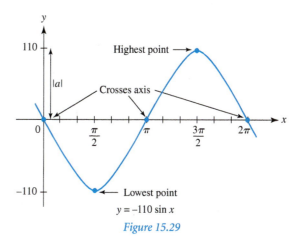

Figure 15.29

Graphing calculators make it easy to graph trigonometric functions. Graphing the function easily allows us more time to explore the nature and behavior of the function.

 Using Technology

Use a graphing calculator to create the graph of the function $y = \sin(x)$. To do this use the following sequence of steps.

1. Begin by putting the calculator in radian mode.

2. Press the Y= key. Enter the expression $\sin(x)$ after the $=$ sign at Y1.

3. Press ZOOM, then #7. This gives you a graph in which the horizontal axis will be in radian measure, taking on values from -2π to 2π.

4. Compare your graph with Figure 15.30.

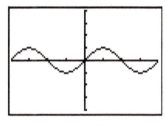

Figure 15.30

We can enter several functions on the same set of axes, all with different values for the **amplitude**. Doing so allows us to explore how changes in that value impact the graph of the sine function.

1. Press the Y= key. Enter the expression $\sin(x)$ after the = sign at Y1.

2. Enter the expression $4\sin(x)$ after the = sign at Y2 and $-2\sin(x)$ after the = sign at Y3. Be sure to use the appropriate negative sign in front of the 2. (Use the (−) key.)

3. Press ZOOM, then #7 to create three graphs on the same set of axes. The graphs will be drawn in the order in which they are entered into your calculator (see Figure 15.31).

4. Compare your graphs with Figure 15.32.

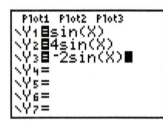

Figure 15.31

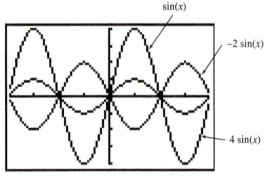

Figure 15.32

The process would be the same for graphing the cosine function.

1. Make sure your calculator is in radian mode.

2. Press the Y= key. Enter the expression $\cos(x)$ after the = sign at Y1.

3. Press ZOOM, then #7. This gives you a graph in which the horizontal axis will be in radian measure, taking on values from -2π to 2π.

4. Compare your graph with Figure 15.33.

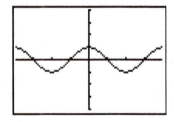

Figure 15.33

The TRACE Key

Once you have created your graph you can use the TRACE key to determine the values of various coordinates on the graph, including the values for the zeros, and the maximums and minimums of the graphs.

15.3 Exercises

In Exercises 1 through 4, use your calculator to complete the following table for the given functions and then plot the resulting graph.

x	$-\pi$	$-\dfrac{3\pi}{4}$	$-\dfrac{\pi}{2}$	$-\dfrac{\pi}{4}$	0	$\dfrac{\pi}{4}$	$\dfrac{\pi}{2}$	$\dfrac{3\pi}{4}$	π	$\dfrac{5\pi}{4}$	$\dfrac{3\pi}{2}$	$\dfrac{7\pi}{4}$	2π	$\dfrac{9\pi}{4}$	$\dfrac{5\pi}{2}$	$\dfrac{11\pi}{4}$	3π
y																	

1. $y = \sin x$ **2.** $y = \cos x$

3. $y = 5 \sin x$ **4.** $y = -3 \cos x$

In Exercises 5 through 8, refer to the graphs shown below and identify the equation of the function displayed. The hash marks on the axes represent one unit each, so that the positive x axis runs from 0 to about 6.3 (or 2π). In each case, the equation is of the form $y = a \sin x$ or $y = a \cos x$.

5. Figure 15.34   **6.** Figure 15.35

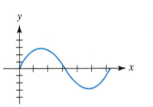

Figure 15.34

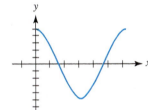

Figure 15.35

7. Figure 15.36   **8.** Figure 15.37

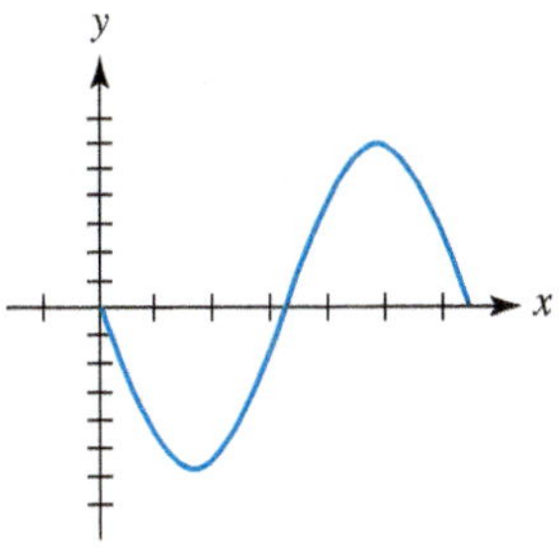

Figure 15.36

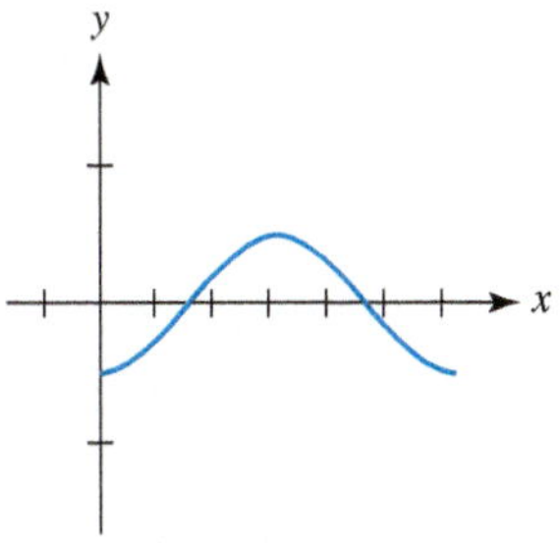

Figure 15.37

In Exercises 9 through 24, sketch the curves of the indicated functions. Use your graphing calculator to check your results.

9. $y = 2 \cos x$ **10.** $y = 4 \cos x$

11. $y = \frac{3}{2} \cos x$ **12.** $y = \frac{9}{4} \cos x$

13. $y = 2 \sin x$ **14.** $y = 3 \sin x$

15. $y = 0.6 \sin x$ **16.** $y = \frac{5}{2} \sin x$

17. $y = -\cos x$ **18.** $y = -4 \cos x$

19. $y = -1.2 \cos x$ **20.** $y = -0.8 \cos x$

21. $y = -\sin x$ **22.** $y = -6 \sin x$

23. $y = -\frac{5}{4} \sin x$ **24.** $y = -0.4 \sin x$

Although units of π are often convenient, we must remember that π is simply a number. Numbers that are not multiples of π can be used as well. In Exercises 25 through 36, plot the indicated graphs by finding the values of y corresponding to the values of 0, 1, 2, 3, 4, 5, 6, and 7 for x. (Remember: The numbers 0, 1, 2, and so forth represent radian measure.) Use your calculator to find the values of y.

25. $y = \cos x$ **26.** $y = 2 \cos x$

27. $y = \sin x$ **28.** $y = 3 \sin x$

29. $y = \frac{2}{3} \sin x$ **30.** $y = \frac{3}{2} \sin x$

31. $y = -2 \cos x$ **32.** $y = -4.5 \cos x$

33. $y = -2.6 \cos x$ **34.** $y = \frac{1}{3} \cos x$

35. $y = -3.5 \sin x$ **36.** $y = -0.8 \sin x$

15.4　Graphs of $y = a \sin bx$ and $y = a \cos bx$

Section Objective

• Find the period for a sine function and a cosine function

In the last section we considered the basic graphs of $y = \sin x$ and $y = \cos x$. We also discussed one variation in these basic graphs and saw the effects changing the amplitude had on the graphs of the form $y = a \sin x$ or $y = a \cos x$. In this section we will consider another variation.

We have noted that the sine and cosine functions are periodic in the sense that they repeat a basic pattern. In fact, all of the equations considered in Section 15.3 have graphs that repeat the pattern every 2π units of x. For such periodic functions, we now find it useful to define the **period** of a function F as the number P where $F(x) = F(x + P)$. Simply stated, *the period P refers to the x distance between any point and the next point at which the same pattern of y values starts repeating.* We might think of the period as the distance (along the x axis) required to get one complete cycle of a repeating pattern.

In Section 15.3 we saw that the amplitude is changed by changing the coefficient preceding the function. For example, $y = \sin x$ and $y = 4 \sin x$ have amplitudes of 1 and 4, respectively. We will now show that *we can change the period of the sine and cosine functions by altering the coefficient of the angle.* To do this, we will now graph $y = \sin 4x$. For this graph we choose suitable values for x, then multiply those values by 4, and then find the sine of the result.

x	0	$\dfrac{\pi}{16}$	$\dfrac{\pi}{8}$	$\dfrac{3\pi}{16}$	$\dfrac{\pi}{4}$	$\dfrac{5\pi}{16}$	$\dfrac{3\pi}{8}$	$\dfrac{7\pi}{16}$	$\dfrac{\pi}{2}$	$\dfrac{9\pi}{16}$	$\dfrac{5\pi}{8}$
$4x$	0	$\dfrac{\pi}{4}$	$\dfrac{\pi}{2}$	$\dfrac{3\pi}{4}$	π	$\dfrac{5\pi}{4}$	$\dfrac{3\pi}{2}$	$\dfrac{7\pi}{4}$	2π	$\dfrac{9\pi}{4}$	$\dfrac{5\pi}{2}$
$y = \sin 4x$	0	0.7	1	0.7	0	-0.7	-1	-0.7	0	0.7	1

Plotting the values of x and y in this table, we get the graph shown in Figure 15.38.

Upon examination of the table and the graph in Figure 15.38, we can see that $y = \sin 4x$ completes one full cycle between $x = 0$ and $x = \pi/2$ so that the period is $\pi/2$. Because $y = \sin x$ has a period of 2π while $y = \sin 4x$ has a period of $\pi/2$, we can see that the 4 had the effect of reducing the period to one-fourth of the period of $y = \sin x$. This suggests the following generalization which is valid: *Both $\sin x$ and $\cos x$ have a period of 2π, whereas $\sin bx$ and $\cos bx$ both have a period of $2\pi/b$.*

We have just seen that $y = \sin x$ and $y = \sin 4x$ have different periods. Note very carefully that *the period is changed by altering the coefficient of the angle.* Both $y = \sin x$ and $y = 4 \sin x$ (see Figure 15.39) have the same period but *different amplitudes*, whereas $y = \sin x$ and $y = \sin 4x$ have the same amplitude but *different periods*.

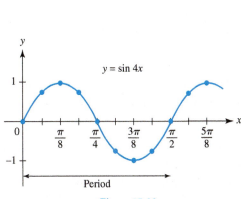

Figure 15.38

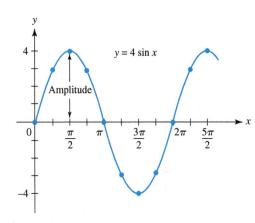

Figure 15.39

Example 1

Find the period of $y = \sin 5x$.

The period of $y = \sin 5x$ is $\dfrac{2\pi}{5}$. This means that the graph of $y = \sin 5x$ will repeat the same cycle every $2\pi/5$ (approximately 1.26) units of x.

Example 2

Find the periods of $y = \cos 3x$, $y = \cos \frac{1}{2}x$, and $y = \cos \pi x$.

The period of $y = \cos 3x$ is $\dfrac{2\pi}{3}$.

The period of $y = \cos \frac{1}{2}x$ is $\dfrac{2\pi}{\frac{1}{2}} = 4\pi$.

The period of $y = \cos \pi x$ is $2\pi/\pi = 2$.

For each of these three cases, the amplitude is 1.

The concepts of amplitude and period can be summarized as follows: *The functions $y = a \sin bx$ and $y = a \cos bx$ both have amplitude $|a|$ and period $2\pi/b$.*

For $y = a \sin bx$ or $y = a \cos bx$:		
amplitude	$\|a\|$	Curve goes as high as $\|a\|$ and as low as $-\|a\|$.
period	$\dfrac{2\pi}{b}$	One full cycle is completed in $\dfrac{2\pi}{b}$ units of x.

These properties are extremely useful in developing sketches, as Examples 3–5 illustrate.

Example 3

A piston is designed to oscillate so that its vertical displacement is described by the equation $y = 4 \sin 2x$, where x is the time (in seconds) and y is the displacement

$$a \qquad b$$

(in centimeters). Sketch the graph of $y = 4 \sin 2x$ for $0 \leq x \leq 2\pi$.

For the given function we know that the amplitude is 4 and the period is $2\pi/2 = \pi$. The amplitude of 4 indicates that the curve can go up to 4 and down to -4. The period of π indicates that a complete cycle will be repeated for every π units of x. The graph shown in Figure 15.40 is a result of *using the amplitude and period plus our knowledge of the basic sine curve.*

These lead us to the following values:

x	0	$\dfrac{\pi}{4}$	$\dfrac{\pi}{2}$	$\dfrac{3\pi}{4}$	π	$\dfrac{5\pi}{4}$	$\dfrac{3\pi}{2}$	$\dfrac{7\pi}{4}$	2π
y	0	4	0	-4	0	4	0	-4	0

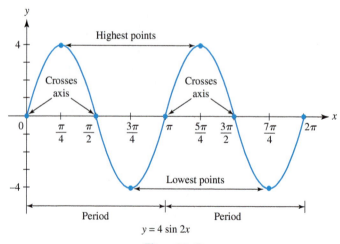

$y = 4 \sin 2x$

Figure 15.40

Example
4

Sketch the graph of $y = -3 \cos 4x$ for $0 \le x \le 2\pi$.

We might begin by recalling the basic pattern of the cosine function. Considering amplitude next, we see that this function will go up to 3 and down to -3. Also, the negative sign will cause the basic curve to be inverted. Finally, the period is $2\pi/4 = \pi/2$, so that a full cycle will be completed every $\pi/2$ units of x. The sketch can therefore be developed by transforming the basic cosine curve as follows: *Invert it, stretch it out vertically and complete the first cycle between $x = 0$ and $x = \pi/2$, then repeat the pattern.* The table of important values that identifies key points on the graph for the first cycle is as follows. The graph is shown in Figure 15.41.

x	0	$\dfrac{\pi}{8}$	$\dfrac{\pi}{4}$	$\dfrac{3\pi}{8}$	$\dfrac{\pi}{2}$
y	-3	0	3	0	-3

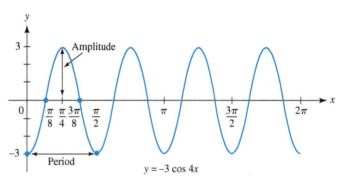

$y = -3 \cos 4x$

Figure 15.41

Example
5

Sketch the function $y = \sin \pi x$ for $0 \le x \le \pi$.

For this function the amplitude is 1 and the period is $2\pi/\pi = 2$. Because the value of the period is not expressed in terms of π, *it will be more convenient to use regular decimal units for x.* The following table lists important values and the graph is shown in Figure 15.42.

x	0	0.5	1	1.5	2	2.5	3
y	0	1	0	-1	0	1	0

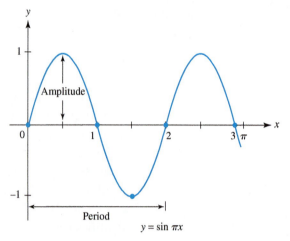

Figure 15.42

15.4 Exercises

In Exercises 1 through 20, find the period of each given function.

1. $y = \sin 3x$

2. $y = 2 \sin 3x$

3. $y = 2 \sin 4x$

4. $y = 3 \sin 10x$

5. $y = 4 \cos 3x$

6. $y = 4 \cos 10x$

7. $y = -2 \sin 6x$

8. $y = -3 \sin 8x$

9. $y = -3 \cos 5x$

10. $y = -4 \cos 6x$

11. $y = 5 \cos 2\pi x$

12. $y = 3 \cos 10\pi x$

13. $y = 5 \sin \frac{1}{2}x$

14. $y = -4 \sin \frac{1}{3}x$

15. $y = -4 \cos \frac{2}{3}x$

16. $y = \frac{1}{3} \cos \frac{1}{4}x$

17. $y = 1.5 \sin \frac{3}{2}\pi x$

18. $y = 0.8 \cos \frac{1}{3}\pi x$

19. $y = 4.5 \cos \pi^2 x$

20. $y = \frac{1}{2} \sin \frac{x}{\pi}$

In Exercises 21 through 40, sketch the graphs of the functions given for Exercises 1 through 20 from 0 to 2π. Use your graphing calculator to check your graph.

In Exercises 41 through 48, the period is given for a function of the form $y = \sin bx$. Write the equation corresponding to the given values of the period.

41. π **42.** $\pi/3$

43. 2   **44.** 3

45. 1   **46.** 6

47. $2\pi/5$ **48.** $3\pi/7$

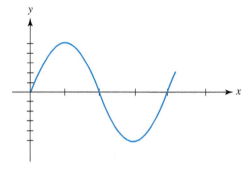

Figure 15.45

In Exercises 49 through 52, refer to the following graphs and identify the equation of the function displayed. The hash marks on the axes represent one unit each. In each case, the equation is of the form $y = a \sin bx$ or $y = a \cos bx$.

52. Figure 15.46

49. Figure 15.43

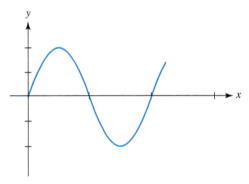

Figure 15.43

50. Figure 15.44

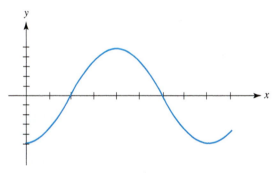

Figure 15.46

In Exercises 53 through 60, sketch one cycle of the indicated graphs. In each case, verify your result by graphing the same function on your graphing calculator.

53. The electric current in a certain alternating-current circuit is given by

$$i = 4 \sin 80\pi t$$

where i is the current in amperes and t is the time in seconds.

54. A generator produces a voltage given by

$$V = 120 \cos 30\pi t$$

where t is the time in seconds.

55. The vertical displacement x of a certain object oscillating at the end of a spring is given by

$$x = 7.5 \cos 3\pi t$$

where x is measured in inches and t in seconds.

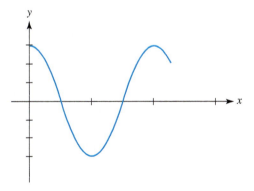

Figure 15.44

51. Figure 15.45

56. The velocity of a piston in a certain engine is given by

$$v = 5.0 \cos 1600\pi t$$

where v is the velocity in centimeters per second and t is the time in seconds.

57. A buoy floats in water. Its vertical motion caused by waves is described by the equation

$$y = 2.37 \sin 0.598t$$

The height y is in feet and the time t is in seconds. Sketch the graph of y versus t and include at least one full cycle of this periodic function.

58. In analyzing the pressure variations caused by the sound wave of a singer hitting the note A, the following equation is determined.

$$y = 0.00001 \sin \frac{8\pi}{3} x$$

The variable y is measured in atmospheres, and x is in meters. Sketch the graph of y versus x and include at least one full cycle of this periodic function.

59. When a tuning fork is activated with a frequency of 60 cycles per second, the pressure p at time t seconds is described by

$$p = \sin 120\pi t$$

Sketch the graph of p versus t and include at least one full cycle of this periodic function.

60. The blade of a saber saw moves vertically up and down at the rate of 18 strokes per second. The vertical displacement of the tip of the blade is described by the equation

$$y = 1.21 \sin 36\pi t$$

where y is in centimeters and t is in seconds. Sketch the graph of y versus t and include at least one full cycle of this periodic function.

15.5 Graphs of $y = a \sin (bx + c)$ and $y = a \cos (bx + c)$

Section Objective

- Find the displacement and phase shift for a sine function and a cosine function

Beginning with the basic graphs of $y = \sin x$ and $y = \cos x$, we have shown the effects caused by changes in amplitude and period. In this section we consider one more variation. In the function $y = a \sin (bx + c)$, c is the **phase angle** and its effect can be seen in Example 1.

Example **1**

Sketch the graph of $y = \sin \left(x + \dfrac{\pi}{2} \right)$.

For this function the amplitude is 1, the period is 2π, but the phase angle is $c = \pi/2$. We first present selected important values in the following table and develop the graph shown in Figure 15.47.

x	$-\dfrac{\pi}{2}$	0	$\dfrac{\pi}{2}$	π	$\dfrac{3\pi}{2}$	2π
y	0	1	0	-1	0	1

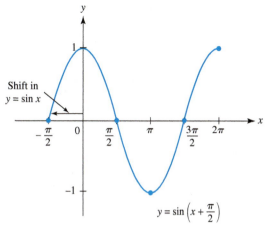

Shift in
$y = \sin x$

$y = \sin \left(x + \dfrac{\pi}{2} \right)$

Figure 15.47

Examining Figure 15.47, we see that the graph of $y = \sin \left(x + \dfrac{\pi}{2} \right)$ is identical to the graph of $y = \sin x$ with the exception that it is shifted $\pi/2$ units to the left. Because the phase angle $c = \pi/2$ is the only variation from $y = \sin x$, it follows that *the phase angle causes the curve to shift horizontally.* From Example 1 it might seem that the amount of shift is $-c$, but it is actually $-\dfrac{c}{b}$, as illustrated in Example 2.

The quantity $-\dfrac{c}{b}$ *is called the* **displacement** *(or* **phase shift***).* Remember, the curve shifts to the left if $-\dfrac{c}{b} < 0$ and to the right if $-\dfrac{c}{b} > 0$.

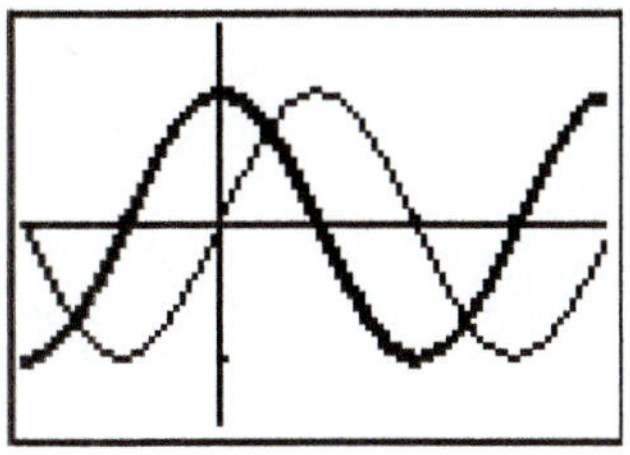

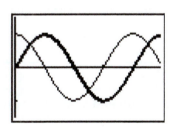

Figure 15.48

Figure 15.49

When we view both $y = \sin x$ and $y = \sin (x + \pi/2)$ on the same set of axes it is easy to see that $\sin (x + \pi/2)$ (shown with the bold line in Figure 15.48) is the graph of $y = \sin x$ shifted to the left by $\pi/2$ units.

Similarly, if we were to graph $y = \cos x$ and $y = \cos (x - \pi/2)$ on the same set of axes, we can see that the graph of $\cos (x - \pi/2)$ is the graph of $y = \cos x$ shifted to the right by $\pi/2$ (see Figure 15.49). The direction of the phase shift will be discussed in Examples 3–5.

Example 2

Sketch the graph of $y = \cos \left(2x + \dfrac{\pi}{4} \right)$.

For this function, $a = 1, b = 2$, and $c = \pi/4$ so that the amplitude is 1, the period is $2\pi/b = 2\pi/2 = \pi$, and the displacement is $-c/b = -(\pi/4)/2 = -\pi/8$.

This is a shift of $\pi/8$ to the *left* because $-c/b < 0$. If we begin a cycle of the curve at $x = -\pi/8$, it will be completed π units to the right of that starting

▶ point, namely $(-\pi/8) + \pi = 7\pi/8$; ***the next cycle then begins at $7\pi/8$,*** and so on. The following table lists some important values and the graph is shown in Figure 15.50.

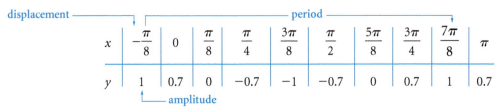

x	$-\dfrac{\pi}{8}$	0	$\dfrac{\pi}{8}$	$\dfrac{\pi}{4}$	$\dfrac{3\pi}{8}$	$\dfrac{\pi}{2}$	$\dfrac{5\pi}{8}$	$\dfrac{3\pi}{4}$	$\dfrac{7\pi}{8}$	π
y	1	0.7	0	-0.7	-1	-0.7	0	0.7	1	0.7

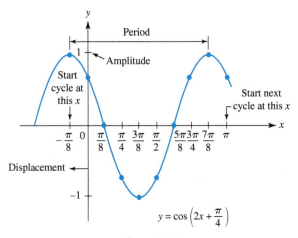

Figure 15.50

After studying Example 2 we can see that we have three important quantities to determine. They are summarized as follows:

For $y = a \sin (bx + c)$ or $y = a \cos (bx + c)$:		
amplitude	$\lvert a \rvert$	Curve goes as high as $\lvert a \rvert$ and as low as $-\lvert a \rvert$.
period	$\dfrac{2\pi}{b}$	One full cycle is completed in $\dfrac{2\pi}{b}$ units of x.
displacement	$-\dfrac{c}{b}$	Curve is displaced $-\dfrac{c}{b}$ units of x

We can find the value of the phase shift or displacement as follows:

$$\text{displacement} = -\frac{c}{b}$$

To determine the direction of the shift:

1. When $-\dfrac{c}{b}$ yields a *positive* result the curve shifts to the *right* by the amount of the displacement.

2. When $-\dfrac{c}{b}$ yields a *negative* result the curve shifts to the *left* by the amount of the displacement.

Here is a helpful hint in identifying which values of x are of enough significance to include in the tables we have been developing: First begin with the displacement value of $-c/b$ and then add one-fourth of the period to get the second x value. To this result add one-fourth of the period to get the third x value, and continue adding one-fourth of the period until at least five points are obtained. This procedure should determine for *one full cycle of the curve those points that are on the x axis or at the highest or lowest points of the curve.* These are important values.

Example 3

Sketch the graph of $y = -3 \sin\left(4x - \dfrac{\pi}{6}\right)$.

where $a = -3$, $b = 4$, $c = -\dfrac{\pi}{6}$.

The amplitude is 3 and the negative sign preceding the 3 has the effect of inverting the curve. The period is $2\pi/4 = \pi/2$. The displacement of

$$-\dfrac{(-\pi/6)}{4} = \dfrac{\pi}{24}$$

means that the graph is shifted to the *right* by an amount of $\pi/24$ units of x. We can consider $x = \pi/24$ to be the start of one cycle of the curve. Because the period is $\pi/2$, one-fourth of the period is

$$\dfrac{1}{4}\left(\dfrac{\pi}{2}\right) = \dfrac{\pi}{8}$$

Adding this to $\pi/24$, we get

$$\dfrac{\pi}{24} + \dfrac{\pi}{8} = \dfrac{\pi}{24} + \dfrac{3\pi}{24} = \dfrac{4\pi}{24} = \dfrac{\pi}{6}$$

which is the next important value of x. Continuing to add $\pi/8$, we get x values of $7\pi/24$, $5\pi/12$, and $13\pi/24$. These values are included in the following table, and the corresponding points are plotted in the graph of Figure 15.51.

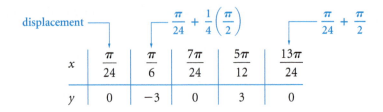

x	$\dfrac{\pi}{24}$	$\dfrac{\pi}{6}$	$\dfrac{7\pi}{24}$	$\dfrac{5\pi}{12}$	$\dfrac{13\pi}{24}$
y	0	-3	0	3	0

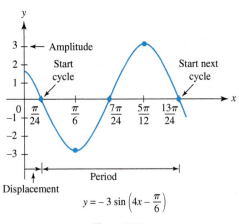

$$y = -3\sin\left(4x - \frac{\pi}{6}\right)$$

Figure 15.51

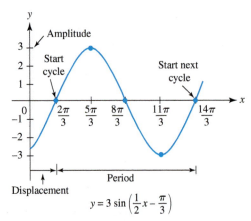

$$y = 3\sin\left(\frac{1}{2}x - \frac{\pi}{3}\right)$$

Figure 15.52

Example 4

Sketch the graph of $y = 3\sin\left(\dfrac{1}{2}x - \dfrac{\pi}{3}\right)$.   with labels a, b, $c = -\dfrac{\pi}{3}$

The amplitude is 3 and the period is $\dfrac{2\pi}{1/2} = 4\pi$. The displacement of $-\dfrac{-\pi/3}{1/2} = \dfrac{2\pi}{3}$ indicates that there is a shift of $\dfrac{2\pi}{3}$ to the right. Using this information and the following table of key values, we get the graph shown in Figure 15.52.

displacement → with $\dfrac{2\pi}{3} + \dfrac{1}{4}(4\pi)$ and $\dfrac{2\pi}{3} + 4\pi$

x	$\dfrac{2\pi}{3}$	$\dfrac{5\pi}{3}$	$\dfrac{8\pi}{3}$	$\dfrac{11\pi}{3}$	$\dfrac{14\pi}{3}$
y	0	3	0	-3	0

Now Try It!

Determine the amplitude, period, and displacement for each of the following functions:

1. $y = 2\sin\left(x - \dfrac{\pi}{6}\right)$

2. $y = -\cos(2x - \pi)$

3. $y = \sin\left(\pi x + \dfrac{\pi}{8}\right)$

4. $y = 40\cos\left(3\pi x - \dfrac{\pi}{5}\right)$

Example 5

The current in a certain alternating-current circuit is described by the equation $y = 15 \sin\left(60\pi x + \dfrac{\pi}{6}\right)$. Sketch the curve.

The amplitude is 15, the period is $\dfrac{2\pi}{60\pi} = \dfrac{1}{30}$, and the displacement is $-\dfrac{(\pi/6)}{60\pi} = -\dfrac{1}{360}$. This represents a shift of 1/360 to the *left*, because the displacement is negative. Because the period and displacement are not expressed in terms of π, we choose values of x that do not involve π. Following the preceding hint, we get the values in the following table:

displacement $\longrightarrow$ $\longleftarrow -\dfrac{1}{360} + \dfrac{1}{4}\left(\dfrac{1}{30}\right)$ $\longleftarrow -\dfrac{1}{360} + \dfrac{1}{30} = \left(\dfrac{11}{360}\right)$

x	$-\dfrac{1}{360}$	$\dfrac{2}{360}$	$\dfrac{5}{360}$	$\dfrac{8}{360}$	$\dfrac{11}{360}$
y	0	15	0	-15	0

Using the values in this table and knowledge of the amplitude, period, and displacement, we get the graph shown in Figure 15.53.

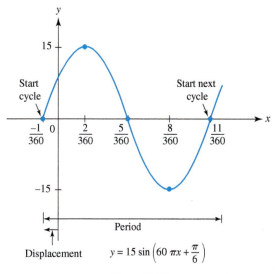

Figure 15.53

In summary:

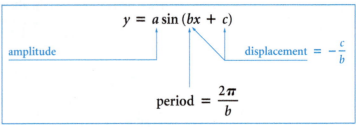

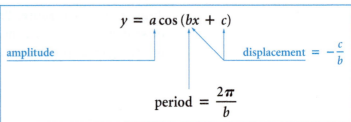

15.5 *Exercises*

In Exercises 1 through 24, determine the amplitude, period, and displacement for each function. Then sketch the graphs of the functions. Use your graphing calculator to verify your work.

1. $y = \sin\left(x + \dfrac{\pi}{3}\right)$

2. $y = 2\sin\left(x - \dfrac{\pi}{3}\right)$

3. $y = \cos\left(x - \dfrac{\pi}{3}\right)$

4. $y = 5\cos\left(x + \dfrac{\pi}{4}\right)$

5. $y = 3\sin\left(2x + \dfrac{\pi}{4}\right)$

6. $y = -\sin\left(2x - \dfrac{\pi}{4}\right)$

7. $y = -4\cos(3x + \pi)$

8. $y = 6\cos\left(2x - \dfrac{\pi}{2}\right)$

9. $y = 2\sin\left(\dfrac{1}{3}x + \dfrac{\pi}{2}\right)$

10. $y = -\dfrac{1}{5}\sin\left(\dfrac{1}{4}x - \dfrac{\pi}{2}\right)$

11. $y = 6\cos\left(\dfrac{2}{3}x + \dfrac{\pi}{3}\right)$

12. $y = \dfrac{1}{6}\cos\left(\dfrac{1}{2}x - \pi\right)$

13. $y = 10\sin(\pi x - 1)$

14. $y = -8\sin(2\pi x + \pi)$

15. $y = \dfrac{3}{2}\cos\left(3\pi x - \dfrac{1}{3}\right)$

16. $y = \dfrac{3}{5}\cos\left(4\pi x + \dfrac{\pi}{5}\right)$

17. $y = -1.2\sin\left(\pi x - \dfrac{\pi}{6}\right)$

18. $y = 0.4\sin\left(2\pi x + \dfrac{1}{4}\right)$

19. $y = 6\cos\left(\pi x - \dfrac{1}{2}\right)$

20. $y = 8\cos(5\pi x + 0.1)$

21. $y = \dfrac{3}{4}\sin(\pi^2 x + \pi)$

22. $y = -\sin\left(4x - \dfrac{2}{\pi}\right)$

23. $y = -\dfrac{5}{2}\cos\left(\pi x - \dfrac{\pi^2}{3}\right)$

24. $y = \pi\cos\left(\dfrac{1}{\pi}x - \dfrac{3}{2}\right)$

In Exercises 25 through 28, sketch the indicated curves.

25. The cross section of a particular water wave is

$$y = 1.8\sin\left(\dfrac{\pi x}{4} - \dfrac{\pi}{6}\right)$$

where x and y are measured in feet. Sketch two cycles of the graph of y versus x.

26. The voltage in a certain alternating-current circuit is given by

$$y = 240 \sin\left(120\pi t + \frac{\pi}{4}\right)$$

where t represents the time in seconds. Sketch three cycles of the curve.

27. An automobile flywheel has a timing mark on its outer edge. The height of the timing mark on the rotating flywheel is given by

$$y = 3.55 \sin\left(x - \frac{\pi}{4}\right)$$

where y is in inches and x is the angle of rotation. Sketch y versus x and include at least one full cycle of this function.

28. For a certain swinging pendulum, if s is the arc length (in meters) measured from the bottom of the path and t is the time in seconds, the motion is described by

$$s = 4 \cos\left(0.9t + \frac{\pi}{2}\right)$$

Sketch s versus t and include at least one full cycle of this function.

Now Try It! Answers

1. $2, \pi, \dfrac{\pi}{6}$ **2.** $1, \pi, \dfrac{\pi}{2}$ **3.** $1, 2, -\dfrac{1}{8}$ **4.** $40, \dfrac{2}{3}, \dfrac{1}{15}$

15.6 The Sine Function as a Function of Time

Section Objective
• Graph the sine function as a function of time

At the start of the chapter we mentioned tidal cycles and blood pressure as two examples of periodic functions. Both of these actions are measured against time. Alternating current (or AC electricity) is also an example of a periodic function and is the type of electricity used in homes. The graph of the alternating current is a sine curve. An oscilloscope is used to measure voltage waves which are displayed on an oscilloscope with *time* on the horizontal axis and *voltage* on the vertical axis.

The sine function $y = a \sin \omega t$ can be created by rotating a vector of length a counterclockwise around a circle with a constant angular velocity of ω, as shown in Figure 15.54. The angle θ through which a body rotates over time is called the *angular displacement* and is defined as the product of the angular velocity ω and elapsed time t. As indicated in Section 15.2, the units of ω are radians per unit of time.

If the length of a given vector is a, then we have $y = a \sin \theta$. Because $\theta = \omega t$ at any moment in time t we have $y = a \sin \omega t$. Notice that in this equation the output y is dependent on t and not on the value of an angle as in the preceding examples. It makes sense that we would define the sine function of many practical applications in terms of time t rather than in terms of an angle measure θ.

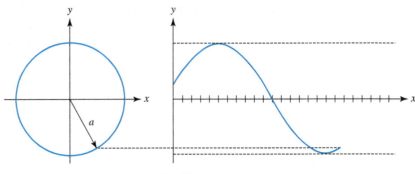

Figure 15.54

Example 1

Given a vector of length 10 rotating with an angular velocity of 15 rad/s we have

$$y = 10 \sin 15t$$

The graph of this function shows the displacement y for a given time t. In this example, note that time t *is the independent variable*. The graph shown in Figure 15.55 represents the displacement over a period of 0.5 seconds.

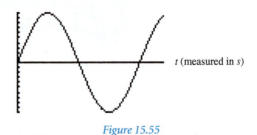

t (measured in s)

Figure 15.55

A sine function as a function of time has an amplitude, period, and phase shift just like the sine functions examined in the previous sections.

Example 2

Given the sine function $y = 375 \sin (55t + \pi/4)$ determine the amplitude, angular velocity, period, and phase shift. Create a graph of this sine function.

a. The amplitude is 375.

b. The angular velocity is 55.

c. The period is $\dfrac{2\pi}{\omega} = \dfrac{2\pi}{55} = 0.1142$ s.

d. The phase shift is $-\dfrac{\pi/4}{55} = -0.0143$ s to the left.

> ### Remember
>
> Given $y = a \sin bx$, x represents an angle measure and b is a coefficient of x with no units attached.
>
> Given $y = a \sin \omega t$, t represents time and ω is an angular velocity measured in rad/s.

Note that the units for the phase shift are in seconds because we are shifting the graph along the horizontal axis.

In Figure 15.56 we can see that the sine function has an amplitude of 375. Each tick mark on the graph represents 0.01 s with the graph completing one complete cycle in approximately 0.11 s. It is also clear that the graph has been shifted about 0.0143 units to the right.

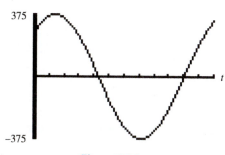

Figure 15.56

Example
3

In alternating current the flow of electrons through a wire alternates direction. This change in direction is caused by the motion of a wire passing through a magnetic field. If the wire is moving in a circular path, with angular velocity ω, the current i in the wire at time t is given by the equation

$$i = I_m \sin(\omega t + \alpha)$$

where I_m is the maximum current attainable and α is the phase angle. If $I_m = 6.00$A, $\omega = 120\pi$ rad/s, and $\alpha = \pi/6$, we can represent the current by the equation

$$i = 6.00 \sin(120\pi t + \pi/6)$$

A graph of this function clearly shows that current takes on both positive and negative values indicating that it is, in fact, alternating the direction in which it is moving.

The **frequency**, f, of electrical signals is the number of times a sine function repeats itself in one second, or *the number of cycles per second*. The unit of measure for frequency is the **hertz** (Hz). Frequencies can be determined in one of two ways. If the period of a sine function is already known, the frequency can be found by simply taking the reciprocal of the period. More often than not we are not given the period of the function. When this is the case the frequency can be found using the formula

$$f = \frac{\omega}{2\pi}$$

We will use both methods in Example 4.

Example
4

In Example 2 we introduced the equation $y = 375 \sin(55t + \pi/4)$ and determined the period of this sine function to be 0.1142 s.

Given

$$f = \frac{1}{period} = \frac{1}{0.1142} = 8.75 \text{ Hz}$$

and using the formula

$$f = \frac{\omega}{2\pi} = \frac{55}{2\pi} = 8.75 \text{ Hz}$$

tells us that 55 rad/s corresponds to 8.75 cycles/s.

15.6 Exercises

In Exercises 1 through 6, sketch two cycles of the graph that has the following information. Use your graphing calculator to verify your graph.

1. $a = 2.40$ cm, $\omega = 2.00$ rad/s

2. $a = 12$ ft, $\omega = 750$ rad/s

3. $a = 18.5$ ft, $f = 0.250$ Hz

4. $a = 5.75$ cm, $f = 18.1$ Hz

5. $a = 500$ mi, $\omega = 3.60$ rad/h, $\alpha = 0$

6. $a = 850$ km, $f = 1.6 \times 10^{-4}$ Hz, $\alpha = \pi/3$

In Exercises 7 and 8, sketch two cylces of the voltage as a function of time for an alternating current circuit in which the voltage e is given by $e = E \cos(\omega t + \alpha)$ for the given values. Use your graphing calculator to verify your graph.

7. $E = 170$ V, $f = 60.0$ Hz, $\alpha = -\pi/3$

8. $E = 80$ V, $\omega = 377$ rad/s, $\alpha = \pi/2$

In Exercises 9 and 10, the air pressure within a plastic container changes above and below the external atmospheric pressure by the formula $p = p_o \sin 2\pi ft$. Sketch two cycles of p for the given values. Use your graphing calculator to verify your graph.

9. $p_o = 2.80$ lb/in^3, $f = 2.30$ Hz

10. $p_o = 45$ kPa, $f = 0.450$ Hz

In Exercises 11 and 12, solve the given problems.

11. An alternating current has the equation $i = 25 \sin(635t - \pi/10)$ where i is measured in amperes, A. Determine the maximum current, period, and frequency for this function.

12. Given an alternating voltage described by the equation $v = 4500 \sin(0.025t + 0.20)$ find the maximum voltage, period, and frequency for the function.

In Exercises 13 through 15, sketch the required graph. Use your graphing calculator to verify your graph.

13. Sketch two cycles of the radio signal $e = 0.014 \cos(2\pi ft + \pi/4)$ (e in volts, f in hertz, and t in seconds) for a station broadcasting with $f = 950$ Hz.

14. Sketch two cycles of the acoustical intensity I of the sound wave for which $I = A \cos(2\pi ft - \alpha)$ given that t is in seconds, $A = 0.027$ W/cm^2, $f = 240$ Hz, and $\alpha = 0.80$.

15. The sinusoidal electromagnetic wave emitted by an antenna in a cellular phone system has a frequency of 7.5×10^9 Hz and an amplitude of 0.0452 V/m. Find the equation representing the wave if it starts at the origin. Sketch two cycles of the graph.

Chapter Summary

Key Terms

radian	periodic	phase shift
arc length	amplitude	frequency
sector	period	hertz
linear velocity	phase angle	
angular velocity	displacement	

Formulas

$\pi \, \text{rad} = 180°$

$1° = \dfrac{\pi}{180} \, \text{rad} = 0.01745 \, \text{rad}$

$1 \, \text{rad} = \dfrac{180°}{\pi} = 57.3°$

$s = r\theta$ (where θ is in radians)   arc length

$A = \dfrac{1}{2}\theta r^2$ (where θ is in radians)   area of sector

$v = \omega r$ (where ω is in radians per unit time)   linear velocity

$y = a \sin(bx + c)$ where $|a|$ = amplitude

$y = a \cos(bx + c)$ $\dfrac{2\pi}{b}$ = period

$y = a \sin \omega t$ $\dfrac{-c}{b}$ = displacement

$f = \dfrac{\omega}{2\pi}$

Review Exercises

In Exercises 1 through 4, express the given angles in terms of π.

In Exercises 5 through 12, the given numbers represent angle measure. Express the measure of each angle in degrees.

1. 63.0°, 137.0°

2. 67.5°, 202.5°

3. 46.0°, 10.0°

4. 318.0°, 284.0°

5. $\dfrac{\pi}{9}, \dfrac{5\pi}{6}$

6. $\dfrac{6\pi}{5}, \dfrac{5\pi}{8}$

7. $\dfrac{7\pi}{18}, \dfrac{9\pi}{20}$

8. $\dfrac{19\pi}{10}, \dfrac{7\pi}{5}$

9. 0.625

10. 1.33

11. 3.45

12. 12.38

In Exercises 13 through 20, express the given angles in radians (not in terms of π).

13. 75.0°

14. 110.0°

15. 340.0°

16. 15.5°

17. 152.5°

18. 215.4°

19. 9.3°

20. 422.0°

In Exercises 21 through 24, find θ for $0 < \theta < 2\pi$.

21. $\sin \theta = 0.6361$

22. $\cos \theta = 0.9925$

23. $\cos \theta = -0.4147$

24. $\tan \theta = 2.087$

In Exercises 25 through 28, the radius and central angle are given for a circle. (a) Find the length of the arc subtended by the central angle. (b) Find the area of the sector.

25. $r = 12.4 \text{ cm}, \theta = 35.7°$

26. $r = 507 \text{ ft}, \theta = 95.6°$

27. $r = 7.24 \text{ in.}, \theta = \dfrac{\pi}{10}$

28. $r = 0.683 \text{ m}, \theta = \dfrac{4\pi}{3}$

In Exercises 29 through 32, an object is moving in a circular path with the given radius and angular velocity. Find the linear velocity.

29. $r = 0.365 \text{ m}, 12.6 \text{ rad/min}$

30. $r = 237 \text{ in.}, 9.59 \text{ rad/min}$

31. $r = 1.25 \text{ m}, 45.0 \text{ r/min}$

32. $r = 82.6 \text{ cm}, 360 \text{ r/min}$

In Exercises 33 through 52, sketch the curves of the given trigonometric functions. Use your graphing calculator to verify your results.

33. $y = \dfrac{5}{2} \cos x$

34. $y = -6 \cos x$

35. $y = -2.5 \sin x$

36. $y = 3.6 \sin x$

37. $y = 3 \cos 4x$

38. $y = 5 \cos \dfrac{1}{2}x$

39. $y = 2 \sin 6x$

40. $y = 5 \sin 2x$

41. $y = \cos \pi x$

42. $y = 5 \cos 4\pi x$

43. $y = 5 \sin 2\pi x$

44. $y = -\sin 3\pi x$

45. $y = -3 \cos \left(4x + \dfrac{\pi}{6} \right)$

46. $y = 2 \cos \left(\dfrac{1}{2}x + \dfrac{\pi}{4} \right)$

47. $y = 3 \sin \left(4x - \dfrac{\pi}{2} \right)$

48. $y = 2 \sin \left(\dfrac{1}{2}x - \dfrac{\pi}{4} \right)$

49. $y = -\cos \left(\pi x + \dfrac{\pi}{3} \right)$

50. $y = 2 \cos (4\pi x - \pi)$

51. $y = 6 \sin \left(4\pi x - \dfrac{\pi}{2} \right)$

52. $y = 5 \sin (2\pi x + \pi)$

In Exercises 53 through 56, refer to the graphs in Figures 15.57 through 15.60 and identify the equation of the function displayed. The hash marks on the axes represent one unit each. In each case, the equation is of the form $y = a \sin bx$ or $y = a \cos bx$.

53. Figure 15.57

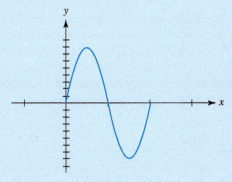

Figure 15.57

54. Figure 15.58

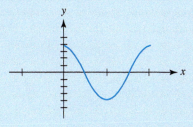

Figure 15.58

55. Figure 15.59

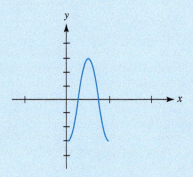

Figure 15.59

56. Figure 15.60

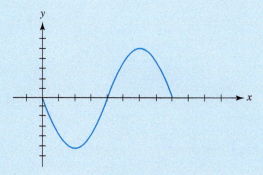

Figure 15.60

In Exercises 57 through 72, solve the given problems.

57. The displacement of a particle moving with simple harmonic motion is given by

$$d = A \cos 5t$$

where A is the maximum displacement and t is the time. Find d given that $A = 12.0$ cm and $t = 0.150$ s.

58. The current in a certain alternating-current circuit is given by

$$i = I \sin 25t$$

where I is the maximum possible current and t is the time. Find i for $t = 0.100$ s and $I = 150$ mA.

59. In constructing a pie chart, a sector of a circle must have an area that is 32.5% of the total area. Find the central angle in degrees. See Figure 15.61.

Figure 15.61

60. A satellite is in orbit at an altitude of 18,600 mi above the surface of the earth, and it makes 1.02 revolutions about the earth each 24.5 h. Find the linear velocity of this satellite. (The radius of the earth is 3960 mi.)

61. A piece of circular filter paper 8.50 cm in diameter is folded so that its effective filtering area is the same as that of a sector with central angle of 230.5°. What is the filtering area? See Figure 15.62.

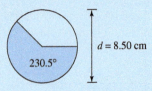

Figure 15.62

62. The armature of a dynamo is 1.25 ft in diameter and is rotating at the rate of 1250 r/min. What is the linear velocity of a point on the outside of the armature?

63. A pulley with a radius of 3.60 in. is belted to another pulley with a radius of 5.75 in. The smaller pulley rotates at 35.0 r/s. What is the angular velocity of the larger pulley? See Figure 15.63.

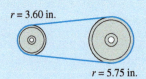

Figure 15.63

64. A lathe is to cut material at the rate of 115 m/min. Calculate the radius of a cylindrical piece that is turned at the rate of 120 r/min.

65. An emery wheel 45.0 cm in diameter has a cutting speed of 1280 m/min. Find the number of revolutions per minute that the wheel makes.

66. A person calculates the distance traveled by a point on the equator as the point travels for exactly 24 h about the earth's center. That person assumed a radius of 3960 mi and that the point makes exactly one complete revolution in 24 h. If it actually requires 24 h 3 min and 55.909 s (mean solar day) to make one complete revolution, find the amount of error in the first result.

67. A simple pendulum is started by giving it a velocity from its equilibrium position. The angle θ between the vertical and the pendulum is given by

$$\theta = a \sin\left(\sqrt{\frac{g}{l}}t\right)$$

where a is the amplitude in radians, $g\,(= 32 \text{ ft/s}^2)$ is the acceleration due to gravity, l is the length of the pendulum in feet, and t is the length of time of the

motion. Sketch two cycles of θ as a function of t for the pendulum whose length is 2.0 ft and $a = 0.1$ rad.

68. The displacement of a certain water wave (height above the calm water level) is given by

$$d = 3.20 \sin 2.50t$$

where d is measured in meters and t in seconds. Find d, given that $t = 1.25$ s.

69. The voltage in a certain alternating-current circuit is given by

$$v = v_{\max} \sin 36t$$

where $v_{\max}$ is the maximum possible voltage and t is the time in seconds. Find v for $t = 0.0050$ s and $v_{\max} = 120$ V.

70. The electric current in a certain circuit is given by

$$i = i_0 \sin\left(\frac{t}{\sqrt{LC}}\right)$$

where i_0 is the initial current, L is an inductance, and C is a capacitance. Sketch two cycles of i as a function of t (in seconds) for the case where $i_0 = 15.0$ A, $L = 1$ H, and $C = 25\ \mu$F.

71. A certain object is oscillating at the end of a spring. See Figure 15.64. The displacement as a function of time is given by the relation

$$y = 2.4 \cos\left(12t + \frac{\pi}{6}\right)$$

where y is measured in meters and t in seconds. Sketch the graph of y versus t.

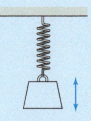

Figure 15.64

72. The current in an alternating-current circuit is described by the equation

$$y = 110 \sin\left(60\pi x - \frac{\pi}{4}\right)$$

Sketch the graph of y versus x.

Chapter Test

1. Change 150° to radian measure with your result written in terms of π.

2. Convert $\dfrac{2\pi}{5}$ to degree measure.

3. Given that 3.572 is the measure of an angle, express the angle in degrees.

4. Given a circle with a radius of $r = 1.50$ and a central angle $\theta = \dfrac{4\pi}{6}$, find the length of the arc subtended by the central angle.

5. Given a circle with a radius of $r = 284$ cm and a central angle $\theta = 46°$, find the area of the sector.

6. An airplane propeller blade is 2.80 feet long and rotates at 2200 revolutions per minute. What is the linear velocity on the tip of the blade?

7. Sketch two cycles of the graph of the function
$$y = 0.5 \cos \frac{\pi}{2}x.$$

8. Sketch two cycles of the graph of the function
$$y = 2 \sin\left(2x - \frac{\pi}{3}\right).$$

9. A wave is traveling in a string. The displacement y (in inches) as a function of time t (in s) from its equilibrium position is given by $y = A \cos\left(\dfrac{2\pi}{T}\right)t$. T is the period (in s) of the motion. If $A = 0.200$ in. and $T = 0.100$ s, sketch two cycles of y as a function of t.

16 Complex Numbers

There are a number of applications in physics and engineering in which the use of complex numbers is helpful.

In electronics, the state of a circuit component is described by two real numbers, the voltage V across the component and the current I flowing through the component. A circuit component may also have a capacitance C and an inductance L that describe its tendency to resist changes in both voltage and current, respectively. These concepts are much better described by complex numbers. Rather than having to describe the component by two different real numbers V and I, it can be described by a single complex number called *impedance,* $z = V + iI$. Inductance and capacitance can also be thought of as having a real and imaginary part and can be described by the single complex number $w = C + iL$. The strength of an electromagnetic

585

field is a second application that is best described as a complex number. Rather than trying to describe an electromagnetic field by two real quantities (electric field strength and magnetic field strength), it is best described as a single complex number, of which the intensity of the electric field and the intensity of the magnetic field are simply the real and imaginary parts.

In this chapter we expand our previous discussions of imaginary numbers to the concept of **complex numbers.** We will examine the basic operations as they apply to complex numbers, and discuss the graphical representation of complex numbers. We conclude the chapter by looking at complex numbers in a special form known as the polar form of a complex number.

16.1 Introduction to Complex Numbers

Section Objectives
- Simplify radicals having negative values under the radical sign
- Evaluate powers of j

> **Note**
>
> In some textbooks you will see the letter i used to represent an imaginary number. In most technical textbooks we use the letter j. This is done to avoid confusion because the letter i is also used to represent electric current.

In Chapter 10 we saw that problems involving the square root of a negative number were resolved by defining the imaginary unit j, where

$$j^2 = -1 \quad \text{or} \quad j = \sqrt{-1}$$

A number in the form of bj (b is a real number) is called an **imaginary number.** For example, the quadratic equation $x^2 + 4 = 0$ has $2j$ and $-2j$ as solutions. We begin with Example 1 that reviews the earlier work done with imaginary numbers.

Example **1**

$$\sqrt{-4} = \sqrt{4j^2} = 2j$$

$$\sqrt{-25} = \sqrt{25j^2} = 5j$$

Also,

$$(8j)^2 = 64j^2 = -64$$
$$(-5j)^2 = 25j^2 = -25$$
$$(-9j)^2 = 81j^2 = -81$$

> **Recall**
>
> $j = \sqrt{-1}$
> $j^2 = -1$
> $j^3 = -j$
> $j^4 = 1$
>
> The powers of j are repetitive with the first four powers repeating themselves.

Example
2

In an alternating-current circuit, the inductive reactance (in ohms) is represented by $\sqrt{-9}$. We can express this value as

$$\sqrt{-9} = \sqrt{9j^2} = 3j$$

The inductive reactance is $3j\ \Omega$.

Example
3

$$(\sqrt{-16})^2 = (\sqrt{16j^2})^2 = (j\sqrt{16})^2$$
$$= 16j^2 = -16$$

In Example 3, it is easy to make the mistake of concluding that

$$(\sqrt{-16})^2 = \sqrt{-16}\ \sqrt{-16} = \sqrt{256} = 16 \quad \text{(wrong!)}$$

The correct result is -16, not 16. In general,

$$\boxed{\sqrt{ab} = \sqrt{a}\ \sqrt{b}} \qquad a \text{ and } b \text{ are positive}$$

but this is valid only for positive values of a and b; it is *not* valid if a or b is a negative number. We must remember that we should never use this property with negative numbers as in $\sqrt{-16}\ \sqrt{-16} = \sqrt{256} = 16$, which is wrong. Much of this difficulty can be avoided if we follow this simple rule: **When working with square roots of negative numbers, express each in terms of *j* immediately before proceeding.** This can be accomplished by applying the following.

$$\boxed{\sqrt{-a} = j\sqrt{a}} \qquad \text{(where } a > 0)$$

Now Try It!

1. Simplify
   a. j^{10} b. j^{19}
   c. j^{30} d. j^{21}
2. Express each number in terms of *j*.
   a. $\sqrt{-49}$ b. $\sqrt{-300}$
   c. $\sqrt{-171}$ d. $\sqrt{-252}$

Example
4

Express $\sqrt{-5}$, $\sqrt{-12}$, and $-\sqrt{-18}$ in terms of *j*.

$$\sqrt{-5} = \sqrt{5j^2} = j\sqrt{5}$$

$$\sqrt{-12} = \sqrt{(4)(3)j^2} = 2j\sqrt{3}$$

$$-\sqrt{-18} = -\sqrt{(9)(2)j^2} = -3j\sqrt{2}$$

When the result contains a factor of *j*, we usually write the *j* before any radical so that it is clear that *j* is not under the radical.

Using real numbers and imaginary numbers, we can define a **complex number** *to be any number that can be expressed in the form of* $a + bj$, where the numbers a and b are both real. The standard form of $a + bj$ is referred to as the **rectangular form of a complex number.** The number a is called the **real part** whereas b is called the **imaginary part.** If $a = 0$, then the number bj is called a **pure imaginary** number. If $b = 0$, then the number a is a **real number.** With these definitions, the real numbers become a subset of the collection of all complex numbers.

Example 5

To express $-7 + \sqrt{-12}$ in the rectangular form of $a + bj$, we proceed as follows.

$$-7 + \sqrt{-12} = -7 + \sqrt{(4)(3)j^2}$$
$$= -7 + 2j\sqrt{3}$$

A complex number has the form $a + bj$, whereas an imaginary number can be written as bj.

Example 6

$3 + 4j$ is a complex number with 3 as the real part and 4 as the imaginary part.

$-5 - 6j$ is a complex number with -5 as the real part and -6 as the imaginary part.

$6 - j$ is a complex number with 6 as the real part and -1 as the imaginary part.

$8j$ is a complex number with 0 as the real part and 8 as the imaginary part; $8j$ is also an imaginary number.

7 is a complex number with 7 as the real part and 0 as the imaginary part; 7 is also a real number.

We conclude this section with a basic definition that will be used later. *The* **conjugate** *of the complex number* $a + bj$ *is the complex number* $a - bj$. To find the conjugate of any complex number, simply change the sign of the imaginary part.

Example 7

The conjugate of $2 - 7j$ is $2 + 7j$.
The conjugate of $-5 + 2j$ is $-5 - 2j$.
The conjugate of $-8 - 3j$ is $-8 + 3j$.

16.1 Exercises

In Exercises 1 through 12, express each number in terms of j.

1. $\sqrt{-16}$ **2.** $\sqrt{-25}$

3. $-\sqrt{-9}$ **4.** $-\sqrt{-36}$

5. $\sqrt{-0.25}$ **6.** $-\sqrt{-0.16}$

7. $\sqrt{-27}$ **8.** $\sqrt{-45}$

9. $-\sqrt{-48}$ **10.** $-\sqrt{-75}$

11. $\sqrt{-0.0004}$ **12.** $-\sqrt{-0.0009}$

In Exercises 13 through 20, express each number in the rectangular form a + bj.

13. $5 + \sqrt{-1}$ **14.** $-3 + \sqrt{-1}$

15. $6 - \sqrt{-4}$ **16.** $-9 - \sqrt{-9}$

17. $-4 + \sqrt{-8}$ **18.** $-8 - \sqrt{-54}$

19. $14 - \sqrt{-63}$ **20.** $-7 + \sqrt{-28}$

In Exercises 21 through 32, simplify each expression. (Be sure to introduce j immediately before proceeding.)

21. $(\sqrt{-9})^2$ **22.** $(\sqrt{-4})^2$

23. $(\sqrt{-6})^2$ **24.** $(\sqrt{-0.5})^2$

25. $\sqrt{(-3)^2}$ **26.** $\sqrt{(-7)^2}$

27. $\sqrt{-3}\sqrt{-4}$ **28.** $\sqrt{-2}\sqrt{-8}$

29. $\sqrt{-3}\sqrt{-12}$ **30.** $\sqrt{-5}\sqrt{-7}$

31. $\sqrt{-3}\sqrt{-11}$ **32.** $\sqrt{-13}\sqrt{-5}$

In Exercises 33 through 44, simplify each expression.

33. j^6 **34.** j^8 **35.** j^9

36. j^{10} **37.** j^{65} **38.** j^{103}

39. j^{402} **40.** j^{604} **41.** $(-j)^{22}$

42. $-j^{43}$ **43.** $-j^{81}$ **44.** $(-j)^{64}$

In Exercises 45 through 52, find the conjugate of each complex number.

45. $3 - 8j$ **46.** 7

47. -4 **48.** $-5 + 4j$

49. $9j$ **50.** $-3j$

51. $-2 - 7j$ **52.** $3 - 8j$

Now Try It! Answers

1. a. -1 **b.** $-j$ **c.** -1 **d.** j

2. a. $7j$ **b.** $10j\sqrt{3}$ **c.** $3j\sqrt{19}$ **d.** $6j\sqrt{7}$

16.2 Basic Operations with Complex Numbers

Section Objective

• Successfully add, subtract, multiply, and divide complex numbers

In this section we consider the basic operations of addition, subtraction, multiplication, and division as they apply to complex numbers. Before defining those operations, we must first establish the rule that is used to determine whether two complex numbers are equal. We stipulate that $a + bj = x + yj$ *if and only if* $a = x$ *and* $b = y$. That is, two complex numbers are equal if and only if the real parts are equal and the imaginary parts are equal. This rule is illustrated in Example 1.

Example 1

If $a + bj = 6 + 3j$, then $a = 6$ and $b = 3$.

If $a + bj = 2 - 5j$, then $a = 2$ and $b = -5$.
If $a + bj = 7$, then $a = 7$ and $b = 0$.
If $a + bj = -8j$, then $a = 0$ and $b = -8$.

We can now proceed to consider the basic operations of arithmetic. We can add (or subtract) two complex numbers by combining the real parts and combining the imaginary parts.

$$(a + bj) + (c + dj) = (a + c) + (b + d)j$$

Example 2

$$(2 + 3j) + (6 + 4j) = (2 + 6) + (3 + 4)j = 8 + 7j$$
$$(6 + j) + (-2 - 8j) = (6 - 2) + (1 - 8)j = 4 - 7j$$
$$(7 + 5j) - (3 + 2j) = (7 - 3) + (5 - 2)j = 4 + 3j$$
$$(-2 + 4j) - (6 - 9j) = (-2 - 6) + (4 - (-9))j = -8 + 13j$$

When multiplying two complex numbers, we can treat them as binomials. Previously we saw that multiplication of two binomials involves multiplying each term of one binomial by each term of the other. When multiplying $a + bj$ and $c + dj$, we get

$$(a + bj)(c + dj) = a(c) + a(dj) + bj(c) + bj(dj)$$
$$= ac + adj + bcj + bdj^2$$
$$= (ac - bd) + (ad + bc)j$$

We summarize this result as follows.

$$(a + bj)(c + dj) = (ac - bd) + (ad + bc)j$$

It is not necessary to memorize the exact form of this equation because it is only meant to show that we multiply two complex numbers by following the same procedure used in the algebraic multiplication of any two binomials.

Example 3

$$(2 + 3j)(6 + 4j) = 12 + 8j + 18j + 12j^2$$
$$= 12 + 8j + 18j + 12(-1)$$
$$= (12 - 12) + (8 + 18)j$$
$$= 26j$$

Example 4

$$(6 + j)(-2 - 8j) = -12 - 48j - 2j - 8j^2$$
$$= -12 - 48j - 2j - 8(-1) \qquad -12 + 8 = -4$$
$$= -4 - 50j$$

Example 5

Multiply the complex number $c + dj$ by its conjugate $c - dj$. We get

$$(c + dj)(c - dj) = c(c) + c(-dj) + dj(c) + dj(-dj)$$

zero

$$= c^2 - cdj + cdj - d^2j^2$$
$$= c^2 - d^2(-1)$$
$$= c^2 + d^2$$

The most important principle that must be remembered for multiplication is that *each term* of the first complex number must be multiplied by *each term* of the second complex number. When adding or subtracting complex numbers, we add or subtract the real parts and then add or subtract the imaginary parts, but multiplication of complex numbers involves more than multiplying the real parts and multiplying the imaginary parts.

When dividing a number (real or complex) by a *complex* number, we multiply the numerator and denominator by the **conjugate** of the denominator. In Example 5 we saw that when the complex number $c + dj$ is multiplied by its conjugate $c - dj$, the result is the real number $c^2 + d^2$. Therefore, multiplication of the numerator and denominator by the conjugate of the denominator will eliminate j from the denominator. We can summarize division as follows.

$$\frac{a + bj}{c + dj} = \frac{(a + bj)(c - dj)}{(c + dj)(c - dj)} = \frac{(ac + bd) + (bc - ad)j}{c^2 + d^2}$$

Example 6

$$\frac{2 + 3j}{3 + 4j} = \frac{(2 + 3j)(3 - 4j)}{(3 + 4j)(3 - 4j)}$$

conjugate

$$= \frac{6 - 8j + 9j - 12j^2}{9 - 12j + 12j - 16j^2}$$

$$= \frac{6 + j - 12(-1)}{9 - 16(-1)}$$

$$= \frac{18 + j}{25}$$

Now Try It!

Perform the indicated operations:

1. $(3 + 7j) + (10 + 2j)$
2. $(4 - 9j) - (9 - 2j)$
3. $(-4 + 2j)(2 - 5j)$
4. $\dfrac{6 - 3j}{4 + 8j}$

The result in Example 6 could be expressed in the standard rectangular form of $a + bj$ as $\frac{18}{25} + \frac{1}{25}j$, but it is usually left as a single fraction.

Example 7

$$\frac{5j}{2 - 7j} = \frac{5j(2 + 7j)}{(2 - 7j)(2 + 7j)} = \frac{10j + 35j^2}{4 + 14j - 14j - 49j^2}$$

$$= \frac{10j + 35(-1)}{4 - 49(-1)}$$

$$= \frac{-35 + 10j}{53}$$

Example 8

$$\frac{3 + 5j}{-4j} = \frac{(3 + 5j)(4j)}{(-4j)(4j)} = \frac{12j + 20j^2}{-16j^2}$$

conjugate

$$= \frac{12j + 20(-1)}{-16(-1)}$$

$$= \frac{-20 + 12j}{16} = \frac{-5 + 3j}{4}$$

reduce

**Example
9**

Express the number

$$\frac{1}{2}\left[(1 + 2j) + \frac{1}{1 + 2j}\right]$$

in the standard rectangular form $a + bj$. Such expressions are used in determining the cross section of an airplane wing.

We can begin by expressing $\dfrac{1}{1 + 2j}$ in standard rectangular form as follows.

$$\frac{1}{1 + 2j} = \frac{1(1 - 2j)}{(1 + 2j)(1 - 2j)} = \frac{1 - 2j}{1 - 2j + 2j - 4j^2}$$

$$= \frac{1 - 2j}{1 - 4(-1)} = \frac{1 - 2j}{5} \quad \text{or} \quad \frac{1}{5} - \frac{2}{5}j$$

The original expression can now be evaluated as

$$\frac{1}{2}\left[(1 + 2j) + \left(\frac{1}{5} - \frac{2}{5}j\right)\right]$$

$$= \frac{1}{2}\left[\left(1 + \frac{1}{5}\right) + \left(2 - \frac{2}{5}\right)j\right]$$

$$= \frac{1}{2}\left[\frac{6}{5} + \frac{8}{5}j\right]$$

$$= \frac{3}{5} + \frac{4}{5}j$$

It is possible to use some scientific and graphing calculators to work through basic operations involving complex numbers. Some calculators include the imaginary unit i as well. (See the note about the use of the imaginary units j and i in the previous section.)

Begin by determining whether your calculator includes an imaginary unit. If you are using the TI-83/84 Plus, you will find the unit i over the decimal point on the bottom row of the calculator (see Figure 16.1).

Figure 16.1

You will probably not use your calculator to add or subtract complex numbers because this process is straightforward.

To Multiply Complex Numbers

Use Example 4 to work through the following steps. To access the imaginary unit i use the keystrokes 2nd $\cdot$.

1. Use the following keystrokes to enter the first complex number: $(6 + 2^{nd} \cdot)$.

2. Press the multiplication key.

3. Use the following keystrokes to enter the second complex number: $(-2 - 8\ 2^{nd} \cdot)$.

4. Press the ENTER key.

5. Compare your result with the screen shot shown in Figure 16.2.

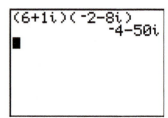

Figure 16.2

To Divide Complex Numbers

Use Example 6 to work through the following steps.

1. Use the following keystrokes to enter the first complex number: $(2 + 3\ 2^{nd} \cdot)$.

2. Press the division key.

3. Use the following keystrokes to enter the second complex number: $(3 + 4\ 2^{nd} \cdot)$.

4. Press the ENTER key.

5. Compare your result with the screen shot shown in Figure 16.3.

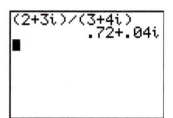

Figure 16.3

You can verify that your result for this problem is equivalent to the result found in your textbook by converting the fractional values to their decimal equivalents.

16.2 *Exercises*

In Exercises 1 through 48, perform the indicated operations and express all answers in the rectangular form $a + bj$. For multiplication and division problems use your calculator to verify your results.

1. $(4 - 6j) + (1 - j)$

2. $(3 - 2j) + (5 + j)$

3. $(7 + j) - (3 - 4j)$

4. $(-9 - 2j) - (3 - 5j)$

5. $(1 + \sqrt{-9}) + (2 - \sqrt{4})$

6. $(3 - \sqrt{-16}) - (-5 + \sqrt{-25})$

7. $(2 - \sqrt{-36}) - (2 + \sqrt{-36})$

8. $(-3 - \sqrt{-64}) + (5 - \sqrt{-49})$

9. $6j + (3 - 8j) + (-1 - 4j)$

10. $(7 - 2j) + (3 - j) + 4j$

11. $\sqrt{-25} + \sqrt{-36} - (3 - 9j)$

12. $(6 - 7j) - \sqrt{-16} - (-5 - 8j)$

13. $5j(6j)$

14. $-3j(2j)$

15. $(2 - j)(2 + j)$

16. $(-3 + 4j)(-3 - 4j)$

17. $(2 - 7j)(3 + 6j)$

18. $(-3 - 9j)(-2 - 8j)$

19. $6j(2j)(-3j)$

20. $-j(-3j)(8j)$

21. $\sqrt{-4}(2 - \sqrt{-9})$

22. $\sqrt{-9}(-3 + \sqrt{-16})$

23. $(2 + \sqrt{-25})(3 - \sqrt{-36})$

24. $(7 - \sqrt{16})(-2 - \sqrt{-9})$

25. $\dfrac{2j}{3 - 4j}$

26. $\dfrac{-3j}{-2 + 5j}$

27. $\dfrac{3 + 8j}{2j}$

28. $\dfrac{5 - 6j}{-3j}$

29. $\dfrac{7 + 3j}{5 + 6j}$

30. $\dfrac{6 - 4j}{9 - 3j}$

31. $\dfrac{-4 + 7j}{1 - 3j}$

32. $\dfrac{-2 + 9j}{-2 - 3j}$

33. $\sqrt{-9} \div (2 + \sqrt{-4})$

34. $\sqrt{-25} \div (4 - \sqrt{-9})$

35. $(2 + \sqrt{-25}) \div (3 - \sqrt{-9})$

36. $(5 + \sqrt{-81}) \div (-2 + \sqrt{-64})$

37. $(1 - j)^2$

38. $(-2 + 3j)^2$

39. $(1 + j)^2 + (2 - 5j)$

40. $(-6 + 2j) - (2 - j)^2$

41. $5j(1 + 3j)^2$

42. $-3j(2 - 7j)^2$

43. $(2 + j)(3 - j) \div (1 - j)$

44. $(3 - 2j)(8 - 3j) \div (2 + 5j)$

45. $6j - (1 + j)(1 + 2j)(1 + 3j)$

46. $-2j + (2 - 3j)(2 + 3j)(3 + 2j)$

47. $(4 - 3j)^2 - (4 + 3j)^2$

48. $(2 - 5j)^2 + (-3 + 4j)^2$

In Exercises 49 through 52, multiply the given complex number by its conjugate.

49. $6 + 8j$

50. $5 - 3j$

51. $-2 + 7j$

52. $-3 - j$

In Exercises 53 through 56, perform the indicated operations and express all answers in the rectangular form of $a + bj$.

53. In an alternating-current circuit, the impedance Z is given by

$$Z = R + j(X_L - X_C)$$

where R is the resistance (in ohms), X_L is the inductive reactance (in ohms), and X_C is the capacitive reactance (in ohms). Find the impedance if $R = 14.7\ \Omega, X_L = 10.2\ \Omega$, and $X_C = 12.3\ \Omega$.

54. In an alternating-current circuit, the impedance Z, the current I, and the voltage V are related by

$$Z = \frac{V}{I}$$

Find the impedance (in ohms) if $V = 26 + 3j$ volts and $I = 6 - 2j$ amperes.

55. In Exercise 54, find the voltage (in volts) if
$Z = -0.20 + 1.3j$ ohms and $I = 4.0 - 3.0j$ amperes.

56. In an alternating-current circuit, two impedances Z_1 and Z_2 are included so that the total impedance Z_T is given by

$$Z_T = \frac{Z_1 Z_2}{Z_1 + Z_2}$$

Find the total impedance if $Z_1 = 2 + 3j$ ohms and $Z_2 = 3 - 4j$ ohms.

Now Try It! Answers

1. $13 + 9j$ **2.** $-5 - 7j$ **3.** $2 + 24j$ **4.** $-\dfrac{3j}{4}$

16.3 Graphical Representation of Complex Numbers

Section Objectives
- Represent a complex number graphically in the complex plane
- Add complex numbers graphically

We represent complex numbers in rectangular form as $a + bj$, where a is the real part and b is the imaginary part. Because each complex number is associated with the pair of numbers a and b, we can represent complex numbers graphically in a rectangular coordinate system. The usual format involves a *horizontal axis called the* **real axis** *and a vertical axis called the* **imaginary axis.** This combination of axes allows us to form the **complex plane.**

Example 1

In Figure 16.4 we depict the complex number $3 + 5j$ as it is graphed in the complex plane.

When constructing the graph of a complex number, we can draw a line from the origin to the point. In this way, we can represent complex numbers in the same way that we represent vectors. In general, *a complex number can be considered to be a vector from the origin to its point in the complex plane.*

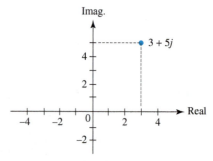

Figure 16.4

Example
2

In Figure 16.5 we show the two complex numbers $3 + 5j$ and $4 + j$. These two complex numbers are shown as vectors.

In Example 2 we graphed the complex numbers $3 + 5j$ and $4 + j$. From Section 16.2 we know that the sum of $3 + 5j$ and $4 + j$ is equal to $7 + 6j$. In Figure 16.3 we show that the vectors $3 + 5j$ and $4 + j$ can be added graphically by following the same parallelogram method used for adding vectors. After representing both complex numbers as vectors, we complete a parallelogram as shown in Figure 16.6. The sum of the two original vectors is represented by a diagonal of the parallelogram. (We always use the diagonal that extends from the origin.)

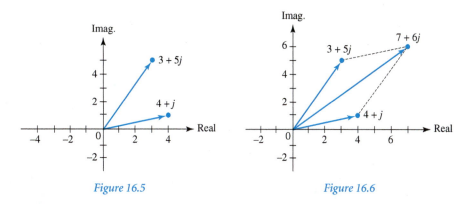

Figure 16.5

Figure 16.6

Steps to Add Complex Numbers Graphically

1. Determine the point in the complex plane that corresponds to the first complex number and draw a line from the origin to this point.

2. Repeat this step for the second complex number.

3. Complete a parallelogram with the lines drawn as adjacent sides.

4. The sum of the two vectors is the resulting fourth vertex of the parallelogram.

The sum of the two original vectors is represented by the diagonal that extends from the origin of the complex plane to the fourth vertex in the parallelogram.

Example
3

Add the complex numbers $4 - j$ and $-3 - 2j$ graphically.

In Figure 16.7 we show the original two vectors with the completed parallelogram. The sum $1 - 3j$ is shown as the vertex of the parallelogram.

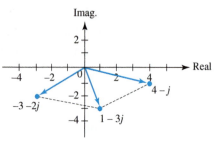

Figure 16.7

Example 4

Subtract $2 - 3j$ from $4j$ graphically.

Subtracting $2 - 3j$ is equivalent to adding $-2 + 3j$. In Figure 16.8 we show the addition of $-2 + 3j$ and $4j$. The sum is $-2 + 7j$. Note that when graphing the complex number $4j$, we consider that number to be in rectangular form as $0 + 4j$ so that the point $(0, 4)$ is plotted in the complex plane.

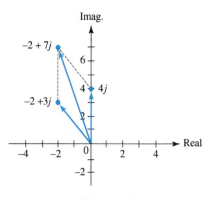

Figure 16.8

Representing complex numbers as vectors has many advantages. This association is helpful in applications. In electronics, for example, the complex plane can be used to represent resistance as the horizontal axis and reactance as the vertical axis so that their sum represents impedance.

Example 5

One common application of complex numbers involves alternating-current circuits. The impedance Z (in ohms) is the sum of the resistance R (in ohms), the capacitive reactance X_C (in ohms), and the inductive capacitance X_L (in ohms). It is standard to represent R along the positive real axis, X_C along the negative real axis, and X_L along the positive imaginary axis. In Figure 16.9, find the impedance Z by adding the given values of R, X_C, and X_L.

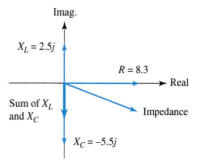

Figure 16.9

We first combine $2.5j$ and $-5.5j$ to get $-3.0j$ as the sum of the inductive reactance and capacitive reactance. The impedance is now represented by $8.3 - 3.0j$ ohms.

16.3 Exercises

In Exercises 1 through 8, graph each complex number as a vector in the complex plane.

1. $3 + 2j$ **2.** $5 - 3j$

3. $3j$ **4.** 4

5. $-2.5 + 3.1j$ **6.** $6.2 - 0.8j$

7. $0.3 - 0.2j$ **8.** $-1.6 + 2.2j$

In Exercises 9 through 28, perform the indicated operations graphically and check the results algebraically.

9. $(4 + 3j) + (1 + j)$

10. $(2 + j) + (3 + 9j)$

11. $5j + (2 + j)$

12. $-2j + (3 + 5j)$

13. $(6 + 2j) - j$

14. $(8 + 3j) - 4j$

15. $(9 + 7j) - (2 + 6j)$

16. $(-5 + 4j) - (3 + 4j)$

17. $(-2 + 3j) + (6 - 4j)$

18. $(-1 - 7j) - (2 - 3j)$

19. $(5 - 6j) - (-4 + 9j)$

20. $(8 + 2j) + (-3 + 5j)$

21. $(2 - j) + (2 + j)$

22. $(1 + 9j) + (8 - 7j)$

23. $(-8 - 4j) - (5 - 6j)$

24. $(3 + 4j) - (3 - 4j)$

25. $(-8 - 7j) - (7 + 4j)$

26. $(-1 + 8j) + (4 + 5j)$

27. $(7 + 4j) + (5 - 8j)$

28. $(9 - 3j) - (6 - 8j)$

In Exercises 29 through 32, graph the given complex number, its negative, and its conjugate on the same set of axes in the complex plane.

29. $2 + 3j$ **30.** $5 - 4j$

31. $-3 + 7j$ **32.** $-6 - j$

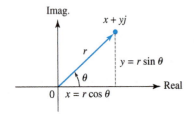

Figure 16.10

16.4 Polar Form of Complex Numbers

Section Objective

• Represent a complex number in polar form

In Section 16.3 we saw that a relationship exists between complex numbers and vectors. We will now use that relationship to develop another way of **expressing** complex numbers. Refer to Figure 16.10, where the general complex number $x + yj$ is depicted in the same way that we depict a vector. That complex number is a distance r from the origin, and the angle θ is in standard position. For any complex number, if we know the magnitude r and the direction θ, then we can locate the correct point in the complex plane.

Recall that $\cos\theta = \dfrac{x}{r}$ and $\sin\theta = \dfrac{y}{r}$. From Figure 16.10 we can use trigonometry and the Pythagorean theorem to get

$$x = r\cos\theta$$

$$y = r\sin\theta$$

$$r^2 = x^2 + y^2$$

$$\tan\theta = \frac{y}{x}$$

We can begin with the rectangular form $x + yj$ of a complex number and then substitute for x and y to get

$$x + yj = r\cos\theta + jr\sin\theta$$

or

$$x + yj = r(\cos\theta + j\sin\theta)$$

The form $r(\cos\theta + j\sin\theta)$ is called the **polar form** *(or* **trigonometric form**) *of a complex number. The length r is called the* **absolute value** *(or the* **modulus**), *and the angle θ is called the* **argument** *of the complex number.* The polar form of $r(\cos\theta + j\sin\theta)$ is sometimes abbreviated as $r\underline{/\theta}$, therefore,

$$r\underline{/\theta} = r(\cos\theta + j\sin\theta)$$

Examples 1 and 2 demonstrate these concepts.

Example

1

$$4(\cos 30° + j \sin 30°) = 4\underline{/30°}$$

$$5\underline{/210°} = 5(\cos 210° + j \sin 210°)$$

$$8 \text{ cis } 93° = 8(\cos 93° + j \sin 93°)$$

Example

2

Represent the complex number $3 + 4j$ graphically; then express it in the polar form $r(\cos \theta + j \sin \theta)$.

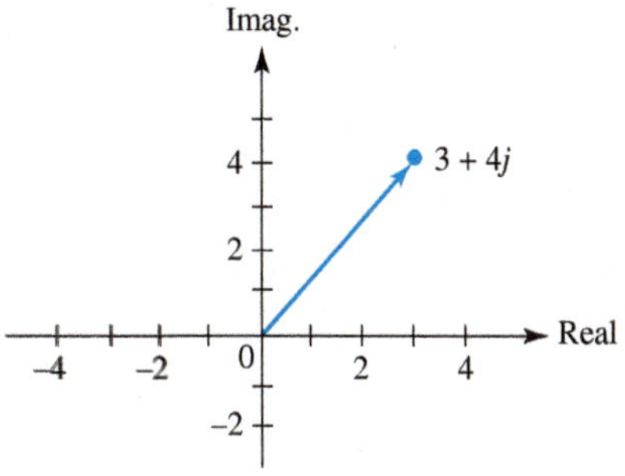

Imag.

3 + 4j

Real

Figure 16.11

From the rectangular form of $3 + 4j$ we see that $x = 3$ and $y = 4$. We plot the point $(3, 4)$ in Figure 16.11, and find the length of r is as follows.

$$r = \sqrt{x^2 + y^2} = \sqrt{3^2 + 4^2} = 5$$

Also, we can find the direction θ by using the definition of tangent as follows.

$$\tan \theta = \frac{y}{x} = \frac{4}{3} = 1.333$$

With $\tan \theta = 1.333$, we can use the INV TAN keys on a calculator to establish that $\theta = 53.1°$. Knowing both r and θ, we can express $3 + 4j$ in polar form as

$$5(\cos 53.1° + j \sin 53.1°)$$

In Figure 16.12 we include the values of r and θ.

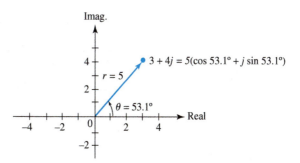

Figure 16.12

In Example 2 we showed how to convert a complex number from the rectangular form $x + yj$ to the polar form $r(\cos \theta + j \sin \theta)$. Some calculators are designed to allow direct conversions between the rectangular and polar forms. See Using Technology at the end of this section for the sequence of keystrokes for converting between the rectangular form and polar form of a complex number.

Example 3

By combining resistance, inductive capacitance, and reactive capacitance, we find that we can represent the impedance in a circuit as $3.14 - 2.07j$ ohms. We will represent that number graphically and then express it in the polar form $r(\cos\theta + j\sin\theta)$.

We first plot the point $x = 3.14$ and $y = -2.07$ in the complex plane. See Figure 16.13. We must now find the values of r and θ so that we can determine the polar form of the given complex number.

$$r = \sqrt{x^2 + y^2} = \sqrt{3.14^2 + (-2.07)^2}$$
$$= 3.76$$

$$\tan\theta = \frac{y}{x} = \frac{-2.07}{3.14} = -0.6592$$

With $\tan\theta = -0.6592$, we use a calculator to determine that $\theta = -33.4°$, and that result agrees with the fourth-quadrant location of the number as it is shown in Figure 16.13.

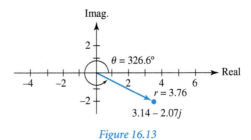

Figure 16.13

Example 4

Express the complex number

$$5(\cos 150.0° + j\sin 150.0°)$$

in rectangular form.

Evaluating $\cos 150.0°$ and $\sin 150.0°$, the number can be expressed in rectangular form as follows.

$$5(\cos 150.0° + j\sin 150.0°) = 5(-0.866 + 0.500j)$$
$$= -4.33 + 2.50j$$

Example 5

Express $12.3\ \underline{/239.4°}$ in rectangular form.

We know that the polar form $12.3\underline{/239.4°}$ is equivalent to the polar form $12.3(\cos 239.4° + j\sin 239.4°)$. Evaluating $\cos 239.4° = -0.509$ and $\sin 239.4° = -0.861$, we get

$$12.3 \underline{/239.4°} = 12.3(\cos 239.4° + j \sin 239.4°)$$
$$= 12.3(-0.509 - 0.861j)$$
$$= -6.26 - 10.6j$$

Example
6

Express $-6.26 - 10.6j$ in polar form.

Since $x = -6.26$ and $y = -10.6$, we can find r.

$$r = \sqrt{x^2 + y^2}$$
$$= \sqrt{(-6.26)^2 + (-10.6)^2} = 12.3$$

Using the definition of tangent we find that

$$\tan \theta = \frac{y}{x} = \frac{-10.6}{-6.26} = 1.693$$

Using a calculator, we would find that $\theta = 59.4°$, but Example 5 and Figure 16.14 both show that 59.4° is *wrong!*

Because $-6.26 - 10.6j$ is in the *third* quadrant, θ must fall between 180° and 270°, so 59.4° is incorrect. Here, the graph of Figure 16.14 helps us to avoid the mistake of simply accepting the value of 59.4° that the calculator would provide. Again, blind use of the calculator will lead to an error; we must *understand* what's happening so that we can adjust the calculator results. However, 59.4° is the reference angle. From Figure 16.14 we can see that the correct value for θ is 239.4°. With $r = 12.3$ and $\theta = 239.4°$, we can express the number in polar form as

$$12.3(\cos 239.4° + j \sin 239.4°)$$

which agrees with the result of Example 5 and Figure 16.14.

Now Try It!

1. Convert
   a. $2 + 2j$ **b.** $3 - 5j$
   to the polar form of a complex number.
2. Convert
   a. $6(\cos 225° + j \sin 225°)$
   b. $6.5(\cos 170° + j \sin 170°)$
   to the rectangular form of a complex number.

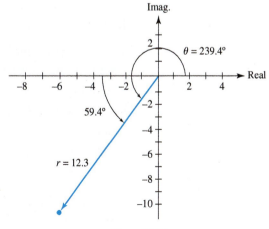

Figure 16.14

The TI-83/84 Plus graphing calculator will allow direct conversion between rectangular and polar forms of a complex number. Locate the ANGLE submenu on your calculator (over the APPS key). To access this submenu press 2nd APPS. You will see a screen similar to the one shown in Figure 16.15.

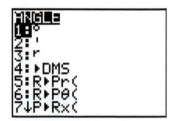

Figure 16.15

To Convert from the Rectangular Form to the Polar Form of a Complex Number

We will use option numbers 5 and 6 in the ANGLE submenu to convert from the rectangular form of a complex number to the polar form of a complex number.

Use Example 2 to work through the following steps. Convert $3 + 4j$ to the polar form of this complex number.

1. Check to make sure that your calculator is in degree mode.

2. Press 2nd APP and choose option #5.

3. Enter the keystrokes 3, 4, and then press ENTER.

4. The calculator returns the value for r as seen in Figure 16.16.

5. Press 2nd APP and choose option #6.

6. Enter the keystrokes 3, 4, and then press ENTER.

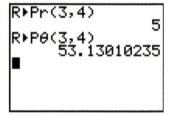

Figure 16.16

7. The calculator returns the value for θ as seen in Figure 16.16.

If your calculator did not return a value of 53.13010235 as shown in Figure 16.16, check to make sure your calculator is in degree mode.

To Convert from the Polar Form to the Rectangular Form of a Complex Number

We will use option numbers 7 and 8 in the ANGLE submenu to convert from the polar form of a complex number to the rectangular form of a complex number.

Use Example 4 to work through the following steps. Convert $5(\cos 150.0° + j \sin 150.0°)$ to the rectangular form of this complex number.

1. Check to make sure that your calculator is in degree mode.

2. Press 2nd APP and choose option #7.

3. Enter the keystrokes 5, 150, and then press ENTER.

4. The calculator returns the value for *x* as seen in Figure 16.17.

5. Press 2ⁿᵈ APP and choose option #8.

6. Enter the keystrokes 5, 150, and then press ENTER.

7. The calculator returns the value for *y* as seen in Figure 16.17.

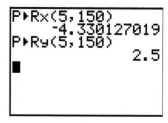

Figure 16.17

Therefore, the rectangular form of this complex number is $-4.33 + 2.50j$.

16.4 *Exercises*

In Exercises 1 through 16, convert each complex number from the given rectangular form to the polar form of $r(\cos\theta + j\sin\theta)$. Use a graph to be sure that θ corresponds to the correct quadrant.

1. $4 + 3j$
2. $-4 + 3j$
3. $4 - 3j$
4. $-4 - 3j$
5. $2.45 - 3.78j$
6. $-4.18 + 1.56j$
7. $3.66 + 5.39j$
8. $-852 - 631j$
9. $-\sqrt{2} - j\sqrt{2}$
10. $\sqrt{3} - j$
11. $-1 + j\sqrt{3}$
12. $1 + j\sqrt{3}$
13. -10
14. 7
15. $6j$
16. $-3j$

In Exercises 17 through 24, express each given complex number in the polar form $r(\cos\theta + j\sin\theta)$.

17. $6.15 \underline{/43.0°}$
18. $256 \underline{/184.6°}$
19. $10.3 \text{ cis } 335.2°$
20. $73.5° \text{ cis } 109.6°$
21. $0.348 \underline{/76.4°}$
22. $8.19 \text{ cis } 27.4°$
23. $56.0 \text{ cis } 212.5°$
24. $403 \text{ cis } 8.3°$

In Exercises 25 through 40, convert each complex number from the given polar form to the rectangular form $x + yj$.

25. $3.00(\cos 70.0° + j\sin 70.0°)$
26. $5.00(\cos 245.0° + j\sin 245.0°)$
27. $1.50(\cos 321.2° + j\sin 321.2°)$
28. $4.50(\cos 255.7° + j\sin 255.7°)$
29. $10(\cos 270° + j\sin 270°)$
30. $20(\cos 0° + j\sin 0°)$
31. $25(\cos 180° + j\sin 180°)$
32. $65(\cos 90° + j\sin 90°)$
33. $50 \underline{/35°}$
34. $75 \underline{/240°}$
35. $12.4 \text{ cis } 300°$
36. $2.93 \text{ cis } 210.0°$
37. $12.48 \underline{/36.25°}$
38. $758.9 \underline{/163.7°}$
39. $2.194 \text{ cis } 235.00°$
40. $5.724 \text{ cis } 123.64°$

In Exercises 41 through 44, solve the given problems.

41. The voltage (in volts) measured on a wire is given as $E = 57.3° \underline{/7.2°}$. Express this number in rectangular form.

42. The current (in microamperes) in a microprocessor circuit is given as $4.25 \underline{/21.5°}$ Express this number in rectangular form.

43. Because complex numbers can be associated with vectors, find the magnitude and direction of a force that is represented by $64.8 - 49.0j$ pounds.

44. Because complex numbers can be associated with vectors, find the magnitude and direction of a displacement vector represented by $0.427 + 0.158j$ millimeters.

Now Try It! Answers

1. a. $2\sqrt{2}(\cos 45° + j\sin 45°)$ **b.** $\sqrt{34}(\cos 301° + j\sin 301°)$
2. a. $-4.24 - 4.24j$ **b.** $3 - 5j$

Chapter Summary ▰◣◥▰◢◣ ▸ ◂▰◢◣◥ ▰◢◣◥ ▰◢◣

Key Terms

complex number   rectangular form of a complex number   complex plane
imaginary number   conjugate   polar form of a complex number

Formulas ◢◣◥ ▰◢◣◥ ▰◢◣◥ ▰◢◣◥ ▰◢◣◥ ▰◢◣

$(a + bj) + (c + dj) = (a + c) + (b + d)j$ **addition and susbtraction**

$(a + bj)(c + dj) = (ac - bd) + (ad + bc)j$ **multiplication**

$\dfrac{a + bj}{c + dj} = \dfrac{(a + bj)(c - dj)}{(c + dj)(c - dj)} = \dfrac{(ac + bd) + (bc - ad)j}{c^2 + d^2}$ **division**

$x = r \cos \theta$

$y = r \sin \theta$

$r^2 = x^2 + y^2$

$\tan \theta = \dfrac{y}{x}$

$x + yj = r(\cos \theta + j \sin \theta)$ **rectangular**

$r\underline{/\theta} = r(\cos \theta + j \sin \theta)$ **polar form**

Review Exercises ▰◢◣◥ ▰◢◣◥ ▰◢◣◥ ▰◢◣◥ ▰◢◣◥ ▰◢

In Exercises 1 through 8, express each number in terms of j.

1. $\sqrt{-64}$ **2.** $-\sqrt{-100}$ **3.** $-\sqrt{-400}$

4. $\sqrt{-81}$ **5.** $\sqrt{-54}$ **6.** $\sqrt{-40}$

7. $-\sqrt{-56}$ **8.** $-\sqrt{-63}$

In Exercises 9 through 12, express each number in the rectangular form $x + yj$.

9. $-6 + \sqrt{-100}$ **10.** $5 - \sqrt{-144}$

11. $3 - \sqrt{-48}$ **12.** $-2 + \sqrt{-27}$

In Exercises 13 through 16, simplify each expression.

13. $(\sqrt{-25})^2$ **14.** $\sqrt{(-8)^2}$

15. $\sqrt{-7}\,\sqrt{-2}$ **16.** $\sqrt{-2}\,\sqrt{-32}$

In Exercises 17 through 20, simplify each expression.

17. j^{14} **18.** $-j^{21}$

19. $(-j)^{40}$ **20.** $-j^{15}$

In Exercises 21 through 32, perform the indicated operations and express all answers in the rectangular form $x + yj$.

21. $(12 + 3j) + (2 + j)$ **22.** $(6 - 7j) - (5 + 4j)$

23. $(-10 - 3j) - (-9 + 4j)$

24. $(-7 + 5j) + (-8 - 4j)$

25. $(3 + 6j)(2 + j)$ **26.** $(10 - 2j)(3 + j)$

27. $(-3 - 4j)(-3 - 5j)$ **28.** $(-8 + 9j)(-5 - 7j)$

29. $(2 - 10j) \div (3 + j)$ **30.** $(-5 + j) \div (4 - j)$

31. $(6 + 2j) \div (-7 - 2j)$ **32.** $(10 + 15j) \div (-5j)$

In Exercises 33 through 36, find the sum of the given number and its conjugate, and find the product of the given number and its conjugate.

33. $6 + 8j$ **34.** $-2 + 5j$

35. $7 - 3j$ **36.** $-4 - 7j$

In Exercises 37 through 40, perform the indicated operations graphically and check the results algebraically.

37. $(1 - j) + (2 + 4j)$ **38.** $3j - (5 - 4j)$

39. $-4j + (-9 - 3j)$ **40.** $(2 + 8j) - (8 - 7j)$

In Exercises 41 through 44, convert each complex number from the given rectangular form to the polar form

$r(\cos \theta + j \sin \theta)$. Use a graph to be sure that θ corresponds to the correct quadrant.

41. $5 + 12j$ **42.** $7.38 - 4.16j$

43. $-18.0 - 12.3j$ **44.** $-62.4 + 87.3j$

In Exercises 45 through 48, convert each complex number from the given polar form to the rectangular form $x + yj$.

45. $15.2(\cos 70.2° + j \sin 70.2°)$

46. $0.912(\cos 216.4° + j \sin 216.4°)$

47. $16.7 \underline{/327.5°}$ **48.** $3.45 \underline{/170.1°}$

In Exercises 49 through 52, answer the given questions or perform the indicated operations.

49. Is $1 + 2j$ a solution to the quadratic equation

$$x^2 - 2x + 5 = 0?$$

Is its conjugate a solution?

50. Show that when the complex number $a + bj$ is multiplied by its conjugate, the product is a real number.

51. In an alternating-current circuit, the voltage V, the impedance Z, and the current I are related by the equation $V = IZ$. Find V (in volts) if $I = 5 - 4j$ amperes and $Z = -3 + 6j$ ohms.

52. In Exercise 51, find the current (in amperes) if $V = 10 + 12j$ volts and $Z = -5 + 8j$ ohms.

Chapter Test

1. Add $(4 - 3j) + (-1 + 4j)$ graphically.

2. Add $(10 - 4j) + (-7 + 8j)$.

3. Multiply $(2 + 4j)(-5 + 3j)$. Express your result in simplest rectangular form.

4. Divide $\dfrac{-4 + j}{3 - 2j}$. Express your result in simplest rectangular form.

5. Express $2 - 7j$ in polar form.

6. Express $2(\cos 225° + j \sin 225°)$ in rectangular form.

7. Determine whether $x = 5 - 2j$ is a solution to the quadratic equation $x^2 - 2x + 4$. Justify your response.

Introduction to Data Analysis

Based on your own experience, you know that it is impossible to open a newspaper or magazine today without finding some reference to statistical results. Many major news stories use statistical analysis or the interpretation of statistical data. Even sports information is presented to us in terms of "stats" for a team or player. And it doesn't stop with the news. Manufacturers use statistical data to determine which products to produce and where to allocate resources. Television networks use statistics to determine what shows will remain in their entertainment line-up. Local, state, and federal government agencies use statistics to make policies. Even the car insurance rates you pay are based on statistical information. Every day we encounter the results of surveys or opinion polls in our newspapers, on the radio, and on television.

Data can be found everywhere. The use of data is growing at incredible rates. Simple examples can be found in opinion polls and in new stories that report on the latest research about the foods you eat. Statistics are used to identify relationships, make predictions, and to communicate information to others. The ability to understand statistical information is extremely important. H. G. Wells, the author of *The War of the Worlds,* wrote "Statistical thinking will one day be as necessary for effective citizenship as the ability to read and write."

This chapter introduces some of the basic concepts of statistics. We will begin by looking at the graphical display of data (such as shown in Figure 17.1). We will also consider some of the more common statistical measures. We will end the chapter with an introduction to probability.

17.1 Creating Pie Charts and Bar Graphs

Section Objectives
- Identify the components of a good graph
- Represent data as a pie chart and a bar graph
- Accurately read and interpret information presented in a pie chart or bar graph

There are many ways to represent data. Earlier in the textbook we considered graphs that used the rectangular coordinate system as a method for representing the relationship between sets of numbers. Sets of data are often presented with other types of graphs as well. We will explore several of these graphs in this section and in Section 17.2, including **pie charts, bar graphs,** and **histograms.**

It is often said that a picture is worth a thousand words. The purpose of any chart or graph is to give a visual summary of the data it is meant to represent. Graphs are generally more informative than looking at a set of data. It is easier to absorb large quantities of information if they are presented in a graph as compared to pages of text. Patterns often emerge in graphs that are not apparent when the data are left in tabular form. Done properly, a good graph conveys a lot of information quickly. We begin by highlighting the characteristics of a good graph: *A good graph is clearly labeled. This includes a title for the graph, a title for each of the axes (including units), and a clear indication as to the scale of the axes.*

The first graph we consider is the **pie chart,** which is a circle graph that represents data in the form of a circle containing wedges. A pie chart is particularly useful for showing the relationship between the various parts of a category to the whole category. This can be seen in Example 1.

Example
1

The following table represents the level of education attained by people aged 25–34 years old in 1998. We wish to display this information with a pie chart.

Education Level	Number of Persons (thousands)
Less than high school	4754
High school graduate	12,568
Some college	11,220
Bachelor's degree	8367
Advanced degree	2444

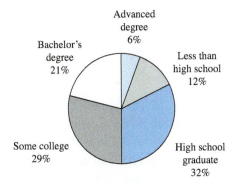

Figure 17.1 Education Levels of 25–34 year olds. (1998)

A *pie chart uses percentages* to compare information. The whole pie is equal to 100%. We begin by finding the total number of persons represented by the table shown (39,353). The percentage of people who had some college education is $\frac{11220}{39353} = 29\%$. To represent this in a pie chart you need to determine how many degrees are represented by 29%. This is done by multiplying 29% by 360°. Therefore, 104°(or 29%) of the circle represents the number of people who had some college education. We can repeat this process in order to create the pie chart in Figure 17.1.

This pie chart conveys information clearly to the reader. The chart title informs us of the information being depicted in the chart. We can see that 88% of the persons surveyed had a high school diploma or higher and that only 12% of those represented by this study did not finish high school.

Computer software packages are available that can be used to create graphs directly from the data given.

Example
2

Representing data with a graph should provide readers with the same information that they would get if they carefully studied the actual data itself. We can examine the title in Figure 17.2 to determine what the pie chart represents (grade distribution). We get a sense for how each of the grading possibilities relates to the total picture. For instance we can see that only 8% of the students did not pass the course.

Another type of graph to consider is the **bar graph.** *This type of graph is valuable for showing the relative size of data.* Bar graphs are useful when comparing results or sets of information. As with all graphs, they allow us to make observations and generalizations about the data quickly. When creating a bar graph we choose a scale suitable for the data and then determine the length of each bar. A bar graph can be drawn with the bars in either a horizontal or vertical alignment. In addition to a title, the axes of the bar graph must also be accurately labeled.

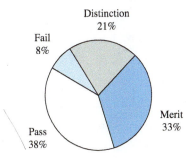

Figure 17.2 Grade Distribution

Example
3

The number of cellular telephone subscribers in the United States is rising quickly as shown in the following table. We can represent this data with a bar graph (see Figure 17.3).

Year	Number of Subscribers (in millions)
1993	14.2
1994	16.9
1995	34
1996	44.5
1997	55.7
1998	69
1999	84.5
2000	110.7
2001	125
2002	142.6
2003	159

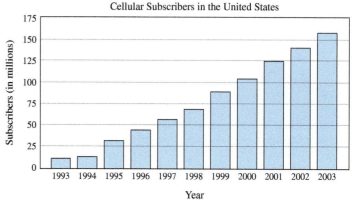

Figure 17.3

Notice that the graph gives a quick visual indication as to the rapid rise in the number of cell phone subscribers over the time period 1993–2003. Also note that the graph contains a title, and clearly labeled horizontal and vertical axes.

Graphs are commonly found in newspapers, magazines, and textbooks. Is it important to be able to create various graphs, and to discuss and analyze the information presented graphically.

Example
4

The graph in Figure 17.4 gives us a snapshot of the enrollment figures for freshman students at a local community college.

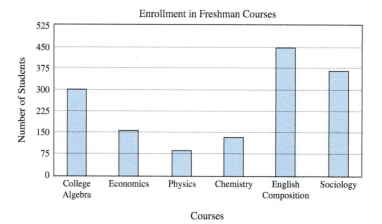

Figure 17.4

In looking at this graph we can easily determine which courses have the most students enrolled and which attract the smallest number of students. We can see that the enrollment in College Algebra is two times larger than the enrollment in Economics. We can also see that the largest number of students are enrolled in English Composition.

17.1 Exercises

In Exercises 1 through 4, create a pie chart to represent the given data.

1. In 1997 there were 92,353 deaths from accidents in the United States. Among these, 42,340 deaths were from motor vehicle accidents, 11,858 from falls, 10,163 from poisoning, 4051 from drowning, 3601 from fires, and the rest for other reasons.

2. Given a gross weekly salary of $750 we find that approximately $187.50 goes towards Federal income tax, $60 is taken out for social security tax, $35 is taken out for health insurance, and $112.50 is taken out for a retirement fund. The balance of $355.00 is the net, or take-home, pay. Represent this information in a pie chart.

3. Seasonal rainfall totals were recorded for a local area with the following results: spring (18 in.), summer (8 in.), fall (10 in.), and winter (20 in.).

4. The top eight gross national incomes by country are shown in the following table. Represent this information as a pie chart.

United States	$9,780,000,000,000
Japan	$4,520,000,000,000
Germany	$1,940,000,000,000
United Kingdom	$1,480,000,000,000
France	$1,380,000,000,000
China	$1,130,000,000,000
Italy	$1,120,000,000,000
Canada	$682,000,000,000

In Exercises 5 through 8, create a bar graph to represent the given data.

5. The number of calories in a one-cup serving of various beverages are: apple juice (125 calories), orange juice (110 calories), cola (104 calories), 2% milk (120 calories), and cocoa (230 calories).

6. The steel production for a particular plant for one year is described as follows: first quarter (25,000 t), second quarter (32,000 t), third quarter (45,000 t), and fourth quarter (28,000 t).

7. In testing a set of electric resistors it is found that (to the nearest ohm) 5 had a resistance of 5 Ω, 8 a resistance of 6 Ω, 17 a resistance of 7 Ω, 4 a resistance of 8 Ω, and 2 a resistance of 9 Ω.

8. The following table gives the water temperature off the North Carolina shore for each month of the year.

Month	Water Temperature (°F)
January	49
February	51
March	55
April	65
May	72
June	79
July	80
August	80
September	77
October	77
November	70
December	50

17.2 **Frequency Tables and Histograms**

Section Objectives
• Organize data into frequency tables
• Create and interpret histograms

Data collection plays an important role in many fields. The data could be test scores, measurements of electric current, lengths of objects, manufacturing specifications for a product, as well as countless other possibilities. Data that have been collected and not yet organized in some way are referred to as **raw data.** We need to be able to organize large amounts of data in a way that makes some sense and will give us useful information.

We begin by looking at **frequency tables** which give us a means of tallying data and listing information based on its corresponding frequency (the number of original scores that fall into a particular class or division). In the process of grouping the data a much clearer overall pattern of the data will emerge. This grouping is done by determining which values belong in a particular group and then tallying the number of values that are within that group. Each group is called a **class,** and the number of values in that class is called the **frequency.** The table that represents this grouping of data is called a **frequency table.** The term *lower class limit* refers to the smallest number that can belong to a class, whereas the *upper class limit* refers to the largest number that can belong to a class.

Example 1

The following data represents the test scores of a particular class:

87, 92, 75, 61, 93, 72, 80, 92, 94, 82, 95, 95, 90

A **frequency table** that would represent the same information would look like the following.

Grades	Tally	Frequency
99–90	⊩⊩ ‖	7
89–80	‖‖	3
79–70	‖	2
69–60	‖	1

In Example 1 we see that each class includes ten values. This is referred to as the **class width.** When creating a frequency table the class width should be the same for each class. Classes should be mutually exclusive, which means that it should be clear which class each value falls in. Notice that the classes in Example 1 were not written as 100–90, 90–80, 80–70, and so on. Had we set the classes up in this way it would not have been clear which class to place the grade of 80. When dividing our data into classes we must decide how many classes to use. In this case the grouping was determined by the grades recorded, but in many cases this may not be as readily apparent. A good rule of thumb is to have between 5 and 15 classes. When choosing the number of classes remember that we are trying to see a pattern in the frequency table created.

To Construct a Frequency Table

1. Determine the number of classes that makes sense given the size of your data set.

2. Determine the width of each class. This should be uniform throughout the frequency table.

3. Create a list of the classes to be used to organize your data.

4. Go back to the data and begin representing each of the values as a tally mark in the appropriate class.

5. Replace the tally marks with the total frequency count for each class.

Example
2

Consider the following information which represents the city driving miles per gallon of automobiles driven in the United States.

19, 30, 23, 31, 23, 19, 23, 17, 20, 22, 24, 26, 25, 17, 18, 17,

18, 19, 20, 27, 29, 28, 35, 36, 20, 24, 33, 35, 26, 29, 30, 32

We begin by determining the number of classes we wish to use. Because there are 32 values in our data set we will use 5 classes. The class width will be 4. Because the lowest value in the data is 17 we can use this as the starting point for the classes we create. After creating all of the classes we complete the following frequency table.

Useful Information

The class width can be determined using the formula

$$class\ width = \frac{range}{number\ of\ classes}.$$

Round off your results to make the class width easy to work with.

The **range** is the difference between the highest value and the lowest value in the set of data.

Classes	Tally	Frequency
17–20	⊞⊞ ⊞⊞ I	11
21–24	⊞⊞ I	6
25–28	⊞⊞	5
29–32	⊞⊞ I	6
33–36	IIII	4

Whereas a bar graph is useful when comparing items with different **qualities**, a **histogram** *is a graph that is useful for displaying the distribution of items with the same quality.* A histogram represents the frequency table by displaying each class of the data as a rectangle. The width of each rectangle represents the *class width* and the height of the rectangle represents the *frequency* of each class.

Example
3

Using the frequency table created in Example 2, which shows the number of cars that get a certain gas mileage in city driving, we create the histogram shown in Figure 17.5.

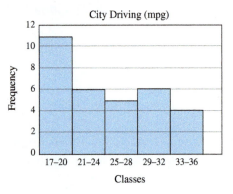

Figure 17.5

Notice that the histogram contains all of the elements of a good graph. In addition, we carefully identify the classes and the frequency for each class. This gives us a good visual representation of the data originally presented.

Example
4

The birth weights of 50 babies are recorded in the following table. We want to organize this data in some way so as to determine any trends in birth **weights.**

5.8	7.4	9.2	7	7.7
7.9	7.8	7.9	7.7	7.2
8.7	7.2	6.1	7.2	8.2
5.9	7	7.8	7.2	8.5
7.4	8.2	9.1	7.3	9.4
8.1	8	7.5	8.7	9
7.6	7.3	7.1	7.6	6.8
7.1	6.9	7	6.7	7.1
7.2	6.9	7	7.9	7
6.4	7.4	7.3	6.4	7.5

We begin by creating a frequency table. Once that is complete we will create a visual representation of this data with a histogram.

To create a frequency table we begin by determining the number of classes that we want to work with. In this example we choose to work with eight classes (although you could choose to work with more or less than eight). We also notice that there is not a wide range between the lowest and highest values in our data. We decide to use a class width of 0.5 because it will also be easy to work with (see Figure 17.6).

Classes	Tally	Frequency
5.6–6	||	2
6.1–6.5	|||	3
6.6–7	卌 ||||	9
7.1–7.5	卌 卌 卌 |	16
7.6–8	卌 卌	10
8.1–8.5	||||	4
8.6–9	|||	3
9.1–9.5	|||	3

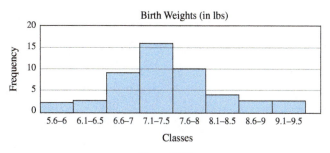

Figure 17.6

— Using Technology

A graphing calculator can be used to display histograms. This involves using both the STAT and STAT PLOT menus, as shown in Figure 17.7.

To Enter Lists of Data

We begin by entering the following data into the calculator:

3, 10, 15, 25, 6, 9, 3, 6, 12, 7, 8, 2, 10, 11, 8, 5, 7, 12, 16, 20

Figure 17.7

1. Press the STAT key to display the Statistics menu.

2. Choose option #1 **Edit.**

3. Use the left cursor to move into the column with the heading L1. Type in the values listed above.

We now are ready to create our histogram.

To Create a Histogram

1. Generally you would begin by entering data into L1. Because we already did this, we move onto step 2.

2. Press 2nd Y= to access the STAT PLOT menu.
   a. Choose option #1 **Plot 1.**
   b. Position the cursor on "ON" and press ENTER.
   c. Move the cursor down to "Type" and select the graph that looks like a histogram. Press ENTER.
   d. Move the cursor down to XList and make sure that L1 is entered. Make sure that Freq has a 1 entered (see Figure 17.8).

3. Press ZOOM 9 or GRAPH to create a histogram.

4. To examine the class limits as well as the frequency associated with each column use the TRACE key. In this case we see that the second column in Figure 17.9 has a frequency of 8 and a lower class limit of 6.6. The upper class limit is less than 11.2.

Figure 17.8

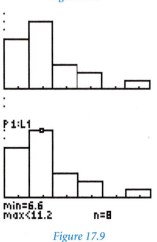

Figure 17.9

17.2 Exercises

In Exercises 1 through 4, create a histogram for the given data.

1. A strobe light was designed to flash every 2.25 s at a certain setting. Sample bulbs were tested with the following results:

Time between flashes (in s)	2.21	2.22	2.23	2.24	2.25	2.26	2.27	2.28	2.29
Number of bulbs	2	7	18	41	56	32	8	3	3

2. In testing the braking system of new cars, the stopping distance required by a car traveling at 70 mi/h was measured in 120 trials. The results are shown in the following table:

Stopping Distance (in ft)	155–159	160–164	165–169	170–174	175–179	180–184	185–189
Times Car Stopped	2	15	32	36	24	10	1

3. Over a period of one week a polymer specialist measured the thickness of the coating of a new substance being tested on PVC pipe to prevent cracking. The specialist recorded the following results:

Thickness (mm)	0.01	0.02	0.03	0.04	0.05	0.06	0.07	0.08	0.09	0.10
Frequency	5	17	23	28	35	42	30	24	10	1

4. Technicians randomly select light bulbs and test them weekly to determine if they will burn the number of hours they are advertised to work. Over a period of ten weeks the number of bulbs that did not meet company standards were recorded in the following chart:

Week	1	2	3	4	5	6	7	8	9	10
Rejected Light bulbs	132	154	90	230	156	176	120	105	125	129

In Exercises 5 through 14, create a frequency table and histogram for the given data.

5. Twenty students were asked to record the total number of hours they spent studying during a given week. This information is given as follows:

   3, 10, 15, 25, 6, 9, 3, 6, 12, 7, 8, 2, 10, 11, 8, 5, 7, 12, 16, 20

6. In testing a computer system, the number of instructions that the system could perform in 1 ns was measured at different points in the program. The number of instructions was recorded as follows:

   19, 21, 22, 25, 22, 20, 18, 21, 20, 19, 22, 21, 19, 23, 21

7. The dosage, in milliroentgens (mR), given by a particular X-ray machine was measured 20 times, with the following readings:

   4.25, 4.36, 3.96, 4.21, 4.37, 4.44, 3.83, 4.27, 4.33, 4.34, 4.15, 3.90, 4.41, 4.51, 4.18, 4.26, 4.29, 4.09, 4.36, 4.23

   Form a histogram with six classes.

8. The price of gasoline was recorded over a 36-month period from January, 2004 to July, 2006.

3.472	3.484	3.447	3.564	3.729	3.64	3.482	3.427	3.531	3.362	3.263	3.131
3.139	3.13	3.241	3.407	3.421	3.404	3.412	3.423	3.422	3.449	3.448	3.394
3.473	3.641	3.748	3.659	3.542	3.514	3.524	3.628	3.728	3.603	3.535	3.494

9. The life of a certain battery (in hours) was measured and recorded with the following results:

   34, 30, 32, 35, 31, 28, 29, 30, 32, 25, 31, 30, 28, 36, 33, 34, 30, 33, 31, 34, 29, 30, 32

10. According to the Energy Information Administration, residential coal fuel consumption (in trillion Btu) is as follows.

    Residential Coal Consumption (Trillion Btu)

93.911	31.114	38.757	49.145	17.451	30.544	1.131	0.887	1.766	1.113
82.126	25.355	40.467	37.313	16.584	30.02	1.035	0.901	11.971	1.93
62.843	25.589	37.218	30.785	15.992	31.791	1.026	0.606	1.363	12.532
58.885	25.749	36.973	39.627	11.548	11.358	0.644	0.78	1.188	1.266
57.461	20.845	30.63	13.981	1.425	0.996	0.724	1.046	1.123	1.358

11. The emissions of greenhouse gases and energy-related carbon dioxide emissions (measured in million metric tons) from residential use of natural gas from 1949 to 2004 can be seen in the following table:

    Energy-Related Carbon Dioxide Emissions—Residential, Natural Gas—1949–2004

55.2	169.6	277.5	238.3	241.0	254.4	266.7	234.4
65.1	177.5	264.0	247.3	234.0	269.1	256.3	241.7
79.8	190.1	258.6	268.8	234.2	263.0	245.6	256.8
87.8	195.1	265.2	284.6	251.2	263.1	250.8	261.5
90.9	205.9	271.8	270.6	260.2	269.2	276.9	125.1
102.1	212.2	259.4	246.6	238.6	259.0	265.5	134.6
113.2	224.9	263.3	256.4	247.3	265.6	159.6	147.3

12. Electric usage is recorded in kilowatt-hours and is often billed in two-month intervals. The following table shows the electric energy consumption (in kWh) for a two-and-a-half-year period.

kWh		
3375	2296	3286
2661	2812	2749
2073	2433	3427
2579	2266	578
2858	3128	3792

13. A heating degree day is the number of degrees that a day's average temperature is below 65°F (18°C), the temperature below which buildings need to be heated. The number of heating degree days (total for a month) for a region in upstate Maine is given in the following table.

Heating Degree Days

2421	1028	116
1841	1967	1457
438	1627	253
15	537	1811
152	26	632

14. The age of female actors at the time each won an Oscar award is given in the following set of data:

22	30	33	34	38	26	41
37	26	28	26	28	37	26
30	29	38	35	27	42	81
62	24	45	34	31	41	42
32	39	24	34	37	35	29
26	24	26	33	30	31	33
31	29	47	49	24	41	35
27	37	41	38	34	33	45
26	30	27	61	60	30	49
27	34	39	21	61	74	39

17.3 Measures of Central Tendency

Section Objective
• Calculate the mean, median, and mode for a given set of data

Tables and graphs give a visual representation of data. It is often useful to describe the data numerically as well. Some of the more common ways to summarize data numerically is with *measures of central tendency* or **measures of center.** These values are those numbers that describe the center of the distribution of a set of data. We will see that different measures of center will generate different values for the same set of data. The measures of center that we will explore include the **arithmetic mean,** the **median,** and the **mode.**

The arithmetic mean is found by adding all of the values in our data set together and then dividing by the number of values in the data set.

$$\text{arithmetic mean} = \frac{\text{sum of all the values}}{\text{number of values in the sample}}$$

This is often referred to as just the *mean.*

The formula for finding the arithmetic mean is

$$\bar{x} = \frac{\sum x}{n}$$

> **Caution**
>
> The *arithmetic mean* is the number most people refer to as the *average* of a set of data.

Some of this notation may not be familiar to you. The symbol $\bar{x}$ represents the mean of a set of data. The symbol Σ is used in mathematics to represent the sum of values. The notation Σx tells us to add up (or sum) all of the values for x, or the values in a set of data. The letter n represents the number of values in the data set. Example 1 demonstrates the use of this formula to find the mean life of batteries.

Example 1

As part of a quality assurance program, the life of a sample of batteries (in hours) was measured and recorded with the following results:

34, 30, 32, 35, 31, 28, 29, 30, 32, 25, 31, 30, 28, 36, 33, 34, 30, 33

Find the arithmetic mean for this set of data.

To find the arithmetic mean we sum the values recorded and divide by the number of data points.

$$\bar{x} = \frac{34 + 30 + 32 + 35 + 31 + 28 + 29 + 30 + 32 + 25 + 31 + 30 + 28 + 36 + 33 + 34 + 30 + 33}{18}$$

$$= \frac{561}{18} = 31.2\,h$$

This tells us that the mean life of the set of batteries tested is approximately 31 hours.

The *median* refers to the value found in the middle of a set of data that has been arranged in order from the lowest value to the highest value. The median tells us that half of the data are above the median and half of the data values are below the median.

The symbol $\tilde{x}$ is used to represent the median. If the number of entries in your data is odd, the median is truly the middle value. If the number of entries in your data is even, the median is the mean of the two middle values.

Example 2

Using the same set of data that we used in Example 1, find the median value for the life of the batteries tested.

Our original data was presented in the following manner:

34, 30, 32, 35, 31, 28, 29, 30, 32, 25, 31, 30, 28, 36, 33, 34, 30, 33

To find the median value we must first sort our data from the lowest value to the highest value. This gives us the following arrangement:

25, 28, 28, 29, 30, 30, 30, 30, 31, 31, 32, 32, 33, 33, 34, 34, 35, 36

middle values

Because there are 18 items the median will be the average of the two middle values. In this case both values are 31, so the median is 31.

Example
3

The selling price for nine homes for a given area is noted in the real estate section of the newspaper. The selling prices are

$342,000 $375,000 $329,000 $338,000 $532,000 $335,000 $350,000
$507,000 $471,000

To find the median price for a home in this area we first need to arrange these prices in order from the lowest price to the highest price:

$329,000 $335,000 $338,000 $342,000 $350,000 $375,000 $471,000
$507,000 $532,000

middle value

Because there are nine values, the fifth value is the median value. In this case the median is $350,000.

The *mode* is that value which occurs most often in a set of data. A set of data may not have a mode or it may have more than one mode.

Example
4

Using our data set from Example 1 we have the following values:

34, 30, 32, 35, 31, 28, 29, 30, 32, 25, 31, 30, 28, 36, 33, 34, 30, 33

In this case we can see that the mode is 30. Four of our data entries have this value. It appears more than any other value in the data set.

Example
5

The following data set gives the length (in inches) of alligators captured in central Florida.

90, 78, 82, 72, 58, 114, 94, 86, 72, 90, 86, 147, 86, 61, 89, 63, 88, 128, 68, 69, 76, 61, 74, 74, 95, 85, 89

Use this data to determine the mean length, the median length, and the mode length of the alligators captured.

The arithmetic mean is $\bar{x} = \dfrac{2275}{27} = 84$ inches.

The median is found after we sort the data from its lowest value to its highest value.

58, 61, 61, 63, 68, 69, 72, 72, 74, 74, 76, 78, 82, 85, 86, 86, 86, 88, 89, 89, 90, 90, 94, 95, 114, 128, 147

Now Try It!

Determine the mean, median, and mode for the following sets of numbers:

1. 3, 6, 4, 2, 5, 4, 7, 6, 3, 4, 6, 4, 5, 7, 3
2. 450, 550, 425, 475, 375, 500, 400, 550, 475, 600, 500, 425, 450, 500
3. 0.48, 0.53, 0.49, 0.45, 0.55, 0.49, 0.55, 0.48, 0.57, 0.51, 0.46

There are 27 values in our data set. The median will be the fourteenth value. In this case the median is 85.

The mode is the value that shows up the most frequently in the data set. With the data arranged in order it is easy to see that the mode is 86 because it appears most frequently in the data set.

The arithmetic mean is 84, the median is 85, and the mode is 86. It is important to note that, for the same set of data, each measure of center is a different value. This is not uncommon.

17.3 Exercises

In each of the following exercises, determine the indicated measure of center.

1. Find the mean, median, and mode for the computer instructions in Exercise 6 in Exercises 17.2.

2. Find the mean, median, and mode for the dosage given by an X-ray machine in Exercise 7 in Exercises 17.2.

3. The snowfall (in inches) from a recent snowstorm was recorded for 18 towns. Find the mean, median, and mode for this data.

   14.6, 12.3, 20.5, 16, 18.5, 15, 9, 9, 16, 14, 14, 13.8, 13.6, 12, 8.5, 26, 23, 20

4. The weekly salaries (in dollars) for the workers in a small factory are as follows:

   $450, $550, $475, $425, $375, $500, $400, $550, $475, $600, $500, $425

   Find the mean, median, and mode of these salaries.

5. Use the following weights (in lb) of the male students enrolled in a college-level mathematics course to determine the mean and the mode.

   144, 216, 220, 262, 144, 162, 156, 180, 125, 132, 204, 180, 135, 144

6. The following test scores for a weekly quiz were recorded and entered into a grade book:

   81 56 90 70 65 90 90 30

   Determine the class mean and median for this quiz.

7. Electric usage is recorded in kilowatt-hours and is often billed in two-month intervals. The following table shows the electric energy consumption (in kWh) for a two-and-a-half-year period.

kWh		
3375	2296	3286
2661	2812	2749
2073	2433	3427
2579	2266	578
2858	3128	3792

Find the mean and median number of kilowatt-hours used per two-month cycle.

8. The price of diamond engagement rings (in dollars) is recorded in the following table. Use this information to determine the mean and median price of an engagement ring.

$5170	$6958	$4426	$1499
$5547	$5885	$6885	$9736
$1859	$6333	$5826	$9859
$7521	$4299	$3670	$1239
$7260	$1589	$7176	$2532
$8139	$6921	$7497	$1100

9. The price of gasoline was recorded over a 36-month period from January, 2004 to July, 2004.

3.472	3.484	3.447	3.564	3.729	3.64	3.482	3.427	3.531	3.362	3.263	3.131
3.139	3.13	3.241	3.407	3.421	3.404	3.412	3.423	3.422	3.449	3.448	3.394
3.473	3.641	3.748	3.659	3.542	3.514	3.524	3.628	3.728	3.603	3.535	3.494

Determine the mean and mode for this set of data.

10. The daily high temperature (in °F) for the month of February was observed and recorded in the following table. Use this information to determine the mean and median daily high temperature.

25	36	37	33
31	37	34	35
33	32	41	38
29	28	40	37
37	43	33	31
35	33	35	38
46	39	33	

11. A heating degree day is the number of degrees that a day's average temperature is below 65°F (18°C), the temperature below which buildings need to be heated. The number of heating degree days over a period of 15 months for upstate Maine is given in the following table.

Heating Degree Days

2421	116	632
1811	15	1627
1028	26	1967
537	152	1811
438	253	1627

Find the average number of heating degree days over this 15-month period.

12. The following table shows the birth weight (in lb) of 35 randomly selected babies.

5.8	7.4	9.2	7.0	8.5	7.6
7.9	7.8	7.9	7.7	9.0	7.1
8.7	7.2	6.1	7.2	7.1	7.2
5.9	7.0	7.8	7.2	7.5	6.4
7.4	8.2	9.1	7.3	7.8	8.2
8.1	8.0	7.5	8.7	7.2	

Find the mean, median, and mode for this data.

13. A test of air pollution in a city gave the following readings of the concentration of sulfur dioxide (in parts per million) for 18 consecutive days:

0.14, 0.18, 0.27, 0.19, 0.15, 0.22, 0.20, 0.18, 0.15
0.17, 0.24, 0.23, 0.22, 0.18, 0.32, 0.26, 0.17, 0.23

Find the mean, median, and mode for this data.

14. In a particular month, the electrical usage (in kWh) of 1000 homes in a certain city was summarized as follows:

Usage	500	600	700	800	900	1000	1100	1200
Number of homes	22	80	106	185	380	122	90	15

Find the mean of the electrical usage.

15. Find the average thickness of the coating in Exercise 3 in Exercises 17.2.

16. Find the mean number of light bulbs that did not meet company standards each week in Exercise 4 in Exercises 17.2.

Now Try It! Answers

1. mean 4.6, median 4, mode 4   **2.** mean 476, median 475, mode 500   **3.** mean 0.48, median 0.49, modes 0.48, 0.49, 0.55

Section Objectives
• Understand the difference between a sample and a population
• Determine the range for a set of data
• Find the standard deviation for a set of data

When working with data we generally collect samples of values and use that information to draw conclusions about a much more complete set of possible values. The term **population** refers to the collection of all elements to be studied whereas **sample** indicates a subset of the population. In the preceding sections we have been working with sample data. We might use that sample data to make some generalizations about the larger group the sample data represents.

In the preceding section we discussed measures of center as a means for describing our data. This information, although useful, does not give us the whole picture. We do not know whether the data are grouped closely or if they contains a large range of values. Measures of spread and variation are those values that tell us how much the data tends to spread out or disperse. When the dispersion is large, the values are widely scattered; when it is small, they are tightly clustered. If the spread is small, the measure of center is more reliable and descriptive of the data than if the data are spread out.

There are several **measures of dispersion.** The two that we will focus on are the range and the standard deviation. The standard deviation tells us to what extent individual values are dispersed or "spread out" around their mean.

Example 1

Suppose there are two students in a class with the following test scores:

Student A   received scores of 74, 72, 74, 80, 77, 83, and 86

Student B   received scores of 58, 74, 100, 74, 81, 77, and 82

These scores are presented in no particular order, meaning this may not be the order in which the students received these scores.

In determining the mean, mode, and median for both of these students we discover that the mean is 78, the median is 77, and the mode is 74. The measures of center for both students are identical.

But are Student A and Student B identical in their ability?

The most basic measure of dispersion is the **range,** which is defined as the difference between the highest and lowest values in our set of data, and is very easy to calculate.

Example 2

In looking back at Student A and Student B we see that

the range for Student A is 86–72 = 14

the range for Student B is 100–58 = 42.

Student B has a larger range (or spread) in test scores than Student A.

A great deal of information is ignored when using the range because we consider only the highest and the lowest value in our data set. The remaining data are ignored. The range can also be easily influenced by the presence of one unusually large or small value in the data set.

The **standard deviation** is generally considered the most important and most useful measure of dispersion. It represents the deviation from the mean and measures the dispersion or variation among scores. As the standard deviation increases the spread between the data gets larger.

The formula for finding the standard deviation is

$$s = \sqrt{\frac{\sum(x - \bar{x})^2}{n - 1}}$$

where x represents the individual scores, $\bar{x}$ represents the mean of the data, and n represents the total number of values in your data set.

Although the formula may look complicated it is not. It is important that you work through the formula in a systematic way and take the time to work through each step carefully.

To Calculate the Standard Deviation

1. Determine the arithmetic mean $\bar{x}$ for your data.

2. Subtract the mean from each value in your data set.

3. Square these differences.

4. Sum the values found in step 3.

5. Divide this sum by $n - 1$.

6. Take the square root of this result.

Example 3

Find the standard deviation for the test scores for both of the students presented in Example 1. Recall the following student scores:

Student A   72, 74, 74, 77, 80, 83, and 86

Student B   58, 74, 74, 77, 81, 82, and 100

Both sets of data had an arithmetic mean of 78.

x	$x - \bar{x}$	$(x - \bar{x})^2$
72	$72 - 78 = -6$	36
74	$74 - 78 = -4$	16
74	$74 - 78 = -4$	16
77	$77 - 78 = -1$	1
80	$80 - 78 = 2$	4
83	$83 - 78 = 5$	25
86	$86 - 78 = 8$	64
		$\sum (x - \bar{x})^2 = 162$

$$s = \sqrt{\frac{\sum (x - \bar{x})^2}{n - 1}} = \sqrt{\frac{162}{6}} = 5.196$$

Therefore, the standard deviation for Student A is 5.196.

If we repeat this process, we find that Student B has a standard deviation of 12.53.

We can see that the data belonging to the student with the larger spread in test scores has a larger standard deviation.

Now Try It!

Determine the standard deviation for the following sets of numbers:

1. 3, 6, 4, 2, 5, 4, 7, 6, 3, 4, 6, 4, 5, 7, 3
2. 450, 550, 425, 475, 375, 500, 400, 550, 475, 600, 500, 425, 450, 500
3. 0.48, 0.53, 0.49, 0.45, 0.55, 0.49, 0.55, 0.48, 0.57, 0.51, 0.46

Using Technology

A graphing calculator can be used to find various statistical values, including the standard deviation.

In a preceding section we entered a list of data in the calculator. Recall that

To Enter Lists of Data

1. Press the STAT key to display the Statistics menu.

2. Choose option #1 **Edit.**

3. Use the left cursor to move into the column with the heading L1. If there are values there from a previous problem you can delete these entries by pressing the DEL key. Enter the test scores for Student A in L1.

4. Press STAT.

5. Use the right arrow key to move to CALC to display the CALC menu.

6. Choose option #1 **1–Var Stats.**

7. Press **ENTER** to display a variety of descriptive statistics. This display shows the mean of the data $\bar{x}$ as well as the sample standard deviation, Sx, and the population standard deviation, σx.

17.4 Exercises

In each of the following exercises, determine the indicated measure of dispersion. Use your calculator to verify your results.

1. In testing a computer system, the number of instructions that the system could perform in 1 ns was measured at different points in the program. The number of instructions was recorded as follows:

   19, 21, 22, 25, 22, 20, 18, 21, 20, 19, 22, 21, 19, 23, 21

   Find the range and standard deviation for this data.

2. A test station was measured for the loudness of the sound of jet aircraft taking off from a certain airport. The decibel (dB) readings were measured to the nearest 5 dB and the readings for the first 17 jets were recorded as follows:

   110, 95, 100, 115, 105, 110, 120, 110, 115, 105, 90, 95, 105, 110, 100, 115, 120

   Find the range and standard deviation for these data.

3. The number of calories per gram of cereal for 16 popular brands is as follows:

   3.7, 3.6, 3.6, 4.0, 4.0, 3.6, 3.8, 3.7, 4.1, 3.9, 3.9, 3.3, 3.5, 3.6, 3.9, 4.0

   Find the range and standard deviation for these data.

4. The height (in inches) of a class of students enrolled in a wellness class were recorded as follows:

   62.5, 64.6, 69.1, 73.9, 67.1, 64.4, 71.1, 71.0, 67.4, 69.3, 64.9, 68.1, 66.5, 67.5, 66.5

   Find the range and standard deviation for these data.

5. Find the standard deviation dosage given by an X-ray machine in Exercise 7 in Exercises 17.2.

6. Find the standard deviation for the price of a gallon of gasoline in Exercise 9 in Exercises 17.3.

7. Find the standard deviation for the concentration of sulfur dioxide in Exercise 13 in Exercises 17.3.

8. Find the range and standard deviation for the number of kilowatt-hours used per two-month cycle in Exercise 7 in Exercises 17.3.

9. Find the range and standard deviation for the number of heating degree days over this 15-month period using Exercise 11 in Exercises 17.3.

10. Fifteen cell phone batteries were randomly selected and tested for the number of minutes the battery went between charges. The number of minutes were as follows:

    350, 420, 220, 60, 95, 110, 300, 600, 500, 370, 230, 485, 140, 95, 540.

    Find the range and standard deviation for the number of hours a cell phone battery can go between charges.

11. Twenty families participated in a school project in which the total number of pounds of garbage generated by a family of four for one week was recorded. This information is as follows.

    Household Garbage in Pounds

10.76	27.9	44.44	24.13
19.96	21.9	45.17	37.6
27.6	21.83	33.07	38.96
38.11	49.27	10.35	17.17
35.54	33.27	44.44	31.6

    Determine the range and standard deviation for the number of pounds of garbage.

12. The following table represents the 2005 baseball salaries for players on the Los Angeles Dodgers and the Florida Marlins.

LA Dodgers		Florida Marlins	
2,000,000	2,150,000	316,000	46,66,667
650,000	7,350,000	300,000	7,500,000
2,500,000	1,000,000	2,400,000	1,100,000
319,500	7,000,000	360,000	400,000
330,000	5,100,000	3,650,000	2,600,000
500,000	330,000	370,000	475,000
351,500	4,500,000	5,166,667	3,700,000
1,300,000	339,000	3,000,000	750,000
13,400,000	316,000	4,000,000	350,000
9,400,000	650,000	750,000	316,000
525,000	332,000	4,435,000	1,500,000
8,000,000	316,000	3,400,000	378,500
327,000	3,500,000	425,000	7,000,000
316,000	9,350,000	1,100,000	
550,000	337,000		

    Determine the average salary and standard deviation for each team.

13. The following data represents the golf scores of two members of the local college golf team for their last 15 rounds of golf.

Golfer 1	Golfer 2
83	90
80	89
77	88
88	73
83	78
78	84
81	73
85	83
82	86
75	80
82	75
83	82
79	85
80	79
81	73

Determine the arithmetic mean, range, and standard deviation of each golfer. Which golfer is the more consistent player?

14. According to the Bureau of Labor Statistics the following table represents the cost of electricity in January and July (price per 500 kWhs). Determine the standard deviation for each month and describe what that tells us about the cost of electricity in January and July (if anything).

Jan	Jul
23.645	
26.698	26.165
31.552	31.513
36.006	35.744
37.184	38.403
38.6	39.507
38.975	42.406
40.223	41.172
40.022	42.138
40.195	42.686
40.828	41.509
41.663	42.825
43.226	43.531
44.501	45.015
46.959	46.862
48.2	48.68
48.874	49.505
48.538	50.552

Now Try It! Answers

1. 1.55   **2.** 62.38   **3.** 0.40

17.5 Probability

Section Objectives
- Understand the basic concept of mathematical probability
- Determine the probability of an event

What is a probability? What does it mean to say that the probability of a fair coin is one half, that there is a 50-50 chance of getting a question right, or that there is a 30% chance of snow in the forecast? We either use or come across phrases like this every day. Mathematicians try to convert statements such as these to their numerical equivalents. Think about some event where the outcome is uncertain.

Examples of such outcomes would be the flip of a coin, or the amount of snow that we get tomorrow. In each case, we cannot know for sure what will happen.

Probability is a numerical measure of the likelihood of the event. It is a number that we attach to an event, say the event of getting a heads when we flip a coin or that we will get over a foot of snow tomorrow. Probability plays a role in many of the decisions we make. Yet most people have a poor understanding and "feel" for what probability really is. We will begin with a few basic definitions.

Definition	Example
An **experiment** is any process by which an outcome is obtained.	The experiment is flipping a coin.
An **outcome** is the result of a single trial of an experiment.	The possible outcomes when flipping a coin are getting a heads or getting a tail.
An **event** is one or more of the outcomes.	One event could be getting a head.
A **sample space** is a list of all possible sample outcomes.	The sample space is the list of all outcomes.
Probability is the measure of how likely an event is.	The probability of a head is $\frac{1}{2}$.

In mathematical notation, the probability $P(A)$ that an event A will occur is defined as follows:

$$P(A) = \frac{the\ number\ of\ ways\ A\ can\ occur}{the\ total\ number\ of\ possible\ outcomes} = \frac{s}{n}$$

This process for finding $P(A)$ assumes that each possible outcome has an equal chance of occurring.

Example 1

Suppose we roll a die and want to find the probability of rolling a 5. The total number of possible outcomes when we roll a single die is 6. There is only one 5 on the die. Therefore, $P(5) = \dfrac{1}{6}$.

Example 2

Suppose we have a drawer filled with 20 resistors, 5 of which have a resistance of 8 Ω. What is the probability of reaching into the drawer and randomly pulling out one of the 8 Ω resistors? The total number of possible outcomes is 20, and there are 5 resistors that will have a resistance of 8 Ω. Therefore, $P(8\ \Omega) = \dfrac{5}{20}$.

A probability is a number between 0 and 1. An *impossible event* has a probability of 0. An event that will *always* occur has a probability of 1. All other events fall somewhere in between with $0 \leq P(A) \leq 1$.

Example 3

The probability of getting a head when we flip a single coin is found to be $P(H) = \dfrac{1}{2} = 0.5$. A probability of 0.5 means that it is as equally likely for the *event to occur* as for the event *to not occur*.

Example 4

If we were to toss two dice we know that 36 possible combinations (or outcomes) of numbers can occur. The sample for such an experiment is shown here:

1,1	1,2	1,3	1,4	1,5	1,6	2,1	2,2	2,3	2,4	2,5	2,6
3,1	3,2	3,3	3,4	3,5	3,6	4,1	4,2	4,3	4,4	4,5	4,6
5,1	5,2	5,3	5,4	5,5	5,6	6,1	6,2	6,3	6,4	6,5	6,6

Find the following probabilities:
P(sum of 2), P(sum of 6), P(sum of 10), and P(sum of 12)

We have determined that there are 36 possible outcomes. A careful inspection of our sample space indicates that there is only one way to get a sum of 2, five ways to get a sum of 6, two ways to get a sum of 10, and one way to get a sum of 12.

Using this information we see that

$$P(\text{sum of }2) = \frac{1}{36} \qquad\qquad P(\text{sum of }6) = \frac{5}{36}$$

$$P(\text{sum of }10) = \frac{3}{36} \qquad\qquad P(\text{sum of }12) = \frac{1}{36}$$

Example 5

Using the information from Example 4, which is more likely to occur: a sum of 6 or a sum of 10; a sum of 2 or a sum of 12?

Since $P(\text{sum of }6) = \dfrac{5}{36}$ and $P(\text{sum of }10) = \dfrac{3}{36}$ we can determine that we have a higher likelihood of rolling two dice and getting a sum of six than we do of rolling two dice and getting a sum of 10.

Now Try It!

Determine the probability for each of the following:

1. The probability that a telephone number will end in an even number.
2. The probability that Christmas will fall on a weekday.
3. The probability that a randomly selected person will have a birthday in July.
4. The probability that a randomly selected battery will be defective if a company reports that, on average, 125 batteries out of each 10,000 produced is defective.

Since $P(\text{sum of }2) = \dfrac{1}{36}$ and $P(\text{sum of }12) = \dfrac{1}{36}$ we can see that we have an equal chance of rolling a sum of 2 or a sum of 12.

We can summarize this observation by noting that

> If $P(A) > P(B)$, then event A is more likely to occur than event B.
>
> If $P(A) = P(B)$, then events A and B are equally likely to occur.

Sometimes we are interested in determining the probability that an event will not occur. We may want to know the probability that it will *not* snow on a particular day. The **complement** of an event, denoted $\overline{A}$, consists of all the outcomes in which event A does not occur.

$$P(\overline{A}) = 1 - P(A)$$

Example 6

Determine the probability of rolling two dice and not getting a sum of 6.

$$P(\text{not sum of }6) = 1 - P(\text{sum of }6)$$

$$= 1 - \frac{5}{36}$$

$$= \frac{31}{36}$$

It is often easier to find a probability using a complement.

17.5 Exercises

In the following exercises, find the indicated probabilities.

1. Government data shows that in a study of 1500 people, 435 were likely to die of heart disease, whereas 345 were likely to die from cancer. Find the probability that a death will occur as a result of heart disease.

2. Use the same information found in Exercise 1 to find the probability that a death will occur as a result of cancer.

3. Use the same information found in Exercise 1 to find the probability that a death will occur as a result of heart disease or cancer.

4. Use the same information found in Exercise 1 to find the probability that a death will occur as a result of some other cause (not heart disease or cancer).

5. At a local high school with a student population of 3300 students it was noted that 1815 of the students were planning to attend college, 990 were planning to attend a two-year technical school, 231 planned to work upon graduation, and the rest were undecided as to what their future plans might be. Determine the probability that a student planned to join the work force at the end of his or her high school education.

6. Use the same information found in Exercise 5 to find the probability that a student will go on to college or a two-year technical school.

7. A couple plans to have three children. There are eight possible combinations of boy and girl babies for three births. Assuming that order does not matter, what is the probability that the couple will have two girls and a boy?

8. Use the same information found in Exercise 7 to find the probability that the couple will have two girls followed by a boy if order does matter.

9. A card is drawn from a standard deck of cards (13 hearts, 13 clubs, 13 diamonds, and 13 spades). What is the probability that a card drawn at random will be a heart?

10. Use the same information found in Exercise 9 to find the probability that a card drawn at random will not be a 2.

11. In a given inventory of 6000 parts it is assumed that 300 of the parts will be defective. What is the probability that a part drawn at random will be defective?

12. At a certain factory with 325 workers, 240 workers exercise regularly, 33 exercise occasionally, and the remainder do not exercise at all. Find the probability that a worker does not exercise at all.

13. A machine produces 55 defective parts for each 3000 parts produced. What is the probability that a randomly selected part will not be defective?

14. In testing a set of electric resistors it is found that (to the nearest ohm) 5 had a resistance of 5 Ω, 8 a resistance of 6 Ω, 17 a resistance of 7 Ω, 4 a resistance of 8 Ω, and 2 a resistance of 9 Ω. What is the probability that a randomly selected resister will have a resistance of 7 Ω?

15. In testing the braking system of new cars, the stopping distance required by a car traveling at 70 mi/h was measured in 120 trials. The results are shown in the following table:

Stopping Distance (in ft)	155–159	160–164	165–169	170–174	175–179	180–184	185–189
Times Car Stopped	2	15	32	36	24	10	1

What is the probability that a car will require a stopping distance between 160 ft and 169 ft?

16. At a recent blood drive the blood types of those who donated blood were recorded. There were 53 donors listed as type O, 32 donors listed as type A, 17 donors listed as type B, and 5 donors listed as type AB. Determine the probability that a randomly selected donor would have blood type AB.

Now Try It! Answers

1. 0.5   **2.** 0.7   **3.** 0.08   **4.** 0.0125

Chapter Summary

Key Terms

pie chart	histogram	standard deviation
bar graph	measures of center	probability
raw data	arithmetic mean	experiment
frequency table	median	outcome
class	mode	event
frequency	measures of dispersion	sample space
class width	range	complement

Key Concepts

• We began this chapter with a brief discussion of the characteristics of a good graph. These included a title for the graph, a title for each of the axes (including units), and a clear indication as to the scale of the axes.

• A pie chart is a circle graph that represents data in the form of a circle.

A pie chart uses percentages to compare information.

• A bar graph is useful when comparing results or sets of information.

• A histogram is useful for displaying the distribution of items.

• Measures of center that were introduced include the arithmetic mean, the median, and the mode. Measures of dispersion that were introduced were the range and the standard deviation.

• Probability is a numerical measure of the likelihood of an event.

Formulas

Arithmetic mean $\bar{x} = \dfrac{\sum x}{n}$

Standard deviation $s = \sqrt{\dfrac{\sum (x - \bar{x})^2}{n - 1}}$

$P(A) = \dfrac{\text{the number of ways A can occur}}{\text{the total number of possible outcomes}} = \dfrac{s}{n}$

Review Exercises

In Exercises 1 and 2, use the information given to create the appropriate graph.

The results of a survey of the housing accommodations most used by students at a large city college are shown in the following table:

Type of Accommodation	Number
House	491
Apartment	543
Town House	387
Dormitory	835

1. Create a pie chart illustrating this information.

2. Create a bar chart using this information.

In Exercises 3 through 9, use the following set of data to answer each question.

1098, 1102, 1101, 1095, 1104, 1097, 1107, 1099
1101, 1098, 1106, 1098, 1104, 1093, 1095, 1102

3. Create a frequency table with five classes and the lowest class limit is 1093.

4. Create a histogram.

5. Determine the mean.

6. Determine the median.

7. Determine the mode.

8. Determine the range.

9. Determine the standard deviation.

In Exercises 10 through 16, use the following information to answer each question.

A medical research team studied the ages of patients who suffered strokes caused by stress. The ages of 34 patients were noted as follows:

29 32 37 44 48 50 59 62 51 31 63 53 32 27 44 63 31
51 54 52 51 26 38 54 46 48 55 46 28 31 36 50 26 50

10. Create a frequency table. Use six classes.

11. Create a histogram.

12. Determine the mean.

13. Determine the median.

14. Determine the mode.

15. Determine the range.

16. Determine the standard deviation.

In Exercises 17 through 23, use the following information to answer each question.

A survey was sent to graduates of the University of South Florida requesting information about jobs and salaries.

The starting salaries of 50 graduates who returned the questionnaire shortly after obtaining a job is given below.

11000	20500	25200	30300
13200	20600	25200	30900
14200	20700	25600	31400
15300	21400	26100	32400
15700	21500	26300	32700
16800	22100	26600	33100
17400	22800	26700	34600
20000	22800	26700	36600
20100	23100	28200	36700
20400	23200	28300	37500
20400	24400	28700	38900
20400	24900	29400	40500
		30000	43900

17. Create a frequency table. Determine the appropriate number of classes.

18. Create a histogram.

19. Determine the mean.

20. Determine the median.

21. Determine the mode.

22. Determine the range.

23. Determine the standard deviation.

24. The U.S. General Accounting Office recently tested the IRS for correctness of answers to taxpayers' questions. For 1733 trials, the IRS was correct 1107 times. Find the probability that a random taxpayer's question will be answered correctly.

25. When Mendel conducted his famous genetics experiments with peas, one sample he used consisted of 428 green peas and 152 yellow peas. Find the probability that a new plant will be a yellow pea plant.

26. A study of 150 randomly selected airline flights showed that 115 arrived on time. What is the probability that a flight will arrive late?

27. A test was given to 300 students. If 55 students received an A, 98 received a B, 60 received a C, and 10 received a D on the test, find the probability of not receiving a passing grade on the test.

Chapter Test

1. A marketing company reported the number of broadcast hours in various categories of radio broadcast. Use the information presented in the following table to construct a pie chart summarizing this data.

Broadcast Category	*Number of Programming Hours*
Music	1943
Lectures	576
News	519
Sports reports	300
Programming for women	143
Miscellaneous	97
TOTAL	*3578*

In Exercises 2 through 10, use the following information to answer each question.

The ages of presidents when they were inaugurated into office is presented in the following table:

Ages of Presidents

57	52	51	42	54	54	56	57	56
65	57	43	69	55	47	51	46	46
55	49	54	50	61	55	57	51	60
54	62	52	51	49	68	56	64	
61	61	64	58	51	55	54	48	

2. Create a frequency table. Use five classes with the lowest class limit at 45.

3. Create a histogram.

4. Determine the mean.

5. Determine the median.

6. Determine the mode.

7. Determine the range.

8. Determine the standard deviation.

9. Determine the probability that a president will be 54 at the time of his or her inauguration.

10. What is the probability that someone will be in his or her 60's when inaugurated as president?

Appendix   *Review of Arithmetic*

A.1   Addition and Subtraction of Whole Numbers

Much of the material in this section should be review. It is important to be skilled in this fundamental material. Finding the solutions to most applied problems in science and technology involves arithmetical computations. To perform these computations, we use the basic arithmetical operations of addition, subtraction, multiplication, and division. These operations are essential to performing computations, and they are fundamental to the development of the various branches of mathematics itself. It is important that these operations be performed accurately and with reasonable speed. Although we assume that you are familiar with these operations, we shall include a brief discussion here for review and reference.

The most fundamental use of numbers is that of counting. *The numbers used for counting, the* **natural numbers,** *are represented by the symbols 1, 2, 3, 4, and so on. When we include* **zero,** *we have the* **whole numbers,** *which are represented by 0, 1, 2, 3, and so on.*

Any of the whole numbers can be written with the use of the 10 symbols 0, 1, 2, 3, 4, 5, 6, 7, 8, and 9 if the actual *position* of a symbol in a given number is properly noted. This important feature used in writing numbers—that of *placing each symbol in a specified position in order to give it a particular meaning—is referred to as* **positional notation.**

Example 1

In the number 3252, read as "three thousand two hundred fifty-two," the left 2 represents the number of hundreds and the right 2 represents the number of ones, because of their respective positions. Even though the symbol is the same, its position gives it a different value. Also, the 3 represents the number of thousands and the 5 represents the number of tens, because of their respective positions.

thousands	hundreds	tens	ones		hundreds	tens	ones
3	2	5	2		3	2	5

In the number 325, the 2 represents the number of tens and the 5 represents the number of ones. In the number 352, the 5 represents the number of tens and the 2 represents the number of ones. Even though the same symbols are used, the different positions result in different values.

The process of finding the total number of objects in two different groups of these objects, without actually counting the objects, is **addition.** *The numbers being added are the* **addends,** *or* **terms,** *and the result is the* **sum.**

When we are adding two whole numbers, we must take into account positional notation and add only those numbers with the same positional value. In this way we are adding like quantities.

Example 2

When we add 46 and 29, we are saying

$$
\begin{array}{r}
4 \text{ tens} + 6 \text{ ones} \\
2 \text{ tens} + 9 \text{ ones} \\
\hline
6 \text{ tens} + 15 \text{ ones}
\end{array}
$$

Since 15 ones = 1 ten and 5 ones, we then have

$$6 \text{ tens} + 1 \text{ ten} + 5 \text{ ones} = 7 \text{ tens} + 5 \text{ ones}$$

We usually perform this addition as follows

$$
\begin{array}{r}
\overset{1}{4}6 \\
+\ 29 \\
\hline
75
\end{array}
\qquad 6 + 9 = 1 \text{ ten} + 5 \text{ ones}
$$

where the 1 shows the number of tens "carried" from the ones column to the tens column.

You should know the basic sums through $9 + 9 = 18$. If you are at all unsure of any of these, write them out so that you can review them. Being able to perform addition accurately and with reasonable speed comes from knowing the basic sums well; this takes practice.

It is also wise to form the habit of checking your work. Several methods of checking addition are available. A simple and effective method is to add the columns in the direction opposite to that used in finding the sum originally.

Example
3

If we find the sum of the indicated numbers by adding the columns downward, we can check the results by adding again, this time upward.

```
    2 2                    2 2
    327                    327
    582                    582
    695     add            695     check
    419                    419
   ─────                  ─────
   2023                   2023
```

The process of **carrying** *a number from one column to the next is necessary whenever the sum of the digits in a column exceeds 9.*

We must often determine how much greater one number is than another. This leads to the operation of **subtraction,** the inverse of addition. *Subtraction consists of reducing the number from which the subtraction is being made (the* **minuend***) by the number being subtracted (the* **subtrahend***). The result is called the* **difference.**

Example
4

If we wish to subtract 29 from 73, we find that for the number of ones involved, we are to reduce 3 by 9. When we consider natural numbers, we cannot perform this operation. However, if we "borrow" 10 ones from the tens of 73, the subtraction amounts to subtracting 2 tens + 9 ones from 6 tens + 13 ones. We can show the subtraction as

$$
\begin{array}{ll}
7 \text{ tens} + 3 \text{ ones} & \quad 6 \text{ tens} + 13 \text{ ones} \\
- \ 2 \text{ tens} + 9 \text{ ones} \quad \text{or} & - \ 2 \text{ tens} + \ \ 9 \text{ ones}
\end{array}
$$

Using the second form, we see that the result is 4 tens + 4 ones, or 44. The usual form of showing the subtraction is

```
                                    73 = 6 tens + 13 ones
                    ⁶
minuend            7¹3 ←
- subtrahend      - 2 9
difference          4 4
```

Here the small 1 shows the number of tens borrowed, and the small 6 shows the remaining tens after the borrowing. Borrowing in subtraction is essentially the opposite of carrying in addition.

We can check subtraction by adding the difference and the subtrahend. The sum should be the minuend. This follows directly from the meaning of subtraction. The check of the preceding subtraction is as follows:

$$
\begin{array}{r}
29 \\
+\ 44 \\
\hline
73
\end{array}
$$

   subtrahend

   + difference

   minuend

In many subtraction problems it is necessary to borrow more than once. It might be necessary to borrow from the tens and then again from the hundreds or thousands. Example 5 illustrates this type of subtraction.

Example 5

The subtraction $8203 - 4659$ is shown as follows:

$$
\begin{array}{r}
7\ 1\ 9 \\
8^1 2^1 0^1 3 \\
-\ 4\ 6\ 5\ 9 \\
\hline
3\ 5\ 4\ 4
\end{array}
$$

$8203 = 7 \text{ thousands} + 11 \text{ hundreds} + 9 \text{ tens} + 13 \text{ ones}$

Here we see that it was necessary to borrow from the tens, although there were initially no tens to borrow from. We had to borrow first from the hundreds and then from the tens. Finally, it was necessary to borrow from the thousands to complete the subtraction in the hundreds column.

Although the additions and subtractions discussed in this section can be easily and quickly performed on a calculator, you must understand these basic arithmetic operations and be able to perform them mentally. A good understanding of all the basic operations is important to comprehending algebra. Only after you understand these operations and can perform them accurately should you consider using a calculator. Many gross errors have been made with calculators because they have been used improperly.

A.1 Exercises

In Exercises 1 through 12, add the given numbers. Be sure to check your work.

1. 36 29 87

2. 45 89 37

3. 627 83 524

4. 433 612 109

5. 446 915 992 67

6. 809 826 278 548

7.	8028	8.	7695	9.	3873	15.	873	16.	921
	4756		4803		9295		− 292		− 224
	4803		986		4082				
	3823		7375		399	17.	8305	18.	2006
					7646		− 7356		− 1197
10.	989	11.	30,964	12.	87,657				
	3216		9,877		93,984	19.	36,047	20.	32,105
	4807		92,286		57,609		− 26,249		− 22,116
	736		5,547		8,726				
	9297		965		92,875	21.	40,165	22.	10,906
							− 9,586		− 9,928

In Exercises 13 through 24, perform the indicated subtractions.
Check your work.

				23.	290,078	24.	872,110
13.	8704	14.	5162		− 194,396		− 682,324
	− 3102		− 2041				

A.2 Multiplication and Division of Whole Numbers

Multiplication *is a short-cut method for doing repeated additions of the same number. The number being multiplied is called the* **multiplicand,** *the number of times it is taken is the* **multiplier,** *and the result is the* **product.** The multiplicand and multiplier are also called factors. The basic notations used to denote multiplication are ×, · , and parentheses.

Example
1

By the expression 3×5 we mean

$$3 \times 5 = 5 + 5 + 5 = 15$$

Here 3 is the multiplier, 5 is the multiplicand, and 15 is the product. The product can also be expressed as

$$3 \cdot 5 = 15 \quad \text{or} \quad (3)(5) = 15$$

To perform multiplication accurately and with reasonable speed, it is necessary, as with addition, to know the basic products through $9 \times 9 = 81$ without hesitation. If you have any doubt about your knowledge of these products, review them and practice until they have been mastered.

The process of multiplication has certain basic properties. One of these, known as the **commutative law,** states that *the order of multiplication does not matter.* Another of these, the **associative law,** deals with the multiplication of more than two factors. It states that *the grouping of the numbers being multiplied (factors) does not matter.* Because multiplication is basically a process of addition, these properties also hold for addition.

Example 2

The commutative and associative laws are as follows.

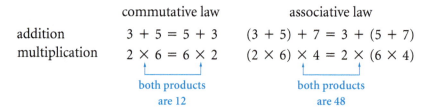

	commutative law	associative law
addition	$3 + 5 = 5 + 3$	$(3 + 5) + 7 = 3 + (5 + 7)$
multiplication	$2 \times 6 = 6 \times 2$	$(2 \times 6) \times 4 = 2 \times (6 \times 4)$

both products are 12   both products are 48

Another important property of numbers that involves multiplication and addition is known as the **distributive law.** This law states that *if the sum of two numbers is multiplied by another given number, then each is multiplied by the given number and the products are added to find the final result.* This may sound complicated, so examine Example 3 closely.

Example 3

Applying the distributive law to the product $(4)(3 + 6)$, we have

$$(4)(3 + 6) = (4)(3) + (4)(6) = 12 + 24 = 36$$

We see that this gives the same result as the product $(4)(9) = 36$, since $3 + 6 = 9$.

Many of us can execute the usual multiplication process even though we might not totally understand the reasons for this procedure. In Example 4 we first illustrate the multiplication procedure and then use the distributive law to explain why it works.

Example 4

When multiplying 27 by 3, the usual procedure is shown as

$$
\begin{array}{r}
\overset{2}{27} \\
\times 3 \\
\hline
81
\end{array}
$$

$3 \times 7 = 2$ tens $+ 1$ one

We multiply 7 by 3 and obtain 21, placing the 1 under the 3 and carrying the 2 into the tens column. The 2 is then added to the product $3 \times 2 = 6$ to obtain the digit 8 which appears in the product 81.

We could have multiplied 27 by 3 by using the distributive law as follows:

$$3(27) = 3(20 + 7)$$
$$= 3(20) + 3(7)$$
$$= 60 + 21 = 60 + (20 + 1)$$
$$= (60 + 20) + 1 = 81$$

Example 5

To find the product of 26 and 124, we would proceed as follows:

$$\begin{array}{r} 124 \\ \times\,26 \\ \hline 744 \\ 248 \\ \hline 3224 \end{array}$$ displaced one place to left

If we now consider the product as

$$(124)(26) = (124)(20 + 6)$$
$$= (124)(20) + (124)(6) = 2480 + 744 = 3224$$

we see that the first line (744) obtained in the multiplication process is also one of the products found by using the distributive law. We then note that the 248 found in the multiplication process is equivalent to the 2480 product found with the distributive law. The final zero is not written, but it is represented because the 248 is displaced one position to the left.

Area of rectangle
= length × width

Length

Width

Figure A.1

There are many applications of multiplication, some of which are demonstrated in the exercises. One of the most basic applications is in determining the **area** *of a* **rectangle,** *which is defined to be the product of the rectangle's* **length** *and* **width** (see Figure A.1).

Example 6

A rectangular plate is used as a ramp for the handicapped. If its length is 4 ft and its width is 6 ft, as shown in Figure A.2, its area is given by

$$\text{Area} = (4\,\text{ft}) \times (6\,\text{ft}) = 24\,\text{ft}^2$$

Here ft^2 is the symbol for square foot.

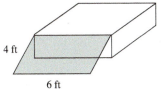

4 ft

6 ft

Figure A.2

Just as subtraction is the inverse process of addition, division is the inverse process of multiplication. Because multiplication can be thought of as a process of repeated addition, we can think of division as a process of repeated subtraction. If we subtract 2 from 10 five times, for example, the result is zero. This means that 10 divided by 2 is 5. Here 10 is the **dividend,** 2 is the **divisor,** and 5 is the **quotient.** A common notation for this is $10 \div 2 = 5$, where $\div$ indicates division.

If the division "comes out even," the product of the quotient and the divisor will be equal to the dividend. This gives us a way to check that the division has been done correctly. It is assumed that you know how to perform the division process, but Example 7 is given for review and also as an explanation of why the division process works.

Example 7

Suppose that we want to divide 3288 by 24. We set up the problem as follows:

$$
\begin{array}{r}
137 \quad \text{quotient} \\
24\,)\overline{3288} \quad \text{dividend} \\
\underline{2400} \quad 24 \times 100 \\
888 \\
\underline{720} \quad 24 \times 30 \\
168 \\
\underline{168} \quad 24 \times 7 \\
\end{array}
$$

divisor

We can now see that

$$3288 = 24 \times 137 = (24 \times 100) + (24 \times 30) + (24 \times 7)$$
$$= 24(100 + 30 + 7)$$

$$
\begin{array}{r}
137 \\
24\,)\overline{3288} \\
\underline{24} \quad 24 \times 1 \\
88 \\
\underline{72} \quad 24 \times 3 \\
168 \\
\underline{168} \quad 24 \times 7 \\
\end{array}
$$

The meanings of the products 2400, 720, and 168 can be seen in this example. Usually, the extra zeros are not written, and only a sufficient number of digits are "brought down." Normally, the division would appear as shown at the left.

In many divisions, the divisor will not divide exactly into the dividend. In these cases, there is a **remainder** in the answer. Example 8 illustrates division with a remainder.

Example 8

Divide 5286 by 25.

$$
\begin{array}{r}
211 \\
25\,)\overline{5286} \\
\underline{50} \quad 25 \times 200 = 5000, \text{ shown as } 25 \times 2 = 50 \\
28 \\
\underline{25} \quad 25 \times 10 = 250, \text{ shown as } 25 \times 1 = 25 \\
36 \\
\underline{25} \quad 25 \times 1 = 25 \\
11 \quad \text{remainder} \\
\end{array}
$$

Thus $5286 = (25 \times 211) + 11$. Because 11 is smaller than the divisor 25, the division process is discontinued and 11 becomes the remainder.

There are many real applications of division. Example 9 illustrates one such application, and others are found in the exercises.

Example 9

$$\begin{array}{r} 305 \\ 27\overline{)8235} \\ \underline{81} \\ 135 \\ \underline{135} \end{array}$$

A certain computer component costs \$27. If a shipment of these components is valued at \$8235, how many components are in the shipment?

Because we have the total value of the shipment and the value of one component, we can find the number of components by dividing \$8235 by \$27, as shown at the left. Two digits are brought down because 27 will not divide into 13. We see from the division that there are 305 components in the shipment.

A.2 *Exercises*

In Exercises 1 through 8, perform the indicated multiplications.

1. 23×458

2. 27×835

3. $(218)(6032)$

4. $(256)(1024)$

5. $\begin{array}{r}1024\\ \times\ 1024\end{array}$

6. $\begin{array}{r}4108\\ \times\ 3897\end{array}$

7. $\begin{array}{r}61547\\ \times\ 3849\end{array}$

8. $\begin{array}{r}78793\\ \times\ 5698\end{array}$

In Exercises 9 through 16, perform the indicated divisions.

9. $3\overline{)732}$

10. $81\overline{)3159}$

11. $54\overline{)17496}$

12. $32\overline{)16256}$

13. $65536 \div 32$

14. $62387 \div 28$

15. $608271 \div 307$

16. $918885 \div 725$

In Exercises 17 through 20, verify the associative law of multiplication by first multiplying the numbers in parentheses and then completing the multiplication on each side.

17. $(17 \times 38) \times 74 = 17 \times (38 \times 74)$

18. $16 \times (312 \times 42) = (16 \times 312) \times 42$

19. $(326 \times 45) \times 217 = 326 \times (45 \times 217)$

20. $52 \times (36 \times 132) = (52 \times 36) \times 132$

In Exercises 21 through 24, perform the indicated multiplications by using the distributive law. Check your answer by adding the numbers within the parentheses first and then complete the multiplication.

21. $15(3 + 7)$

22. $14(8 + 12)$

23. $628(29 + 86)$

24. $4159(387 + 832)$

Exercises 25 through 32 give the lengths and widths of certain rectangles. Determine their areas. (In Exercise 29, cm is the symbol for centimeter.)

25. Length = 20 mi, width = 8 mi

26. Length = 44 ft, width = 24 ft

27. Length = 17 in, width = 14 in.

28. Length = 682 ft, width = 273 ft

29. Length = 296 cm, width = 35 cm

30. Length = 543 yd, width = 274 ft

31. Length = 18 in, width = 2 ft

32. Length = 34 in, width = 3 ft

In Exercises 33 through 40, solve the given problems. Where appropriate, the following units of measurement are designated by the given symbols: miles per hour—mi/h, hour—h, revolution—r, square foot—ft^2, square meter—m^2, meter—m, millimeter—mm, square millimeter—mm^2, square inch—in^2.

33. A jet plane averages a speed of 595 mi/h for 17 h. How far does it travel?

34. A certain gear makes 78 r/min (revolutions per minute). How many revolutions does it make in an hour?

35. What is the cost of carpeting a rectangular room 8 yd by 5 yd if the total cost is $32 per square yard? See Figure A.3.

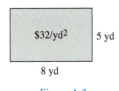

$32/yd²   5 yd

8 yd

Figure A.3

36. Suppose that the area of a rectangle is 61,884 ft^2 and the length is 573 ft. Find the width. See Figure A.4.

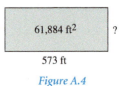

61,884 ft²   ?

573 ft

Figure A.4

37. A surveyor is partitioning a parcel of land into 1-acre rectangular plots. One acre is 43,560 ft^2. If one such plot is 132 ft wide, what is its length?

38. Suppose that the area of a rectangle is 56,056 m^2 and the width is 98 m. What is the length?

39. If a car traveled 234 mi on 13 gal of gasoline, what was its gas consumption in miles per gallon?

40. If a class of 27 students allots $945 for computer usage, what allotment does each individual student have?

A.3 Fractions

Mathematics can be applied to many situations in which only whole numbers are necessary. However, there are many other cases in which parts of a quantity or less than the total of a group must be used. An industrial plant might have a total energy cost that is the sum of fuel and electricity expenses. If we consider only the fuel cost, we are considering only a fractional part of the total energy cost.

Considerations such as these lead us to fractions and the basic operations with them. In general, *a **fraction** is the indicated division of one number by another.* It is also possible to interpret this definition as a certain number of equal parts of a given unit or given group.

Example 1

The fraction $\frac{5}{8}$ is the indicated division of 5 by 8. In this fraction, 5 is the **numerator** and 8 is the **denominator.**

$$\frac{5}{8} \xleftarrow{} \text{numerator}$$
$$\phantom{\frac{5}{8}} \xleftarrow{} \text{denominator}$$

It is also possible to interpret the fraction $\frac{5}{8}$ as referring to 5 of 8 equal parts of a whole (see Figure A.5). Here the line segment is the given unit, of which 5 parts of 8 are being considered. That is, the line segment $\overline{AB}$ is $\frac{5}{8}$ of the line segment $\overline{AC}$.

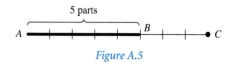

Figure A.5

If, in a group of eight batteries, five are 6-V batteries, we could say that $\frac{5}{8}$ of the group are 6-V batteries.

If the numerator of a fraction is numerically less than the denominator, the fraction is called a **proper fraction.** *If the numerator equals or is numerically greater than the denominator, the fraction is called an* **improper fraction.** Because an improper fraction in which the numerator is greater than the denominator represents a number numerically greater than 1, it is often convenient to use a **mixed number**—*a whole number plus a proper fraction*—to represent the same number. Consider Example 2.

Example 2

The fraction $\frac{4}{9}$ is a proper fraction, whereas $\frac{9}{4}$ is an improper fraction.

$$\text{Proper } \frac{4}{9} \begin{array}{l} \text{numerator} \\ \text{less than} \\ \text{denominator} \end{array} \qquad \text{Improper } \frac{9}{4} \begin{array}{l} \text{numerator} \\ \text{greater than} \\ \text{denominator} \end{array}$$

An improper fraction in which the numerator equals the denominator is equal to the whole number 1. Also, an improper fraction with a denominator of 1 equals the number in the numerator. Each of these can be seen to be valid if we think of the fraction in terms of division.

Example 3

The fraction $\frac{3}{3}$ equals 1, because $3 \div 3 = 1$. In the same way, $\frac{7}{7} = 1$ and $\frac{73}{73} = 1$. This is true because any number (except zero) divided by itself is 1.

$$\begin{array}{l} \text{numerator} \\ \text{equals} \\ \text{denominator} \end{array} \frac{3}{3} = 1 \qquad \frac{3}{1} = 3$$

$$\text{denominator} = 1$$

The fraction $\frac{3}{1}$ equals 3, because $3 \div 1 = 3$. In the same way, $\frac{7}{1} = 7$ and $\frac{73}{1} = 73$. This is true because any number divided by 1 equals that number.

To convert an improper fraction to a mixed number, divide the **numerator by** the denominator. The number of times the denominator divides evenly is the whole-number part of the mixed number. The remainder obtained is **the nu-**merator of the fractional part of the mixed number, and the denominator of this fraction is the same as the denominator of the improper fraction.

Example 4

To convert $\frac{9}{4}$ to a mixed number, we divide 9 by 4. The result is 2 with a remainder of 1. Thus $\frac{9}{4} = 2\frac{1}{4}$.

Converting $\frac{73}{14}$ to a mixed number, we divide 73 by 14, obtaining 5 with a remainder of 3. Thus $\frac{73}{14} = 5\frac{3}{14}$ as illustrated below.

$$\frac{73}{14} \longrightarrow 14\overline{)73} \longrightarrow \begin{array}{r} 5 \\ 14\overline{)73} \\ \underline{70} \\ 3 \end{array} \longrightarrow 5\frac{3}{14}$$

To convert a mixed number to an improper fraction, multiply the **whole num-**ber of the mixed number by the denominator of the fraction. Add this **result to the** numerator of the fraction. Place this result over the denominator of the **fraction.**

Example 5

The selling price of a stock is listed as $2\frac{1}{4}$ (dollars). Write $2\frac{1}{4}$ as an improper fraction.

To convert $2\frac{1}{4}$ to an improper fraction, we multiply 2 by 4, obtaining **8. This is** added to 1, obtaining 9, which is now the numerator of the improper fraction. Thus,

$$2\frac{1}{4} = \frac{(2 \times 4) + 1}{4} = \frac{8 + 1}{4} = \frac{9}{4}$$

A.3 **Exercises**

In Exercises 1 through 4, write fractions for the indicated selected parts.

1. Five equal parts from among nine such parts.

2. Four equal parts from among eleven such parts.

3. One part from among seven such equal parts.

4. Eight equal parts from among eight such parts.

In Exercises 5 through 8, write fractions equal to each indicated quotient.

5. $7 \div 13; 3 \div 16$

6. $1 \div 3; 11 \div 17$

7. $9 \div 8; 1 \div 12$

8. $23 \div 4; 2 \div 35$

In Exercises 9 through 12, write the whole number that equals the given fraction.

9. $\dfrac{6}{6}, \dfrac{6}{1}$ 10. $\dfrac{19}{19}, \dfrac{19}{1}$ 11. $\dfrac{32}{1}, \dfrac{32}{32}$ 12. $\dfrac{503}{1}, \dfrac{503}{503}$

In Exercises 13 through 20, convert the given improper fractions to mixed numbers.

13. $\dfrac{5}{3}$ 14. $\dfrac{18}{5}$ 15. $\dfrac{64}{13}$ 16. $\dfrac{55}{32}$

17. $\dfrac{278}{75}$ 18. $\dfrac{315}{32}$ 19. $\dfrac{1329}{25}$ 20. $\dfrac{4376}{118}$

In Exercises 21 through 28, convert the given mixed numbers to improper fractions.

21. $3\dfrac{2}{5}$ 22. $6\dfrac{3}{4}$ 23. $9\dfrac{7}{8}$ 24. $12\dfrac{3}{5}$

25. $17\dfrac{2}{13}$ 26. $34\dfrac{3}{8}$ 27. $105\dfrac{3}{4}$ 28. $235\dfrac{16}{25}$

In Exercises 29 through 36, solve the given problems.

29. A case of oil contains 24 cans. If 7 cans are removed, write a fraction for the part of the case that remains.

30. Write a fraction that indicates the cars with automatic transmission from a group of cars of which seven have automatic transmission and five have standard transmission.

31. Write a fraction that indicates the 60-W electric light bulbs from a container that has ten 60-W bulbs and thirteen 100-W bulbs.

32. The breakeven point for a certain production crew is found to be 48 units per hour. If 55 units are produced each hour, what fraction of the breakeven point is the current production level?

33. A standard sheet of typing paper is $\dfrac{17}{2}$ in. wide. Express this width as a mixed number.

34. A cylinder holds $\dfrac{39}{4}$ oz of fuel. Express this capacity as a mixed number.

35. The snowfall from a storm was $3\dfrac{7}{10}$ in. deep. Express the snowfall as an improper fraction.

36. A fuel gauge gives the indication shown in Figure A.6. Write a fraction that indicates the part of the tank that contains fuel.

Figure A.6

A.4 **Equivalent Fractions**

We have considered the basic meanings of various types of fractions. Now we shall determine how a fraction can be equivalent to other fractions. Determining equality of two or more fractions is very important in working with fractions and performing the basic operations on them.

Consider the area of rectangle *A* as shown in Figure A.7. It has been divided into four equal parts, one of which is shaded. Thus the fraction $\frac{1}{4}$ can be used to represent the shaded part. Rectangle *B* is the same as rectangle *A*, but it has been divided again by a line through the middle. Now the shaded area consists of 2 of 8 equal parts, and the fraction $\frac{2}{8}$ can be used to represent the shaded area. However, the shaded areas of the two rectangles are the same, which means that $\frac{1}{4} = \frac{2}{8}$. Because the two fractions represent the same fractional part, they must be equal.

The fraction $\frac{2}{8}$ could be obtained by doubling both the numerator and the denominator of the fraction $\frac{1}{4}$. Also, we can obtain the fraction $\frac{1}{4}$ by dividing both the numerator and the denominator of the fraction $\frac{2}{8}$ by 2. These two fractions are known as **equivalent fractions.** In general: *Two fractions are equivalent*

Rectangle *A*

Rectangle *B*

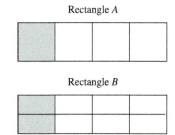

Figure A.7

(equal) if the numerator and the denominator of one of the fractions can be multiplied, or divided, by the same number (not zero) to obtain the other fraction.

Example 1

The fraction $\frac{18}{24}$ can be obtained by multiplying the numerator and the denominator of the fraction $\frac{3}{4}$ by 6. That is,

$$\frac{3}{4} = \frac{3 \times 6}{4 \times 6} = \frac{18}{24}$$

Therefore $\frac{3}{4}$ and $\frac{18}{24}$ are equivalent fractions.

If we multiply the numerator and the denominator of $\frac{3}{4}$ by 3, we obtain the fraction $\frac{9}{12}$. Thus $\frac{3}{4}$ and $\frac{9}{12}$ are equivalent, which means that $\frac{3}{4} = \frac{9}{12}$. Other fractions equivalent to $\frac{3}{4}$ can be obtained by multiplying the numerator and the denominator by other nonzero numbers.

Example 2

Following are equivalent fractions that have been obtained by performing the indicated operation on the numerator and the denominator of the left-hand fraction.

$\frac{2}{4} = \frac{4}{8}$ multiplication by 2 $\frac{2}{3} = \frac{14}{21}$ multiplication by 7

$\frac{5}{8} = \frac{20}{32}$ multiplication by 4 $\frac{5}{8} = \frac{45}{72}$ multiplication by 9

$\frac{28}{100} = \frac{7}{25}$ division by 4 $\frac{12}{54} = \frac{2}{9}$ division by 6

$\frac{121}{154} = \frac{11}{14}$ division by 11 $\frac{156}{84} = \frac{13}{7}$ division by 12

In changing one fraction to an equivalent fraction, we say that *the fraction has been converted to* **higher terms** *if the resulting numerator and denominator are larger. A fraction is said to be* **reduced** *if it is changed to a fraction in which the numerator and denominator are smaller.*

One of the most important operations performed on a fraction is reducing it to **lowest terms** or **simplest form.** *This means that the resulting numerator and denominator are not both evenly divisible by any whole number other than 1.*

Example 3

The fraction $\frac{10}{24}$ is not in lowest terms, because both the numerator and the denominator are divisible by 2. To reduce $\frac{10}{24}$ to lowest terms we divide the numerator and the denominator by 2 and obtain $\frac{5}{12}$. Because 5 and 12 are not both divisible by any whole number other than 1, the lowest term of $\frac{10}{24}$ is $\frac{5}{12}$. Therefore $\frac{10}{24} = \frac{5}{12}$.

Example
4

The fraction $\frac{42}{54}$ is not in simplest form. Dividing both 42 and 54 by 2, we obtain the following:

$$\frac{42}{54} = \frac{42 \div 2}{54 \div 2} = \frac{21}{27}$$

▶ However, we note that 21 and 27 are both evenly divisible by 3. Although $\frac{21}{27}$ is a reduced form of $\frac{42}{54}$, *it is not the simplest form.* Dividing 21 and 27 by 3, we get

$$\frac{21}{27} = \frac{21 \div 3}{27 \div 3} = \frac{7}{9}$$

Now 7 and 9 are not both evenly divisible by any whole number other than 1. Thus $\frac{7}{9}$ is the simplest form of $\frac{42}{54}$. (If we had originally noted that 42 and 54 were both divisible by 6, we would have obtained $\frac{7}{9}$ directly.)

Reducing an improper fraction to lowest terms is done in precisely the same way as with a proper fraction. The final form should be left as an improper fraction unless specifically noted otherwise.

When reducing a fraction to lowest terms, we may find that the largest number that will divide into both the numerator and the denominator is not obvious, as illustrated in Example 4.

For purposes of reducing fractions with large numerators and denominators, as well as performing the basic operations on fractions, it is convenient to determine the various factors of a whole number. We recall that each of the numbers being multiplied together is a **factor** of a given product. Here the whole number under consideration is the product in question. Thus *the process is that of* **factoring** *whole numbers, which means we shall be determining those numbers that, when multiplied together, give the whole number.*

Example
5

The whole number 10 is equal to 10×1. In general, any number can be considered to be the product of itself and 1. Thus 10 and 1 are factors of 10. However, we also note that $10 = 2 \times 5$, which means that 2 and 5 are also factors of 10. Therefore, the number 10 has 1, 2, 5, and 10 as factors—each of these whole numbers can be multiplied by another whole number to obtain 10.

Those factors that are the most important are called **prime numbers.** *A prime number is a whole number that is evenly divisible only by itself and 1.* Neither 0 nor 1 is considered to be a prime number.

Example
6

2 is prime, because it is divisible evenly only by 2 and 1.

3 is prime, because it is divisible evenly only by 3 and 1.

4 is not prime, because $4 = 2 \times 2$. That is, 4 is divisible by a whole number other than itself or 1.

5 is prime, because it is divisible evenly only by 5 and 1.

6 is not prime, because $6 = 2 \times 3$.

Other prime numbers are 7, 11, 13, 17, and 19.

Other numbers that are not prime are 8 ($8 = 2 \times 2 \times 2$), 9 ($9 = 3 \times 3$), 10 ($10 = 5 \times 2$), 12 ($12 = 2 \times 2 \times 3$), 14 ($14 = 2 \times 7$), 15 ($15 = 3 \times 5$), 16 ($16 = 2 \times 2 \times 2 \times 2$), and 18 ($18 = 2 \times 3 \times 3$).

Because the most useful factorization of a whole number consists of prime factors, *we shall factor all whole numbers until only prime factors are present. When the original number has been factored so that all of the factors are prime numbers, we say that the number has been* **factored completely.**

Example
7

In factoring 18, we may note that $18 = 2 \times 9$. However, 9 is not prime, so we further factor 9 as 3×3. Therefore $18 = 2 \times 3 \times 3$. Because all factors are now prime, the factorization is complete. We also note that the prime factor 3 appears twice.

prime
factors

$$18 = 2 \times 3 \times 3$$

In factoring 34, we note that 34 is even so that 2 must be a factor. This leads to the factorization $34 = 2 \times 17$, which is the complete factorization, because 17 is prime.

In factoring 55, we note that 55 ends in 5 so that 5 must be a factor. We can express 55 as 5×11 and, because 5 and 11 are both prime, we say that 55 has been factored completely.

In factoring 91, we divide by the prime numbers. The first prime number to divide evenly is 7 so that $91 = 7 \times 13$, which is the complete factorization because 7 and 13 are both prime.

To reduce a fraction by use of factoring, first factor both the numerator and the denominator into prime factors. At this point any factors that are common to both the numerator and the denominator can be determined by inspection. Any such factors can then be divided out.

Example 8

To reduce the fraction $\frac{65}{78}$, first factor 65 and 78 into prime factors. This gives

$$\frac{65}{78} = \frac{5 \times 13}{2 \times 3 \times 13} \quad \longleftarrow \text{prime factors}$$

We note that the factor 13 appears in both the numerator and the denominator so that both can be divided by 13. Dividing both numerator and denominator by 13 we get

$$\frac{65}{78} = \frac{65 \div 13}{78 \div 13} = \frac{5}{6}$$

After we have factored the numerator and the denominator, it often happens that more than one common factor appears in each. When this happens, every factor common to both the numerator and the denominator should be canceled. *It must be remembered, however, that* **cancellation is actually a process of dividing both the numerator and the denominator by the common factor.**

Example 9

In reducing the fraction $\frac{21}{70}$, we first factor both 21 and 70 to get

$$\frac{21}{70} = \frac{3 \times 7}{2 \times 5 \times 7} \quad \longleftarrow \text{prime factors}$$

Note that the factor 7 appears in both the numerator and the denominator. We can do the division by crossing out the 7's. Performing this cancellation, we get

$$\frac{21}{70} = \frac{3 \times \overset{1}{\cancel{7}}}{2 \times 5 \times \underset{1}{\cancel{7}}} = \frac{3}{10} \qquad \text{Cancel common prime factors}$$

We must remember that we really divided by 7, leaving quotients of 3 and 10.
In reducing the fraction $\frac{140}{56}$, we get the following:

$$\frac{140}{56} = \frac{\overset{1}{\cancel{2}} \times \overset{1}{\cancel{2}} \times 5 \times \overset{1}{\cancel{7}}}{\underset{1}{\cancel{2}} \times \underset{1}{\cancel{2}} \times 2 \times \underset{1}{\cancel{7}}} = \frac{5}{2}$$

Example
10

In reducing $\frac{15}{75}$, we factor the numerator and the denominator as follows:

$$\frac{15}{75} = \frac{3 \times 5}{3 \times 5 \times 5}$$

In canceling, note that both the 3 and the 5 of the numerator are to be canceled. Because this amounts to dividing the numerator by 15, we get

$$\frac{15}{75} = \frac{\overset{1}{\cancel{3}} \times \overset{1}{\cancel{5}}}{\underset{1}{\cancel{3}} \times \underset{1}{\cancel{5}} \times 5} = \frac{1}{5} \quad \longleftarrow \text{All prime factors canceled}$$

Example
11

In reducing $\frac{72}{6}$ we factor the numerator and denominator to get

$$\frac{72}{6} = \frac{2 \times 2 \times 2 \times 3 \times 3}{2 \times 3} = \frac{12}{1} = 12 \quad \longleftarrow \text{All prime factors canceled}$$

Instead of expressing the result as $\frac{12}{1}$, we write it simply as 12.

A.4   Exercises

In Exercises 1 through 12, determine fractions equivalent to the given fractions by performing the indicated operation on the numerator and the denominator.

1. $\frac{3}{7}$ (multiply by 2)

2. $\frac{5}{9}$ (multiply by 3)

3. $\frac{16}{20}$ (divide by 4)

4. $\frac{15}{125}$ (divide by 5)

5. $\frac{4}{13}$ (multiply by 6)

6. $\frac{8}{15}$ (multiply by 11)

7. $\frac{60}{156}$ (divide by 12)

8. $\frac{140}{42}$ (divide by 7)

9. $\frac{13}{25}$ (multiply by 7)

10. $\frac{17}{15}$ (multiply by 12)

11. $\frac{1024}{64}$ (divide by 32)

12. $\frac{289}{340}$ (divide by 17)

In Exercises 13 through 24, reduce each of the given fractions to lowest terms.

13. $\frac{4}{8}$

14. $\frac{12}{18}$

15. $\frac{15}{10}$

16. $\frac{21}{6}$

17. $\frac{20}{25}$

18. $\frac{21}{28}$

19. $\frac{27}{45}$

20. $\frac{16}{40}$

21. $\frac{24}{60}$

22. $\frac{55}{66}$

23. $\frac{30}{75}$

24. $\frac{9}{78}$

In Exercises 25 through 36, find the value of the missing numbers.

25. $\frac{1}{3} = \frac{?}{12}$

26. $\frac{5}{8} = \frac{?}{16}$

27. $\frac{2}{3} = \frac{10}{?}$

28. $\frac{7}{8} = \frac{21}{?}$

29. $\frac{12}{64} = \frac{?}{16}$

30. $\frac{27}{36} = \frac{?}{16}$

31. $\dfrac{?}{6} = \dfrac{2}{3}$ **32.** $\dfrac{?}{18} = \dfrac{4}{9}$ **33.** $\dfrac{3}{?} = \dfrac{6}{80}$

34. $\dfrac{5}{?} = \dfrac{30}{18}$ **35.** $\dfrac{13}{?} = \dfrac{52}{64}$ **36.** $\dfrac{7}{?} = \dfrac{21}{192}$

In Exercises 37 through 48, factor the given numbers into their prime factors.

37. 20   **38.** 28   **39.** 16

40. 32   **41.** 36   **42.** 44

43. 48   **44.** 52   **45.** 57

46. 84   **47.** 105   **48.** 132

In Exercises 49 through 60, reduce each fraction to lowest terms by factoring the numerator and the denominator into prime factors and then canceling any common factors.

49. $\dfrac{30}{35}$ **50.** $\dfrac{28}{63}$ **51.** $\dfrac{24}{30}$

52. $\dfrac{45}{75}$ **53.** $\dfrac{56}{24}$ **54.** $\dfrac{99}{27}$

55. $\dfrac{52}{78}$ **56.** $\dfrac{70}{84}$ **57.** $\dfrac{17}{68}$

58. $\dfrac{19}{57}$ **59.** $\dfrac{63}{105}$ **60.** $\dfrac{78}{117}$

In Exercises 61 through 69, solve the given problems. Reduce all fractions to lowest terms.

61. How would a $\frac{12}{16}$-in. wrench be labeled if all wrenches in a set are labeled with fractions reduced to lowest terms?

62. A $\frac{3}{8}$-in. hole must be drilled for a bolt so that a voltage regulator can be mounted in a truck. If the drill bits are measured in 64ths of an inch, which bit should be used?

63. One road has 32 homes and 14 of them use solar energy to heat water. What fraction of the homes on this road heat their hot water with solar energy?

64. An electric current is split between two wires. If 6 A pass through the first wire and 14 A pass through the second wire, what fraction of the total current passes through the first wire?

65. A chemist mixes 18 mL of sulfuric acid with 27 mL of water. What fraction of the mixture is pure acid?

66. A small business acquires a laptop computer at a cost of $4500. Two years later this same laptop computer is sold for $2700. What fraction of the original cost is lost?

67. A business spends approximately $2400 for each computer it orders. If $2000 is spent on computer hardware and $400 is spent on software, determine the fractional amount spent in both of these categories.

68. On a recent algebra quiz of 50 problems a student counted 38 questions right and 12 questions wrong. What fractional part of the test was correct? Write your result as an equivalent fraction with a denominator of 100.

69. In testing 750 products as they come from an assembly line, it was discovered that 25 items were defective. What fractional number of products was defective? Express your result in lowest terms.

A.5 Addition and Subtraction of Fractions

We can think of a fraction as a number of equal parts of a given unit or group so that each of these parts or units is equal. Consequently, *fractions in which the denominators are equal are called* **like** *fractions.*

Example 1

The fractions $\frac{3}{8}$, $\frac{5}{8}$, and $\frac{11}{8}$ are like fractions because all have denominators of 8. With 3 eighths, 5 eighths, and 11 eighths, we have like quantities because they are all eighths. The fractions $\frac{1}{3}$ and $\frac{2}{6}$ are not like fractions because the denominators are different. They are equivalent fractions, but they are not like fractions.

We add like fractions by placing the sum of the numerators over the denominator that is common to them. In the same way, when we subtract fractions we place the difference of numerators over the common denominator. These procedures are illustrated in Example 2.

Example 2

sum of numerators

a. $\dfrac{2}{9} + \dfrac{5}{9} = \dfrac{2 + 5}{9} = \dfrac{7}{9}$ ← common denominator

b. $\dfrac{3}{8} + \dfrac{1}{8} = \dfrac{3 + 1}{8} = \dfrac{4}{8} = \dfrac{1}{2}$

c. $\dfrac{7}{12} + \dfrac{1}{12} - \dfrac{5}{12} = \dfrac{7 + 1 - 5}{12} = \dfrac{8 - 5}{12} = \dfrac{3}{12} = \dfrac{1}{4}$

d. $\dfrac{5}{11} - \dfrac{3}{11} = \dfrac{5 - 3}{11} = \dfrac{2}{11}$

e. $2\dfrac{1}{7} + \dfrac{3}{7} = \dfrac{15}{7} + \dfrac{3}{7} = \dfrac{15 + 3}{7} = \dfrac{18}{7}$

In Example 2(e) the mixed number is first changed to an improper fraction; then the addition is performed. Although this is not the only procedure that can be followed, it is standard. In 2(c), where more than two fractions are being combined, we add and subtract from left to right.

   If the fractions being added or subtracted do not have the same denominators, it is necessary to convert them so that all denominators are equal. This is consistent with the requirement that we can add or subtract only like quantities.

Example 3

To perform the addition $\frac{1}{3} + \frac{2}{9}$, we must first change the fraction $\frac{1}{3}$ to its equivalent form $\frac{3}{9}$. When this is done we have

$$\dfrac{1}{3} + \dfrac{2}{9} = \dfrac{3}{9} + \dfrac{2}{9} = \dfrac{3 + 2}{9} = \dfrac{5}{9}$$

denominators must be equal in order to add

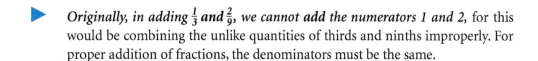

> *Originally, in adding $\frac{1}{3}$ and $\frac{2}{9}$, we cannot **add** the numerators 1 and 2,* for this would be combining the unlike quantities of thirds and ninths improperly. For proper addition of fractions, the denominators must be the same.

Although any proper common denominator can be used for adding fractions, there are distinct advantages if this denominator is the least possible value. Example 4 illustrates this.

Example 4

If we wish to add $\frac{5}{12}$ and $\frac{7}{8}$, we might note that both fractions can be converted to fractions with a common denominator of 96. Therefore, converting these fractions, we have

$$\frac{5}{12} = \frac{40}{96} \quad \text{and} \quad \frac{7}{8} = \frac{84}{96}$$

Adding $\frac{5}{12}$ and $\frac{7}{8}$ is equivalent to adding $\frac{40}{96}$ and $\frac{84}{96}$, or

$$\frac{5}{12} + \frac{7}{8} = \frac{40}{96} + \frac{84}{96} = \frac{124}{96}$$

This final fraction, $\frac{124}{96}$, should be reduced to its simplest form. This is done by dividing both the numerator and denominator by 4.

$$\frac{5}{12} + \frac{7}{8} = \frac{40}{96} + \frac{84}{96} = \frac{124}{96} = \frac{31}{24}$$

If we had first recognized that a denominator of 24 could be used, the addition would be

$$\frac{5}{12} + \frac{7}{8} = \frac{10}{24} + \frac{21}{24} = \frac{31}{24}$$

In this case, the conversions are simpler and the final result is already in its simplest form. In other cases, the result might not be in simplest form, but this use of a smaller denominator does simplify the process.

In Example 4 we saw that either 96 or 24 could be used as a common denominator. Actually, many other possibilities such as 48, 72, and 120 also exist. However, 24 is the smallest of all of the possibilities. It is the **lowest common denominator** for these fractions. In general: *The lowest common denominator of a set of fractions is the smallest number that is evenly divisible by all denominators of the set.* The advantages of using the lowest common denominator are illustrated in Example 4. The conversions to higher terms are simpler, as is the simplification of the result. Therefore, when adding or subtracting fractions, we shall always use the lowest common denominator. We shall now see how the lowest common denominator is determined.

In many cases the lowest common denominator can be determined by inspection. For example, if we are adding the fractions $\frac{1}{2}$ and $\frac{3}{8}$, we can easily determine that the lowest common denominator is 8. When adding $\frac{1}{2}$ and $\frac{2}{3}$, the lowest common denominator is clearly 6. However, if we cannot readily determine the lowest common denominator by observation, we need a systematic method that we can follow.

When we discussed equivalent fractions, we saw how whole numbers are factored into their prime factors. We can also determine the lowest common denominator of a set of fractions by factoring the denominators into their prime factors. Therefore: *After each denominator has been factored into its prime factors, **the lowest common denominator is the product of all the different prime factors, each taken the greatest number of times it appears in any one of the denominators.***

The procedure for finding the lowest common denominator can be summarized as follows.

1. List all denominators.

2. Next to each denominator, write the number as a product of its prime factors.

3. The lowest common denominator is found by multiplying the prime factors from step 2. For any prime factor that is repeated within one or more denominators, repeat it as a factor in the lowest common denominator. For each denominator, find the number of times that the prime factor occurs, then select the largest of these numbers of occurrences. That is the number of times the factor is repeated in the lowest common denominator.

Example 5

When we want to find the sum $\frac{5}{12} + \frac{7}{8}$, we must find the prime factors of 12 and 8. Because $12 = 2 \times 2 \times 3$ and $8 = 2 \times 2 \times 2$, we see that the only different prime factors are 2 and 3. *Because 2 occurs twice in 12 and three times in 8, it is taken three times in the least common denominator. Because 3 occurs once in 12 and does not occur in 8, it is taken once.* The lowest common denominator is

$$2 \times 2 \times 2 \times 3 = 24$$

This means that $\quad \dfrac{5}{12} + \dfrac{7}{8} = \dfrac{10}{24} + \dfrac{21}{24} = \dfrac{31}{24}$

Convert to fractions with lowest common denominator

Compare this result with Example 4.

Example
6

Add: $\frac{5}{18} + \frac{3}{14}$

First we determine the prime factors of 18 and 14 as follows.

$$18 = 2 \times 3 \times 3, \qquad 14 = 2 \times 7$$

▶ We now see that the prime factors for the lowest common denominator are 2, 3, and 7. The greatest number of times each appears is *once for 2 (once each in 18 and 14),* twice for 3 (in 18), and once for 7 (in 14). The lowest common denominator is

$$2 \times 3 \times 3 \times 7 = 126$$

Now converting to fractions with 126 as the denominator and adding, we have

$$\frac{5}{18} + \frac{3}{14} = \frac{35}{126} + \frac{27}{126} = \frac{62}{126} = \frac{31}{63}$$

▶ Here we see that *the initial result of $\frac{62}{126}$ is not in lowest terms and must be reduced* by cancelling the common factor of 2.

Example
7

Combine: $\frac{2}{15} + \frac{11}{27} - \frac{7}{50}$.

First we determine the prime factors of 15, 27, and 50 as follows.

$$15 = 3 \times 5, \quad 27 = 3 \times 3 \times 3, \quad 50 = 2 \times 5 \times 5$$

We now observe that the prime factors for the lowest common denominator are 2, 3, and 5. The greatest number of times each appears is once for 2 (in 50), three times for 3 (in 27), and twice for 5 (in 50). The lowest common denominator can now be expressed as

$$2 \times 3 \times 3 \times 3 \times 5 \times 5 = 1350$$

We can now proceed to combine the fractions as follows.

$$\frac{2}{15} + \frac{11}{27} - \frac{7}{50} = \frac{2(90)}{15(90)} + \frac{11(50)}{27(50)} - \frac{7(27)}{50(27)}$$

$$= \frac{180}{1350} + \frac{550}{1350} - \frac{189}{1350} = \frac{180 + 550 - 189}{1350}$$

$$= \frac{730 - 189}{1350} = \frac{541}{1350}$$

▶ We know that *the only prime factors of 1350 are 2, 3, and 5. Because none of these divides evenly into 541, the result is in simplest form.*

Sometimes a whole number is involved in addition or subtraction with fractions. In such cases we express the whole number as an improper fraction with the proper common denominator of the fractions. (Note that this procedure is the same as the procedure for changing a mixed number to an improper fraction.)

Example 8

Combine: $3 - \frac{16}{21} + \frac{4}{9}$.

We first determine the lowest common denominator of the fractions. Factoring the denominators, we have $21 = 3 \times 7$ and $9 = 3 \times 3$. The lowest common denominator is $3 \times 3 \times 7 = 63$. Now, since $3 = \frac{3}{1}$, we can change 3 into an improper fraction with a denominator of 63 by multiplying both the numerator and denominator by 63. We get

$$3 - \frac{16}{21} + \frac{4}{9} = \frac{3(63)}{1(63)} - \frac{16(3)}{21(3)} + \frac{4(7)}{9(7)}$$

$$= \frac{189}{63} - \frac{48}{63} + \frac{28}{63} = \frac{189 - 48 + 28}{6}$$

$$= \frac{169}{63}$$

The result is in simplest form.

Example 9

One section of an aircraft firewall is $\frac{1}{2}$ in. thick. That thickness is increased by a $\frac{3}{8}$-in. washer and a $\frac{5}{16}$-in. bolt. See Figure A.8. What is the total combined thickness?

Because we must add the three fractions, we begin by determining the lowest common denominator. Factoring the denominators, we get the three results 2, $2 \times 2 \times 2$, and $2 \times 2 \times 2 \times 2$, so the lowest common denominator is $2 \times 2 \times 2 \times 2 = 16$. We then change each fraction to an equivalent fraction with a denominator of 16.

$$\frac{1}{2} + \frac{3}{8} + \frac{5}{16} = \frac{1(8)}{2(8)} + \frac{3(2)}{8(2)} + \frac{5}{16}$$

$$= \frac{8}{16} + \frac{6}{16} + \frac{5}{16} = \frac{19}{16} \text{ in.}$$

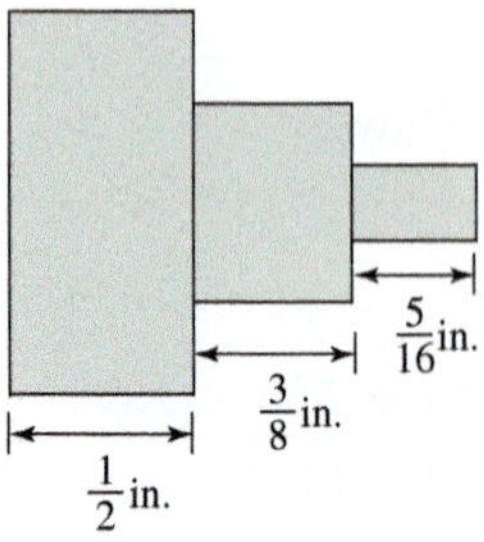

$\frac{5}{16}$in.

$\frac{3}{8}$ in.

$\frac{1}{2}$ in.

Figure A.8

The total combined thickness is $\frac{19}{16}$ in. (or $1\frac{3}{16}$ in.).

A.5 Exercises

In Exercises 1 through 8, find the lowest common denominator, assuming that the given numbers are denominators of fractions to be added or subtracted.

1. 2, 4

2. 2, 3

3. 6, 8

4. 6, 9

5. 8, 12, 18

6. 6, 10, 14

7. 10, 12, 25

8. 22, 24, 33

In Exercises 9 through 36, perform the indicated additions or subtractions, expressing each result in simplest form. In Exercises 23 through 36, use factoring to determine the lowest common denominator.

9. $\dfrac{1}{5} + \dfrac{3}{5}$

10. $\dfrac{2}{11} + \dfrac{5}{11}$

11. $\dfrac{5}{7} - \dfrac{3}{7}$

12. $\dfrac{4}{5} - \dfrac{1}{5}$

13. $\dfrac{1}{2} + \dfrac{1}{4}$

14. $\dfrac{1}{5} + \dfrac{2}{15}$

15. $\dfrac{23}{24} - \dfrac{5}{6}$

16. $\dfrac{2}{3} - \dfrac{1}{6}$

17. $\dfrac{1}{3} + \dfrac{3}{4}$

18. $\dfrac{3}{5} + \dfrac{5}{6}$

19. $\dfrac{1}{2} - \dfrac{2}{5}$

20. $\dfrac{2}{7} - \dfrac{1}{9}$

21. $5 - \dfrac{8}{3}$

22. $3 - \dfrac{11}{4}$

23. $\dfrac{11}{20} + \dfrac{5}{8}$

24. $\dfrac{2}{9} + \dfrac{5}{21}$

25. $\dfrac{5}{6} - \dfrac{3}{26}$

26. $\dfrac{19}{28} - \dfrac{5}{24}$

27. $3\dfrac{5}{6} - \dfrac{3}{8}$

28. $2\dfrac{4}{9} + \dfrac{5}{12}$

29. $\dfrac{4}{7} + \dfrac{1}{3} - \dfrac{3}{14}$

30. $\dfrac{1}{2} + \dfrac{5}{6} - \dfrac{9}{22}$

31. $\dfrac{3}{8} + \dfrac{7}{12} + \dfrac{1}{18}$

32. $3\dfrac{2}{9} + \dfrac{7}{15} - \dfrac{4}{25}$

33. $\dfrac{1}{3} + 1\dfrac{9}{14} - \dfrac{5}{21}$

34. $\dfrac{26}{27} - \dfrac{7}{18} + \dfrac{1}{10}$

35. $\dfrac{3}{4} + \dfrac{13}{30} - \dfrac{7}{12} + 2$

36. $5 - \dfrac{4}{25} + \dfrac{4}{5} + \dfrac{3}{35}$

In Exercises 37 through 44, solve the given problems.

37. One section of an aircraft firewall is $\frac{1}{4}$ in. thick. That thickness is increased by a $\frac{1}{16}$-in. washer and a $\frac{3}{16}$-in. bolt. What is the total combined thickness?

38. In an electric circuit, three resistors are connected in series. The total resistance is found by adding the individual resistances. What is the sum of $3\frac{3}{4}$ Ω, $2\frac{1}{2}$ Ω, and $3\frac{1}{4}$ Ω?

39. An engine coolant mixture consists of $3\frac{3}{4}$ gal of ethylene glycol and $4\frac{1}{5}$ gal of water. What is the total amount of coolant?

40. A ramp for the disabled is to be bordered with a special reflective tape. What length of tape is needed if the ramp is a rectangular plate $4\frac{5}{12}$ ft wide and $8\frac{11}{12}$ ft long?

41. A weather station records precipitation of $\frac{1}{4}$ in., $1\frac{1}{12}$ in., and $\frac{3}{8}$ in. on three successive days. What is the total amount of precipitation?

42. A technician experiments with the conducting properties of an alloy. If the alloy is $\frac{2}{3}$ copper, $\frac{1}{5}$ gold, and the remaining part is silver, what fraction is silver?

43. A cement truck delivers $34\frac{1}{3}$ cubic yards of cement to one building site and $17\frac{1}{3}$ cubic yards to a second site. If a total of $43\frac{1}{2}$ cubic yards of cement is used, how much is left over?

44. A home heating oil storage tank holds 275 gal of heating oil when full. During a recent oil delivery it was noted that $165\frac{5}{6}$ gal of oil were delivered. How much oil was in the oil storage tank at the time of the delivery?

A.6   Multiplication and Division of Fractions

To develop the procedure for multiplying one fraction by another, we shall consider the area of a rectangle. See Figure A.9. If the rectangle is 7 in. long and 3 in. wide, the area is 21 in.^{2}. Let us now mark both the length and the width at 1-in. intervals and divide the area into 21 equal parts, as shown. Each square has an area of 1 in.^{2}. If we now find the area of a section that is 5 in. long and 2 in. wide, it is 10 in.^{2}. This is equivalent to finding the area of a smaller rectangle whose length is $\frac{5}{7}$ of the length of the original rectangle and whose width is $\frac{2}{3}$ of the width of the original rectangle. We note that the resulting area is $\frac{10}{21}$ of the area of the original rectangle. Because we find area by multiplying length by width, we have

$$\frac{5}{7} \times \frac{2}{3} = \frac{10}{21}$$

This example suggests that multiplication is accomplished by multiplying the numerators and multiplying the denominators. In general: *The product of two fractions is the fraction whose numerator is the product of the numerators and whose denominator is the product of the denominators.* This can be seen in Example 1.

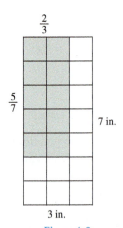

$\frac{2}{3}$

$\frac{5}{7}$

7 in.

3 in.

Figure A.9

Example 1

$$\frac{5}{7} \times \frac{2}{3} = \frac{5 \times 2}{7 \times 3} = \frac{10}{21} \longleftarrow \text{product of numerators}$$
$$\longleftarrow \text{product of denominators}$$

$$\frac{2}{5} \times \frac{8}{9} = \frac{2 \times 8}{5 \times 9} = \frac{16}{45}$$

$$\frac{9}{4} \times \frac{7}{2} = \frac{9 \times 7}{4 \times 2} = \frac{63}{8}$$

$$\frac{3}{14} \times \frac{15}{4} = \frac{3 \times 15}{14 \times 4} = \frac{45}{56}$$

If the resulting fraction is not in simplest form, it should be reduced to this form. We can do this by multiplying the numerators and denominators and reducing the resulting fraction, but this approach can lead to much more arithmetic than is necessary. Because a fraction is reduced to its lowest terms by dividing both the numerator and denominator by the same number, we can first divide any factor present in both the numerator and denominator before performing the multiplication. In Example 2, compare the two methods used for finding the product.

Example 2

In finding the product $\frac{24}{7} \times \frac{17}{32}$, we can multiply directly to get

$$\frac{24}{7} \times \frac{17}{32} = \frac{408}{224}$$

We can now reduce this result by dividing both the numerator and denominator by 8, giving

$$\frac{24}{7} \times \frac{17}{32} = \frac{408}{224} = \frac{51}{28}$$

However, if we indicate the multiplication only as

$$\frac{24}{7} \times \frac{17}{32} = \frac{24 \times 17}{7 \times 32}$$

we note that both 24 and 32 are divisible by 8. If we perform this division before we multiply the factors in the numerator, we have

$$\frac{24}{7} \times \frac{17}{32} = \frac{24 \times 17}{7 \times 32} = \frac{3 \times 17}{7 \times 4} = \frac{51}{28}$$

▶ We obtain the same result, but the arithmetic operations are simpler and the numbers involved are smaller. Consequently, we should *divide out any factors that are common to both the numerator and denominator before we actually multiply the numerators and denominators.*

Example 3

$$\frac{4}{5} \times \frac{3}{8} = \frac{4 \times 3}{5 \times 8} = \frac{1 \times 3}{5 \times 2} = \frac{3}{10} \qquad \text{divide 4 and 8 by 4}$$

$$\frac{18}{25} \times \frac{4}{27} = \frac{18 \times 4}{25 \times 27} = \frac{2 \times 4}{25 \times 3} = \frac{8}{75} \qquad \text{divide 18 and 27 by 9}$$

$$\frac{16}{15} \times \frac{5}{12} = \frac{16 \times 5}{15 \times 12} = \frac{4 \times 1}{3 \times 3} = \frac{4}{9} \qquad \text{divide 16 and 12 by 4 and divide 5 and 15 by 5}$$

$$\frac{30}{7} \times \frac{28}{33} = \frac{30 \times 28}{7 \times 33} = \frac{10 \times 4}{1 \times 11} = \frac{40}{11} \qquad \text{divide 30 and 33 by 3 and divide 28 and 7 by 7}$$

▶ In many cases a factor that is common to both the numerator and the denominator of the resulting fraction is fairly obvious after the multiplication has been indicated. If such a factor is not evident, *the resulting numerator and denominator can be factored completely to determine the common factors.*

To multiply a whole number by a fraction, we can write the whole number as a fraction with 1 as the denominator. The result is the same if the whole number is multiplied by the numerator of the fraction. *If a mixed number is to be multiplied by another number,* **the mixed number must first be converted to an improper fraction.**

Example 4

$$4 \times \frac{3}{7} = \frac{4}{1} \times \frac{3}{7} = \frac{4 \times 3}{1 \times 7} = \frac{12}{7}$$

$$9 \times \frac{5}{12} = \frac{9 \times 5}{12} = \frac{3 \times 5}{4} = \frac{15}{4}$$

$$2\frac{1}{3} \times \frac{3}{5} = \frac{7}{3} \times \frac{3}{5} = \frac{7 \times 3}{3 \times 5} = \frac{7 \times 1}{1 \times 5} = \frac{7}{5}$$

$$6 \times 3\frac{2}{5} = 6 \times \frac{17}{5} = \frac{6 \times 17}{5} = \frac{102}{5}$$

If more than two fractions are to be multiplied, the resulting numerator is the product of the numerators and the resulting denominator is the product of the denominators.

Example 5

$$\frac{2}{5} \times \frac{3}{7} \times \frac{4}{11} = \frac{2 \times 3 \times 4}{5 \times 7 \times 11} = \frac{24}{385}$$

$$\frac{3}{7} \times \frac{4}{8} \times \frac{5}{13} = \frac{3 \times 4 \times 5}{7 \times 8 \times 13} = \frac{3 \times 1 \times 5}{7 \times 2 \times 13} = \frac{15}{182}$$

$$7 \times \frac{5}{28} \times \frac{2}{3} = \frac{7 \times 5 \times 2}{28 \times 3} = \frac{1 \times 5 \times 1}{2 \times 3} = \frac{5}{6}$$

$$\frac{25}{12} \times \frac{8}{15} \times \frac{3}{10} = \frac{25 \times 8 \times 3}{12 \times 15 \times 10} = \frac{5 \times 2 \times 3}{3 \times 3 \times 10}$$

$$= \frac{1 \times 1 \times 1}{3 \times 1 \times 1} = \frac{1}{3}$$

We will now illustrate the usual procedure used when dividing by a fraction. *To divide a number by a fraction, invert the divisor and multiply,* as in Example 6.

Example
6

$$\dfrac{\dfrac{3}{7}}{\dfrac{4}{9}} = \dfrac{3}{7} \times \dfrac{9}{4} = \dfrac{3 \times 9}{7 \times 4} = \dfrac{27}{28}$$

invert and multiply

$$\dfrac{8}{5} \div \dfrac{4}{15} = \dfrac{8}{5} \times \dfrac{15}{4} = \dfrac{8 \times 15}{5 \times 4} = \dfrac{2 \times 3}{1 \times 1} = 6$$

invert and multiply

$$\dfrac{6}{\dfrac{2}{7}} = 6 \times \dfrac{7}{2} = \dfrac{6 \times 7}{2} = \dfrac{3 \times 7}{1} = 21$$

$$3 \div \dfrac{4}{7} = 3 \times \dfrac{7}{4} = \dfrac{3 \times 7}{4} = \dfrac{21}{4}$$

$$\dfrac{\dfrac{2}{5}}{\dfrac{4}{1}} = \dfrac{\dfrac{2}{5}}{4} = \dfrac{2}{5} \times \dfrac{1}{4} = \dfrac{2 \times 1}{5 \times 4} = \dfrac{1 \times 1}{5 \times 2} = \dfrac{1}{10}$$

$$\dfrac{2}{3} \div 6 = \dfrac{2}{3} \div \dfrac{6}{1} = \dfrac{2}{3} \times \dfrac{1}{6} = \dfrac{2 \times 1}{3 \times 6} = \dfrac{1 \times 1}{3 \times 3} = \dfrac{1}{9}$$

The method used in Example 6 can be justified by the fact that we can express a quotient like $\frac{3}{7} \div \frac{4}{9}$ as a fraction with a numerator of $\frac{3}{7}$ and a denominator of $\frac{4}{9}$. We could then proceed as follows.

$$\dfrac{\dfrac{3}{7}}{\dfrac{4}{9}} = \dfrac{\dfrac{3}{7} \times \dfrac{9}{4}}{\dfrac{4}{9} \times \dfrac{9}{4}} = \dfrac{\dfrac{3 \times 9}{7 \times 4}}{\dfrac{4 \times 9}{9 \times 4}} = \dfrac{\dfrac{27}{28}}{\dfrac{1 \times 1}{1 \times 1}} = \dfrac{\dfrac{27}{28}}{1} = \dfrac{27}{28}$$

final product
is $\frac{3}{7} \times \frac{9}{4} = \frac{27}{28}$

multiply numerator
and denominator by $\frac{9}{4}$

We have just seen that when dividing by a fraction it is necessary to invert the divisor. In general, *the **reciprocal** of a number is 1 divided by that number.* In following this definition we find that *the reciprocal of a fraction is the fraction with the numerator and denominator switched.*

Example
7

The reciprocal of 8 is $\frac{1}{8}$. The reciprocal of $\frac{1}{3}$ is

$$\dfrac{1}{\dfrac{1}{3}} = 1 \times \dfrac{3}{1} = 3$$

The reciprocal of $\frac{7}{12}$ is

$$\frac{\frac{1}{7}}{\frac{7}{12}} = 1 \times \frac{12}{7} = \frac{12}{7} \longleftarrow \text{numerator and denominator of } \frac{7}{12} \text{ inverted}$$

Finally, we consider the case in which the numerator or denominator, or both, of a fraction is itself the sum or difference of fractions. Such a fraction is called a **complex fraction.** *To simplify a complex fraction,* **we must first perform the additions or subtractions.**

Example 8

perform addition first

$$\frac{\frac{1}{2} + \frac{3}{4}}{8} = \frac{\frac{2}{4} + \frac{3}{4}}{8} = \frac{\frac{5}{4}}{8} = \frac{5}{4} \times \frac{1}{8} = \frac{5}{32}$$

then perform division

Example 9

perform addition and subtraction first

$$\frac{\frac{1}{5} + \frac{7}{10}}{\frac{2}{6} - \frac{1}{8}} = \frac{\frac{2}{10} + \frac{7}{10}}{\frac{8}{24} - \frac{3}{24}} = \frac{\frac{2 + 7}{10}}{\frac{8 - 3}{24}} = \frac{\frac{9}{10}}{\frac{5}{24}} = \frac{9}{10} \times \frac{24}{5}$$

then perform division

$$= \frac{9 \times 12}{5 \times 5} = \frac{108}{25}$$

A.6 Exercises

In Exercises 1 through 36, perform the indicated multiplications and divisions.

1. $\dfrac{2}{7} \times \dfrac{3}{11}$

2. $\dfrac{7}{8} \times \dfrac{3}{5}$

3. $3 \times \dfrac{2}{5}$

4. $\dfrac{2}{7} \times 1\dfrac{2}{3}$

5. $\dfrac{7}{8} \div \dfrac{5}{6}$

6. $\dfrac{7}{4} \div \dfrac{2}{5}$

7. $\dfrac{5}{9} \div 3$

8. $2\dfrac{1}{2} \div \dfrac{8}{3}$

9. $\dfrac{8}{9} \times \dfrac{5}{16}$

10. $\dfrac{5}{12} \times \dfrac{3}{7}$

11. $2\frac{1}{3} \times \frac{9}{14}$

12. $\frac{22}{25} \times \frac{15}{33}$

13. $\frac{8}{15} \div \frac{12}{35}$

14. $\frac{21}{44} \div \frac{28}{33}$

15. $\frac{8}{17} \div 4$

16. $\frac{39}{35} \div \frac{13}{21}$

17. $\frac{3}{5} \times \frac{15}{7} \times \frac{14}{9}$

18. $\frac{2}{7} \times \frac{19}{4} \times \frac{21}{38}$

19. $\left(\frac{3}{4} \times \frac{28}{27}\right) \div \frac{35}{6}$

20. $\left(\frac{2}{3} \div \frac{7}{9}\right) \times \frac{14}{15}$

21. $\left(\frac{6}{11} \div \frac{12}{13}\right) \div \frac{26}{121}$

22. $\frac{18}{25} \div \left(\frac{17}{5} \div \frac{34}{15}\right)$

23. $\frac{9}{16} \times \left(\frac{1}{2} + \frac{1}{4}\right)$

24. $\left(\frac{7}{15} - \frac{3}{10}\right) \times \frac{12}{7}$

25. $6 \div \left(\frac{1}{14} + \frac{3}{4}\right)$

26. $\left(\frac{9}{16} - \frac{1}{6}\right) \div 4\frac{3}{4}$

27. $\dfrac{\frac{1}{2} + \frac{1}{3}}{\frac{2}{3}}$

28. $\dfrac{\frac{15}{14}}{\frac{7}{8} - \frac{1}{4}}$

29. $\dfrac{\frac{1}{3} + \frac{5}{6}}{\frac{5}{12} - \frac{1}{4}}$

30. $\dfrac{\frac{7}{18} - \frac{2}{9}}{2 - \frac{4}{15}}$

31. $\dfrac{\frac{4}{21} + \frac{5}{9}}{\frac{3}{4} + \frac{9}{14}}$

32. $\dfrac{\frac{2}{6} + \frac{7}{15}}{\frac{7}{10} - \frac{1}{4}}$

33. $\dfrac{\frac{3}{4} + \frac{1}{5}}{\frac{9}{10} - \frac{5}{6}}$

34. $\dfrac{\frac{13}{64} - \frac{1}{8}}{\frac{13}{16} - \frac{2}{5}}$

35. $\dfrac{\frac{26}{11} - 2}{\frac{3}{5} + \frac{2}{3}}$

36. $\dfrac{\frac{25}{8} - \frac{7}{3}}{4 - \frac{3}{8}}$

In Exercises 37 through 44, find the reciprocals of the given numbers.

37. 5; 13

38. 2; 9

39. $\frac{1}{2}; \frac{1}{5}$

40. $\frac{2}{9}; \frac{3}{7}$

41. $5\frac{1}{3}; 3\frac{1}{2}$

42. $3\frac{1}{4}; 9\frac{1}{32}$

43. $5\frac{1}{16}; 7\frac{1}{10}$

44. $12\frac{1}{10}; 15\frac{1}{3}$

In Exercises 45 through 52, solve the problems. (Where appropriate, h is the symbol for hour.)

45. A machinist makes 16 shearing pins in 3 h. How many will be made in $2\frac{1}{4}$ h?

46. If 30 carbon-zinc flashlight batteries are connected in series, the total voltage is found by adding the individual voltages. What is the sum of the voltages of 30 batteries, each of which is $1\frac{1}{2}$ V?

47. If a civil engineer can survey $2\frac{1}{4}$ acres in $6\frac{1}{2}$ h, what is his rate of surveying?

48. A car's cooling system contains $14\frac{2}{3}$ qt, of which $\frac{3}{5}$ is alcohol. How many quarts of alcohol are in the system?

49. The acceleration due to gravity on Mars is about $\frac{2}{5}$ of that on the earth. If the acceleration due to gravity on the earth is $32\frac{1}{5}$ ft/s^2, what is it on Mars?

50. An employee for a computer company earns an annual salary of $52,500. After taxes the employee takes home $\frac{3}{5}$ of the annual salary. How much does the employee bring home each year?

51. How many pieces of copper wire $\frac{3}{4}$ ft long can be cut from a coil of wire that measures 24 ft in length?

52. A small engine parts company has budgeted 300 h of labor during a given month. If $\frac{3}{5}$ of the hours are for assembly and $\frac{1}{4}$ of the hours are for product testing, how many hours are left over for packing parts to be shipped to customers?

A.7 Decimals

In discussing the basic operations on fractions in the last few sections, we have seen how to express and combine numbers that represent parts of quantities. In many applied situations, fractions are quite useful. However, there are many times in scientific work when a different way of expressing parts of quantities is more convenient. Measurements such as meter readings and distances are often expressed in terms of whole numbers and **decimal** parts.

Fractions whose denominators are 10, 100, 1000, and so forth are called **decimal fractions.** For example, $\frac{7}{10}$ and $\frac{193}{10000}$ are decimal fractions. Making further use of positional notation, as introduced in Section A.1, we place a **decimal point** to the right of the units digit and let the first position to the right stand for the number of tenths, the digit in the second position to the right stand for the number of hundredths, and so on. *Numbers written in this form are called* **decimals.**

Example 1

The meaning of the decimal 6352.1879 is illustrated in Figure A.10. We can also show the meaning of this decimal number as

$$6(1000) + 3(100) + 5(10) + 2(1) + \frac{1}{10} + \frac{8}{100} + \frac{7}{1000} + \frac{9}{10,000}$$

In this form we can easily see the relationship between the decimal and decimal fraction.

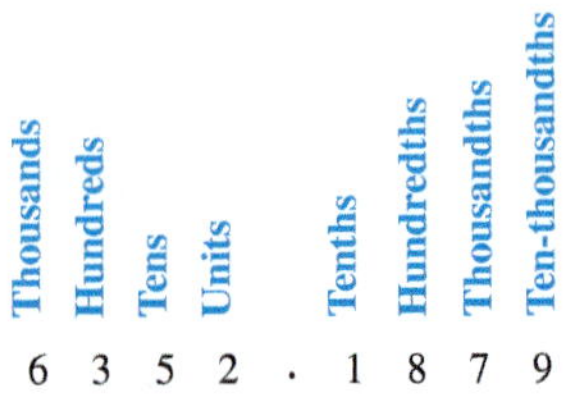

Figure A.10

Because a fraction is the indicated division of one number by another, *we can change a fraction to an equivalent decimal by division.* We place a decimal point and additional zeros to the right of the units position of the numerator and then perform the division.

Example 2

A stock has a listed value of $\frac{5}{8}$ (dollars). Express that value in decimal form.

To change $\frac{5}{8}$ into an equivalent decimal, we perform the division shown at the left. Therefore $\frac{5}{8} = 0.625$. (It is common practice to place a zero to the left of the decimal point in a decimal less than 1. It clarifies that the decimal point is properly positioned.)

$$\begin{array}{r} 0.625 \\ 8)\overline{5.000} \\ \underline{4\,8} \\ 20 \\ \underline{16} \\ 40 \\ \underline{40} \end{array}$$

It often happens that in converting a fraction to a decimal, the division does not come out even, regardless of the number of places to the right of the decimal point. When this occurs, the most useful decimal form is one that is *approximated* by **rounding off** the result of the division. *When we round off a decimal to a required accuracy, we want the decimal that is the closest approximation to the original but with the specified number of decimal positions.*

Example 3

Because $\frac{3}{7} = 0.428\ldots$ (the three dots indicate that the division can be continued), we can round off the result to one decimal place (the nearest tenth) as $\frac{3}{7} = 0.4$, because 0.4 is the decimal closest to $\frac{3}{7}$, with the tenth position as the last position that is written.

Rounded off to two decimal places (the nearest hundredth), $0.428\ldots = 0.43$. Note that when we rounded off the number to hundredths, we increased the hundredths position by one.

These examples suggest that when rounding off a decimal to a specified number of positions after the decimal point, the following rules apply:

1. *If the digit to the right of the round-off place is 5 or more, increase the digit in the round-off place by 1. If the digit to the right of the round-off place is 4 or less, do not change the digit in the round-off place.*

2. *Delete all digits to the right of the round-off digit.*

This procedure is illustrated in Example 4.

Example 4

hundredths
(4 or less) delete

$0.862 = 0.86$ (to two decimal places or hundredths)

hundredths

$0.867 = 0.87$ (to two decimal places or hundredths)

(5 or more) delete
and add 1 to 6

$0.09326 = 0.093$ (to three decimal places or thousandths)

(4 or less) delete
and delete the 6

$0.09326 = 0.0933$ (to four decimal places or ten thousandths)

(5 or more) delete
and add 1 to 2

Increasing the digit in the round-off place by 1 may cause a change in some of the digits to its left. For example, if 0.0598 is rounded to three decimal places, the result is 0.060.

To change a decimal to a fractional form, we write it in its decimal fraction form and then reduce to lowest terms. Consider the illustrations in Example 5.

Example 5

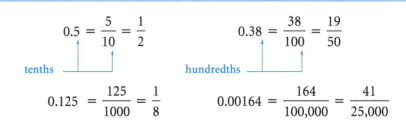

$$0.5 = \frac{5}{10} = \frac{1}{2} \qquad\qquad 0.38 = \frac{38}{100} = \frac{19}{50}$$

tenths ⎏   hundredths ⎏

$$0.125 = \frac{125}{1000} = \frac{1}{8} \qquad\qquad 0.00164 = \frac{164}{100,000} = \frac{41}{25,000}$$

When we consider the addition and subtraction of decimals, we see that the principles are the same as those for the whole numbers. Because only like quantities can be added or subtracted, we must then add tenths to tenths, hundredths to hundredths, and so on. Therefore, when we add and subtract decimals, we align the decimal points and perform the addition or subtraction as we did with the whole numbers.

Example 6

$$
\begin{array}{r}
326.49 \\
98.362 \\
+\ 5937.8 \\
\hline
6362.652
\end{array}
\qquad\qquad
\begin{array}{r}
7862.472 \\
-\ 794.56 \\
\hline
7067.912
\end{array}
$$

To develop the method for multiplying one decimal by another, we express each decimal in its decimal fraction form and then perform the multiplication. An observation of the results leads to the standard method. Consider Example 7.

Example 7

To multiply 0.053 by 3.4, we express the multiplication as

$$(0.053)(3.4) = \frac{53}{1000} \cdot \frac{34}{10} = \frac{1802}{10,000} = 0.1802$$

We note that there are three decimal places (to the right of the decimal point) in the first of the numbers being multiplied, one decimal place in the second number, and four in the final result. Also, the numerator in the product is the product of 53 and 34 and is not affected by the denominators. In general, the positioning of the decimal point in the numbers being multiplied depends only on the denominators.

Considering the results of Example 7, when we multiply one decimal by another, *the number of decimal places (to the right of the decimal point) in the product is the sum of the number of decimal places in the numbers being multiplied.* It is not necessary to line up decimal points as we do in addition and subtraction. The numbers are multiplied as if they were natural numbers, and the decimal point is then properly positioned in the product.

Example
8

5.307 3 places	0.004
× 2.63 2 places	× 0.02
15921	0008
31842	0000
10614	0000
13.95741 5 places	0.00008

As in the other basic operations, the division of one decimal by another is very similar to the division of whole numbers. *Before the division is actually performed, however, the decimal point in the divisor is moved to the right a sufficient number of places to make the divisor a whole number. The decimal point in the dividend is then moved the same number of places.* (This is equivalent to multiplying the numerator and denominator of the fractional form of the division by the same number.) The decimal point in the quotient is directly above that of the dividend.

Example
9

A circuit has a voltage level of 74.362 V and its resistance is 3.26 Ω. The current (in amperes) can be found by dividing 74.362 by 3.26. Find the quotient.

We perform the division as follows.

$$
\begin{array}{r}
22.8 \\
3.26.)\overline{74.36.2} \\
\end{array}
$$

decimal point moved two places to right

$$
\begin{array}{r}
65\ 2 \\
\hline
9\ 16 \\
6\ 52 \\
\hline
2\ 642 \\
2\ 608 \\
\hline
34
\end{array}
$$

The result is expressed as 22.8 A.

A.7 Exercises

In Exercises 1 through 4, write the given decimals as a sum of 1's, 10's, 100's, and so on, and decimal fractions, as in Example 1.

1. 47.3   **2.** 29.26   **3.** 429.486   **4.** 5230.3727

In Exercises 5 through 12, round off the given decimals as indicated.

5. 27.32   (tenths)

6. 404.878   (hundredths)

7. 57.544   (two decimal places)

8. 6.3833   (three decimal places)

9. 8.0327   (hundredths)

10. 0.4063   (hundredths)

11. 17.3846   (tenths)

12. 86.30241   (thousandths)

In Exercises 13 through 20, change the given fractions into equivalent decimals. Where necessary round off the result to the accuracy indicated.

13. $\dfrac{2}{5}$

14. $\dfrac{7}{16}$

15. $\dfrac{4}{19}$ (hundredths)

16. $\dfrac{11}{23}$ (hundredths)

17. $\dfrac{47}{27}$ (tenths)

18. $\dfrac{882}{67}$ (tenths)

19. $\dfrac{362}{725}$ (thousandths)

20. $\dfrac{29}{426}$ (thousandths)

In Exercises 21 through 28, change the given decimals into equivalent fractions in simplest form.

21. 0.8

22. 0.002

23. 0.45

24. 0.075

25. 5.34

26. 17.6

27. 0.0252

28. 0.0084

In Exercises 29 through 44, perform the indicated operations.

29. $3.26 + 18.941 + 9.094$

30. $18.87 + 8.9 + 182.825$

31. $18.046 + 1924.3 + 874.91$

32. $0.046 + 19.35 + 186.6942$

33. $18.623 - 9.86$

34. $2.057 - 1.388$

35. $0.03106 - 0.00478$

36. $0.8694 - 0.0996$

37. $(2.36)(5.932)$

38. $(37.4)(8.207)$

39. $(0.27)(36.6)$

40. $(0.0805)(1.006)$

41. $5.6(3.72 + 18.6)$

42. $0.042(3.072 + 92.23)$

43. $6.75(0.107 - 0.089)$

44. $0.375(4.70 - 2.92)$

In Exercises 45 through 52, perform the indicated divisions, rounding off as indicated.

45. $32.6 \div 2.6$ (tenths)

46. $37.9 \div 41.6$ (tenths)

47. $192 \div 0.65$ (tenths)

48. $132 \div 2.35$ (tenths)

49. $\dfrac{96.288}{18.5}$ (hundredths)

50. $\dfrac{96.7}{0.826}$ (hundredths)

51. $\dfrac{6.238}{13.5}$ (hundredths)

52. $\dfrac{82.75}{103.6}$ (hundredths)

In Exercises 53 through 64, solve the given problems. (Where appropriate, h is the symbol for hour.)

53. A power supply line to a house is carrying 117.6 V. One day later, a reading of 118.2 V is obtained. By what amount did the voltage change?

54. The diameter of a tree is measured as 2.94 ft. One year later the same diameter is measured as 3.11 ft. By what amount did the diameter increase?

55. If an employee is given $416.34 for expenses but spends only $238.85, how much is left?

56. Three electrical resistors have resistances of 13.2 Ω, 6.9 Ω, and 8.4 Ω, respectively. What is the sum of these resistances?

57. If a tank has 318.62 L of fuel and 82.74 L are removed, how much remains? (The symbol L denotes liter.)

58. While using a calculator, a surveyor obtains an indicated length of 8.25 ft. First express that value as a mixed number; then express that distance in feet and inches.

59. A satellite circling the earth travels at 16,500 mi/h. If one orbit takes 1.58 h, how far does it travel in one orbit?

60. A machine is rented for $16.80 per hour for the first 8 h. The cost is $1\frac{1}{2}$ times the regular rate for any use past the 8 h limit. If the machine is used for $14\frac{1}{2}$ h, what is the total cost?

61. High speeds are often compared to the speed of sound. To determine the Mach number for a speed, we divide the given speed by the speed of sound (742 mi/h). Determine, to the nearest tenth, the Mach number of a jet that is traveling at 1340 mi/h.

62. A recycling plant pays $0.425 for each pound of recyclable paper. How much does a plant pay for 5000 pounds of recyclable paper?

63. An average monthly electric bill in upstate New York is $157.00. The average cost for electricity in the same area is $0.139 per kilowatt-hour. How many kilowatt-hours does the average customer use in this area?

64. During a certain chemical process it is necessary to use 2.75 liters of a mixture that costs $20.95 per liter. Determine the cost of the mixture needed for this chemical process.

A.8 Percent

Prior to this section we have used fractions and decimals for representing parts of a unit or quantity. We have seen that decimals have advantages: We use the same basic procedures for their basic operations as whole numbers, and they are more easily compared in size than fractions. In this section we consider the concept of **percent.**

The word "percent" means "per hundred" or hundredths, so *percent represents a decimal fraction with a denominator of 100.* The familiar % symbol is used to denote percent.

Example 1

3% means $\frac{3}{100}$ or $3(0.01) = 0.03$. Also:

$$25\% = \frac{25}{100} = 0.25, \ 300\% = \frac{300}{100} = 3.00, \ 0.4\% = \frac{0.4}{100} = 0.004$$

Percent is very convenient, because we need consider only hundredths. With fractions we might use halves, fifths, tenths, and so on, and comparing fractions is difficult unless the denominators are the same. With decimals we use tenths, hundredths, thousandths, and so on. Comparisons of percents are easy, because only one denominator is used.

Example 2

Suppose you are told that $\frac{3}{20}$ of brand A fuses are defective and $\frac{4}{25}$ of brand B fuses are defective. You would first have to determine that $\frac{3}{20} = \frac{15}{100}$ and $\frac{4}{25} = \frac{16}{100}$ before you could compare. However, if you are told that 15% of brand A fuses are defective and 16% of brand B fuses are defective, the comparison is easy.

Using the meaning of percent allows us to convert percents to decimals, decimals to percents, percents to fractions, and fractions to percents. Examples 3 and 4 illustrate these conversions.

Example 3

$$5\% = \frac{5}{100} = 0.05 \qquad \text{(percent to decimal)}$$

$$132\% = \frac{132}{100} = 1.32 \qquad \text{(percent to decimal)}$$

$$0.2\% = \frac{0.2}{100} = 0.002 \qquad \text{(percent to decimal)}$$

two places to left

$$0.45 = \frac{45}{100} = 45\% \qquad \text{(decimal to percent)}$$

$$0.3 = \frac{3}{10} = \frac{30}{100} = 30\% \qquad \text{(decimal to percent)}$$

$$0.826 = \frac{826}{1000} = \frac{82.6}{100} = 82.6\% \qquad \text{(decimal to percent)}$$

two places to right

Note: To convert a percent to a decimal or a decimal to a percent, the following rules apply.

1. *To change a percent to a decimal, move the decimal point two places to the left and omit the % sign.*

2. *To change a decimal to a percent, move the decimal point two places to the right and attach the % sign.*

Example 4

$$20\% = \frac{20}{100} = \frac{1}{5} \qquad \text{(percent to fraction)}$$

$$0.6\% = \frac{0.6}{100} = \frac{6}{1000} = \frac{3}{500} \qquad \text{(percent to fraction)}$$

$$\frac{3}{8} = 0.375 = 37.5\% \qquad \text{(fraction to percent)}$$

two places to right

$$\frac{15}{12} = 1.25 = 125\% \qquad \text{(fraction to percent)}$$

two places to right

From Example 4 we see that *to change a percent to its equivalent fractional form we first write the percent in its decimal fraction form.* This is done by writing the number of the percent over 100. *This fraction is then reduced to lowest terms. To change a fraction to a percent, we first change the fraction to its decimal form by dividing the numerator by the denominator. The resulting decimal is then changed to a percent and rounded off if necessary.*

There are many applications of percent. Bank interest, income taxes, sales taxes, sales records of corporations, unemployment records, numerous statistics in sports, efficiency ratings of machines, compositions of alloys—these are only a few of the many uses of percent.

The number for which we are finding the percent is the **base,** *the percent expressed as a decimal is the* **rate,** *and the product of the rate and the base is the* **percentage.** Thus,

$$\boxed{\textbf{Percentage } = \text{ rate} \times \text{ base}}$$

This formula is known as the **percentage formula.** In using it we must be careful to identify the base and percentage properly. We must also remember that the rate is the percent expressed as a decimal.

Example
5

A state has a sales tax of $5\frac{1}{2}$%. What is the sales tax on a word-processing device costing $4000?

Here we must recognize that we are to find 5.5% of $4000. This means that 0.055 is the rate, $4000 is the base, and the tax is the percentage to be determined. Thus,

$$\text{percentage} = \text{rate} \times \text{base}$$
$$\text{Tax} = 0.055 \times \$4000 = \$220$$

Therefore, the sales tax on $4000 is $220.

Many situations arise in which the percentage and base are known and the rate is to be found. In general,

$$\boxed{\text{Rate} = \frac{\text{percentage}}{\text{base}}}$$

is an alternative way of writing the percentage formula when it is the rate we seek.

In these percentage, rate, and base problems, it is helpful to remember that the base is the total amount, the rate corresponds to the percent, and the percentage is the amount found when you take a percent of the base.

Percentage = *rate* × *base*

Base: Total amount

Percent: Rate

Percentage: Amount found by multiplying the rate and base

Example
6

A manufacturer bought a machine for $28,000 and later sold it for $12,000. By what percent did the machine decrease in value?

▶ **The machine dropped in value by $16,000 and this amount represents the percentage. Because the value dropped from the original amount of $28,000, that is the base.** We now calculate

$$\text{Rate} = \frac{\overset{\text{percentage}}{16,000}}{\underset{\text{base}}{28,000}} = \underset{\text{rate}}{0.571} \qquad \text{(rounded off)}$$

We conclude that the machine decreased 57.1% in value.

Another problem arises when the percentage and rate are known and the base is to be determined. In general,

$$\text{Base} = \frac{\text{percentage}}{\text{rate}}$$

is a third way we can express the percentage formula when it is the base that is to be found.

Example 7

A fuel mixture is 40% alcohol. One shipment of the mixture contains 300 gal of alcohol. How many gallons are in the shipment?

▶ **The percentage is 300 gal, because there are 300 gal of alcohol in the mixture. The rate is 0.40, because 40% of the mixture is alcohol. Thus, the base is to be found.** This means

$$\text{Total number of gallons} = \frac{\overset{\text{percentage}}{\downarrow}300}{\underset{\underset{\text{rate}}{\uparrow}}{0.40}} = \underset{\underset{\text{base}}{\uparrow}}{750} \text{ gal}$$

Thus, there are 750 gal in the shipment.

A.8 Exercises

In Exercises 1 through 8, change the given percent to equivalent decimals.

1. 8%

2. 78%

3. 236%

4. 482%

5. 0.3%

6. 0.082%

7. 5.6%

8. 10.3%

In Exercises 9 through 16, change the given decimals to percent.

9. 0.27

10. 0.09

11. 3.21

12. 21.6

13. 0.0064

14. 0.0007

15. 7

16. 8.3

In Exercises 17 through 24, change the given percent to equivalent fractions.

17. 30%

18. 48%

19. 2.5%

20. 0.8%

21. 120%

22. 0.036%

23. 0.57%

24. 0.14%

In Exercises 25 through 32, change the given fractions to percent. Round off to the nearest tenth of a percent where necessary.

25. $\dfrac{3}{5}$

26. $\dfrac{7}{20}$

27. $\dfrac{4}{7}$

28. $\dfrac{16}{11}$

29. $\dfrac{8}{35}$

30. $\dfrac{18}{29}$

31. $\dfrac{8}{3}$

32. $\dfrac{9}{7}$

In Exercises 33 through 62, solve the given problems.

33. Find 20% of 65.

34. Find 2.6% of 230.

35. What is 0.52% of 1020?

36. What is 126% of 300?

37. What percent of 72 is 18?

38. What percent of 250 is 5?

39. 3.6 is what percent of 48?

40. 0.14 is what percent of 3.5?

41. 25 is 50% of what number?

42. 3.6 is 25% of what number?

43. If 1.75% of a number is 7, what is the number?

44. If 226% of a number is 3.7, what is the number? (Round off to tenths.)

45. A software package retails for $300, but it is sold at a 15% discount. What is the discounted price?

46. The sales tax in a certain state is 6%. What is the tax in this state on tools with a total price of $378?

47. A company pays taxes of $27,200 on earnings of $85,000. What percent of earnings is paid in taxes?

48. An incandescent light bulb converts 4% of its input electrical energy into light. The remaining energy is given off as waste heat. Of 52,400 J of energy supplied, how much is actually *wasted* by such a bulb?

49. A machine rents for $16.80 per hour. If that cost is increased by 5%, what is the new cost per hour?

50. A factory uses a solar system to heat 42% of the hot water it needs. In one day, there is a need for 6500 gal of hot water. How many gallons were heated by the solar system?

51. A laptop costing $1995 is discounted by 15% for educational institutions. What does a college pay for one such unit?

52. The flue gas of a domestic oil burner is 84.7% nitrogen when measured by volume. What is the volume of nitrogen in 872,000 cm^3 of flue gas? (The symbol cm^3 denotes cubic centimeter.)

53. An alloy weighs a total of 640 lb, and it includes 224 lb of zinc. What percent of the alloy is zinc?

54. If $730 is deposited in a company credit union account, and $65.70 interest is earned for a one-year term, what is the annual rate of interest (in percent)?

55. If 2890 g of ocean water is found to contain 9.1 g of salt, what percent of ocean water is salt?

56. A company manager sets a production goal of 850 tires for one particular day, but only 793 tires were produced. What percent of the goal was met?

57. A particular light bulb is labeled 60 watts, but actual tests show that the electric power level is really 58.2 W. What percent of the advertised wattage was actually achieved by this bulb?

58. A sales representative earns $400 per week plus 12% commission on all sales he makes. What are his earnings for a week in which he sells $2600 worth of merchandise?

59. If a sales tax of 6% amounts to $34.80, what is the original cost of the goods purchased?

60. The efficiency of a motor is defined as power output divided by power input, usually expressed in percent. What is the efficiency of an electric motor whose power input is 850 W and whose power output is 561 W?

61. The effective value of an alternating current is 70.7% of its maximum value. What is the maximum value (to the nearest ampere) of a current whose effective value is 8 A?

62. In one hour, a solar cell can convert 110.6 Btu of solar energy to 4.95 Btu of electrical energy. The efficiency of the conversion is the energy output divided by the energy input, expressed in percent. Find the efficiency.

Answers to Chapter Exercises

Exercises 1.1 (ODD)

1.

3.
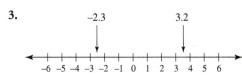

5.
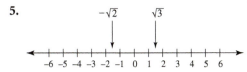

7.
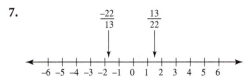

9. 2   **11.** −1   **13.** 6 > 2

15. 0 < 4   **17.** −3 > −7   **19.** −7 < −5

21. $\sqrt{5}$ > 2.2   **23.** |6| = |−6| **25.** |6| = 6 |−6| = 6

27. $\left|-\dfrac{6}{7}\right| = \dfrac{6}{7}$ $\left|\dfrac{8}{5}\right| = \dfrac{8}{5}$

29. −30   **31.** −8

33. Answers will vary depending on current year.

35. −30°C < −5°C   **37.** −2 volts > −5 volts

39. a. −890 feet, −1425 feet
    b. the submarine at −1425 ft depth

41. a. 100 volts   **b.** −200 volts

Exercises 1.2 (ODD)

1. 11   **3.** −15   **5.** 3

7. −8   **9.** 7   **11.** −3

13. −10   **15.** 16   **17.** 1

19. −6   **21.** −8   **23.** 1

25. 6   **27.** −17   **29.** −2

31. −6   **33. a.** −5°C   **b.** −15°C

35. 17 amperes   **37.** 5000   **39.** −5

41. 11   **43.** 12   **45.** 44

Exercises 1.3 (ODD)

1. −63   **3.** −84   **5.** 30

7. 0   **9.** −56   **11.** 30

13. −168   **15.** −8   **17.** −3

19. −11   **21.** 15   **23.** 0

25. undefined   **27.** −16   **29.** 3

31. 3   **33.** −5   **35.** negative

37. negative   **39.** positive   **41.** zero

43. 5 seconds   **45.** 5 days   **47.** 18

49. −56   **51.** 21   **53.** undefined

55. −9   **57.** 2

Exercises 1.4 (ODD)

1. 8^3 **3.** 2^4 **5.** 3^5

7. 10^5 **9.** 8×8

11. $-(3 \times 3 \times 3 \times 3 \times 3 \times 3)$

13. $7 \times 7 \times 7 \times 7 \times 7 \times 7 \times 7 \times 7$

15. $5 \times 5 \times 5 \times 5 \times 5 \times 5$

17. −243   **19.** 64   **21.** 0.09

23. 42.875   **25.** 4   **27.** 11

29. −4   **31.** 2   **33.** 0.4

35. 0.3   **37.** 108   **39.** −4500

41. 891   **43.** 0.637   **45.** 70,000 joules

47. $\dfrac{15}{4}$ seconds   **49.** $10,737,418.24   **51.** 484 feet

53. 125.340   **55.** 27 inches

Exercises 1.5 (ODD)

1. 20   **3.** −14   **5.** −49

7. −1   **9.** −69   **11.** −5

13. −106   **15.** 13   **17.** 14

19. $\dfrac{20}{3}$ **21.** −5   **23.** 108

25. −4   **27.** 36   **29.** −1000 feet

31. −78.0 meters   **33.** 46.2 gallons   **35.** 2480 gallons

37. 3   **39.** −311.94   **41.** −663

Exercises 1.6 (ODD)

1. 4,000,000   **3.** 0.08   **5.** 2.17

7. 0.00365   **9.** 3×10^3 **11.** 7.6×10^{-2}

13. 7.04×10^{-1} **15.** 9.21   **17.** 5.3×10^{-5}

19. 2.01×10^9 **21.** 1.55×10^8 **23.** 9.30×10^{-3}

25. 4.719×10^3 **27.** 2.80×10^2 **29.** 1.12064×10^5

31. 4.26×10^1 **33.** 6.86×10^7 **35.** 9.1×10^{-28}

37. 50,000 pounds   **39.** 3.6×10^8 square kilometers

41. 0.0000000000016 watts   **43.** 7.1×10^8 years

45. 6×10^{-19} joules   **47.** 6.06×10^7 hertz

49. 2.715×10^{-6}**51.** 6×10^{-5} meters

53. 6,500,000,000; 6.5×10^9

Exercises 1.7 (ODD)

1. 6 hr   **3.** 600 liters   **5.** 70

7. 0 ft   **9.** 136

Chapter 1—Review Exercises

1. −2   **2.** −8   **3.** −13

4. 4   **5.** −63   **6.** 96

7. 9   **8.** −8   **9.** 10

10. 13   **11.** 19   **12.** −19

13. 96   **14.** −28   **15.** −60

16. −360   **17.** 10   **18.** 4

19. −6   **20.** −6   **21.** 256

22. −343   **23.** 5   **24.** 8

25. 0   **26.** −9   **27.** −7

28. −17   **29.** 10   **30.** −16

31. −16   **32.** −15   **33.** −4

34. 69   **35.** $\dfrac{20}{3}$ **36.** −13

37. −5   **38.** 2.05×10^5 **39.** 9.805×10^9

40. 4.005×10^3 **41.** 3.5×10^{-4} **42.** 7.5×10^{-7}

43. 7.07×10^{-1} **44.** 47000   **45.** 302,000,000

46. 831.9   **47.** 0.00187   **48.** 0.0000000077

49. 0.0000000000091   **50.** 3.64×10^6

51. 4.1006×10^6 **52.** 5.572×10^{-1} **53.** 2.555×10^{-3}

54. 1.16×10^{17} **55.** 5 quarts   **56.** −3000

57. 26 feet   **58.** −4   **59.** −0.2%

60. $440   **61.** 1650   **62.** 1.039

63. −7° C   **64.** 75,000,000   **65.** ≈ 487 seconds

66. Hydrogen atom by 2.4888×10^{-23}

Chapter 1—Test

1. −8   **2.** 5   **3.** −11

4. 24   **5.** 4   **6.** $-\dfrac{1}{8}$

7. 3   **8.** −28   **9.** −2

10. 0.47   **11.** 44735318.5   **12.** 11.8

13. 1.7×10^{-12} **14.** 5.67×10^6 **15.** 0.000000035

16. 12400   **17.** 5.945×10^{-2} **18.** 4.599×10^4

19. 1.4514×10^{-8} **20.** 3.125×10^8 **21.** 43 yard gain

22. 66.652 coulombs   **23.** 12 hours

Exercises 2.1 (ODD)

1. 28,800 s   **3.** 900 in.   **5.** 1360 oz

7. 60 mph   **9.** ≈95 ft/s   **11.** $21.63 per hour

13. ≈4.5 h   **15.** ≈25,227 mph

Exercises 2.2 (ODD)

1. mA   1 mA = 0.001 A   **3.** kV   1 kV = 1000 V

5. kW   1 kW = 1000 W   **7.** ML   1 ML = 1,000,000 L

9. megavolt   1 MV = 1,000,000 V

11. microsecond $1\mu s = 0.000001$ s

13. centivolt   1 cV = 0.01 V

15. nanoampere   1 nA = 0.000000001 A

17. a. m^2 **b.** ft^2 **19. a.** m/s   **b.** ft/s

21. 4 s   **23.** 8 m   **25.** 4 cs

27. $3\,\mu s$ **29.** 8 cg   **31.** 4 kg

33. mm^2 **35.** m/gal (or mpg)

37. l/s   **39.** lb/ft^2 **41.** $kg \cdot m/s^2$

43. $A \cdot s$

Exercises 2.3 (ODD)

1. 10   **3.** 63,360   **5.** 288 in.^{2}

7. 30.48 cm   **9.** 220 qt   **11.** 23.7 L

13. 52,000 cm^2 **15.** 9.05 ft^3 **17.** 50 mph

19. 3.8 L   **21.** 1129 ft/s   **23.** 5520 kg/m^3

25. ≈0.25 mi   **27.** ≈91.4 m   **29.** ≈1.6 lb/ft^3

Exercises 2.4 (ODD)

1. exact   **3.** approximate   **5.** approximate

7. 1 cm, 1mm are approximate; $2.80 is exact

9. 3; 4   **11.** 4; 4   **13.** 3; 3   **15.** 4; 5

17. 3.763 is more precise and more accurate

19. 0.01 is more precise; 30.8 is more accurate

21. 0.1 and 78.0 are equally precise; 78.0 is more accurate

23. 0.004 is more precise; both are equally accurate

25. a. 5.71   **b.** 5.7   **27. a.** 6.93   **b.** 6.9

29. a. 4100   **b.** 4100   **31. a.** 46,800   **b.** 47,000

33. a. 501   **b.** 500   **35. a.** 0.215   **b.** 0.22

37. 128.25 ft, 128.35 ft   **39.** 81.5 L, 82.5 L

41. 0.1733 qt   **43.** 91.4 m   **45.** 327 ft

47. 0.0373 W

Chapter 2—Review Exercises

1. 4; 4   **2.** 1; 2   **3.** 3; 4   **4.** 4; 4

5. 7.32 is more precise; both are equally accurate

6. 80.0 is more precise and more accurate

7. both are equally precise; 207.31 is more accurate

8. 0.0021 is more precise; 98.568 is more accurate

9. a. 98.5   **b.** 98   **10. a.** 2.73   **b.** 2.7

11. a. 60,500   **b.** 61,000   **12. a.** 220,000   **b.** 220,000

13. a. 673   **b.** 670   **14. a.** 69,000   **b.** 69,000

15. a. 0.700   **b.** 0.70   **16. a.** 4940   **b.** 4900

17. $1\,\mu g = 0.000001$ g

18. 1 cA = 0.01 A   **19.** 1 ks = 1000 s

20. 1 MV = 1,000,000 V   **21.** 0.385 cm^3

22. 4750 cm^2 **23.** 13 cm   **24.** 1.18 m

25. 35.62 pt   **26.** 1,850,000 cm   **27.** 764.64 L

28. 1.992 km   **29.** 740 in^3 **30.** 0.197 ft^3

31. 1.10 m/s   **32.** 95100 cm/h

33. 0.1855 in. and 0.1865 in.   **34.** 1.45 V and 1.55 V

35. 21.84 g   **36.** 0.37 m   **37.** .02 m

38. 5.97 in.   **39.** 0.67 kg/L   **40.** 51.8 mph

41. 0.000287 in^2 **42.** 0.5 s

Chapter 2—Test

1. a. 3   **b.** 3   **c.** 3

2. a. 4.36   **b.** 0.00616   **c.** 105

3. a. kilovolt; 1 kV = 1000 V
   b. milliamper; 1 mA = 0.001 A
   c. millimeter; 1 mm = 0.001 m

4. a. 100,000 cm   **b.** 12.3 lb

5. 15 gal   **6.** 0.54 L/s   **7.** 1.6 lb/ft^3

8. The Empire State Building is 431 m tall; therefore it is taller than the Eiffel Tower.

9. 212°F   **10. a.** 0.135 km/L   **b.** 92 L per passenger

Exercises 3.1 (ODD)

1. b, c **3.** $7, p, q, r$ **5.** i, i, R

7. a, b, c, c, c **9.** 6   **11.** 2π

13. $4\pi e$ **15.** mw^2 **17.** $x = 4y$

19. $m = 10c$ **21.** $V = 7.48lwd$ **23.** $d = \dfrac{1}{2}gt^2$

25. $N = 5280x$ **27.** $N = 9xy$ **29.** $N = 24n$

31. $C = 4\,cs$ **33.** 1500 mi   **35.** 1.5 V

37. 720 gal   **39.** 81.5 ft   **41.** 3456

43. $64.80

Exercises 3.2 (ODD)

1. $x^2, 4xy, -7x$ **3.** $12, -5xy, 7x, \dfrac{-x}{8}$

5. $3x$ and $2x$ **7.** x and $5x$

9. $-8mn$ and $-mn$ **11.** $6(x - y)$ and $-3(x - y)$

13. $6a(a - x)$ **15.** $x^2(a - x)(a + x)$

17. $\dfrac{2}{5a}$ **19.** $\dfrac{6}{a - b}$

21. $K = 328$ **23.** $t = 2$

25. $L = 50.0 - x$ **27.** $A = 2x^2 + 4lx$

29. $m = \dfrac{1}{2}(s + l)$ **31.** $E = \dfrac{I - P}{I}$

33. 22.6 ft   **35.** 77.8 ft^2

37. $1000   **39.** 0.65

41. 9.741 in.

Exercises 3.3 (ODD)

1. $6x + y$ **3.** $7a - 4b^2$

5. $5s + 3t$ **7.** $2a + 3$

9. $-4 - 4x$ **11.** $2s - 1$

13. $-9 + 4y$ **15.** $10x + 1$

17. $2s + 2$ **19.** $6x - 12$

21. $7t - 5x - 5p^2 + 9$ **23.** $2a^2 + 3x + 1$

25. $7 + 2t$ **27.** $4x + 5a$

29. $8x$ **31.** $2x + 28$

33. $2R + 30$ **35.** $2M - 40$

37. 80 ft   **39.** $56 - 8x$

Exercises 3.4 (ODD)

1. x^{10} **3.** y^8 **5.** t^{10}

7. n^{14} **9.** $a^2x^4b^2$ **11.** $-a^5t^{10}$

13. $-84r^2s^2t^3$ **15.** $-8s^3t^9x^3$ **17.** $2a^2 + 6ax$

19. $6a^2x - 3a^4$ **21.** $-2s^2tx + 2st^3y$

23. $-3x^3y^2 + 9ax^2y^7$ **25.** $x^2 - 4x + 3$

27. $s^2 + s - 6$ **29.** $2x^2 + x - 1$

31. $10v^2 + 17v + 3$ **33.** $a^2 - 3ax + 2x^2$

35. $6a^2 + ac - 2c^2$ **37.** $2x^2 + 18x - 5tx - 45t$

39. $4a^2 - 81p^2y^2$ **41.** $2a^3 - 5a^2 - 13a - 5$

43. $a^2 + 2axy - 4ax - 2x^2y + 3x^2$

45. $x^2 - 4x + 4$ **47.** $x^2 + 4xy + 4y^2$

49. $x^3 - 3x^2 - 10x + 24$ **51.** $x^3 + 3x^2 + 3x + 1$

53. a. F $(2x)^3 = 8x^3$ **b.** F $(x^2y)^3 = x^6y^3$ **c.** T

55. a. T   **b.** T   **c.** F $(xy)^6 = x^6y^6$

57. a. F $(a + b)^2 = a^2 + 2ab + b^2$
   b. F $(x - 3)^2 = x^2 - 6x + 9$
   c. T

59. a. F $(t + 1)^2 = t^2 + 2t + 1$ **b.** T   **c.** T

61. $x^2 + x - 6 \text{ ft}^2$ **63.** $200x^2 + 1600x + 3000$

65. $(13)(7) = 100 - 9$ **67.** $nr_2 - nr_1 - r_2 + r_1$
$91 = 91$

Exercises 3.5 (ODD)

1. x^3 **3.** a **5.** $4n^3$

7. $-x^2y^2$ **9.** $\dfrac{-x}{4r^2}$ **11.** $\dfrac{3as}{5d^3}$

13. $6b + 5$ **15.** $3m - 1$ **17.** $-a^2x^2 + ax$

19. $y^2 - 2xy^3$ **21.** $bc - ab^3c^4 - 2a$

23. $-ab^2 + 2a^2b^3 + b + 1$ **25.** $x - 3$

27. $2x + 1$ **29.** $4x - 3 + \dfrac{2}{2x + 3}$

31. $2x^2 - 3x - 4$ **33.** $2x^2 - x + 3$

35. $x^3 + x^2 + x + 1$ **37.** $5(x - 1)$

39. $2x^3 - x^2 + 2x + 1$ **41.** $a - 3b$

43. $x^2 - 4xy - 2y^2$

45. a. F $x^8 \div x^2 = x^6$ **b.** F $r \div r^3 = \dfrac{1}{r^2}$

 **c.** F $x^6 \div x^3 = x^3$

47. a. $\dfrac{x^2 + y^2}{x} = x + \dfrac{y^2}{x}$ **b.** T   **c.** T

49. a. T   **b.** T   **c.** $\dfrac{6x^2 - 8}{x^2} = 6 - \dfrac{8}{x^2}$

51. $6r - 4 + \dfrac{18}{r + 2}$ **53.** $3x - 16 \text{ mi/h}$

Chapter 3—Review Exercises

1. $-7a$ and $5a$ **2.** $3ax^2$ and $-6ax^2$

3. $5(a - b)$ and $-7(a - b)$ **4.** $8x^2$ and $-2x^2$

5. $7a - 4b$ **6.** $3a^2b + 5ab$ **7.** $12ax - 18bx$

8. $-24a + 15ab$ **9.** $3x + y$ **10.** $x^2 + 6xy$

11. $3a - x$ **12.** $3x - 3s$ **13.** $9x - 13y$

14. $-7s$ **15.** $-4 + 2n$ **16.** -12

17. $8y$ **18.** $-2x + 6r$ **19.** $-6a^3b^6$

20. $8s^4t^6$ **21.** $56x^4y^9z^8$ **22.** $-12x^5y^7$

23. $27a^3b^6$ **24.** $16a^{16}c^4$ **25.** $16x^8y^4z^{12}$

26. $-x^{15}y^{10}z^{20}$ **27.** $-4ax^3$ **28.** $\dfrac{3s^2}{4r^3t}$

29. $\dfrac{5x}{y^4z^3}$ **30.** $\dfrac{8b^4}{3a^6c^4}$ **31.** $8 - 5x$

32. $2y - 2$ **33.** $5x - 2a$ **34.** $3y - 4x - 5xy$

35. $4x - 7$ **36.** $9a - 6$ **37.** $5b - 10$

38. $4x - 5y$ **39.** $2x^5 - 6x^3$ **40.** $3s^7 - 2s^4$

41. $-2a^3x + 2a^2t$ **42.** $-9a^2j^5 + 12a^3j - 3a^3j^2$

43. $2x^2 + x - 21$ **44.** $3x^2 + 13x - 10$

45. $6a^2 - 11ab - 10b^2$ **46.** $-10x^2 - y^2 + 7xy$

47. $x^3 + 1$ **48.** $2x^3 - 5x^2 - 5x + 6$

49. $-2x^3 + 6x^2 + 4x - 16$

50. $-3x^2y + 6xy^2 - 9qxy + 3qx - 6qy + 9q^2$

51. $-2y^3 + 3x^3$ **52.** $3a^2b^2 - 4b^3$

53. $h - 3j^2 - 6h^3j^3$ **54.** $3f^2g - 4fk^4 + 6k^2$

55. $(x + 4)$ **56.** $(3x - 5)$

57. $x^2 - x + 1 + \dfrac{-11}{2x + 3}$

58. $2x^2 - 3x + 1 + \dfrac{2}{3x + 5}$

59. $x^2 - x + 1$ **60.** $2x^2 + x + 1$

61. $2x^2 - x - 4$ **62.** $3x^2 + 2x + 2$

63. $x + 2y$ **64.** $2b + 3a^2$

65. $x^2 + x + 1$ **66.** $x - 3$

67. 42   **68.** 27   **69.** 98

70. -602 **71.** 218,448   **72.** -192

73. $1.71C + 33.6$ **74.** $2T_1 - 2T_2$ **75.** $-4x^2 + 8x$

76. $x - 11$ **77.** $36x - 4x^2$ **78.** $80t - 32t^2$

79. $lh + 2atlh + a^2t^2lh$ **80.** $8xd$

81. $2x^2 + 20x + 50, 450$ pounds   **82.** $10x + 4$

Chapter 3—Test

1. $-10a$ and $9a$ **2.** $5ax^2$ and $-16ax^2$

3. $10a - 4b$ **4.** $-19a^2b$

5. $12a^2 - 18a^2b - 12ab$ **6.** $12a^3b^6$

7. $-8x^6y^3z^6$

8. $\dfrac{-4s^4}{ad}$

9. $6x^2 - 13xy - 28y^2$

10. $-xy - 3y^4 + 6xy^2$

11. $2x^2 + 4x + 2 + \dfrac{3}{2x - 1}$

12. 36

13. $2Vr - Va - Vb$

14. $4t - 4h - 2t^2 - 4th - 2h^2$

15. $\dfrac{r}{k} - \dfrac{2h^2}{kr} + \dfrac{h^2v^2}{k^2}$

Exercises 4.1 (ODD)

1. 2

3. 3

5. 7

7. 35

9. -18

11. -4

13. 9

15. -2

17. 3

19. -6

21. -1

23. 8

25. 1

27. $\dfrac{1}{2}$

29. $\dfrac{1}{3}$

31. 2

33. 9

35. 32

37. 3

39. 68

41. -2

43. $\dfrac{25}{19}$

45. the first has no solution, the second is an identity

47. a, d

49. 120°C

51. 750 gal

Exercises 4.2 (ODD)

1. $\dfrac{N}{A - s}$

3. $\dfrac{R_1L_2}{L_1}$

5. $v_2 - at$

7. $\dfrac{PV}{R}$

9. $\dfrac{Id^2}{5300E}$

11. $\dfrac{yd}{ml}$

13. $180 - A - C$

15. $\dfrac{RM}{CV}$

17. $\dfrac{MD_m}{D_p}$

19. $\dfrac{L - 3.14r_1 - 2d}{3.14}$

21. $\dfrac{L - L_0}{L_0t}$

23. $\dfrac{p - p_a + dgy_1}{dg}$

25. $\dfrac{n_1A + n_2A - n_2p_2}{n_1}$

27. $\dfrac{f_su - fu}{f}$

29. $a - bc$

31. $\dfrac{f - 3y}{a}$

33. $\dfrac{3y - 2ay}{2a}$

35. $2b + 4$

37. $\dfrac{x_1 - x_2 - 3a}{a}$

39. $2R_3 - R_1$

41. $\dfrac{3y + 6 - 7ay}{7a}$

43. $\dfrac{x - 3a - ax}{a}$

45. $3A - a - b$

47. $\dfrac{E - Ir}{I}$

49. $\dfrac{I - xr_1}{x + 1000}$

51. $\dfrac{C - x}{7}$

53. 10.6 m

55. 2.17 ft^2

Exercises 4.3 (ODD)

1. $x > 12$

3. $x < 5$

5. $x > 18$

7. $x < -6$

9. $x \leq 4$

11. $x \leq -1$

13. $x < -4$

15. $x < -2$

17. $x > -1$

19. $x \geq 6$

21. $2000 \leq M \leq 1{,}000{,}000$

23. $1490 \leq v \geq 1610$

25. $62.6 \leq p \leq 67.2$

27. greater than 40

29. $4 < x < 9$

Exercises 4.4 (ODD)

1. 220 Ω

3. 16 ft

5. 9 ft

7. 150 m by 200 m

9. 10 Gbytes and 55 Gbytes

11. $59.67

13. 800 gal, 1200 gal, 2400 gal

15. 4 h

17. 3600 ft/s

19. 12.5 lb and 37.5 lb

21. 30 t

23. $4000 federal, $800 state

25. $2\dfrac{2}{7}$ qt

27. 344 m

29. $11,000

Exercises 4.5 (ODD)

1. $\dfrac{12}{5} \quad \dfrac{3}{23}$

3. $\dfrac{7}{1} \quad \dfrac{1}{6}$

5. $\dfrac{2}{3} \quad \dfrac{2}{3}$

7. $\dfrac{2}{11} \dfrac{2}{7}$ **9.** $\dfrac{15}{4}$ **11.** $\dfrac{3}{10}$

13. $\dfrac{1}{6}$ **15.** $\dfrac{4}{9}$ **17.** 2 m/s

19. $\dfrac{2}{9}$ lb/ft^3 **21.** $\dfrac{5}{4}$ **23.** $\dfrac{6}{7}$

25. 5 **27.** 9 **29.** 14

31. $\dfrac{35}{2}$ **33.** $\dfrac{1}{4}$ **35.** 4540 g

37. $\dfrac{100}{3}$ in. **39.** 180 mg, 100 mg **41.** 37 h

43. 9.3 mm **45.** $\dfrac{9}{5}$ **47.** $y = kt$

49. $y = ks^2$ **51.** $t = \dfrac{k}{y}$ **53.** $y = kst$

55. $y = \dfrac{ks}{t}$ **57.** $x = \dfrac{kyz}{t^2}$ **59.** $y = 5s$

61. $u = \dfrac{272}{d^2}$ **63.** $y = \dfrac{9x}{t}$ **65.** 16

67. $\dfrac{32}{5}$ **69.** 2700 **71.** 0.5

73. $E = 23I$ **75.** $F = \dfrac{kQ_1Q_2}{s^2}$ **77.** 35.4 hp

79. 0.116 Ω **81.** 240 r/min

Chapter 4—Review Exercises

1. 18 **2.** −3 **3.** 9

4. $\dfrac{1}{2}$ **5.** −4 **6.** 9

7. 4 **8.** 3 **9.** −3

10. −2 **11.** $\dfrac{1}{2}$ **12.** $11\dfrac{1}{2}$

13. $R_3 = R - R_1 - R_2$ **14.** $g = \dfrac{wa}{F}$

15. $s_2 = \dfrac{ms_1}{r}$ **16.** $R = \dfrac{P}{I^2}$

17. $n = \dfrac{d_m}{A} + 1$ **18.** $T_1 = -\dfrac{T_2(w - q)}{q}$

19. $M_1 = \dfrac{PT - M_2V_2}{V_1}$ **20.** $v_s = \dfrac{f_s u - fu}{f}$

21. $H = \dfrac{wL}{R(w + L)}$ **22.** $r = \dfrac{-uR_o - AR_o}{A}$

23. $T = \dfrac{W + H_1 - H_2}{S_1 - S_2}$ **24.** $C = \dfrac{2p + dv^2 + 2dW}{2d}$

25. $y = \dfrac{2a + ax}{3}$ **26.** $a = \dfrac{bx + bc - c^2}{cx}$

27. $c = \dfrac{ax + ab - bx}{b}$ **28.** $y = \dfrac{2a - 3}{a}$

29. $b = \dfrac{6 - x - a^2}{a}$ **30.** $r_3 = \dfrac{2r_1r_2}{-3}$

31. $x = \dfrac{2 - 3a}{6}$ **32.** $a = \dfrac{2x + 30}{x}$

33. $x < 9$ **34.** $x < -2$ **35.** $x \geq 5$

36. $x \geq 3$ **37.** $x < -3$ **38.** $x < 4$

39. $x < -4$ **40.** $x < 1$ **41.** $x < -2$

42. $x < -4$ **43.** $x \leq 3$ **44.** $x \leq 6$

45. $\dfrac{2}{3}$ **46.** $\dfrac{8}{15}$ **47.** $\dfrac{6}{1}$

48. $\dfrac{1}{4}$ **49.** $\dfrac{8}{3}$ **50.** $\dfrac{17}{3}$

51. 4.5 **52.** 12 **53.** 4

54. 8 **55.** 4 **56.** $-2\dfrac{1}{2}$

57. $\dfrac{7}{3}$ **58.** $\dfrac{175}{28}$ **59.** $\dfrac{21}{2}$

60. $\dfrac{65}{3}$ **61.** $y = 6x$ **62.** $s = 15t^2$

63. $m = \dfrac{15}{\sqrt{r}}$ **64.** $v = \dfrac{24}{z^3}$ **65.** 5

66. $\dfrac{243}{8} = 30.375$ **67.** $\dfrac{243}{8} = 30.375$ **68.** 43.2

69. 54 in. by 36 in. **70.** 15 m **71.** \$50, \$150

72. 36 ft by 24 ft   **73.** 95   **74.** 48 ft

75. $3200, $7700, $10,700   **76.** 3 mA, 9 mA

77. 1850 km/h, 2150 km/h   **78.** 4.5 mi

79. 10 L   **80.** 3.6 mL   **81.** $9000

82. $19,500   **83.** 20 lb   **84.** 3.51 s

Chapter 4—Test

1. -4.5 **2.** $\dfrac{47}{8}$ **3.** $-\dfrac{19}{5}$

4. $Z = \dfrac{R}{n^2}$ **5.** $R = \dfrac{V - Ir}{I}$ **6.** $J = \dfrac{f - 2B}{2B}$

7. $x > 6$ **8.** $x > \dfrac{-5}{3}$ **9.** $\dfrac{19}{3}$

10. $\dfrac{8}{1}$ **11.** 75   **12.** $130.50, $259.50

13. 9.2 lb   **14.** $P = 30,000t$ **15.** 2163 Ω

Exercises 5.1 (ODD)

1. y dependent variable; x independent variable

3. p dependent variable; V independent variable

5. multiplication

7. square the value of the independent variable

9. $f(x) = 5 - x$ **11.** $f(t) = t^2 - 3t$

13. $0, 3$ **15.** $7, -5$ **17.** $-2, -\dfrac{5}{4}$

19. $-1, 2.91$ **21.** $-1, 8$ **23.** $2, 2$

25. $\dfrac{1}{2}$, undefined   **27.** $a^2 - 2a^4$, $\dfrac{a - 2}{a^2}$

29. function   **31.** not a function

33. answers will vary   **35.** $x \geq -7$

37. x can be any real number, except x cannot equal 0

39. $V(e) = e^3$ **41.** $C(P) = P + 0.5P$ **43.** $i(v) = \dfrac{V}{5}$

45. a. $C(h) = \$12 + 6(h - 6)$
     b. h is the independent variable
       c is the dependent variable
     c. $h \geq 0$

Exercises 5.2 (ODD)

1. A(2, 1) B(−2, 3)   **3.** E(4, 0) F(−2, 1)

5. I(1, 5.5) J(3, 5.5)   **7.** M(−9.5, 0) N(−9.5, 2)

9.

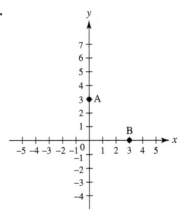

11.

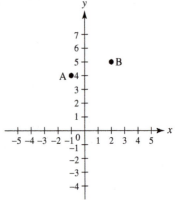

13.

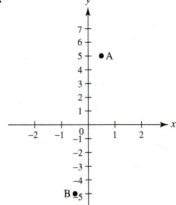

15.

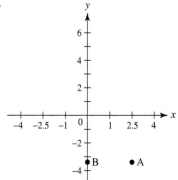

17. I, II   **19.** III, II   **21.** 0

23. I and III   **25.** I

Exercises 5.3 (ODD)

1.

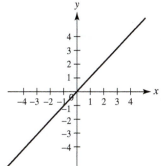

3.

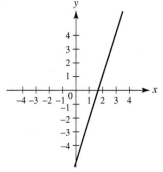

5.

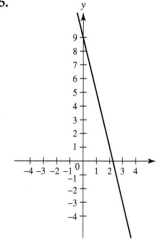

7.

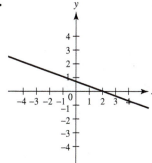

9.

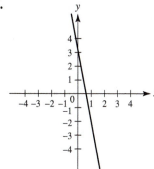

11.

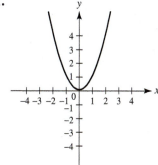

13.

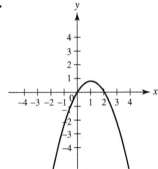

15.

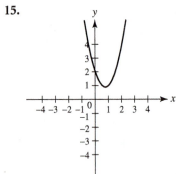

17.

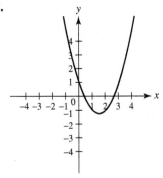

19.

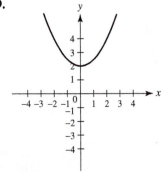

21.

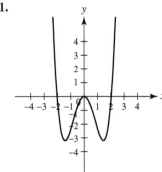

23.

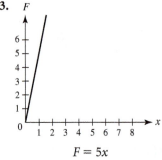

$$F = 5x$$

25.

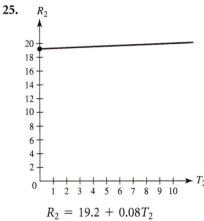

$$R_2 = 19.2 + 0.08T_2$$

27.

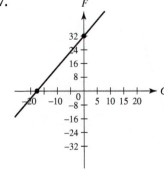

29.

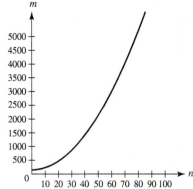

31.

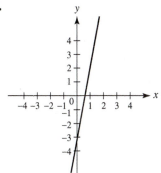

33.

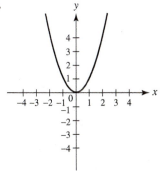

35.

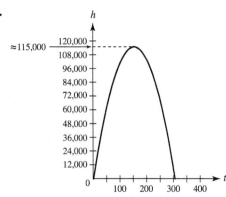

Exercises 5.4 (ODD)

1. 2   **3.** -1 **5.** 1

7. $\dfrac{11}{-7}$ **9.** $\dfrac{.4}{2.5}$ **11.** undefined

13. 0   **15.** $\dfrac{1}{2}$

17. $(4, 0)$ and $(0, 2)$ **19.** $(4, 0)$ and $(0, -10)$

21. $(-18, 0)$ and $(0, 4.5)$ **23.** $\left(-\dfrac{8}{5}, 0\right)$ and $(0, 2)$

25. $m = -2$ and $(0, 1)$ **27.** $m = 1$ and $(0, 4)$

29. $m = -2$ and $\left(0, \dfrac{7}{-2}\right)$ **31.** $m = \dfrac{-5}{8}$ and $\left(0, \dfrac{10}{8}\right)$

Exercises 5.5 (ODD)

1.

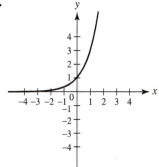

3.

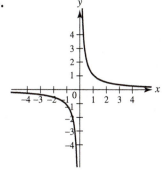

5.

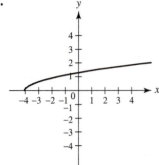

13.

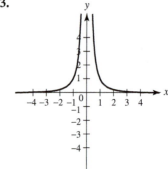

7.

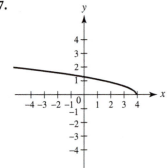

15.

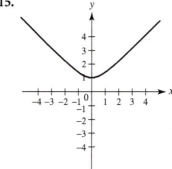

9.

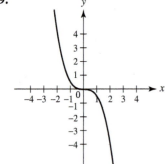

17.

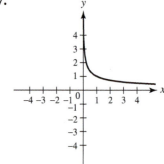

11.

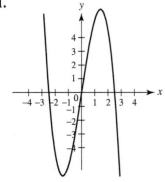

19.

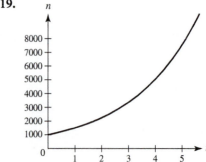

21.

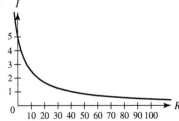

Exercises 5.7 (ODD)

1.

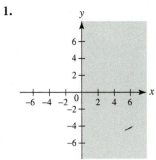

23.

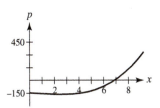

3.

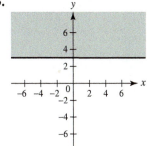

25.

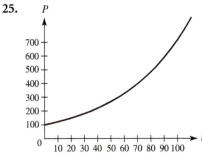

5.

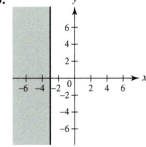

Exercises 5.6 (ODD)

1. 1.4, −1.2   **3.** 9, −8   **5.** 6.5, 11.6

7. 9.2, 1.7   **9.** 8.9, 2.1   **11.** −2.1, 5.0

13. 0.7   **15.** −1.2, 1.7   **17.** 1.7

19. 4.2   **21.** 3.5   **23.** 1.5

25. 0.0055 C, 0.0001 C

27. 35 g, 70 g   **29.** 28 lb   **31.** 2.1 in.

7.

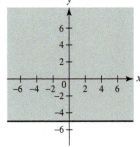

9.

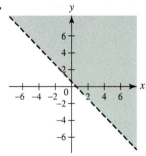

17.

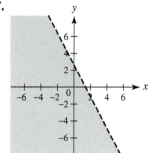

11.

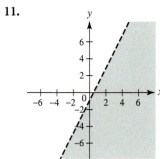

19.

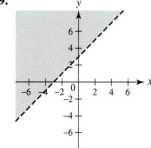

13.

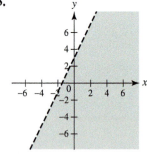

21.

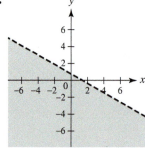

15.

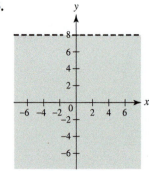

23.

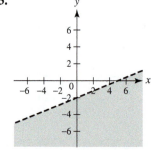

25.

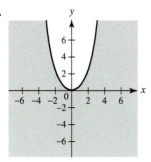

Chapter 5—Review Exercises

1. $3, 2$

2. $-3, 3$

3. $4, \dfrac{5}{3}$

4. $-22, -10$

5. $3, -\dfrac{1}{2}$

6. $-9, -16$

7. $12, -3.36$

8. $4, 7 + v - 4v^2$

9. $0, 8$

10. $20, -15$

11. $1, 5$

12. $\dfrac{2}{3}, 2$

27.

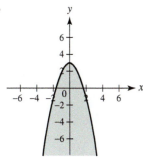

29.

31.

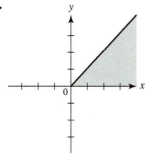

21.

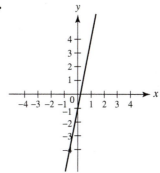

22.

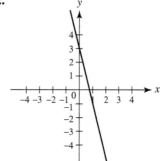

23.

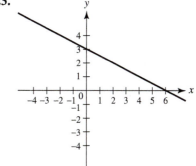

24.

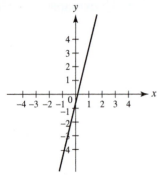

25.

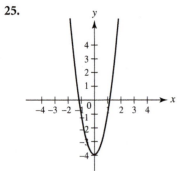

26.

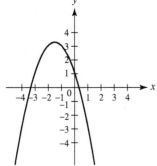

27.

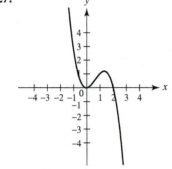

28.

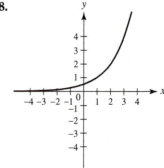

29.

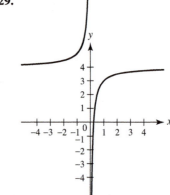

30.

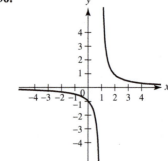

31.

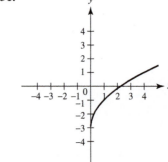

32.

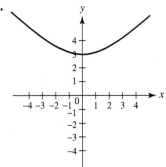

33. $\dfrac{-1}{7}$ **34.** $\dfrac{16}{7}$ **35.** $\dfrac{11}{-7}$

36. 0 **37.** $\dfrac{-5}{1}$ **38.** undefined

39. $(4, 0)$ and $(0, 8)$ **40.** $\left(\dfrac{3}{2}, 0\right)$ and $(0, 9)$

41. $\left(\dfrac{9}{2}, 0\right)$ and $(0, -6)$ **42.** $(5, 0)$ and $(0, -1)$

43. $(-13, 0)$ and $(0, 6.5)$ **44.** $m = -1, (0, 2)$

45. $m = -2, (0, 0)$ **46.** $m = 2, (0, -14)$

47. $m = 1, (0, 4)$

48.

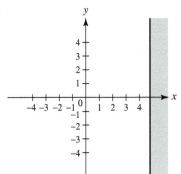

49.

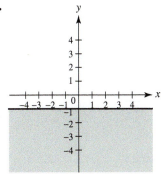

50.

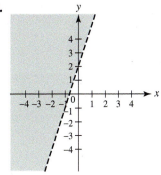

51.

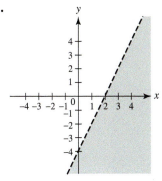

52.

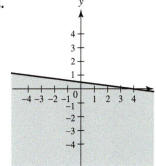

53.

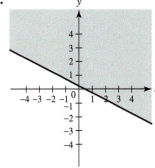

54.

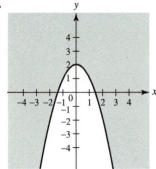

55.

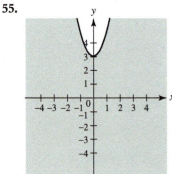

56.

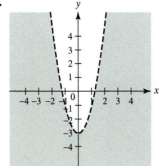

57.

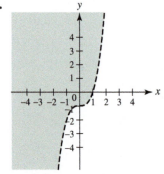

58. 1.3

59. -1.8

60. $-\dfrac{2}{3}, 1.0$

61. $0.7, -7.7$

62. 0.6

63. $T = 0.06\,C$

64. $I = 200 + 0.03S$

65. $H = 240\,I^2$

66. $V = 2A$

67. $700, $1700, 2400

68. 4.0 mi/s, 4.2 mi/s, 4.9 mi/s

69. 101 ft, 103 ft, 112 ft

70. 67 m

71. $F = \dfrac{9}{5}C + 32$ $F = C$ at $-40°$

72. the projectile travels 2500 ft

Chapter 5—TEST

1. a. $-7, 2, -10$ **b.** $33, 26$ **c.** 18

2. a.

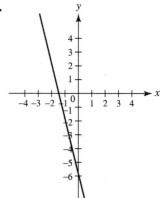

b.

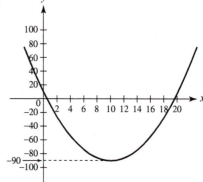

3. $\dfrac{-1}{2}$ **4.** $(2, 0)$ and $(0, -4)$ **5.** $m = \dfrac{4}{3}, (0, 2)$

6. a.

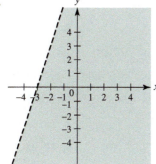

b.

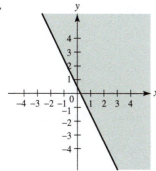

7. a. $\dfrac{-3}{2}$ **b.** -5.16 and 1.16

8. a.

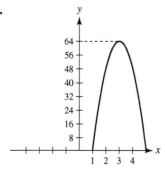

    b. 1 s and 5 s   **c.** 64 ft

9. a. $900, $4400, $3000   **b.** 53 items

Exercises 6.1 (ODD)

1. 26°   **3.** 104°   **5.** 56°24′

7. 136°27′   **9.** 156.25°   **11.** 67.1°

13. ∠EBD, ∠DBC   **15.** ∠ABD

17. ∠ABE, ∠EBC   **19.** ∠ABD = 125°

21. △ABD   **23.** 54°

25. AB = 5 in.   **27.** ∠A and ∠C, ∠B and ∠D

29. 8 cm   **31.** 3 in.

33. 4 cm   **43. a.** yes   **b.** no

45. rhombus   **47.** 54°

49. 94 km   **51.** west

Exercises 6.2 (ODD)

1. 12 ft   **3.** 123 mm   **5.** 896 m

7. 84.8 in.   **9.** 193.8 mm   **11.** 57.1 in.

13. 2.6 m   **15.** 0.60 mi   **17.** 168 cm

19. 230.6 ft   **21.** 94.9 cm   **23.** 21.2 in.

25. 36 ft   **27.** 23.0 cm   **29.** 45.36 in.

31. 24.22 m   **33.** $p = 2s + a$ **35.** $p = 2r + \pi r$

37. $p = 5s$ **39.** $p = 2a + b_1 + b_2$

41. $69.30   **43.** $\approx$ 24900 mi   **45.** 892 ft

47. 25 ft   **49.** 55 in.   **51.** 1428 ft

53. 26.00 in.   **55.** 8.89 cm   **57.** 4.58 cm

Exercises 6.3 (ODD)

1. 2700 cm^2 **3.** 170 in.^{2} **5.** 57.8 in.^{2}

7. 6.45 in.^{2} **9.** 2450 mm^2 **11.** 24.4 cm^2

13. 106 in.^{2} **15.** 40 ft^2 **17.** 0.240 m^2

19. 340 cm^2 **21.** 238 ft^2 **23.** 15,700 cm^2

25. 130 cm^2 **27.** 12 ft^2 **29.** 0.000258 km^2

31. 31.0 m^2 **33.** 0.718 ft^2 **35.** 1692 ft^2

37. 21.65 in.^{2} **39.** 1.27 ft^2

41. a. 36 in. **b.** 60 in.^{2} **43. a.** 60 cm   **b.** 225 cm^2

45. a. 30 in. **b.** 30 in.^{2} **47. a.** 26.6 in.   **b.** 35 in.^{2}

49. $179.20   **51.** 2 gal   **53.** 122,000 m^2

55. 1070 cm^2 **57.** $12,900   **59.** 9 times greater

61. a. 203 ft^2 **b.** $474

Exercises 6.4 (ODD)

1. 20,100 mm^3 **3.** 0.08 cm^3 **5.** 4600 m^3

7. 1,500,000 cm^3 **9.** 3400 cm^3 **11.** 0.512 in.^{3}

13. 0.01 m^3 **15.** 7156 cm^3 **17.** 18,850 cm^3

19. 2544 ft^3 **21.** 14,730,000 mm^3 **23.** 54,000 ft^3

25. 958 mm^3 **27.** 1,466 ft^3 **29.** 27,200 cm^3

31. 85,200 cm^3 **33.** 14,700 m^3 **35.** 21,000 in.^{3}

37. 0.735 mm^3 **39.** 1728 in.^{3} **41.** 1,000,000 cm^3

43. 1300 ft^3 **45.** 15.9   **47.** no

49. 122,674   **51.** 3,330,000 yd^3 **53.** 23,500 lb

55. 4.71 m^3 **57.** 311 in.^{3}

Chapter 6—Review Exercises

1. 37°30′   **2.** 43°57′   **3.** 12°33′

4. 45°16′   **5.** 63.5°   **6.** 82.33°

7. 105.9°   **8.** 215.75°   **9.** 40.2 in.

10. 158.1 cm   **11.** 25.5 mm   **12.** 1.29 yd

13. 27.2 m   **14.** 60.8 in.   **15.** 1798 ft

16. 28.0 m   **17.** 278 cm   **18.** 400 in.

19. 26.7 ft   **20.** 119 cm   **21.** 153 mm

22. 1.68 mi   **23.** 15.5 mm   **24.** 25.6 ft

25. 224 cm^2 **26.** 8840 in.^{2} **27.** 3.08 yd^2

28. 15,000 mm^2 **29.** 4.80 cm^2 **30.** 1460 ft^2

31. 3.06 in.^{2} **32.** 398 m^2 **33.** 0.001 km^2

34. 432 in.^{2} **35.** 35.5 in.^{2} **36.** 20,700 cm^2

37. 140,000 mm^2 **38.** 0.290 ft^2 **39.** 3.30 m^2

40. 9.47 ft^2 **41.** 3.60 m^3 **42.** 3750 in.^{3}

43. 42.9 yd^3 **44.** 10,600,000 mm^3 **45.** 678 cm^3

46. 125 ft^3 **47.** 27.4 yd^3 **48.** 4.032 m^3

49. 140 ft, 208 ft   **50.** 204 m^2 **51.** 30,700 lb

52. 15 cm   **53.** 508 mm   **54.** 26,200 mi

55. 128,000 lb   **56.** 8400 ft^2, 13,100 ft^2

57. 8.77 ft^2 **58.** 116 cm^3

59. 150 gal   **60.** 50 acres

Chapter 6—Test

1. 45°42′   **2.** 62.58°

3. a. 80.4 in.   **b.** 79.7 cm   **c.** 48.7 in.

4. a. 3.06 ft^2 **b.** 7605 mm^2 **c.** 14 yd^2

5. a. 9770 in^3 **b.** 160,400 ft^3 **c.** 5650 mm^3

6. 12 cm   **7.** 26200 mi

8. 30.4 ft^2 **9.** 49,800,000 ft^3

10. 2,350,000 ft^3

Exercises 7.1 (ODD)

1. $x = 3, y = 0$ **3.** $x = 2, y = 1$

5. $r = 4, x = -3$ **7.** $x = 2, y = 2$

9. $x = -1, y = 2$ **11.** $x = -2, y = 1.5$

13. $a = -1.9, b = -0.5$ **15.** inconsistent

17. $p = 2.4, q = 1.1$ **19.** $x = 1.9, y = -2.2$

21. $x = 5.9, y = -0.2$ **23.** $x = 0.8, y = 3.6$

25. inconsistent   **27.** dependent

29. dependent   **31.** inconsistent

33. $x = 95, y = 25$

35. $g = \$6.00$ per hour, $d = \$4.00$ per hour

37. $T_1 = 12.2$ lb, $T_2 = 14.3$ lb

39. 8.5 ft for compact cars, 11 ft for full-size cars

Exercises 7.2 (ODD)

1. $x = 5, y = 8$ **3.** $x = 6, y = 4$

5. $x = \dfrac{2}{5}, y = \dfrac{3}{5}$ **7.** $x = -1, y = 2$

9. $x = 2, y = 1$ **11.** $x = 0, y = 0$

13. $x = 6, y = 0$ **15.** $u = -\dfrac{6}{17}, k = \dfrac{20}{17}$

17. $x = \dfrac{81}{10}, y = \dfrac{63}{10}$ **19.** $x = -2, y = -1$

21. $r = 4, s = -2$ **23.** $x = \dfrac{32}{31}, k = -\dfrac{69}{31}$

25. $x = 1000$ km/h, $y = 200$ km/h

27. $r_1 = 1.25$ gal/h, $r_2 = 0.75$ gal/h

29. $V_1 = 40$ V, $V_2 = 20$ V **31.** $b = 24$ ft, $r = 13$ ft

Exercises 7.3 (ODD)

1. $x = 5, y = 2$ **3.** $x = 2, y = 1$

5. $m = 4, n = 8$ **7.** $d = -1, t = 4$

9. $x = 1, n = 5$ **11.** $x = 9, y = 2$

13. inconsistent   **15.** $a = 1, b = 2$

17. $p = \dfrac{3}{2}, q = -\dfrac{1}{12}$ **19.** $x = -14, y = -3$

21. $m = -4, n = 6$ **23.** $x = 1, y = 1$

25. $V_1 = 6.0$ V, $V_2 = 1.5$ V   **27.** $x = \dfrac{1}{16}, y = 1$

29. 650 math, 500 verbal

31. $120 for graphing calculators, $25 for scientific calculators

Exercises 7.4 (ODD)

1. -2 **3.** 31   **5.** 0

7. -2 **9.** 6   **11.** 52

13. $x = 1, y = 5$ **15.** $x = 2, y = 3$

17. $s = 4, t = 3$ **19.** $v_1 = 5, v_2 = 3$

21. $x = -6, y = 7$ **23.** $x = 12, y = 9$

25. $x = -12, y = -10$ **27.** $x = 3, y = \dfrac{1}{2}$

29. $x = 0.2, y = 0.3$ **31.** $x = \dfrac{1}{2}, y = \dfrac{1}{4}$

33. $R_1 = 24{,}000\ \Omega$, $R_2 = 8000\ \Omega$

35. $x = 5.3, y = 10.7$ **37.** $f = $4.50, b = 0.75

39. $R_1 = 40.2\ \Omega$, $R_2 = 15.5\ \Omega$

41. $x = 1.43, y = -0.653$

43. $x = 0.006088, y = 0.001498$

Exercises 7.5 (ODD)

1. 8 A, -3 A   **3.** 75 h, 30 h

5. 8 mm, 21 mm   **7.** $8000 at 4%, $200 at 5%

9. 90,000 at 80%, 60,000 at 70% **11.** 42%

13. 16, 48   **15.** 80 g, 40 g

17. 210 cell phones, 110 radar detectors

19. 100 breadboards, 60 testers

Chapter 7—Review Exercises

1. $x = \dfrac{12}{5}, y = \dfrac{6}{5}$ **2.** $x = 6, y = -6$

3. $x = 4, y = 4$ **4.** $x = 4, y = 4$

5. $x = 4, y = 4$ **6.** $x = -\dfrac{6}{11}, y = -\dfrac{21}{11}$

7. $p = \dfrac{18}{5}, q = \dfrac{2}{5}$ **8.** $m = \dfrac{7}{2}, n = \dfrac{1}{2}$

9. $u = 1, v = -1$ **10.** $x = 2, y = -3$

11. $a = 12, b = -3$ **12.** $p = -\dfrac{7}{2}, x = -2$

13. $x = -9, y = 11$ **14.** $h = \dfrac{6}{5}, y = \dfrac{5}{3}$

15. $y = 1, z = -\dfrac{1}{2}$ **16.** $x = -\dfrac{3}{17}, y = \dfrac{8}{17}$

17. dependent   **18.** inconsistent

19. $x = -\dfrac{7}{3}, y = 2$ **20.** $x = -\dfrac{1}{4}, y = -8$

21. $x = \dfrac{43}{19}, y = -\dfrac{22}{19}$ **22.** $x = \dfrac{30}{59}, y = \dfrac{7}{59}$

23. $s = \dfrac{94}{107}, t = -\dfrac{22}{107}$ **24.** $x = -\dfrac{47}{191}, y = -\dfrac{301}{191}$

25. $x = 100, y = -1$ **26.** $x = \dfrac{1400}{43}, y = \dfrac{220}{43}$

27. $x = \dfrac{119}{201}, y = -\dfrac{59}{201}$ **28.** $r = \dfrac{3}{2}, t = \dfrac{1}{3}$

29. $x = 12, y = 24$ **30.** $x = \dfrac{180}{23}, y = \dfrac{132}{23}$

31. $r = \dfrac{1}{3}, s = \dfrac{1}{5}$ **32.** $x = \dfrac{1}{4}, y = \dfrac{1}{2}$

33. $x = 1, y = 0.3$ **34.** $x = 6, y = -4$

35. $x = 1.6, y = -1.2$

36. $x = 1.7, y = -2.3$

37. $u = 2, v = -6$

38. $r = 4, s = 0.5$

39. inconsistent

40. $m = 2.6, n = 1.5$

41. -2

42. -28

43. -7

44. -159

45. $x = -7, y = 2$

46. $x = 1, y = -8$

47. $x = \dfrac{1}{2}, y = 4$

48. $x = 4, y = -6$

49. $i_1 = -\dfrac{3}{22}A, i_2 = -\dfrac{39}{110}A$ **50.** $v_0 = 6$ ft/s, $a = 10$ ft/s^2

51. $m = 417$ mol/h, $n = 331$ mol/h

52. $T_2 = 25$ lb, $T_3 = 43.3$ lb

53. 45 spots/h, 30 spots/h   **54.** 48, 72

55. 72 refunds, 112 additional payments

56. $300 fixed costs, $0.25 per booklet

57. 45 mi/h, 55 mi/h   **58.** 175 km/h, 25 km/h

59. 750 mL of 5% solution, 250 mL of 25% solution

60. 31.7 mL, 63.3 mL

Chapter 7—Test

1. $x = 2.2, y = 1.3$

2. $x = 7, y = -2$

3. $x = 2, y = -4$

4. 19

5. $x = 2, y = 1$

6. $I_1 = 2.4\,\Omega, I_2 = 6.2\,\Omega$

7. $58 for 2-inch deck sets, $75 for 5/4-inch deck sets

Exercises 8.1 (ODD)

1. $5(x + y)$

3. $7(a^2 - 2bc)$

5. $a(a + 2)$

7. $2x(x - 2)$

9. $3(ab - c)$

11. $2p(2 - 3q)$

13. $3y^2(1 - 3z)$

15. $abx(1 - xy)$

17. $6x(1 - 3y)$

19. $3ab(a + 3)$

21. $acf(abc - 4)$

23. $ax^2y^2(x + y)$

25. $2(x + y - z)$

27. $5(x^2 + 3xy - 4y^3)$

29. $2x(3x + 2y - 4)$

31. $4pq(3q - 2 - 7q^2)$

33. $7a^2b^2(5ab^2c^2 + 2b^3c^3 - 3a)$

35. $3a(2ab - 1 + 3b^2 - 4ab^2)$ **37.** $(x^2 + 2)(x + 3)$

39. $(3x^2 + 5)(x - 5)$ **41.** $(x + 3)(y + 1)$

43. $nR\left[\dfrac{T_2}{T_1} - 1\right]$ **45.** $2wh(7 + 2w)$

47. $2x(x^2 - 3x + 5)$

49. a. yes   **b.** no, $2x(x - 4)$

51. a. yes   **b.** no, $12\,ay(xy - 3)$

53. a. yes   **b.** no   **c.** no

55. a. no   **b.** yes   **c.** no

Exercises 8.2 (ODD)

1. $(x + 1)(x + 2)$ **3.** $(x + 4)(x - 3)$

5. $(y - 5)(y + 1)$ **7.** $(x + 5)(x + 5)$

9. $(x + 7)(x + 2)$ **11.** $(x + 6)(x - 7)$

13. $(x + 4)(x + 8)$ **15.** $(x + 1)(x + 1)$

17. $(x - 5)(x - 3)$ **19.** $(x + 7)(x - 2)$

21. $(x + 6)(x + 2)$ **23.** $(x - 10)(x + 4)$

25. $(x - 12)(x + 9)$ **27.** $(x + 1)(x + 1)$

29. $(x - 4)(x - 4)$ **31.** $N(r + 1)(r + 1)$

33. $P(R + 1)(R + 1)$

Exercises 8.3 (ODD)

1. $(2q + 1)(q + 5)$ **3.** $(3x + 1)(x - 3)$

5. $(5c - 1)(c + 7)$ **7.** prime

9. $(2s - 3t)(s - 5t)$ **11.** $(5x + 2)(x + 3)$

13. $(2x - 3)(2x - 1)$ **15.** $(6q + 1)(2q + 3)$

17. $(6t - 5u)(t + 2u)$ **19.** $(4x - 3)(2x + 3)$

21. $(4x - 3)(x + 6)$ **23.** $(4n - 5)(2n + 3)$

25. $2(x - 3)(x - 8)$ **27.** $2(2x - 3z)(x + 2z)$

29. $2x(x + 1)(x + 2)$ **31.** $a(5x - y)(2x + 5y)$

33. $3a(x + 5)(x - 3)$ **35.** $7a^3(2x + 1)(x - 1)$

37. $(2x + 1)^2$ **39.** $(3x - 1)^2$

41. $2(p - 4)(p - 50)$ **43.** $2(x - 4)(x - 8)$

45. $4x(x - 6)^2$ **47.** $3x(2x - 1)(x - 2)$

49. $4(4t + 1)(t - 8)$

Exercises 8.4 (ODD)

1. $(a + 1)(a - 1)$ **3.** $(t + 3)(t - 3)$

5. $(4 + x)(4 - x)$ **7.** $(2x^2 + y)(2x^2 - y)$

9. $(10 + a^2b)(10 - a^2b)$ **11.** $(ab + y)(ab - y)$

13. $(9x^2 + 2y^3)(9x^2 - 2y^3)$ **15.** $5(x^2 + 3)(x^2 - 3)$

17. $4(x + 5y)(x - 5y)$ **19.** $(x^2 + 1)(x + 1)(x - 1)$

21. $4(x^2 + 9y^2)$ **23.** $3s^2(t^4 + 4)$

25. $(5 + x + y)(5 - x - y)$ **27.** $4xy$

29. $a(x + y + 1)(x + y - 1)$

Exercises 8.5 (ODD)

1. $(a - 1)(a^2 + a + 1)$

3. $(t + 2)(t^2 - 2t + 4)$

5. $(1 - x)(1 + x + x^2)$

7. $(2x + 3a)(4x^2 - 6ax + 9a^2)$

9. $(2x^2 - y)(4x^2 + 2x^2y + y^2)$

11. $(ax - y^2)(a^2x^2 + axy^2 + y^4)$

13. $8x(x + 1)(x^2 - x + 1)$

15. $ax^2(1 - y)(1 + y + y^2)$

17. $2k(R_1 + 2R_2)(R_1^2 - 2R_1R_2 + 4R_2^2)$

19. $2(3t_1^2 - t_2)(9t_1^4 + 3t_1^2t_2 + t_2^2)$

21. $(5s^2 - 4t^3)(25s^4 + 20s^2t^3 + 16t^6)$

23. $(ab + c^5)(a^2b^2 - abc^5 + c^{10})$

25. $(1 - ax)(1 + ax)(1 + ax + a^2x^2)(1 - ax + a^2x^2)$

27. $(1 - x - y)(1 + x + y + (x + y)^2)$

29. $N(x - y)(x^2 + xy + y^2)$

31. $(2x + 3)(4x^2 - 6x + 9)$

33. $at^3(t - 1)(t^2 + t + 1)$

Chapter 8—Review Exercises

1. $5(a - c)$ **2.** $4(r + 2s)$

3. $3a(a + 2)$ **4.** $2t^2(3t - 4)$

5. $4ab(3a + 1)$ **6.** $5ty(3t^2 - 2y)$

7. $8stu^2(1 - 3s^2)$ **8.** $4xyz^4(4x - 1)$

9. $(2x + y)(2x - y)$ **10.** $(p + 3uv)(p - 3uv)$

11. $(4y^2 + x)(4y^2 - x)$ **12.** $(rst + 2x)(rst - 2x)$

13. $(x + 1)(x + 1)$ **14.** $(x - 3)(x + 1)$

15. $(x - 1)(x - 6)$ **16.** $(x + 9)(x - 7)$

17. $a(x^2 + 3ax - a^2)$ **18.** $3r^2t(6 - 3rt - 2t^2)$

19. $2nm(m^2 - 2nm + 3n^2)$ **20.** $8y^2(1 + 3yz - 4z^4)$

21. $4t^2(p^3 - 3t^2 - 1 + a)$

22. $11rst^2(2rs - 11 - 2r^2 + 3r^3s)$

23. $2xy^3(1 - 7x + 8y - 3x^2y^2)$

24. $3st^2(1 - 2s^2tu - 4u + 3t)$

25. $(4rs + 3y)(4rs - 3y)$

26. $(7r^2t^2 + y^3)(7r^2t^2 - y^3)$ **27.** prime

28. $(a + b + c)(a + b - c)$

29. $(2x + 7)(x + 1)$ **30.** $(3y - 5)(y + 2)$

31. $(5s + 2)(s - 1)$ **32.** $(3a - 1)(a + 7)$

33. $(7t + 1)(2t - 3)$ **34.** $(5x - 4)(x + 1)$

35. $(3x + 1)(3x + 1)$ **36.** $(4r - 5)(2r + 3)$

37. $(x + y)(x + 2y)$ **38.** $(3a - 4b)(2a - 3b)$

39. $(5c - d)(2c + 5d)$ **40.** $(2p - 3q)(2p - 3q)$

41. $(8x + 7)(11x - 12)$ **42.** $(4y + 7)(4y + 7)$

43. $2(x + 3y)(x - 3y)$ **44.** $4(rt + 3pq)(rt - 3pq)$

45. $8x^4y^2(xy + 2)(xy - 2)$ **46.** $3mn(m^2 + 3n)(m^2 - 3n)$

47. $3a(x - 3)(x + 4)$ **48.** $2x(9c + 5)(2c - 3)$

49. $3r(6r - 13s)(3r + 4s)$ **50.** $4c^2(2x + 9)(x + 2)$

51. $16y^3(y - 1)(y - 3)$ **52.** $a^2(9u - 2)(2u + 3)$

53. $5(x^2 + 5)(x^2 - 5)$

54. $4(a^4 + 4)(a^2 + 2)(a^2 - 2)$

55. $(4x^2 + 1)(2x + 1)(2x - 1)$

56. $(x^4 + 1)(x^2 + 1)(x - 1)(x + 1)$

57. $(x + 3)(x^2 - 3x + 9)$

58. $(t - 4)(t^2 + 4t + 16)$

59. $(2x + 1)(4x^2 - 2x + 1)$

60. $8(x^2 - y)(x^4 + x^2y + y^2)$

61. $axy(x - y)(x^2 + x^2y + y^2)$

62. $ab(b + a)(b^2 - ba + a^2)$

63. $i(R_1 + R_2 + R_3)$ **64.** $16t(4 - t)$

65. $P(N + 2)$ **66.** $k(R + r)(R - r)$

67. $k(D + 2r)(D - 2r)$ **68.** $P(V_2 - V_1)(C + R)$

69. $b(x + y)(x - y)$ **70.** $(T - 10)(T + 530)$

Chapter 8 — Test

1. $6(x - 4y)$ **2.** $3xy(4 + y + 2xy)$

3. $(x + 3)(y - 1)$ **4.** $(x + 3)(x + 8)$

5. $(7x - 2)(x - 1)$ **6.** $(2x + 5)(2x - 5)$

7. $8(x - 2y)(x^2 + 2xy + 4y^2)$ **8.** $x(x + 3)(x^2 - 3x + 9)$

9. $(R - 100)(R - 300)$ **10.** $16(x - 6)(x - 1)$

Exercises 9.1 (ODD)

1. $\dfrac{4}{7}$ **3.** $\dfrac{2a}{3a^2}$ **5.** $\dfrac{2}{x + 1}$

7. $\dfrac{2x - 1}{x + 1}$ **9.** $\dfrac{3(x - 2)}{x + 2}$ **11.** $\dfrac{-(2 + x)}{x - 3}$

13. $\dfrac{1}{3}$ **15.** $\dfrac{ab}{4}$ **17.** $\dfrac{8}{9}$

19. $\dfrac{2x - 1}{x - 2}$ **21.** $\dfrac{(x + 1)(x + 2)}{2(x + 3)}$ **23.** $\dfrac{x + 1}{x - 1}$

25. $\dfrac{x}{x + 2}$ **27.** $\dfrac{3x - 2}{4x + 3}$ **29.** $\dfrac{x + 3y}{3y}$

31. $\dfrac{5 - x}{2 + x}$ **33.** $-3x$ **35.** $\dfrac{2 - x}{3 + x}$

37. $\dfrac{3}{7}$

Exercises 9.2 (ODD)

1. $\dfrac{1}{8n}$, $13s$ **3.** $\dfrac{3b}{a}$, $\dfrac{a}{3b}$

5. $\dfrac{x - y}{x + y}$, $\dfrac{x^2}{x^2 + y^2}$ **7.** $\dfrac{a + b}{a}$, $-\dfrac{V}{IR}$

9. $\dfrac{4}{9t}$ **11.** $\dfrac{2a}{15}$ **13.** $\dfrac{rt}{12}$

15. $\dfrac{81}{256}$ **17.** $\dfrac{a^{10}}{32x^5}$ **19.** $\dfrac{a^3x^9}{b^6}$

21. $\dfrac{26}{35cx}$ **23.** $\dfrac{24mx}{7}$ **25.** $\dfrac{y}{45x}$

27. $\dfrac{1}{2a^2b^5}$ **29.** $\dfrac{a + 3b}{a + b}$ **31.** $\dfrac{5x(x - y)}{6}$

33. $\dfrac{x + 1}{x + 2}$ **35.** $\dfrac{(x + 3)(x - 3)}{(x - 2)(x - 4)}$

37. $\dfrac{5b - 2}{10}$ **39.** $\dfrac{3(a - b)}{(a - 2b)(a + b)}$

41. $\dfrac{(s + 2)(s + 7)}{(s - 12)(s + 11)}$ **43.** $\dfrac{3.2(x + 2)}{x}$

45. $\dfrac{1 - 2n + n^2}{1 + 2n + n^2}$

Exercises 9.3 (ODD)

1. 18 **3.** 36 **5.** $12a$

7. $40t$ **9.** $90y$ **11.** $9x^2$

13. $8x^2$ **15.** $420ax$ **17.** $375ax^2$

19. $75a^3$ **21.** $96a^3b^3$ **23.** $15a^2$

25. $60a^2cx^3$ **27.** $8x(x - 1)$ **29.** $3a(a + 3)$

31. $6x(x - y)(x + y)$ **33.** $2(x - 1)^2(x - 2)$

35. $2(2t + 3)(t - 4)(t^2 + 5t + 3)$

37. $(x - 3)(x + 2)(x + 3)^2$

39. $x^2(2x - 1)(x + 1)(x - 1)$

Exercises 9.4 (ODD)

1. $\dfrac{20}{36a} - \dfrac{21}{36a}$ **3.** $\dfrac{5b}{abx} + \dfrac{a}{abx} - \dfrac{4bx}{abx}$

5. $\dfrac{8x}{2x^2(x-1)} - \dfrac{3}{2x^2(x-1)}$

7. $\dfrac{x(x+2)^2}{2(x-2)(x+2)^2} + \dfrac{10(x+2)}{2(x-2)(x+2)^2} - \dfrac{6x(x-2)}{2(x-2)(x+2)^2}$

9. $\dfrac{19}{10x}$

11. $\dfrac{2b-5a}{3ab}$

13. $\dfrac{6x+3}{x^2}$

15. $\dfrac{26b-25}{40b}$

17. $\dfrac{8y-b}{by^2}$

19. $\dfrac{9+2x^2}{3x^3y}$

21. $\dfrac{2xy+5x^2-3}{x^2y}$

23. $\dfrac{42yz-15xz+2xy}{12xyz}$

25. $\dfrac{4(2a-1)}{(a-2)(a+2)}$

27. $\dfrac{2x^2+3x+9}{4(x-3)(x+3)}$

29. $\dfrac{3x^2-17x+14}{(2-3x)(2+3x)}$

31. $\dfrac{4+15x-5x^2}{(x-2)(x+2)}$

33. $\dfrac{2x^2+3x-125}{3(x+5)(x-5)}$

35. $\dfrac{26.500x^2+5.080x^3+0.004}{x^2}$

37. $\dfrac{P^2-2gm^2rM}{2mr^2}$

39. $\dfrac{R_1+R_2}{R_1R_2}$

41. $\dfrac{u-u_d}{u_du}$

Exercises 9.5 (ODD)

1. 4

3. -2

5. 10

7. $-\dfrac{3}{4}$

9. $1-3b$

11. $4a+2$

13. $\dfrac{2a-4}{3a^2}$

15. $-\dfrac{16}{7b}$

17. 2

19. $\dfrac{7}{3}$

21. $\dfrac{3}{4}$

23. no solution

25. $\dfrac{52}{11}$

27. $\dfrac{pf}{p-f}$

29. $\dfrac{WT_1+Q_1T_1}{Q_1}$

31. \$2000

33. 1.8 mi

35. 240 h

Chapter 9—Review Exercises

1. $\dfrac{3rt^4}{s^3}$

2. $\dfrac{-z}{6y^3}$

3. $\dfrac{a}{3bc^2}$

4. $\dfrac{4x^2y}{z^2}$

5. $\dfrac{4}{x-2y}$

6. $\dfrac{a}{x+y}$

7. $\dfrac{p+q}{3+2p^2}$

8. $\dfrac{4a}{5b}$

9. $\dfrac{a}{2b}$

10. $\dfrac{p+1}{p+3}$

11. $\dfrac{3x+y}{2x-y}$

12. $\dfrac{2(y-5)}{3-2y}$

13. 18

14. $10x^2$

15. $12t^2$

16. abt^3

17. $48b^2t$

18. aby^2z^4

19. $20x^2(x-2)$

20. $(x+5)(x-3)(x+3)$

21. $\dfrac{10a}{3x^2}$

22. $\dfrac{b^2c^4}{4}$

23. $\dfrac{15y}{4x}$

24. $\dfrac{q}{3p^2}$

25. $\dfrac{6}{a}$

26. $\dfrac{b^2}{ac^2}$

27. $\dfrac{2bu}{av}$

28. $\dfrac{40m^3}{81}$

29. $\dfrac{10b-3a}{5a^2b}$

30. $\dfrac{9x^2-10y^2}{12xy}$

31. $\dfrac{10cd+c-6}{2c^2d}$

32. $\dfrac{3x^2+18x-2a}{12x^3}$

33. $\dfrac{8}{27}$

34. $\dfrac{I^3}{R^3t^3}$

35. $\dfrac{a^4x^4}{81y^8}$

36. $\dfrac{32x^5}{z^{15}}$

37. $\dfrac{2}{x(x+1)}$

38. $\dfrac{1}{2a(a-3)}$

39. $\dfrac{x-5}{4}$

40. $\dfrac{3(x-3)(x-4)(x+3)}{2(x-1)(x-5)}$

41. $a(a-1)$

42. $\dfrac{4x}{x-1}$

43. $(3x+2y)(y-2x)$

44. $\dfrac{3r+s}{4(r-2s)}$

45. $\dfrac{5x+9}{x^2(x+3)}$

46. $\dfrac{8-3a}{2a(a-2)}$

47. $\dfrac{(x-1)(x-3)}{(x-2)^2}$

48. $\dfrac{3x-6y-7}{(3x+y)(x-2y)}$

49. $\dfrac{-(2x+3)^2}{(2x+5)(x-3)(x+3)}$

50. $\dfrac{3x^2-2x}{2(x+4)(x-2)}$

51. $\dfrac{-2x^3+9x^2-43x+15}{x(x+5)(x-5)(2x-1)}$

52. $\dfrac{25x+32}{8(x+2)(x-2)}$

53. $\dfrac{2}{5}$

54. 1

55. $\dfrac{(x+1)(x-3)}{(x-2)(x+3)}$

56. $\dfrac{3s + 2}{s^2(s + 1)}$ **57.** 9   **58.** $\dfrac{34}{63}$

59. $\dfrac{a + 6}{4(b - a)}$ **60.** $\dfrac{5by}{4b - 5}$ **61.** no solution

62. no solution   **63.** $-\dfrac{8}{21}$ **64.** 6

65. $\dfrac{3r - h}{12r^3}$ **66.** $\dfrac{24dL^2s - ds^3}{24L^3}$ **67.** $\dfrac{Z_2 - Z_1}{Z_1 + Z_2}$

68. $Prt + P$ **69.** $\dfrac{k(h - U)}{hu}$ **70.** $\dfrac{2akM}{av^2 + kM}$

71. $150,000   **72.** 3.2 min

73. $37,500 and $125,000   **74.** 6.5 h

Chapter 9—Test

1. $\dfrac{4m}{5n}$ **2.** $\dfrac{b}{c}$

3. $\dfrac{3a}{a + 2}$ **4.** $\dfrac{20}{3bc}$

5. $\dfrac{1}{(x + 1)(2x + 3)}$ **6.** $\dfrac{a + x}{2}$

7. $\dfrac{-2x^3 - 5x - 3}{4x^2(x - 1)}$

8. $\dfrac{11x^3 + 36x^2 - 9x - 2}{(x + 1)(x + 2)(x + 5)(x - 1)}$

9. 2   **10.** 260 h

Exercises 10.1 (ODD)

1. $\dfrac{1}{t^5}$ **3.** $\dfrac{1}{x^4}$ **5.** x^3

7. R_1^3 **9.** $\dfrac{3}{c^2}$ **11.** $\dfrac{c}{3}$

13. 1   **15.** 1   **17.** 5

19. $9y^2$ **21.** 3^6 **23.** 6

25. $\dfrac{1}{ax}$ **27.** $\dfrac{2}{c^8}$ **29.** $\dfrac{y^2}{x^6}$

31. $\dfrac{x^2}{125}$ **33.** $\dfrac{s}{t^2}$ **35.** $\dfrac{x^4}{8y^4}$

37. $\dfrac{b^7}{9a}$ **39.** $\dfrac{4}{25a^2b^2}$ **41.** $\dfrac{y^5}{x^4}$

43. $\dfrac{a^5c^2}{18}$ **45.** $\dfrac{1}{R_1} + \dfrac{1}{R_2}$ **47.** $g \times cm^{-3}; m \times s^{-2}$

49. 362 Btu/h $\times$ ft^2

Exercises 10.2 (ODD)

1. $\sqrt{5}$ **3.** $\sqrt[4]{a}$ **5.** $\sqrt[5]{x^3}$

7. $\sqrt[3]{R^7}$ **9.** $a^{\frac{1}{3}}$ **11.** $x^{\frac{1}{2}}$

13. $x^{\frac{2}{3}}$ **15.** $b^{\frac{8}{5}}$ **17.** 3

19. 2   **21.** 8   **23.** 16

25. 27   **27.** 4   **29.** $\dfrac{1}{6}$

31. $\dfrac{1}{2}$ **33.** $\dfrac{1}{2}$ **35.** 18

37. 2   **39.** $a^{\frac{3}{2}}$ **41.** $a^{\frac{3}{4}}b$

43. $x^{\frac{29}{15}}$ **45.** $x^{\frac{5}{6}}$ **47.** $k = x^{5/2} + y^{5/2}$

49. $R = \dfrac{\sqrt{(1 + D_1^2)^3}}{D_2}$ **51.** 4.72×10^{22}

Exercises 10.3 (ODD)

1. 7   **3.** -12 **5.** 0.4

7. -0.2 **9.** 20   **11.** -40

13. 2   **15.** -2 **17.** -5

19. 0.5   **21.** 2   **23.** 3

25. 2   **27.** $2j$ **29.** $-20j$

31. $0.7j$ **33.** -25 **35.** -20

37. -12 **39.** 64   **41.** 6.00 ft

43. $\dfrac{9}{5}$ **45.** 0.1

Exercises 10.4 (ODD)

1. $\dfrac{\sqrt{2}}{2}$ **3.** $\dfrac{2\sqrt{5}}{5}$ **5.** $\dfrac{\sqrt{a}}{a}$

7. $\dfrac{\sqrt{ab}}{b}$ **9.** $\dfrac{\sqrt{15}}{5}$ **11.** $\dfrac{a\sqrt{3a}}{3}$

13. $2\sqrt{3}$ **15.** $2\sqrt{7}$ **17.** $3\sqrt{5}$

19. $5\sqrt{6}$ **21.** $7\sqrt{3}$ **23.** $9\sqrt{3}$

25. $c\sqrt{a}$ **27.** $ab\sqrt{a}$ **29.** $2ac\sqrt{bc}$

31. $4x^2z^2\sqrt{5yz}$ **33.** $\dfrac{b\sqrt{3a}}{6}$ **35.** $\dfrac{x\sqrt{10y}}{5a^4}$

37. $3\sqrt[3]{2}$ **39.** $2a\sqrt[3]{a}$ **41.** $2a^2\sqrt[4]{a}$

43. $3a^2\sqrt[4]{3a^3}$ **45.** $3a^2x^3\sqrt[4]{2a^2}$ **47.** $2rs^2t^2\sqrt[7]{2t^2}$

49. $\dfrac{2\sqrt{3.14A}}{3.14}$ **51.** 80 ft, 160 ft **53.** $V = \dfrac{k\sqrt[3]{PW^2}}{W}$

55. $d = \dfrac{k\sqrt[3]{16JC^2}}{C}$

Exercises 10.5 (ODD)

1. $2\sqrt{7}$ **3.** $4\sqrt{7} + \sqrt{5}$ **5.** $5\sqrt{3}$

7. $9\sqrt{10}$ **9.** 10 **11.** $5\sqrt{3}$

13. $9\sqrt{2}$ **15.** $\sqrt{7}$ **17.** $-\sqrt{2} - 4\sqrt{3}$

19. $4\sqrt{a}$ **21.** $(3 + 4a)\sqrt{2a}$

23. $7a\sqrt{2} - 2\sqrt{3a}$ **25.** $\sqrt{21} - 9\sqrt{2}$

27. 26 **29.** $a\sqrt{b} + 3a\sqrt{c}$ **31.** $1 + \sqrt{6}$

33. $-17 - 3\sqrt{15}$ **35.** $2a - \sqrt{ac} - 15c$

37. $7 + 4\sqrt{3}$ **39.** $\dfrac{\sqrt{6} + 2}{2}$ **41.** 7

43. $\dfrac{7 - \sqrt{21}}{4}$ **45.** $\dfrac{4 + 3\sqrt{2}}{2}$ **47.** $\dfrac{a - 2\sqrt{ab}}{a - 4b}$

49. $\dfrac{R_1\sqrt{R_2} - R_2\sqrt{R_1}}{R_1 - R_2}$ **51.** $\dfrac{13 - 2\sqrt{30}}{7}$

Exercises 10.6 (ODD)

1. 12 **3.** -4 **5.** -6

7. 4 **9.** -11 **11.** 3

13. 1 **15.** no solution

17. $L = \dfrac{1}{(2\pi f)^2 C}$ **19.** $A = \pi r^2$ area of a circle

21. $h = \dfrac{v_0^2 - v^2}{2g}$ **23.** $R = \dfrac{v^2}{usg}$

25. $h = \dfrac{\sqrt{S^2 - \pi^2 r^4}}{\pi r}$

Chapter 10—Review Exercises

1. $\dfrac{1}{10}$ **2.** $\dfrac{1}{16}$ **3.** 9

4. 4 **5.** $\dfrac{1}{6}$ **6.** 64

7. 13 **8.** -30 **9.** 5

10. 5 **11.** $\dfrac{1}{4}$ **12.** $\dfrac{2}{5}$

13. $-\dfrac{3}{11}$ **14.** $-\dfrac{12}{13}$ **15.** 10

16. 10 **17.** 343 **18.** 1331

19. 128 **20.** $\dfrac{8}{9}$ **21.** 625

22. $9j$ **23.** $12j$ **24.** $-0.8j$

25. $-0.1j$ **26.** $\dfrac{3b}{a^2}$ **27.** $\dfrac{2x}{y}$

28. $\dfrac{m^4}{n^2}$ **29.** $\dfrac{2rt^5}{s}$ **30.** $\dfrac{2x^5}{3y^3}$

31. $\dfrac{c^4}{4a^2b^2}$ **32.** $a^{7/12}$ **33.** $x^{13/15}$

34. $a^{7/6}$ **35.** $\dfrac{1}{b^{1/2}}$ **36.** $\dfrac{x^{1/2}}{y}$

37. $\dfrac{2y^{1/2}}{x}$ **38.** $s^{2/3}t^{7/3}$ **39.** $\dfrac{8c^{17/10}}{a}$

40. $2\sqrt{11}$ **41.** $3\sqrt{3}$ **42.** $6\sqrt{2}$

43. $3\sqrt{6}$ **44.** $8\sqrt{2}$ **45.** $2\sqrt{31}$

46. $2\sqrt[3]{5}$ **47.** $3\sqrt[3]{4}$ **48.** $\dfrac{\sqrt{11}}{11}$

49. $\dfrac{2\sqrt{7}}{7}$ **50.** $2a$ **51.** $2\sqrt{7a}$

52. $5b\sqrt{5c}$

53. $3b\sqrt{10b}$

54. $\dfrac{\sqrt{6a}}{a}$

55. $\dfrac{\sqrt{3}}{a}$

56. $\dfrac{2\sqrt{21a}}{3a}$

57. $\dfrac{200\sqrt{10ab}}{ab}$

58. $2a\sqrt[3]{2}$

59. $3y\sqrt[3]{3x^2y}$

60. $-\sqrt{7}$

61. $-6\sqrt{5}$

62. $5\sqrt{7}-2\sqrt{6}$

63. $(6 - 3a)\sqrt{2}$

64. $(4a - 7)\sqrt{5a}$

65. $-6\sqrt{3}$

66. $18\sqrt{2}$

67. $a\sqrt{b} - 3\sqrt{5ab}$

68. $b\sqrt{a} - 3a\sqrt{b}$

69. $27 - 5\sqrt{30}$

70. $2a - 3b - 5\sqrt{ab}$

71. $2ab - \sqrt{abc} - c$

72. $\dfrac{2 + \sqrt{10}}{3}$

73. $5\sqrt{2} - 7$

74. $\dfrac{9 - \sqrt{55}}{13}$

75. $\dfrac{r^2 + 4R^2}{2\pi R}$

76. $v = \dfrac{\sqrt{Ed}}{d}$

77. $v = \dfrac{\sqrt{2emV}}{m}$

78. $\dfrac{21}{8}$ in.

79. $\dfrac{\sqrt{3}}{3}$

80. 0.182 ft

81. 1963°C

82. 34

83. 2

84. -7

85. 4

86. 243

Chapter 10—Test

1. $-\sqrt{5}$

2. 4000

3. $3a^2b\sqrt{3b}$

4. $\dfrac{2b^{7/4}}{a^{3/2}}$

5. $2x - 6\sqrt{2xy} + 9y$

6. $\dfrac{3\sqrt{x} - 2x}{2x}$

7. $\dfrac{10x + 5\sqrt{5x} + 2}{5x - 4}$

8. $\dfrac{49}{11}$

9. a. $v = k\left(\dfrac{P}{W}\right)^{1/3}$

b. $v = \dfrac{k\sqrt[3]{PW^2}}{W}$

10. $\dfrac{\sqrt{2\omega rn}}{2n}$

Exercises 11.1 (ODD)

1. $x^2 - 7x - 4 = 0$ with $a = 1, b = -7, c = -4$

3. not a quadratic

5. $x^2 + 4x + 4 = 0$ with $a = 1, b = 4, c = 4$

7. $7x^2 - x = 0$ with $a = 7, b = -1, c = 0$

9. not a quadratic

11. $-6x^2 + 3x - 1 = 0$ with $a = -6, b = +3, c = -1$

13. 2, 3

15. 2

17. $-1, 2$

19. $\dfrac{1}{2}, 1$

21. $3, -4$

23. none of the values

25. $-5x^2 + 8x + 132 = 0$

27. $605 = 500(1 + r)^2$ becomes $500r^2 + 1000r - 105 = 0$

Exercises 11.2 (ODD)

1. $3, -3$

3. $-2, 1$

5. $-2, 5$

7. $-2, \dfrac{1}{3}$

9. $-2, \dfrac{1}{2}$

11. $-\dfrac{5}{2}, \dfrac{1}{3}$

13. $-\dfrac{1}{2}, \dfrac{2}{3}$

15. $\dfrac{1}{5}, 4$

17. $-1, \dfrac{9}{2}$

19. $-2, -2$

21. $0, 8$

23. $-\dfrac{7}{2}, \dfrac{1}{2}$

25. $\dfrac{7}{2}, \dfrac{7}{2}$

27. $0, 7$

29. $-\dfrac{1}{3}, 3$

31. $-2a, 2a$

33. 2, 17

35. 10 m by 25 m

37. 2.5 s

39. 1, 16

41. 9 s

Exercises 11.3 (ODD)

1. $\pm\sqrt{13} - 2$

3. no solution

5. $-2, -1$

7. $\pm\sqrt{13} - 3$

9. $-5, 3$

11. $\dfrac{-3 \pm \sqrt{21}}{3}$

13. $\pm\sqrt{\dfrac{7}{16}} + \dfrac{1}{4}$

15. $-3, \dfrac{1}{2}$

17. $\dfrac{\pm\sqrt{33} + 3}{6}$

19. $-\dfrac{1}{3}, -\dfrac{1}{3}$

21. 5°C, 15°C

Exercises 11.4 (ODD)

1. $3, -1$

3. $-3, -\frac{1}{2}$

5. $\dfrac{-5 \pm \sqrt{13}}{2}$

7. $2 \pm \sqrt{6}$

9. $\dfrac{1}{2}, \dfrac{3}{2}$

11. $\dfrac{4}{3}, -\dfrac{4}{3}$

13. $-\dfrac{1}{2}, \dfrac{7}{2}$

15. $-1 + 2j, -1 - 2j$

17. $\dfrac{1 \pm \sqrt{33}}{2}$

19. $\dfrac{-1 \pm \sqrt{33}}{4}$

21. $0, 7$

23. $\dfrac{3 \pm j\sqrt{55}}{4}$

25. $-\dfrac{2}{3}, 4$

27. $-\dfrac{1}{3a}, -\dfrac{3}{2a}$

29. $16, 17$

31. 12 cm by 17 cm

33. 0.382 atm

35. 64.9 m

37. 17

Exercises 11.5 (ODD)

1. opens up, minimum

3. opens down, maximum

5. opens up, minimum

7.

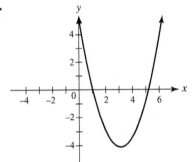

vertex $(3, -4)$
y-intercept $(0, 5)$

9.

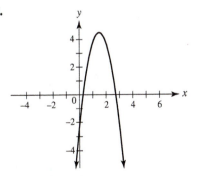

vertex $\left(\dfrac{5}{3}, \dfrac{13}{3}\right)$
y-intercept $(0, -4)$

11.

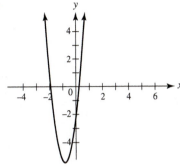

vertex $(-2, -13)$
y-intercept $(0, -5)$

13.

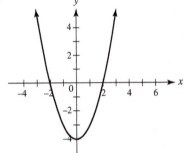

vertex $(0, -4)$
y-intercept $(0, -4)$
x-intercepts $(-2, 0)$ and $(2, 0)$

15.

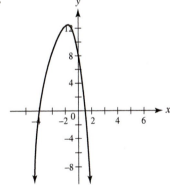

vertex $(-1.5, 12.5)$
y-intercept $(0, 8)$
x-intercepts $(-4, 0)$ and $(1, 0)$

17.

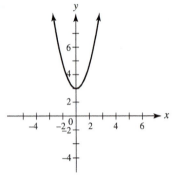

axis of symmetry $x = 0$
y-intercept $(0, 3)$

19.

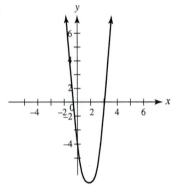

axis of symmetry $x = \dfrac{7}{4}$

y-intercept $(0, -4)$

21.

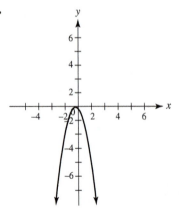

axis of symmetry $x = -\dfrac{1}{6}$

y-intercept $(0, 0)$

23.

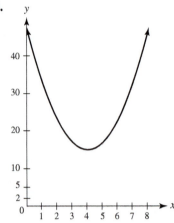

x window $[0, 8]$
y window $[0, 50]$

25. 64 ft

Chapter 11—Review Exercises

1. $-3, -4$ **2.** $8, -1$ **3.** $6, -\dfrac{1}{2}$

4. $-\dfrac{1}{2}, \dfrac{3}{2}$ **5.** $-\dfrac{1}{6}, 6$ **6.** $0, \dfrac{9}{8}$

7. $\dfrac{3}{4}, \dfrac{3}{4}$ **8.** $\dfrac{7}{2}, \dfrac{7}{2}$ **9.** $0, -\dfrac{7}{5}$

10. $\dfrac{1}{9}, 6$ **11.** $-11, 10$ **12.** $-20, \dfrac{1}{2}$

13. $3, -8$ **14.** $7, -3$ **15.** $-3, -3$

16. $1, \dfrac{3}{2}$ **17.** $\dfrac{5}{6}, 1$ **18.** $-4, \dfrac{4}{7}$

19. $1, 1$ **20.** $0, -8$ **21.** $-2 \pm \sqrt{2}$

22. $\dfrac{-5 \pm \sqrt{33}}{4}$ **23.** $3 \pm \sqrt{15}$ **24.** $\dfrac{5 \pm \sqrt{85}}{10}$

25. $-6, 5$ **26.** $\dfrac{1 \pm \sqrt{33}}{4}$ **27.** $1 \pm \sqrt{6}$

28. $\dfrac{2 \pm \sqrt{7}}{2}$ **29.** $1, -9$ **30.** $\dfrac{-3 \pm \sqrt{29}}{2}$

31. $\dfrac{1 \pm \sqrt{11}}{2}$ **32.** $2, 2$ **33.** $1 \pm 2\sqrt{2}$

34. $-1, \dfrac{5}{2}$ **35.** $\dfrac{1 \pm \sqrt{41}}{5}$ **36.** $\dfrac{2}{3}, -2$

37.

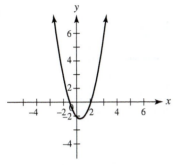

vertex $\left(\dfrac{1}{4}, -\dfrac{9}{8}\right)$

y-intercept $(0, -1)$

38.

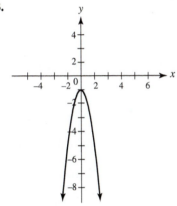

vertex $(0, -1)$

y-intercept $(0, -1)$

39.

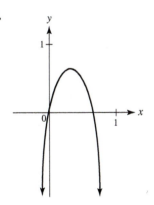

vertex $\left(\dfrac{1}{6}, \dfrac{1}{12}\right)$

y-intercept $(0, 0)$

40.

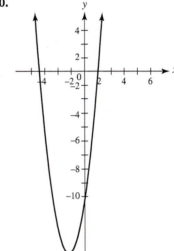

vertex $(-2, -18)$

y-intercept $(0, -10)$

41. $-1.69, 1.19$ **42.** no real roots

43. no real roots   **44.** $-0.43, 1.23$

45. 27   **46.** 20, 21   **47.** 4 in., 5 in.

48. 7 m   **49.** 1.18   **50.** 6 s

51. 5, 20   **52.** 7 mm   **53.** 33 psi

54. 50 items for a maximum profit of \$4950

Chapter 11—Test

1. $-2 \pm \sqrt{2}$ **2.** $\dfrac{1}{2}, 4$ **3.** $\dfrac{2 \pm \sqrt{7}}{2}$

4. $\dfrac{6}{5}$ **5.** $3 \pm 3\sqrt{2}$

6.

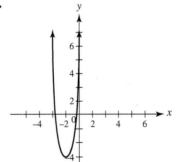

opens up

vertex $(-2, -3)$

y-intercept $(0, 5)$

7.

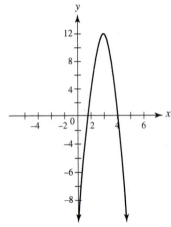

$x = 0.83$ and 4.84

8. at $t = 2$ **9.** 20 ft by 26 ft

23.

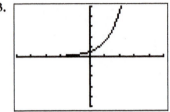

25.

27. $303.88   **29.** 4.92 mg   **31.** 632 million

Exercise 12.1 (ODD)

1. exponential function

3. exponential function

5. not an exponential function

7. not an exponential function

9. 3   **11.** π

13. increasing   **15.** increasing

17. $\dfrac{1}{16}$ **19.** 18.6

21.

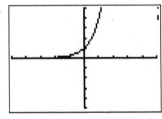

Exercise 12.2 (ODD)

1. $\log 100 = 2$ **3.** $\log 0.01 = -2$

5. $\log 2884 = 3.4600$ **7.** $\log 0.0003594 = -3.4444$

9. $\log_2 1024 = 10$ **11.** $\log_6 216 = 3$

13. $10^1 = 10$ **15.** $10^3 = 1000$

17. $10^{-2} = 0.01$ **19.** $10^{2.7536} = 567$

21. $2^3 = 8$ **23.** $3^5 = 243$

25. 4   **27.** -3

29. 6   **31.** -3

33. 4   **35.** 3

37. 0.8609   **39.** 3.4150

41. 2.3181   **43.** -2.4461

45. 2   **47.** $-\dfrac{1}{2}$

49. 343   **51.** 256

53. 9   **55.** $\dfrac{1}{64}$

57.

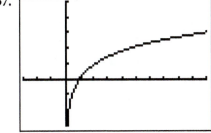

59.

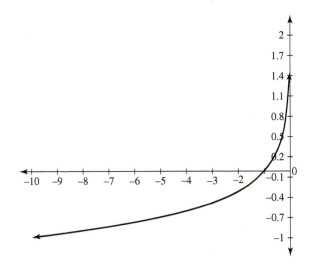

61. 8.4   **63.** 20 dB

Exercise 12.3 (ODD)

1. $\log a + \log b$ **3.** $\log x + \log y + \log z$

5. $\log x - \log 3$ **7.** $5 \log a$

9. $(\log 2 + \log a + \log c) - (\log 3 + 2 \log b)$

11. $\frac{1}{2} \log x - 2 \log a$ **13.** $\log (ac)$

15. $\log \left(\frac{9}{3} \right)$ **17.** $\log x^2 \sqrt{x}$

19. $\log(2^2 n^3)$ **21.** $\log 1 - \log 32 = -1.51$

23. $2.4 \log(3) = 1.15$ **25.** $\frac{1}{4} \log 27 = 0.36$

27. 1.88   **29.** 6.4   **31.** 20

33. 17 dB   **35.** 6560

37. 4.56×10^{192} **39.** 3.27×10^{150}

41. 1.34×10^{154} **43.** 3.76×10^{414}

Exercise 12.4 (ODD)

1. 2.1282   **3.** -0.1054 **5.** 5.2983

7. -6.9078 **9.** 9.90   **11.** 304.2

13. 28.1   **15.** 5.46×10^{18} **17.** 2.7080

19. 4.8282   **21.** 4.46   **23.** 9.9%

25. 8.5%   **27.** 2.2%   **29.** 4.987

Exercise 12.5 (ODD)

1. 4   **3.** -0.748 **5.** -2.864

7. 1.43   **9.** -0.757 **11.** 1.95

13. 2.71   **15.** $\frac{1}{8}$ **17.** 1.6

19. 250   **21.** 64.5   **23.** 1.42

25. 1.67 min   **27.** 104.8 mi   **29.** 31.2%

Exercise 12.6 (ODD)

1.

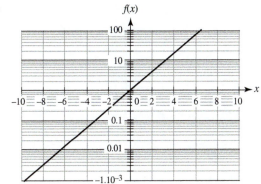

3.

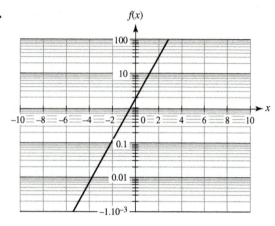

5.

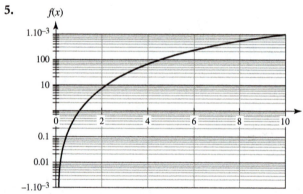

7.

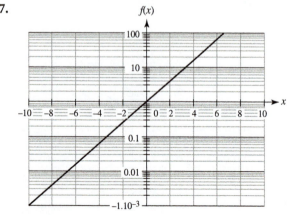

9.

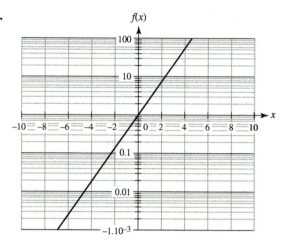

11.

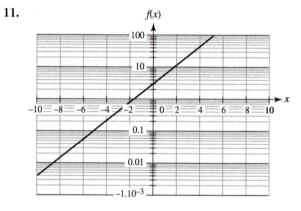

13.

15.

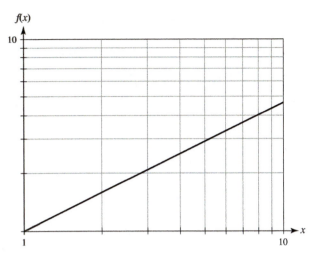

17.

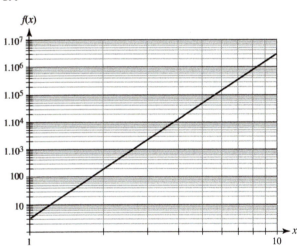

19.

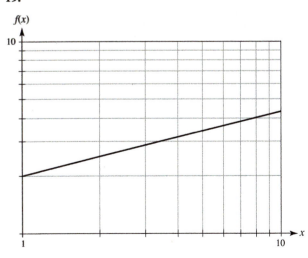

21.

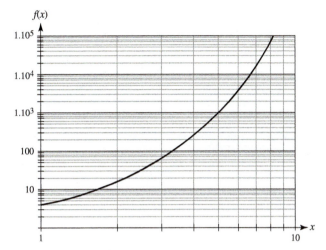

23.

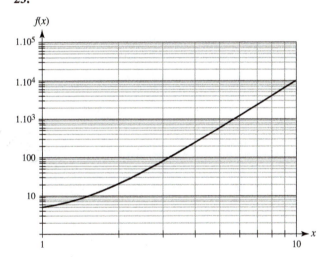

25.

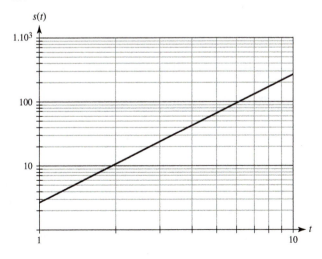

27. a.

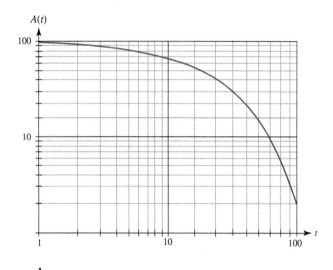

b.

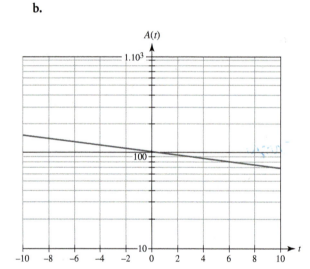

Chapter 12 Review Exercises

1. $\log 10{,}000 = 4$

2. $\log 0.001 = -3$

3. $\log_6 1296 = 4$

4. $\log_5 0.008 = -3$

5. $\log 10 = 1$

6. $\log 17.0 = 1.23$

7. $\log 0.00288 = -2.54$

8. $\log 2.5 = 0.4$

9. $2^7 = 128$

10. $3^4 = 81$

11. $10^5 = 100{,}000$

12. $10^{-1} = 0.1$

13. $e^{3.0} = 20$

14. $e^{-1.4} = 0.25$

15. $4^5 = 1024$

16. $5^{-2} = 0.04$

17. $\log 2.07 + \log 3.45$

18. $\log 8.75 + \log 0.205$

19. $\log x + \log y$

20. $\log 3 + \log z$

21. $\log 2 + \log a + \log b + \log c$

22. $\log 7.14 - \log 9310$

23. $\log 89.15 - \log 9.176$

24. $(\log x + \log z) - \log 3$

25. $(\log a + \log b) - (\log c + \log d)$

26. $3 \log 6.184$

27. $0.3 \log 1.034$

28. $\log 2 + 4 \log x$

29. $5 \log a + 2 \log b$

30. $\dfrac{1}{3} \log 1.17$

31. $\dfrac{1}{3} \log 0.9006$

32. $\log 6.12 + \dfrac{1}{2} \log 128$

33. $\dfrac{1}{2} \log 86000 - \log 45.8$

34. $\dfrac{1}{5}(\log x + \log y)$

35. $\dfrac{1}{2}(\log 2 + \log x) - 2 \log a$

36. $\dfrac{1}{3}(\log 5 + \log y) - \log 7$

37. 1.4110

38. 2.4849

39. -0.51

40. -2.3

41. 1.7917

42. 2.0793

43. 0.4055

44. 3.4655

45. $\log(4c)$

46. $\log \dfrac{3x}{7}$

47. $\log \sqrt{\dfrac{7}{x}}$

48. $\log(7x)(5y)$

49. $-\log y^2 \sqrt{2x}$

50. $\log x^3 y$

51. $\log(2x)^2(y^3)$

52. $\log \dfrac{\sqrt{a}}{5^2}$

53. $\log(x^2 y)^2$

54. $\dfrac{\log x}{\log 4y}$

55. $\log \left(\dfrac{y}{x^3}\right)^2$

56. $\dfrac{\log x^2}{\log 7y}$

57. 0.8

58. 1.25

59. 4.3

60. 8.75

61. 2

62. -0.2

63. 48

64. 1

65.

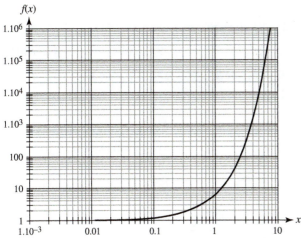

66.

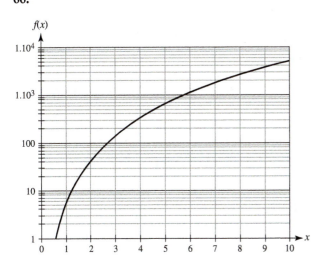

67.

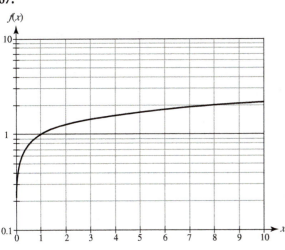

68.

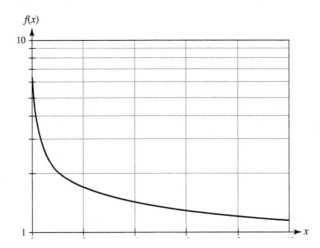

69. 4.6   **70.** 160,000   **71.** 140 dB

72. 1370   **73.** 0.000392   **74.** 0.0000001

75. 15.8°C   **76.** 10.8 km/s   **77.** 6.16 atm

78. 20008 s

Chapter 12 Test

1. −0.01   **2.** 2.05   **3.** 1.43

4. 14.15   **5.** −0.16   **6.** 0.9

7.

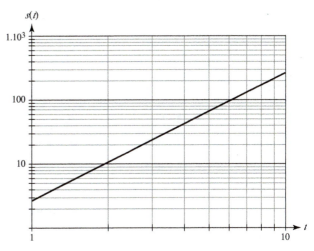

8. $(\log 4 + 5 \log a) - \log 9$ **9.** $8\frac{2}{3}$ yr

Exercise 13.1(ODD)

1. 53°   3. 21°   5. ∠BEC or ∠CED

7. ∠AEB and ∠BEC
   ∠BEC and ∠CED
   ∠BED and ∠AEB
   ∠AEC and ∠CED

9. ∠CBE and ∠EBA   11. ∠CBD   13. 25°

15. 115°   17. 40°   19. 62°

21. ∠1 and ∠5
    ∠3 and ∠4

23. ∠1 and ∠3 OR ∠4 and ∠5   25. 50°

27. 130°   29. 58°   31. 148°

33. 40°   35. 50°   37. 40°

39. 100°

Exercise 13.2 (ODD)

1. 56°   3. 48°   5. 68°

7. 80°   9. 60°   11. 120°

13. 32°   15. 76°   17. AE, GD

19. GC, BC   21. ∠BOG   23. BG, GC

25. 60°   27. 30°   29. 110°

31. 35°   33. 3   35. 5

37. 40°   39. 135° each   41. 117°

Exercise 13.3 (ODD)

1. 5   3. 17   5. 8

7. 4.9   9. 10.6   11. 28.3

13. 59.9   15. 40.9   17. 2.67

19. 39.1   21. 5.66 cm   23. 12.0 m

25. 7.9 m   27. 521 ft   29. 36.8 ft

31. 18.9 Ω   33. 894 lb   35. 8.42 m

37. 162 ft   39. 56.1 cm

Exercise 13.4 (ODD)

1. 20°, 100°, 60°   3. 16.5, 19.5   5. 40°, 65°, 75°

7. 12.7, 70°, 70°   9. similar   11. neither

13. ∠E, side EF   15. ∠U, side ST   17. 10

19. 12   21. 21   23. 20

25. 16   27. 8

29.   31. 3.3 cm, 4.0 cm

33. ∠XKY = ∠NKF, ∠KXY = ∠KNF, ∠XYK = ∠NFK

35. 8.0 in.   37. 6.7 ft

39. 350 cm   41. $2\frac{1}{4}$ in.

43. 1530 km, 1800 km, 760 km   45. 5.3 ft

Exercise 13.5 (ODD)

1. $\dfrac{9}{41}, \dfrac{9}{40}, \dfrac{9}{41}$ 3. $\dfrac{9}{40}, \dfrac{41}{40}, \dfrac{40}{9}$

5. $\dfrac{1}{2}, 2, \sqrt{3}$ 7. $\dfrac{\sqrt{3}}{3}, \dfrac{1}{2}, \dfrac{2\sqrt{3}}{3}$

9. 0.624, 1.25, 1.60   11. 0.782, 0.782, 0.798

13. $\dfrac{4}{5}, \dfrac{3}{4}$ 15. $\dfrac{7}{24}, \dfrac{24}{25}$ 17. $\dfrac{\sqrt{2}}{2}, 1$

19. 1.4, 0.73   21. 0.63, 1.3   23. 0.679, 1.47

25. $\dfrac{\sqrt{2}}{2}$ 27. 1.4   29. 0.491

31. 1.04   33. 0.5000   35. 1.41

37. cot B, sec B

39. $\sin A = \dfrac{a}{c}, \cos A = \dfrac{b}{c}, \tan A = \dfrac{a}{b}$

    $\csc A = \dfrac{c}{a}, \sec A = \dfrac{c}{b}, \cot A = \dfrac{b}{a}$

41. 1

43. $\dfrac{5}{13}, \dfrac{12}{13}, \dfrac{5}{12}$

$\dfrac{5}{13} \div \dfrac{12}{13} = \dfrac{5}{12}$

Exercise 13.6 (ODD)

1. 0.8480   **3.** 0.4557   **5.** 1.500

7. 0.9178   **9.** 0.2382   **11.** 0.7046

13. 4.222   **15.** 2.475   **17.** 32°

19. 18.9°   **21.** 60.5°   **23.** 61.9°

25. 52.7°   **27.** 18.0°   **29.** 84.3°

31. 58.2°   **33.** 1.206   **35.** 1.179

37. 53.1°   **39.** 40.9°   **41.** 34.6°

43. 79.4°   **45.** 13 V   **47.** 126.5 m

49. 1.56   **51.** 23.9 m

Exercise 13.7 (ODD)

1. $\angle B = 60°, b = 20.8, c = 24$

3. $\angle A = 33.7°, a = 12.5, b = 18.7$

5. $\angle B = 13.2°, b = 3.06, a = 13.0$

7. $\angle A = 28.8°, \angle B = 61.2°, b = 1.18$

9. $\angle A = 82.6°, \angle B = 7.4°, a = 44.6$

11. $\angle A = 64.2°, \angle B = 25.8°, b = 4.71$

13. $\angle B = 83°, a = 1.88, c = 15.4$

15. $\angle A = 33.5°, \angle B = 56.5°, c = 118$

17. 29.1 ft   **19.** 304 cm   **21.** 318 cm

23. 85.2°   **25.** 4000 ft   **27.** 6.8°

29. 1640 ft   **31.** 243 km   **33.** 4707 ft

35. 22.9°

Chapter 13 Review Exercises

1. 61°   **2.** 151°   **3.** 90°

4. 60°   **5.** 27°   **6.** $\angle C = \angle D = 41°$

7. $\angle CBE$ **8.** $\angle ABE, \angle EBD$ **9.** $\angle EBD$

10. $\angle EBC$ **11.** $\angle 2$ and $\angle 5$ **12.** $\angle 1$ and $\angle 3$

13. 180°   **14.** 180°   **15.** 65°

16. 155°   **17.** 32°   **18.** 148°

19. 52°   **20.** 128°   **21.** 132°

22. 132°   **23.** 36°   **24.** 62°

25. 50°   **26.** 80°   **27.** 120°

28. 160°   **29.** 40°   **30.** 80°

31. 25°   **32.** 65°   **33.** 65°

34. 90°   **35.** 41   **36.** 50

37. 42   **38.** 33   **39.** 7.36

40. 281   **41.** 21.1   **42.** 0.414

43. 7.5   **44.** 2.4   **45.** 0.735

46. 0.482   **47.** 0.130   **48.** 0.794

49. 0.116   **50.** 2.01   **51.** 1.16

52. 0.251   **53.** 30.0°   **54.** 75.0°

55. 45.0°   **56.** 89.0°   **57.** 3.0°

58. 55.5°   **59.** 56.7°   **60.** 42.8°

61. $\dfrac{23}{41}, \dfrac{23}{41}$ **62.** $\dfrac{34}{23}, \dfrac{34}{41}$ **63.** $\dfrac{41}{23}, \dfrac{23}{34}$

64. $\dfrac{34}{23}, \dfrac{41}{34}$ **65.** 0.385, 0.417   **66.** 1.67, 1.33

67. 3.78, 0.967   **68.** 0.845, 1.18   **69.** 0.461, 0.418

70. 1.09, 2.51   **71.** 0.898, 2.04   **72.** 0.736, 1.09

73. 38.0°   **74.** 28.4°   **75.** 13.1°

76. 71.0°   **77.** 28.4°   **78.** 23.2°

79. 53.4°   **80.** 70.0°

81. $\angle B = 69.0°, a = 2.48, b = 6.47$

82. $\angle A = 72.0°, c = 0.379, b = 0.117$

83. $\angle A = 57.3°, a = 71.5, c = 85$

84. $\angle A = 82.6°, a = 1870, b = 243$

85. $\angle A = 59.1°, \angle B = 30.9°, c = 10.1$

86. $\angle A = 32.3°, \angle B = 57.7°, c = 0.0142$

87. $\angle A = 63.9°, \angle B = 26.1°, b = 47.5$

88. $\angle A = 71.2°, \angle B = 18.8°, a = 51.1$

89. $\angle B = 52.75°, a = 8397, b = 11,040$

90. $\angle A = 25.667°, a = 2540, c = 5860$

91. $\angle A = 88.9125°, \angle B = 1.08748°, a = 112.483$

92. $\angle A = 89.98°, \angle B = 0.02°, a = 112$

93. 21.1 ft   **94.** 510 m

95. 56.1 m   **96.** 9.9 ft, 14.9 ft, 17.9 ft

97. 22.5°, 67.5°   **98.** 30°, 60°, 90°

99. 5.0 ft   **100.** 20 ft   **101.** 5.03 cm

102. 1600 cm   **103.** 1.50 m   **104.** 0.614 μm

105. 55.1 mA   **106.** 30.6°   **107.** 1210 ft-lb

108. 35.9°   **109.** 77.5 ft   **110.** 940 ft

111. 26.6°, 63.4°   **112.** 139 m   **113.** 28.1°

114. 4.9°   **115.** 6430 ft   **116.** 575 ft

117. 16.1°   **118.** 2290 m   **119.** 4051 mi

120. 20.8 in.   **121.** no, yes   **122.** yes, no

Chapter 13 Test

1. $\angle EBD$ and $\angle DBC$ **2.** $\angle ABD$

3. $\angle EBD = 25°$ **4.** $\angle ABD = 115°$

5. $\angle EBD$ **6.** 62°   **7.** 118°

8. 28°   **9.** 152°   **10.** 8

11. a. 0.9092   **b.** 49.03°

12. $\dfrac{2}{\sqrt{5}}$ **13.** 0.7728

14. $\angle B = 53°, a = 39.8\, c = 66.1$

15. 67 ft

Exercise 14.1 (ODD)

1.

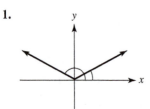

3.

(graph)

5. $\dfrac{12}{13}, \dfrac{12}{5}$ **7.** $-\dfrac{\sqrt{13}}{3}, \dfrac{2}{\sqrt{13}}$

9. $1, -\dfrac{1}{\sqrt{2}}$ **11.** $\dfrac{3}{5}, -\dfrac{4}{5}$

13. 0.316, 0.333   **15.** $-0.413, 2.62$

17. $-0.772, -1.57$ **19.** $-1.85, 0.842$

21. positive, negative, negative

23. negative, negative, negative

25. positive, positive, positive

27. negative, negative, negative

29. IV   **31.** I   **33.** II

35. II   **37.** II   **39.** II

41. II   **43.** lies on positive y axis

45. $\sin 15°, -\cos 50°$ **47.** $-\cos 27°, -\csc 70°$

49. $-\tan 18°, \sec 10°$ **51.** $-\cot 70°, \tan 60°$

53. -0.5446 **55.** -10.78

57. -1.090 **59.** -0.9265

61. 3.732   **63.** -1.718 **65.** 0.9759

67. 2.778   **69.** -0.5358 **71.** 2.414

73. 238.0°, 302.0°   **75.** 66.4°, 293.6°

77. 62.2°, 242.2°   **79.** 15.8°, 195.8°

81. 232.0°, 308.0°   **83.** 219.3°, 320.7°

85. 167.8°, 192.2°   **87.** 178.0°, 182.0°

89. 334°   **91.** 129.8°   **93.** 119.6°

95. 189.2°   **97.** 85.2°   **99.** 354 lb

101. 334   **103.** 2.4°

Exercise 14.2 (ODD)

1. $\angle C = 72.6°, b = 4.52, c = 7.23$

3. $\angle C = 109.0°, a = 1390, b = 1300$

5. $\angle A = 149.7°, \angle C = 9.6°, a = 221$

7. $\angle B = 8.5°, \angle C = 28.3°, c = 0.735$

9. $\angle A = 99.4°, b = 55.1, c = 24.4$

11. $\angle A = 68.1°, a = 552, c = 537$

13. $\angle A_1 = 61.5°, \angle C_1 = 70.4°, c_1 = 28.1$
$\angle A_2 = 118.5°, \angle C_2 = 13.4°, c_2 = 6.89$

15. $\angle A_1 = 107.3°, \angle C_1 = 41.3°, a_1 = 1060$
$\angle A_2 = 9.9°, \angle C_2 = 138.7°, a_2 = 191$

17. $\angle B = 68.5°, \angle C = 42.4°, b = 93.8$

19. no solution

21. $\angle B = 60.0°, \angle C = 90.0°, b = 173$

23. no solution   **25.** 15.6 in., 27.2 in.

27. 21,000 m   **29.** 19.7 km

31. 29,000 km

Exercise 14.3 (ODD)

1. $\angle A = 55.3°, \angle B = 37.2°, c = 27.1$

3. $\angle A = 9.9°, \angle C = 111.2°, b = 38,600$

5. $\angle A = 70.9°, \angle B = 11.1°, c = 1580$

7. $\angle A = 18.2°, \angle B = 22.2°, \angle C = 139.6°$

9. $\angle A = 42.3°, \angle B = 30.3°, \angle C = 107.4°$

11. $\angle A = 51.5°, \angle B = 35.1°, \angle C = 93.4°$

13. $\angle A = 46.1°, \angle B = 109.2°, c = 138$

15. $\angle A = 132.4°, \angle C = 10.3°, b = 4.20$

17. $\angle A = 39.9°, \angle B = 56.8°, \angle C = 83.4°$

19. $\angle A = 48.6°, \angle B = 102.3°, \angle C = 29.1°$

21. $\angle A = 44.37°, \angle B = 60.51°, \angle C = 75.12°$

23. $\angle B = 4.05°, \angle C = 166.30°, a = 24.25$

25. 1290 m   **27.** 96.2 cm

29. 19.8°   **31.** 30.0°, 60.0°, 90.0°

Exercise 14.4 (ODD)

1. a. scalar because only the magnitude is given
    b. vector because it has both magnitude and direction

3. a. vector because it has both magnitude and direction
    b. scalar because only the magnitude is given

5. a. vector because it has both magnitude and direction
    b. scalar because only the magnitude is given

7. a. scalar because only the magnitude is given
    b. vector because it has both magnitude and direction

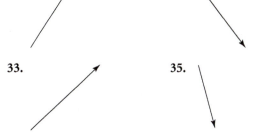

37. 335 ft   **39.** 1300 lb

41. 3200 lb   **43.** 18 m/s

Exercise 14.5 (ODD)

5. $A_x = 25.4, A_y = 12.9$

7. $A_x = -8.60, A_y = 57.6$

9. $A_x = 3340, A_y = -5590$

11. $A_x = -701, A_y = -229$

13. $A_x = -1344, A_y = -3212$

15. $A_x = -0.1306, A_y = 0.8636$

17. $A_x = 18.2, A_y = 18.2$

19. $A_x = 3.45, A_y = 5.32$

21. $A_x = -26.7, A_y = -217$

23. $A_x = -969, A_y = -969$

25. $A_x = 52.4, A_y = -69.7$

27. $A_x = -0.03233, A_y = -0.04370$

29. $A_x = 25815, A_y = -1714.6$

31. $A_x = -13.378, A_y = 31.981$

33. 33.3 mi east, 14.8 mi north

35. 0.013 along the x axis and 0.020 along the y axis

37. 3.72 A, 39.1°

Exercise 14.6 (ODD)

1. $R = 5.39, \theta = 21.8°$ **3.** $R = 1460, \theta = 59.2°$

5. $R = 38.3, \theta = 25.2°$ **7.** $R = 25.49, \theta = 61.8°$

9. $R = 10.7, \theta = 23.7°$ **11.** $R = 276, \theta = 55.3°$

13. $R = 115, \theta = 102.0°$ **15.** $R = 121, \theta = 272.1°$

17. $R = 10, \theta = 36.9°$ **19.** $R = 61.1, \theta = 116.8°$

21. $R = 1560, \theta = 201.6°$ **23.** $R = 7675.1, \theta = 175.07°$

25. $R = 13, \theta = 293°$ **27.** $R = 28.9, \theta = 327.5°$

29. $R = 4352, \theta = 321.0°$ **31.** $R = 0.321, \theta = 193.7°$

33. 264 lb, 36.7° angle   **35.** 46.0 mi, 34.3° south of east

37. 3990 km/h, 5.0°   **39.** 53.9 lb, $\theta = 68.2°$

41. 130 V, 6.9° from V_1 **43.** yes, to the right

Chapter 14 Review Exercises

1. $\sin \theta = \dfrac{3}{5}, \cos \theta = \dfrac{4}{5}, \tan \theta = \dfrac{3}{4},$
$\csc \theta = \dfrac{5}{3}, \sec \theta = \dfrac{5}{4}, \cot \theta = \dfrac{4}{3}$

2. $\sin \theta = \dfrac{12}{13}, \cos \theta = -\dfrac{5}{13}, \tan \theta = -\dfrac{12}{5},$
$\csc \theta = \dfrac{13}{12}, \sec \theta = -\dfrac{13}{5}, \cot \theta = -\dfrac{5}{12}$

3. $\sin \theta = -0.275, \cos \theta = 0.962, \tan \theta = -0.286,$
$\csc \theta = -3.64, \sec \theta = 1.04, \cot \theta = -3.5$

4. $\sin \theta = -0.832, \cos \theta = -0.555, \tan \theta = 1.5,$
$\csc \theta = -1.2, \sec \theta = -1.8, \cot \theta = 0.667$

5. II   **6.** III

7. IV   **8.** IV

9. $-\cos 48°, \tan 14°$ **10.** $-\sin 63°, -\cot 42°$

11. $-\sin 71°, \sec 15°$ **12.** $-\cos 77°, -\csc 80°$

13. -0.4540 **14.** 0.6561

15. -0.3057 **16.** 0.1405

17. -1.082 **18.** -9.113

19. -0.5764 **20.** 0.9219

21. -2.552 **22.** -0.3502

23. 4.230   **24.** -0.9914

25. 0.6820   **26.** -1.600

27. 1.003   **28.** -1.001

29. 37.0°, 217.0°   **30.** 212.0°, 328.0°

31. 114.9°, 245.1°   **32.** 151.3°, 331.3°

33. 27.4°, 152.6°   **34.** 45.6°, 314.4°

35. 189.4°, 350.6°   **36.** 74.1°, 285.9°

37. 155.0°, 335.0°   **38.** 8.3°, 188.3°

39. 56.3°, 123.7°   **40.** 112.4°, 236.6°

41. $\angle C = 71.7°, b = 120, c = 130$

42. $\angle B = 58.5°, b = 40.0, c = 45.0$

43. $\angle A = 21.2°, b = 128, c = 43.1$

44. $\angle C = 20.0°, a = 136, b = 191$

45. $\angle A = 34.8°, \angle B = 53.5°, c = 5.60$

46. $\angle B = 56.0°, \angle B = 39.2°, a = 4590$

47. $\angle A = 59.8°, \angle C = 58.2°, b = 289$

48. $\angle B = 115.8°, \angle C = 6.6°, a = 856$

49. $\angle A_1 = 60.6°, \angle C_1 = 65.1°, a_1 = 17.5,$
$\angle A_2 = 10.8°, \angle C_2 = 114.9°, a_2 = 3.75$

50. $\angle B = 25.9°, \angle C = 5.2°, c = 158$

51. $\angle B_1 = 68.5°, \angle C_1 = 60.5°, c_1 = 73.4,$
$\angle B_2 = 111.5°, \angle C_2 = 17.5°, c_2 = 25.3$

52. $\angle B_1 = 33.0°, \angle C_1 = 127.8°, c_1 = 2020,$
$\angle B_2 = 147.0°, \angle C_2 = 13.8°, c_2 = 608$

53. 62.4°, 83.3°, 34.3°   **54.** 41.7°, 39.0°, 99.3°

55. 10.5°, 36.4°, 133.1°   **56.** 32.0°, 121.9°, 26.1°

57. 41.1°, 32.8°, 431   **58.** 40.8°, 50.7°, 0.558

59. a. scalar because only the magnitude is given
b. vector because both the magnitude and direction are given

60. a. vector because both the magnitude and direction are given
b. scalar because only the magnitude is given

61. a. scalar because only the magnitude is given
b. vector because both the magnitude and direction are given

62. a. vector because both the magnitude and direction are given
b. scalar because only the magnitude is given

63.   **64.**   **65.**
66.   **67.**   **68.**
69.   **70.**
71.   **72.**
73.   **74.**

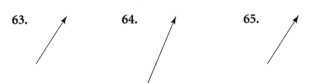
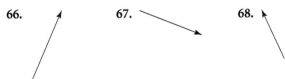

75. $A_x = 6.77, A_y = 2.73$ **76.** $A_x = 15.1, A_y = 44.0$

77. $A_x = -1380, A_y = 778$

78. $A_x = 0.448, A_y = -0.788$

79. $A_x = 15.9, A_y = 6.50$ **80.** $A_x = -85.6, A_y = 93.8$

81. $A_x = 8.903, A_y = -13.87$

82. $A_x = -0.111, A_y = -0.890$

83. 35 mi/h west, 9.3 mi/h north

84. 254 mi/h west, 92.3 mi/h south

85. 431 lb downward, 349 lb horizontally

86. 2.4 lb horizontally, 27 lb vertically

87. $R = 38.8, \theta = 27.2°$ **88.** $R = 1060, \theta = 62.4°$

89. $R = 5530, \theta = 81.7°$ **90.** $R = 1588.6, \theta = 41.2°$

91. $R = 27.66, \theta = 40.0°$ **92.** $R = 753, \theta = 23.4°$

93. $R = 1.10, \theta = 48.0°$ **94.** $R = 52.1, \theta = 125.5°$

95. $R = 47, \theta = 314°$ **96.** $R = 7.18, \theta = 74.8°$

97. $R = 64.8, \theta = 68.7°$ **98.** $R = 656, \theta = 269.6°$

99. 65.4°, 45.5°, 69.1°   **100.** 24.0°, 123.5°, 32.5°

101. -115 V   **102.** -15.0 cm

103. 585 m   **104.** 83.8°

105. 2.2°   **106.** 83.3°, 52.6°, 44.0°

107. 281 m   **108.** 14.8 m   **109.** 2787 ft

110. 201 km   **111.** 113 lb

112. $A_x = -4510$ lb, $A_y = -32,100$ lb

113. 348 km/h horizontally, 2220 km/h vertically

114. 367 ft/s horizontally, 711 ft/s vertically

115. 232 lb

116. 353 mi/h horizontally, 140 mi/h vertically

117. $R = 1470$ lb, $\theta = 54.7°$ below the horizontal force

118. $R = 13,500$ lb, $\theta = 45.8$ above the horizontal

119. $R = 88.3$ lb, $\theta = 8.93°$ **120.** $R = 1180$ lb, $\theta = 24.7°$

121. 32 lb   **122.** 374 lb

123. $R = 356$ N, $\theta = 29.6°$ **124.** 61.0 mi, 35.0° north of east

Chapter 14 Test

1. $\sin \theta = \dfrac{6}{10}, \cos \theta = \dfrac{8}{10}, \tan \theta = \dfrac{6}{8},$
$\csc \theta = \dfrac{10}{6}, \sec \theta = \dfrac{10}{8}, \cot \theta = \dfrac{8}{6}$

2. II

3. $\angle C = 64°, b = 181, c = 175$

4. $\angle A = 67.0°, \angle B = 50.9°, \angle C = 62.1°$

5. $A_x = 57, A_y = 31$ **6.** $R = 320.6, \theta = 113.4°$

7. 27,300 km

8. distance from A to C is 30.8 m, distance from B to C is 85.6 m

9. 6.60 lb, 29.5°   **10.** 24 m, 36.5° north of east

43. 7.086   **45.** 0.85730   **47.** -1.145

49. -9.657 **51.** 3.796   **53.** 1.047, 5.236

55. 1.9116, 5.0532   **57.** 0.4500, 2.692   **59.** 4.470, 4.954

61. 162 V   **63.** 33.4 m   **65.** $1620°, 9\pi$ rad

67. 146°   **69.** 13.8°   **71.** 1.3090

73. 0.1745   **75.** -0.315209 **77.** 0.000421

Exercise 15.1 (ODD)

1. $\dfrac{2\pi}{9}, \dfrac{4\pi}{45}$ **3.** $\dfrac{11\pi}{36}, \dfrac{11\pi}{6}$ **5.** $\dfrac{\pi}{6}, \dfrac{3\pi}{4}$

7. $\dfrac{35\pi}{36}, \dfrac{7\pi}{6}$ **9.** 120°, 36°   **11.** 15°, 225°

13. 168°, 140°   **15.** 35°, 135°   **17.** 0.802

19. 3.31   **21.** 4.86   **23.** 3.18

25. 46°   **27.** 143°   **29.** 186°

31. 710°   **33.** 0.5000   **35.** -1.732

37. 0.8674   **39.** 1.197   **41.** 1.556

Exercise 15.2 (ODD)

1. a. 7.34 cm   **b.** 13.4 cm^2

3. a. 355 mm   **b.** 73,000 mm^2

5. a. 0.620 ft   **b.** 0.735 ft^2

7. a. 26.9 in.   **b.** 88.0 in.^{2}

9. 70.6 cm/s   **11.** 6920 ft/min   **13.** 37.7 cm

15. 679 cm^2 **17.** 1.62   **19.** 87.8 ft^2

21. 0.114 m   **23.** 26.4 ft   **25.** 59 rad/s

27. 153,000 cm/min   **29.** 134 ft^2

31. 0.00000485   **33.** 80,400,000 mi

Exercise 15.3 (ODD)

1.

x	$-\pi$	$-\dfrac{3\pi}{4}$	$-\dfrac{\pi}{2}$	$-\dfrac{\pi}{4}$	0	$\dfrac{\pi}{4}$	$\dfrac{\pi}{2}$	$\dfrac{3\pi}{4}$	π	$\dfrac{5\pi}{4}$	$\dfrac{3\pi}{2}$	$\dfrac{7\pi}{4}$	2π	$\dfrac{9\pi}{4}$	$\dfrac{5\pi}{2}$	$\dfrac{11\pi}{4}$	3π
y	0	-0.7	-1	-0.7	0	0.7	1	0.7	0	-0.7	-1	-0.7	0	0.7	1	0.7	0

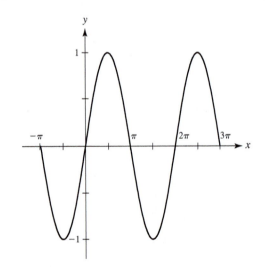

3.

x	$-\pi$	$-\dfrac{3\pi}{4}$	$-\dfrac{\pi}{2}$	$-\dfrac{\pi}{4}$	0	$\dfrac{\pi}{4}$	$\dfrac{\pi}{2}$	$\dfrac{3\pi}{4}$	π	$\dfrac{5\pi}{4}$	$\dfrac{3\pi}{2}$	$\dfrac{7\pi}{4}$	2π	$\dfrac{9\pi}{4}$	$\dfrac{5\pi}{2}$	$\dfrac{11\pi}{4}$	3π
y	0	-3.5	-5	-3.5	0	3.5	5	3.5	0	-3.5	-5	-3.5	0	3.5	5	3.5	0

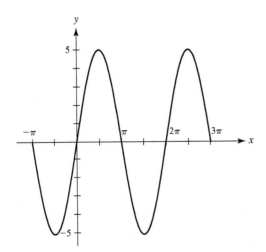

5. $y = 3\sin(x)$ **7.** $y = -6\sin(x)$ **13.**

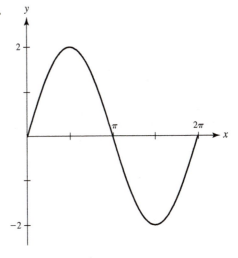

9.

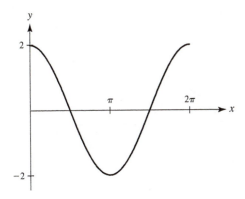

11.

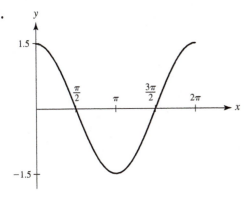

15.

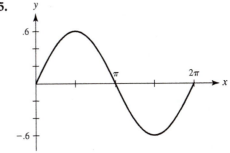

17.

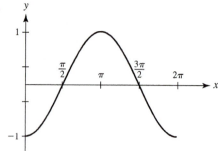

19.

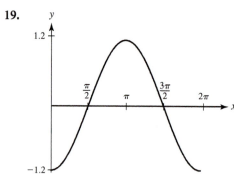

21.

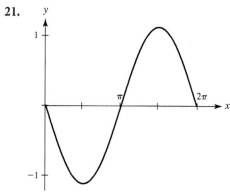

23.

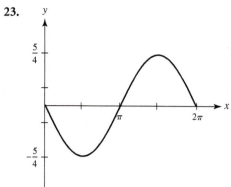

25.

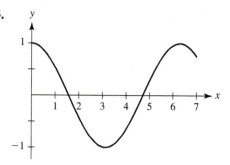

27.

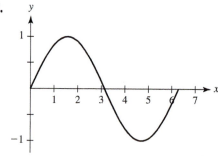

29.

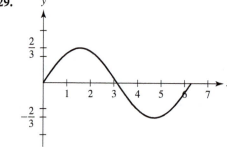

31.

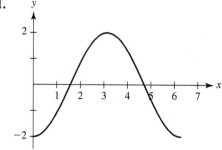

33.

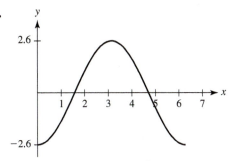

35.

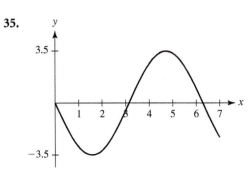

23.

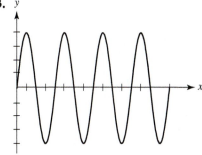

25.

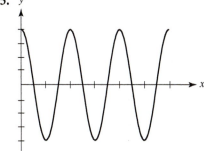

27.

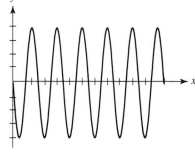

Exercise 15.4 (ODD)

1. $\dfrac{2\pi}{3}$

3. $\dfrac{\pi}{2}$

5. $\dfrac{2\pi}{3}$

7. $\dfrac{\pi}{3}$

9. $\dfrac{2\pi}{5}$

11. 1

13. 4π

15. 3π

17. $\dfrac{4}{3}$

19. $\dfrac{2}{\pi}$

21.

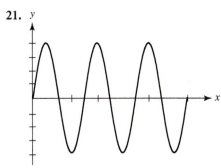

29.

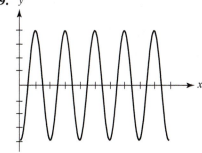

31.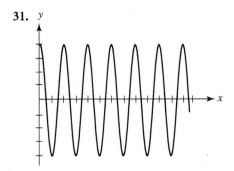

33.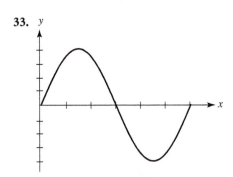

graphed from 0 to 4π

35.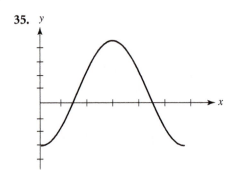

graphed from 0 to 3π

37.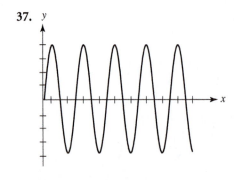

39.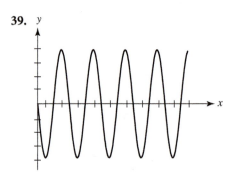

graphed from 0 to π

41. $y = \sin 2x$ **43.** $y = \sin \pi x$

45. $y = \sin 2\pi x$ **47.** $y = \sin 5x$

49. $y = 2 \sin \pi x$ **51.** $y = 5 \sin \dfrac{\pi}{2}x$

53.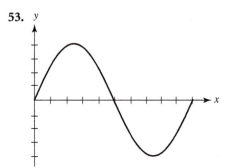

graphed from 0 to 0.025

55.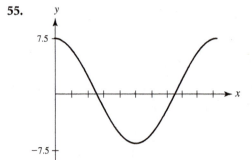

graphed from 0 to $\dfrac{2}{3}$

57.

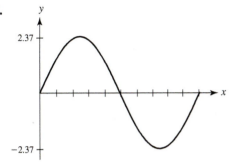

graphed from 0 to 10.5

59.

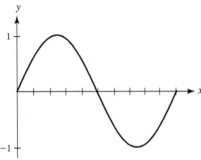

graphed from 0 to $\dfrac{1}{60}$

Exercise 15.5 (ODD)

1. $1, 2\pi, -\dfrac{\pi}{3}$

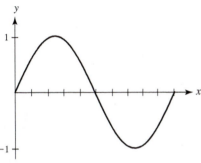

3. $1, 2\pi, \dfrac{\pi}{3}$

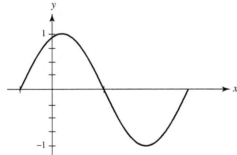

5. $3, \pi, -\dfrac{\pi}{8}$

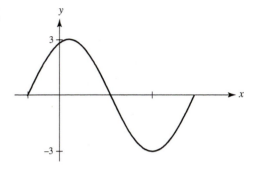

7. $4, \dfrac{2\pi}{3}, -\dfrac{\pi}{3}$

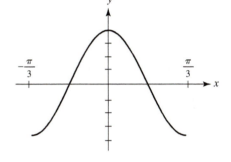

9. $2, 6\pi, -\dfrac{3\pi}{2}$

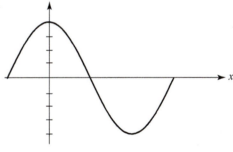

11. $6, 3\pi, -\dfrac{\pi}{2}$

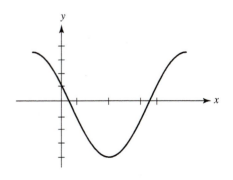

13. $10, 2, \dfrac{1}{\pi}$

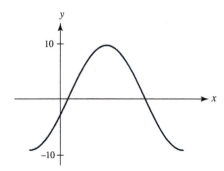

21. $\dfrac{3}{4}, \dfrac{2}{\pi}, -\dfrac{1}{\pi}$

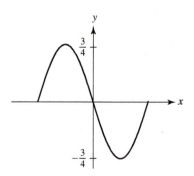

15. $\dfrac{3}{2}, \dfrac{2}{3}, \dfrac{1}{9\pi}$

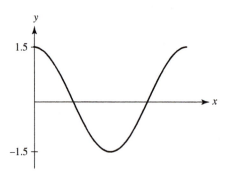

23. $\dfrac{5}{2}, 2, \dfrac{\pi}{3}$

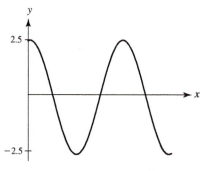

17. $1.2, 2, \dfrac{1}{6}$

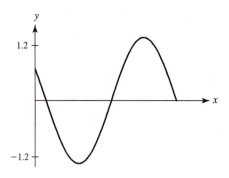

25.

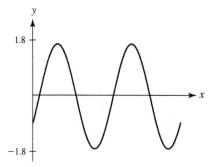

19. $6, 2, \dfrac{1}{2\pi}$

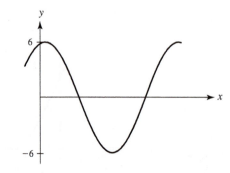

27.

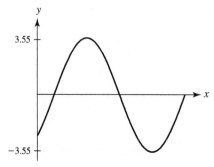

Exercise 15.6 (ODD)

1.

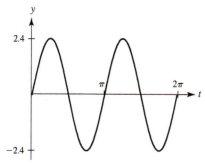

3.

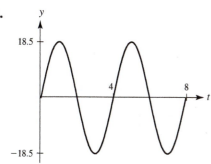

5.

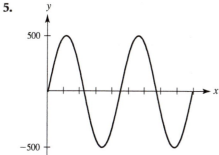

7.

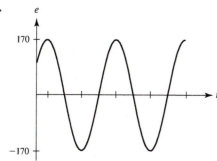

9.

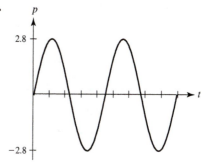

11. 25 A, 0.010 s, 101.1 Hz

13.

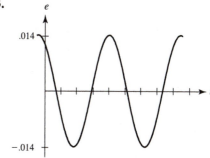

15. The equation should be $y = 0.045 \sin (1.5 \times 10^{10}\pi t)$

The graph should look like this.

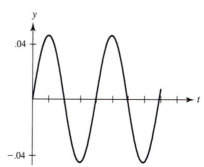

Chapter 15 Review Exercises

1. $\dfrac{7\pi}{20}, \dfrac{137\pi}{180}$ **2.** $\dfrac{3\pi}{8}, \dfrac{9\pi}{8}$ **3.** $\dfrac{23\pi}{90}, \dfrac{\pi}{18}$

4. $\dfrac{53\pi}{30}, \dfrac{71\pi}{45}$ **5.** $20°, 150°$ **6.** $216°, 112.5°$

7. $70°, 81°$ **8.** $342°, 252°$ **9.** $35.8°$

10. $76.2°$ **11.** $197.7°$ **12.** $709.3°$

13. 1.31 **14.** 1.920 **15.** 5.934

16. 0.271   **17.** 2.662   **18.** 3.759

19. 0.16   **20.** 7.365

21. 0.6894, 2.452   **22.** 0.1226, 6.161

23. 1.998, 4.285   **24.** 1.124, 4.266

25. a. 7.73 cm   **b.** 47.9 cm^2

26. a. 846 ft   **b.** 214,000 ft^2

27. a. 2.27 in.   **b.** 8.23 in.^{2}

28. a. 2.86 m   **b.** 0.977 m^2

29. 4.60 m/min   **30.** 2270 in./min

31. 353 m/min   **32.** 187,000 cm/min

33.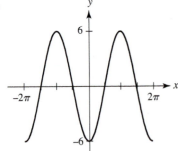

34.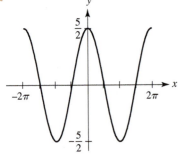

35.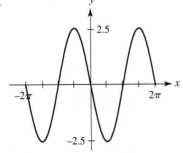

36.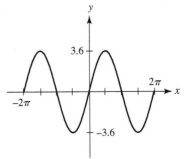

37.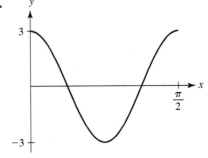

38.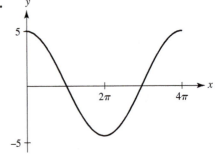

39.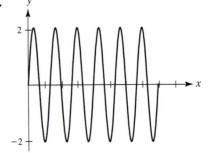

40.

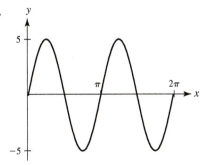

41.

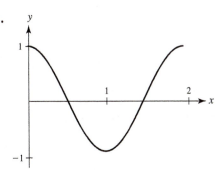

42.

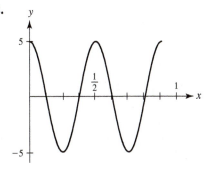

43.

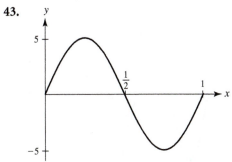

44.

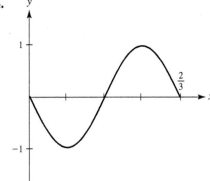

45.

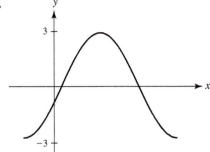

46.

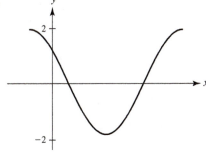

47.

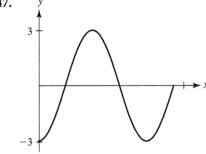

48.

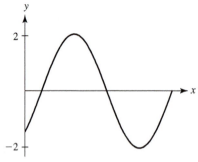

49.

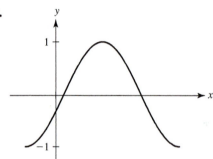

50.

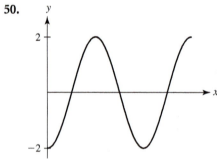

51.

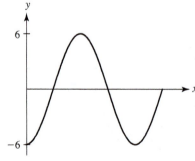

52.

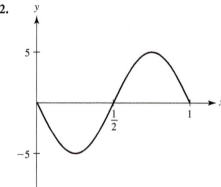

53. $y = 8 \sin (\pi x)$

54. $y = 4 \cos (\pi x)$

55. $y = -3 \cos (2\pi x)$

56. $y = -5 \sin \left(\dfrac{\pi}{4} x \right)$

57. 8.78 cm

58. 89.8 mA

59. 117.0° or 2.04 rad

60. 5900 mi/h

61. 36.3 cm^2

62. 4900 ft/min

63. 138 rad/s or 21.9 r/s

64. 0.163 m

65. 905 r/min

66. 67.8 mi

67.

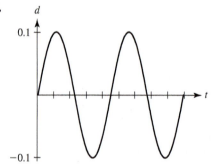

68. 0.053 m

69. 21 V

70.

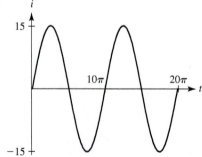

71.

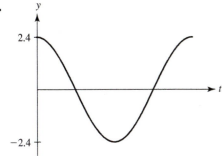

72.

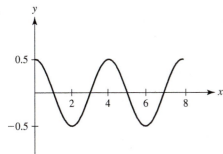

Chapter 15 Test

1. $\dfrac{5\pi}{6}$ **2.** 72°   **3.** 205°

4. 3.14   **5.** 32400 cm^2 **6.** 38,700 ft/min

7.

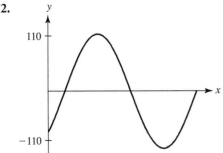

8.

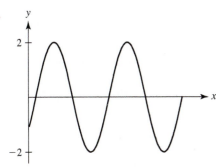

9.

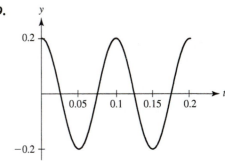

Exercise 16.1 (ODD)

1. $4j$	**3.** $-3j$	**5.** $0.5j$
7. $3j\sqrt{3}$	**9.** $-4j\sqrt{3}$	**11.** $0.02j$
13. $5 + j$	**15.** $6 - 2j$	**17.** $-4 + 2j\sqrt{2}$
19. $14 - 3j\sqrt{7}$	**21.** -9	**23.** -6
25. 3	**27.** $-2\sqrt{3}$	**29.** -6
31. $-\sqrt{33}$	**33.** -1	**35.** j
37. j	**39.** -1	**41.** -1
43. $-j$	**45.** $3 + 8j$	**47.** -4
49. $-9j$	**51.** $-2 + 7j$	

Exercise 16.2 (ODD)

1. $5 - 7j$	**3.** $4 + 5j$	**5.** $1 + 3j$
7. $-12j$	**9.** $2 - 6j$	**11.** $-3 + 20j$
13. -30	**15.** 5	**17.** $48 - 9j$
19. $36j$	**21.** $6 + 4j$	**23.** $36 + 3j$
25. $\dfrac{-8 + 6j}{25}$	**27.** $\dfrac{8 - 3j}{2}$	**29.** $\dfrac{53 - 27j}{61}$
31. $\dfrac{-5 - j}{2}$	**33.** $\dfrac{3 + 3j}{4}$	**35.** $\dfrac{-3 + 7j}{6}$
37. $-2j$	**39.** $2 - 3j$	**41.** $-30 - 40j$
43. $3 + 4j$	**45.** $10 + 6j$	**47.** $-48j$
49. 100	**51.** 53	
53. $14.7 - 2.1j$ ohms	**55.** $3.1 + 5.8j$ volts	

Exercise 16.3 (ODD)

1.

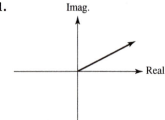

3.

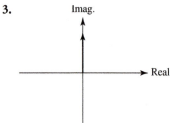

5.

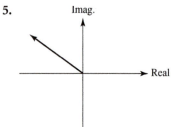

7.

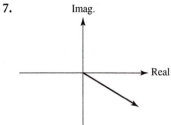

9. $5 + 4j$ **11.** $2 + 6j$

13. $6 + j$ **15.** $7 + j$

17. $4 - j$ **19.** $9 - 15j$

21. 4   **23.** $-13 + 2j$

25. $-15 - 11j$ **27.** $12 - 4j$

29.

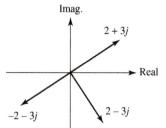

31.

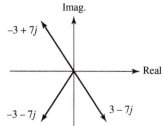

Exercise 16.4 (ODD)

1. $5(\cos 37° + j \sin 37°)$ **3.** $5(\cos 323° + j \sin 323°)$

5. $4.50(\cos 302.9° + j \sin 302.9°)$

7. $6.52(\cos 55.8° + j \sin 55.8°)$

9. $2(\cos 225° + j \sin 225°)$ **11.** $2(\cos 120° + j \sin 120°)$

13. $10(\cos 180° + j \sin 180°)$ **15.** $6(\cos 90° + j \sin 90°)$

17. $6.15(\cos 43.0° + j \sin 43.0°)$

19. $10.3(\cos 335.2° + j \sin 335.2°)$

21. $0.348(\cos 76.4° + j \sin 76.4°)$

23. $56.0(\cos 212.5° + j \sin 212.5°)$

25. $1.03 + 2.82j$ **27.** $1.17 - 0.940j$

29. $-10j$ **31.** -25

33. $41 + 29j$ **35.** $6.2 - 11j$

37. $10.06 + 7.380j$ **39.** $-1.258 - 1.797j$

41. $56.8 + 7.18j$ volts   **43.** 81.2 lb, $322.9°$

Chapter 16—Review Exercises

1. $8j$ **2.** $-10j$ **3.** $-20j$

4. $9j$ **5.** $3j\sqrt{6}$ **6.** $2j\sqrt{10}$

7. $-2j\sqrt{14}$ **8.** $-3j\sqrt{7}$ **9.** $-6 + 10j$

10. $5 - 12j$ **11.** $3 - 4j\sqrt{3}$ **12.** $-2 + 3j\sqrt{3}$

13. -25

14. 8

15. $-\sqrt{14}$

16. -8

17. -1

18. $-j$

19. 1

20. j

21. $14 + 4j$

22. $1 - 11j$

23. $-1 - 7j$

24. $-15 + j$

25. $15j$

25. $32 + 4j$

27. $-11 + 27j$

28. $103 + 11j$

29. $\dfrac{-2 - 16j}{5}$

30. $\dfrac{-21 - j}{17}$

31. $\dfrac{-46 - 2j}{53}$

32. $-3 + 2j$

33. $12; 100$

34. $-4; 29$

35. $14; 58$

36. $-8; 65$

37. $3 + 3j$

38. $-5 + 7j$

39. $-9 - 7j$

40. $-6 + 15j$

41. $13(\cos 67° + j \sin 67°)$

42. $8.47(\cos 330.6° + j \sin 330.6°)$

43. $21.8(\cos 214.3° + j \sin 214.3°)$

44. $107(\cos 125.6° + j \sin 125.6°)$

45. $5.15 + 14.3j$

46. $-0.734 - 0.541j$

47. $14.1 - 8.97j$

48. $-3.40 + 0.593j$

49. yes; yes

50. $(a + bj)(a - bj) = a^2 - abj + abj - b^2j^2 = a^2 + b^2$

51. $9 + 42j$ volts

52. $\dfrac{46 - 140j}{89}$ amperes

Chapter 16—Test

1.

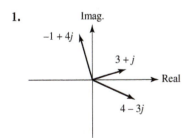

2. $3 + 4j$

3. $-22 - 14j$

4. $-1.07 - 0.38j$

5. $7.28(\cos 286° + j \sin 286°)$

6. $-1.41 - 1.41j$

7. Not a solution. Using the quadratic formula to solve $x^2 - 2x + 4$ we find the solution to be $1 \pm j\sqrt{3}$.

Exercises 17.1 (ODD)

1.

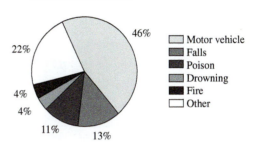

Deaths from Accidents

3.

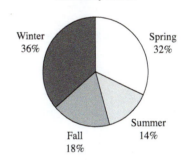

Rainfall by Season

5.

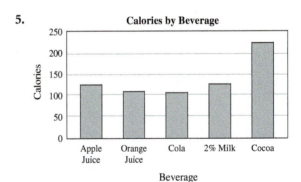

7.

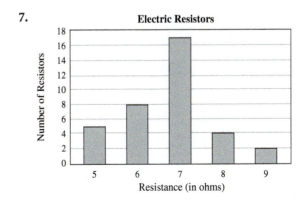

Exercises 17.2 (ODD)

1.

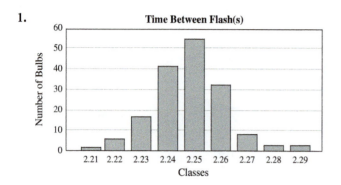

3.

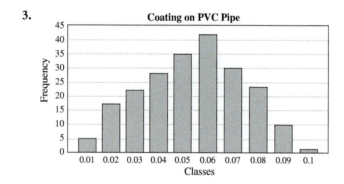

5. a. *Hours Studying*   *Number of Students*

Hours Studying	Number of Students
1–5	4
6–10	9
11–15	4
16–20	2
21–25	1

b.

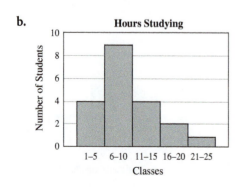

7. a. *X-ray Dosage (in mR)*   *Frequency*

X-ray Dosage (in mR)	Frequency
3.80–3.92	2
3.93–4.05	1
4.06–4.18	3
4.19–4.31	6
4.32–4.44	7
4.45–4.57	1

b.

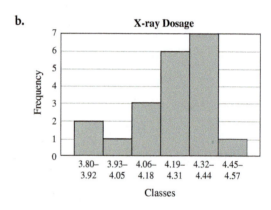

9. a. *Life Expectancy (in hours)*   *Number of Batteries*

Life Expectancy (in hours)	Number of Batteries
25–26	1
27–28	2
29–30	7
31–32	6
33–34	5
35–36	2

b.

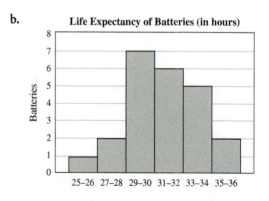

11.

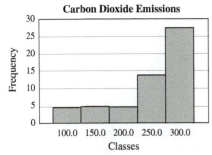

Carbon Dioxide Emissions

13.

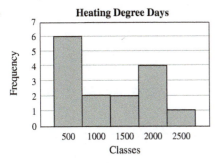

Heating Degree Days

Exercises 17.3 (ODD)

1. mean 21, median 21, mode 21

3. mean 15.3, median 14.3, mode 9, 14, 16

5. mean 172, mode 144

7. mean 2688, median 2749

9. mean 3.470, no mode

11. 964 heating degree days

13. mean 0.21, median 0.195, mode 0.18

15. 0.05 mm

Exercises 17.4 (ODD)

1. range 7, standard deviation 1.81

3. range 0.8, standard deviation 0.222

5. 0.18

7. 0.05

9. range 2406, standard deviation 825

11. range 38.9, standard deviation 11.39

13. Golfer 1: mean 81, range 13, standard deviation 3.2
Golfer 2: mean 81, range 17, standard deviation 5.9
Golfer 1 is the more consistent golfer.

Exercises 17.5 (ODD)

1. 0.29   **3.** 0.52   **5.** 0.07   **7.** 0.375

9. 0.25   **11.** 0.05   **13.** 0.98   **15.** 0.39

Chapter 17—Review Exercises

1.

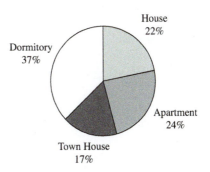

College Housing Accommodations

2.

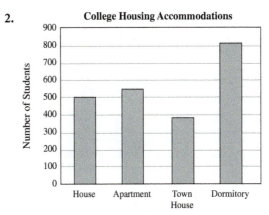

College Housing Accommodations

3.

Classes	Frequency
1093–1095	3
1096–1098	4
1099–1101	3
1102–1104	4
1105–1107	2

4.

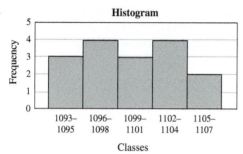
Histogram

5. 1100   **6.** 1100

7. 1098   **8.** 14

9. 4.066

10.

Classes	Frequency
25–31	8
32–38	5
39–45	2
46–52	11
53–59	5
60–66	3

11.

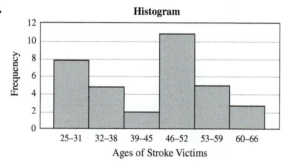
Histogram

12. 44   **13.** 47   **14.** 31, 51

15. 37   **16.** 11.54

17.

Classes	Frequency
11000–17999	7
18000–24999	17
25000–31999	16
32000–38999	8
39000–45999	2

18.

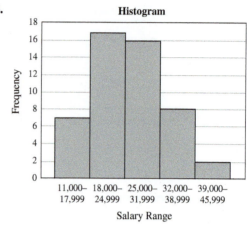
Histogram

19. 25,708   **20.** 25,200   **21.** 20,400

22. 32,900   **23.** 7337   **24.** 0.64

25. 0.26   **26.** 0.23   **27.** 0.26

Chapter 17—Test

1.

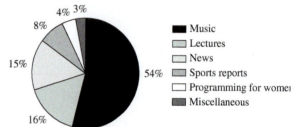
Broadcast Hours

Music
Lectures
News
Sports reports
Programming for women
Miscellaneous

2.

Classes	Frequency
45–49	8
50–54	13
55–59	12
60–64	7
65–69	3

3.

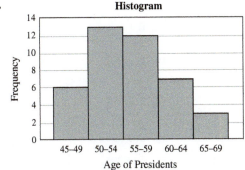

4. 54

5. 55

6. 51, 54

7. 27

8. 6.2

9. $P(54) = 0.12$

10. $P(60s) = 0.23$

Exercises A.1 (ODD)

1. 152

3. 1234

5. 2420

7. 21,410

9. 25,295

11. 139,639

13. 5602

15. 581

17. 949

19. 9,798

21. 30,579

23. 95,682

Exercises A.2 (ODD)

1. 10,534

3. 1,314,976

5. 1,048,576

7. 236,894,403

9. 244

11. 324

13. 2048

15. 1981 R104

17. 47,804

19. 3,183,390

21. 150

23. 72,220

25. 160 mi^2

27. 238 in.^{2}

29. 10,360 cm^2

31. 432 in.^{2}

33. 10,115 mi

35. $1280

37. 330 ft

39. 18 mi/gal

Exercises A.3 (ODD)

1. $\frac{5}{9}$

3. $\frac{1}{7}$

5. $\frac{7}{13}, \frac{3}{16}$

7. $\frac{9}{8}, \frac{1}{12}$

9. 1, 6

11. 32, 1

13. $1\frac{2}{3}$

15. $4\frac{12}{13}$

17. $3\frac{53}{75}$

19. $53\frac{4}{25}$

21. $\frac{17}{5}$

23. $\frac{79}{8}$

25. $\frac{223}{13}$

27. $\frac{423}{4}$

29. $\frac{17}{24}$

31. $\frac{10}{23}$

33. $8\frac{1}{2}$ in.

35. $\frac{37}{10}$ in.

Exercises A.4 (ODD)

1. $\frac{6}{14}$

3. $\frac{4}{5}$

5. $\frac{24}{78}$

7. $\frac{5}{13}$

9. $\frac{91}{175}$

11. $\frac{32}{2}$

13. $\frac{1}{2}$

15. $\frac{3}{2}$

17. $\frac{4}{5}$

19. $\frac{3}{5}$

21. $\frac{2}{5}$

23. $\frac{2}{5}$

25. 4

27. 15

29. 3

31. 4

33. 40

35. 16

37. $2 \times 2 \times 5$

39. $2 \times 2 \times 2 \times 2$

41. $2 \times 2 \times 3 \times 3$

43. $2 \times 2 \times 2 \times 2 \times 3$

45. 3×19

47. $3 \times 5 \times 7$

49. $\frac{6}{7}$

51. $\frac{4}{5}$

53. $\frac{7}{3}$

55. $\frac{2}{3}$

57. $\frac{1}{4}$

59. $\frac{3}{5}$

61. $\frac{3}{4}$ in.

63. $\frac{7}{16}$

65. $\frac{2}{5}$

67. $\frac{5}{6}, \frac{1}{6}$

69. $\frac{1}{30}$

Exercises A.5 (ODD)

1. 4

3. 24

5. 72

7. 300

9. $\frac{4}{5}$

11. $\frac{2}{7}$

13. $\frac{3}{4}$

15. $\frac{1}{8}$

17. $\frac{13}{12}$

19. $\dfrac{1}{10}$ **21.** $\dfrac{7}{3}$ **23.** $\dfrac{47}{40}$

25. $\dfrac{28}{39}$ **27.** $\dfrac{83}{24}$ **29.** $\dfrac{29}{42}$

31. $\dfrac{73}{72}$ **33.** $\dfrac{73}{42}$ **35.** $\dfrac{13}{5}$

37. $\dfrac{1}{2}$ in. **39.** $7\dfrac{19}{20}$ gal **41.** $1\dfrac{17}{24}$ in.

43. $8\dfrac{1}{6}$ cu yd

Exercises A.6 (ODD)

1. $\dfrac{6}{77}$ **3.** $\dfrac{6}{5}$ **5.** $\dfrac{21}{20}$

7. $\dfrac{5}{27}$ **9.** $\dfrac{5}{18}$ **11.** $\dfrac{3}{2}$

13. $\dfrac{14}{9}$ **15.** $\dfrac{2}{17}$ **17.** 2

19. $\dfrac{2}{15}$ **21.** $\dfrac{11}{4}$ **23.** $\dfrac{27}{64}$

25. $\dfrac{168}{23}$ **27.** $\dfrac{5}{4}$ **29.** 7

31. $\dfrac{188}{351}$ **33.** $\dfrac{57}{4}$ **35.** $\dfrac{60}{209}$

37. $\dfrac{1}{5}, \dfrac{1}{13}$ **39.** 2, 5 **41.** $\dfrac{3}{16}, \dfrac{2}{7}$

43. $\dfrac{16}{81}, \dfrac{10}{71}$ **45.** 12 **47.** $\dfrac{9}{26}$ acre/h

49. $12\dfrac{22}{25}$ ft/s^2 **51.** 32

Exercises A.7 (ODD)

1. $4(10) + 7(1) + \dfrac{3}{10}$

3. $4(100) + 2(10) + 9(1) + \dfrac{4}{10} + \dfrac{8}{100} + \dfrac{6}{1000}$

5. 27.3 **7.** 57.54 **9.** 8.03

11. 17.4 **13.** 0.4 **15.** 0.21

17. 1.7 **19.** 0.499 **21.** $\dfrac{4}{5}$

23. $\dfrac{9}{20}$ **25.** $\dfrac{267}{50}$ **27.** $\dfrac{63}{2500}$

29. 31.295 **31.** 2817.256 **33.** 8.763

35. 0.02628 **37.** 13.99952 **39.** 9.882

41. 124.992 **43.** 0.1215 **45.** 12.5

47. 295.4 **49.** 5.20 **51.** 0.46

53. 0.6 V **55.** \$177.49 **57.** 235.88 L

59. 26,070 mi **61.** 1.8

63. 1130 kilowatt-hours

Exercises A.8 (ODD)

1. 0.08 **3.** 2.36 **5.** 0.003

7. 0.056 **9.** 27% **11.** 321%

13. 0.64% **15.** 700% **17.** $\dfrac{3}{10}$

19. $\dfrac{1}{40}$ **21.** $\dfrac{6}{5}$ **23.** $\dfrac{57}{10,000}$

25. 60% **27.** 57.1% **29.** 22.9%

31. 266.7% **33.** 13 **35.** 5.304

37. 25% **39.** 7.5% **41.** 50

43. 400 **45.** \$255 **47.** 32%

49. \$17.64 **51.** \$1695.75 **53.** 35%

55. 0.3% **57.** 97% **59.** \$580

61. 11 A

Index

Algebra

Exponents and Radicals

$a^m \times a^n = a^{m+n}$

$\dfrac{a^m}{a^n} = a^{m-n} \quad a \neq 0$

$(a^m)^n = a^{mn}$

$(ab)^n = a^n b^n$

$\left(\dfrac{a}{b}\right)^n = \dfrac{a^n}{b^n} \quad b \neq 0$

$a^0 = 1 \quad a \neq 0$

$a^{-n} = \dfrac{1}{a^n} \quad a \neq 0$

$a^{m/n} = \sqrt[n]{a^m} = (\sqrt[n]{a})^m$

$\sqrt[n]{ab} = \sqrt[n]{a}\sqrt[n]{b}$

Special Products

$a(x + y) = ax + ay$

$(x + y)(x - y) = x^2 - y^2$

$(x + y)^2 = x^2 + 2xy + y^2$

$(x - y)^2 = x^2 - 2xy + y^2$

Quadratic Equation and Formula

$ax^2 + bx + c = 0 \quad a \neq 0$

$x = \dfrac{-b \pm \sqrt{b^2 - 4ac}}{2a}$

Properties of Logarithms

$\log_b x + \log_b y = \log_b xy$

$\log_b x - \log_b y = \log_b\left(\dfrac{x}{y}\right)$

$n \log_b x = \log_b (x^n)$

Complex Numbers

$\sqrt{-a} = j\sqrt{a} \quad (a > 0)$

$x + yj = r(\cos\theta + j\sin\theta)$
$\qquad = re^{j\theta} = r\angle\theta$

Variation

Direct variation: $y = kx$

Inverse variation: $y = k/x$

Geometric Formulas

Triangle

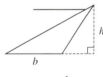

$A = \dfrac{1}{2}bh$

Circle

$A = \pi r^2$
$c = 2\pi r$

Parallelogram

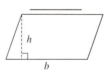

$A = bh$

Trapezoid

$A = \dfrac{1}{2}h(b_1 + b_2)$

Pythagorean Theorem

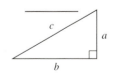

$c^2 = a^2 + b^2$

Cylinder

$V = \pi r^2 h$
$A = 2\pi rh + 2\pi r^2$

Cone

$V = \dfrac{1}{3}\pi r^2 h$
$A = \pi rs + \pi r^2$

Sphere

$V = \dfrac{4}{3}\pi r^3$
$A = 4\pi r^2$

Metric and English Conversion

Length

English	Metric
1 inch (in.) =	2.54 cm
1 foot (ft) =	30.5 cm
1 yard (yd) =	91.4 cm
1 mile (mi) =	1610 m
1 mi =	1.61 km
0.0394 in. =	1 mm
0.394 in. =	1 cm
39.4 in. =	1 m
3.28 ft =	1 m
1.09 yd =	1 m
0.621 mi =	1 km

Weight

English	Metric
1 ounce (oz) =	28.3 g
1 pound (lb) =	454 g
1 lb =	0.454 kg
0.0353 oz =	1 g
0.00220 lb =	1 g
2.20 lb =	1 kg

Capacity

English	Metric
1 gallon (gal) =	3.79 L
1 quart (qt) =	0.946 L
0.264 gal =	1 L
1.06 qt =	1 L
1 metric ton =	1000 kg

Area

English

$1 \text{ ft}^2 = 144 \text{ in.}^2$
$1 \text{ yd}^2 = 9 \text{ ft}^2$
$1 \text{ rd}^2 = 30.25 \text{ yd}^2$
$1 \text{ acre} = 160 \text{ rd}^2$
$= 4840 \text{ yd}^2$
$= 43{,}560 \text{ ft}^2$
$1 \text{ mi}^2 = 640 \text{ acres}$
$= 1 \text{ section}$

Metric

$1 \text{ m}^2 = 10{,}000 \text{ cm}^2 \text{ or } 10^4 \text{ cm}^2$
$= 1{,}000{,}000 \text{ mm}^2 \text{ or } 10^6 \text{ mm}^2$
$1 \text{ cm}^2 = 100 \text{ mm}^2$
$= 0.0001 \text{ m}^2$
$1 \text{ km}^2 = 1{,}000{,}000 \text{ m}^2$
$1 \text{ hectare (ha)} = 10{,}000 \text{ m}^2$
$= 1 \text{ hm}^2$

English → Metric

$1 \text{ in.}^2 = 6.45 \text{ cm}^2$
$= 645 \text{ mm}^2$
$1 \text{ ft}^2 = 929 \text{ cm}^2$
$= 0.0929 \text{ m}^2$
$1 \text{ yd}^2 = 8361 \text{ cm}^2$
$= 0.8361 \text{ m}^2$
$1 \text{ rd}^2 = 25.3 \text{ m}^2$
$1 \text{ acre} = 4047 \text{ m}^2$
$= 0.004047 \text{ km}^2$
$= 0.4047 \text{ ha}$
$1 \text{ mi}^2 = 2.59 \text{ km}^2$

Metric → English

$1 \text{ m}^2 = 10.76 \text{ ft}^2$
$= 1550 \text{ in.}^2$
$= 0.0395 \text{ rd}^2$
$= 1.196 \text{ yd}^2$
$1 \text{ cm}^2 = 0.155 \text{ in.}^2$
$1 \text{ km}^2 = 247 \text{ acres}$
$= 1.08 \times 10^7 \text{ ft}^2$
$= 0.386 \text{ mi}^2$
$1 \text{ ha} = 2.47 \text{ acres}$

Volume

English

$1 \text{ ft}^3 = 1728 \text{ in.}^3$
$1 \text{ yd}^3 = 27 \text{ ft}^3$

Metric

$1 \text{ m}^3 = 10^6 \text{ cm}^3$
$1 \text{ cm}^3 = 10^{-6} \text{ m}^3$
$= 10^3 \text{ mm}^3$
$1 \text{ cm}^3 = 1 \text{ mL}$

English → Metric

$1 \text{ in.}^3 = 16.39 \text{ cm}^3$
$1 \text{ ft}^3 = 28{,}317 \text{ cm}^3$
$= 0.028317 \text{ m}^3$
$1 \text{ yd}^3 = 0.7646 \text{ m}^3$

Metric → English

$1 \text{ cm}^3 = 0.06102 \text{ in.}^3$
$1 \text{ m}^3 = 35.3 \text{ ft}^3$
$= 1.31 \text{ yd}^3$